MW00717666

"Why do engineers prefer GLP study guides?"

Some comments about the GLP approach...

"In your FE book you cut to the chase and don't embellish it. Yours is the only review that doesn't scare the pants off people. It's exactly what I want for teaching my review course."

—John Thorington

"I passed the first time I took the test. Potter's book is a great, great review. It's right to the point and doesn't include extra stuff you don't need. All I did was work through the book. My friends took review courses and didn't use your book—they failed! Do you have a PE review? I don't want to take the exam until you do!"

—Dana, calling to back-order our new PE Civil Review

"I used your FE book and thought it was far superior to any other. I took the exam and did great. Now I need your PE Review!"

—Tom Roach, Indiana Midwest Steel

"Your FE book kept straight to the point, was very informative and more than adequate to pass the exam. I'm calling to see if you have a review for the PE exam."

—Burton Mills, Fruehauf Company

"When you're working 50 hours a week, you don't have a lot of time to prepare. The Potter FE book is condensed and focuses on everything that is on the exam. It has a good strategy: it concentrates on the basics—just what you need to pass. It doesn't overwhelm you."

—James Alt, graduate of University of Wisconsin

"We have been using your FE review manual for several years with great success. Congratulations on a job well done!"

—Ted Huddleston, review course coordinator, University of S. Alabama

"The FE Reference is very concise. It does a good job with respect to the time that you have to study, getting you through the material for the test. It's right where it should be in terms of length and depth of review. It doesn't overwhelm those studying for the exam. I agree with the approach you took. It's a very useful book."

—Dr. Packard, University of California

"Several of the engineers I sold a copy of the FE book to passed the exam after trying and failing repeatedly. They all said "What is it with this book?" They didn't even attend my course and they passed with the book alone."

—Shahin Mansour, S&R Engineering Foundation

International Standard Book Number 1-881018-12-1

All comments and inquiries should be addressed to:
Great Lakes Press—Customer Service
PO Box 172
Grover, MO 63040-0172
Phone (314) 273-6016
www.glpbooks.com glpbooks@aol.com

Library of Congress
Cataloging-in-Publication Data

Printed in the USA by Braun-Brumfield, Inc., Ann Arbor, Michigan.

10 9 8 7 6 5 4 3 2 1

...from the Professors who know it best...

Principles & Practice of Mechanical Engineering

The most efficient and authoritative review book for the PE License Exam

Editor: Merle C. Potter, PhD, PE
Professor, Michigan State University

Authors:	Frank Hatfield, PhD, PE	Engineering Economics
	George E. Mase, PhD	Mechanics
	Merle C. Potter, PhD, PE	Thermodynamics
	Charles W. Radcliffe, PhD	Dynamics
	Clark J. Radcliffe, PhD	Design
	Bassem H. Ramadan, PhD	HVAC
	Craig W. Somerton, PhD	Power Plants
	David C. Wiggert, PhD, PE	Fluid Flow
	Keith Woodbury, PhD	Heat Transfer

The authors are professors at Michigan State University, with the exception of K. Woodbury, of The University of Alabama and C. W. Radcliffe, from The University of California-Berkeley

published by:

Great Lakes Press

P.O. Box 483

Okemos, MI 48805-0483

Customer Service (314) 273-6016

www.glpbooks.com glpbooks@aol.com

Table of Contents

Preface

Introduction

CHAPTER 1 Mathematics

CHAPTER 2 Statics

CHAPTER 8 Stress Analysis

CHAPTER 9 Vibration

CHAPTER 10 Machine Design

CHAPTER 11 Control Systems

CHAPTER 12 Heat Transfer

CHAPTER 13 Heating Ventilation, Air Conditioning and Refrigeration

CHAPTER 16 Engineering Economics

Index

Preface

This book was written to provide the most efficient review possible for the P.E./Mechanical, Principles and Practice of Engineering Examination. Nine currently lecturing professors from major universities have done us the courtesy of delivering this review. We feel that a professor who lectures in a specialized area of engineering is the person most qualified to write a review of the pertinent material from that area. Until now, concise materials from lecturing professors have not been available to those wishing to prepare for the PE. You are now getting ready to use the result of the efforts of our distinctive team; we feel that it is an authoritative, highly efficient review. Credit is also due Jeff Potter, Michelle Gruender and John Gruender for their diligence and hard work in bringing the strengths of this unique manual to the engineering community.

A passing grade on the FE exam is the first step for all engineers in the registration process. The second exam, The Principles and Practice of Engineering Examination, is presented in each major area of engineering. It may be taken after varying amounts of engineering apprenticeship, depending on state requirements. To practice engineering in some states an engineer must pass an examination in the particular area in which the engineer wishes to practice. Consequently, most states require each engineer to specify the particular area of engineering for testing; this allows for reciprocity in the event that an engineer moves from one state to another.

Background and Format of this PE Review

We published a review book in 1984 for the FE (EIT) exam and by 1994 over 100,000 engineers had used this review in their preparation. Many of the users of our FE Review have requested a review book for the PE exam. This PE/Mechanical Review is the second of a series we plan to publish.

As with our FE Review, we include detailed reviews, numerous examples, and problems with full solutions in one book. Since the PE exams include both essay design problems (partial credit) and multiple-choice problems, we have included both types of problems at the end of the chapters. This should provide sufficient practice for the actual exam. There seems to be an excellent correlation between an engineer's performance on the machine-graded, multiple-choice problems and that same engineer's performance on the partial-credit essay problems graded by professional engineers. Since a machine avoids the subjectivity of humans, there is a tendency to move in the direction of more multiple-choice problems. It's worth noting that a multiple-choice question sometimes helps lead the test-taker through a tough problem that may otherwise be quite formidable.

We have included in this review the major subjects encountered in most design problems that mechanical engineers are required to solve. You may find types of problems on the exam that we have not included in this review book. We have not attempted to include all areas of review, only the major ones. Of the numerous problems that can be selected, there will always be several that are thoroughly reviewed in this book. A good understanding of these subjects should ensure passing.

The problems on the exam may be presented using SI units or english units. If you are familiar with english units only, you may want to convert the given information to english units and then convert your answer back to SI units. We have presented problems using both sets of units, and have included at the end of the introduction some conversions between the two systems.

A Little Strategic Advice

Make sure you're proficient in the basics—math, statics, dynamics, solid mechanics, fluid mechanics and thermodynamics—these basic subjects are used in most mechanical engineering design problems; they form the first chapters in this review book. Then select those areas in which you're most familiar and emphasize those areas in your review. Because you may not find four problems in the morning session and four problems in the afternoon session in one or two areas, you will probably have to work problems in three or perhaps four areas. So, to be safe, you should familiarize yourself with several areas.

The PE exam may also contain some economics problems, so we have added a chapter on engineering economics as Chapter 16. An economics problem may be one of the essay problems of the morning session. Some of the multiple choice questions of the afternoon session may require economic analysis.

Use your time wisely in your review—start early—perhaps several months early. Sign up for a review course if there's one in your area; that'll force you to review. Get a good night's sleep before the examination, and we here at Great Lakes Press wish you well!

Dr. Merle C. Potter
Okemos, Michigan

P.S. Your comments are a vital part of our user-oriented publishing method and are greatly appreciated. With this in mind, note that a book has its most errors in the first printing of its first edition even though we've proofed and reviewed each chapter several times. Please fill out and return the postpaid comment and errata cards included in the back.

Introduction

The NCEES

The National Council of Examiners for Engineering and Surveying (NCEES) develops both the FE/EIT exam and the various Principles and Practice of Engineering examinations (the PE exams). The PE exams are given in a variety of major areas including Civil (Sanitary/Structural), Mechanical, Electrical, and Chemical Engineering. The NCEES develops the examinations using standard guidelines that allow for maximum fairness and the highest quality. Experienced testing specialists and professional engineers from across the nation including those from private consulting, government, industry and education, prepare examinations which are valid measures of competency.

Applications

Each state establishes the education and experience level required for eligibility to take the PE exam. Application procedures and fee requirements vary from state to state. Sufficient time must be allotted for the application process. Consult the Board of the state in which you desire registration for the details of the registration process and an application for the PE exam. The address of each state is given in a table that follows.

Major Area

Some states permit an engineer to practice only in the major area of the PE exam that was passed, while other states allow engineers to practice in all areas of engineering by passing a PE exam in only one major area. The Board of a state must be consulted for the requirements in that particular state. To allow for reciprocity, a particular state may require that the PE exam be taken in one major area only.

When and Where

The PE exam is offered twice a year: in mid-April and late October. The exact dates are selected by each state Board. The location may also change year to year. Be sure and consult the state Board for the exact time, date and location of the exam.

Content of Exam

The PE exam is an eight-hour examination consisting of a morning session and an afternoon session. The four-hour morning session consists of ten essay type problems and the four-hour afternoon session consists of ten multiple-choice type problems. Four problems are to be solved in each session. This provides about one hour for each problem.

The essay problems are similar to the long problems given in an undergraduate hour-long exam in a design course. The situation is described in detail and the expected response is carefully outlined. Partial credit is awarded if mistakes or errors in judgment are made.

The multiple-choice problems each consist of a particular situation that is again described in detail, followed by ten multiple-choice questions. There is no partial credit, however, guessing is permitted. There is some advantage in this type of problem in that the ten questions may lead a candidate through the steps to a successful solution, even though the problem appears unsolvable without the "hints" contained in the wording of the ten multiple-choice questions.

Scoring

The four problems selected in the morning session are each worth ten points for a maximum score of forty points, with partial credit awarded on each problem. Scores of 0, 2, 4, 6, 8 or 10 points are earned on each problem with 6 being the score of minimum competency.

The four multiple-choice problems selected in the afternoon session are also worth ten points each for a maximum score of forty points. Each of the ten questions in a particular problem is worth one point with 6 again representing minimum competency.

A combined score of 48 is usually needed to pass the exam. This is the score of minimum competency established by the NCEES. Much effort is made through workshops composed of practicing engineers to assure the quality and fairness of the exam. The engineers are representative of the profession regarding gender, ethnic background, geography and area of practice.

Minimum competency is defined by NCEES as:

> "The lowest level of knowledge at which a person can practice professional engineering in such a manner that will safeguard life, health and property and promote the public welfare."

The NCEES, with the aide of committees of professional engineers, continues to develop examinations that assure minimum competency of all who pass.

The Examination Day

For the morning session proctors distribute an exam booklet and a solution pamphlet to each candidate. Detailed instructions are given before the booklets are to be opened. Your name must be recorded on the front cover. In addition, your board code and identification number, the number of the problem you are solving and the number of pages in the solution must be recorded in the upper right corner on each page of the solution pamphlet. You must also write the numbers of the problems that you solved on the front of the solutions pamphlet. If you do not enter this information in the appropriate places, errors may result in the scoring process.

There is no partial credit for the machine-scored answers of the multiple choice questions of the afternoon session. As in all machine-scored answers, you must use a #2 pencil to avoid any improper scanning by the machine. Remember to answer all the questions, even if you must guess, since there is no penalty for a wrong answer.

If you complete either session of the examination with more than 30 minutes remaining in the session, you will be free to leave. If there is less than 30 minutes remaining in the session, you must wait until the end of the session to permit an orderly collection of exam materials and to avoid disruptions of candidates. Give full attention to the proctors' instructions at the end of each session as well as at the beginning.

What to Bring

The PE exam is open-book. You may bring review books, text books, handbooks, and other bound material into the exam. You should probably predetermine the areas in which you will most likely select problems to solve. The handbooks that are used for design purposes in those areas (at most 4 areas) should be part of your selection. Select one handbook, if one exists, for each area. Use the handbook you're most familiar with even when a new edition has just been released. A good, battery-operated, silent, non-printing calculator that you can easily operate is a must. Don't make last minute selections of materials. Work with your selected materials all through the review process. Familiarity is very important.

HP-48G—the ideal PE Exam Calculator

You really should use a silent, pre-programmed calculator at this exam. A calculator is essential when solving many problems. In fact, with the exception of a couple states, the premier engineering calculator, the HP-48G, is allowed into the exam (check with your state board). This calculator is a hand computer which has hundreds of basic equations and constants preprogrammed. If you need extra help with exam calculations, the HP-48G will be very useful. We at GLP offer this calculator and its higher-powered downloadable relative, the GX, for sale at substantial discounts.

We now also offer a 150-pg. manual, "JumpStart the HP-48G", to guide you through the steps to using this powerful calculator effectively. The manufacturer's manual is known to be difficult to use for even basic operations. Our engineering-oriented manual will help you make the most of HP-48 calculators in the PE exam room. Call us at 1-800-837-0201 for ordering or info.

Special Requirements

If you require special accommodations due to some condition that would put you at a disadvantage, you should consult your state board. Special arrangement can be made if you communicate such requirements well in advance of the exam.

What's in this Review

This book presents an efficient review of the major subject areas covered on the PE exam and also serves as an excellent desk reference.

Select those chapters that are related to the subject area of the PE exam that you are most familiar with. Select at least four of the major design areas:

- Kinematics and Dynamics of Machines
- Stress Analysis and Structures
- Vibration
- Machine Design
- Control Systems
- Heat Transfer
- HVAC

- Power Plants
- Fluid Flow

To help you in your overall preparation, this book contains:

- Succinct reviews of both theoretical and practical aspects of the subject areas
- Tables, charts and figures pertinent to the subjects
- Example problems with detailed solutions illustrating important concepts
- Practice problems similar to those that may be encountered on the PE exam
- Solutions to all practice problems

Why Our PE Review Stands Apart

We at Great Lakes Press developed this reference with the cooperation of a team of professorial colleagues in response to the existing prep situation. The kind of concise material that an engineer naturally desires to use was needed. We were also motivated by the fact that over the years, we've received hundreds of requests for a PE title from users of our FE review. So we recruited popular lecturers from the university campus and prepared this study guide. The design of this review is to deliver an adequate preparation without excess. Our goal as publisher is always to act as an advocate for the test-taker in our interactions with engineering departments, associations, review courses and registration governing bodies. We are dedicated to making the entire licensing and registration process as reasonable as possible.

How To Become A Professional Engineer

To become registered as an engineer, a state may require that you:

1. graduate from an ABET-accredited engineering program
2. pass the *Fundamentals of Engineering* (FE) exam
3. pass the *Principles and Practice of Engineering* (PE) exam after several years of engineering experience

Requirements vary from state to state, so you should obtain local guidelines and follow them carefully.

Why Get an Engineering License?

Registration is necessary if an engineer works as a consultant, and is highly recommended in certain industries—especially when one is, or hopes to be, in a management position.

Handling Multiple Choice Problems

The afternoon portion of the PE exam consists of 4 multiple choice problems, each having 10 questions. For those of us with less immediate experience with this sort of test question, a few points of strategic advice will make for much less confusion and frustration, and help you maximize your score for any given amount of preparation.

- Sometimes the best way to find the right answer is to look for the wrong ones and cross them out. On questions which are difficult for you, wrong answers are often much easier to find than right ones!
- Answers are seldom given with more than three significant figures, and may be given with two significant figures. The choice **closest** to your own solution should be selected.

- Use the **process of elimination** when *guessing* at an answer. If only one answer is negative and four answers are positive, eliminate the one odd answer from your guess. Also, when in doubt, work backwards and eliminate those answers that you believe are untrue until you are left with only 2 or 3 and then guess. By using a combination of methods, you greatly improve your odds of answering correctly.
- **Do not leave any answers blank** on your answer key. A guess *cannot* hurt you, only help you. Your score is based on the number of questions you answer correctly. An incorrect answer does not harm your score.
- **Place a question mark** beside choices you are uncertain of, but seem correct. If time prevents you from re-working that problem, you will have at least identified your best guess.
- **Be sure to make a best guess** if you have spent some time on a problem. Then circle the problem number, so you can go back to it later if time permits. But your guess is most "educated"—and most likely to be correct—if you make it *immediately* while working on the problem.

State Boards of Registration

All State Boards of Registration administer the National Council of Engineering Examiners and Surveyors (NCEES) uniform examination. The dates of the exams cover a span of three days in mid-April and three days in late October. The specific dates are selected by each State Board. To be accepted to take the PE exam, an applicant must apply well in advance. For information regarding the specific requirements in your state, contact your state board's office. Any comments relating to the exam should be addressed to the Executive Director of NCEES, POB 1686, Clemson, SC 29633-1686, ph: (803) 654-6824.

State Boards of Registration Addresses and Phone Numbers

ALABAMA: State Board of Registration for Professional Engineers and Land Surveyors, P. O. Box 304451, Montgomery 36130-4451. Executive Secretary, Telephone: (205) 242-5568.

ALASKA: State Board of Registration for Architects, Engineers and Land Surveyors, Pouch D, Juneau 99811. Licensing Examiner, Telephone: (907) 465-2540.

ARIZONA: State Board of Technical Registration, 1951 W. Camelback Rd., Suite 250, Phoenix 85015. Executive Director, Telephone: (602) 255-4053.

ARKANSAS: State Board of Registration for Professional Engineers and Land Surveyors, P. O. Box 2541, Little Rock 72203. Secretary-Treasurer, Telephone: (501) 324-9085.

CALIFORNIA: State Board of Registration for Professional Engineers and Land Surveyors, 2535 Capitol Oaks Dr., Sacramento 95853. Executive Secretary, Telephone: (916) 263-2222.

COLORADO: State Board of Registration for Professional Engineers and Professional Land Surveyors, 1860 Broadway, Suite 1370, Denver 80202. Program Administrator, Telephone: (303) 894-7788.

CONNECTICUT: State Board of Registration for Professional Engineers and Land Surveyors, The State Office Building, Room G-3A, 165 Capitol Avenue, Hartford 06106. Administrator, Ph: (203) 566-3290.

DELAWARE: Delaware Association of Professional Engineers, 2005 Concord Pike, Wilmington 19803. Executive Secretary, Telephone: (302) 577-6500.

DISTRICT OF COLUMBIA: Board of Registration for Professional Engineers, 614 H Street, N.W., Room 910, Washington 20001. Executive Secretary, Telephone: (202) 727-7468.

FLORIDA: Department of Business and Professional Regulations, Board of Professional Engineers, Northwood Centre, 1940 N. Monroe Street, Tallahassee 32399-0755. Executive Director, Telephone: (904) 488-9912.

GEORGIA: State Board of Registration for Professional Engineers and Land Surveyors, 166 Pryor Street, SW, Atlanta 30303-3465. Executive Director, Telephone: (404) 656-3926.

GUAM: Territorial Board of Registration for Professional Engineers, Architects and Land Surveyors, Department of Public Works, Government of Guam, P. 0. Box 2950, Agana 96911. Chairman, Telephone: (671) 646-3115.

HAWAII: State Board of Registration for Professional Engineers, Architects, Land Surveyors and Landscape Architects, P. 0. Box 3469, (1010 Richards Street), Honolulu 96801. Executive Secretary, Telephone: (808) 586-2693.

IDAHO: Board of Professional Engineers and Land Surveyors, 600 S. Orchard, Suite A, Boise 83705. Executive Secretary, Telephone: (208) 334-3860.

ILLINOIS: Department of Registration and Education, Professional Engineers' Examining Committee, 320 West Washington, 3rd Floor, Springfield 62786. Unit Manager, Telephone: (217) 782-0458.

INDIANA: Indiana Professional Licensing Agency, 302 W. Washington St., Room E034, Indianapolis 46204. Executive Director, Telephone: (317) 232-2980.

IOWA: State Board of Engineering Examiners, Capitol Complex, 1918 S.E. Hulsizer, Ankeny 50021. Executive Secretary, Telephone: (515) 281-5602.

KANSAS: State Board of Technical Professions, 900 Jackson, Room 507, Topeka 66612. Executive Secretary, Telephone: (913) 296-3053.

KENTUCKY: State Board of Registration for Professional Engineers and Land Surveyors, 160 Democrat Dr., Frankfort 40601. Executive Director, Telephone: (502) 573-2680.

LOUISIANA: State Board of Registration for Professional Engineers and Land Surveyors, 1055 St. Charles Avenue, Suite 415, New Orleans 70130. Executive Secretary, Telephone: (504) 568-8450.

MAINE: State Board of Registration for Professional Engineers, State House, Station 92, Augusta 04333. Secretary, Telephone: (207) 289-3236.

MARYLAND: State Board of Registration for Professional Engineers, 501 St. Paul Place, Room 902, Baltimore 21202. Executive Secretary, Telephone: (410) 659-6322.

MASSACHUSETTS: State Board of Registration of Professional Engineers and of Land Surveyors, Room 1512, Leverett Saltonstall Building, 100 Cambridge Street, Boston 02202. Secretary, Ph: (617) 727-9957.

MICHIGAN: Board of Professional Engineers, P. O. Box 30018, (611 West Ottawa), Lansing 48909. Administrative Secretary, Telephone: (517) 335-1669.

MINNESOTA: State Board of Registration for Architects, Engineers, Land Surveyors and Landscape Architects, 133 7th St. E., St. Paul 55101. Executive Secretary, Ph: (612) 296-2388.

MISSISSIPPI: State Board of Registration for Professional Engineers and Land Surveyors P. 0. Box 3, Jackson 39205. Executive Director, Telephone: (601) 359-6160.

MISSOURI: Board of Architects, Professional Engineers and Land Surveyors, P. O. Box 184, Jefferson City 65102. Executive Director, Telephone: (314) 751-0047.

MONTANA: State Board of Professional Engineers and Land Surveyors, Department of Commerce, 111 N. Jackson, POB 200513, Helena 59620-0513. Administrative Secretary, Telephone: (406) 444-4285.

NEBRASKA: State Board of Examiners for Professional Engineers and Architects, P. 0. Box 94751, Lincoln 68509. Executive Director, Telephone: (402) 471-2021.

NEVADA: State Board of Registered Professional Engineers and Land Surveyors, 1755 East Plum Lane, Ste. 135, Reno 89502. Executive Secretary, Telephone: (702) 688-1231.

NEW HAMPSHIRE: State Board of Professional Engineers, 57 Regional Drive, Concord 03301. Executive Secretary, Telephone: (603) 271-2219.

NEW JERSEY: State Board of Professional Engineers and Land Surveyors, P.O. Box 45015, Newark 07101. Executive Secretary-Director, Telephone: (201) 504-6460.

NEW MEXICO: State Board of Registration for Professional Engineers and Land Surveyors, 1010 Marquez Pl., Santa Fe 87501. Secretary, Telephone: (505) 827-7561.

NEW YORK: State Board for Engineering and Land Surveying, The State Education Department, Cultural Education Center, Madison Avenue, Albany 12230. Executive Secretary, Telephone: (518) 474-3846.

NORTH CAROLINA: State Board of Professional Engineers and Surveyors, 3620 Six Forks Rd., Suite 300, Raleigh 27609. Executive Secretary, Telephone (919) 781-9499

NORTH DAKOTA: State Board of Registration for Professional Engineers and Land Surveyors, P. 0. Box 1357, Bismarck 58502. Executive Secretary, Telephone: (701) 258-0786.

OHIO: State Board of Registration for Professional Engineers and Surveyors, 77 S. High St., 16th Fl., Columbus 43266-0314. Executive Secretary, Telephone: (614) 466-8999.

OKLAHOMA: State Board of Registration for Professional Engineers and Land Surveyors, Oklahoma Engineering Center, Room 120, 201 N.E. 27th Street, Oklahoma City, 73105. Executive Secretary, Telephone: (405) 521-2874.

OREGON: State Board of Engineering Examiners, Department of Commerce, 750 Front St., NE, Suite 240, Salem 97310. Executive Secretary, Telephone: (503) 378-4180.

PENNSYLVANIA: State Registration Board for Professional Engineers, P. O. Box 2649, Harrisburg 17120-264g. Administrative Assistant, Telephone: (800) 877-3926.

PUERTO RICO: Board of Examiners of Engineers, Architects, and Surveyors, Box 3271, San Juan 00904. Director, Examining Boards, Telephone: (809) 722-2122.

RHODE ISLAND: State Board of Registration for Professional Engineers and Land Surveyors, 10 Orms St., Suite 324, Providence 02904. Administrative Assistant, Telephone: (401) 277-2565.

SOUTH CAROLINA: State Board of Registration for Professional Engineers and Land Surveyors, P.O. Drawer 50408, Columbia 29250. Agency Director, Ph: (803) 734-9166.

SOUTH DAKOTA: State Commission of Engineering, Architectural Examiners and Land Surveying, 2040 West Main Street, Suite 304, Rapid City 57702-2447. Executive Secretary, Telephone: (605) 394-2510.

TENNESSEE: State Board of Architectural and Engineering Examiners, 3rd Fl. Volunteer Plaza, 500 James Robertson Pkwy, Nashville 37243-1142. Administrator, Telephone: (615) 741-3221.

TEXAS: State Board of Registration for Professional Engineers, P.O. Drawer 18329, Austin 78760-8329. Executive Director, Telephone: (512) 440-7723.

UTAH: Division of Occupational and Professional Licensing, P.O. Box 45805, Salt Lake City 84145-0805. Director, Ph: (801) 530-6628.

VERMONT: State Board of Registration for Professional Engineers, Division of Licensing and Registration, Pavilion Building, Montpelier 05602. Executive Secretary, Telephone: (802) 828-2875.

VIRGINIA: State Board of Architects, Professional Engineers, Land Surveyors and Certified Landscape Architects, 3600 West Broad St., Seaboard Building, 5th Floor, Richmond 23230-4917. Assistant Director, Telephone: (804) 367-8506.

VIRGIN ISLANDS: Board for Architects, Engineers and Land Surveyors, Bldg 1, Sub-Base, Rm 205, St. Thomas 00801. Secretary, Telephone: (809) 774-3130.

WASHINGTON: State Board of Registration for Professional Engineers and Land Surveyors, P.O. Box 9649, Olympia 98504. Executive Secretary, Telephone: (206) 753-3634.

WEST VIRGINIA: State Board of Registration for Professional Engineers, 608 Union Building, Charleston 25301. Executive Director, Telephone: (304) 348-3554.

WISCONSIN: State Examining Board of Professional Engineers, P.O. Box 8935, Madison 53708. Administrator, Telephone: (608) 266-1397.

WYOMING: State Board of Examining Engineers, Herschler Building, Room 4135, Cheyenne 82002. Secretary-Accountant, Telephone: (307) 777-6156.

English and SI Units

The following tables presenting SI (Systems International) units and the conversion of English units to SI units conclude this introduction.

SI Prefixes

Multiplication Factor	*Prefix*	*Symbol*
10^{15}	peta	P
10^{12}	tera	T
10^{9}	giga	G
10^{6}	mega	M
10^{3}	kilo	k
10^{-1}	deci	d
10^{-2}	centi	c
10^{-3}	mili	m
10^{-6}	micro	μ
10^{-9}	nano	n
10^{-12}	pico	p
10^{-15}	femto	f

SI Base Units

Quantity	*Name*	*Symbol*
length	meter	m
mass	kilogram	kg
time	second	s
electric current	ampere	A
temperature	kelvin	K
amount of substance	mole	mol
luminous intensity	candela	cd

SI Derived Units

Quantity	*Name*	*Symbol*	*In Terms of Other Units*
area	square meter		m^2
volume	cubic meter		m^3
velocity	meter per second		m/s
acceleration	meter per second squared		m/s^2
density	kilogram per cubic meter		kg/m^3
specific volume	cubic meter per kilogram		m^3/kg
frequency	hertz	Hz	s^{-1}
force	newton	N	$m \cdot kg/s^2$
pressure, stress	pascal	Pa	$kg/(m \cdot s^2)$
energy, work, heat	joule	J	$N \cdot m$
power	watt	W	J/s
electric charge	coulomb	C	$A \cdot s$
electric potential	volt	V	W/A
capacitance	farad	F	C/V
electric resistance	ohm	Ω	V/A
conductance	siemens	S	A/V
magnetic flux	weber	Wb	$V \cdot s$
inductance	henry	H	Wb/A
viscosity	pascal second		$Pa \cdot s$
moment (torque)	meter newton		$N \cdot m$
heat flux	watt per square meter		W/m^2
entropy	joule per kelvin		J/K
specific heat	joule per kilogram-kelvin		$J/(kg \cdot K)$
conductivity	watt per meter-kelvin		$W/(m \cdot K)$

Conversion Factors to SI Units

English	*SI*	*SI Symbol*	*To Convert from English to SI Multiply by*
	Area		
square inch	square centimeter	cm^2	6.452
square foot	square meter	m^2	0.09290
acre	hectare	ha	0.4047
	Length		
inch	centimeter	cm	2.54
foot	meter	m	0.3048
mile	kilometer	km	1.6093
	Volume		
cubic inch	cubic centimeter	cm^3	16.387
cubic foot	cubic meter	m^3	0.02832
gallon	cubic meter	m^3	0.003785
gallon	liter	L	3.785
	Mass		
pound mass	kilogram	kg	0.4536
slug	kilogram	kg	14.59
	Force		
pound	newton	N	4.448
kip(1000 lb)	newton	N	4448
	Density		
pound/cubic foot	kilogram/cubic meter	kg/m^3	16.02
pound/cubic foot	grams/liter	g/L	16.02
	Work, Energy, Heat		
foot-pound	joule	J	1.356
Btu	joule	J	1055
Btu	kilowatt-hour	kWh	0.000293
therm	kilowatt-hour	kWh	29.3

Conversion Factors to SI Units (continued)

English	SI	SI Symbol	To Convert from English to SI Multiply by
	Power, Heat, Rate		
horsepower	watt	W	745.7
foot pound/sec	watt	W	1.356
Btu/hour	watt	W	0.2931
Btu/hour-ft^2-°F	watt/meter squared-°C	W/m^2 · °C	5.678
tons of refrig.	kilowatts	kW	3.517
	Pressure		
pound/square inch	kilopascal	kPa	6.895
pound/square foot	kilopascal	kPa	0.04788
inches of H_20	kilopascal	kPa	0.2486
inches of Hg	kilopascal	kPa	3.374
one atmosphere	kilopascal	kPa	101.3
	Temperature		
Fahrenheit	Celsius	°C	5 (°F-32)/9
Fahrenheit	kelvin	K	5 (°F+460)/9
	Velocity		
foot/second	meter/second	m/s	0.3048
mile/hour	meter/second	m/s	0.4470
mile/hour	kilometer/hour	km/h	1.609
	Acceleration		
foot/second squared	meter/second squared	m/s^2	0.3048
	Torque		
pound-foot	newton-meter	N · m	1.356
pound-inch	newton-meter	N · m	0.1130
	Viscosity, Kinematic Viscosity		
pound-sec/square foot	newton-sec/square meter	N · s/m^2	47.88
square foot/second	square meter/second	m^2/s	0.09290
	Flow Rate		
cubic foot/minute	cubic meter/second	m^3/s	0.0004719
cubic foot/minute	liter/second	L/s	0.4719
	Frequency		
cycles/second	hertz	Hz	1.00

Conversion Factors

Length

1 cm	= 0.3937 in
1 m	= 3.281 ft
1 yd	= 3 ft
1 mi	= 5280 ft
1 mi	= 1760 yd
1 km	= 3281 ft

Area

1 cm^2	= 0.155 in^2
1 m^2	= 10.76 ft^2
1 ha	= 10^4 m^2
1 acre	= 100 m^2
1 acre	= 4047 m^2
1 acre	= 43,560 ft^2

Volume

1 ft^3	= 28.32 L
1 L	= 0.03531 ft^3
1 L	= 0.2642 gal
1 m^3	= 264.2 gal
1 ft^3	= 7.481 gal
1 m^3	= 35.31 ft^3
1 acre-ft	= 43,560 ft^3
1 m^3	= 1000 L

Velocity

1 m/s	= 3.281 ft/s
1 mph	= 1.467 ft/s
1 mph	= 0.8684 knot
1 knot	= 1.688 ft/s
1 km/h	= 0.2778 m/s
1 km/h	= 0.6214 mph

Force

1 lb	= 4.448 x 10^5 dyne
1 lb	= 32.17 pdl
1 lb	= 0.4536 kg
1 N	= 10^5 dyne
1 N	= 0.2248 lb
1 kip	= 1000 lb

Mass

1 oz	= 28.35 g
1 lb	= 0.4536 kg
1 kg	= 2.205 lb
1 slug	= 14.59 kg
1 slug	= 32.17 lb

Work and Heat

1 Btu	= 778.2 ft-lb
1 Btu	= 1055 J
1 Cal	= 3.088 ft-lb
1 J	= 10^7 ergs
1 kJ	= 0.9478 ft-lb
1 Btu	= 0.2929 W · hr
1 ton	= 12,000 Btu/hr
1 kWh	= 3414 Btu
1 quad	= 10^{15} Btu
1 therm	= 10^5 Btu

Power

1 Hp	= 550 ft-lb/s
1 HP	= 33,000 ft-lb/min
1 Hp	= 0.7067 Btu/s
1 Hp	= 2545 Btu/hr
1 Hp	= 745.7 W
1 W	= 3.414 Btu/hr
1 kW	= 1.341 Hp

Volume Flow Rate

1 cfm	= 7.481 gal/min
1 cfm	= 0.4719 L/s
1 m^3/s	= 35.31 ft^3/s
1 m^3/s	= 2119 cfm
1 gal/min	= 0.1337 cfm

Torque

1 N · m	= 10^7 dyne · cm
1 N · m	= 0.7376 lb-ft
1 N · m	= 10 197 g · cm
1 lb-ft	= 1.356 N · m

Viscosity

1 lb-s/ft^2	= 478 poise
1 poise	= 1 g/cm · s
1 N · s/m^2	= 0.02089 lb-s/ft^2

Pressure

1 atm	= 14.7 psi
1 atm	= 29.92 in Hg
1 atm	= 33.93 ft H_20
1 atm	= 1.013 bar
1 atm	= 1.033 kg/cm^2
1 atm	= 101.3 kPa
1 psi	= 2.036 in Hg
1 psi	= 6.895 kPa
1 psi	= 68 950 dyne/cm^2
1 ft H_20	= 0.4331 psi
1 kPa	= 0.145 psi

Mathematics

by Merle C. Potter

The engineer uses mathematics as a tool to help solve the problems encountered in the analysis and design of physical systems. We will review those parts of mathematics that are used fairly often by the engineer and which may appear on the exam. The topics include: algebra, trigonometry, analytic geometry, linear algebra (matrices), calculus, differential equations, and probability and statistics. The review here is intended to be brief and not exhaustive. The majority of the questions on the exam will be based on the material included in this chapter. There may be a few questions, however, that will require information not covered here; to cover all possible points would not be in the spirit of an efficient review.

1.1 Algebra

It is assumed that the reader is familiar with most of the rules and laws of algebra as applied to both real and complex numbers. We will review some of the more important of these and illustrate several with examples. The three basic rules are:

commutative law:	$a + b = b + a$	$ab = ba$	**(1.1.1)**
distributive law:	$a(b + c) = ab + ac$		**(1.1.2)**
associative law:	$a + (b + c) = (a + b) + c$	$a(bc) = (ab)c$	**(1.1.3)**

Exponents

Laws of exponents are used in many manipulations. For positive x and y we use

$$x^{-a} = \frac{1}{x^a}$$

$$x^a x^b = x^{a+b}$$

$$(xy)^a = x^a y^a$$
$$x^{ab} = \left(x^a\right)^b \tag{1.1.4}$$

Logarithms

Logarithms are actually exponents. For example if $b^x = y$ then $x = \log_b y$; that is, the exponent x is equal to the logarithm of y to the base b. Most engineering applications involve common logs which have a base of 10, written as log y, or natural logs which have a base of e (e = 2.7183 ...), written as ln y. If any other base is used it will be so designated, such as $\log_5 y$.

Remember, logarithms of numbers less than one are negative, the logarithm of one is zero, and logarithms of numbers greater that one are positive. The following identities are often useful when manipulating logarithms:

$$\begin{aligned} \ln x^a &= a \ln x \\ \ln(xy) &= \ln x + \ln y \\ \ln(x/y) &= \ln x - \ln y \\ \ln x &= 2.303 \log x \\ \log_b b &= 1 \\ \ln 1 &= 0 \\ \ln e^a &= a \end{aligned} \tag{1.1.5}$$

The Quadratic Formula and the Binomial Theorem

We often encounter the quadratic equation $ax^2 + bx + c = 0$ when solving engineering problems. The *quadratic formula* provides its solution; it is

$$x = \frac{-b \pm \sqrt{b^2 - 4ac}}{2a}. \tag{1.1.6}$$

If $b^2 < 4ac$, the two roots are complex numbers. Cubic and higher order equations are most often solved by trial and error.

The *binomial theorem* is used to expand an algebraic expression of the form $(a + x)^n$. It is

$$(a+x)^n = a^n + na^{n-1}x + \frac{n(n-1)}{2!}a^{n-2}x^2 + \cdots. \tag{1.1.7}$$

If n is a positive integer, the expansion contains (n + 1) terms. If it is a negative integer or a fraction, an infinite series expansion results.

Partial Fractions

A rational fraction $P(x) / Q(x)$, where $P(x)$ and $Q(x)$ are polynomials, can be resolved into partial fractions for the following cases.

Case 1: $Q(x)$ factors into n different linear terms,

$$Q(x) = (x - a_1)(x - a_2)\ldots(x - a_n).$$

Then

$$\frac{P(x)}{Q(x)} = \sum_{i=1}^{n} \frac{A_i}{x - a_i}. \tag{1.1.8}$$

Case 2: $Q(x)$ factors into n identical terms,

$$Q(x) = (x - a)^n.$$

Then

$$\frac{P(x)}{Q(x)} = \sum_{i=1}^{n} \frac{A_i}{(x-a)^i}. \tag{1.1.9}$$

Case 3: $Q(x)$ factors into n different quadratic terms,

$$Q(x) = \left(x^2 + a_1 x + b_1\right)\left(x^2 + a_2 x + b_2\right)\ldots\left(x^2 + a_n x + b_n\right).$$

Then

$$\frac{P(x)}{Q(x)} = \sum_{i=1}^{n} \frac{A_i\, x + B_i}{x^2 + a_i\, x + b_i}. \tag{1.1.10}$$

Case 4: $Q(x)$ factors into n identical quadratic terms,

$$Q(x) = \left(x^2 + ax + b\right)^n.$$

Then

$$\frac{P(x)}{Q(x)} = \sum_{i=1}^{n} \frac{A_i\, x + B_i}{\left(x^2 + a\, x + b\right)^i}. \tag{1.1.11}$$

Case 5: $Q(x)$ factors into a combination of the above. The partial fractions are the obvious ones from the appropriate expansions above.

Example 1.1

The temperature at a point in a body is given by $T(t) = 100e^{-0.02t}$. At what value of t does $T = 20$?

Solution. The equation takes the form

$$20 = 100e^{-0.02t}$$
$$0.2 = e^{-0.02t}.$$

Take the natural logarithm of both sides and obtain

$$\ln 0.2 = \ln e^{-0.02t}.$$

Using a calculator, we find

$$-1.6094 = -0.02t.$$
$$\therefore t = 80.47.$$

Example 1.2

Find an expansion for $(9+x)^{1/2}$.

Solution. Using the binomial theorem, Eq. 1.1.7, we have

$$(9+x)^{1/2} = 3\left(1 + \frac{x}{9}\right)^{1/2}$$
$$= 3\left[1 + \frac{1}{2}\left(\frac{x}{9}\right) + \frac{1/2(-1/2)}{2}\left(\frac{x}{9}\right)^2 + \frac{1/2(-1/2)(-3/2)}{3\cdot 2}\left(\frac{x}{9}\right)^3 + \cdots\right]$$
$$= 3 + \frac{x}{6} - \frac{x^2}{216} + \frac{x^3}{3888} + \cdots.$$

Note: We factored out $9^{1/2} = 3$ so that in Eq. 1.1.7 $a = 1$; this simplifies the expansion.

Example 1.3

Resolve $\frac{x^2+2}{x^4+4x^3+x^2}$ into partial fractions.

Solution. The denominator is factored into

$$x^4+4x^3+x^2 = x^2\left(x^2+4x+1\right).$$

Using Cases 2 and 3 there results

$$\frac{x^2+2}{x^4+4x^3+x^2} = \frac{A_1}{x}+\frac{A_2}{x^2}+\frac{A_3x+B_3}{x^2+4x+1}.$$

This can be written as

$$\frac{x^2+2}{x^4+4x^3+x^2} = \frac{A_1x\left(x^2+4x+1\right)+A_2\left(x^2+4x+1\right)+\left(A_3x+B_3\right)x^2}{x^2\left(x^2+4x+1\right)}$$

$$= \frac{(A_1+A_3)x^3+(4A_1+A_2+B_3)x^2+(A_1+4A_2)x+A_2}{x^2\left(x^2+4x+1\right)}.$$

The numerators on both sides must be equal. Equating the coefficients of the various powers of x provides us with the four equations:

$$A_1+A_3=0$$

$$4A_1+A_2+B_3=1$$

$$A_1+4A_2=0$$

$$A_2=2.$$

These are solved quite easily to give $A_2=2$, $A_1=-8$, $A_3=8$, $B_3=31$. Finally,

$$\frac{x^2+2}{x^4+4x^3+x^2} = \frac{8}{x}+\frac{2}{x^2}+\frac{8x+31}{x^2+4x+1}.$$

1.2 Trigonometry

The primary functions in trigonometry involve the ratios between the sides of a right triangle. Referring to the right triangle in Fig. 1.1, the functions are defined by

$$\sin\theta = \frac{y}{r}, \qquad \cos\theta = \frac{x}{r}, \qquad \tan\theta = \frac{y}{x}. \tag{1.2.1}$$

In addition, there are three other functions that find occasional use, namely,

$$\cot\theta = \frac{x}{y}, \qquad \sec\theta = \frac{r}{x}, \qquad \csc\theta = \frac{r}{y}. \tag{1.2.2}$$

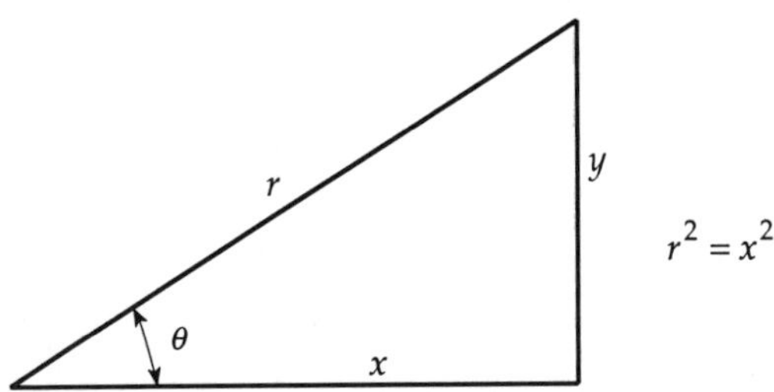

Figure 1.1 A right triangle.

The trig functions $\sin\theta$ and $\cos\theta$ are periodic functions with a period of 2π. Fig. 1.2 shows a plot of the three primary functions.

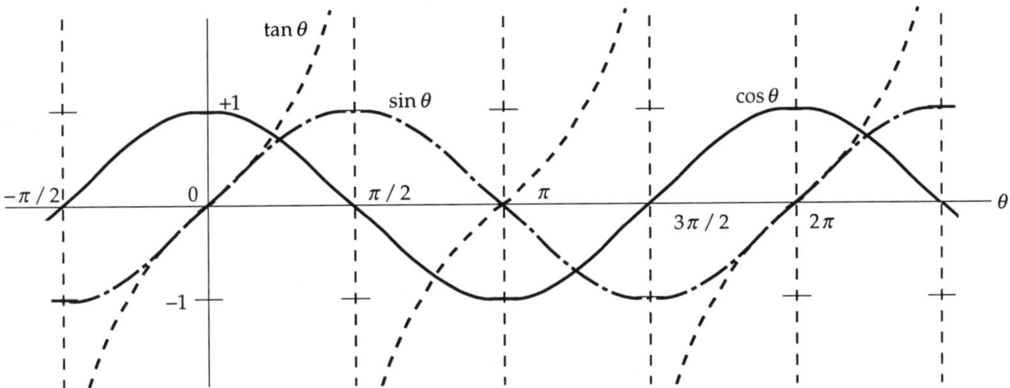

Figure 1.2 The Trig Functions.

In the above relationships, the angle θ is usually given in radians for mathematical equations. It is possible, however, to express the angle in degrees; if that is done it may be necessary to relate degrees to radians. This can be done by remembering that there are 2π radians in 360°. Hence, we multiply radians by $(180/\pi)$ to obtain degrees, or multiply degrees by $(\pi/180)$ to obtain radians. A calculator may use either degrees or radians for an input angle.

Most problems involving trigonometry can be solved using a few fundamental identities. They are

$$\sin^2\theta + \cos^2\theta = 1 \tag{1.2.3}$$

$$\sin 2\theta = 2\sin\theta\cos\theta \tag{1.2.4}$$

$$\cos 2\theta = \cos^2\theta - \sin^2\theta \tag{1.2.5}$$

$$\sin(\alpha \pm \beta) = \sin\alpha\cos\beta \pm \sin\beta\cos\alpha \tag{1.2.6}$$

$$\cos(\alpha \pm \beta) = \cos\alpha\cos\beta \mp \sin\alpha\sin\beta. \tag{1.2.7}$$

A general triangle may be encountered such as that shown in Fig. 1.3. For this triangle we may use the following equations:

$$\text{law of sines: } \frac{\sin\alpha}{a} = \frac{\sin\beta}{b} = \frac{\sin\gamma}{c} \tag{1.2.8}$$

$$\text{law of cosines: } a^2 = b^2 + c^2 - 2bc\cos\alpha \tag{1.2.9}$$

Note that if $\gamma = 90°$, the law of cosines becomes the *Pythagorean Theorem*

$$c^2 = a^2 + b^2. \tag{1.2.10}$$

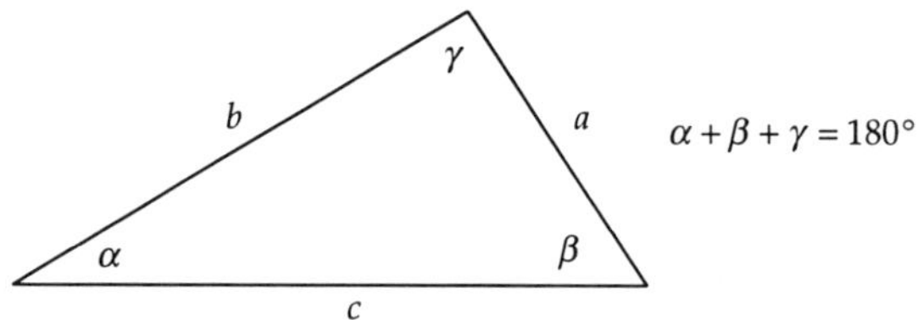

Figure 1.3 A general triangle.

The hyperbolic trig functions also find occasional use. They are defined by

$$\sinh x = \frac{e^x - e^{-x}}{2}, \quad \cosh x = \frac{e^x + e^{-x}}{2}, \quad \tanh x = \frac{\sinh x}{\cosh x}. \tag{1.2.11}$$

Useful identities follow:

$$\cosh^2 x - \sinh^2 x = 1 \tag{1.2.12}$$

$$\sinh(x \pm y) = \sinh x \cosh y \pm \cosh x \sinh y \tag{1.2.13}$$

$$\cosh(x \pm y) = \cosh x \cosh y \pm \sinh x \sinh y \tag{1.2.14}$$

The values of the primary trig functions of certain angles are listed in Table 1.1.

TABLE 1.1 Functions of Certain Angles.

	0	30°	45°	60°	90°	135°	180°	270°	360°
$\sin\theta$	0	1/2	$\sqrt{2}/2$	$\sqrt{3}/2$	1	$\sqrt{2}/2$	0	−1	0
$\cos\theta$	1	$\sqrt{3}/2$	$\sqrt{2}/2$	1/2	0	$-\sqrt{2}/2$	−1	0	1
$\tan\theta$	0	$1/\sqrt{3}$	1	$\sqrt{3}$	∞	−1	0	$-\infty$	0

Example 1.4

Express $\cos^2\theta$ as a function of $\cos 2\theta$.

Solution. Substitute Eq. 1.2.3 into Eq. 1.2.5 and obtain

$$\begin{aligned}\cos 2\theta &= \cos^2\theta - \left(1-\cos^2\theta\right)\\ &= 2\cos^2\theta - 1.\end{aligned}$$

There results

$$\cos^2\theta = \frac{1}{2}(1+\cos 2\theta).$$

Example 1.5

If $\sin\theta = x$, what is $\tan\theta$?

Solution. Think of $x = x/1$. Thus, the hypotenuse of an imaginary triangle is of length unity and the side opposite θ is of length x. The adjacent side is of length $\sqrt{1-x^2}$. Hence,

$$\tan\theta = \frac{x}{\sqrt{1-x^2}}$$

Example 1.6

An airplane leaves Lansing flying due southwest at 300 km/hr, and a second leaves Lansing at the same time flying due west at 500 km/hr. How far apart are the airplanes after 2 hours?

Solution. After 2 hours, the respective distances from Lansing are 600 km and 1000 km. A sketch is quite helpful. The distance d that the two airplanes are apart is found using the law of cosines:

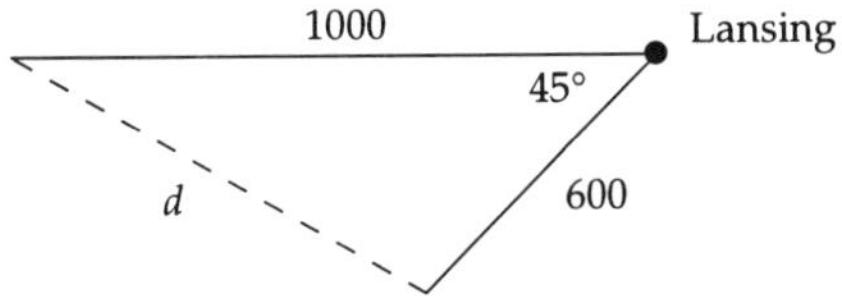

$$\begin{aligned}d^2 &= 1000^2 + 600^2 - 2\times 1000\times 600\cos 45^\circ\\ &= 511470.\end{aligned}$$

$$\therefore d = 715.2 \text{ km}.$$

1.3 Geometry

A regular polygon with n sides has a vertex angle (the central angle subtended by one side) of $2\pi/n$. The included angle between two successive sides is given by $\pi(n-2)/n$.

Some common geometric shapes are displayed in Fig. 1.4.

The equation of a straight line can be written in the general form

$$Ax + By + C = 0. \tag{1.3.1}$$

There are three particular forms that this equation can take. They are:

Point-slope: $$y - y_1 = m(x - x_1) \tag{1.3.2}$$

Slope-intercept: $$y = mx + b \tag{1.3.3}$$

Two-intercept: $$\frac{x}{a} + \frac{y}{b} = 1. \tag{1.3.4}$$

In the above equations m is the slope, (x_1, y_1) a point on the line, "a" the x-intercept, and "b" the y-intercept (see Fig. 1.5). The perpendicular distance d from the point (x_3, y_3) to the line $Ax + By + C = 0$ is given by (see Fig. 1.5).

$$d = \frac{|Ax_3 + By_3 + C|}{\sqrt{A^2 + B^2}}. \tag{1.3.5}$$

The equation of a plane surface is given as

$$Ax + By + Cz + D = 0. \tag{1.3.6}$$

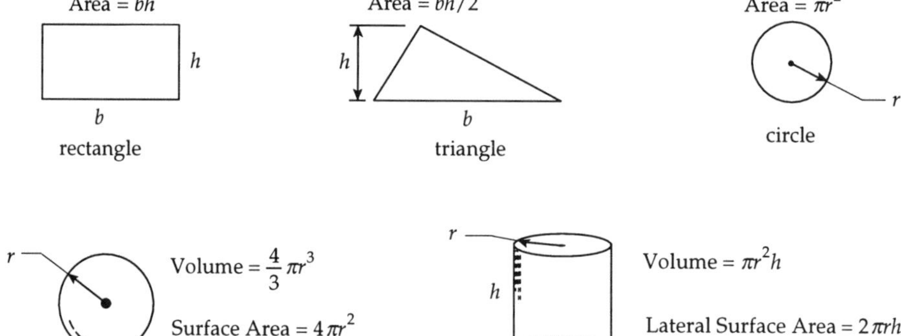

Figure 1.4 Common geometric shapes.

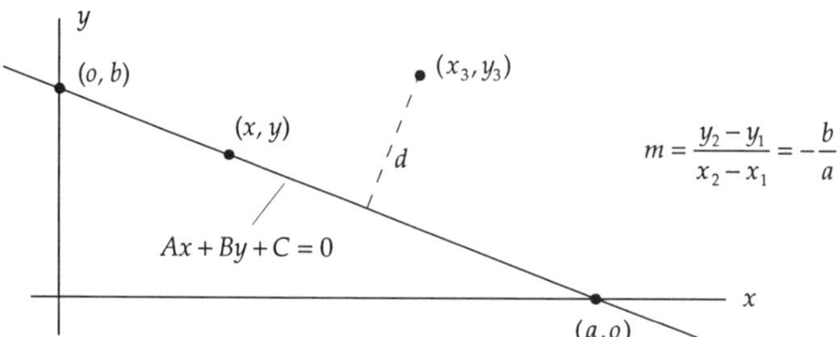

Figure 1.5 A straight line.

The general equation of second degree

$$Ax^2 + 2Bxy + Cy^2 + 2Dx + 2Ey + F = 0 \tag{1.3.7}$$

represents a set of geometric shapes called *conic sections.* They are classified as follows:

$$\begin{aligned} &\text{ellipse:} && B^2 - AC < 0 (\text{circle: } B = 0,\ A = C) \\ &\text{parabola:} && B^2 - AC = 0 \\ &\text{hyperbola:} && B^2 - AC > 0 \end{aligned} \tag{1.3.8}$$

If $A = B = C = 0$, the equation represents a line in the xy-plane, not a parabola. Let's consider each in detail.

Circle: The circle is a special case of an ellipse with $A = C$. Its general form can be expressed as

$$(x-a)^2 + (y-b)^2 = r^2 \tag{1.3.9}$$

where its center is at (a, b) and r is the radius.

Ellipse: The sum of the distances from the two foci, F, to any point on an ellipse is a constant. For an ellipse centered at the origin

$$\frac{x^2}{a^2} + \frac{y^2}{b^2} = 1 \tag{1.3.10}$$

where a and b are the semi-major and semi-minor axes. The foci are at $(\pm\ c,0)$ where $c^2 = a^2 - b^2$. See Fig. 1.6a.

Parabola: The locus of points on a parabola are equidistant from the focus and a line (the directrix). If the vertex is at the origin and the parabola opens to the right, it is written as

$$y^2 = 2px \tag{1.3.11}$$

where the focus is at $(p/2, 0)$ and the directrix is at $x = -p/2$. See Fig. 1.6b. For a parabola opening to the left, simply change the sign of p. For a parabola opening upward or downward, interchange x and y.

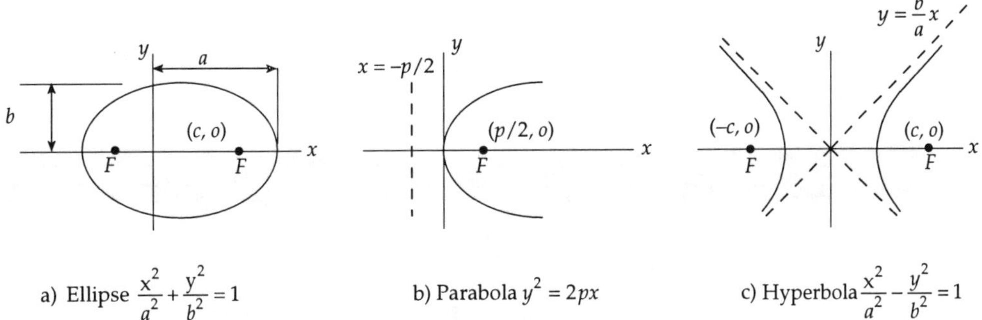

a) Ellipse $\frac{x^2}{a^2} + \frac{y^2}{b^2} = 1$ b) Parabola $y^2 = 2px$ c) Hyperbola $\frac{x^2}{a^2} - \frac{y^2}{b^2} = 1$

Figure 1.6 The three conic sections.

Hyperbola: The difference of the distances from the foci to any point on a hyperbola is a constant. For a hyperbola centered at the origin opening left and right, the equation can be written as

$$\frac{x^2}{a^2} - \frac{y^2}{b^2} = 1. \tag{1.3.12}$$

The lines to which the hyperbola is asymptotic are asymptotes:

$$y = \pm\frac{b}{a}x. \tag{1.3.13}$$

If the asymptotes are perpendicular, a rectangular hyperbola results. If the asymptotes are the x and y axes, the equation can be written as

$$xy = \pm k^2. \tag{1.3.14}$$

Finally, in our review of geometry, we will present three other coordinate systems often used in engineering analysis. They are the polar (r, θ) coordinate system, the cylindrical (r, θ, z) coordinate system, and the spherical (r, θ, ϕ) coordinate system. The polar coordinate system is restricted to a plane:

$$x = r\cos\theta, \quad y = r\sin\theta. \tag{1.3.15}$$

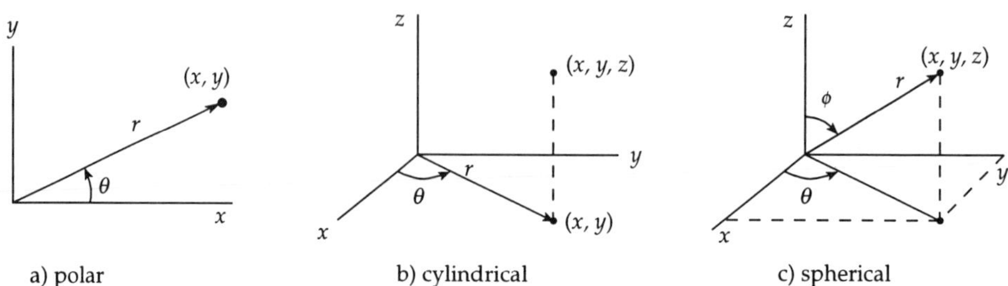

Figure 1.7 The polar, cylindrical and spherical coordinate systems.

For the cylindrical coordinate system

$$x = r\cos\theta, \quad y = r\sin\theta, \quad z = z. \tag{1.3.16}$$

And, for the spherical coordinate system

$$x = r\sin\phi\cos\theta, \quad y = r\sin\phi\sin\theta, \\ z = r\cos\phi. \tag{1.3.17}$$

Example 1.7

What conic section is represented by $2x^2 - 4xy + 5x = 10$?

Solution. Comparing this with the general form Eq. 1.3.7, we see that

$$A = 2, \quad B = -2, \quad C = 0.$$

Thus, $B^2 - AC = 4$, which is greater than zero. Hence, the conic section is a hyperbola.

Example 1.8

Write the general form of the equation of a parabola, vertex at (2, 4), opening upward, with directrix at $y = 2$.

Solution. The equation of the parabola (see Eq. 1.3.11) can be written as

$$(x - x_1)^2 = 2p(y - y_1)$$

where we have interchanged x and y so that the parabola opens upward. For this example, $x_1 = 2$, $y_1 = 4$ and $p = 4$ ($p/2$ is the distance from the vertex to the directrix). Hence, the equation is

$$(x - 2)^2 = 2(4)(y - 4),$$

or, in general form,

$$x^2 - 4x - 8y + 36 = 0.$$

Example 1.9

Express the rectangular coordinates (3, 4, 5) in cylindrical coordinates and spherical coordinates.

Solution. In cylindrical coordinates

$$r = \sqrt{x^2+y^2} = \sqrt{3^2+4^2} = 5,$$

$$\theta = \tan^{-1} y/x = \tan^{-1} 4/3 = 0.927 \text{ rad}.$$

Thus, in cylindrical coordinates, the point is located at (5, 0.927, 5).

In spherical coordinates

$$r = \sqrt{x^2+y^2+z^2} = \sqrt{3^2+4^2+5^2} = 7.071,$$

$$\phi = \cos^{-1} z/r = \cos^{-1} 5/7.071 = 0.785 \text{ rad},$$

$$\theta = \tan^{-1} y/x = \tan^{-1} 4/3 = 0.927 \text{ rad}.$$

Finally, in spherical coordinates, the point is located at (7.071, 0.927, 0.785).

1.4 Complex Numbers

A complex number consists of a real part x and an imaginary part y, written as $x + iy$, where $i = \sqrt{-1}$. (In electrical engineering, however, it is common to let $j = \sqrt{-1}$ since i represents current.) In real number theory, the square root of a negative number does not exist; in complex number theory, we would write $\sqrt{-4} = \sqrt{4(-1)} = 2i$. The complex number may be plotted using the real x-axis and the imaginary y-axis, as shown in Fig. 1.8.

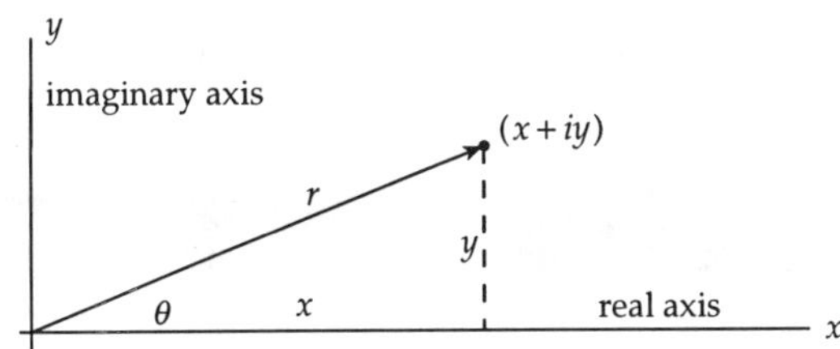

Figure 1.8 The complex number.

It is often useful to express a complex number in polar form as (both $re^{i\theta}$ and $r\angle\theta$ are polar forms)

$$x + iy = re^{i\theta} \quad \text{or} \quad r\angle\theta \tag{1.4.1}$$

where we use *Euler's equation*

$$e^{i\theta} = \cos\theta + i\sin\theta \tag{1.4.2}$$

to verify the relations

$$x = r\cos\theta, \quad y = r\sin\theta. \tag{1.4.3}$$

Note that $e^{i\theta} = e^{i(\theta+2n\pi)}$ where n is an integer. This simply adds 360° (2π radians) to θ and hence in Fig. 1.8 $re^{i\theta}$ and $re^{i(\theta+2n\pi)}$ represent the identical point.

Multiplication and division are accomplished with either form:

$$\begin{aligned}(a+ib)(c+id) &= ac - bd + i(ad+bc)\\ &= r_1e^{i\theta_1}r_2e^{i\theta_2} = r_1r_2e^{i(\theta_1+\theta_2)}\\ &= r_1r_2 \angle\theta_1+\theta_2\end{aligned} \quad \textbf{(1.4.4)}$$

$$\begin{aligned}\frac{a+ib}{c+id} &= \frac{a+ib}{c+id}\frac{c-id}{c-id} = \frac{(a+ib)(c-id)}{c^2+d^2}\\ &= \frac{r_1}{r_2}e^{i(\theta_1-\theta_2)} = \frac{r_1}{r_2}\angle\theta_1-\theta_2\end{aligned} \quad \textbf{(1.4.5)}$$

It is usually easier to find powers and roots of complex numbers using the polar form:

$$(x+iy)^k = r^k e^{ik\theta}, \quad (x+iy)^{1/k} = r^{1/k}e^{i\theta/k} \quad \textbf{(1.4.6)}$$

When finding roots, more than one root results by using $e^{i\theta}$ and $e^{i(\theta+2n\pi)}$. An example illustrates. Remember, in mathematical equations we usually express θ in radians; however, when displaying the angle as $\angle\theta$ it is expressed in degrees.

Using Euler's equation we can show that

$$\sin\theta = \frac{e^{i\theta}-e^{-i\theta}}{2i}, \qquad \cos\theta = \frac{e^{i\theta}+e^{-i\theta}}{2} \quad \textbf{(1.4.7)}$$

Example 1.10

Divide (3 + 4i) by (4 + 3i).

Solution. We perform the division as follows:

$$\frac{3+4i}{4+3i} = \frac{3+4i}{4+3i}\cdot\frac{4-3i}{4-3i} = \frac{12+16i-9i+12}{16+9} = \frac{24+7i}{25} = 0.96+0.28i.$$

Note that we multiplied the numerator and the denominator by the *complex conjugate* of the denominator. A complex conjugate is formed simply by changing the sign of the imaginary part.

Example 1.11

Find $(2+3i)^4$.

Solution. This can be done using the polar form. Hence,

$$r = \sqrt{2^2+3^2} = \sqrt{13}, \quad \theta = \tan^{-1}3/2 = 0.9828 \text{ rad}.$$

We normally express θ in radians. The complex number, in polar form, is

$$2+3i = \sqrt{13}\,e^{0.9828i}.$$

Thus,

$$(2+3i)^4 = \left(\sqrt{13}\right)^4 e^{4(0.9828i)} = 169\,e^{3.9312i}.$$

Converting back to rectangular form we have

$$169\,e^{3.9312i} = 169(\cos 3.9312 + i\sin 3.9312)$$

$$= \quad 169(-0.7041 - 0.7101i) = -119 - 120i.$$

Alternatively, we recognize that this could be expressed in polar form as $169\,\angle 225.2^\circ$ or $\angle -134.8^\circ$, where a positive angle is measured counterclockwise.

Example 1.12

Find the three roots of $2 + 3i$.

Solution. We express the complex number (see Example 1.11) in polar form as

$$2 + 3i = \sqrt{13}\,e^{0.9828i}.$$

Since the trig functions are periodic, we know that

$$\sin\theta = \sin(\theta + 2\pi) = \sin(\theta + 4\pi)$$

$$\cos\theta = \cos(\theta + 2\pi) = \cos(\theta + 4\pi).$$

Thus, in addition to the first form, we have

$$2 + 3i = \sqrt{13}\,e^{(0.9828+2\pi)i} = \sqrt{13}\,e^{(0.9828+4\pi)i}.$$

Taking the one-third root of each form, we find the three roots to be

$$(2+3i)^{1/3} = \left(\sqrt{13}\right)^{1/3} e^{0.3276i}$$

$$= \quad 1.533(0.9468 + 0.3218i) = 1.452 + 0.4935i \quad \text{or} \quad 1.533\,\angle 18.77^\circ.$$

$$(2+3i)^{1/3} = \left(\sqrt{13}\right)^{1/3} e^{2.422i}$$

$$= \quad 1.533(-0.7521 + 0.6591i) = -1.153 + 0.010i \quad \text{or} \quad 1.533\,\angle 138.77^\circ.$$

$$(2+3i)^{1/3} = \left(\sqrt{13}\right)^{1/3} e^{4.516i}$$

$$= \quad 1.533(-0.1951 - 0.9808i) = -0.2991 - 1.504i \quad \text{or} \quad 1.533\,\angle 258.77^\circ.$$

If we added 6π to the angle we would be repeating the first root, so obviously this is not done.

1.5 Linear Algebra

The primary objective in linear algebra is to find the solution to a set of n linear algebraic equations for n unknowns. To do this we must learn how to manipulate a matrix, a rectangular array of quantities arranged into rows and columns.

An $m \times n$ matrix has m rows (the horizontal lines) and n columns (the vertical lines). An $m \times n$ matrix multiplied by an $n \times s$ matrix produces an $m \times s$ matrix. When multiplying two matrices the columns of the first matrix must equal the rows of the second. Their product is a third matrix:

$$\left[c_{ij}\right] = \sum_{k=1}^{n}\left[a_{ik}\right]\left[b_{kj}\right] \tag{1.5.1}$$

We are primarily interested in square matrices since we usually have the same number of equations as unknowns, such as

$$\begin{aligned}
a_{11}x_1 + a_{12}x_2 + a_{13}x_3 + a_{14}x_4 &= r_1\\
a_{21}x_1 + a_{22}x_2 + a_{23}x_3 + a_{24}x_4 &= r_2\\
a_{31}x_1 + a_{32}x_2 + a_{33}x_3 + a_{34}x_4 &= r_3\\
a_{41}x_1 + a_{42}x_2 + a_{43}x_3 + a_{44}x_4 &= r_4.
\end{aligned} \tag{1.5.2}$$

In matrix form this can be written as

$$\left[a_{ij}\right]\left[x_j\right] = \left[r_i\right] \quad \text{or} \quad \mathbf{Ax} = \mathbf{r} \tag{1.5.3}$$

where $\left[x_j\right]$ and $\left[r_i\right]$ are column matrices. (A column matrix is often referred to as a *vector*.) The coefficient matrix $\left[a_{ij}\right]$ and the column matrix $\left[r_i\right]$ are assumed to be known quantities. The solution $\left[x_j\right]$ is expressed as

$$\left[x_j\right] = \left[a_{ij}\right]^{-1}\left[r_i\right] \quad \text{or} \quad \mathbf{x} = \mathbf{A}^{-1}\mathbf{r} \tag{1.5.4}$$

where $[a_{ij}]^{-1}$ is the *inverse* matrix of $\left[a_{ij}\right]$. It is defined as

$$\left[a_{ij}\right]^{-1} = \frac{\left[a_{ij}\right]^{+}}{\left|a_{ij}\right|} \quad \text{or} \quad \mathbf{A}^{-1} = \frac{\mathbf{A}^{+}}{|\mathbf{A}|} \tag{1.5.5}$$

where $[a_{ij}]^{+}$ is the *adjoint* matrix and $\left|a_{ij}\right|$ is the *determinant* of $\left[a_{ij}\right]$. Let us review how the determinant and the adjoint are evaluated.

In general, the determinant may be found using the *cofactor* A_{ij} of the element a_{ij}. The cofactor is defined to be $(-1)^{i+j}$ times the *minor*, the determinant obtained by deleting the ith row and the jth column. The determinant is then

$$\left|a_{ij}\right| = \sum_{j=1}^{n} a_{ij}\, A_{ij}\,, \tag{1.5.6}$$

where i is any value from 1 to n. Recall that the third-order determinant can be evaluated by writing the first two columns after the determinant and then summing the products of the elements of the diagonals, using negative signs with the diagonals sloping upward.

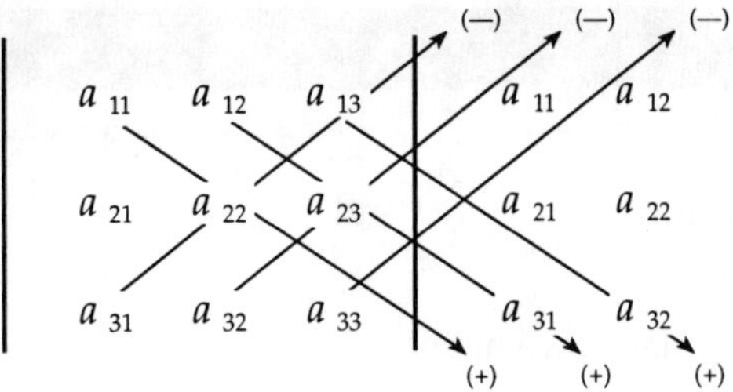

The elements of the adjoint $[a_{ij}]^+$ are the cofactors A_{ij} of the elements a_{ij}; for the matrix $[a_{ij}]$ of Eq. 1.5.2 we have

$$\left[a_{ij}\right]^+ = \begin{bmatrix} A_{11} & A_{21} & A_{31} & A_{41} \\ A_{12} & A_{22} & A_{32} & A_{42} \\ A_{13} & A_{23} & A_{33} & A_{43} \\ A_{14} & A_{24} & A_{34} & A_{44} \end{bmatrix} \tag{1.5.7}$$

Note that A_{ij} takes the position of a_{ji}.

Finally, the solution $[x_j]$ of Eq. 1.5.4 results if we multiply the square matrix $\left[a_{ij}\right]^{-1}$ by the column matrix $\left[r_i\right]$. In general, we multiply the elements in each left-hand matrix row by the elements in each right-hand matrix column, add the products, and place the sum at the location where the row and column intersect. The following examples will illustrate.

Before we work some examples though, we should point out that the above matrix presentation can also be presented as *Cramer's rule*, which states that the solution element x_n can be expressed as

$$x_n = \frac{\left|b_{ij}\right|}{\left|a_{ij}\right|} \tag{1.5.8}$$

where $\left|b_{ij}\right|$ is formed by replacing the nth column of $\left|a_{ij}\right|$ with the elements of the column matrix $\left[r_i\right]$.

Notes: If the system of equations is homogeneous, i.e., $r_i = 0$, a solution may exist if $\left|a_{ij}\right| = 0$. If the determinant of a matrix is zero, that matrix is *singular* and its inverse does not exist.

In the solution of a system of first-order differential equations, we encounter the matrix equation

$$(\mathbf{A} - \lambda\mathbf{I})\mathbf{x} = \mathbf{O} \tag{1.5.9}$$

The scalar λ is the *eigenvalue* and the vector **x** is the *eigenvector* associated with the eigenvalue. The matrix **I** is called the *unit matrix*, which for a 2x2 matrix is $\begin{bmatrix} 1 & 0 \\ 0 & 1 \end{bmatrix}$. If **A** is 2x2, λ has two distinct values, and if **A** is 3x3 it has three distinct values. Since $r_i = 0$ in Eq. 1.5.9, the scalar equation

$$\left|\mathbf{A} - \lambda\mathbf{I}\right| = 0 \tag{1.5.10}$$

provides the equation that yields the eigenvalues. Then Eq. 1.5.9 is solved with each eigenvalue to give the eigenvectors. We often say that λ represents the eigenvalues of the matrix **A**.

Example 1.13

Calculate the determinants of $\begin{bmatrix} 2 & -3 \\ 1 & 4 \end{bmatrix}$ and $\begin{bmatrix} 2 & 3 & 0 \\ 1 & 4 & -2 \\ 0 & 3 & 5 \end{bmatrix}$.

Solution. For the first matrix we have

$$\begin{vmatrix} 2 & -3 \\ 1 & 4 \end{vmatrix} = 2\times 4 - 1(-3) = 11.$$

The second matrix is set up as follows:

$$\left|\begin{matrix} 2 & 3 & 0 \\ 1 & 4 & -2 \\ 0 & 3 & 5 \end{matrix}\right|\begin{matrix} 2 & 3 \\ 1 & 4 \\ 0 & 3 \end{matrix} = 40+0+0-0-(-12)-15 = 37.$$

(Diagonals running down to the right are marked (+), (+), (+); diagonals running up to the right are marked (−), (−), (−).)

Example 1.14

Expanding with cofactors, evaluate the determinant $D = \begin{vmatrix} 1 & 0 & -2 \\ -1 & 2 & 0 \\ 1 & 2 & 1 \end{vmatrix}$.

Solution. Choosing the first row ($i = 1$ in Eq. 1.5.6) we have

$$D = 1\begin{vmatrix} 2 & 0 \\ 2 & 1 \end{vmatrix}(-1)^2 - 0\begin{vmatrix} -1 & 0 \\ 1 & 1 \end{vmatrix}(-1)^3 + (-2)\begin{vmatrix} -1 & 2 \\ 1 & 2 \end{vmatrix}(-1)^4$$
$$= 2 + 0 - 2(-4) = 10.$$

Note: If any two columns (or rows) are a multiple of each other, a determinant is zero.

Example 1.15

Find the adjoint of the matrix if

$$[a_{ij}] = [b_{in}][c_{nj}] \quad \text{where } [b_{in}] = \begin{bmatrix} 0 & 0 & 1 \\ 4 & 1 & -2 \\ -1 & 1 & 0 \end{bmatrix} \text{ and } [c_{nj}] = \begin{bmatrix} 0 & 0 & -1 \\ 1 & 2 & 0 \\ 1 & 0 & -2 \end{bmatrix}.$$

Solution. First, multiply the two matrices using Eq. 1.5.1. If we desire a_{23} we use the second row of $[b_{in}]$ and the third column of $[c_{nj}]$ so that

$$a_{23} = b_{21}c_{13} + b_{22}c_{23} + b_{23}c_{33} = 4(-1) + 1(0) + (-2)(-2) = 0.$$

Doing this for all elements of $[a_{ij}]$ we find

$$[a_{ij}] = \begin{bmatrix} 1 & 0 & -2 \\ -1 & 2 & 0 \\ 1 & 2 & 1 \end{bmatrix}.$$

The cofactor of each element of $[a_{ij}]$ must be determined. The cofactor is found by multiplying $(-1)^{i+j}$ times the determinant formed by deleting the ith row and the jth column. They are found to be

$$A_{11} = 2, \quad A_{12} = 1, \quad A_{13} = -4$$

$$A_{21} = -4, \quad A_{22} = 3, \quad A_{23} = -2$$

$$A_{31} = 4, \quad A_{32} = 2, \quad A_{33} = 2.$$

The adjoint is then

$$[a_{ij}]^+ = [A_{ji}] = \begin{bmatrix} 2 & -4 & 4 \\ 1 & 3 & 2 \\ -4 & -2 & 2 \end{bmatrix}.$$

Note: The matrix $[A_{ji}]$ is called the *transpose* of $[A_{ij}]$, i.e., $[A_{ji}] = [A_{ij}]^T$.

Example 1.16

Find the inverse of the matrix $[a_{ij}] = \begin{bmatrix} 1 & 0 & -2 \\ -1 & 2 & 0 \\ 1 & 2 & 1 \end{bmatrix}$.

Solution. The inverse is defined to be the adjoint matrix divided by the determinant $|a_{ij}|$. Hence, the inverse is (see Examples 1.14 and 1.15)

$$[a_{ij}]^{-1} = \frac{1}{10}\begin{bmatrix} 2 & -4 & 4 \\ 1 & 3 & 2 \\ -4 & -2 & 2 \end{bmatrix} = \begin{bmatrix} 0.2 & -0.4 & 0.4 \\ 0.1 & 0.3 & 0.2 \\ -0.4 & -0.2 & 0.2 \end{bmatrix}.$$

Example 1.17

Find the solution to

$$\begin{aligned} x_1 \qquad\quad - 2x_3 &= 2 \\ -x_1 + 2x_2 \qquad\quad &= 0 \\ x_1 + 2x_2 + x_3 &= -4 \end{aligned} \quad .$$

Solution. The solution matrix is (see Example 1.14 and 1.15 or use $[a]^{-1}$ fromExample 1.16)

$$[x_j] = [a_{ij}]^{-1}[r_i] = \frac{[a_{ij}]^+}{|a_{ij}|}[r_i]$$

$$= \frac{1}{10}\begin{bmatrix} 2 & -4 & 4 \\ 1 & 3 & 2 \\ -4 & -2 & 2 \end{bmatrix}\begin{bmatrix} 2 \\ 0 \\ -4 \end{bmatrix}.$$

First, let's multiply the two matrices; they are multiplied row by column as follows:

$$\begin{aligned} 2\cdot 2+(-4)\cdot 0+4\cdot(-4) &= -12 \\ 1\cdot 2+3\cdot 0+2\cdot(-4) &= -6 \\ -4\cdot 2-2\cdot 0+2\cdot(-4) &= -16 \end{aligned}$$

The solution vector is then

$$[x_i] = \frac{1}{10}\begin{bmatrix} -12 \\ -6 \\ -16 \end{bmatrix} = \begin{bmatrix} -1.2 \\ -0.6 \\ -1.6 \end{bmatrix}.$$

In component form, the solution is

$$x_1 = -1.2, \qquad x_2 = -0.6, \qquad x_3 = -1.6.$$

Example 1.18

Use Cramer's rule and solve

$$\begin{aligned} x_1 \quad\quad\quad - 2x_3 &= 2 \\ -x_1 + 2x_2 \quad\quad\quad &= 0 \\ x_1 + 2x_2 + x_3 &= -4 \end{aligned}$$

Solution. The solution is found (see Example 1.14) by evaluating the ratios as follows:

$$x_1 = \frac{\begin{vmatrix} 2 & 0 & -2 \\ 0 & 2 & 0 \\ -4 & 2 & 1 \end{vmatrix}}{D} = \frac{-12}{10} = -1.2 \qquad x_2 = \frac{\begin{vmatrix} 1 & 2 & -2 \\ -1 & 0 & 0 \\ 1 & -4 & 1 \end{vmatrix}}{D} = \frac{-6}{10} = -0.6$$

$$x_3 = \frac{\begin{vmatrix} 1 & 0 & 2 \\ -1 & 2 & 0 \\ 1 & 2 & -4 \end{vmatrix}}{D} = \frac{-16}{10} = -1.6$$

where

$$D = \begin{vmatrix} 1 & 0 & -2 \\ -1 & 2 & 0 \\ 1 & 2 & 1 \end{vmatrix} = 10.$$

Note that the numerator is the determinant formed by replacing the ith column with right-hand side elements r_i when solving for x_i.

Example 1.19

Find the eignevalues of $\mathbf{A} = \begin{bmatrix} 4 & 0 & 2 \\ 0 & 8 & 0 \\ 3 & 0 & 5 \end{bmatrix}$ and one of its eigenvectors.

Solution. To find the eigenvalues of the matrix, we form the equation

$$|\mathbf{A} - \lambda \mathbf{I}| = 0 \qquad \text{or} \qquad \begin{vmatrix} 4-\lambda & 0 & 2 \\ 0 & 8-\lambda & 0 \\ 3 & 0 & 5-\lambda \end{vmatrix} = 0$$

Expanding the determinant using cofactors we have

$$(8-\lambda)\begin{vmatrix} 4-\lambda & 2 \\ 3 & 5-\lambda \end{vmatrix} = (8-\lambda)[(4-\lambda)(5-\lambda)-6] = (8-\lambda)[\lambda^2 - 9\lambda + 14] = 0 \ .$$

or

$$(8-\lambda)(\lambda-7)(\lambda-2) = 0.$$

The eigenvalues are then $\lambda = 8, 7, 2$.
To find an eigenvector, use $\lambda = 2$. Then Eq. 1.5.9 is

$$(\mathbf{A} - 2\mathbf{I})\mathbf{x} = \mathbf{0} \qquad \text{or} \qquad \begin{bmatrix} 2 & 0 & 2 \\ 0 & 6 & 0 \\ 3 & 0 & 3 \end{bmatrix}\begin{bmatrix} x_1 \\ x_2 \\ x_3 \end{bmatrix} = \begin{bmatrix} 0 \\ 0 \\ 0 \end{bmatrix}$$

Write out the three equations:

$$\begin{aligned} 2x_1 + 2x_3 &= 0 \\ 6x_2 &= 0 \\ 3x_1 + 3x_3 &= 0 \end{aligned}$$

The only solution is

$$\begin{aligned} x_2 &= 0 \\ x_1 &= -x_3 \end{aligned}$$

An eigenvector is then

$$\mathbf{x} = \begin{bmatrix} 1 \\ 0 \\ -1 \end{bmatrix}$$

Note that eigenvectors are not unique. We simply select $x_1 = 1$ so that $x_3 = -1$. We could have selected $x_1 = -1$ so that $x_3 = 1$, an equally valid result. Or we could select x_1, x_2, and x_3 so that a unit vector results.

1.6 Calculus

Differentiation

The slope of a curve $y = f(x)$ is the ratio of the change in y to the change in x as the change in x becomes infinitesimally small. This is the first derivative, written as

$$\frac{dy}{dx} = \lim_{\Delta x \to 0} \frac{\Delta y}{\Delta x}. \quad (1.6.1)$$

This may be written using abbreviated notation as

$$\frac{dy}{dx} = Dy = y' = \dot{y}. \quad (1.6.2)$$

The second derivative is written as

$$\frac{d^2y}{dx^2} = D^2y = y'' = \ddot{y}, \quad (1.6.3)$$

and is defined by

$$\frac{d^2y}{dx^2} = \lim_{\Delta x \to 0} \frac{\Delta y'}{\Delta x}. \quad (1.6.4)$$

Some derivative formulas, where f and g are functions of x, and k is constant, are given below.

$$\begin{aligned}
\frac{dk}{dx} &= 0 \\
\frac{d\left(k\,x^n\right)}{dx} &= k\,n\,x^{n-1} \\
\frac{d}{dx}(f+g) &= f' + g' \\
\frac{df^n}{dx} &= n\,f^{n-1}\,f' \\
\frac{d}{dx}(fg) &= fg' + gf' \\
\frac{d}{dx}(\ln x) &= \frac{1}{x} \\
\frac{d}{dx}\left(e^{kx}\right) &= ke^{kx} \\
\frac{d}{dx}(\sin x) &= \cos x \\
\frac{d}{dx}(\cos x) &= -\sin x
\end{aligned} \quad (1.6.5)$$

If a function f depends on more than one variable, partial derivatives are used. If $z = f(x, y)$, then $\partial z / \partial x$ is the derivative of z with respect to x holding y constant. It would represent the slope of a line tangent to the surface in a plane of constant y.

Maxima and Minima

Derivatives are used to locate points of inflection, maxima, and minima. Note the following:

$$\begin{aligned} f'(x) &= 0 \text{ at a maximum or a minimum.} \\ f''(x) &= 0 \text{ at an inflection point.} \\ f''(x) &> 0 \text{ at a minimum.} \\ f''(x) &< 0 \text{ at a maximum.} \end{aligned}$$

An inflection point always exists between a maximum and a minimum.

L'Hospital's Rule

Differentiation is also useful in establishing the limit of $f(x)/g(x)$ as $x \to a$ if $f(a)$ and $g(a)$ are both zero or $\pm\infty$. *L'Hospital's rule* is used in such cases and is as follows:

$$\lim_{x\to a}\frac{f(x)}{g(x)} = \lim_{x\to a}\frac{f'(x)}{g'(x)} = \lim_{x\to a}\frac{f''(x)}{g''(x)}. \quad \textbf{(1.6.6)}$$

Taylor's Series

Derivatives are used to expand a continuous function as a power series around $x = a$. Taylor's series is as follows:

$$f(x) = f(a) + (x-a)f'(a) + (x-a)^2 f''(a)/2! + \cdots. \quad \textbf{(1.6.7)}$$

This series is often used to express a function as a polynomial near a point $x = a$ providing the series can be truncated after a few terms. Using Taylor's series we can show that (expanding about $a = 0$):

$$\begin{aligned} \sin x &= x - x^3/3! + x^5/5! - \cdots \\ \cos x &= 1 - x^2/2! + x^4/4! - \cdots \\ \ln(1+x) &= x - x^2/2 + x^3/3 - \cdots \\ \frac{1}{1-x} &= 1 + x + x^2 + \cdots \\ e^x &= 1 + x + x^2/2! + x^3/3! + \cdots \end{aligned} \quad \textbf{(1.6.8)}$$

When expanding with $a = 0$, the Taylor series may be called a *Maclaurin series*.

Integration

The inverse of differentiation is the process called integration. If a curve is given by $y = f(x)$, then the area under the curve from $x = a$ to $x = b$ is given by

$$A = \int_a^b y\,dx. \quad \textbf{(1.6.9)}$$

The length of the curve between the two points is expressed as

$$L = \int_a^b \left(1 + y'^2\right)^{1/2} dx. \quad \textbf{(1.6.10)}$$

Volumes of various objects are also found by an appropriate integration.

If the integral has limits, it is a *definite integral*; if it does not have limits, it is an *indefinite integral* and a constant is always added. Some common indefinite integrals follow:

$$\int dx = x + C.$$

$$\int cy dx = c\int y dx$$

$$\int x^n dx = \frac{x^{n+1}}{n+1} + C \quad n \neq -1$$

$$\int x^{-1} dx = \ln x + C$$

$$\int e^{ax} dx = \frac{1}{a} e^{ax} + C \tag{1.6.11}$$

$$\int \sin x \, dx = -\cos x + C$$

$$\int \cos x \, dx = \sin x + C$$

$$\int \cos^2 x \, dx = \frac{x}{2} + \frac{1}{4}\sin 2x + C$$

$$\int u \, dv = uv - \int v \, du.$$

This last integral is often referred to as "integration by parts." If the integrand (the coefficients of the differential) is not one of the above, then in the last integral, $\int v \, du$ may in fact be integrable. An example will illustrate.

Example 1.20

Find the slope of $y = x^2 + \sin x$ at $x = 0.5$.

Solution. The derivative is the slope:

$$y'(x) = 2x + \cos x.$$

At $x = 0.5$ the slope is

$$y'(0.5) = 2 \cdot 0.5 + \cos 0.5 = 1.878.$$

Example 1.21

Find $\frac{d}{dx}(\tan x)$.

Solution. Writing $\tan x = \sin x / \cos x = f(x) \cdot g(x)$ we find

$$\frac{d}{dx}(\tan x) = \frac{1}{\cos x}\frac{d}{dx}(\sin x) + \sin x \frac{d}{dx}(\cos x)^{-1}$$

$$= \frac{\cos x}{\cos x} + \frac{\sin^2 x}{\cos^2 x} = 1 + \tan^2 x$$

$$= \frac{\cos^2 x + \sin^2 x}{\cos^2 x} = \frac{1}{\cos^2 x} = \sec^2 x.$$

Either expression is acceptable although the latter is usually used.

Example 1.22

Locate the maximum and minimum points of the function $y(x) = x^3 - 12x - 9$ and evaluate y at those points.

Solution. The derivative is

$$y'(x) = 3x^2 - 12.$$

The points at which $y'(x) = 0$ are at

$$x = 2, -2.$$

At these two points the extrema are:

$$y_{\min} = (2)^3 - 12 \cdot 2 - 9 = -25$$

$$y_{\max} = (-2)^3 - 12(-2) - 9 = 7.$$

Let us check the second derivative. At the two points we have:

$$y''(2) = 6 \cdot 2 = 12$$

$$y''(-2) = 6 \cdot (-2) = -12$$

Obviously, the point $x = 2$ is a minimum since its second derivative is positive there.

Example 1.23

Find the limit as $x \to 0$ of $\sin x / x$.

Solution. If we let $x = 0$ we are faced with the ratio of 0/0, an indeterminate quantity. Hence, we use L'Hospital's rule and differentiate both numerator and denominator to obtain

$$\lim_{x \to 0} \frac{\sin x}{x} = \lim_{x \to 0} \frac{\cos x}{1}.$$

Now, we let $x = 0$ and find

$$\lim_{x \to 0} \frac{\sin x}{x} = \frac{1}{1} = 1.$$

Example 1.24

Verify that $\sin x = x - x^3/3! + x^5/5! - \cdots$.

Solution. We expand in a Taylor's series about $x = 0$:

$$f(x) = f(0) + x f'(0) + x^2 f''(0)/2! + \cdots.$$

Letting $f(x) = \sin x$ so that $f' = \cos x$, $f'' = -\sin x$, etc., there results

$$\sin x = 0 + x(1) + x^2(0/2!) + x^3(-1)/3! + \cdots$$

$$= x - x^3/3! + x^5/5! - \cdots.$$

Example 1.25

Find the area of the shaded area in the figure.

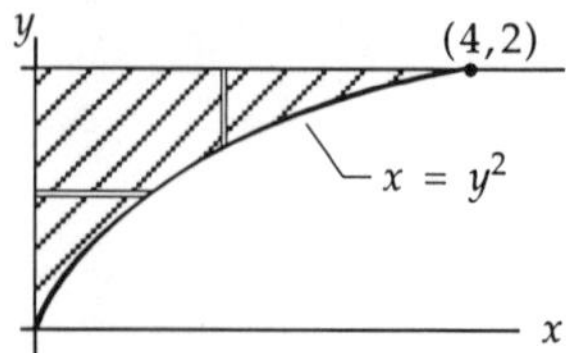

Solution. We can find this area by using either a horizontal strip or a vertical strip. We will use both. First, for a horizontal strip:

$$\begin{aligned} A &= \int_0^2 x dy \\ &= \int_0^2 y^2 dy = \left.\frac{y^3}{3}\right|_0^2 = 8/3. \end{aligned}$$

Using a vertical strip we have

$$\begin{aligned} A &= \int_0^4 (2-y)\, dx\,, \\ &= \int_0^4 \left(2-x^{1/2}\right) dx = \left[2x - \frac{2}{3}x^{3/2}\right]_0^4 = 8 - \frac{2}{3}8 = 8/3. \end{aligned}$$

Either technique is acceptable. The first appears to be the simpler one.

Example 1.26

Find the volume enclosed by rotating the shaded area of Example 1.25 about the y-axis.

Solution. If we rotate the horizontal strip about the y-axis we will obtain a disc with volume

$$dV = \pi x^2 dy.$$

This can be integrated to give the volume, which is

$$\begin{aligned} V &= \int_0^2 \pi x^2 dy \\ &= \pi \int_0^2 y^4 dy = \left.\frac{\pi y^5}{5}\right|_0^2 = \frac{32\pi}{5}. \end{aligned}$$

Now, let us rotate the vertical strip about the y-axis to form a cylinder with volume

$$dV = 2\pi x(2-y)dx.$$

This can be integrated to yield

$$
\begin{aligned}
V &= \int_0^4 2\pi x(2-y)dx \\
&= \int_0^4 2\pi x\left(2-x^{1/2}\right)dx = 2\pi\left[x^2 - \frac{2x^{5/2}}{5}\right]_0^4 = \frac{32\pi}{5}.
\end{aligned}
$$

Again, the horizontal strip is simpler.

Example 1.27

Show that $\int xe^x dx = (x-1)e^x + C$.

Solution. Let's attempt the last integral of (1.6.11). Define the following:

$$u = x, \quad dv = e^x dx$$

Then,

$$du = dx, \quad v = \int e^x dx = e^x$$

and we find that

$$
\begin{aligned}
\int xe^x dx &= x e^x - \int e^x dx \\
&= x e^x - e^x + C = (x-1)e^x + C.
\end{aligned}
$$

1.7 Differential Equations

A differential equation is *linear* if no term contains the dependent variable or any of its derivatives to a power other than one (terms that do not contain the dependent variable are not considered in the test of linearity). For example,

$$y'' + 2xy' - y\sin x = 3x^2 \tag{1.7.1}$$

is a linear differential equation. The dependent variable is y and the independent variable is x. If a term contained y'^2, or $y^{1/2}$ or $\sin y$ the equation would be nonlinear. We use either primes or dots to denote a derivative, i.e., $\dot{y} = y' = dy/dx$.

A differential equation is *homogeneous* if all of its terms contain the dependent variable. Eq. 1.7.1 is *nonhomogeneous* because of the term $3x^2$.

The *order* of a differential equation is established by its highest order derivative. Eq. 1.7.1 is a second order differential equation.

The general solution of a differential equation involves a number of arbitrary constants equal to the order of the equation. If conditions are specified, the arbitrary constants may be calculated.

First Order

A first order differential equation is *separable* if it can be expressed as

$$M(x)\,dx + N(y)\,dy = 0. \tag{1.7.2}$$

The solution follows by integrating each of the terms. This equation is, in general, nonlinear.

If a first order differential equation is not separable, we will consider only those that are linear with a constant coefficient, written as

$$y'+ay = f(x) \tag{1.7.3}$$

For the homogeneous equation, $f(x) = 0$, it has the solution

$$y(x) = Ce^{-ax} \tag{1.7.4}$$

The constant C would be determined from an imposed condition

$$y(0) = C \tag{1.7.5}$$

Second Order

The general form of a second order, linear, homogeneous differential equation with constant coefficients is

$$y'' + 2ay' + by = 0. \tag{1.7.6}$$

To find a solution we assume $y = e^{rx}$ and solve the *characteristic equation*

$$r^2 + 2ar + b = 0. \tag{1.7.7}$$

Let r_1 and r_2 be its roots. If $r_1 \neq r_2$ and both are real, the general solution is

$$y(x) = C_1 e^{r_1 x} + C_2 e^{r_2 x}. \tag{1.7.8}$$

If $r_1 = r_2$, the general solution is

$$y(x) = C_1 e^{r_1 x} + C_2 xe^{r_2 x}. \tag{1.7.9}$$

Finally, if $r_1 = \alpha + i\beta$ and $r_2 = \alpha - i\beta$, the general solution is

$$y(x) = (C_1 \sin \beta x + C_2 \cos \beta x)\, e^{-ax} \tag{1.7.10}$$

where $\beta = \sqrt{b - a^2}$. and $\alpha = -a$.

Linear, Nonhomogeneous, with Constant Coefficients

If Eq. 1.7.6 were nonhomogeneous, it would be written as

$$y'' + 2ay' + by = f(x). \tag{1.7.11}$$

The general solution is found by finding the solution $y_h(x)$ to the homogeneous equation (simply let the right-hand side be zero and solve the equation as in Section 1.7.2) and adding to it a particular solution $y_p(x)$ found by using Table 1.2. This is expressed as

$$y(x) = y_h(x) + y_p(x). \tag{1.7.12}$$

TABLE 1.2 Particular Solutions

f(x)	$y_p(x)$	provisions
A	C	
$Ax + B$	$Cx + D$	
e^{Ax}	Ce^{Ax}	if r_1 or $r_2 \neq a$
	Cxe^{Ax}	if r_1 or $r_2 = a$
$B \sin Ax$	$C \sin Ax + D \cos Ax$	if $r_{1,2} \neq \pm ai$
	$Cx \sin Ax + Dx \cos Ax$	if $r_{1,2} = \pm ai$
$B \cos Ax$	(same as above)	

Example 1.28

Find the general solution to $2xy' + y^2 = 2y$.

Solution. Write the differential equation as

$$2x\frac{dy}{dx} = 2y - y^2$$

This is separable and written as

$$\frac{2dy}{y(2-y)} = \frac{dx}{x}$$

This is rewritten as

$$\frac{dy}{y} + \frac{dy}{2-y} = \frac{dx}{x}$$

Integrating provides a solution:

$$\ln y - \ln(2-y) = \ln x - \ln C$$

Rewrite this as

$$\ln\frac{Cy}{2-y} = \ln x$$

Finally, the solution is best written as

$$\frac{Cy}{2-y} = x \quad \text{or} \quad y(x) = \frac{2x}{C+x}$$

Example 1.29

Find the solution to

$$y' + 2y = 4x, \qquad y(0) = 2$$

Solution. We will use the method outlined in Section 1.7.2. The characteristic equation of the homogeneous equation is

$$r + 2 = 0. \qquad \therefore r = -2$$

The homogeneous solution is then

$$y_h(x) = Ce^{-2x}$$

The particular solution is assumed to be of the form

$$y_p(x) = Ax + B$$

Substituting this into the original differential equation gives

$$A + 2(Ax + B) = 4x. \qquad \therefore A = 2, \quad B = -1$$

The solution is then

$$\begin{aligned} y(x) &= y_h(x) + y_p(x) \\ &= Ce^{-2x} + 2x - 1 \end{aligned}$$

Using the given condition

$$y(0) = 2 = C - 1. \quad \therefore C = 3$$

Finally,

$$y(x) = 3e^{-2x} + 2x - 1$$

Example 1.30

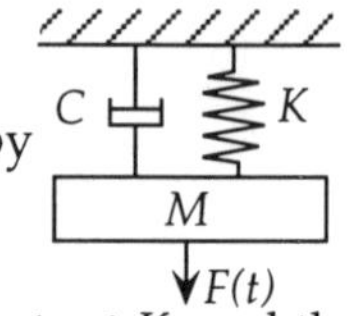

The motion of a simple spring-mass system is represented by

$$M\ddot{y} + C\dot{y} + Ky = F(t)$$

where the mass M, the damping coefficient C, the spring constant K, and the forcing function $F(t)$ have the appropriate units. Find the general solution if $M = 2$, $C = 0$, $K = 50$, and $F(t) = 0$.

Solution. The differential equation simplifies to $2\ddot{y} + 50y = 0$.

The characteristic equation is then

$$2r^2 + 50 = 0$$

$$\therefore r_1 = 5i, \quad r_2 = -5i$$

The solution is then (see Eq. 1.7.10)

$$y(t) = C_1 \sin 5t + C_2 \cos 5t$$

This situation with zero damping represents simple *harmonic motion.*

Note that we have used dots to represent the time derivative $dy/dt = \dot{y}$. Also, note that the coefficient of t in the undamped system is the *natural frequency* $\omega_n = \sqrt{K/M}$; the frequency in hertz is $f = \omega_n/2\pi$; and the *period* is $\tau = f^{-1} = 2\pi/\omega_n$. For this problem $\omega_n = 5$ rad/s, $f = 5/2\pi$ Hz, and $\tau = 2\pi/5$ seconds.

Example 1.31

In the equation of Ex. 1.30, let $M = 2$, $C = 12$, $K = 50$, and $F(t) = 60 \sin 5t$. Find the general solution.

Solution. The differential equation is

$$2\ddot{y} + 12\dot{y} + 50y = 60 \sin 5t.$$

The characteristic equation of the homogeneous differential equation is found by letting $y = e^{rx}$:

$$2r^2 + 12r + 50 = 0.$$

$$\therefore r_1 = -3 + 4i, \quad r_2 = -3 - 4i.$$

The homogeneous solution is (see Eq. 1.7.10)

$$y_h(t) = e^{-3t}(C_1 \sin 4t + C_2 \cos 4t).$$

The particular solution is found by assuming that

$$y_p(t) = A\sin 5t + B\cos 5t\,.$$

Substitute this into the original differential equation:

$$2[-25A\sin 5t - 25B\cos 5t] + 12[5A\cos 5t - 5B\sin 5t] + 50[A\sin 5t + B\cos 5t] = 60\sin 5t$$

Equating coefficients of sin terms and then cos terms:

$$-50A - 60B + 50A = 60. \quad \therefore B = -1$$

$$-50B + 60A + 50B = 0. \quad \therefore A = 0\,.$$

The two solutions are superposed as in Eq. 1.7.12 to give

$$y(t) = (C_1\sin 4t + C_2\cos 4t)\,e^{-3t} - \cos 5t\,.$$

Laplace Transforms

The solution of nonhomogeneous, constant coefficient, linear differential equations can be obtained quite easily using Laplace transforms. The initial conditions must be given and with them we obtain the solution directly; we do not superpose homogeneous and particular solutions.

The *Laplace transform F(s)* of *f(t)* is defined as

$$F(s) = \int_0^\infty f(t)e^{-st}dt\,. \qquad \textbf{(1.7.13)}$$

We also use the notation

$$L(f) = F(s) \qquad \textbf{(1.7.14)}$$

where $L(f)$ is the Laplace transform of *f(t)*.

By applying Eq. 1.7.13 to a variety of functions, including derivatives, we can obtain the Laplace transforms presented in Table 1.3.

Table 1.3 Laplace Transforms

f(t)	*F(s)*	*f(t)*	*F(s)*
(t), Impulse at $t = 0$	1	$\dfrac{d^n f(t)}{dt^n}$	$s^n F(s) - \sum_{m=0}^{n-1} s^{n-m-1}\dfrac{d^m f(0)}{dt^m}$
$u(t)$, Step at $t = 0$	$1/s$		
$t\,u(t)$, Ramp at $t = 0$	$1/s^2$	$\int_0^t f(\tau)d\tau$	$(1/s)F(s)$
$e^{-\alpha t}$	$1/(s+\alpha)$	$\int_0^t f(t-\tau)h(\tau)d\tau$	$F(s)H(s)$
$te^{-\alpha t}$	$1/(s+\alpha)^2$	$\lim_{t\to\infty} f(t)$	$\lim_{s\to 0} sF(s)$
$e^{-\alpha t}\sin\beta t$	$\beta/\left[(s+\alpha)^2+\beta^2\right]$	$\lim_{t\to 0} f(t)$	$\lim_{s\to\infty} sF(s)$
$e^{-\alpha t}\cos\beta t$	$(s+\alpha)/\left[(s+\alpha)^2+\beta^2\right]$		
$t\sin\beta t$	$2bs/\left(s^2+\beta^2\right)^2$		
$t\cos\beta t$	$\left(s^2-\beta^2\right)/\left(s^2+\beta^2\right)^2$		

We can now use Table 1.3 to solve ordinary differential equations. We can write:

$$\begin{aligned} \mathrm{L}\{y'(t)\} &= sY(s) - y(0) \\ \mathrm{L}\{y''(t)\} &= s^2Y(s) - sy(0) - y'(0) \end{aligned} \qquad \textbf{(1.7.15)}$$

If we take the Laplace transform of Eq. 1.7.11, replacing the independent variable x with t, we find the *subsidiary* equation:

$$s^2Y(s) + 2asY(s) + bY(s) - sy(0) - y'(0) - y(0) = F(s) \qquad \textbf{(1.7.16)}$$

This can be rearranged as

$$Y(s)\frac{sy(0) + y'(0) + y(0)}{s^2 + 2as + b} + \frac{F(s)}{s^2 + 2as + b} \qquad \textbf{(1.7.17)}$$

The solution $y(t)$ is then the inverse Laplace transform

$$y(t) = \mathrm{L}^{-1}\{Y(s)\} \qquad \textbf{(1.7.18)}$$

To find the inverse Laplace transform we find the inverse functions of the fractions on the right-hand side of Eq. 1.7.17. Section 1.1.4 on partial fractions will aid in this process.

Example 1.32

Solve the differential equation $2y'' + 12y' + 50y = 60\sin 5t$ if $y(0) = 0$ and $y'(0) = 10$.

Solution. We will find the solution using Laplace transforms. The Laplace transform of our differential equation is (first divide by 2 so that the coefficient of y'' is 1)

$$s^2Y(s) + 6sY(s) + 25Y(s) - s\,\cancelto{0}{y(0)} - 10 - \cancelto{0}{y(0)} = \frac{30\,(5)}{s^2 + 5^2}$$

where we have used $\alpha = 0$ and $\beta = 5$ in the Laplace transform of $e^{-\alpha t}\sin\beta t$. Next, we see, with some algebra, that

$$\begin{aligned} Y(s) &= \frac{10}{s^2 + 6s + 25} + \frac{150}{(s^2 + 25)(s^2 + 6s + 25)} \\ &= \frac{10}{s^2 + 6s + 25} + \frac{-s}{s^2 + 25} + \frac{s + 6}{s^2 + 6s + 25} \\ &= \frac{13}{(s + 3)^2 + 16} - \frac{s}{s^2 + 25} + \frac{s + 3}{(s + 3)^2 + 16}. \end{aligned}$$

The inverse Laplace transform is found using Table 1.3 to be:

$$\begin{aligned} y(t) &= \frac{13}{4}e^{-3t}\sin 4t - \cos 5t + e^{-3t}\cos 4t \\ &= \left(\frac{13}{4}\sin 4t + \cos 4t\right)e^{-3t} - \cos 5t \end{aligned}$$

Compare this solution to that of Example 1.31. Also, check and make sure that those initial conditions are satisfied.

1.8 Vectors

There are two vector multiplications. The first, the *scalar product,* or *dot product,* is the scalar defined by

$$\mathbf{A}\cdot\mathbf{B} = AB\cos\theta \tag{1.8.1}$$

where θ is the angle between the two vectors, as shown in Fig. 1.9, and A and B are the magnitudes of the two vectors. In a rectangular coordinate system the dot product becomes

$$\begin{aligned}\mathbf{A}\cdot\mathbf{B} &= \left(A_x\mathbf{i}+A_y\mathbf{j}+A_z\mathbf{k}\right)\cdot\left(B_x\mathbf{i}+B_y\mathbf{j}+B_z\mathbf{k}\right)\\ &= A_xB_x + A_yB_y + A_zB_z.\end{aligned} \tag{1.8.2}$$

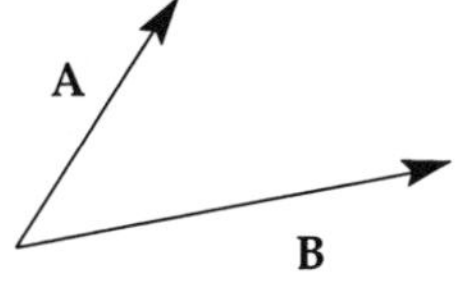

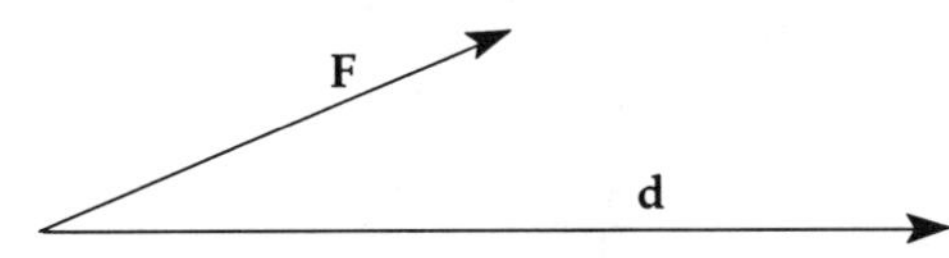

Figure 1.9 Vectors.

The scalar quantity *work* can be defined using the dot product:

$$w = \mathbf{F}\cdot\mathbf{d} \tag{1.8.3}$$

where **d** is the directed distance moved by the force.

The second, the *vector product,* or *cross product,* of the two vectors **A** and **B** is a vector defined by

$$\mathbf{C} = \mathbf{A}\times\mathbf{B} \tag{1.8.4}$$

where the magnitude of $\mathbf{C}$ is given by

$$C = AB\sin\theta. \tag{1.8.5}$$

The vector **C** acts in a direction perpendicular to the plane of **A** and **B** so that the three vectors form a right-handed set of vectors. (If the fingers curl **A** into **B**, the thumb points in the direction of **C**.) In a rectangular coordinate system the cross product is

$$\mathbf{A}\times\mathbf{B} = \begin{vmatrix}\mathbf{i} & \mathbf{j} & \mathbf{k}\\ A_x & A_y & A_z\\ B_x & B_y & B_z\end{vmatrix} = \left(A_yB_z - A_zB_y\right)\mathbf{i} + (A_zB_x - A_xB_z)\mathbf{j} + \left(A_xB_y - A_yB_x\right)\mathbf{k}. \tag{1.8.6}$$

The magnitude of **C** is the area of the parallelogram with sides **A** and **B**.

The volume of the parallelepiped with sides **A**, **B**, and **C** is the scalar triple product given by

$$\mathbf{A}\times\mathbf{B}\cdot\mathbf{C} = \begin{vmatrix}A_x & A_y & A_z\\ B_x & B_y & B_z\\ C_x & C_y & C_z\end{vmatrix}. \tag{1.8.7}$$

Using the above definition we can produce the following identities:

$$
\begin{aligned}
&\mathbf{i}\cdot\mathbf{i}=\mathbf{j}\cdot\mathbf{j}=\mathbf{k}\cdot\mathbf{k}=1\\
&\mathbf{i}\times\mathbf{i}=\mathbf{j}\times\mathbf{j}=\mathbf{k}\times\mathbf{k}=0\\
&\mathbf{A}\cdot\mathbf{B}=0 \quad \text{if} \quad \mathbf{A}\perp\mathbf{B}\\
&\mathbf{A}\cdot\mathbf{B}=AB \quad \text{if} \quad \mathbf{A}\|\mathbf{B}\\
&\mathbf{i}\cdot\mathbf{j}=\mathbf{j}\cdot\mathbf{k}=\mathbf{k}\cdot\mathbf{i}=0\\
&\mathbf{i}\times\mathbf{j}=\mathbf{k},\ \ \mathbf{j}\times\mathbf{k}=\mathbf{i},\ \ \mathbf{k}\times\mathbf{i}=\mathbf{j}\\
&\mathbf{A}\times\mathbf{B}=0 \quad \text{if} \quad \mathbf{A}\|\mathbf{B}
\end{aligned} \tag{1.8.8}
$$

Example 1.33

Find the projection of **A** on **B** if $\mathbf{A}=12\mathbf{i}-18\mathbf{j}+6\mathbf{k}$ and $\mathbf{B}=2\mathbf{i}-4\mathbf{j}+4\mathbf{k}$.

Solution. Find the unit vector $\mathbf{i}_\mathbf{B}$ in the direction of B:

$$\mathbf{i}_\mathbf{B}=\frac{\mathbf{B}}{B}=\frac{2\mathbf{i}-4\mathbf{j}+4\mathbf{k}}{\sqrt{2^2+4^2+4^2}}=\frac{1}{6}(2\mathbf{i}-4\mathbf{j}+4\mathbf{k}),$$

The projection of **A** on **B** is then

$$
\begin{aligned}
\mathbf{A}\cdot\mathbf{i}_\mathbf{B} &=(12\mathbf{i}-18\mathbf{j}+6\mathbf{k})\cdot\frac{1}{6}(2\mathbf{i}-4\mathbf{j}+4\mathbf{k})\\
&=4+12+4=20.
\end{aligned}
$$

Example 1.34

Find the area of a parallelogram with two sides identified by vectors from the origin to the points (3,4) and (8,0).

Solution. The two vectors are represented by

$$\mathbf{A}=3\mathbf{i}+4\mathbf{j}, \qquad \mathbf{B}=8\mathbf{i}.$$

The area of the parallelogram is then

$$|\mathbf{A}\times\mathbf{B}|=|(3\mathbf{i}+4\mathbf{j})\times 8\mathbf{i}|=32$$

since $\mathbf{i}\times\mathbf{i}=0$ and $\mathbf{j}\times\mathbf{i}=-\mathbf{k}$.

Example 1.35

Find a unit vector perpendicular to the plane that contains both $\mathbf{A}=\mathbf{i}-2\mathbf{j}+3\mathbf{k}$ and $\mathbf{B}=\mathbf{i}+2\mathbf{j}-\mathbf{k}$.

Solution. The vector $\mathbf{C}=\mathbf{A}\times\mathbf{B}$ is perpendicular to the plane of A and B. Using Eq. 1.8.6,

$$
\begin{aligned}
\mathbf{C} &= [-2(-1)-3\times 2]\mathbf{i}+[3\times 1-1(-1)]\mathbf{j}+[1\times 2-(-2)\times 1]\mathbf{k}\\
&= -4\mathbf{i}+4\mathbf{j}+4\mathbf{k}.
\end{aligned}
$$

A unit vector in the direction of **C** is the desired unit vector:

$$\mathbf{i}_c=\frac{-4\mathbf{i}+4\mathbf{j}+4\mathbf{k}}{\sqrt{4^2+4^2+4^2}}=\frac{1}{\sqrt{3}}(-\mathbf{i}+\mathbf{j}+\mathbf{k}).$$

The Gradient

We define a vector differential operator ∇, referred to as *del*, to be

$$\nabla = \frac{\partial}{\partial x}\mathbf{i} + \frac{\partial}{\partial y}\mathbf{j} + \frac{\partial}{\partial z}\mathbf{k} \tag{1.8.9}$$

The *gradient* of a scalar function ϕ is

$$\nabla\phi = \frac{\partial\phi}{\partial x}\mathbf{i} + \frac{\partial\phi}{\partial y}\mathbf{j} + \frac{\partial\phi}{\partial z}\mathbf{k}. \tag{1.8.10}$$

It is a vector quantity that points in the direction of the maximum rate of change of ϕ; hence, it is normal to a constant ϕ surface.

The *divergence* of a vector function $\mathbf{u}$ is the scalar quantity

$$\nabla\cdot\mathbf{u} = \frac{\partial u_x}{\partial x} + \frac{\partial u_y}{\partial y} + \frac{\partial u_z}{\partial z}. \tag{1.8.11}$$

The *curl* of a vector function $\mathbf{u}$ is the vector quantity

$$\nabla\times\mathbf{u} = \left(\frac{\partial u_z}{\partial y} - \frac{\partial u_y}{\partial z}\right)\mathbf{i} + \left(\frac{\partial u_x}{\partial z} - \frac{\partial u_z}{\partial x}\right)\mathbf{j} + \left(\frac{\partial u_y}{\partial x} - \frac{\partial u_x}{\partial y}\right)\mathbf{k}. \tag{1.8.12}$$

The *Laplacian* of a scalar function is, using $\nabla\cdot\nabla = \nabla^2$,

$$\nabla^2\phi = \frac{\partial^2\phi}{\partial x^2} + \frac{\partial^2\phi}{\partial y^2} + \frac{\partial^2\phi}{\partial z^2} \tag{1.8.13}$$

and *Laplace's equation* is

$$\nabla^2\phi = 0. \tag{1.8.14}$$

A vector function $\mathbf{u}$ is a *conservative* (or *irrotational*) vector function if it is given by the gradient of a scalar function ϕ, i.e.,

$$\mathbf{u} = \nabla\phi. \tag{1.8.15}$$

The curl of a conservative vector function is always zero. A vector field is *solenoidal* if its divergence is zero.

Example 1.36

Find a unit vector in the direction of the maximum rate of change of $\phi = x^2 - 2yz + xy$ at the point (2, –1, –1).

Solution. The gradient of ϕ is

$$\begin{aligned}\nabla\phi &= \frac{\partial\phi}{\partial x}\mathbf{i} + \frac{\partial\phi}{\partial y}\mathbf{j} + \frac{\partial\phi}{\partial z}\mathbf{k} \\ &= (2x+y)\mathbf{i} + (-2z+x)\mathbf{j} + (-2y)\mathbf{k} = 3\mathbf{i} + 4\mathbf{j} + 2\mathbf{k}.\end{aligned}$$

A unit vector in the direction of $\nabla\phi$ is

$$\mathbf{i}_n = \frac{\nabla\phi}{|\nabla\phi|} = \frac{3\mathbf{i}+4\mathbf{j}+2\mathbf{k}}{\sqrt{3^2+4^2+2^2}} = \frac{1}{\sqrt{29}}(3\mathbf{i}+4\mathbf{j}+2\mathbf{k}).$$

Example 1.37

Given a vector function $\mathbf{u} = x^2\mathbf{i} + 2yz\mathbf{j} + y^2\mathbf{k}$, find a) the divergence, b) the curl, and c) its associated scalar function.

Solution. a) the divergence is the scalar function

$$\begin{aligned}\nabla \cdot \mathbf{u} &= \frac{\partial u_x}{\partial x} + \frac{\partial u_y}{\partial y} + \frac{\partial u_z}{\partial z} \\ &= 2x + 2z + 0 = 2(x+z).\end{aligned}$$

b) The curl is the vector function

$$\begin{aligned}\nabla \times \mathbf{u} &= \left(\frac{\partial u_z}{\partial y} - \frac{\partial u_y}{\partial z}\right)\mathbf{i} + \left(\frac{\partial u_x}{\partial z} - \frac{\partial u_z}{\partial x}\right)\mathbf{j} + \left(\frac{\partial u_y}{\partial x} - \frac{\partial u_x}{\partial y}\right)\mathbf{k} \\ &= (2y - 2y)\mathbf{i} + (0-0)\mathbf{j} + (0-0)\mathbf{k} = 0,\end{aligned}$$

c) We know that an associated scalar function exists since the curl is everywhere zero. It is found by equating the scalar components of Eq. 1.8.15:

$$u_x = \frac{\partial \phi}{\partial x} \Rightarrow x^2 = \frac{\partial \phi}{\partial x}. \quad \therefore \phi = \frac{x^3}{3} + f(y,z).$$

$$u_y = 2yz = \frac{\partial \phi}{\partial y} = \frac{\partial}{\partial y}\left[\frac{x^3}{3} + f(y,z)\right] = \frac{\partial f}{\partial y}.$$

$$\therefore \frac{\partial f}{\partial y} = 2yz \quad \text{and} \quad f(y,z) = y^2 z + g(z).$$

$$u_z = y^2 = \frac{\partial \phi}{\partial z} = \frac{\partial}{\partial z}\left[\frac{x^3}{3} + y^2 z + g(z)\right] = y^2 + \frac{dg}{dz}.$$

$$\therefore \frac{dg}{dz} = 0 \qquad \text{so that} \quad g(z) = c.$$

Finally,

$$\phi = \frac{x^3}{3} + y^2 z + c.$$

Practice Problems

ALGEBRA

1.1 A growth curve is given by $A = 10\,e^{2t}$. At what value of t is A = 100?

a) 5.261 b) 3.070 c) 1.151 d) 0.726 e) 0.531

1.2 If $\ln x = 3.2$, what is x?

a) 18.65 b) 24.53 c) 31.83 d) 64.58 e) 126.7

1.3 If $\ln_5 x = -1.8$, find x.

a) 0.00483 b) 0.0169 c) 0.0552 d) 0.0783 e) 0.1786

1.4 One root of the equation $3x^2 - 2x - 2 = 0$ is

a) 1.215 b) 1.064 c) 0.937 d) 0.826 e) 0.549

1.5 $\sqrt{4+x}$ can be written as the series

a) $2 - x/4 + x^2/64 + \cdots$
b) $2 + x/8 - x^2/128 + \cdots$
c) $2 - x^2/4 - x^4/64 + \cdots$
d) $2 + x^2/8 + x^4/128 + \cdots$
e) $2 + x/4 - x^2/64 + \cdots$

1.6 Resolve $\dfrac{2}{x\left(x^2 - 3x + 2\right)}$ into partial fractions.

a) $1/x + 1/(x-2) - 2/(x-1)$
b) $1/x - 2/(x-2) + 1/(x-1)$
c) $2/x - 1/(x-2) - 2/(x-1)$
d) $-1/x + 2/(x-2) + 1/(x-1)$
e) $-1/x - 2/(x-2) + 1/(x-1)$

1.7 Express $\dfrac{4}{x^2\left(x^2 - 4x + 4\right)}$ as the sum of fractions.

a) $1/x + 1/(x-2)^2 - 1/(x-2)$
b) $1/x + 1/x^2 - 1/(x-2) + 1/(x-2)^2$
c) $1/x^2 + 1/(x-2)^2$
d) $1/x + 1/x^2 + 1/(x-2) + 1/(x-2)^2$
e) $1/x^2 - 1/(x-2) + 1/(x-2)^2$

1.8 A germ population has a growth curve of $Ae^{0.4t}$. At what value of t does its original value double?

a) 9.682 b) 7.733 c) 4.672 d) 1.733 e) 0.5641

TRIGONOMETRY

1.9 If $\sin\theta = 0.7$, what is $\tan\theta$?

a) 0.98 b) 0.94 c) 0.88 d) 0.85 e) 0.81

1.10 If the short leg of a right triangle is 5 units long and the long leg is 7 units long, find the angle opposite the short leg, in degrees.

a) 26.3 b) 28.9 c) 31.2 d) 33.8 e) 35.5

1.11 The expression $\tan\theta\sec\theta\left(1-\sin^2\theta\right)/\cos\theta$ simplifies to

a) $\sin\theta$ b) $\cos\theta$ c) $\tan\theta$ d) $\sec\theta$ e) $\csc\theta$

1.12 A triangle has sides of length 2, 3 and 4. What angle, in radians, is opposite the side of length 3?

a) 0.55 b) 0.61 c) 0.76 d) 0.81 e) 0.95

1.13 The length of a lake is to be determined. A distance of 850 m is measured from one end to a point x on the shore. A distance of 732 m is measured from x to the other end. If an angle of 154° is measured between the two lines connecting x, what is the length of the lake?

a) 1542 b) 1421 c) 1368 d) 1261 e) 1050

1.14 Express $2\sin^2\theta$ as a function of $\cos 2\theta$.

a) $\cos 2\theta - 1$ b) $\cos 2\theta + 1$ c) $\cos 2\theta + 2$ d) $2 - \cos 2\theta$ e) $1 - \cos 2\theta$

GEOMETRY

1.15 The included angle between two successive sides of a regular eight-sided polygon is

a) 150° b) 135° c) 120° d) 75° e) 45°

1.16 A large 15-m-dia cylindrical tank that sits on the ground is to be painted. If one liter of paint covers 10 m^2, how many liters are required if it is 10 m high? (Include the top.)

a) 65 b) 53 c) 47 d) 38 e) 29

1.17 The equation of a line that has a slope of –2 and intercepts the x-axis at $x = 2$ is

a) $y + 2x = 4$
b) $y - 2x = 4$
c) $y + 2x = -4$
d) $2y + x = 2$
e) $2y - x = -2$

1.18 The equation of a line that intercepts the x-axis at $x = 4$ and the y-axis at $y = -6$ is

a) $2x - 3y = 12$
b) $3x - 2y = 12$
c) $2x + 3y = 12$
d) $3x + 2y = 12$
e) $3y - 2x = 12$

1.19 The shortest distance from the line $3x - 4y = 3$ to the point (6, 8) is

a) 4.8 b) 4.2 c) 3.8 d) 3.4 e) 2.6

1.20 The equation $x^2 + 4xy + 4y^2 + 2x = 10$ represents which conic section?

a) circle b) ellipse c) parabola d) hyperbola e) plane

1.21 The x- and y-axes are the asymptotes of a hyperbola that passes through the point (2, 2). Its equation is

a) $x^2 - y^2 = 0$ b) $xy = 4$ c) $y^2 - x^2 = 0$ d) $x^2 + y^2 = 4$ e) $x^2 y = 8$

1.22 A 100-m-long track is to be built 50 m wide. If it is to be elliptical, what equation could describe it?

a) $50x^2 + 100y^2 = 1000$
b) $2x^2 + y^2 = 250$
c) $4x^2 + y^2 = 2500$
d) $x^2 + 2y^2 = 250$
e) $x^2 + 4y^2 = 10000$

1.23 The cylindrical coordinates (5, 30°, 12) are expressed in spherical coordinates as

a) (13, 30°, 67.4°)
b) (13, 30°, 22.6°)
c) (15, 52.6°, 22.6°)
d) (15, 52.6°, –22.6°)
e) (13, 30°, –67.4°)

1.24 The equation of a 4-m-radius sphere using cylindrical coordinates is

a) $x^2 + y^2 + z^2 = 16$
b) $r^2 = 16$
c) $r^2 + z^2 = 16$
d) $x^2 + y^2 = 16$
e) $r^2 + y^2 = 16$

COMPLEX NUMBERS

1.25 Divide $3 - i$ by $1 + i$.

a) $1 - 2i$ b) $1 + 2i$ c) $2 - i$ d) $2 + i$ e) $2 + 2i$

1.26 Find $(1 + i)^6$.

a) $1 + i$ b) $1 - i$ c) $8i$ d) $-8i$ e) $-1 - i$

1.27 Find the first root of $(1+i)^{1/5}$.

a) $0.168 + 1.06i$
b) $1.06 + 0.168i$
c) $1.06 - 0.168i$
d) $0.168 - 1.06i$
e) $-1.06 - 0.168i$

1.28 Express $(3+2i)\,e^{2it} + (3-2i)\,e^{-2it}$ in terms of trigonometric functions.

a) $3\cos 2t - 4\sin 2t$
b) $3\cos 2t - 2\sin 2t$
c) $6\cos 2t - 4\sin 2t$
d) $3\sin 2t + 2\sin 2t$
e) $6\cos 2t + 4\sin 2t$

1.29 Subtract $5e^{0.2i}$ from $6e^{2.3i}$. Express the answer in polar form.

a) $9.56\angle -21.4°$
b) $3.59\angle 104.5°$
c) $3.59\angle 255.5°$
d) $3.59\angle 201.4°$
e) $9.56\angle 158.6°$

LINEAR ALGEBRA

1.30 Find the value of the determinant $\begin{vmatrix} 3 & 2 & 1 \\ 0 & -1 & -1 \\ 2 & 0 & 2 \end{vmatrix}$.

a) 8 b) 4 c) 0 d) –4 e) –8

1.31 Evaluate the determinant $\begin{vmatrix} 1 & 0 & 1 & 1 \\ 2 & -1 & 0 & 1 \\ 0 & 0 & 2 & 0 \\ 3 & 2 & 1 & 1 \end{vmatrix}$.

a) 8 b) 4 c) 0 d) –4 e) –8

1.32 The cofactor A_{21} of the determinant of Prob. 1.30 is

a) –5 b) –4 c) 3 d) 4 e) 5

1.33 The cofactor A_{34} of the determinant of Prob. 1.31 is

a) 4 b) 6 c) 0 d) –4 e) –6

1.34 Find the adjoint matrix of $\begin{vmatrix} 1 & -4 \\ 0 & 2 \end{vmatrix}$.

a) $\begin{bmatrix} 4 & 2 \\ 0 & 1 \end{bmatrix}$ b) $\begin{bmatrix} 1 & 0 \\ 4 & 2 \end{bmatrix}$ c) $\begin{bmatrix} 2 & 4 \\ 1 & 0 \end{bmatrix}$ d) $\begin{bmatrix} 2 & 4 \\ 0 & 1 \end{bmatrix}$ e) $\begin{bmatrix} 1 & 4 \\ 0 & 2 \end{bmatrix}$

1.35 The inverse matrix of $\begin{bmatrix} 2 & 3 \\ 1 & 1 \end{bmatrix}$ is

a) $\begin{bmatrix} -1 & 3 \\ 1 & -2 \end{bmatrix}$ b) $\begin{bmatrix} 1 & -1 \\ -3 & 2 \end{bmatrix}$ c) $\begin{bmatrix} -1 & 1 \\ -3 & 2 \end{bmatrix}$ d) $\begin{bmatrix} -2 & 3 \\ 1 & -1 \end{bmatrix}$ e) $\begin{bmatrix} 2 & 3 \\ -1 & -1 \end{bmatrix}$

1.36 Calculate $\begin{bmatrix} 2 & -1 \\ 3 & 2 \end{bmatrix}\begin{bmatrix} 2 \\ 1 \end{bmatrix}$.

a) $\begin{bmatrix} 8 \\ 3 \end{bmatrix}$ b) $\begin{bmatrix} 3 \\ 8 \end{bmatrix}$ c) $\begin{bmatrix} -3 \\ -8 \end{bmatrix}$ d) $[3,8]$ e) $[8,3]$

1.37 Determine $\begin{bmatrix} 1 & 2 \\ 2 & 1 \end{bmatrix}\begin{bmatrix} -1 & 0 \\ 1 & 2 \end{bmatrix}$.

a) $\begin{bmatrix} 1 & 4 \\ -1 & 2 \end{bmatrix}$ b) $\begin{bmatrix} 1 & -1 \\ 4 & 2 \end{bmatrix}$ c) $\begin{bmatrix} 1 \\ -1 \end{bmatrix}$ d) $\begin{bmatrix} 4 \\ 2 \end{bmatrix}$ e) $[1,4]$

1.38 Solve for $[x_i]$.

$$\begin{aligned} 3x_1 + 2x_2 \quad\quad &= -2 \\ x_1 - x_2 + x_3 &= 0 \\ 4x_1 \quad\quad + 2x_3 &= 4 \end{aligned}$$

a) $\begin{bmatrix} 2 \\ 4 \\ -6 \end{bmatrix}$ b) $\begin{bmatrix} -2 \\ 4 \\ 12 \end{bmatrix}$ c) $\begin{bmatrix} 2 \\ 8 \\ 4 \end{bmatrix}$ d) $\begin{bmatrix} -6 \\ 8 \\ 14 \end{bmatrix}$ e) $\begin{bmatrix} 6 \\ 4 \\ 3 \end{bmatrix}$

1.39 Find the eigenvalues of $\begin{bmatrix} 1 & 2 \\ 3 & 2 \end{bmatrix}$

a) 4, –1 b) 4, 1 c) 1, –4 d) 3, 2 e) 1, 2

1.40 Find an eigenvector of $\begin{bmatrix} 1 & 2 \\ 3 & 2 \end{bmatrix}$

a) $\begin{bmatrix} 4 \\ -1 \end{bmatrix}$ b) $\begin{bmatrix} 2 \\ 3 \end{bmatrix}$ c) $\begin{bmatrix} 1 \\ 4 \end{bmatrix}$ d) $\begin{bmatrix} -3 \\ 2 \end{bmatrix}$ e) $\begin{bmatrix} 1 \\ 2 \end{bmatrix}$

CALCULUS

1.41 The slope of the curve $y = 2x^3 - 3x$ at $x = 1$ is

a) 3 b) 5 c) 6 d) 8 e) 9

1.42 If $y = \ln x + e^x \sin x$, find dy/dx at $x = 1$.

a) 1.23 b) 3.68 c) 4.76 d) 6.12 e) 8.35

1.43 At what value of x does a maximum of $y = x^3 - 3x$ occur?

a) 2 b) 1 c) 0 d) –1 e) –2

1.44 Where does an inflection point occur for $y = x^3 - 3x$?

a) 2 b) 1 c) 0 d) –1 e) –2

1.45 Evaluate $\lim\limits_{x\to\infty} \dfrac{2x^2 - x}{x^2 + x}$.

a) 2 b) 1 c) 0 d) –1 e) –2

1.46 If a quantity η and its derivatives η' and η'' are known at a point, its approximate value at a small distance h is

a) $\eta + h^2\eta''/2$ b) $\eta + h\eta/2 + h^2\eta''$ c) $\eta + h\eta' + h^2\eta''/2$

d) $\eta + h\eta' + h^2\eta''$ e) $\eta + h\eta'$

1.47 Find an approximation to $e^x \sin x$ for small x.

a) $x - x^2 + x^3$ b) $x + x^2 + x^3/3$ c) $x - x^2/2 + x^3/6$

d) $x + x^2 - x^3/6$ e) $x + x^2 + x^3/2$

1.48 Find the area between the y-axis and $y = x^2$ from $y = 4$ to $y = 9$.

a) 29/3 b) 32/3 c) 34/3 d) 38/3 e) 43/3

1.49 The area contained between $4x = y^2$ and $4y = x^2$ is

a) 10/3 b) 11/3 c) 13/3 d) 14/3 e) 16/3

1.50 Rotate the shaded area of Example 1.25 about the x-axis. What volume is formed?

a) 4π b) 6π c) 8π d) 10π e) 12π

1.51 Evaluate $\int_0^2 \left(e^x + \sin x\right) dx$.

a) 7.81 b) 6.21 c) 5.92 d) 5.61 e) 4.21

1.52 Evaluate $2\int_0^1 e^x \sin x \, dx$.

a) 1.82 b) 1.94 c) 2.05 d) 2.16 e) 2.22

1.53 Derive an expression $\int x \cos x \, dx$.

a) $x \cos x - \sin x + C$ b) $-x \cos x + \sin x + C$ c) $x \sin x - \cos x + C$
d) $x \cos x + \sin x + C$ e) $x \sin x + \cos x + C$

DIFFERENTIAL EQUATIONS

1.54 The differential equation $y'' + x^2 y' + y + 2 = 0$ is

a) linear and homogeneous.
b) linear and nonhomogeneous.
c) nonlinear and homogeneous.
d) nonlinear and nonhomogeneous.
e) not a differential equation.

1.55 Given: $y' + 2xy = 0$, $y(0) = 2$. Find : $y(2)$.

a) 0.0366 b) 0.127 c) 0.936 d) 2.36 e) 27.3

1.56 Given: $y' + 2x = 0$, $y(0) = 1$. Find: $y(10)$.

a) –100 b) –99 c) –91 d) –86 e) –54

1.57 A spring-mass system is represented by $2\ddot{y} + \dot{y} + 50y = 0$. What frequency, in hertz, is contained in the solution?

a) 0.79 b) 1.56 c) 2.18 d) 3.76 e) 4.99

1.58 Find the solution to $\ddot{y} + 16y = 0$.

a) $C_1 \cos 4t + C_2 \sin 4t$
b) $C_1 e^{4t} + C_2 e^{-4t}$
c) $C_1 e^{4t} + C_2 t e^{4t}$
d) $C_1 \cos 4t + C_2 t \cos 4t$
e) $C_1 \sin 4t + C_2 t \sin 4t$

1.59 Find the solution to $\ddot{y}+8\dot{y}+16y=0$.

a) $C_1\cos 4t + C_2\sin 4t$
b) $C_1 t\cos 4t + C_2\cos 4t$
c) $C_1\sin 4t + C_2 t\sin 4t$
d) $C_1e^{4t} + C_2te^{4t}$
e) $C_1e^{-4t} + C_2te^{-4t}$

1.60 Solve the equation $\ddot{y}-5\dot{y}+6y=4e^t$.

a) $C_1e^{2t}+C_2e^{3t}$
b) $C_1e^{-2t}+C_2e^{-3t}+2e^t$
c) $C_1e^{2t}+C_2te^{3t}+2e^t$
d) $C_1e^{2t}+C_2e^{3t}+2e^t$
e) $C_1e^{2t}+C_2e^{3t}+C_3e^t$

1.61 Solve $\ddot{y}+16y=8\sin 4t$.

a) $c_1\sin 4t+c_2\cos 4t-t\cos 4t$
b) $c_1\sin 4t+c_2\cos 4t+t\cos 4t$
c) $c_1\sin 4t+c_2\cos 4t-\cos 4t$
d) $c_1\sin 4t+c_2\cos 4t+\cos 4t$
e) $c_1\sin 4t+c_2\cos 4t-\sin 4t$

1.62 Solve $\ddot{y}+5\dot{y}+6y=0$ if $y(0)=2$ and $\dot{y}(0)=0$. Use Laplace transforms.

a) $2e^{-3t}-3e^{-2t}$
b) $2e^{3t}-3e^{2t}$
c) $4e^{-3t}-6e^{-2t}$
d) $-4e^{-3t}+6e^{-2t}$
e) $4e^{3t}-6e^{2t}$

1.63 Solve $\ddot{y}+16y=120te^{-t}$ if $y(0)=0$ and $\dot{y}(0)=12$.

a) $8te^{-t}+\sin 4t$
b) $4te^{t}+2\sin 4t$
c) $6te^{-t}+6\sin t$
d) $6te^{t}+6\sin t$
e) $12te^{-t}$

VECTOR ANALYSIS

1.64 Given: $\mathbf{A}=3\mathbf{i}-6\mathbf{j}+2\mathbf{k}$, $\mathbf{B}=10\mathbf{i}+4\mathbf{j}-6\mathbf{k}$. Find: $\mathbf{A}\cdot\mathbf{B}$.

a) $-6\mathbf{i}$ b) 6 c) –6 d) $30\mathbf{i}-24\mathbf{j}-12\mathbf{k}$ e) $28\mathbf{i}+38\mathbf{j}+72\mathbf{k}$

1.65 Given: $\mathbf{A}=2\mathbf{i}-5\mathbf{k}$, $\mathbf{B}=\mathbf{j}$. Find: $\mathbf{A}\times\mathbf{B}$.

a) 0 b) $5\mathbf{i}+2\mathbf{k}$ c) –3 d) $-3\mathbf{k}$ e) 7

1.66 Find the projection of **A** in the direction of **B** if $\mathbf{A}=14\mathbf{i}-7\mathbf{j}$ and $\mathbf{B}=6\mathbf{i}+3\mathbf{j}-2\mathbf{k}$.

a) –9 b) $12\mathbf{i}-3\mathbf{j}$ c) 0 d) 15 e) 9

1.67 The equation of a plane perpendicular to and passing through the end of the vector $\mathbf{A}=2\mathbf{i}-4\mathbf{j}+6\mathbf{k}$ is given by $2x-4y+6z=k$ where k is

a) 56 b) 24 c) 0 d) –8 e) –16

1.68 Estimate the area of the parallelogram with sides $\mathbf{A} = 2\mathbf{i}+3\mathbf{j}$ and $\mathbf{B} = 4\mathbf{i}-6\mathbf{j}+5\mathbf{k}$.

a) 32 b) 30 c) 26 d) 20 e) 10

1.69 Find a unit vector normal to the surface $x^2+3y^2-3z=12$ at the point (1,1,0).

a) $\frac{\mathbf{i}+\mathbf{j}}{\sqrt{2}}$ b) $\frac{2\mathbf{i}-3\mathbf{j}}{\sqrt{13}}$ c) $\frac{\mathbf{i}+3\mathbf{j}-3\mathbf{k}}{\sqrt{19}}$ d) $\frac{\mathbf{i}+3\mathbf{j}}{\sqrt{10}}$ e) $\frac{2\mathbf{i}+6\mathbf{j}-3\mathbf{k}}{7}$

1.70 The divergence of $\mathbf{u} = x^2\mathbf{i}+y^2\mathbf{j}+z^2\mathbf{k}$ at the point (1,1,1) is

a) $2\mathbf{i}+2\mathbf{j}+2\mathbf{k}$ b) $\mathbf{i}+\mathbf{j}+\mathbf{k}$ c) 3 d) 6 e) 0

1.71 The curl of $\mathbf{u} = x^2\mathbf{i}+y^2\mathbf{j}+z^2\mathbf{k}$ at the point (1,1,1) is

a) $2\mathbf{i}+2\mathbf{j}+2\mathbf{k}$ b) $\mathbf{i}+\mathbf{j}+\mathbf{k}$ c) 3 d) 6 e) 0

1.72 The vector function $\mathbf{u} = yz\mathbf{i}+xz\mathbf{j}+xy\mathbf{k}$ is

a) conservative and solenoidal
b) conservative only
c) solenoidal only
d) neither conservative nor solenoidal
e) undefined at the origin

Solutions to Practice Problems

1.1 c) $100 = 10e^{2t}. \quad \therefore \ln e^{2t} = \ln 10. \quad \therefore 2t = 2.303 \quad \therefore t = 1.151$

1.2 b) $\ln x = 3.2. \quad \therefore e^{3.2} = x. \quad \therefore x = 24.53$

1.3 c) $\ln_5 x = -1.8. \quad \therefore 5^{-1.8} = x. \quad \therefore x = 0.0552$

1.4 a) $x = \dfrac{-(-2)\pm\sqrt{2^2-4(3)(-2)}}{3\cdot 2} = 1.215$

1.5 e) $(4+x)^{1/2} = 4^{1/2}+\frac{1}{2}4^{-1/2}x+\dfrac{\frac{1}{2}\left(1-\frac{1}{2}\right)}{2}4^{-3/2}x^2+\cdots \quad = 2+x/4-x^2/64+\cdots$

1.6 a) $\dfrac{2}{x\left(x^2-3x+2\right)} = \dfrac{A_1}{x}+\dfrac{A_2}{x-2}+\dfrac{A_3}{x-1} = \dfrac{A_1\left(x^2-3x+2\right)+A_2\left(x^2-x\right)+A_3\left(x^2-2x\right)}{x\left(x^2-3x+2\right)}$

$$\therefore \quad \left.\begin{aligned} A_1+A_2+A_3 &= 0 \\ -3A_1-A_2-2A_3 &= 0 \\ 2A_1 &= 2 \end{aligned}\right\} \qquad \left.\begin{aligned} A_1 &= 1 \\ A_2+A_3 &= -1 \\ -A_2-2A_3 &= 3 \end{aligned}\right\} \qquad \begin{aligned} A_3 &= -2 \\ A_2 &= 1 \end{aligned}$$

1.7 b) $\dfrac{4}{x^2\left(x^2-4x+4\right)}=\dfrac{A_1}{x}+\dfrac{A_2}{x^2}+\dfrac{A_3}{x-2}+\dfrac{A_4}{(x-2)^2}$

$$=\frac{A_1\left(x^3-4x^2+4x\right)+A_2\left(x^2-4x+4\right)+A_3\left(x^3-2x^2\right)+A_4x^2}{x^2(x-2)^2}$$

$$\therefore \quad \left.\begin{aligned} A_1+A_3&=0\\ -4A_1+A_2-2A_3+A_4&=0\\ 4A_1-4A_2&=0\\ 4A_2&=4\end{aligned}\right\} \qquad \begin{aligned} A_2&=1\\ A_1&=1\\ A_3&=-1\\ A_4&=1\end{aligned}$$

1.8 d) at $t=0$, population $=A$. $\therefore 2A=Ae^{0.4t}$. $\ln 2=0.4t$. $\therefore t=1.733$

1.9 a) $\sin\theta=0.7$ $\therefore \theta=44.43°$. $\tan 44.43°=0.980$

1.10 e) $\tan\theta=5/7$. $\therefore \theta=35.54°$

1.11 c) $\tan\theta\sec\theta\left(1-\sin^2\theta\right)/\cos\theta=\tan\theta\dfrac{1}{\cos\theta}\cos^2\theta\dfrac{1}{\cos\theta}=\tan\theta$

1.12 d) $3^2=4^2+2^2-2\cdot 2\cdot 4\cos\theta$. $\therefore \cos\theta=0.6875$. $\theta=46.6°$. $\therefore$ rad $=0.813$

1.13 a) $L^2=850^2+732^2-2\cdot 850\cdot 732\cos 154°$. $\therefore L=1542$ m

1.14 e) $\cos 2\theta=\cos^2\theta-\sin^2\theta=1-\sin^2\theta-\sin^2\theta=1-2\sin^2\theta$. $\therefore 2\sin^2\theta=1-\cos 2\theta$

1.15 b) $\theta=\pi(n-2)/n=\pi(8-2)/8$ radians. $\dfrac{6\pi}{8}\times\dfrac{180}{\pi}=135°$

1.16 a) $\text{Area}=\text{Area}_{top}+\text{Area}_{sides}=\pi R^2+\pi DL$

$$=\pi\times 7.5^2+\pi\times 15\times 10=648\text{ m}^2.\quad 648\div 10\approx 65$$

1.17 a) $y=mx+b$. $y=-2x+b$. $0=-2(2)+b$. $\therefore b=4$. $\therefore y=-2x+4$

1.18 b) $y=mx+b$. $0=4m+b$. $-6=b$. $\therefore m=3/2$. $\therefore y=3x/2-6$ or $2y=3x-12$

1.19 d) $3x-4y-3=0$. $A=3, B=-4$. $d=\dfrac{|3\times 6-4\times 8-3|}{\sqrt{3^2+(-4)^2}}=3.4$

1.20 c) $B^2-AC=2^2-1\times 4=0$. $\therefore$ parabola

1.21 b) $xy=\pm k^2=4$

1.22 c) $\dfrac{x^2}{25^2}+\dfrac{y^2}{50^2}=1$. $\therefore 4x^2+y^2=2500$

1.23 b) $x=r\cos\theta=5\times 0.866=4.33$. $y=r\sin\theta=5\times 0.5=2.5$

Spherical coordinates:

$$r=\sqrt{4.33^2+2.5^2+12^2}=13. \quad \phi=\cos^{-1}\frac{z}{r}=\cos^{-1}\frac{12}{13}=22.6°.$$

$$\theta=\tan^{-1}\frac{y}{x}=\tan^{-1}\frac{2.5}{4.33}. \quad \therefore \theta=30°.$$

1.24 c) a) is rectangular coordinates. b) is spherical coordinates.

1.25 a) $\frac{3-i}{1+i}=\frac{3-i}{1+i}\frac{1-i}{1-i}=\frac{3-1-4i}{1-(-1)}=\frac{1}{2}(2-4i)=1-2i$

1.26 d) $1+i=re^{i\theta}$. $r=\sqrt{1^2+1^2}=\sqrt{2}$. $\theta=\tan^{-1}\frac{1}{1}=\pi/4$ rad. $\therefore 1+i=\sqrt{2}e^{i\pi/4}$.

$\therefore (1+i)^6=\left(\sqrt{2}\right)^6 e^{i\pi 3/2}=8\left(\cos\frac{3\pi}{2}+i\sin\frac{3\pi}{2}\right)=-8i$. Note: angles are in radians.

1.27 b) $1+i=\sqrt{2}e^{i\pi/4}$. $\therefore (1+i)^{1/5}=1.414^{1/5}e^{i\pi/20}=1.072\left(\cos\frac{\pi}{20}+i\sin\frac{\pi}{20}\right)$

$$=1.06+0.168i$$

1.28 c) $(3+2i)(\cos 2t+i\sin 2t)+(3-2i)(\cos 2t-i\sin 2t)=6\cos 2t+4i(i\sin 2t)$

$$=6\cos 2t-4\sin 2t$$

1.29 e) $6(\cos 2.3+i\sin 2.3)-5(\cos 0.2+i\sin 0.2)=6(-0.666+0.746i)-5(0.98+0.199i)$

$$=-8.90+3.48i$$

$$=9.56\angle 158.6°$$

1.30 e) $\begin{vmatrix}3 & 2 & 1\\ 0 & -1 & -1\\ 2 & 0 & 2\end{vmatrix}=-6-4+2=-8$

1.31 a) $\begin{vmatrix}1 & 0 & 1 & 1\\ 2 & -1 & 0 & 1\\ 0 & 0 & 2 & 0\\ 3 & 2 & 1 & 1\end{vmatrix}=2\begin{vmatrix}1 & 0 & 1\\ 2 & -1 & 1\\ 3 & 2 & 1\end{vmatrix}=2(-1+4+3-2)=8$

Note: Expand using the third row.

1.32 b) $(-1)^3\begin{vmatrix}2 & 1\\ 0 & 2\end{vmatrix}=-4$

1.33 e) $(-1)^7\begin{vmatrix}1 & 0 & 1\\ 2 & -1 & 0\\ 3 & 2 & 1\end{vmatrix}=-(-1+4+3)=-6$

1.34 d) $\left[a_{ij}\right]^+=\begin{bmatrix}A_{11} & A_{21}\\ A_{12} & A_{22}\end{bmatrix}=\begin{bmatrix}2 & 4\\ 0 & 1\end{bmatrix}$

1.35 a) $\left[a_{ij}\right]^{-1}=\frac{\left[a_{ij}\right]^+}{\left|a_{ij}\right|}=\frac{\begin{bmatrix}1 & -3\\ -1 & 2\end{bmatrix}}{-1}=\begin{bmatrix}-1 & 3\\ 1 & -2\end{bmatrix}$

1.36 b) $\begin{bmatrix} 2 & -1 \\ 3 & 2 \end{bmatrix}\begin{bmatrix} 2 \\ 1 \end{bmatrix} = \begin{bmatrix} 4-1 \\ 6+2 \end{bmatrix} = \begin{bmatrix} 3 \\ 8 \end{bmatrix}$

1.37 a) $\begin{bmatrix} 1 & 2 \\ 2 & 1 \end{bmatrix}\begin{bmatrix} -1 & 0 \\ 1 & 2 \end{bmatrix} = \begin{bmatrix} 1 & 4 \\ -1 & 2 \end{bmatrix}$

1.38 d) $\left[a_{ij}\right] = \begin{bmatrix} 3 & 2 & 0 \\ 1 & -1 & 1 \\ 4 & 0 & 2 \end{bmatrix}$. $\left[a_{ij}\right]^+ = \begin{bmatrix} -2 & -4 & 2 \\ 2 & 6 & -3 \\ 4 & 8 & -5 \end{bmatrix}$. $\left|a_{ij}\right| = -2.$

$\therefore \left[a_{ij}\right]^{-1} = \dfrac{\left[a_{ij}\right]^+}{\left|a_{ij}\right|} = \begin{bmatrix} 1 & 2 & -1 \\ -1 & -3 & 3/2 \\ -2 & -4 & 5/2 \end{bmatrix}$. $\left[x_j\right] = \left[a_{ij}\right]^{-1}\left[r_i\right] = \left[a_{ij}\right]^{-1}\begin{bmatrix} -2 \\ 0 \\ 4 \end{bmatrix} = \begin{bmatrix} -6 \\ 8 \\ 14 \end{bmatrix}$

1.39 a) $\begin{vmatrix} 1-\lambda & 2 \\ 3 & 2-\lambda \end{vmatrix} = \lambda^2 - 3\lambda + 2 - 6 = \lambda^2 - 3\lambda - 4 = 0.$

$(\lambda-4)(\lambda+1) = 0. \quad \therefore \ \lambda = 4, -1$

1.40 b) Use $\lambda = 4$. $\begin{bmatrix} -3 & 2 \\ 3 & -2 \end{bmatrix}\begin{bmatrix} x_1 \\ x_2 \end{bmatrix} = 0$.

$-3x_1 + 2x_2 = 0.$

$3x_1 - 2x_2 = 0.$ $\quad \therefore \ \mathbf{x} = \begin{bmatrix} 2 \\ 3 \end{bmatrix}$

Any multiple is also an eigenvector, e.g., $\mathbf{x} = \begin{bmatrix} 1 \\ 3/2 \end{bmatrix}$ or $\mathbf{x} = \begin{bmatrix} 2/\sqrt{13} \\ 3/\sqrt{13} \end{bmatrix}$

1.41 a) $\dfrac{dy}{dx} = 6x^2 - 3 = 6(1)^2 - 3 = 3$

1.42 c) $\dfrac{dy}{dx} = \dfrac{1}{x} + e^x \cos x + e^x \sin x = 1 + e\cos 1 + e \sin 1 = 4.76.$ $\quad (\cos 1 = \cos 57.3°)$

1.43 d) $\dfrac{dy}{dx} = 3x^2 - 3 = 0. \quad \therefore x^2 = 1. \quad \therefore x = \pm 1. \qquad \dfrac{d^2y}{dx^2} = 6x. \quad \therefore x = -1$ is a maximum.

1.44 c) $y' = 3x^2 - 3. \quad y'' = 6x. \quad \therefore x = 0$ is inflection.

1.45 a) $\lim\limits_{x\to\infty} \dfrac{2x^2 - x}{x^2 + x} = \lim\limits_{x\to\infty} \dfrac{4x-1}{2x+1} = \lim\limits_{x\to\infty} \dfrac{4}{2} = 2$

1.46 c) $\eta(x+h) = \eta + h\eta' + \dfrac{h^2}{2}\eta''$

1.47 b) $e^x \sin x = \left(1 + x + \dfrac{x^2}{2}\right)\left(x - \dfrac{x^3}{6}\right) = x + x^2 + \dfrac{x^3}{2} - \dfrac{x^3}{6} = x + x^2 + x^3/3$

1.48 d) Area $= \int_4^9 x\,dy = \int_4^9 y^{1/2}dy = \frac{2}{3}(27-8) = 12\frac{2}{3}$

1.49 e) $\text{Area} = \int_0^4 (x_2 - x_1)dy = \int_0^4 \left(2y^{1/2} - \frac{y^2}{4}\right)dy$

(4, 4)

$= 2 \times \frac{2}{3} \times 8 - \frac{1}{12} \times 64 = 16/3$

1.50 c) $V = \int_0^2 2\pi y\, x dy = 2\pi \int_0^2 y^3 dy = 2\pi \times \frac{2^4}{4} = 8\pi$

1.51 a) $\int_0^2 \left(e^x + \sin x\right)dx = e^x - \cos x\Big|_0^2 = e^2 - 1 - \cos 2 + 1 = 7.81$

1.52 a) $\int_0^1 e^x \sin x\, dx = e^x \sin x\Big|_0^1 - \int_0^1 e^x \cos x\, dx$

$u = \sin x \quad dv = e^x dx$

$du = \cos x dx \quad v = e^x$

1st integral

$u = \cos x \quad dv = e^x dx$

$v = e^x \quad du = -\sin x dx$

2nd integral

$\therefore \int_0^1 e^x \sin x\, dx = e \sin 1 - \left[e^x \cos x\Big|_0^1 + \int_0^1 e^x \sin x\, dx\right]$

$\therefore 2\int_0^1 e^x \sin x\, dx = e \sin 1 - e \cos 1 + 1 \times 1 = 1.819. \quad \therefore \int_0^1 e^x \sin x\, dx = 0.909$

1.53 e) $\int x \cos x\, dx = x \sin x - \int \sin x\, dx = x \sin x + \cos x + C$

$u = x \quad dv = \cos x dx$

$du = dx \quad v = \sin x$

1st integral

1.54 b) linear and nonhomogeneous. The term (+2) makes it nonhomogeneous. $x^2 y'$ is linear.

1.55 a) $\frac{dy}{dx} = -2xy. \quad \frac{dy}{y} = -2xdx. \quad \therefore \ln y = -x^2 + C. \quad \ln 2 = 0 + C. \quad \therefore C = \ln 2.$

$y(2) = \exp\left(-2^2 + \ln 2\right) = 0.0366$

1.56 b) $\frac{dy}{dx} = -2x. \quad dy = -2xdx. \quad \therefore y = -x^2 + C. \quad 1 = 0 + C. \quad \therefore C = 1.$

$y(10) = -10^2 + 1 = -99$

1.57 a) $2m^2 + m + 50 = 0. \quad \therefore m = \frac{-1 \pm \sqrt{1 - 400}}{4} = -\frac{1}{4} \pm 4.99i.$

$\therefore y(t) = e^{-t/4}\left(A \cos 4.99t + B \sin 4.99t\right).$

$\therefore \omega = 4.99 \text{ rad/s}. \quad \therefore f = \frac{\omega}{2\pi} = \frac{4.99}{2\pi} = 0.794 \text{ hertz}$

1.58 a) $m^2 + 16 = 0. \quad \therefore m = \pm 4i. \quad \therefore y(t) = C_1 \cos 4t + C_2 \sin 4t$

1.59 e) $m^2+8m+16=0.\quad (m+4)^2=0.\quad m=-4,-4.\quad \therefore y(t)=C_1e^{-4t}+C_2te^{-4t}$

1.60 d) $m^2-5m+6=0.\quad m=\dfrac{5\pm\sqrt{25-24}}{2}=3,\ 2.\quad \therefore y_h=C_1e^{3t}+C_2e^{2t}.$

Assume $y_p=Ae^t$. Then $Ae^t-5Ae^t+6Ae^t=4e^t.\quad \therefore A=2.$

1.61 a) homogeneous: $m^2+16=0.\quad \therefore m=\pm 4i.\quad \therefore y_h(t)=C_1\sin 4t+C_2\cos 4t$

particular: $y_p=At\cos 4t.$ (This is resonance.) $\dot{y}_p=A\cos 4t-4At\sin 4t$

$\therefore -4A\sin 4t-4A\sin 4t-16At\cos 4t+16At\cos 4t=8\sin 4t$

$\therefore -8A=8.\quad A=-1.\quad \therefore y=y_h+y_p=C_1\sin 4t+C_2\cos 4t-t\cos 4t$

1.62 d) $s^2Y-2s+5sY-10+6Y=0.\quad Y=\dfrac{2s+10}{s^2+5s+6}=\dfrac{-4}{s+3}+\dfrac{6}{s+2}.$

$\therefore y(t)=-4e^{-3t}+6e^{-2t}$

1.63 a) $s^2Y-12+16Y=\dfrac{120}{(s+1)^2}.\quad Y=\dfrac{12}{s^2+16}+\dfrac{120}{(s+1)^2(s^2+16)}$

$$=\frac{12}{s^2+16}+\frac{8}{(s+1)^2}-\frac{8}{s^2+16}.$$

$\therefore y(t)=\sin 4t+8te^{-t}$

1.64 c) $\mathbf{A}\cdot\mathbf{B}=3\cdot 10+(-6)\cdot 4+2(-6)=-6$

1.65 b) $\mathbf{A}\times\mathbf{B}=(2\mathbf{i}-5\mathbf{k})\times\mathbf{j}=2\mathbf{i}\times\mathbf{j}-5\mathbf{k}\times\mathbf{j}=2\mathbf{k}-5(-\mathbf{i})=5\mathbf{i}+2\mathbf{k}$

1.66 e) $\mathbf{i}_B=(6\mathbf{i}+3\mathbf{j}-2\mathbf{k})/\sqrt{6^2+3^2+2^2}=\frac{1}{7}(6\mathbf{i}+3\mathbf{j}-2\mathbf{k})$

$\mathbf{A}\cdot\mathbf{i}_B=[14\cdot 6-7(3)]/7=12-3=9$

1.67 a) $[(x\mathbf{i}+y\mathbf{j}+z\mathbf{k})-(2\mathbf{i}-4\mathbf{j}+6\mathbf{k})]\cdot(2\mathbf{i}-4\mathbf{j}+6\mathbf{k})=0.$

$2(x-2)-4(y+4)+6(z-6)=0.\quad \therefore 2x-4y+6z=56$

(x, y, z)

1.68 b) $|\mathbf{A}\times\mathbf{B}|=|15\mathbf{i}-10\mathbf{j}-24\mathbf{k}|=\sqrt{901}\cong 30$

1.69 e) $\nabla\phi=2x\mathbf{i}+6y\mathbf{j}-3\mathbf{k}=2\mathbf{i}+6\mathbf{j}-3\mathbf{k}.\quad |\nabla\phi|=7.\quad \therefore \mathbf{i}_n=\nabla\phi/|\nabla\phi|=(2\mathbf{i}+6\mathbf{j}-3\mathbf{k})/7$

1.70 d) $\nabla\cdot u=2x+2y+2z=2+2+2=6$

1.71 e) $\nabla\times\mathbf{u}=(0-0)\mathbf{i}+(0-0)\mathbf{j}+(0-0)\mathbf{k}=0$

1.72 a) $\nabla\cdot\mathbf{u}=0+0+0=0.\quad \therefore$ solenoidal. $\nabla\times\mathbf{u}=(x-x)\mathbf{i}+(y-y)\mathbf{j}+(z-z)\mathbf{k}=0.$

$\therefore$ conservative and solenoidal.

Statics

by George E. Mase

Statics is concerned primarily with the equilibrium of bodies subjected to force systems. Also, traditionally in engineering statics we consider centroids, center of gravity and moments of inertia.

2.1 Forces, Moments and Resultants

A force is the manifestation of the action of one body upon another. Forces arise from the direct action of two bodies in contact with one another, or from the "action at a distance" of one body upon another as occurs with gravitational and magnetic forces. We classify forces as either *body forces* which act (and are distributed) throughout the volume of the body, or as *surface forces*, which act over a surface portion (either external or internal) of the body. If the surface over which the force system acts is very small we usually assume localization at a specific point in the surface and speak of a *concentrated force* at that point.

Mathematically, forces are represented by *vectors*. Geometrically, a vector is a directed line segment having a head and a tail, i.e., an arrow. The length of the arrow corresponds to the magnitude of the force, its orientation defines the line of action, and the direction of the arrow (tail to head) gives the sense of the force.

Systems of concentrated forces are listed as *concurrent* when all of the forces act, or could act, at a single point; otherwise they are termed *non-concurrent* systems. Parallel force systems are in this second group. Also, force systems are often described as two-dimensional (acting in a single plane), or three-dimensional (spatial systems).

In addition to the "push or pull" effect on the point at which it acts, a force creates a *moment* about other points of the body. Conceptually, a moment may be thought of as a torque. Mathematically, the moment of the force **F** with respect to point A when the force acts at point B is defined by the vector cross product

$$\mathbf{M}_A = \mathbf{r}_{B/A} \times \mathbf{F} \tag{2.1.1}$$

where $\mathbf{r}_{B/A}$ is the position vector of B relative to A as shown by Fig. 2.1. Actually, it is easily shown that

$$\mathbf{M}_A = \mathbf{r}_{Q/A} \times \mathbf{F} = \mathbf{r}_{B/A} \times \mathbf{F} \tag{2.1.2}$$

where Q is any point on the line of action of **F**, as shown in Fig. 2.1.

A moment must always be designated with respect to a specific point, and is represented by a vector perpendicular to the plane of **r** and **F**. Moment vectors are denoted by double-headed arrows to distinguish them from force vectors. The component of the moment vector in the direction of any axis (line) passing through A is said to be the moment of the force about that axis. If the direction of the axis is defined by the unit vector $\boldsymbol{\lambda}$, the moment component, a scalar along that axis, is given by

$$M_\lambda = \mathbf{r} \times \mathbf{F} \cdot \boldsymbol{\lambda}\,. \tag{2.1.3}$$

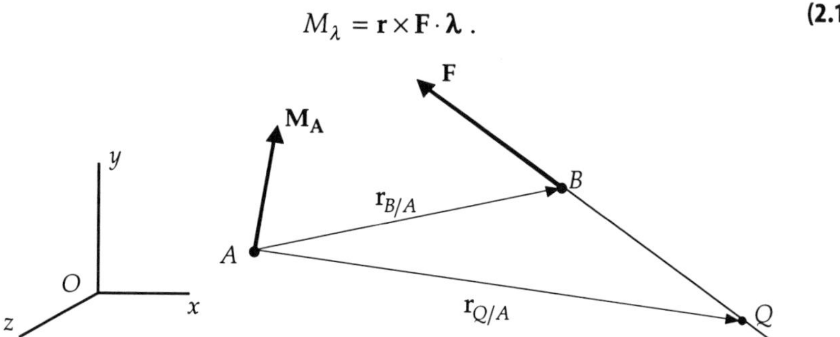

Figure 2.1 Moment of F about *A*.

The "turning effect" about the hinge axis of a door due to the application of a force to the doorknob is an example. If $\boldsymbol{\lambda} = \mathbf{k}$ in Eq. 2.1.3, the moment component is labeled M_z, the scalar moment about the z-axis, and we find that

$$M_z = xF_y - yF_x \tag{2.1.4}$$

where x and y are the components of $\mathbf{r}_{B/A}$ of Fig. 2.1. Similarly, for $\boldsymbol{\lambda} = \mathbf{j}$ and $\boldsymbol{\lambda} = \mathbf{i}$, respectively,

$$M_y = zF_x - xF_z \tag{2.1.5}$$

$$M_x = yF_z - zF_y\,. \tag{2.1.6}$$

Furthermore, it may be shown from Eq. 2.1.4 that the scalar moment about any point A in the plane of **r** and **F** is given by

$$M_A = Fd \tag{2.1.7}$$

where d is the perpendicular distance from A to the line of action of **F**.

The *resultant* of a system of forces is the equivalent force and moment of the total system at any point. Thus, the resultant of a concurrent force system is a single force acting at the point of concurrency, and being the vector sum of the individual forces. In contrast, the resultant of a non-concurrent system depends upon the point at which it is determined, and in general consists of a resultant force and resultant moment. Actually, the resultant force (the vector sum of the individual forces) is the same at all points, but the resultant moment will vary (in magnitude and direction) from point to point.

The resultant of a pair of equal, but oppositely directed parallel forces, known as a *couple*, is simply a moment having a magnitude Fd where F is the magnitude of the forces and d the perpendicular distance between their lines of action. Figure 2.2 illustrates equivalent couples and the curly symbol often used.

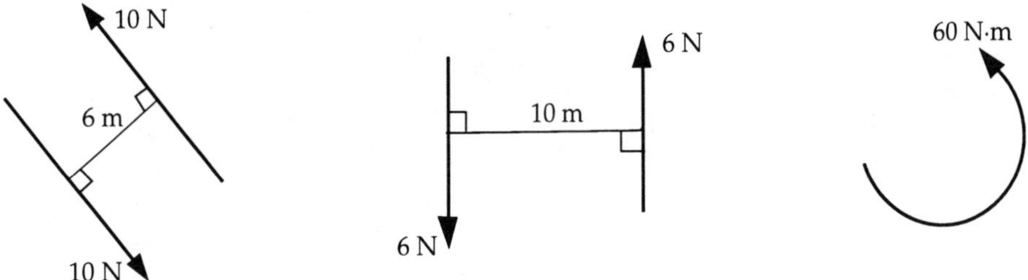

Figure 2.2 Equivalent couples.

Example 2.1

Determine the resultant force for the (a) plane, and (b) space concurrent systems shown.

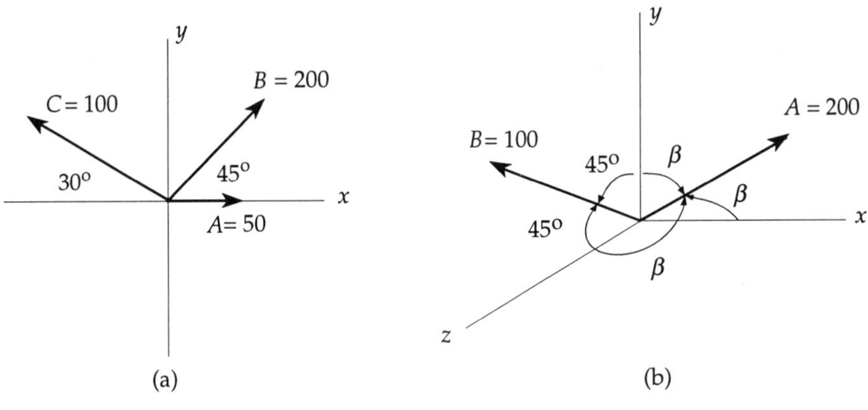

Solution. (a) $\mathbf{A} = 50\mathbf{i}$, $\quad \mathbf{B} = 141.4\mathbf{i} + 141.4\mathbf{j}$ $\quad \mathbf{C} = -86.6\mathbf{i} + 50\mathbf{j}$

$$\mathbf{R} = \mathbf{A} + \mathbf{B} + \mathbf{C} = 104.8\mathbf{i} + 191.4\mathbf{j}$$

(b) $\mathbf{A} = 200(\mathbf{i} + \mathbf{j} + \mathbf{k})/\sqrt{3}$, $\quad \mathbf{B} = 70.7\mathbf{j} + 70.7\mathbf{k}$

$$\mathbf{R} = 115.47\mathbf{i} + 186.17\mathbf{j} + 186.17\mathbf{k}$$

Example 2.2

Determine the moment of force **F**

(a) with respect to the origin. (b) with respect to point Q.

(c) about the axis OQ. (d) about the x-axis. (e) about the y-axis.

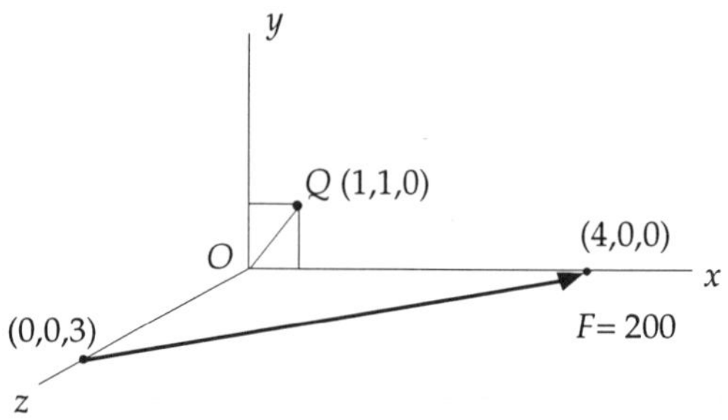

Solution. $\mathbf{F} = 0.8(200)\mathbf{i} - 0.6(200)\mathbf{k} = 160\mathbf{i} - 120\mathbf{k}$.

(a) $\mathrm{M}_O = 4\mathrm{i} \times (160\mathrm{i} - 120\mathrm{k}) = 3\mathrm{k} \times (160\mathrm{i} - 120\mathrm{k}) = 480\mathrm{j}$.

(b) $\mathrm{M}_Q = (3\mathrm{i} - \mathrm{j}) \times (160\mathrm{i} - 120\mathrm{k}) = (-\mathrm{i} - \mathrm{j} + 3\mathrm{k}) \times (160\mathrm{i} - 120\mathrm{k}) = 120\mathrm{i} + 360\mathrm{j} + 160\mathrm{k}$.

(c) $M_{OQ} = (120\mathrm{i} + 360\mathrm{j} + 160\mathrm{k}) \cdot (\mathrm{i} + \mathrm{j}) / \sqrt{2}$

$= (120 + 360)/\sqrt{2} = 480/\sqrt{2} = 339.5$.

(d) $M_x = 0$ (force F intersects the x-axis).

(e) $M_y = Fd = 200(3\cos\beta) = 200(2.4) = 480\ \mathrm{N \cdot m}$.

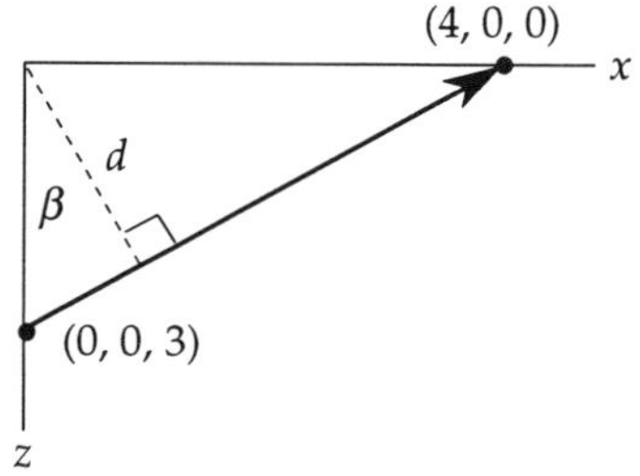

2.2 Equilibrium

If the system of forces acting on a body is one whose resultant is absolutely zero (vector sum of all forces is zero, and the resultant moment of the forces about every point is zero) the body is in *equilibrium*. Mathematically, equilibrium requires the equations

$$\sum \mathbf{F} = 0, \qquad \sum \mathbf{M}_A = 0 \tag{2.2.1}$$

to be simultaneously satisfied, with A arbitrary. These two vector equations are equivalent to the six scalar equations

$$\sum F_x = 0 \qquad \sum M_x = 0 \tag{2.2.2}$$
$$\sum F_y = 0 \qquad \sum M_y = 0$$
$$\sum F_z = 0 \qquad \sum M_z = 0$$

which must hold at every point A for any orientation of the xyz-axes. The moment components in Eq. 2.2.2 are the coordinate axes components of $\mathbf{M}_A$. If the forces are concurrent and their vector sum is zero, the sum of moments about every point will be satisfied automatically.

If all the forces act in a single plane, say the xy-plane, one of the above force equations, and two of the moment equations are satisfied identically, so that equilibrium requires only

$$\sum F_x = 0, \qquad \sum F_y = 0, \qquad \sum M_z = 0. \tag{2.2.3}$$

In this case we can solve for only three unknowns instead of six as when Eqs. 2.2.2 are required.

The solution for unknown forces and moments in equilibrium problems rests firmly upon the construction of a good *free body diagram*, abbreviated FBD, from which the detailed equations 2.2.2 or 2.2.3 may be obtained. A free body diagram is a neat sketch of the body (or of any appropriate portion of it) showing all forces and moments acting on the body, together with all important linear and angular dimensions.

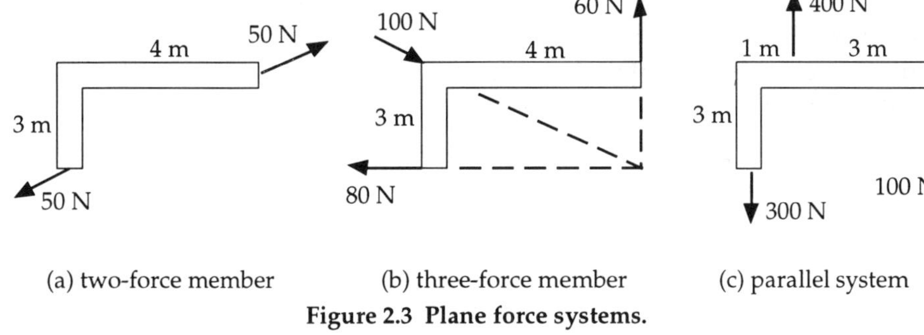

(a) two-force member (b) three-force member (c) parallel system

Figure 2.3 Plane force systems.

A body in equilibrium under the action of two forces only is called a *two-force member*, and the two forces must be equal in magnitude and oppositely directed along the line joining their points of application. If a body is in equilibrium under the action of three forces (a *three-force member*) those forces must be coplanar, and concurrent (unless they form a parallel system). Examples are shown in Fig. 2.3.

A knowledge of the possible reaction forces and moments at various supports is essential in preparing a correct free body diagram. Several of the basic reactions are illustrated in Fig. 2.4 showing a block of concrete subjected to a horizontal pull P, Fig. 2.4a, and a cantilever beam carrying both a distributed and concentrated load, Fig. 2.4b. The correct FBDs are to the right of the sketches.

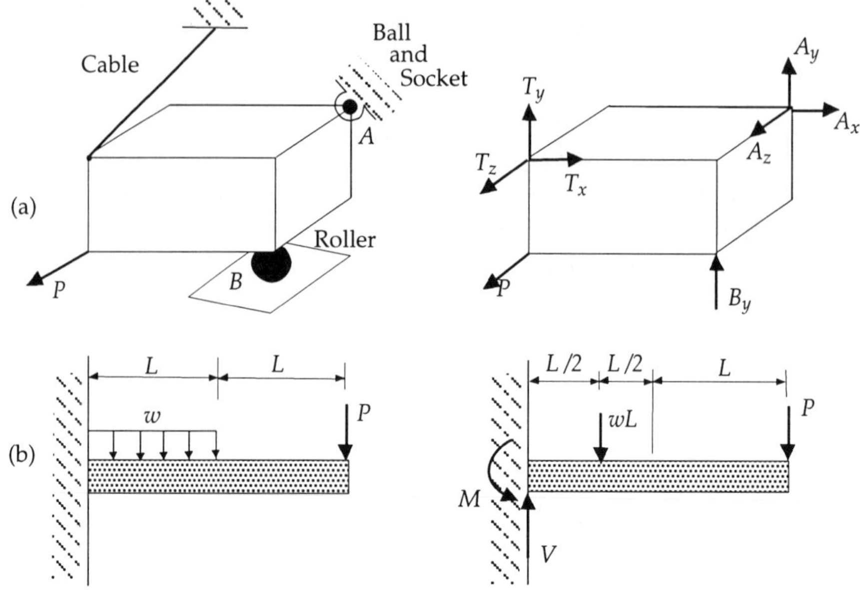

Figure 2.4 Free body diagrams.

Example 2.3

Determine the tension in the two cables supporting the 700 N block.

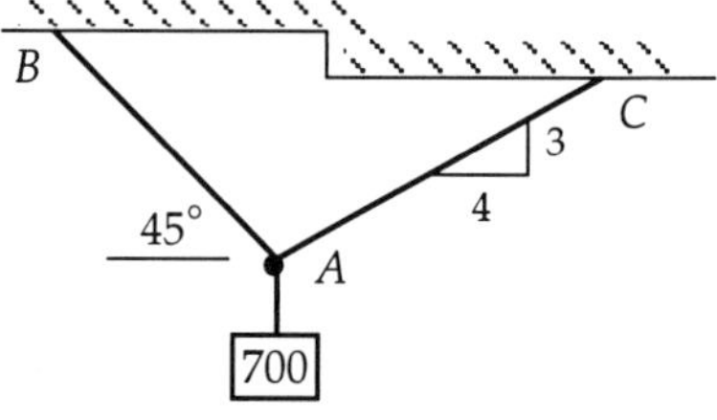

Solution. Construct the FBD of junction A of the cables. Sum forces in x and y directions:

$$\sum F_x = -0.707T_{AB} + 0.8T_{AC} = 0$$

$$\sum F_y = 0.707T_{AB} + 0.6T_{AC} - 700 = 0.$$

Solve, simultaneously, and find

$$T_{AC} = 700/1.4 = 500 \text{ N}$$

$$T_{AB} = 0.8(500)/0.707 = 565.8 \text{ N}.$$

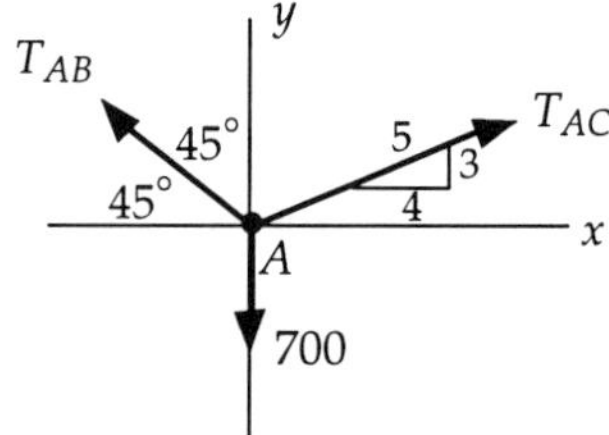

Alternative solution. Draw the force polygon (vector sum of forces) which must close for equilibrium. Determine angles, and use law of sines:

$$\frac{700}{\sin 81.87^\circ} = \frac{T_{AB}}{\sin 53.1^\circ} = \frac{T_{AC}}{\sin 45^\circ}$$

$$T_{AC} = \frac{700(0.707)}{(0.99)} = 500 \text{ N}$$

$$T_{AB} = \frac{700(0.8)}{(0.99)} = 565.8 \text{ N}.$$

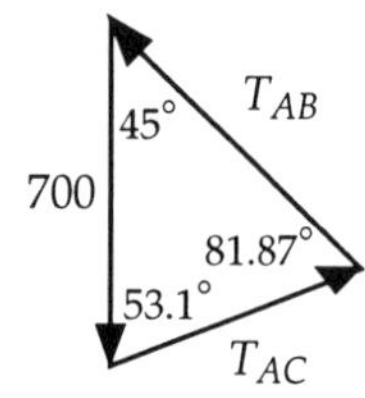

Example 2.4

A 12 m bar weighing 140 N is hinged to a vertical wall at A, and supported by the cable BC. Determine the tension in the cable together with the horizontal and vertical components of the force reaction at A.

Solution. Construct the FBD showing force components at A and B. Write the equilibrium equations and solve:

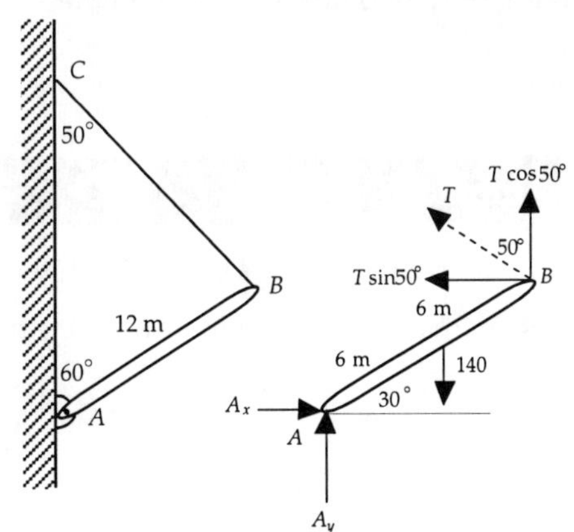

$$\sum M_A = 6T\sin 50^\circ + 6\sqrt{3}T\cos 50^\circ - 140(6)\sqrt{3}/2 = 0$$

$$\therefore T = 64.5 \text{ N}$$

$$\sum F_x = A_x - T\sin 50^\circ = A_x - 64.5(0.766) = 0$$

$$\therefore A_x = 49.4 \text{ N}$$

$$\sum F_y = A_y + T\cos 50^\circ - 140 = A_y + 64.5(0.643) - 140 = 0$$

$$\therefore A_y = 98.5 \text{ N}.$$

2.3 Trusses and Frames

Simple pin-connected trusses and plane frames provide us with elementary examples of structures that may be solved by the equilibrium concepts of statics.

The classic truss problem resembles the one-lane country bridge, as shown schematically in Fig. 2.5a. All members are assumed to be two-force members and are therefore in simple (axial) tension or compression. All loads are assumed to act at the joints (labeled A, B, C, etc.) where the members are pinned together. External reactions such as A_x, A_y and E_y may be determined as a non-concurrent force problem from a FBD of the entire truss. Following that, the internal forces in the members themselves may be determined from a FBD of each joint in turn

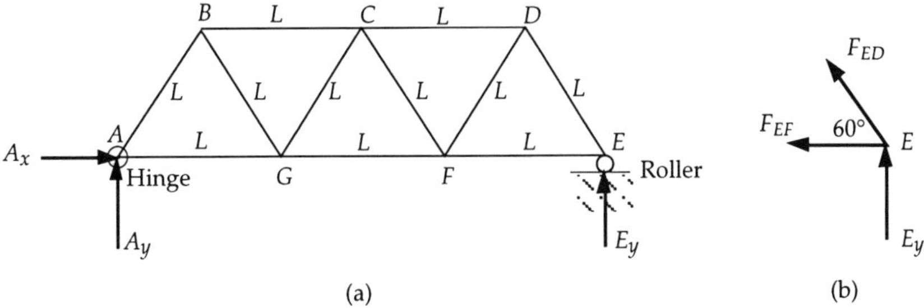

Figure 2.5 Simple truss.

(method of joints) starting, for example, with joint E of the truss as shown in Fig. 2.5b. Thus, we solve a sequence of concurrent force problems at successive joints having only two unknowns. As noted by Fig. 2.5b we may assume the unknown internal forces such as F_{ED}, F_{EF}, etc., to be tension. A negative result indicates compression.

Example 2.5

Using the right sub-truss of Fig 2.6 determine the forces in members CD, CF and FG.

Solution. Summing moments about pin F of the FBD of the right sub-truss,

$$\sum M_F = F_{DC}(\sqrt{3}L/2) + 10P(L) - 6P(L/2) = 0.$$

$$\therefore F_{DC} = -8.085P \text{ (comp)}.$$

Summing vertical forces on the sub-truss,

$$\sum F_y = 10P - 3P - 6P + 0.866F_{FC} = 0.$$

$$\therefore F_{FC} = -1.156P \text{ (comp)}.$$

Summing horizontal forces,

$$\sum F_x = -F_{FG} + 8.085P + 1.156P/2 = 0.$$

$$\therefore F_{FG} = 8.663P \text{ (tens)}.$$

Note: These results agree with those determined by the method of joints in Example 2.6

Example 2.6

Determine the forces in the members of the pin-connected truss loaded as shown below. All members have length L.

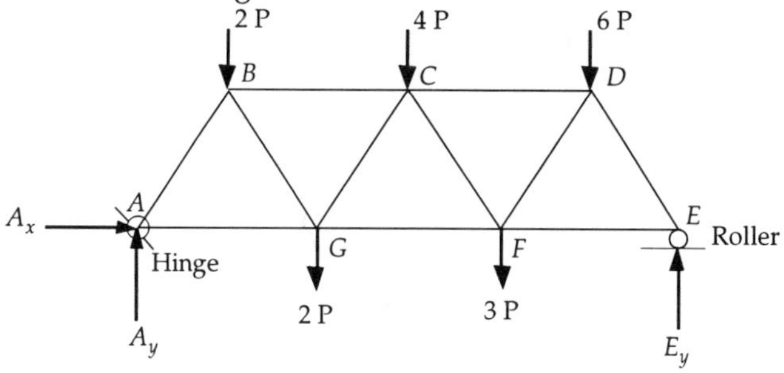

Solution. Summing moments about pin A we solve for the reaction at roller E (counterclockwise moments are positive):

$$\sum M_A = 3LE_y - 2PL - 3P(2L) - 2P(L/2) - 4P(3L/2) - 6P(5L/2) = 0.$$

$$\therefore E_y = 10P.$$

Also,

$$\sum M_E = -3LA_y + 3P(L) + 2P(2L) + 6P(L/2) + 4P(3L/2) + 2P(5L/2) = 0.$$

$$\therefore A_y = 7P.$$

Summing Horizontal forces,

$$\sum F_x = A_x = 0.$$

Consider joint E (see FBD at right),

$$\sum F_y = T_{ED} \sin 60^\circ + 10P = 0$$

$$\therefore T_{ED} = -11.55P \text{ (comp).}$$

$$\sum F_x = -T_{EF} - T_{ED} \cos 60^\circ = 0$$

$$\therefore T_{EF} = 5.775P \text{ (tens).}$$

Consider next joint D (see FBD at right),

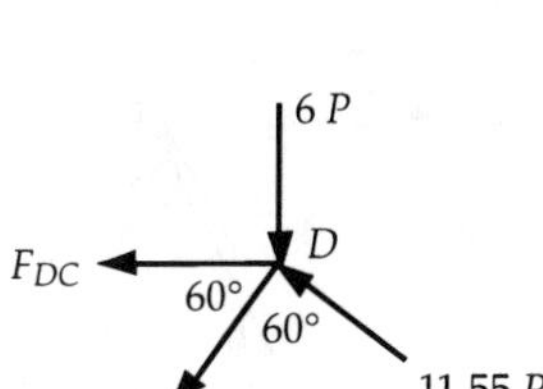

$$\sum F_y = -6P - 0.866F_{DF} + 0.866(11.55P) = 0.$$

$$\therefore F_{DF} = 4.62P \text{ (tens).}$$

$$\sum F_x = -F_{DC} - 0.5(11.55P) - 0.5(4.62P) = 0$$

$$\therefore F_{DC} = -8.085P \text{ (comp).}$$

Consider next joint F (see FBD at right),

$$\sum F_y = -3P + 0.866(4.62P) + 0.866F_{FC} = 0.$$

$$\therefore F_{FC} = -1.156P \text{ (comp).}$$

$$\sum F_x = -F_{FG} - 0.5(-1.156P) + 0.5(4.62P) + 5.775P = 0$$

$$\therefore F_{FG} = 8.663P \text{ (tens).}$$

Note: This example should be completed by considering joint C next, then joint G, and so on.

If the internal forces in only a few selected members are required, the *method of sections* may be used. For example, to obtain only the forces in members CD, CF and GF of the above truss, we "section" it into two portions by cutting across those members as shown in Fig. 2.6. Each portion becomes a sub-truss, and the internal forces of the sectioned members become external reactions of the two sub-trusses. Both force $(\sum F = 0)$ and moment $(\sum M_A = 0)$ equations are useful in this method.

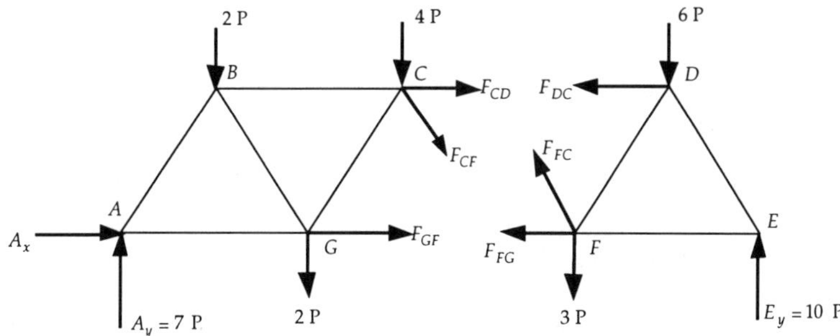

Figure 2.6 Sectioned truss.

A plane frame is a structure that consists of both two-force and three-force members, or even four-force members, etc. Loads may act at any location on the frame. The problem is to determine the components of the reactions at all pins of the frame. This usually requires not only a FBD of the entire frame, but also a FBD of each member. We illustrate by the following example.

Example 2.7

For the frame shown, determine the horizontal and vertical components of the reactions at all pins.

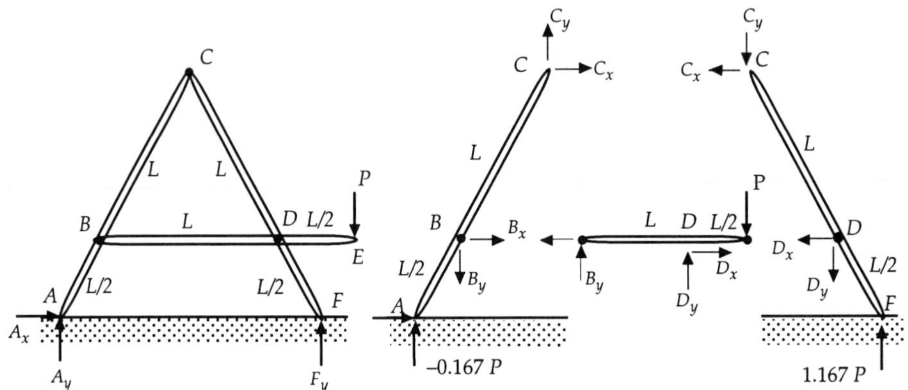

Solution. From a FBD of entire frame,

$$\sum F_x = A_x = 0. \qquad \therefore A_x = 0.$$
$$\sum M_A = (1.5L)F_y - (1.75L)P = 0. \qquad \therefore F_y = 1.167P$$
$$\sum F_y = A_y + 1.167P - P = 0. \qquad \therefore A_y = -0.167P$$

From a FBD of a member BDE,

$$\sum M_D = -(L)B_y - (L/2)P = 0. \qquad \therefore B_y = -0.5P.$$
$$\sum F_y = -0.5P + D_y - P = 0. \qquad \therefore D_y = 1.5P.$$

Transfer the vertical components at B and D of members BDE to members ABC and CDF by changing directions (action and reaction principle). From a FBD of member ABC,

$$\sum F_y = -0.167P + 0.5P + C_y = 0. \qquad \therefore C_y = -0.333P$$
$$\sum M_B = -(\sqrt{3}L/2)C_x - (L/2)0.333P + (L/4)0.167P = 0. \qquad \therefore C_x = -0.144P$$
$$\sum F_x = B_x - 0.144P = 0. \qquad \therefore B_x = 0.144P$$

Transfer horizontal components B_x and C_x to members BDE and CDF. From a FBD of member BDE,

$$\sum F_x = -0.144P + D_x = 0. \qquad \therefore D_x = 0.144P.$$

From a FBD of member CDF,

$$\sum F_y = 0.333P + D_y + 1.167P = 0. \qquad \therefore D_y = -1.5P.$$

Completed FBDs of all members are shown on the next page.

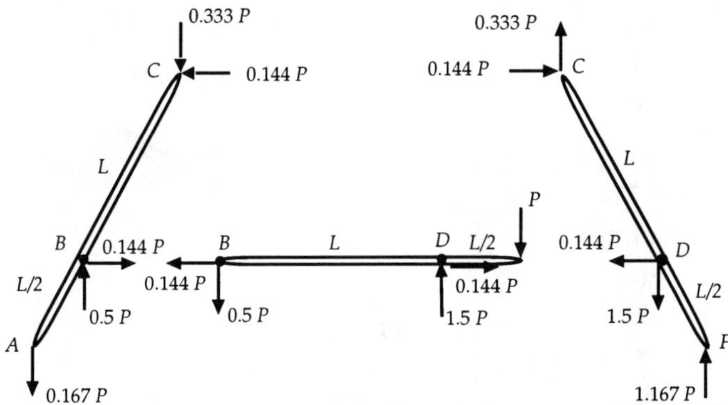

2.4 Friction

Consider a block of weight W at rest on a dry, rough, horizontal plane, Fig. 2.7a. Let a horizontal force P act to the right on the block, Fig. 2.7b. A *friction force* F is developed at the surface of contact between the block and plane, and maintains equilibrium ($F = P$) as long as $P < \mu_s N = \mu_s W$, where μ_s is the *static coefficient of friction*. If P is increased until $P = P^* = \mu_s N = \mu_s W$ (Fig. 2.7c), the equilibrium state is on the verge of collapse, and motion to the right is impending. In summary, dry *Coulomb friction* described here is governed by

$$F \leq \mu_s N. \tag{2.4.1}$$

If relative motion between the block and the plane occurs (Fig. 2.7d), the friction force is given by

$$F = \mu_k N \tag{2.4.2}$$

where μ_k is the *kinetic coefficient of friction*, slightly less in value than μ_s, with both coefficients having a range $0 \leq \mu \leq 1$.

At impending motion, the equality holds in Eq. 2.4.1 and the resultant R of F and N makes an angle ϕ with the normal N, called the *friction angle.* From Fig. 2.8a it is clear that

$$\phi = \tan^{-1} \mu_s. \tag{2.4.3}$$

Also, it may be readily shown that a block will remain at rest (in equilibrium) on a rough inclined plane, Fig. 2.8b, as long as $\beta < \phi$ for the surfaces of contact.

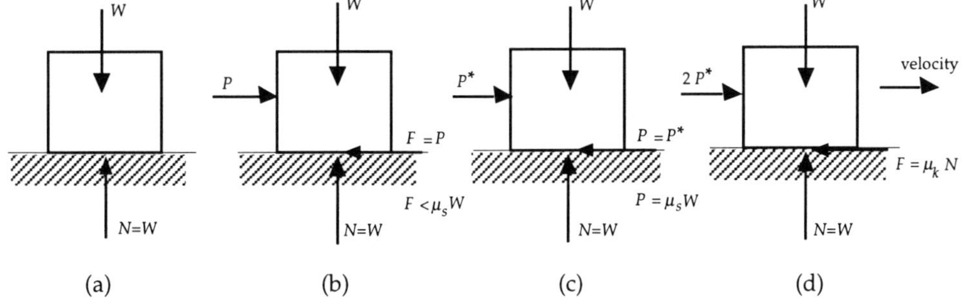

Figure 2.7 Friction forces.

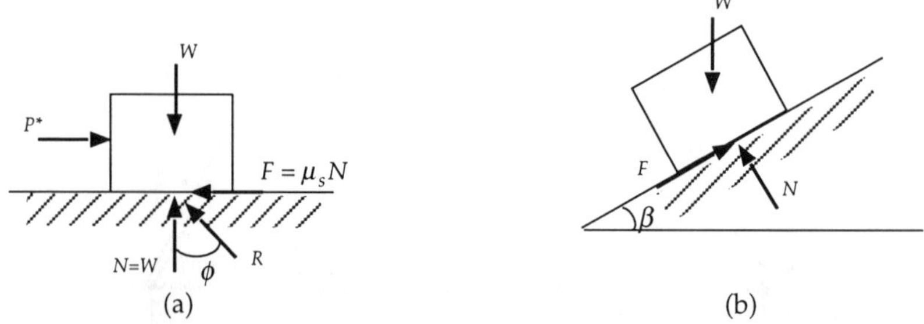

Figure 2.8 Friction angle.

Example 2.8

Determine the horizontal force P required to cause impending motion of the 50 kg block, (a) up the plane, (b) down the plane, if $\mu_s = 0.6$ between the block and plane.

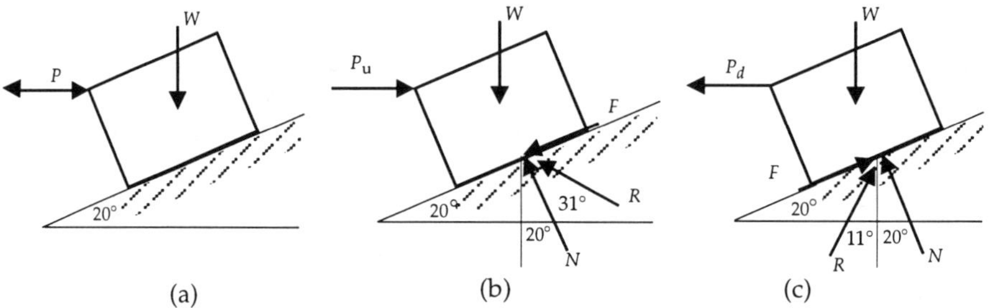

Solution. Note that, $\phi = \tan^{-1}(0.6) = 31°$ so that the block would remain in place if undisturbed.

(a) At impending motion up the plane, Fig. b, the resultant R of the friction force F and normal force N makes an angle of 51° to the right of the vertical and from the FBD of the block,

$$\sum F_x = P_u - R\sin 51^\circ = 0$$

$$\sum F_y = R\cos 51^\circ - 50 \times 9.81 = 0.$$

Solving these equations,

$$R = 779 \text{ N}, \qquad P_u = 606 \text{ N}.$$

(b) At impending motion down the plane, R makes an angle of 11° to the left of the vertical, Fig. c, so that now

$$\sum F_x = -P_d + R\sin 11° = 0$$

$$\sum F_y = R\cos 11^\circ - 50 \times 9.81 = 0.$$

Therefore

$$R = 499 \text{ N}, \qquad P_d = 95.2 \text{ N}.$$

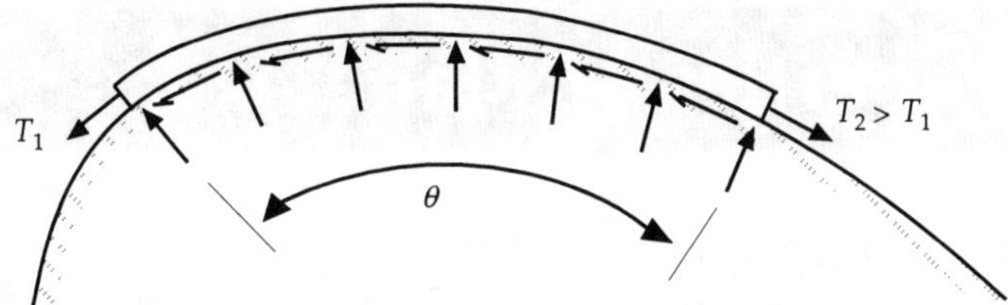

Figure 2.9 A belt with friction.

If a belt, or rope, is pressed firmly against some portion of a rough stationary curved surface, and pulled in one direction or the other, the tension in the belt will increase in the direction of pull due to the frictional resistance between the belt and surface, as shown in Fig. 2.9. It may be shown that

$$T_2 = T_1 e^{\mu_s \theta} \qquad T_2 > T_1 \tag{2.4.4}$$

where θ is the angle of contact, in radians, and μ_s is the static coefficient of friction.

Example 2.9

A 100 kg block rests on a 30° rough inclined plane ($\mu_s = 0.4$) and is attached by a rope to a mass m in the arrangement shown. If the static coefficient of friction between the rope and the circular support is 0.25, determine the maximum m that can be supported without slipping.

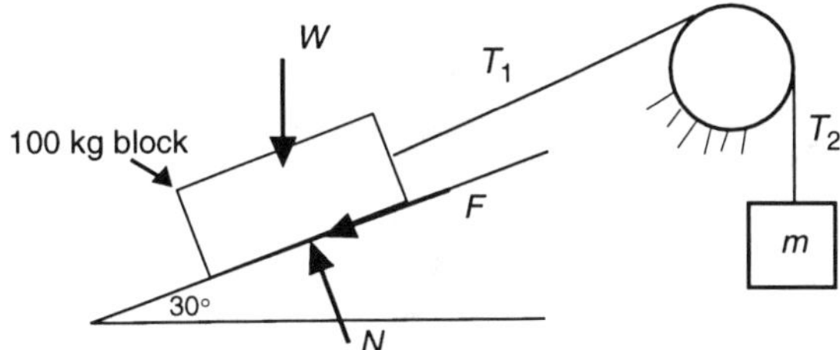

Solution. Summing forces perpendicular to the plane we determine

$$N = 981\cos 30° = 849.5 \text{ N}.$$

Thus,

$$F = 0.4(849.5) = 339.8 \text{ N}$$

at impending motion, and by summing forces along the plane

$$T_1 = 339.8 + 981\sin 30° = 830.3 \text{ N}.$$

For the circular support we have,

$$T_2 = T_1 e^{(0.25)(120\times\pi/180)} = 830.3(1.69) = 1403 \text{ N}$$

and

$$m = 1403/9.81 = 143 \text{ kg}.$$

2.5 Properties of Plane Areas

Associated with every plane area A (in the xy-plane, for example) there is a point C, known as the *centroid*, whose coordinates $\bar{x}$ and $\bar{y}$ are defined by the integrals

$$\bar{x} = \frac{\int_A x dA}{\int_A dA} \qquad \bar{y} = \frac{\int_A y dA}{\int_A dA} \tag{2.5.1}$$

where dA is differential element of area having coordinates x and y as shown in Fig. 2.10a.

Although the integrals in Eq. 2.5.1 may be evaluated by a double integration using $dA = dxdy$, in practice it is often advantageous to calculate centroidal coordinates by a single integration using either a horizontal or vertical strip for dA (Fig. 2.10b and Fig. 2.10c) for which Eq. 2.5.1 may be expressed in the form

$$\bar{x} = \frac{\int_A x_e dA}{\int_A dA} \qquad \bar{y} = \frac{\int_A y_e dA}{\int_A dA} \tag{2.5.2}$$

where x_e and y_e are the coordinates of the centroids of the strip elements as shown.

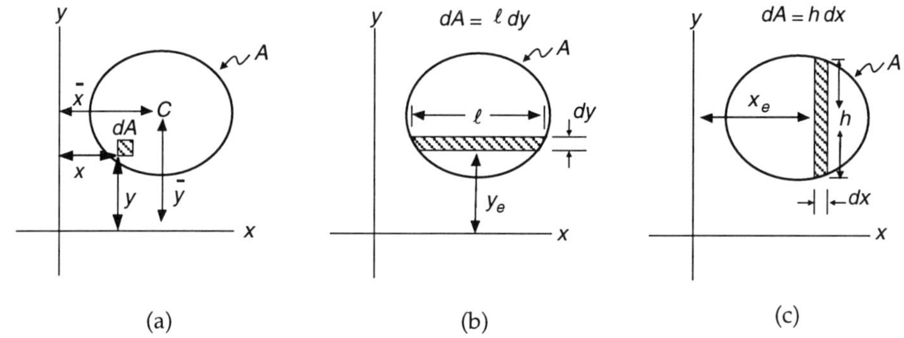

Figure 2.10 Plane area centroid.

The *plane moments of inertia* I_x and I_y of A with respect to the x and y axes, respectively, are defined by

$$I_x = \int_A y^2 dA, \qquad I_y = \int_A x^2 dA \tag{2.5.3}$$

where in integrating for I_x we use the horizontal strip for dA, and for I_y the vertical strip. The *polar moment of inertia* with respect to the origin O is defined by

$$J_O = \int_A r^2 dA = \int_A (x^2 + y^2) dA = I_x + I_y. \tag{2.5.4}$$

All moments of inertia may be expressed in terms of their respective *radii of gyration*. Thus,

$$I_x = r_x{}^2 A, \qquad I_y = r_y{}^2 A, \qquad J_O = r_O{}^2 A. \tag{2.5.5}$$

The *product of inertia* of A is defined with respect to a pair of perpendicular axes. For the coordinate axes we have

$$I_{xy} = \int_A xy\,dA \tag{2.5.6}$$

which normally must be evaluated by a double integration over A. If either one (or both) of the reference axes is an axis of symmetry the product of inertia is zero relative to that pair of axes.

The *transfer theorem* establishes a relationship between the moment of inertia about an arbitrary axis and the moment of inertia about a parallel axis passing through the centroid C. Thus

$$I_P = I_C + Ad^2, \qquad J_P = J_C + Ad^2 \tag{2.5.7}$$

where the subscript C indicates the centroidal moment of inertia, subscript P indicates the moment of inertia about the parallel axis, A is the area, and d the distance separating the two axes. Similarly, for products of inertia,

$$I_{x_P y_P} = I_{x_C y_C} + Ax_1 y_1 \tag{2.5.8}$$

where x_1 and y_1 are the distances between axes x and x_C, and y and y_C, respectively.

Example 2.10

For the shaded area shown by the sketch, determine $\overline{x}$, $\overline{y}$, I_x, I_y, J_O, I_{xy}, $(I_x)_C$, $(I_y)_C$, J_C and $I_{x_C y_C}$. Units are mm.

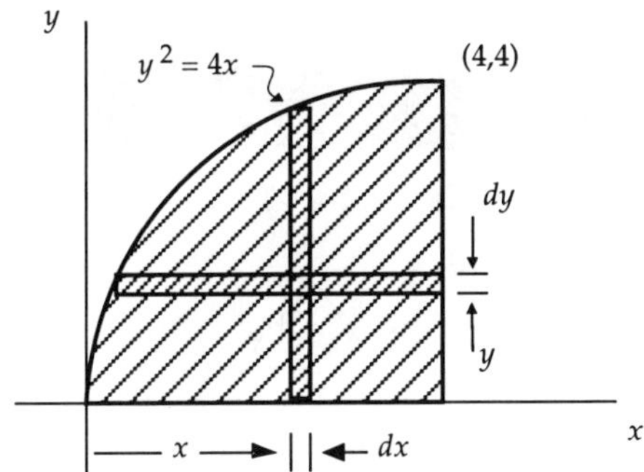

Solution. First we calculate A using the vertical strip, $dA = ydx = 2\sqrt{x}dx$:

$$A = \int_0^4 2\sqrt{x}dx = 32/3 \text{ mm}^2.$$

Thus, from Eq. 2.5.1, with $dA = 2\sqrt{x}dx$,

$$\overline{x} = \frac{3}{32}\int_0^4 x(2\sqrt{x})dx = 2.4 \text{ mm}$$

and, using the horizontal strip $dA = (4 - y^2/4)dy$,

$$\overline{y} = \frac{3}{32}\int_0^4 y(4 - y^2/4)dy = 1.5 \text{ mm}.$$

From Eq. 2.5.3, with $dA = (4 - y^2/4)dy$,

$$I_x = \int_0^4 y^2(4 - y^2/4)dy = 34.13 \text{ mm}^4$$

and, using $dA = 2\sqrt{x}dx$,

$$I_y = \int_0^4 x^2(2\sqrt{x})dx = 73.14 \text{ mm}^4.$$

From Eq. 2.5.4

$$J_O = I_x + I_y = 34.13 + 73.14 = 107.27 \text{ mm}^4.$$

From Eq. 2.5.6, with $dA = dxdy$,

$$I_{xy} = \int_0^4 \int_0^{2\sqrt{x}} xy\,dydx = \int_0^4 \left[\frac{xy^2}{2}\right]_0^{2\sqrt{x}} dx = \int_0^4 2x^2 dx = 42.67 \text{ mm}^4.$$

From Eq. 2.5.7,

$$(I_x)_C = I_x - Ad^2 = 34.13 - 10.67(1.5)^2 = 10.13 \text{ mm}^4$$

$$(I_y)_C = I_y - Ad^2 = 73.14 - 10.67(2.4)^2 = 11.68 \text{ mm}^4$$

$$J_C = J_O - Ad^2 = 107.27 - 10.67\left[(1.5)^2 + (2.4)^2\right] = 21.80 \text{ mm}^4.$$

Note that moments of inertia are always minimum about a centroidal axis. Finally, from Eq. 2.5.8,

$$I_{x_C y_C} = I_{xy} - A(-1.5)(-2.4) = 42.67 - 38.40 = 4.27 \text{ mm}^4$$

Example 2.11

Determine I_x and J_O for the circular sector shown below.

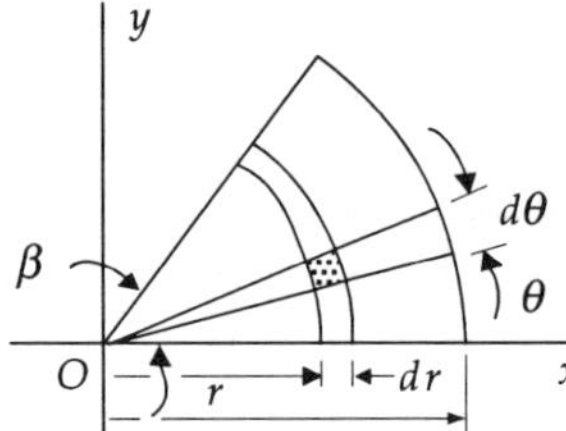

Solution. Using polar coordinates with $dA = rdrd\theta$ and $y = r\sin\theta$ in Eq. 2.5.3:

$$I_x = \int_0^\beta \int_0^a (r\sin\theta)^2 rdrd\theta = \frac{a^4}{4}\left[\frac{\beta}{2} - \frac{\sin 2\beta}{4}\right].$$

When $\beta = \pi/2$,

$$I_x = \pi a^4/16.$$

From Eq. 2.5.4,

$$J_O = \int_0^\beta \int_0^a r^3 drd\theta = \frac{a^4\beta}{4}.$$

When $\beta = \pi/2$,

$$J_O = \pi a^4/8.$$

TABLE 2.1. Properties of Areas.

Shape	Dimensions	Centroid	Inertia
Rectangle		$\bar{x} = b/2$ $\bar{y} = h/2$	$I_C = bh^3/12$ $I_x = bh^3/3$ $I_y = hb^3/3$
Triangle		$\bar{y} = h/3$	$I_C = bh^3/36$ $I_x = bh^3/12$
Circle		$\bar{x} = 0$ $\bar{y} = 0$	$I_x = \pi r^4/4$ $J = \pi d^4/32$
Quarter Circle		$\bar{y} = 4r/3\pi$	$I_x = \pi r^4/16$
Half Circle		$\bar{y} = 4r/3\pi$	$I_x = \pi r^4/8$

The properties of common areas may be determined by integration. A brief list is given in Table 2.1. Using data from Table 2.1, we may calculate centroids and moments of inertia of *composite areas* made up of combinations of two or more (including cutouts) of the common areas. Thus,

$$\bar{x} = \frac{\sum_{i=1}^{N} x_i A_i}{\sum_{i=1}^{N} A_i}, \qquad \bar{y} = \frac{\sum_{i=1}^{N} y_i A_i}{\sum_{i=1}^{N} A_i} \tag{2.5.9}$$

where N is the number of areas, and x_i is the centroidal distance for area A_i. Likewise, for moments of inertia

$$I = \sum_{i=1}^{N} I_i = I_1 + I_2 + \ldots + I_N. \tag{2.5.10}$$

An example illustrates.

Example 2.12

Determine the centroidal coordinates, and I_x and I_y for the composite area shown.

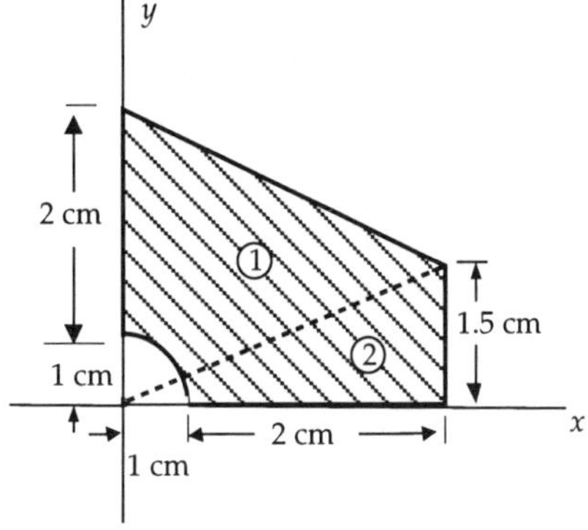

Solution. Decompose the composite into two triangular areas 1 and 2, and the negative quarter circular area 3:

$$A = A_1 + A_2 + A_3 = 4.5 + 2.25 - \pi/4 = 5.97 \text{ cm}^2.$$

$$A\bar{x} = x_1A_1 + x_2A_2 + x_3A_3$$
$$= (1)(4.5) + (2)(2.25) + (4/3\pi)(-\pi/4) = 8.67.$$
$$\therefore \bar{x} = 8.67/5.97 = 1.45 \text{ cm}.$$

$$A\bar{y} = y_1 \cdot A_1 + y_2A_2 + y_3A_3$$
$$= (1.5)(4.5) + (0.5)(2.25) + (4/3\pi)(-\pi/4) = 7.54.$$
$$\therefore \bar{y} = 7.54/5.97 = 1.26 \text{ cm}.$$

$$I_x = I_{1x} + I_{2x} + I_{3x}$$
$$= \left\{2\left[3(1.5)^3/36\right] + (4.5)(1.5)^2\right\} + 3(1.5)^3/12 - \pi/16 = 11.33 \text{ cm}^4,$$

where we have used the parallel-axis theorem to obtain I_{1x}. Finally,

$$I_y = I_{1y} + I_{2y} + I_{3y}$$
$$= 3(3)^3/12 + 1.5(3)^3/4 - \pi/16 = 16.68 \text{ cm}^4.$$

2.6 Properties of Masses and Volumes

The coordinates of the *center of gravity* G of an arbitrary mass m occupying a volume V of space are defined by

$$x_G = \frac{\int_V x\rho dV}{\int_V \rho dV}, \qquad y_G = \frac{\int_V y\rho dV}{\int_V \rho dV}, \qquad z_G = \frac{\int_V z\rho dV}{\int_V \rho dV} \tag{2.6.1}$$

where ρ is the mass density, dV is the differential element of volume, and x, y and z are the coordinates of dV, as shown in Fig. 2.11.

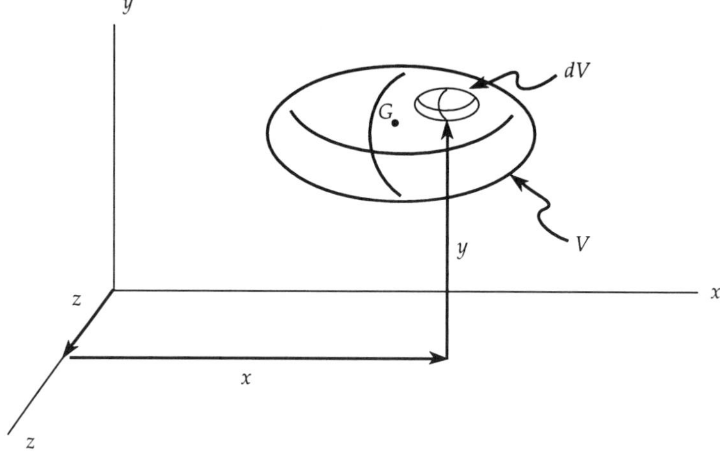

Figure 2.11 Center of gravity of an arbitrary mass.

The density may be a function of the space variables, $\rho = \rho(x,y,z)$. If the density is constant throughout the volume, the integrals in Eq. 2.6.1 reduce to

$$x_C = \frac{\int_V x dV}{V}, \qquad y_C = \frac{\int_V y dV}{V}, \qquad z_C = \frac{\int_V z dV}{V} \tag{2.6.2}$$

which defines the coordinates of the centroid C of volume V. If ρ is constant, G and C coincide for a given body. As with areas, if an axis of symmetry exists for the volume, C is on that axis. The coordinates of G and C are readily calculated for geometries having an axis of revolution.

Example 2.13

Let the area of Example 2.10 be rotated about the x-axis to form a solid (volume) of revolution. Determine (a) G for the solid if $\rho = \rho_o x$ where ρ_o is a constant, (b) C for the volume.

Solution. Since there is symmetry about the x-axis, and since ρ is at most a function of x, we have

$$y_G = z_G = y_C = z_C = 0$$

(a) Let the element of mass be a thin disk for which $\rho dV = \rho_o x \pi y^2 dx$. Therefore,

$$x_G = \frac{\int_0^4 \rho_o \pi x^2 (4x) dx}{\int_0^4 \rho_o \pi x (4x) dx} = \frac{4 \times 4^4/4}{4 \times 4^3/3} = 3.00 \text{ mm.}$$

(b) For the same element with $\rho = \rho_o$

$$x_C = \frac{\int_0^4 \rho_o \pi x (4x) dx}{\int_0^4 \rho_o \pi (4x) dx} = \frac{4 \times 4^3/3}{4 \times 4^2/2} = 2.67 \text{ mm.}$$

Because the topic of *mass moments of inertia* properly belongs in the subject of dynamics, and is not a factor in statics, we present only a brief comment or two here for comparison with area moments of inertia. The mass moment of inertia of the three-dimensional body in Fig. 2.11 about any axis is defined by the integral

$$I = \int_V r^2 \rho dV = \int_V r^2 dm \tag{2.6.3}$$

where r is the perpendicular distance of the mass element dm from the axis. With respect to the coordinate axes, Eq. 2.6.3 may be specialized to yield

$$I_x = \int_V (y^2 + z^2) dm \tag{2.6.4}$$

$$I_y = \int_V (z^2 + x^2) dm \tag{2.6.5}$$

$$I_z = \int_V (x^2 + y^2) dm. \tag{2.6.6}$$

There is also a parallel-axis theorem for mass moments of inertia, namely,

$$I_P = I_G + md^2 \tag{2.6.7}$$

with d being the distance between the center of gravity axis and the parallel axis of interest. Several mass moments of inertia are presented in Table 2.2.

TABLE 2.2 Mass Moments of Inertia.

Shape	Dimensions	Moment of Inertia
Slender rod	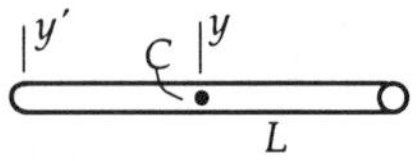	$I_y = mL^2/12$ $I_{y'} = mL^2/3$
Circular cylinder	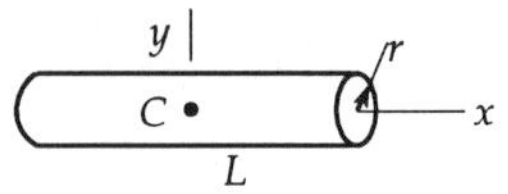	$I_x = mr^2/2$ $I_y = m(L^2 + 3r^2)/12$
Disk	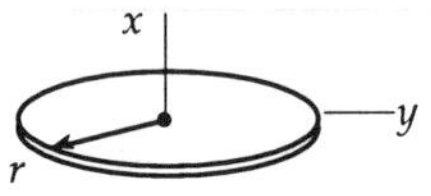	$I_x = mr^2/2$ $I_y = mr^2/4$
Rectangular parallelpiped	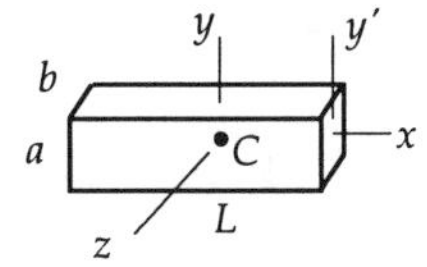	$I_x = m(a^2 + b^2)/12$ $I_y = m(L^2 + b^2)/12$ $I_z = m(L^2 + a^2)/12$ $I_{y'} = m(4L^2 + b^2)/12$
Sphere	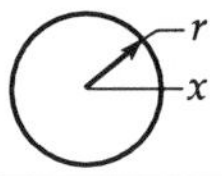	$I_x = 2mr^2/5$

Practice Problems

(If you choose to work only a few problems, select those with a star.)

GENERAL

*2.1 Find the component of the vector $\mathbf{A} = 15\mathbf{i} - 9\mathbf{j} + 15\mathbf{k}$ in the direction of $\mathbf{B} = \mathbf{i} - 2\mathbf{j} - 2\mathbf{k}$.

a) 1 b) 3 c) 5 d) 7 e) 9

2.2 Find the magnitude of the resultant of $\mathbf{A} = 2\mathbf{i} + 5\mathbf{j}$, $\mathbf{B} = 6\mathbf{i} - 7\mathbf{k}$, and $\mathbf{C} = 2\mathbf{i} - 6\mathbf{j} + 10\mathbf{k}$.

a) 8.2 b) 9.3 c) 10.5 d) 11.7 e) 12.8

*2.3 Determine the moment about the y-axis of the force $\mathbf{F} = 200\mathbf{i} + 400\mathbf{j}$ acting at (4, –6, 4).

a) 0 b) 200 c) 400 d) 600 e) 800

2.4 What total moment do the two forces $\mathbf{F}_1 = 50\mathbf{i} - 40\mathbf{k}$ and $\mathbf{F}_2 = 60\mathbf{j} + 80\mathbf{k}$ acting at (2, 0, –4) and (–4, 2, 0), respectively, produce about the x-axis?

a) 0 b) 80 c) 160 d) 240 e) 320

*2.5 The force system shown may be referred to as being

a) concurrent
b) coplanar
c) parallel
d) two-dimensional
e) non-concurrent, non-coplanar

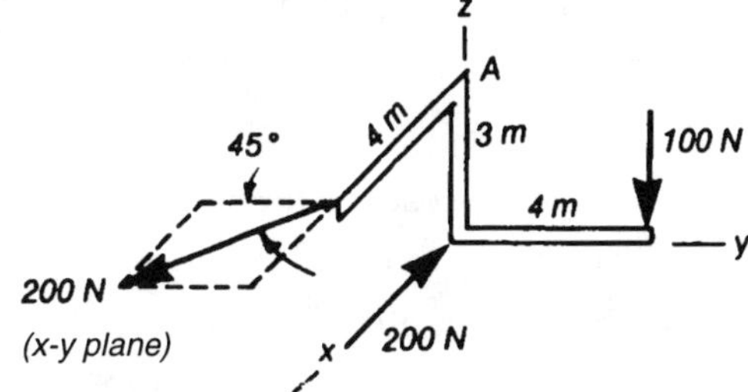

EQUILIBRIUM

2.6 If equilibrium exists due to a rigid support at A in the figure of Prob. 2.5, what reactive force must exist at A?

a) $-59\mathbf{i} - 141\mathbf{j} + 10\mathbf{k}$
b) $59\mathbf{i} + 141\mathbf{j} + 100\mathbf{k}$
c) $341\mathbf{i} - 141\mathbf{j} - 100\mathbf{k}$
d) $341\mathbf{i} + 141\mathbf{j} - 100\mathbf{k}$
e) $59\mathbf{i} - 141\mathbf{j} + 100\mathbf{k}$

*2.7 If equilibrium exists on the object in Prob. 2.5, what reactive moment must exist at the rigid support A?

a) $600\mathbf{i} + 400\mathbf{j} + 564\mathbf{k}$
b) $400\mathbf{i} + 564\mathbf{k}$
c) $400\mathbf{i} - 600\mathbf{j} + 564\mathbf{k}$
d) $400\mathbf{i} - 600\mathbf{j}$
e) $-600\mathbf{j} + 564\mathbf{j}$

*2.8 If three nonparallel forces hold a rigid body in equilibrium, they must
- a) be equal in magnitude.
- b) be concurrent.
- c) be non-concurrent.
- d) form an equilateral triangle.
- e) be colinear.

*2.9 A truss member
- a) is a two-force body.
- b) is a three-force body.
- c) resists forces in compression only.
- d) may resist three concurrent forces.
- e) resists lateral forces.

*2.10 Find the magnitude of the reaction at support *B*.

a) 400 d) 700
b) 500 e) 800
c) 600

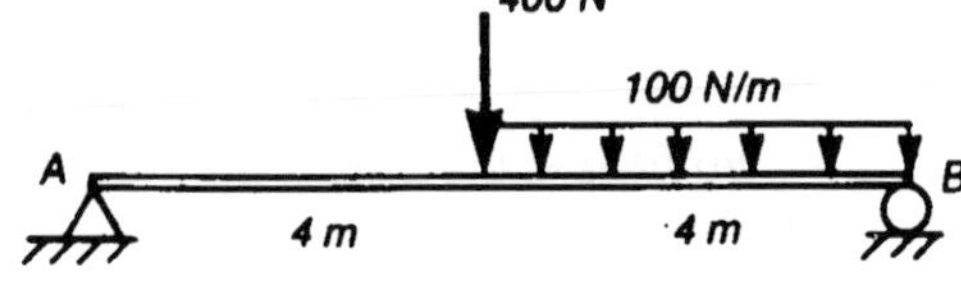

2.11 What moment *M* exists at support *A*?

a) 5600 d) 4000
b) 5000 e) 3600
c) 4400

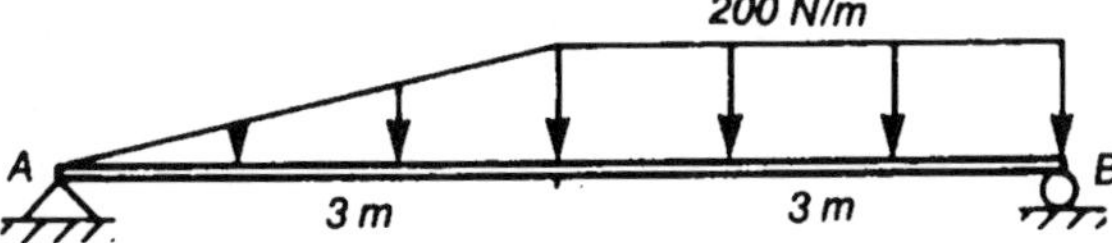

2.12 Calculate the reactive force at support *A*.

a) 250 d) 550
b) 350 e) 650
c) 450

2.13 Find the support moment at *A*.

a) 66 d) 99
b) 77 e) 111
c) 88

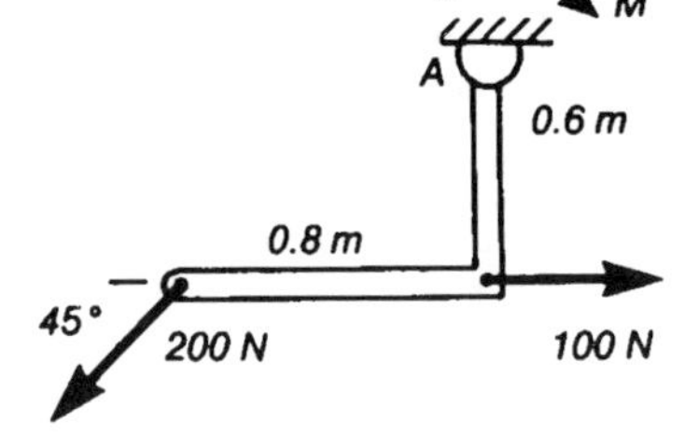

*2.14 To ensure equilibrium, what couple must be applied to this member?

a) 283 cw d) 400 ccw
b) 283 ccw e) 0
c) 400 cw

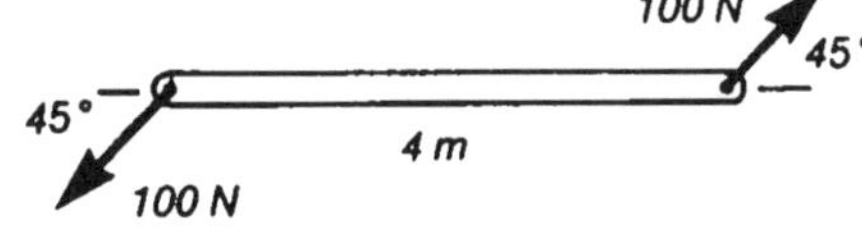

*2.15 Calculate the magnitude of the equilibrating force at A for the three-force body shown.

a) 217 d) 385
b) 287 e) 492
c) 343

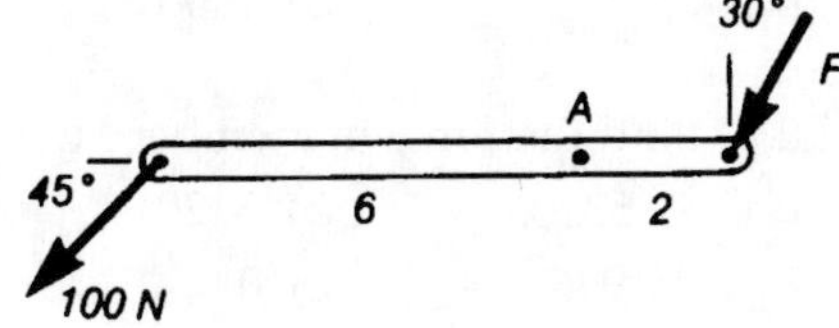

2.16 Find the magnitude of the equilibrating force at point A.

a) 187 d) 99
b) 142 e) 84
c) 114

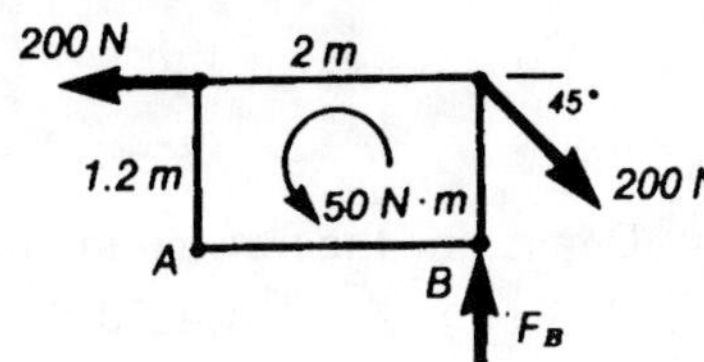

TRUSSES AND FRAMES

2.17 Find F_{DE} if all angles are equal.

a) 121 d) 176
b) 142 e) 189
c) 163

*2.18 Find F_{DE}.

a) 0 d) 2500
b) 1000 e) 5000
c) 2000

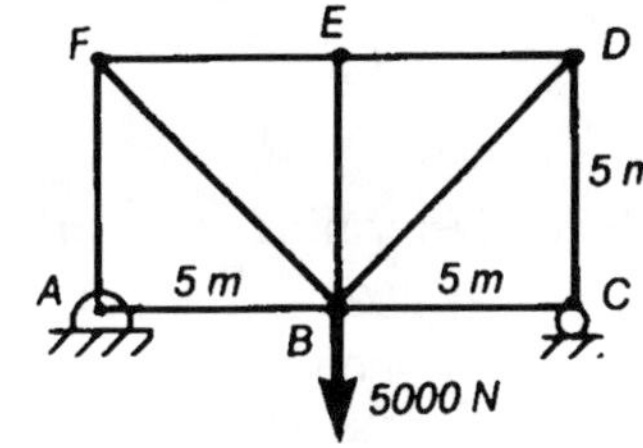

2.19 What is the force in member DE?

a) 1532 d) 1946
b) 1768 e) 2231
c) 1895

*2.20 Calculate F_{FB} in the truss of Problem 2.19.

a) 0 b) 932 c) 1561 d) 1732 e) 1887

2.21 Find the force in member IC.

a) 0
b) 1000
c) 1250
d) 1500
e) 2000

2.22 What force exists in member BC in the truss of Prob. 2.21?

a) 0 b) 1000 c) 2000 d) 3000 e) 4000

2.23 Find the force in member FC.

a) 5320 b) 3420 c) 2560
d) 936 e) 0

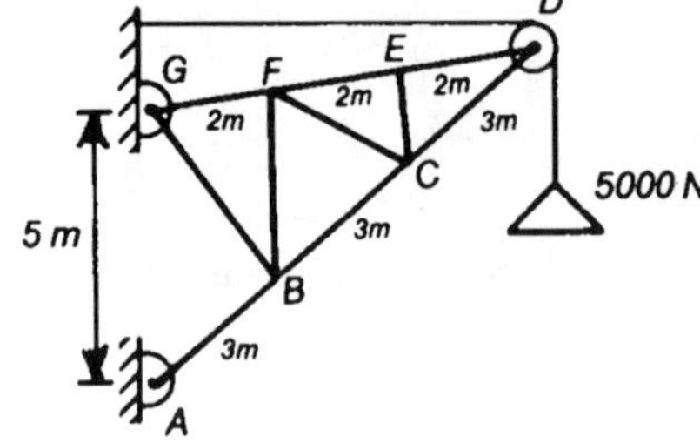

2.24 Determine the force in member BC in the truss of Prob. 2.23.

a) 3560 b) 4230 c) 4960 d) 5230 e) 5820

*2.25 Find the magnitude of the reactive force at support A.

a) 1400 b) 1300 c) 1200
d) 1100 e) 1000

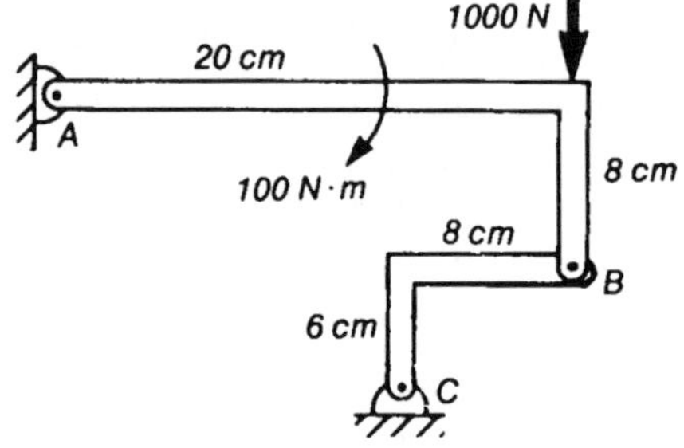

2.26 Determine the distributed force intensity w, in N/m, for equilibrium to exist.

a) 2000 b) 4000 c) 6000
d) 8000 e) 10 000

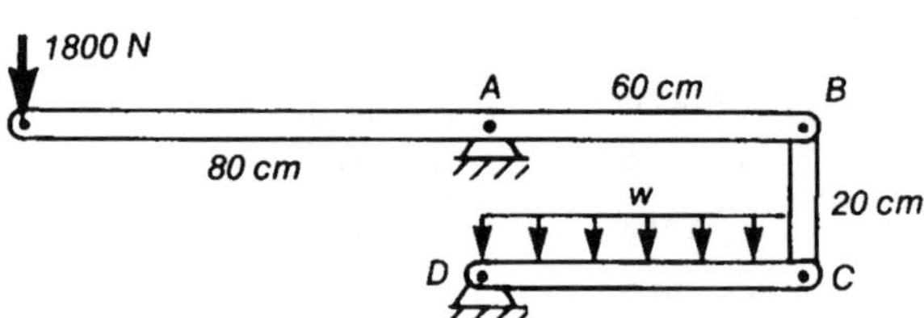

*2.27 Find the magnitude of the reactive force at support A.

a) 2580 b) 2670 c) 279
d) 2880 e) 2960

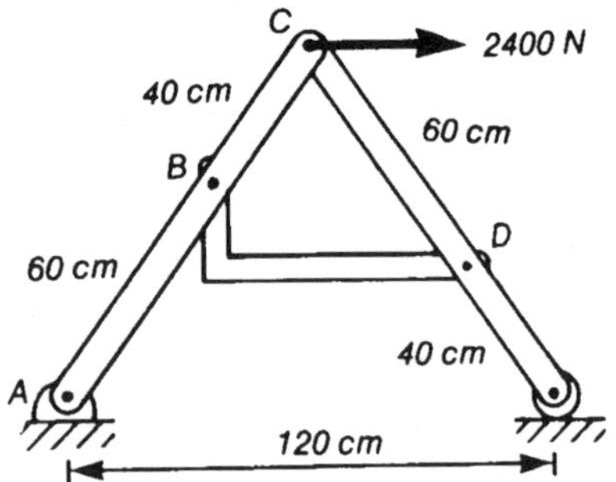

2.28 Calculate the magnitude of the force in member BD of Prob. 2.27.

a) 2590 b) 2670 c) 2790 d) 2880 e) 2960

FRICTION

*2.29 What force, in Newtons, will cause impending motion up the plane?

a) 731 b) 821 c) 973
d) 1102 e) 1245

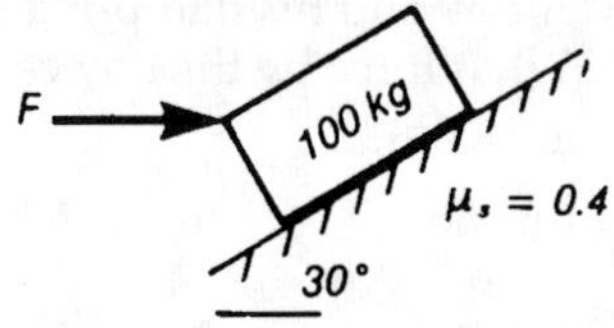

*2.30 What is the maximum force F, in Newtons, that can be applied without causing motion to impend?

a) 184 b) 294 c) 316
d) 346 e) 392

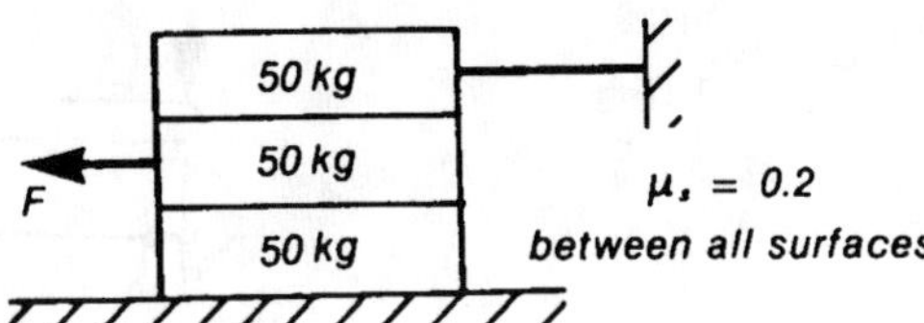

*2.31 Only the rear wheels provide braking. At what angle θ will the car slide if $\mu_s = 0.6$?

a) 10 b) 12 c) 16
d) 20 e) 24

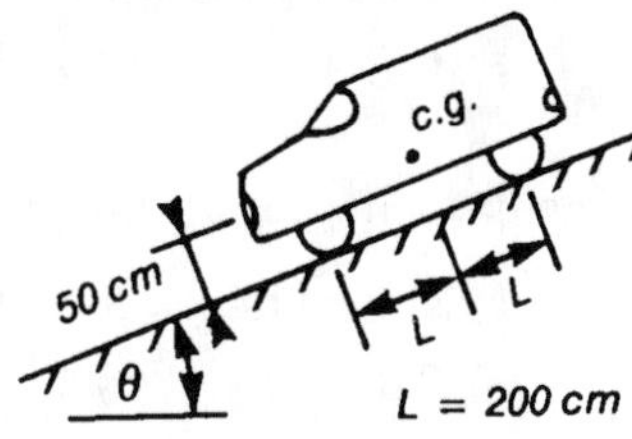

2.32 Find the minimum h value at which tipping will occur.

a) 8 b) 10 c) 12
d) 14 e) 16

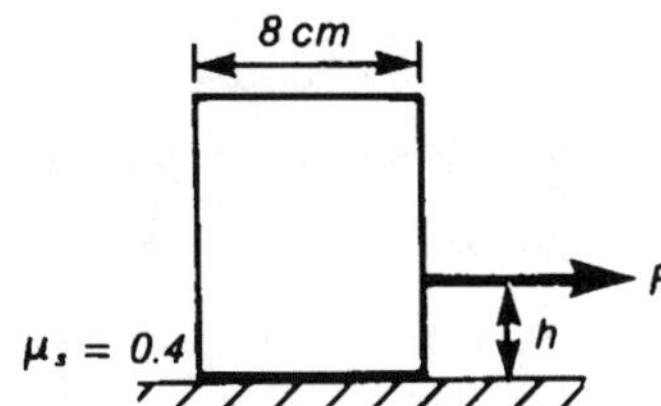

*2.33 What force F, in Newtons, will cause impending motion?

a) 240 b) 260 c) 280
d) 300 e) 320

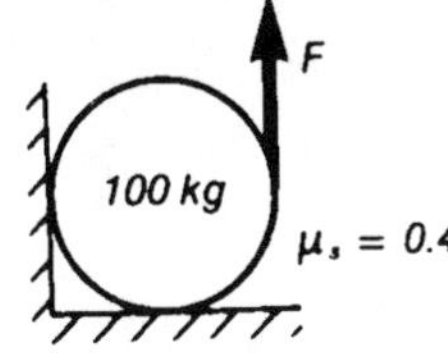

2.34 The angle θ at which the ladder is about to slip is

a) 50 b) 46 c) 42
d) 38 e) 34

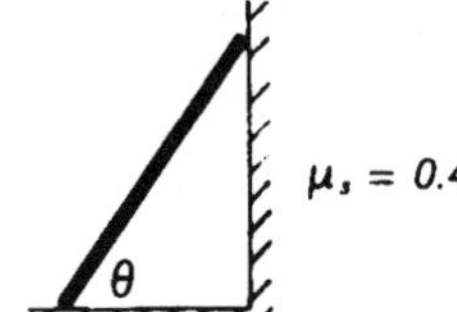

*2.35 A boy and his dad put a rope around a tree and stand side by side. What force by the boy can resist a force of 800 N by his dad? Use $\mu_s = 0.5$.

a) 166 b) 192 c) 231 d) 246 e) 297

2.36 What moment, in N
cause
impending motion?

a) 88
b) 99
c) 110
d) 121
e) 146

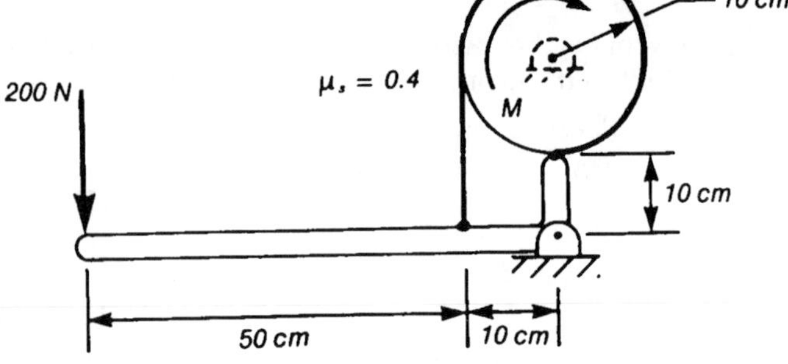

2.37 A 12-m-long rope is draped over a horizontal cylinder of 1.2-m-diameter so that both ends hang free. What is the length of the longer end at impending motion? Use $\mu_s = 0.5$.

a) 6.98 b) 7.65 c) 7.92 d) 8.37 e) 8.83

CENTROIDS AND MOMENTS OF INERTIA

*2.38 Find the x-coordinate of the centroid of the area bounded by the x-axis, the line $x = 3$, and the parabola $y = x^2$.

a) 2.0 b) 2.15 c) 2.20 d) 2.25 e) 2.30

2.39 What is the y-coordinate of the centroid of the area of Prob. 2.38?

a) 2.70 b) 2.65 c) 2.60 d) 2.55 e) 2.50

2.40 Calculate the x-coordinate of the centroid of the area enclosed by the parabolas $y = x^2$ and $x = y^2$.

a) 0.43 b) 0.44 c) 0.45 d) 0.46 e) 0.47

2.41 Find the y-component of the centroid of the area shown.

a) 3.35 b) 3.40 c) 3.45
d) 3.50 e) 3.55

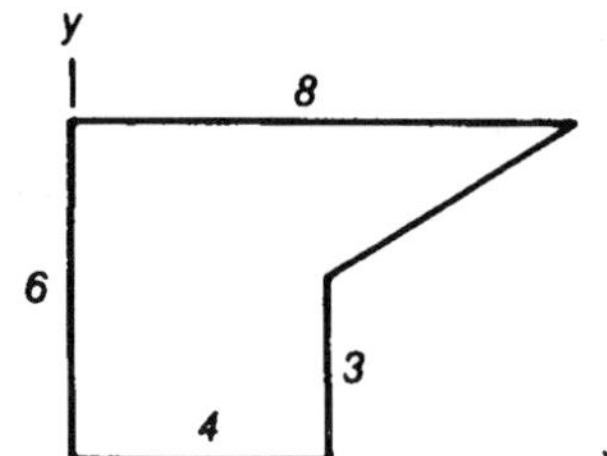

*2.42 Calculate the y-component of the centroid of the area shown.

a) 3.52 b) 3.56 c) 3.60
d) 3.64 e) 3.68

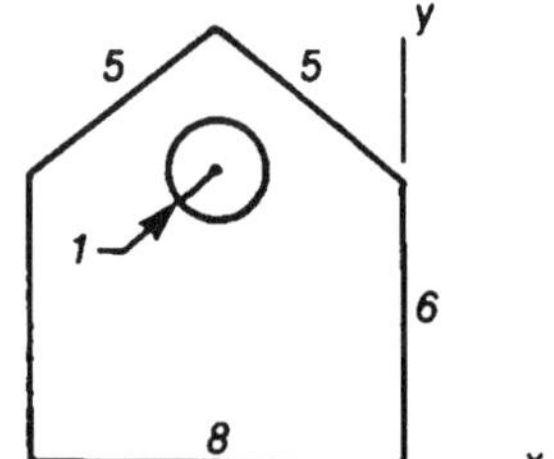

2.43 Find the x-component of the center of gravity of the three objects.

a) 2.33
b) 2.42
c) 2.84
d) 3.22
e) 3.64

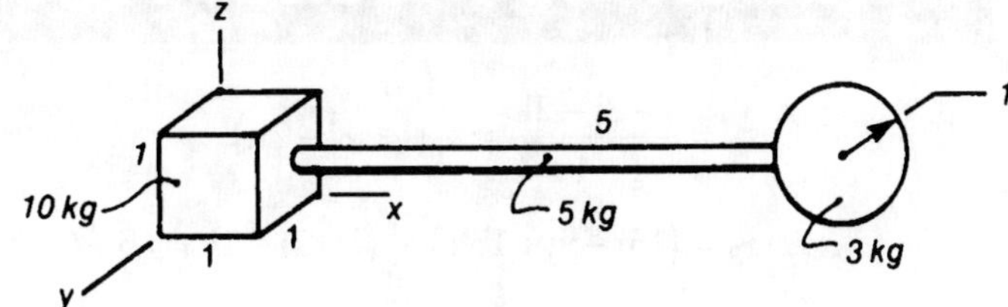

*2.44 Calculate the moment of inertia about the x-axis of the area of Prob. 2.32.

a) 94 b) 104 c) 112 d) 124 e) 132

2.45 What is I_x for the area of Prob. 2.42?

a) 736 b) 842 c) 936 d) 982 e) 1056

*2.46 Find I_y for the symmetrical area shown.

a) 4267 b) 4036 c) 3827
d) 3652 e) 3421

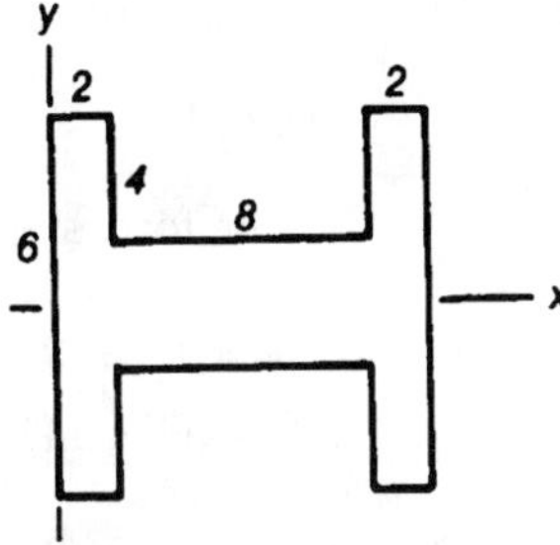

2.47 Determine the mass moment of inertia of a cube with edges of length b, about an axis passing through an edge.

a) $2mb^2/3$ b) $mb^2/6$ c) $3mb^2/2$ d) $mb^2/2$ e) $3mb^2/4$

*2.48 Find the mass moment of inertia about the x-axis if the mass of the rods per unit length is 1.0 kg/m.

a) 224 b) 268 c) 336
d) 386 e) 432

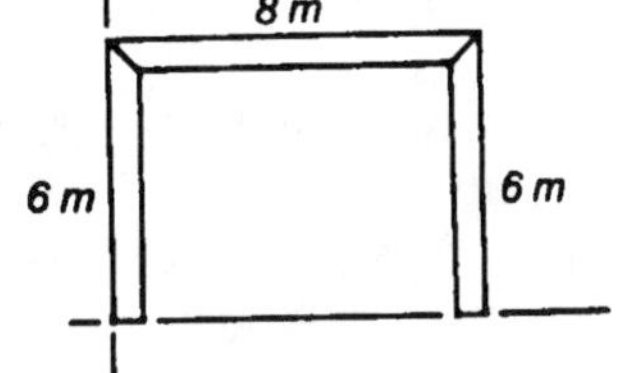

Solutions to Practice Problems

2.1 a) $\mathbf{i}_B = \dfrac{\mathbf{i}-2\mathbf{j}-2\mathbf{k}}{\sqrt{1+4+4}} = \frac{1}{3}(\mathbf{i}-2\mathbf{j}-2\mathbf{k})$

$\mathbf{A}\cdot\mathbf{i}_B = (15\mathbf{i}-9\mathbf{j}+15\mathbf{k})\cdot\frac{1}{3}(\mathbf{i}-2\mathbf{j}-2\mathbf{k}) = 5+6-10 = 1$

2.2 c) $\mathbf{A}+\mathbf{B}+\mathbf{C} = (2\mathbf{i}+5\mathbf{j})+(6\mathbf{i}-7\mathbf{k})+(2\mathbf{i}-6\mathbf{j}+10\mathbf{k}) = 10\mathbf{i}-\mathbf{j}+3\mathbf{k}$

$\text{magnitude} = \sqrt{10^2+1^2+3^2} = 10.49$

2.3 e) $\mathbf{M} = \mathbf{r}\times\mathbf{F} = (4\mathbf{i}-6\mathbf{j}+4\mathbf{k})\times(200\mathbf{i}+400\mathbf{j}).\quad M_y = 4\times 200 = 800\quad$ since $\mathbf{k}\times\mathbf{i} = \mathbf{j}$

2.4 c) $\mathbf{M} = \mathbf{r}_1\times\mathbf{F}_1+\mathbf{r}_2\times\mathbf{F}_2 = (2\mathbf{i}-4\mathbf{k})\times(50\mathbf{i}-40\mathbf{k})+(-4\mathbf{i}+2\mathbf{j})\times(60\mathbf{j}+80\mathbf{k})$

$M_x = 2\times 80 = 160\quad$ since $\mathbf{j}\times\mathbf{k} = \mathbf{i}$

2.5 e) Concurrent $\Rightarrow$ all pass through a point.

Coplanar $\Rightarrow$ all in the same plane.

The forces are three - dimensional.

2.6 b) $\sum\mathbf{F} = 0.\quad\therefore\mathbf{R}+141\mathbf{i}-141\mathbf{j}-200\mathbf{i}-100\mathbf{k} = 0$

$\therefore\mathbf{R} = 59\mathbf{i}+141\mathbf{j}+100\mathbf{k}$

2.7 c) $\sum\mathbf{M} = 0.\quad\therefore\mathbf{M}_A+(4\mathbf{j}-3\mathbf{k})\times(-100\mathbf{k})-3\mathbf{k}\times(-200\mathbf{i})+4\mathbf{i}\times(141\mathbf{i}-141\mathbf{j}) = 0$

$\therefore\mathbf{M}_A = 400\mathbf{i}-600\mathbf{j}+564\mathbf{k}$

2.8 b) They must be concurrent, otherwise a resultant moment would occur.

2.9 a) It is a two-force body.

2.10 b) $\sum M_A = 0.\quad F_B\times 8 = 400\times 4+400\times 6.\quad\therefore F_B = 500\text{ N}$

2.11 a) $M_A = 400\times 8+400\times 6 = 5600\text{ N}\cdot\text{m}$

2.12 b) $\sum M_B = 0.\quad 6F_A = 4\times 300+600\times 3/2.\quad\therefore F_A = 350\text{ N}$

2.13 c) $M_A = 0.6\times 100-141\times 0.6+141\times 0.8 = 88.2$

2.14 a) $M = 100\sin 45°\times 4 = 282.8\text{ cw}$

2.15 c) $\sum M_A = 0. \quad \therefore 6 \times 70.7 = 2 \times 0.866 F_1. \quad \therefore F_1 = 245$

$\sum F_x = 0. \quad \therefore -70.7 - 245 \times 0.5 + F_{Ax} = 0. \quad \therefore F_{Ax} = 193$

$\sum F_y = 0. \quad \therefore -70.7 - 245 \times 0.866 + F_{Ay} = 0. \quad \therefore F_{Ay} = 283$

$\therefore F_A = \sqrt{F_{Ax}^2 + F_{Ay}^2} = \sqrt{193^2 + 283^2} = 343$

2.16 e) $\sum M_A = 0. \quad \therefore 2F_B + 1.2 \times 200 - 141.4 \times 2 - 141.4 \times 1.2 + 50 = 0. \quad \therefore F_B = 81.2$

$\sum F_x = 0. \quad \therefore F_{Ax} - 200 + 141.4 = 0. \quad \therefore F_{Ax} = 58.6$

$\sum F_y = 0. \quad \therefore F_{Ay} + 81.2 - 141.4 = 0. \quad \therefore F_{Ay} = 60.2$

$\therefore F_A = \sqrt{F_{Ax}^2 + F_{Ay}^2} = \sqrt{58.6^2 + 60.2^2} = 84.0$

2.17 e) $\sum M_A = 0. \quad \therefore 500\ell + 200 \times 0.866\ell - F_C \times 2\ell = 0. \quad \therefore F_C = 337$

$0.866 F_{DC} = 337. \quad \therefore F_{DC} = 389$

$0.866 \times 389 = 0.866 F_{BD}. \quad \therefore F_{BD} = 389$

$-F_{DE} + 200 - 389 \times 0.5 - 389 \times 0.5 = 0. \quad \therefore F_{DE} = -189$

2.18 d) $\sum M_A = 0. \quad \therefore 5 \times 5000 = 10 \times F_C. \quad \therefore F_C = 2500 \quad \therefore F_{DC} = 2500$

$0.707 F_{BD} = 2500. \quad \therefore F_{BD} = 3536$

$0.707 \times 3536 = F_{DE}. \quad \therefore F_{DE} = 2500$

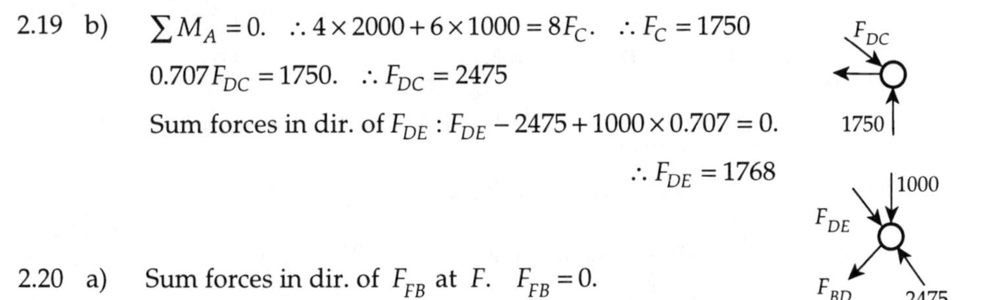

2.19 b) $\sum M_A = 0. \quad \therefore 4 \times 2000 + 6 \times 1000 = 8F_C. \quad \therefore F_C = 1750$

$0.707 F_{DC} = 1750. \quad \therefore F_{DC} = 2475$

Sum forces in dir. of F_{DE} : $F_{DE} - 2475 + 1000 \times 0.707 = 0.$

$\therefore F_{DE} = 1768$

2.20 a) Sum forces in dir. of F_{FB} at F. $F_{FB} = 0.$

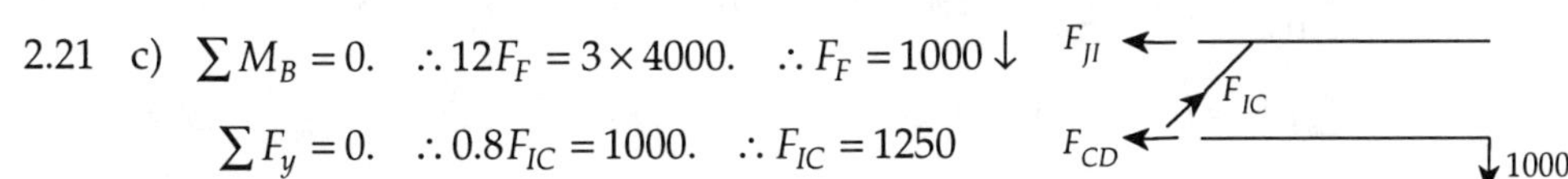

2.21 c) $\sum M_B = 0. \quad \therefore 12F_F = 3 \times 4000. \quad \therefore F_F = 1000 \downarrow$

$\sum F_y = 0. \quad \therefore 0.8 F_{IC} = 1000. \quad \therefore F_{IC} = 1250$

2.22 d) Cut vertically through link KA. Then $F_{KA} = 5000.$
Obviously, $F_{AL} = 0. \quad \therefore F_{AB} = 3000. \quad \therefore F_{BC} = 3000$

2.23 e) At E we see that $F_{EC} = 0. \quad \therefore$ At C, $F_{FC} = 0$

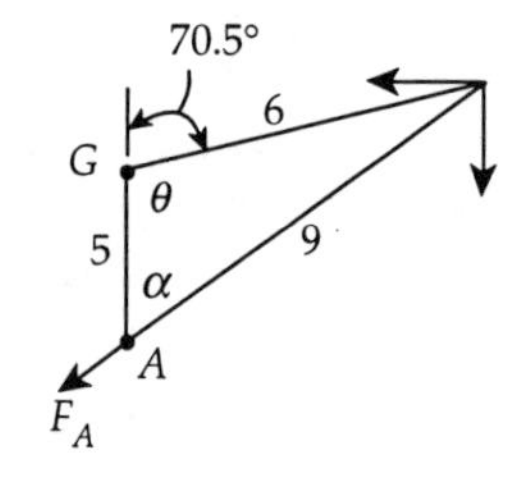

2.24 e) $9^2 = 6^2 + 5^2 - 2 \times 5 \times 6 \cos\theta$. $\therefore \theta = 109.5°$

$6^2 = 9^2 + 5^2 - 2 \times 9 \times 5 \cos\alpha$. $\therefore \alpha = 38.9°$

From pts E, C, F, B we see that $F_{EC} = F_{FC} = F_{FB} = F_{GB} = 0$.

Also, $F_A = F_{BC}$. $\sum M_G = 0$.

$\therefore 5 \times F_A \sin 38.9° + 5000 \times 6 \sin 70.5° = 5000 \times 6 \cos 70.5°$.

$\therefore F_A = -5817$

2.25 b) Recognize that link BC is a two-force member. $\sum M_A = 0$.

$\therefore 0.2 \times 1000 + 100 = 0.08 \times F_{BC} \times 0.8 + 0.2 \times F_{BC} \times 0.6$

$\therefore F_{BC} = 1630$. $A_x = 1630 \times 0.8 = 1304$. $A_y = 1630 \times 0.6 - 1000 = -22$

$\therefore F_A = \sqrt{1304^2 + 22^2} = 1304$ N

1000
A_x
100 N•m
A_y
F_{BC}

2.26 d) $1800 \times 0.8 = 0.6 F_{BC}$. $\therefore F_{BC} = 2400$

$0.3 \times 0.6w = 0.6 \times 2400$. $\therefore w = 8000$

.6w
F_{BC}
D

2.27 d) $\sum M_A = 0$. $\therefore 1.2 F_E = 0.8 \times 2400$. $\therefore F_E = 1600$

$\therefore A_x = 2400$. $A_y = 1600$

$\therefore F_A = \sqrt{2400^2 + 1600^2} = 2884$ N

2400
100
60°
A_x
A_y
F_E

2.28 a) Link BD is a two-force member. $\therefore$ the force acts from D to B. Hence, the angles are found.

$120^2 = 100^2 + 100^2 - 2 \times 100 \times 100 \cos\beta$. $\therefore \beta = 73.7°$

$\overline{BD}^2 = 60^2 + 40^2 - 2 \times 60 \times 40 \cos 73.7°$. $\therefore BD = 62.1$

$\dfrac{62.1}{\sin 73.7°} = \dfrac{40}{\sin\phi}$. $\therefore \phi = 38.2°$. $\alpha = (180 - 73.7)/2 = 53.2°$

$\sum M_C = 0$. $1600 \times 100 \cos 53.2° = 60 \times F_{BD} \sin 38.2°$. $\therefore F_{BD} = 2587$

C
β
ϕ
F_{BD}
α
160

2.29 e) $\sum F_y = 0$. $N \times 0.866 - 980 - 0.4N \times 0.5 = 0$. $\therefore N = 1471$

$\sum F_x = 0$. $F = 1471 \times 0.5 + 0.4 \times 1471 \times 0.866 = 1245$

y
F
0.4 N
N
980

2.30 b) $N_1 = 490.\quad N_2 = 980.\quad \therefore F = 0.2(490 + 980) = 294$

2.31 c) $\sum M_{\text{front wheel}} = 0.\quad \therefore 400N_2 - W\cos\theta \times 200 + W\sin\theta \times 50 = 0$

$\sum F_x = 0.\quad \therefore 0.6N_2 = W\sin\theta.$

$\therefore 400(W\sin\theta)/0.6 + 50W\sin\theta = 200W\cos\theta$

$\therefore \dfrac{\sin\theta}{\cos\theta} = \dfrac{200}{716.7} = \tan\theta.\quad \therefore \theta = 15.6°$

2.32 b) If $h < h_{\text{min}}$ then sliding occurs, and $F_f = 0.4N$.

If $h > h_{\text{min}}$ tipping occurs and $F_f < 0.4N$.

When $h = h_{\text{min}}$, $F_f = 0.4N = 0.4W = F$.

$\sum M_A = 0.\quad \therefore 4W = hF = h \times 0.4W.\quad \therefore h = 10 \text{ cm}$

2.33 e) $\sum F_x = 0.\quad \therefore N_2 = 0.4N_1.$ Also, $W = 980$

$\sum M_A = 0.\quad \therefore W \cdot r = (N_1 + 0.4N_1 + 2 \times 0.4N_2)r.$

$\therefore N_1 = 0.5814W = 570$

$\sum F_y = 0.\quad \therefore F = 980 - 570 - 0.16 \times 570 = 319$

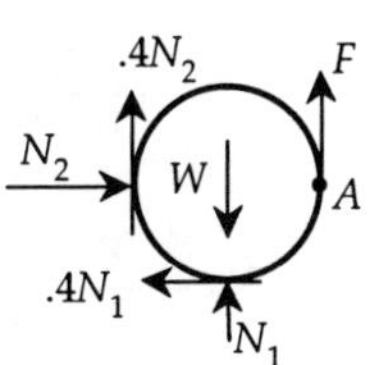

2.34 b) $\sum F_x = 0.\quad \therefore N_2 = 0.4N_1.\quad \sum F_y = 0.\quad \therefore N_1 + 0.4N_2 = W.$

$\therefore N_2 = 0.345W$

$\sum M_A = 0.\quad \therefore \dfrac{L}{2} \times W\cos\theta = N_2 \times L\sin\theta + 0.4N_2 \times L\cos\theta.$

This gives $\tan\theta = 1.049.\quad \therefore \theta = 46.4°$

2.35 a) $F_B = F_D e^{-\mu\theta} = 800e^{-0.5\pi} = 166\,\text{N}$

2.36 a) $\sum M_A = 0.\quad \therefore 200 \times 0.6 = 0.1 \times T_1 + 0.1 \times T_2.$

$T_1 = T_2 e^{0.4\times 3\pi/2} = 6.59T_2.$ Thus, $T_2 = 158$ and $T_1 = 1042.$

$\sum M_{\text{center}} = 0.\quad \therefore M = 0.1 \times (1042 - 158) = 88.4\ \text{N}\cdot\text{m}$

2.37 d) Let h = long end. m = mass/unit length. Then,

$(12 - 1.88 - h)mge^{0.5\pi} = hmg.\quad \therefore h = 8.38\ \text{m}$

2.38 d) $\bar{x} = \dfrac{\int_0^3 xydx}{\int_0^3 ydx} = \dfrac{\int_0^3 x^3dx}{\int_0^3 x^2dx} = \dfrac{3^4/4}{3^3/3} = 2.25$

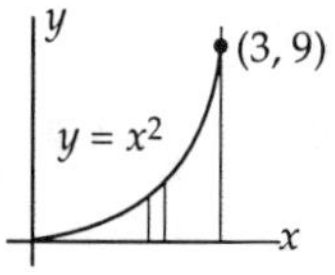

2.39 a) $\bar{y} = \dfrac{\int_0^3 \frac{y}{2} ydx}{\int_0^3 ydx} = \dfrac{\frac{1}{2}\int_0^3 x^4dx}{\int_0^3 x^2dx} = \dfrac{3^5/10}{3^3/3} = 2.7$

2.40 c) $\bar{x} = \dfrac{\int_0^1 \left(\sqrt{x} - x^2\right)xdx}{\int_0^1 \left(\sqrt{x} - x^2\right)dx} = \dfrac{\frac{1}{5/2} - \frac{1}{4}}{\frac{1}{3/2} - \frac{1}{3}} = 0.45$

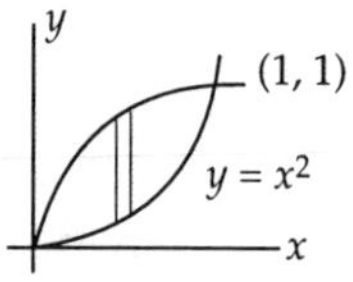

2.41 b) $\bar{y} = \dfrac{24 \times 3 + 6 \times 5}{6 \times 4 + 4 \times 3/2} = 3.4$

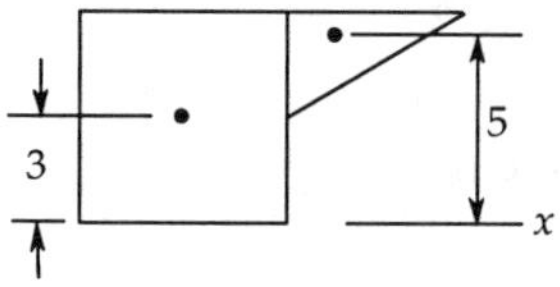

2.42 e) $\bar{y} = \dfrac{48 \times 3 + 12 \times 7 - \pi \times 6}{8 \times 6 + 3 \times 4 - \pi} = 3.68$

2.43 b) $\bar{x} = \dfrac{10 \times \frac{1}{2} + 5 \times 3.5 + 3 \times 7}{10 + 5 + 3} = 2.42$

2.44 b) $I_x = \int_0^3 y^3\, dx/3 = \int_0^3 x^6\, dx/3 = 3^7/21 = 104.1.$

With a horizontal strip: $I_x = \int_0^9 y^2(3-x)dy = \int_0^9 y^2\left(3 - \sqrt{y}\right)dy = 9^3 - \dfrac{9^{7/2}}{7/2} = 104.1$

2.45 e) $I_x = 8 \times 6^3/3 + \left(8 \times 3^3/36 + 12 \times 7^2\right) - \left(\pi \times 1^4/4 + \pi \times 6^2\right) = 1056$

2.46 a) $I_y = 12 \times 12^3/3 - \left(8 \times 8^3/12 + 64 \times 6^2\right) = 4267.$ Or, alternatively:

$I_y = 8 \times 2^3/3 + 4 \times 12^3/3 + 8 \times 2^3/12 + 16 \times 11^2 = 4267$

2.47 a) $I_{\text{edge}} = I_{\text{c.g.}} + Md^2 = \frac{1}{12}M\left(b^2 + b^2\right) + M\frac{b^2}{2} = \frac{2}{3}Mb^2$

2.48 e) $I_x = \frac{1}{3}(6m) \times 6^2 \times 2 + 8m \times 6^2 = 432$ with $m = 1$

Dynamics

by Merle C. Potter

Overview

Dynamics is separated into two major divisions: *kinematics*, which is a study of motion without reference to the forces causing the motion, and *kinetics*, which relates the forces on bodies to their resulting motions. Newton's laws of motion are necessary in relating forces to motions; they are:

1st law: A particle remains at rest or continues to move in a straight line with a constant velocity if no unbalanced force acts on it.

2nd law: The acceleration of a particle is proportional to the force acting on it and inversely proportional to the particle mass; the direction of acceleration is the same as the force direction.

3rd law: The forces of action and reaction between contacting bodies are equal in magnitude, opposite in direction, and colinear.

Law of gravitation: The force of attraction between two bodies is proportional to the product of their masses and inversely proportional to the square of the distance between their centers.

3.1 Kinematics

In kinematics we will consider three different kinds of particle motion: rectilinear motion, angular motion, and curvilinear motion. These will be followed by a review of motion of rigid bodies.

Rectilinear Motion

In rectilinear motion of a particle in which the particle moves in a straight line, the acceleration a, the velocity v, and the displacement s are related by

$$a = \frac{dv}{dt}, \qquad v = \frac{ds}{dt}, \qquad a = \frac{d^2s}{dt^2} = v\frac{dv}{ds}. \tag{3.1.1}$$

If the acceleration is a known function of time, the above can be integrated to give $v(t)$ and $s(t)$. For the important case of constant acceleration, integration yields

$$\begin{aligned} v &= v_o + at \\ s &= v_o t + at^2/2 \\ v^2 &= v_o^2 + 2as \end{aligned} \tag{3.1.2}$$

where at $t = 0$, $v = v_o$ and $s_o = 0$.

Angular Motion

Angular displacement is the angle θ that a line makes with a fixed axis, usually the positive x-axis. Counter clockwise motion is assumed to be positive, as shown in Fig. 3.1. The angular acceleration α, theangular velocity ω, and θ are related by

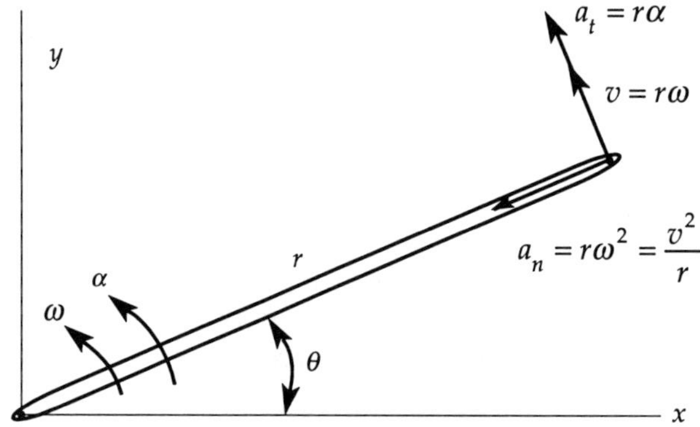

Figure 3.1 Angular motion.

$$\alpha = \frac{d\omega}{dt}, \qquad \omega = \frac{d\theta}{dt}, \qquad \alpha = \omega\frac{d\omega}{d\theta} = \frac{d^2\theta}{dt^2}. \tag{3.1.3}$$

If α is a constant, integration of these equations gives

$$\begin{aligned} \omega &= \omega_o + \alpha t \\ \theta &= \omega_o t + \alpha t^2/2 \\ \omega^2 &= \omega_o^2 + 2\alpha\theta \end{aligned} \tag{3.1.4}$$

where we have assumed that $\omega = \omega_o$ and $\theta_o = 0$ at $t = 0$.

Curvilinear Motion

When a particle moves on a plane curve as shown in Fig. 3.2, the motion may be described in terms of coordinates along the normal n and the tangent t to the curve at the instantaneous position of the particle.

The acceleration is the vector sum of the normal acceleration a_n and the tangential acceleration a_t. These components are given as

$$a_n = \frac{v^2}{r}, \qquad a_t = \frac{dv}{dt} \tag{3.1.5}$$

where r is the radius of curvature and v is the magnitude of the velocity. The velocity is always tangential to the curve so no subscript is necessary to identify the velocity.

It should be noted that rigid body traveling without rotation can be treated as particle motion.

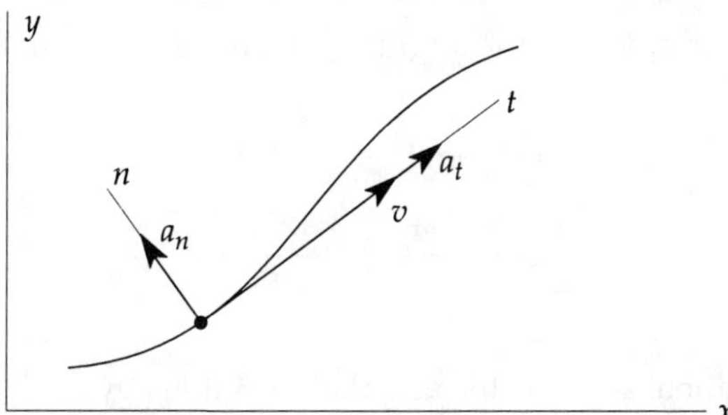

Figure 3.2 Motion on a plane curve.

Example 3.1

The velocity of a particle undergoing rectilinear motion is $v(t) = 3t^2 + 10t$ m/s. Find the acceleration and the displacement at t = 10 s, if $s_o = 0$ at $t = 0$.

Solution. The acceleration is found to be

$$a = \frac{dv}{dt}$$

$$= 6t + 10 = 6 \times 10 + 10 = 70 \text{ m/s}^2.$$

The displacement is found by integration

$$s = \int_0^{10} v\, dt$$

$$= \int_0^{10} (3t^2 + 10t)dt = t^3 + 5t^2 \Big|_0^{10} = 1500 \text{ m}.$$

Example 3.2

An automobile skids to a stop 60 m after its brakes are applied while traveling 25 m/s. What is its acceleration?

Solution. We use the relationship

$$v^2 = v_o{}^2 + 2as.$$

Letting $v = 0$, we find,

$$a = -\frac{v_o{}^2}{2s}$$

$$= -\frac{25^2}{2 \times 60} = -5.21 \text{ m/s}^2.$$

Example 3.3

A wheel, rotating at 100 rad/s ccw (counterclockwise), is subjected to an angular acceleration of 20 rad/s^2 cw. Find the total number of revolutions (cw plus ccw) through which the wheel rotates in 8 seconds.

Solution. The time at which the angular velocity is zero is found as follows:

$$\cancelto{0}{\omega} = \omega_o + \alpha t$$

$$\therefore t = -\frac{\omega_o}{\alpha} = -\frac{100}{-20} = 5 \text{ s}.$$

After three additional seconds the angular velocity is by

$$\omega = \cancelto{0}{\omega_o} + \alpha t$$
$$= -20 \times 3 = -60 \text{ rad / s}.$$

The angular displacement from 0 to 5 s is

$$\theta = \omega_o t + \alpha t^2/2$$
$$= 100 \times 5 - 20 \times 5^2/2 = 250 \text{ rad}.$$

During the next 3 s, the angular displacement is

$$\theta = \alpha t^2/2$$
$$= -20 \times 3^2/2 = -90 \text{ rad}.$$

The total number of revolutions rotated is

$$\theta = (250 + 90)/2\pi = 54.1 \text{ rev}.$$

Example 3.4

Consider idealized projectile motion (no air drag) in which $a_x = 0$ and $a_y = -g$. Find expressions for the range R and the maximum height H in terms of v_o and θ.

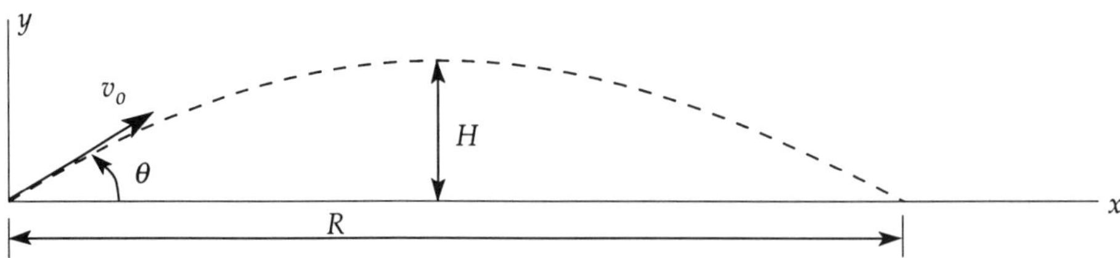

Solution. Using Eq. 3.1.2 for constant acceleration we have the point $(R, 0)$:

$$\cancelto{0}{y} = (v_o \sin\theta)t - gt^2/2.$$
$$\therefore t = 2v_o \sin\theta/g.$$

From the x-component equation:

$$x = (v_o \cos\theta)t + \cancelto{0}{a_x} t^2/2.$$
$$\therefore R = (v_o \cos\theta)2v_o \sin\theta / g = v_o^2 \sin 2\theta / g.$$

Obviously, the maximum height occurs when the time is one-half that which yields the range R. Hence,

$$H = (v_o \sin\theta)\frac{v_o \sin\theta}{g} - \frac{g}{2}\left(\frac{v_o \sin\theta}{g}\right)^2$$
$$= \frac{v_o^2}{2g}\sin^2\theta.$$

Note: The maximum R for a given v_o occurs when $\sin 2\theta = 1$, which means $\theta = 45°$ for R_{max}.

Example 3.5

It is desired that the normal acceleration of a satellite be 9.6 m/s^2 at an elevation of 200 km. What should be the velocity be for a circular orbit? The radius of the earth is 6400 km.

Solution. The normal acceleration, which points toward the center of the earth, is

$$a_n = \frac{v^2}{r}.$$
$$\therefore v = \sqrt{a_n r}$$
$$= \sqrt{9.6 \times (6400 + 200) \times 1000} = 7960 \text{ m/s}.$$

The normal acceleration is essentially the value of gravity near the earth's surface. Gravity varies only slightly if the elevation is small with respect to the earth's radius.

Rigid Body Motion

The motion of a rigid body can be described using the relative velocity and relative acceleration equations

$$\mathbf{v}_A = \mathbf{v}_B + \mathbf{v}_{A/B}$$
$$\mathbf{a}_A = \mathbf{a}_B + \mathbf{a}_{A/B} \quad \textbf{(3.1.6)}$$

where the velocity $\mathbf{v}_{A/B}$ is the velocity of point A with respect to point B and the acceleration $\mathbf{a}_{A/B}$ is the acceleration of point A with respect to point B. If points A and B are on the same rigid body then point A must move perpendicular to the line AB and

$$v_{A/B} = r\omega \tag{3.1.7}$$
$$\left(a_{A/B}\right)_n = r\omega^2$$
$$\left(a_{A/B}\right)_t = r\alpha$$

where ω is the angular velocity, α is the angular acceleration of the body, and r is the length of $\overline{AB}$. If point A is located on a body which moves with a constant velocity v relative to a coincident point B which is located on a second body rotating with an angular velocity ω (see Fig. 3.3), the acceleration of A with respect to B is called the *Coriolis acceleration*, given by

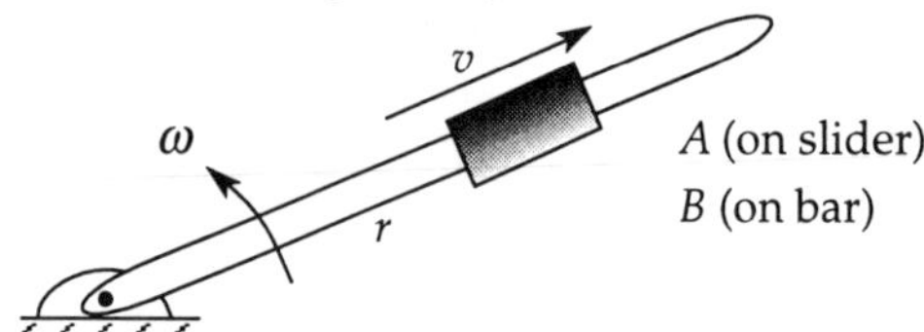

Figure 3.3 Coriolis acceleration.

We note that the Coriolis acceleration acts normal to both vectors $\boldsymbol{\omega}$ (use the right-hand rule) and $\mathbf{v}$. Thus it acts normal to the arm.

A final note regarding the instant center of zero velocity (a point which is often off the body, that has zero velocity): if such a point B can be located, then the magnitude of the velocity of point A is simply $r\omega$.

$$\mathbf{a}_{A/B} = 2\boldsymbol{\omega} \times \mathbf{v}. \tag{3.1.8}$$

Example 3.6

Find the magnitude of the velocity and acceleration of point A.

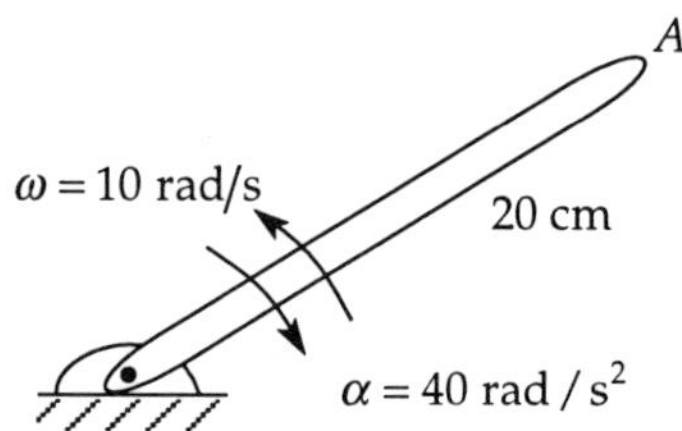

Solution. The velocity is found to be

$$v_A = r\omega$$
$$= 0.2 \times 10 = 2 \text{ m/s}.$$

The acceleration components are

$$a_n = r\omega^2 = 0.2 \times 10^2 = 20 \text{ m/s}^2$$

$$a_t = r\alpha = 0.2 \times 40 = 8 \text{ m/s}^2.$$

Thus,

$$a = \sqrt{a_n^2 + a_t^2} = \sqrt{20^2 + 8^2} = 21.5 \text{ m/s}^2.$$

Example 3.7

Find the velocity of C and ω_{BC}.

Solution. The velocity of B is normal to $\overline{AB}$ and is equal to

$$
\begin{aligned}
v_B &= r\omega \\
&= 0.4 \times 100 = 40 \text{ m/s}.
\end{aligned}
$$

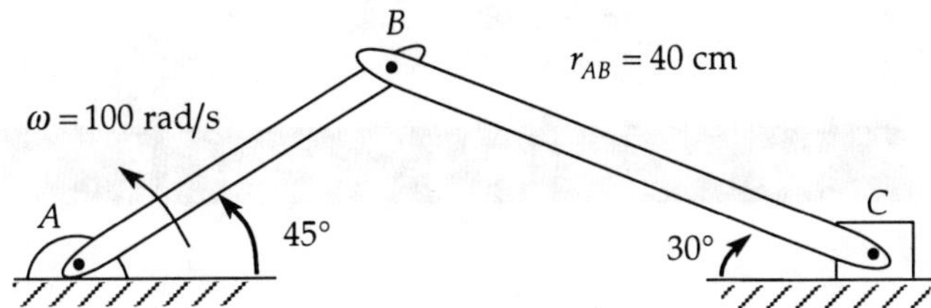

To find V_C we use the relative motion equation

$$\mathbf{v}_C = \mathbf{v}_B + \mathbf{v}_{C/B}.$$

The velocity of C must be horizontal and $\mathbf{v}_{C/B}$ must be normal to $\overline{BC}$. This can be displayed in a velocity polygon as follows:

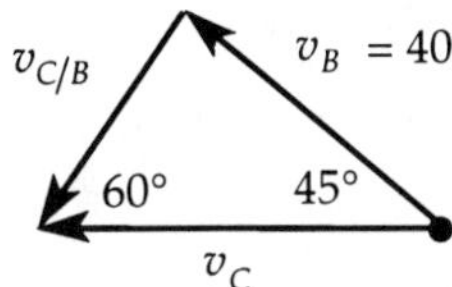

From the velocity polygon we use some simple trigonometry and find v_C to be to the left with magnitude

$$v_C = 44.6 \text{ m/s}.$$

The angular velocity of $\overline{BC}$ is found to be

$$
\begin{aligned}
\omega_{BC} &= v_{C/B}/r_{BC} \\
&= 32.7 / 0.5656 = 57.8 \text{ rad/s cw}.
\end{aligned}
$$

Example 3.8

Find the acceleration of C in Example 3.7, assuming $\alpha_{AB} = 0$.

Solution. The acceleration of B is

$$
\begin{aligned}
(a_B)_n &= r\omega^2 \\
&= 0.4 \times 100^2 = 4000 \text{ m/s}^2.
\end{aligned}
$$

The relative acceleration equation

$$\mathbf{a}_C = \mathbf{a}_B + \mathbf{a}_{C/B}$$

can be displayed in an acceleration polygon, realizing that $\mathbf{a}_C$ must be horizontal, and $\mathbf{a}_{C/B}$ has both normal and tangential components. We find $(a_{C/B})_n$ to be

$$\left(a_{C/B}\right)_n = r_{BC}\omega_{BC}^2$$
$$= 0.5656 \times 57.8^2 = 1890 \text{ m/s}^2.$$

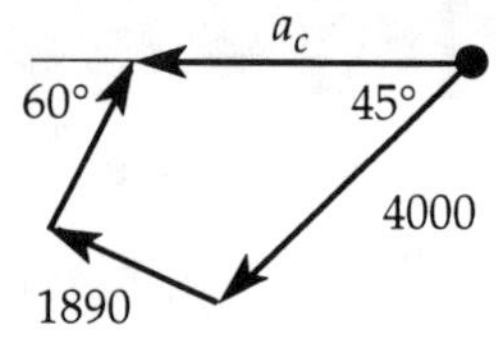

The acceleration polygon is shown. From the polygon we can find a_C to be to the left with magnitude

$$a_C = 3380 \text{ m/s}^2.$$

Example 3.9

Find v_A and a_A.

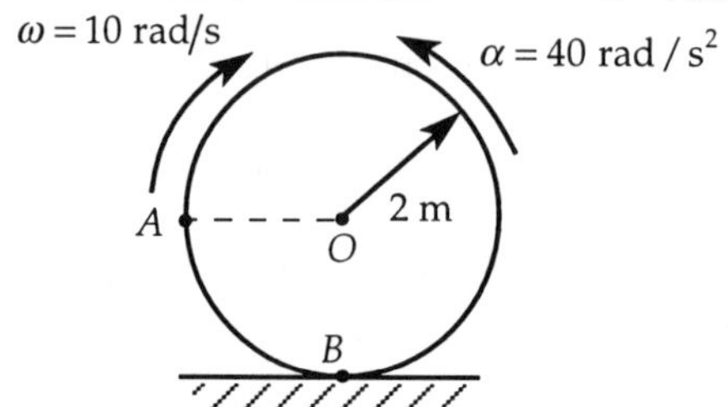

Solution. At this instant the wheel rotates about B. Hence, we can find v_A by using

$$v_A = r_{AB}\omega$$
$$= \frac{2}{\sin 45^\circ} \times 10 = 28.3 \text{ m/s}.$$

To find the acceleration we relate A to the center O and use

$$\mathbf{a}_A = \mathbf{a}_O + \mathbf{a}_{A/O}.$$

The acceleration polygon follows using

$$a_O = r\alpha = 2 \times 40 = 80 \text{ m/s}^2$$
$$(a_{A/O})_t = r\alpha = 2 \times 40 = 80 \text{ m/s}^2$$
$$(a_{A/O})_n = r\omega^2 = 2 \times 10^2 = 200 \text{ m/s}^2.$$

From the polygon we find a_A to be $a_A = 144 \text{ m/s}^2$.

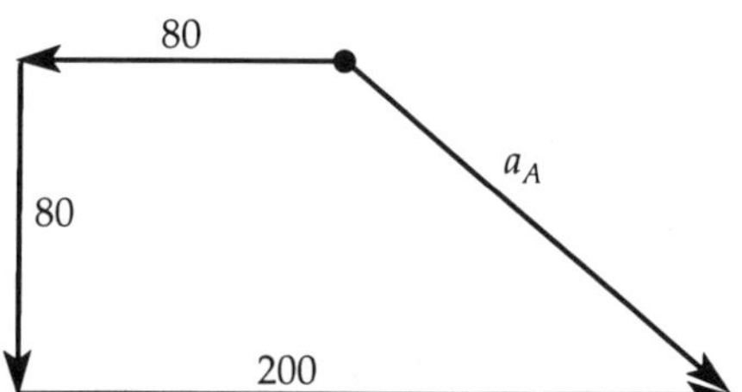

Example 3.10

Calculate a_A where A is on the slider.

Solution. The relative acceleration equation is

$$\mathbf{a}_A = \mathbf{a}_B + \mathbf{a}_{A/B}$$

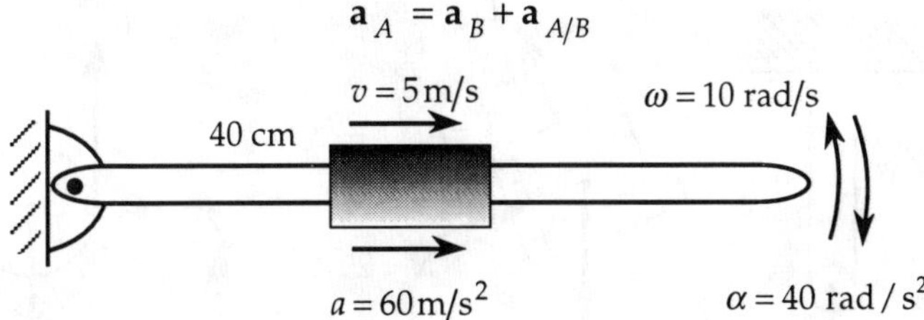

where point B is on the arm, coincident with point A. We know that

$$(a_B)_n = r\omega^2 = 0.4 \times 10^2 = 40 \ \text{m/s}^2$$
$$(a_B)_t = r\alpha = 0.4 \times 40 = 16 \ \text{m/s}^2$$
$$(a_{A/B})_t = 60 \ \text{m/s}^2$$
$$(a_{A/B})_C = 2\omega v = 2 \times 10 \times 5 = 100 \ \text{m/s}^2.$$

The acceleration polygon appears as follows:

The acceleration a_A is found to be $a_A = 86.4 \ \text{m/s}^2$.

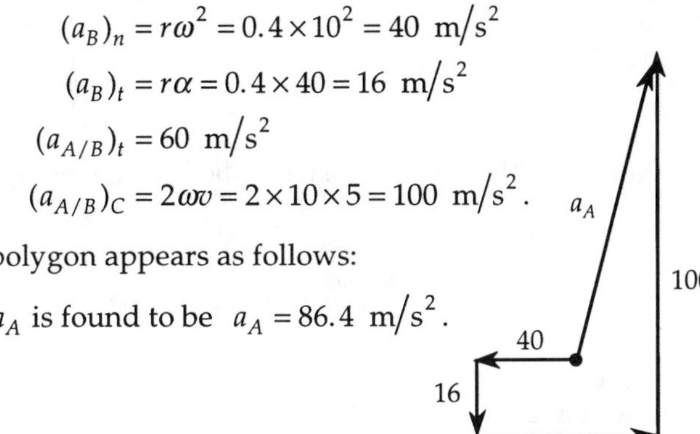

3.2 Kinetics

To relate the force acting on a body to the motion of that body we use Newton's laws of motion. Newton's 2nd law is used in the form

$$\sum \mathbf{F} = m\mathbf{a} \qquad \textbf{(3.2.1)}$$

where the mass of the body is assumed to be constant and $\mathbf{a}$ is the acceleration of the center of mass (center of gravity) if the body is rotating. We also require

$$\sum M = I\alpha \qquad \textbf{(3.2.2)}$$

where the moments must be summed about an axis passing through the center of mass. The mass moment of inertia I is often found by using the radius of gyration k and the relation $I = mk^2$.

The gravitational attractive force between one body and another is given by

$$F = K\frac{m_1 m_2}{r^2} \qquad \textbf{(3.2.3)}$$

where $K = 6.67 \times 10^{-11} \ \text{N}\cdot\text{m}^2/\text{kg}^2$.

Note: Since metric units are used in the above relations, mass must be measured in kilograms. The weight is related to the mass by

$$W = mg \qquad \textbf{(3.2.4)}$$

where we will use $g = 9.8 \ \text{m/s}^2$, unless otherwise stated.

Example 3.11

Find the tension in the rope and the distance the 600 kg mass moves in 3 seconds. The mass starts from rest and the mass of the pulleys is negligible.

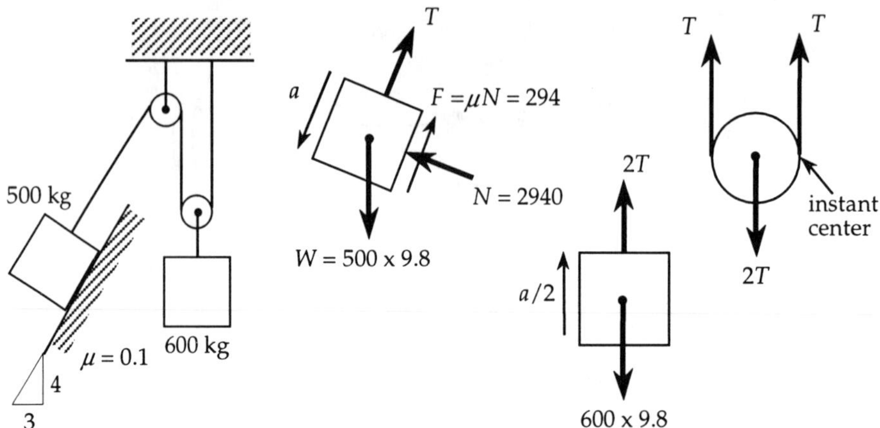

Solution. Applying Newton's 2nd law to the 500 kg mass gives

$$\sum F = ma$$

$$0.8 \times 500 \times 9.8 - 294 - T = 500\ a.$$

By studying the lower pulley we observe the 600 kg mass to be accelerating at $a/2$. Hence, we have

$$\sum F = ma$$

$$2T - 600 \times 9.8 = 600 \times a/2.$$

Solving the above equations simultaneously results in

$$a = 1.055\ \text{m/s}^2, \qquad T = 3100\ \text{N}.$$

The distance the 600 kg mass moves is

$$s = \frac{1}{2}\frac{a}{2}t^2$$

$$= \frac{1}{2} \times \frac{1.055}{2} \times 3^2 = 2.37\ \text{m}.$$

Example 3.12

Find the tension in the string if at the position shown, $v = 4$ m/s. Calculate the angular acceleration.

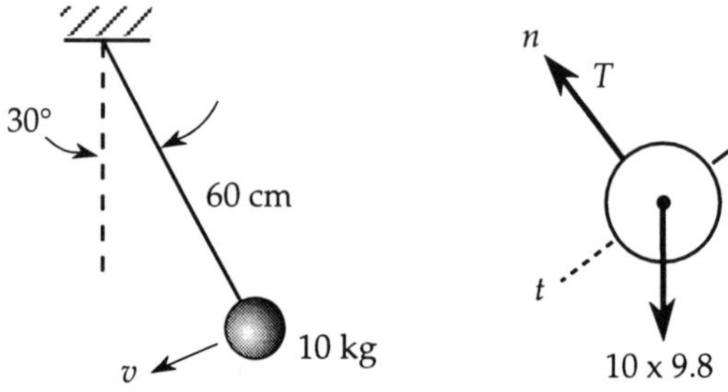

Solution. Sum forces in the normal direction and obtain

$$\sum F_n = ma_n = mv^2 / r$$

$$T - 10 \times 9.8 \cos 30^\circ = 10 \times 4^2 / 0.6.$$

$$\therefore \ T = 352 \text{ N}.$$

Sum forces in the tangential direction and find

$$\sum F_t = ma_t = mr\alpha$$

$$10 \times 9.8 \sin 30^\circ = 10 \times 0.6\alpha$$

$$\therefore \ \alpha = 8.17 \ \text{rad/s}^2.$$

Example 3.13

Find the angular acceleration of the 60 kg cylindrical pulley, and the tension in the rope.

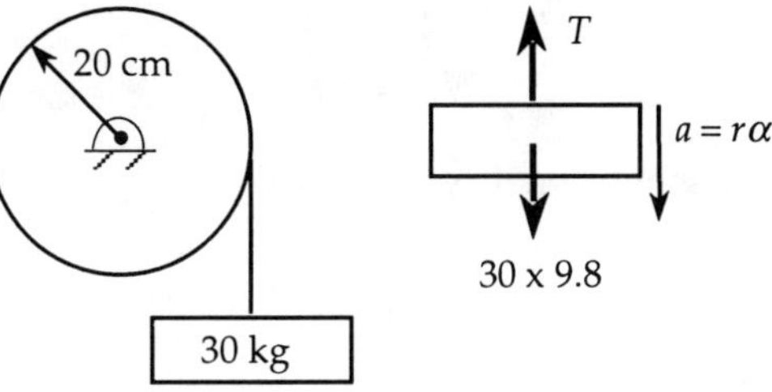

Solution. Summing forces on the 30 kg mass gives

$$\sum F = ma$$

$$-T + 30 \times 9.8 = 30 \times 0.2\alpha.$$

Summing moments about the center of the pulley yields

$$\sum M = I\alpha$$

$$T \times 0.2 = \frac{1}{2} \times 60 \times 0.2^2 \alpha$$

where $I = mr^2/2$ for a cylinder (see Table 2.2). A simultaneous solution results in

$$\alpha = 24.5 \ \text{rad/s}^2, \qquad T = 147 \text{ N}.$$

3.3 Work and Energy

Work is defined to be the dot product between a force and the distance it moves, that is,

$$W = \int \mathbf{F} \cdot d\mathbf{s}, \tag{3.3.1}$$

or, if the force is constant,

$$W = \mathbf{F} \cdot \Delta \mathbf{s}. \tag{3.3.2}$$

For a rotating body the work is

$$W = M\Delta\theta. \tag{3.3.3}$$

The *work-energy equation*, which results from integrating Newton's 2nd law, Eqs. 3.2.1 and 3.2.2, states that the net work done on a body (or several connected bodies) equals the change in energy of the body (or several bodies). This is expressed as

$$W_{net} = \Delta E \tag{3.3.4}$$

where E represents the kinetic energy, given by

$$E = \frac{1}{2}mv^2 \tag{3.3.5}$$

for a translating body, and

$$E = \frac{1}{2}I\omega^2 \tag{3.3.6}$$

for a rotating body. For a translating and rotating body, refer v and I to the mass center. Potential energy can be realized by allowing the body forces to do work; or they can be incorporated in the ΔE-term by using the potential energy as

$$E_p = mgh \tag{3.3.7}$$

where h is the distance above a selected datum.

By applying Eq. 3.3.1 to a spring, the work necessary to compress that spring a distance x is

$$W = \frac{1}{2}Kx^2. \tag{3.3.8}$$

The quantity $Kx^2/2$ can be considered the potential energy stored in the spring.

Example 3.14

Neglecting friction, find v of the slider when it hits B if it starts from rest at A.

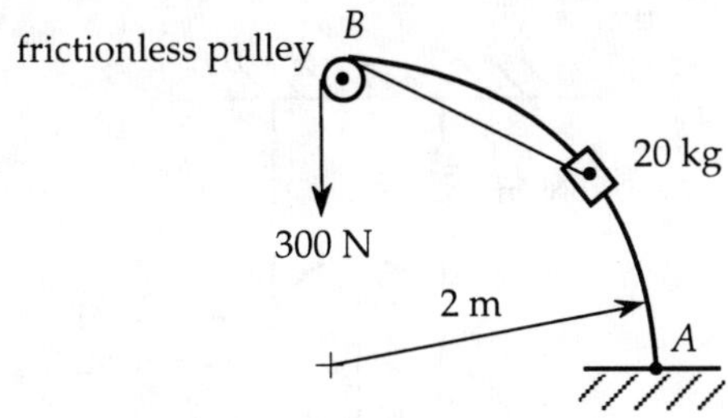

Solution. The distance the force F moves is

$$s = \frac{2}{\sin 45^\circ} = 2.828 \text{ m}.$$

The work-energy equation is written as

$$W_{net} = \frac{1}{2}mv^2$$

$$300 \times 2.828 - (20 \times 9.8) \times 2 = \frac{1}{2} \times 20v^2.$$

$$\therefore v = 6.76 \text{ m/s}.$$

Note that the body force does negative work since the motion is up and the body force acts down.

Example 3.15

Neglect friction and estimate the angular velocity of the cylinder after the mass falls 2 m from rest.

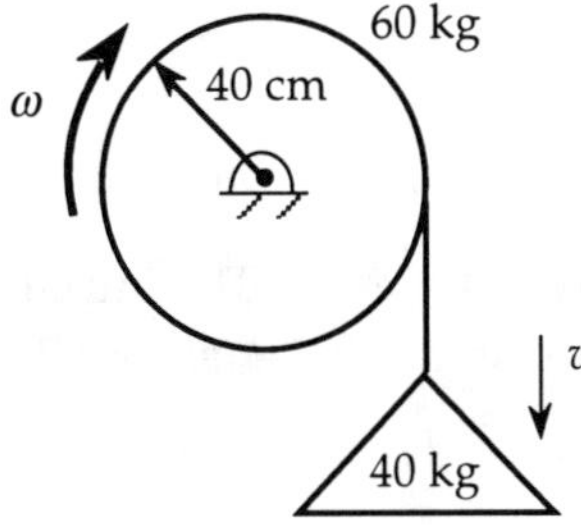

Solution. The work-energy equation provides

$$W_{net} = \frac{1}{2}mv^2 + \frac{1}{2}I\omega^2$$

$$(40 \times 9.8) \times 2 = \frac{1}{2} \times 40 \times (0.4\omega)^2 + \frac{1}{2}(\frac{1}{2} \times 60 \times 0.4^2)\omega^2.$$

$$\therefore \omega = 11.83 \text{ rad/s}.$$

Example 3.16

What is the velocity of the 40 kg mass after it falls 20 cm from rest? The spring is initially stretched 10 cm.

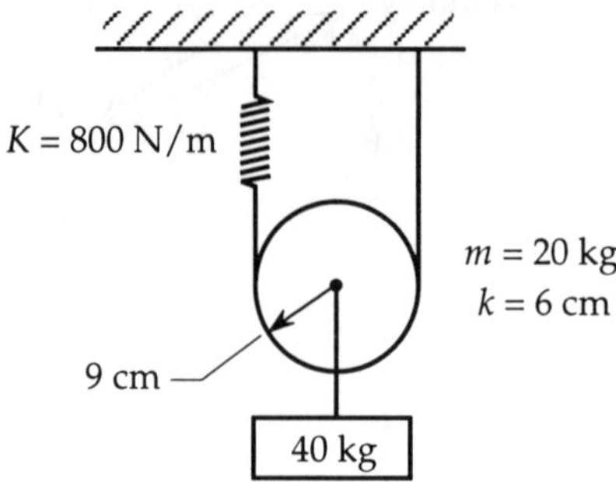

Solution. The spring will stretch an additional 40 cm. Thus, the work-energy equation results in

$$W_{net} = \frac{1}{2}mv^2 + \frac{1}{2}I\omega^2,$$

$$(20+40)\times 9.8\times 0.2 - \frac{1}{2}800\left(0.5^2 - 0.1^2\right) = \frac{1}{2}(20+40)v^2 + \frac{1}{2}\left(20\times 0.06^2\right)\left(\frac{v}{0.09}\right)^2.$$

$\therefore v = 0.792\ \text{m/s}.$

3.4 Impulse and Momentum

The impulse-momentum equations also result from integrating Newton's 2nd law. *Impulse* is defined for linear and rotating bodies, respectively, as

$$i_\ell = \int F dt$$
$$i_r = \int M dt. \quad \textbf{(3.4.1)}$$

Momentum is velocity multiplied by mass. The impulse-momentum equations for a constant force and moment take the form (see Eqs. 3.2.1 and 3.2.2)

$$F\Delta t = m\Delta v$$
$$M\Delta t = I\Delta\omega. \quad \textbf{(3.4.2)}$$

Objects impacting each other with no external forces acting experience a conservation of momentum. The *coefficient of restitution* e is used in such problems. It is defined as

$$e = \frac{\text{relative separation velocity}}{\text{relative approach velocity}}. \quad \textbf{(3.4.3)}$$

If $e = 1$, the collision is *elastic* with no energy loss. If $e = 0$, the collision is *plastic* with maximum energy loss.

Example 3.17

Neglecting friction, estimate the angular velocity of the cylinder after 2 seconds if the motion starts from rest.

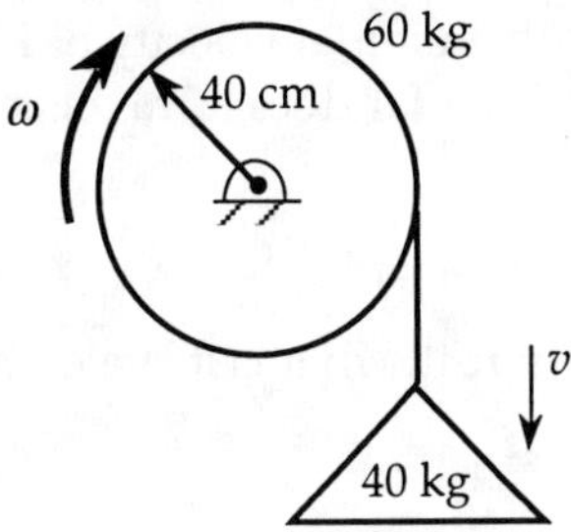

Solution. The impulse-momentum equation is used as follows:

$$M\Delta t = I\Delta\omega + m\Delta v \times r$$

$$0.4 \times (40 \times 9.8) \times 2 = \frac{1}{2} \times 60 \times 0.4^2 \omega + 40 \times 0.4\omega \times 0.4$$

$$\therefore \omega = 28.0 \text{ rad/s}.$$

Example 3.18

Find v' and θ if the coefficient of restitution is 0.8.

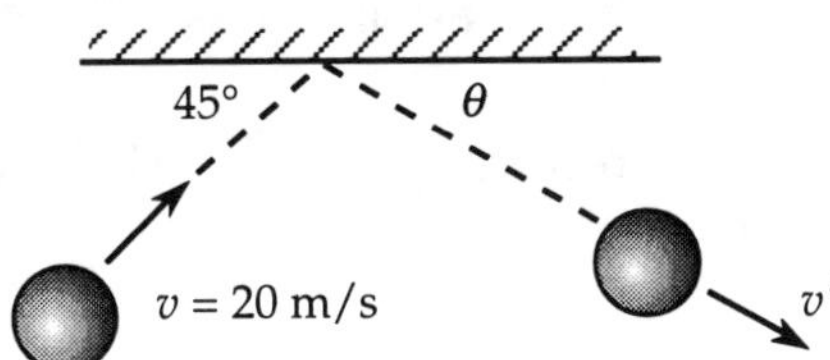

Solution. The coefficient of restitution is based on the normal components of velocity. Thus,

$$e = \frac{v' \sin\theta}{v \sin 45^\circ}.$$

$$\therefore v' \sin\theta = 0.8 \times 20 \times 0.707 = 11.31.$$

The tangential velocity component remains unchanged so that

$$v' \cos\theta = 20 \cos 45^\circ = 14.14.$$

Simultaneous solution of the above results in

$$v' = 18.11 \text{ m/s}, \qquad \theta = 38.65^\circ.$$

Practice Problems

(If you choose to work only a few problems, select those with a star.)

RECTILINEAR MOTION

*3.1 An object is moving with an initial velocity of 20 m/s. If it is decelerating at 5 m/s^2 how far does it travel, in meters, before it stops?
a) 10 b) 20 c) 30 d) 40 e) 50

3.2 If the particle starts from rest, what is its velocity, in m/s, at $t = 4$ s?

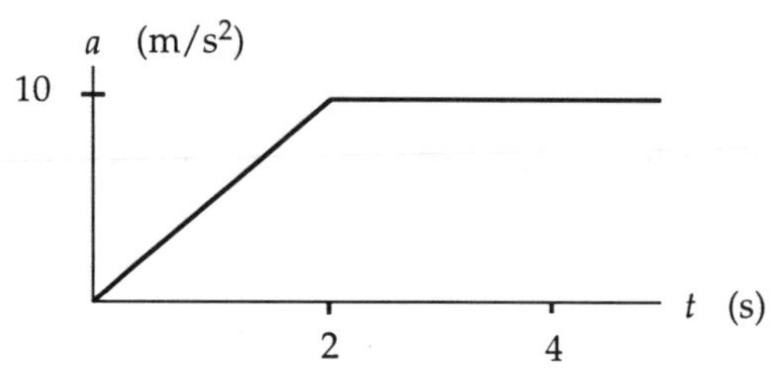

a) 10 b) 20 c) 30 d) 40 e) 50

*3.3 A projectile is shot straight up with $v_o = 40$ m/s. After how many seconds will it return if drag is neglected?
a) 4 b) 6 c) 8 d) 10 e) 12

*3.4 An automobile is traveling at 25 m/s. It takes 0.3 s to apply the brakes after which the deceleration is 6.0 m/s^2. How far does the automobile travel, in meters, before it stops?
a) 40 b) 45 c) 50 d) 55 e) 60

ANGULAR MOTION

3.5 A wheel accelerates from rest with $\alpha = 6 \text{ rad/s}^2$. How many revolutions are experienced in 4 s?
a) 7.64 b) 9.82 c) 12.36 d) 25.6 e) 38.4

*3.6 A 2-m-long shaft rotates about one end at 20 rad/s. It begins to accelerates with $\alpha = 10 \text{ rad/s}^2$. After how long, in seconds, will the velocity of the free end reach 100 m/s?
a) 7 b) 6 c) 5 d) 4 e) 3

CURVILINEAR MOTION

*3.7 A roller-coaster reaches a velocity of 20 m/s at a location where the radius of curvature is 40 m. Calculate the acceleration, in m/s^2.
a) 8 b) 9 c) 10 d) 11 e) 12

3.8 A bucket full of water is to be rotated in the vertical plane. What minimum angular velocity, in rad/s, is necessary to keep the water inside if the rotating arm is 120 cm?
a) 5.31 b) 4.26 c) 3.86 d) 3.15 e) 2.86

3.9 An automobile is accelerating at 5 m/s^2 on a straight road on a hill where the radius of curvature of the hill is 200 m. What is the magnitude of the total acceleration when the car's speed is 30 m/s?
a) 5 b) 5.46 c) 6.04 d) 6.73 e) 7.93

3.10 A particle experiences the displacement shown. What is its velocity at $t = 1$ s?

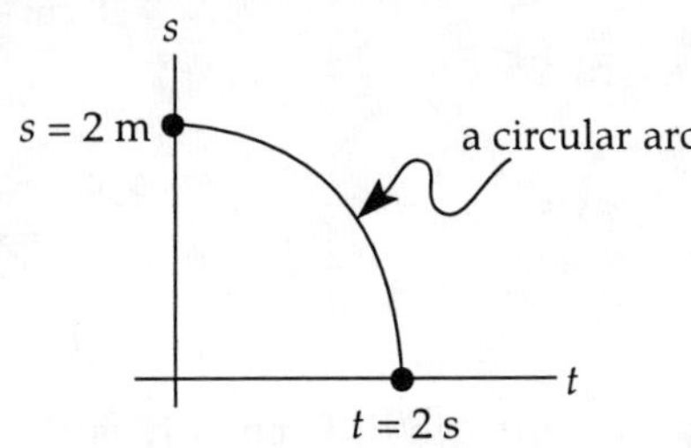

a) –1.24 b) – 1.0 c) –0.92 d) –0.76 e) – 0.58

3.11 Neglecting the change of gravity with elevation, estimate the speed a satellite must have to orbit the earth at an elevation of 100 km. Earth's radius = 6400 km.

a) 4000 b) 6000 c) 8000 d) 10 000 e) 12 000

PROJECTILE MOTION

3.12 Find an expression for the maximum range of a projectile with initial velocity v_0 at angle θ with the horizontal.

a) $\frac{v_0^2}{g}\sin 2\theta$ b) $\frac{v_0^2}{2g}\sin^2\theta$ c) $\frac{v_0^2}{2g}\sin\theta$

d) $\frac{v_0^2}{g}\cos\theta$ e) $\frac{v_0^2}{2g}\sin\theta\cos\theta$

*3.13 Find the maximum height, in meters.

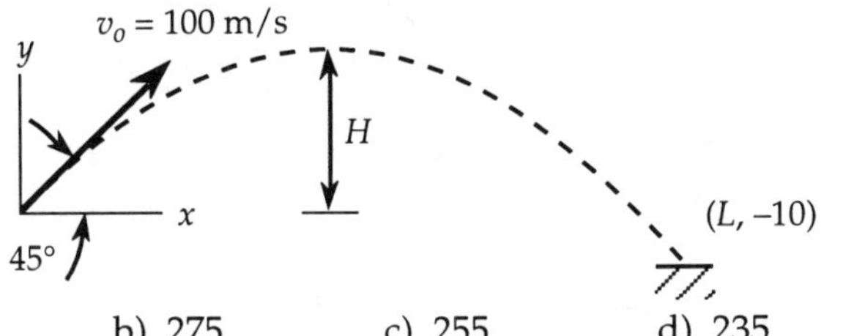

a) 295 b) 275 c) 255 d) 235 e) 215

3.14 Calculate the time, in seconds, it takes the projectile of Problem 3.13 to reach the low point.

a) 14.6 b) 12.2 c) 11.0 d) 10.2 e) 8.31

3.15 What is the distance L, in meters, in Problem 3.13?

a) 530 b) 730 c) 930 d) 1030 e) 1330

RIGID BODY MOTION

*3.16 The acceleration of the center O is given by

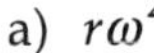

a) $r\omega^2$ b) v^2/r c) 0

d) ω^2/r e) rv^2

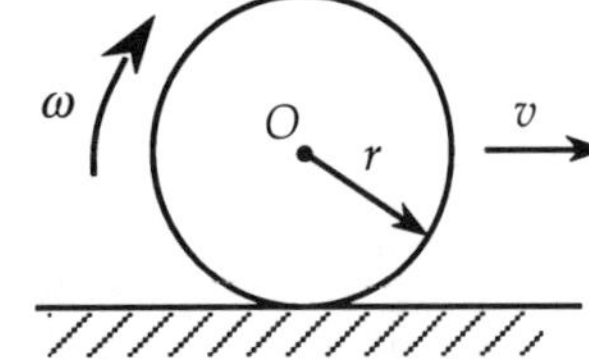

3.17 The spool rolls without slipping. Find v_o in m/s.

a) 30 b) 25 c) 20

d) 10 e) 5

*3.18 The acceleration a_A, in m/s^2, in Problem 3.17 is

a) 1000 b) 800 c) 600 d) 400 e) 200

*3.19 If the acceleration of B is $60\mathbf{i} - 20\mathbf{j}$ m/s^2, find $\mathbf{a}_A$ in m/s^2.

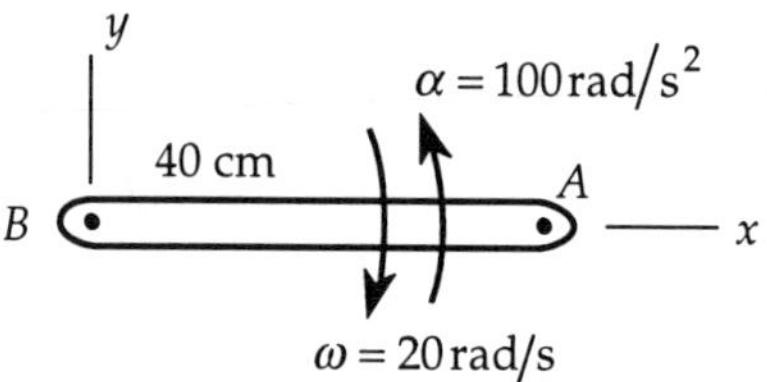

a) $220\mathbf{i} + 60\mathbf{j}$

b) $100\mathbf{i} + 20\mathbf{j}$

c) $-100\mathbf{i} + 20\mathbf{j}$

d) $-100\mathbf{i} - 20\mathbf{j}$

e) $200\mathbf{i} + 20\mathbf{j}$

*3.20 Find ω_{AB} in rad/s.

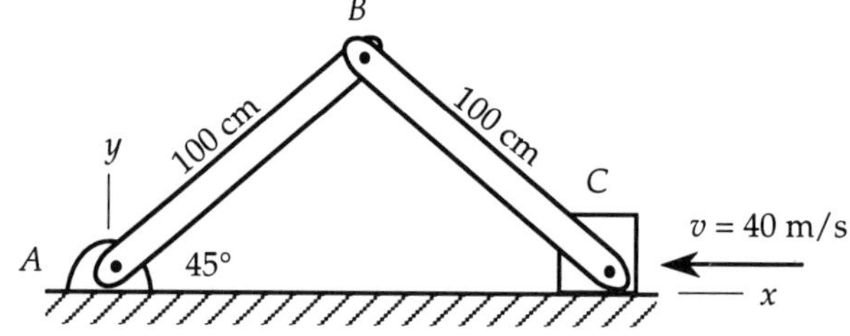

a) 56.6 ccw

b) 56.6 cw

c) 34.1 ccw

d) 28.3 ccw

e) 28.3 cw

*3.21 What is ω_{BC}, in rad/s, for the linkage of Problem 3.20?

a) 56.6 ccw b) 56.6 cw c) 34.1 ccw d) 28.3 ccw e) 28.3 cw

3.22 Determine α_{AB}, in rad/s^2, if $a_C = 0$, in Problem 3.20.

a) 800 cw b) 800 ccw c) 1160 cw d) 3200 cw e) 3200 ccw

3.23 What is $\mathbf{a}_B$, in m/s^2, if $a_C = 0$, in Problem 3.20?

a) 1130 $\mathbf{i}$ b) −1130 $\mathbf{i}$ c) 1130 $\mathbf{j}$ d) −1130 $\mathbf{j}$ e) 1130 $\mathbf{k}$

3.24 Find ω_{AB} in rad/s.

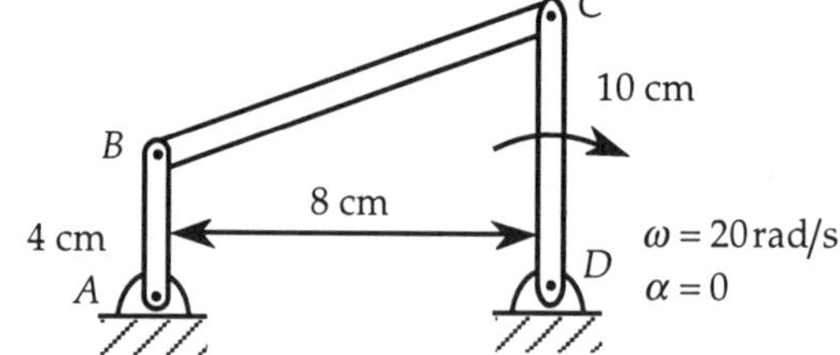

a) 20 ccw

b) 20 cw

c) 10 ccw

d) 10 cw

e) 0

3.25 Find α_{AB}, in rad/s^2, if $a_C = 0$, in Problem 3.24.

a) 750 cw b) 750 ccw c) 400 cw d) 1000 cw e) 1000 ccw

*3.26 What is the acceleration, in m/s^2, of the bead if it is 10 cm from the center?

a) $20\,\mathbf{i} + 40\,\mathbf{j}$
b) $-40\,\mathbf{i} - 20\,\mathbf{j}$
c) $-40\,\mathbf{i} + 40\,\mathbf{j}$
d) $20\,\mathbf{i} - 40\,\mathbf{j}$
e) $40\,\mathbf{i} + 20\,\mathbf{j}$

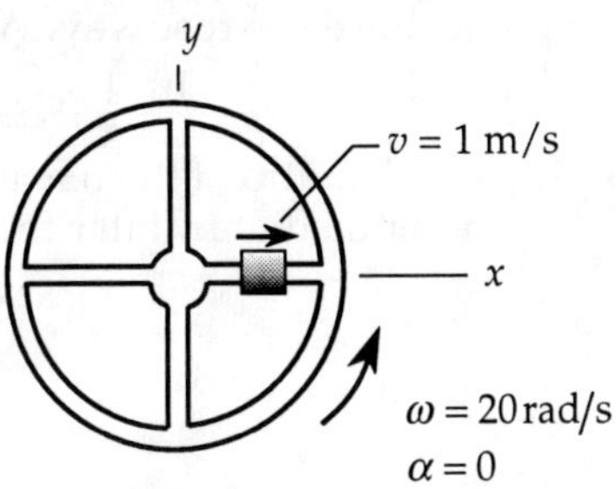

3.27 A boy moves toward the center of a merry-go-round at 4 m/s. If the merry-go-round rotates at 5 rev/min., calculate the acceleration component of the boy normal to the radius.

a) 6.7 b) 6.1 c) 5.8 d) 5.2 e) 4.2

KINETICS

*3.28 What is a_A in m/s^2?

a) 2.09
b) 1.85
c) 1.63
d) 1.47
e) 1.22

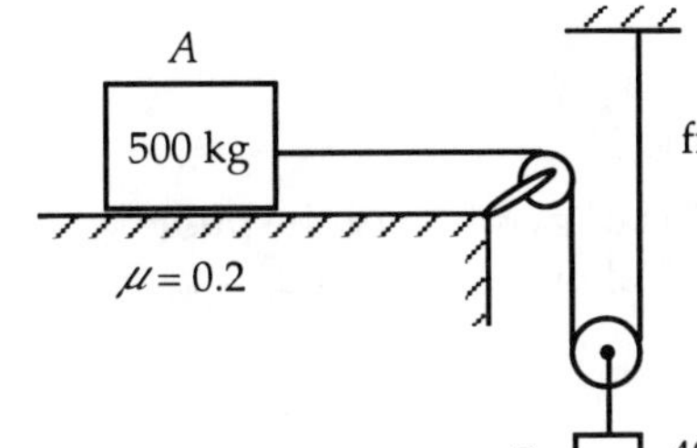

*3.29 How far, in meters, will the weight move in 10 s, if released from rest?

a) 350 b) 300 c) 250
d) 200 e) 150

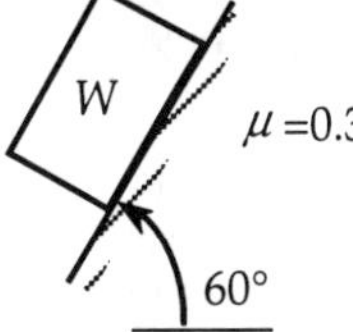

*3.30 At what angle, in degrees, should a road be slanted to prevent an automobile traveling at 25 m/s from tending to slip? The radius of curvature is 200 m.

a) 22 b) 20 c) 18 d) 16 e) 14

3.31 A satellite orbits the earth 200 km above the surface. What speed, in m/s, is necessary for a circular orbit? The radius of the earth is 6400 km and $g = 3.2\ m/s^2$.

a) 7800 b) 7200 c) 6600 d) 6000 e) 5400

3.32 Determine the mass of the earth, in kg, if the radius of the earth is 6400 km.

a) 6×10^{22} b) 6×10^{23} c) 6×10^{24} d) 6×10^{25} e) 6×10^{26}

3.33 The coefficient of sliding friction between rubber and asphalt is about 0.6. What minimum distance, in meters, can an automobile slide on a horizontal surface if it is traveling at 25 m/s?

a) 38 b) 43 c) 48 d) 53 e) 58

*3.34 The center of mass is 30 cm in front of the rear wheel of a motorcycle and 80 cm above the roadway. What maximum acceleration, in m/s^2, is possible?

a) 4.5 b) 4.3 c) 4.1 d) 3.9 e) 3.7

3.35 Find the ratio of the tension in the wire before and immediately after the string is cut.

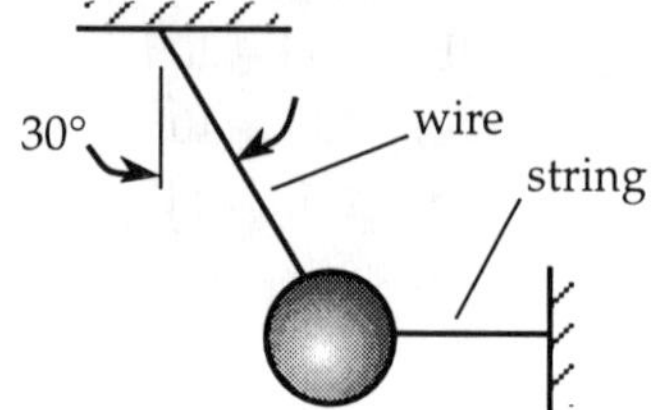

a) $2/\sqrt{3}$ b) $4/3$ c) $2/3$

d) $3/4$ e) $\sqrt{3}/2$

*3.36 Find the force, in kN, on the front wheels if $a = 2\ \text{m/s}^2$. The center of mass is at *G*.

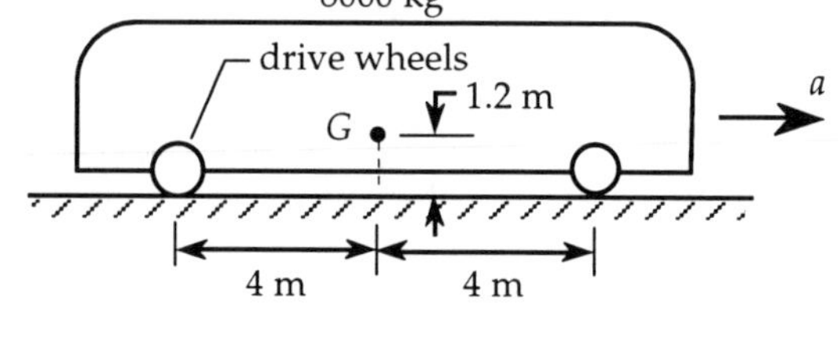

a) 58.2
b) 47.3
c) 41.6
d) 36.8
e) 22.8

3.37 The radius of gyration of the pulley is 10 cm. Calculate α in rad/s^2.

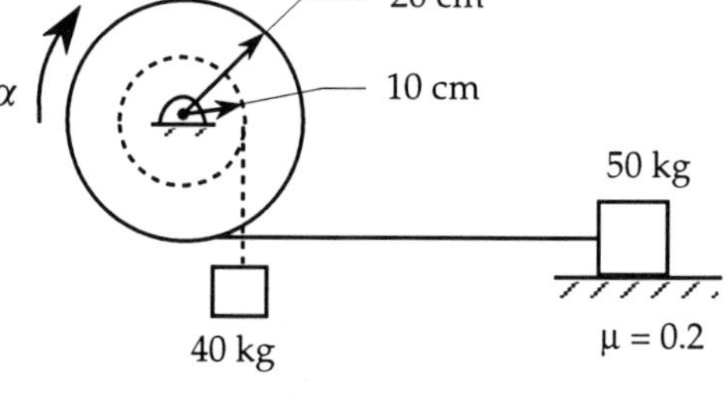

a) 8.52
b) 7.26
c) 6.58
d) 5.32
e) 4.69

3.38 What is the force at *O* immediately after the string is cut?

a) $\frac{mg}{2}$ b) $\frac{mg}{3}$ c) $\frac{mg}{10}$

d) $\frac{mg}{5}$ e) 0

O ℓ

$I_o = \frac{1}{3}m\ell^2$

<u>**Questions 3.39–3.41:**</u>

The force $F(t)$ acts on a 2 kg mass that moves in a straight line on a frictionless plane. It starts from rest at $x = 0$ at $t = 0$.

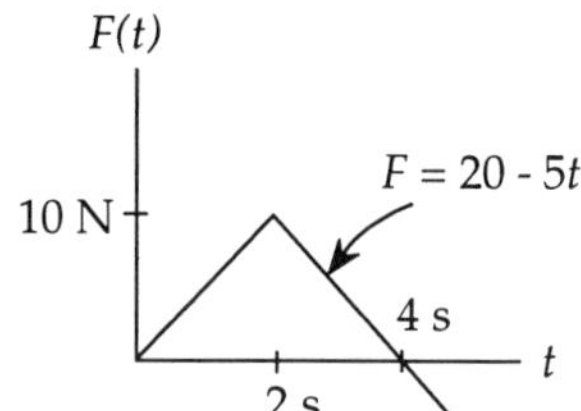

3.39 What is the displacement x after 2 seconds?

a) 5 b) 5/3 c) 3 d) 10/3 e) 2

3.40 What is the velocity after 4 seconds?

a) 5 b) 6 c) 8 d) 9 e) 10

3.41 How long does it take for the mass to reach its maximum displacement from $x = 0$?

a) 2 s b) 4 s c) 6.83 s d) 7.24 s e) 8.61 s

WORK ENERGY

3.42 Find the velocity, in m/s, after the mass moves 10 m if it starts from rest.

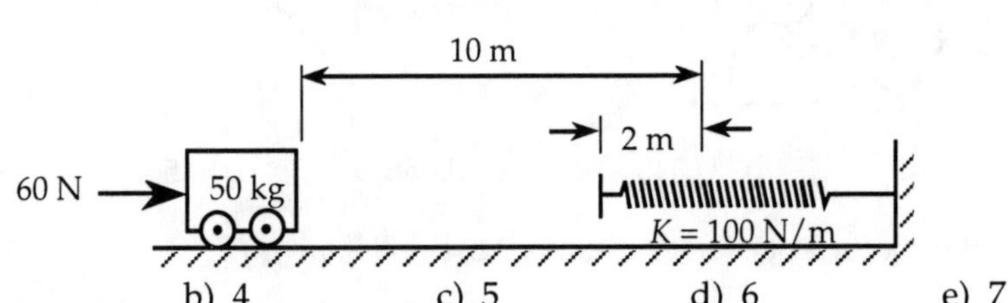

a) 3 b) 4 c) 5 d) 6 e) 7

*3.43 If the force acts through 4 m, what is the angular velocity, in rad/s, of the solid cylinder? Assume no slip and the cylinder starts from rest.

a) 18 b) 16 c) 14

d) 12 e) 10

100 N
20 cm
100 kg
$I_o = \frac{1}{2}mr^2$

*3.44 The spring is initially free. Calculate the velocity, in m/s, of the 2 kg mass after it falls 40 cm. It starts from rest.

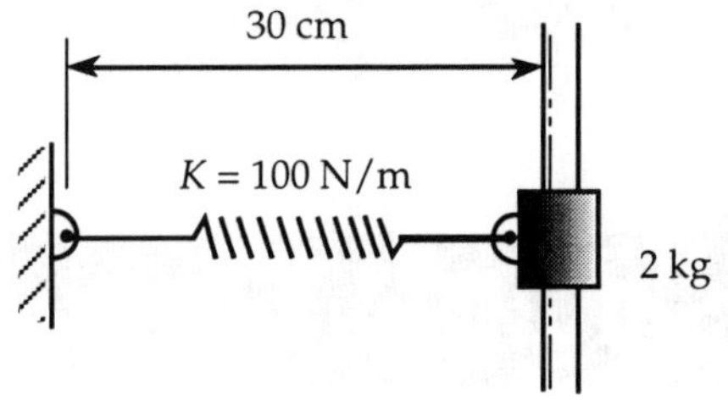

a) 4.62
b) 3.84
c) 2.96
d) 2.42
e) 1.95

3.45 Find the velocity, in m/s, of the end of the 10 kg bar as it passes A. The free spring length is 30 cm. The moment of inertia of a bar about its mass center is $m\ell^2/12$.

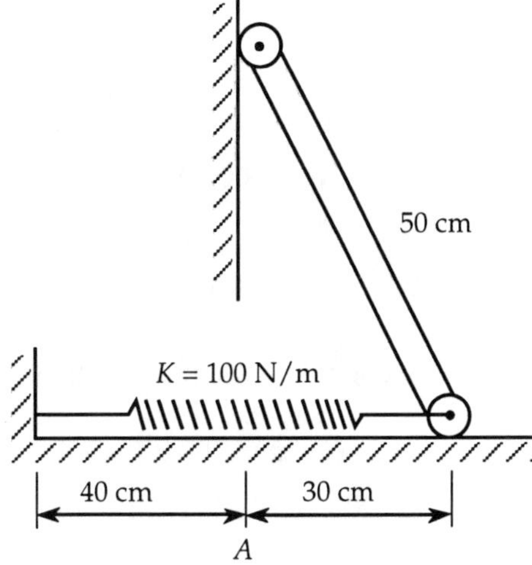

a) 5.2
b) 4.6
c) 3.5
d) 2.4
e) 1.2

IMPULSE-MOMENTUM

*3.46 If the force in Problem 3.43 acts for 4 seconds, find the angular velocity, in rad/s, assuming no slip. The cylinder starts from rest.

a) 2.4 b) 5.2 c) 8.6 d) 10.2 e) 26.7

*3.47 Find the velocity, in m/s, of a 100 kg mass at $t = 2$ s. The coefficient of friction is 0.2.

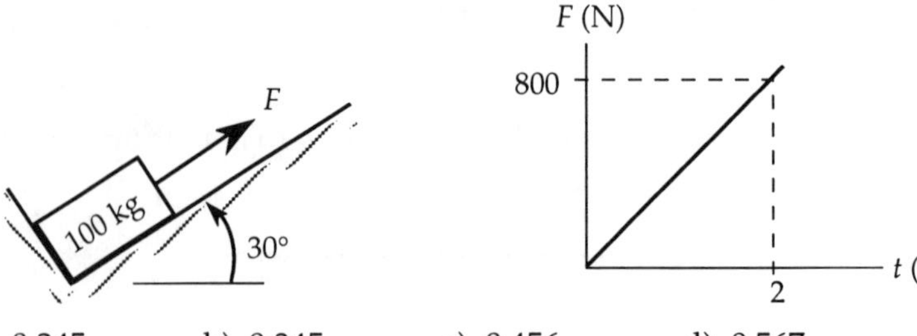

a) 0.245 b) 0.345 c) 0.456 d) 0.567 e) 0.678

*3.48 If the coefficient of restitution is 0.8, find v'_B in m/s.

a) 16
b) 13
c) 11
d) 9
e) 7

10 m/s

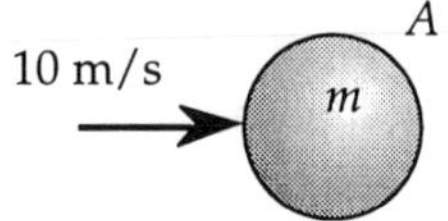

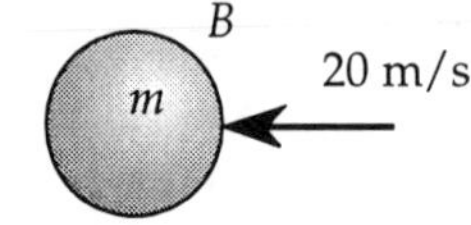

20 m/s

*3.49 Calculate the energy lost, in joules, in the collision of Problem 3.48 if $m = 2$ kg.

a) 200 b) 180 c) 160 d) 140 e) 120

Solutions to Practice Problems

3.1 d) $v^2 = v_o^2 + 2as.\quad 0 = 20^2 - 2\times 5s \quad \therefore\ s = 40\text{ m}$

3.2 c) $a = 5t \quad t \le 2 \quad v = \int a dt = \int_0^2 5t dt + \int_2^4 10 dt = 5\frac{2^2}{2} + 10(4-2) = 30\text{ m/s}$

3.3 c) $v = v_o + at.\quad 0 = 40 - 9.8t \quad \therefore\ t = 4.08\text{ s}.\quad \therefore\ t_{\text{total}} = 2t = 8.16\text{ s}$

3.4 e) $\Delta s = v\Delta t = 25\times 0.3 = 7.5\text{ m}$

$v^2 = v_o^2 + 2as.\quad 0 = 25^2 - 2\times 6s \quad \therefore\ s = 52.1\text{ m}$

$\therefore\ s_{\text{total}} = 7.5 + 52.1 = 59.6\text{ m}$

3.5 a) $\theta = \alpha t^2/2 = 6\times 4^2/2 = 48\text{ rad}.\quad 48/2\pi = 7.64\text{ rev}.$

3.6 e) $\omega = \omega_o + \alpha t = 20 + 10t.\quad \omega = v/L = 100/2 = 50.$

$\therefore\ 20 + 10t = 50.\quad \therefore\ t = 3\text{ s}.$

3.7 c) $a = \frac{v^2}{r} = \frac{20^2}{40} = 10\text{ m/s}^2$

3.8 e) $\frac{v^2}{r} = g.\quad \frac{v^2}{1.20} = 9.8.\quad \therefore\ v = 3.43\text{ m/s}.$

$v = r\omega \quad \therefore\ \omega = 3.43/1.2 = 2.86\text{ rad/s}$

3.9 d) $a_t = 5. \quad a_n = \frac{v^2}{R} = \frac{30^2}{200} = 4.5. \quad a = \sqrt{a_t^2 + a_n^2} = \sqrt{5^2 + 4.5^2} = 6.73 \text{ m/s}^2.$

3.10 e) The equation of the circle is

$s^2 + t^2 = 4. \quad 2s\frac{ds}{dt} + 2t = 0. \quad \therefore \frac{ds}{dt} = -\frac{2t}{2s} = -\frac{t}{s} = -\frac{1}{\sqrt{4-1}} = -0.577 \text{ m/s}.$

3.11 c) $g = \frac{V^2}{R}. \quad 9.8 = \frac{V^2}{6\,400\,000}. \quad \therefore V = 7920 \text{ m/s}.$

3.12 a) Because of symmetry, the time is twice that to reach the maximum height.

Let $v_y = 0$: $gt = v_0 \sin\theta. \quad \therefore t = 2\left(\frac{v_0}{g}\sin\theta\right).$

$\therefore x_{max} = v_0\left(\frac{2v_0}{g}\sin\theta\right)\cos\theta = \frac{v_0^2}{g}\sin 2\theta.$

3.13 c) $H = v_0^2 \sin^2\theta / 2g = 100^2 \times .707^2 2 \times 9.8 = 255 \text{ m}$

3.14 a) $y = v_o t \sin\theta - gt^2/2. \quad -10 = 100t \times .707 - 9.8t^2/2. \quad \therefore t = 14.6 \text{ s}$

3.15 d) $x = v_o t \cos\theta = 100 \times 14.6 \times 0.707 = 1032 \text{ m}$

3.16 c) The acceleration of the point of contact is v^2/r or $r\omega^2$. The acceleration of the center is 0.

3.17 c) Motion is about the point of contact. Thus,
$\omega = 10/.02 = 50 \text{ rad/s}. \quad v = r\omega = 0.4 \times 50 = 20 \text{ m/s}.$

3.18 a) $a = v^2/r = 20^2/0.4 = 1000 \text{ m/s}^2$

3.19 c) $\mathbf{a}_A = \mathbf{a}_B + \mathbf{a}_{A/B}$

$= 60\mathbf{i} - 20\mathbf{j} - .4 \times 20^2\mathbf{i} + .4 \times 100\mathbf{j} = -100\mathbf{i} + 20\mathbf{j}$

3.20 d) $v_B = 40\cos 45° = 28.3 = r\omega_{AB} = 1.0\omega_{AB}$

$\therefore \omega_{AB} = 28.3 \text{ ccw}.$

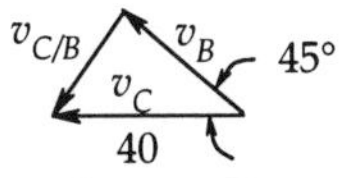

3.21 e) $v_{C/B} = 40\sin 45° = 28.3 = r_{BC}\omega_{BC} = 1.0\omega_{BC}. \quad \therefore \omega_{BC} = 28.3 \text{ cw}$

3.22 a) $\mathbf{a}_C = \mathbf{a}_B + \mathbf{a}_{C/B}$

$r_{AB}\omega_{AB}^2 = 1.0 \times 28.3^2 = 800$

$r_{BC}\omega_{BC}^2 = 1.0 \times 28.3^2 = 800$

Then, $r_{AB}\alpha_{AB} = 800. \quad \therefore \alpha_{AB} = 800 \text{ cw}$

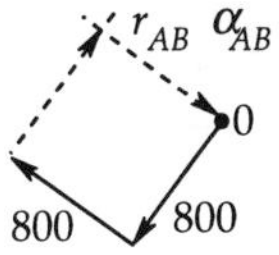

3.23 d) $\mathbf{a}_B = \overset{0}{\cancel{\mathbf{a}_A}} + \mathbf{a}_{B/A}$

$r\omega^2 = 1.0 \times 28.3^2 = 800$

$r\alpha = 1.0 \times 800 = 800$

$\therefore\ \mathbf{a}_B = -1130j$

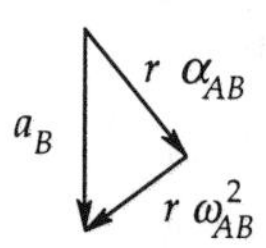

3.24 e) $\mathrm{v}_A = \mathrm{v}_B + \mathrm{v}_{A/B}$. $v_{A/B} = r\omega_{AB} \perp \overline{AB}$. But, v_A and v_B are both horizontal. Thus, $\mathrm{v}_{A/B} = 0$ and $\omega_{AB} = 0$.

3.25 b) $v_A = v_B = 20 \times 0.1 = 0.04\,\omega_A$.

$\therefore\ \omega_A = 50.\quad r_B\omega_B^2 = .1 \times 20^2 = 40$

$r_A\omega_A^2 = .04 \times 50^2 = 100.$

$\therefore\ r_{AB}\alpha_{AB} = \dfrac{60}{0.8} = 75.$

But, $r_{AB} = 0.1.\quad \therefore\ \alpha_{AB} = 750$ ccw.

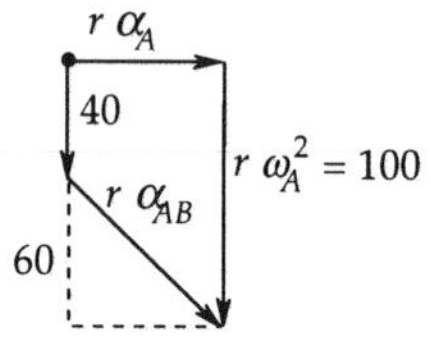

$\vec{a}_A = \vec{a}_B + \vec{a}_{A/B}$

3.26 c) $\mathbf{a} = -r\omega^2\mathbf{i} + 2\omega v\mathbf{j} = -0.1 \times 20^2\mathbf{i} + 2 \times 20 \times 1.0\mathbf{j} = -40\mathbf{i} + 40\mathbf{j}$

3.27 e) $a_{\text{Coriolis}} = 2v\omega = 2 \times 4 \times \dfrac{5 \times 2\pi}{60} = 4.19\ \text{m/s}^2.$

3.28 c) $a_A = 2a_B$(from small pulley above B)

$400 \times 9.8 - 2T = 400\frac{a_A}{2}$ (body B)

$T - 0.2 \times 500 \times 9.8 = 500\ a_A$ (body A), $\therefore\ a_A = 1.63\ \text{m/s}^2$

3.29 a) $W\cos 60° - W\mu\sin 60° = \dfrac{W}{g}a.$

$\therefore\ a = 9.8(0.866 - 0.3 \times 0.5) = 7.02\ \text{m/s}^2.\quad \therefore\ s = at^2/2 = 7.02 \times 10^2/2 = 351\ \text{m}$

3.30 c) $W\sin\theta = \dfrac{W}{g}\dfrac{v^2}{r}\cos\theta \qquad \tan\theta = \dfrac{25^2}{200 \times 9.8}.\quad \therefore\ \theta = 17.7°$

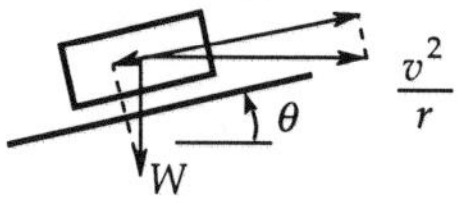

3.31 a) $W = \dfrac{W}{g}\dfrac{v^2}{r}.\quad \therefore\ v = \sqrt{9.2 \times 6\ 600\ 000} = 7790\ \text{m/s}$

3.32 c) $F = k\dfrac{m_1m_2}{r^2}.\quad W = k\dfrac{m_eW/g}{r^2}.$ (Let W be your weight!)

$\therefore\ m_e = r^2g/k = 6\,400\,000^2 \times 9.8/6.67 \times 10^{-11} = 6 \times 10^{24}\ \text{kg}$

3.33 d) $v^2 = v_o^2 + 2as$. $-0.6W = \frac{W}{g}a$. $\therefore a = -.06g$. $0 = 25^2 - 2\times.6\times9.8s$.

$\therefore s = 53.2$ m

3.34 e) $\Sigma M = 0$

For maximum accel. the force on the front wheel = 0.

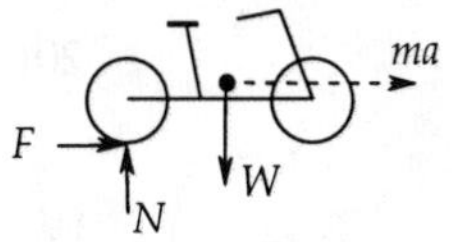

Thus, $30W = 80\ ma$. $\therefore a = \frac{30\times9.8}{80} = 3.68\ \text{m/s}^2$

ma
F
N
W

3.35 b) Before: $\Sigma F_y = 0$. $T\cos30° = W$. $\therefore T = \frac{W}{0.866}$

After: $F = 0$, $\Sigma F_n = ma_n$. $n \perp t$. $\therefore$ $T = W\cos\ = 0.866\ W$.

ratio = $\frac{\text{before}}{\text{after}} = \frac{1}{0.866\times0.866} = 1.33$

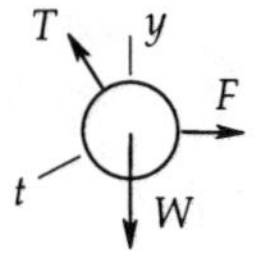

T
y
F
t
W

3.36 d) Take moments about the back wheels.

$4W - 8N_2 = 1.2\frac{W}{g}\times2$. $\therefore N_2 = 0.469W = .469\times8000\times9.8 = 36\ 800$ N

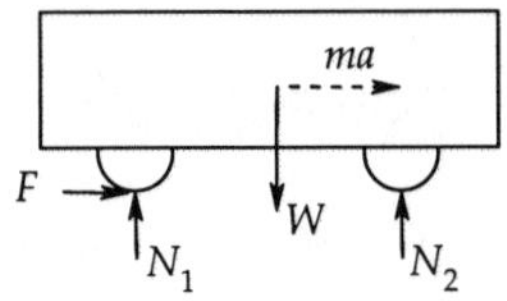

ma
F
W
N_1
N_2

3.37 b) Acceleration of block = 0.2α. Thus, $T = 50\times0.2\alpha + 50\times9.8\times0.2$.

$\Sigma M = I\alpha$. Use $I = mk^2$. $(40\times9.8 - 40\alpha)\times0.1 - (10\alpha + 9.8)\times0.2 = 30\times.1^2\alpha$.

$\therefore \alpha = 7.26\ \text{rad/s}^2$

3.38 c) $\Sigma M_o = I_o\alpha$. $mg\times\frac{\ell}{2} = \frac{1}{3}m\ell^2\alpha$. $\therefore \alpha = \frac{3g}{2\ell}$.

$\Sigma F_y = m\bar{a}$ ($\bar{a}$ is acc. of mass center).

$mg - F_o = m\frac{\ell}{2}\alpha = m\frac{3}{4}g$. $\therefore F_o = \frac{mg}{4}$.

3.39 d) $F = 5t$ for $t < 2$ s. $\therefore 5t = 2\frac{dv}{dt}$ or $\int_0^t 5t\,dt = \int_0^v 2dv$. $\therefore v = \frac{5}{4}t^2$.

$v = \frac{dx}{dt} = \frac{5}{4}t^2$. $\therefore \int_0^x dx = \frac{5}{4}\int_0^2 t^2dt$ and $x = \frac{5}{4}\times\frac{2^3}{3} = \frac{10}{3}$ m.

3.40 e) $v = \frac{5}{4}t^2 = \frac{5\times2^2}{4} = 5$ m/s after 2 seconds. For $t > 2$, $20 - 5t = 2\frac{dv}{dt}$.

$\therefore \int_2^4 (20 - 5t)dt = 2\int_5^v dv$. $20\times(4-2) - \frac{5}{2}(16-4) = 2(v-5)$. $\therefore v = 10$ m/s.

3.41 c) From Problem 3.40,

$$\int_2^t (20-5t)dt = 2\int_5^v dv \quad \text{or} \quad 20(t-2) - \frac{5}{2}\left(t^2-4\right) = 2(v-5).$$

The mass reaches its maximum displacement when $v = 0$. Set $v = 0$ and solve for t:

$$20t - 40 - \frac{5}{2}t^2 + 10 = -10. \quad \therefore t^2 - 8t + 8 = 0. \quad \therefore t = \frac{8 \pm \sqrt{64-32}}{2} = 6.83 \text{ s}.$$

3.42 b) $W_{net} = \Delta E = \frac{1}{2}mv^2. \quad 60\times 10 - \frac{1}{2}\times 100\times 2^2 = \frac{1}{2}50v^2. \quad \therefore v = 4 \text{ m/s}$

3.43 d) $W_{net} = \Delta KE = \frac{1}{2}mv^2 + \frac{1}{2}I\omega^2. \quad v = r\omega.$

$$\therefore\ 100\times 4 = \frac{1}{2}\times 100(0.2\omega)^2 + \frac{1}{2}\left(\frac{1}{2}\times 100\times .2^2\right)\omega^2. \quad \therefore \omega = 11.55 \text{ rad/s}$$

3.44 d) $W_{net} = \Delta KE = \frac{1}{2}mv^2. \quad \Delta x = 50 - 30 = 20.$

$$\therefore\ 2\times 9.8\times 0.4 - \frac{1}{2}\times 100\times .2^2 = \frac{1}{2}\times 2v^2. \quad \therefore v = 2.42 \text{ m/s}$$

3.45 e) $W_{net} = \Delta KE = \frac{1}{2}mv^2 + \frac{1}{2}I\omega^2.$

$$\frac{1}{2}100\left(.4^2 - .1^2\right) - 10\times 9.8(.25 - .20) = \frac{1}{2}10\left(\frac{v}{2}\right)^2 + \frac{1}{2}\left(10\times .5^2/12\right)\left(\frac{v}{.5}\right)^2$$

where $\overline{v} = v/2$ and $\omega = v/.5$. The above gives $v = 1.25$ m/s.

3.46 b) $\Sigma M_C \Delta t = I_C\omega$ where C is the point of contact. $I_C = I_o + mr^2 = \frac{3}{2}mr^2.$

$$100\times 0.4\times 4 = \frac{3}{2}\times 100\times .2^2\omega. \quad \therefore \omega = 26.7 \text{ rad/s}.$$

3.47 a) Integrate Eq. 3.2.1 as follows: $\int F dt = m\Delta v. \quad F = 400t.$

$F_f = \mu N = 0.2\times 100\times 0.866\times 9.8 = 170$

$400t = 100\times 0.5\times 9.8 + 170. \quad \therefore t = 1.65$ s when motion initiates.

$$\int_{1.65}^{2.0} 400t dt - (170 + 50\times 9.8)\times 0.35 = 100v. \quad \therefore v = 0.245 \text{ m/s}$$

3.48 e) $10\,m - 20\,m = v'_A m + v'_B m. \quad \therefore\ v'_A + v'_B = -10$

$0.8 = \frac{v'_B - v'_A}{20+10}. \quad \therefore v'_B - v'_A = 24.$ Simultaneous solutions yields $v'_B = 7$ m/s,

$v'_A = -17$ m/s.

3.49 c) $\Delta KE = \frac{1}{2}mv_A^2 + \frac{1}{2}mv_B^2 - \frac{1}{2}mv_A'^2 - \frac{1}{2}mv_B'^2 = \frac{2}{2}\left(10^2 + 20^2\right) - \frac{2}{2}\left(7^2 + 17^2\right) = 162 \text{ J}.$

Mechanics of Materials

by George E. Mase

The mechanics of materials is one of a number of names given to the study of deformable solids subjected to applied forces and moments. The foundations of this subject reside in three basic topics:

1. internal equilibrium (stress concepts)
2. geometry of deformation (strain concepts)
3. mechanical and thermal properties (by which stress and strain are related)

Additionally, we assume homogeneity (properties are independent of position) and isotropy (absence of directional properties) in the materials considered.

4.1 Stress and Strain, Elastic Behavior

Consider a prismatic bar of length L and a cross sectional area A situated along the x-axis as shown in Fig. 4.1a. Let the bar be subjected to a constant axial force P applied at the centroids of the end faces so as to stretch the bar by an amount δ, Fig. 4.1b.

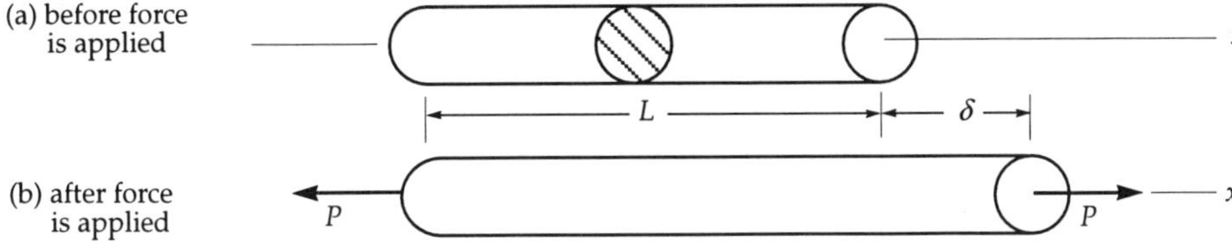

Figure 4.1 Normal stress and strain.

We define the *longitudinal*, or *normal strain* ε_x by the ratio

$$\varepsilon_x = \delta/L. \tag{4.1.1}$$

Strain is dimensionless, having the units of m/m, or in/in, etc. Normal strains are positive if due to elongation, negative if the result is shortening. The stress

$$\sigma_x = P/A \tag{4.1.2}$$

is the *normal*, or *axial stress* in the bar. Stress has units of N/m^2, lbs/in^2 (psi) or kips/in^2 (ksi). One Newton per square meter is called a *Pascal*, abbreviated Pa. Note that the stress and strain defined here are averages, constant over the length of the bar, and uniform over its cross section. By contrast, for a tapered bar hanging from the ceiling under its own weight, the stress and strain would vary along the bar. Also, it is natural that a positive longitudinal strain will be accompanied by a negative lateral strain. Indeed, this ratio

$$\nu = -\frac{\text{lateral strain}}{\text{longitudinal strain}} \tag{4.1.3a}$$

is called *Poisson's ratio*, an important property of a given material. Thus, for the bar in Fig. 4.1,

$$\varepsilon_y = \varepsilon_z = -\nu\varepsilon_x. \tag{4.1.3b}$$

If the bar in Fig. 4.1 is made of a *linear elastic* material, its axial stress and strain are related by the formula, often called *Hooke's law*,

$$\sigma_x = E\varepsilon_x \tag{4.1.4}$$

where E is a material constant called *Young's modulus*, or the *modulus of elasticity*. The units of E are the same as those of stress. By inserting Eqs. 4.1.1 and 4.1.2 into Eq. 4.1.4 and solving for δ we obtain the useful formula

$$\delta = \frac{PL}{AE}. \tag{4.1.5}$$

Stress is always accompanied by strain, but strain may occur without stress. In particular, a temperature change in an unconstrained bar will cause it to expand (or shrink) inducing a thermal deformation

$$\delta_t = \alpha L(T - T_o) \tag{4.1.6}$$

where α is the *coefficient of thermal expansion*, and $(T - T_o)$ the temperature change. Typical units of α are meters per meter per degree Celsius ($^\circ\text{C}^{-1}$). Important properties of several materials are listed in Table 4.1.

Next consider a material cube subjected to a pair of equilibrating couples acting in the plane of the faces of the cube as shown pictorially in Fig. 4.2a, and schematically in Fig. 4.2b. For a cube whose faces have an area A we define the *shear stress* in a plane parallel to those on which the forces act as

$$\tau_{xy} = F/A. \tag{4.1.7}$$

Table 4.1 Average Material Properties.

	Modulus of Elasticity E		Shear Modulus G		Poisson's Ratio ν	Density ρ		Coefficient of Thermal Expansion α	
	$\times 10^6$ kPa	$\times 10^6$ psi	$\times 10^6$ kPa	$\times 10^6$ psi		kg/m^3	lb/ft^3	$\times 10^{-6}$ °C^{-1}	$\times 10^{-6}$ °F^{-1}
steel	210	30	83	12	.28	7850	490	11.7	6.5
aluminum	70	10	27	3.9	.33	2770	173	23.0	12.8
magnesium	45	6.5	17	2.4	.35	1790	112	26.1	14.5
cast iron	140	20	55	8	.27	7080	442	10.1	5.6
titanium	106	15.4	40	6	.34	4520	282	8.8	4.9
brass	100	15	40	6	.33	8410	525	21.2	11.8
concrete	20	3	—	—	—	2400	150	11.2	6.2

If the material of the cube is linearly elastic, the top will be displaced relative to the bottom as shown in Fig. 4.2c. The angle γ_{xy} measures the *shear strain* of the cube, and since for elastic behavior this angle is very small, we define the shear strain as

$$\gamma_{xy} \approx \tan \gamma_{xy} = \Delta x/h. \tag{4.1.8}$$

Also, for elastic behavior,

$$\tau_{xy} = G\gamma_{xy} \tag{4.1.9}$$

where G is the *shear modulus*, or *modulus of rigidity*, having the units of Pa or psi. It is related to E and ν by

$$G = \frac{E}{2(1+\nu)}\ . \tag{4.1.10}$$

There are only two independent material properties in an isotropic, elastic solid. These properties may depend on position as in a heat-treated steel.

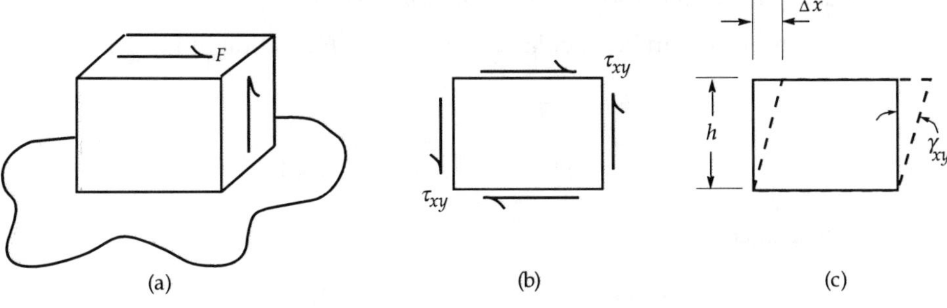

Figure 4.2 Shear stress and strain.

Example 4.1

A 2 cm × 2 cm square aluminum bar *AB* supported by 1.25 cm diameter steel cable *BC* carries a 7000 N load in the arrangement shown. Determine the stresses in the steel and in the aluminum. Also, calculate the elongation of the cable and the shortening of the bar.

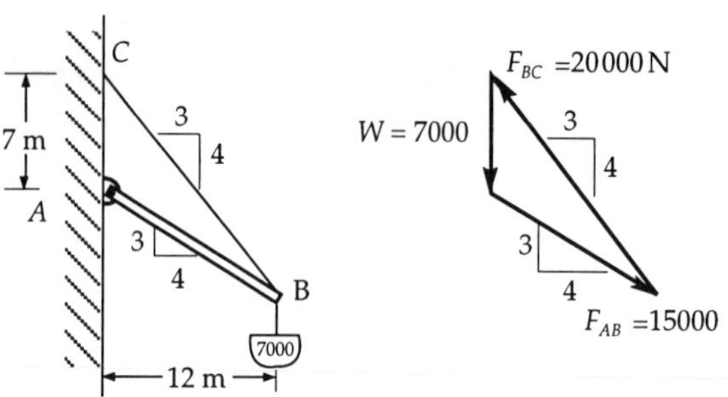

Solution. As shown by the force polygon for the equilibrium, $F_{AB} = -15\ 000$ N and $F_{BC} = 20\ 000$ N. Thus there results

$$\sigma_{al} = \frac{P}{A} = -\frac{15\ 000}{0.02 \times 0.02} = -37.5 \times 10^6 \text{ Pa} \quad \text{or} \quad -37.5 \text{ MPa}.$$

$$\sigma_{st} = \frac{P}{A} = \frac{20\ 000}{\pi(0.0125)^2/4} = 163 \times 10^6 \text{ Pa} \quad \text{or} \quad 163 \text{ MPa}.$$

From Table 4.1 and Eq. 4.1.5,

$$\delta_{al} = \frac{PL}{AE} = \frac{-15\ 000(15)}{0.02 \times 0.02 \times 70 \times 10^9} = -0.00804 \text{ m} \quad \text{or} \quad -8.04 \text{ mm}.$$

$$\delta_{st} = \frac{PL}{AE} = -\frac{20\ 000(20)}{(\pi \times 0.0125^2/4) \times 210 \times 10^9} = 0.0155 \text{ m} \quad \text{or} \quad 15.5 \text{ mm}.$$

Example 4.2

A 5-cm-dia, 80-cm-long steel bar is restrained from moving. If its temperature is increased 100°C, what compressive stress is induced?

Solution. The strain can be calculated using the deformation of Eq. 4.1.6:

$$\varepsilon = \frac{\delta_t}{L} = \alpha(T - T_o)$$
$$= 11.7 \times 10^{-6} \times 100 = 11.7 \times 10^{-4}.$$

Hence, the induced stress is

$$\sigma = \varepsilon E = 11.7 \times 10^{-4} \times 210 \times 10^6 = 246\ 000 \text{ kPa}.$$

Example 4.3

The steel block ($G = 83 \times 10^6$ kPa) is welded securely to a horizontal platen and subjected to 1000 kN horizontal force as shown. Determine the shear stress in a typical horizontal plane of the block, and the horizontal displacement of the top edge *AB*.

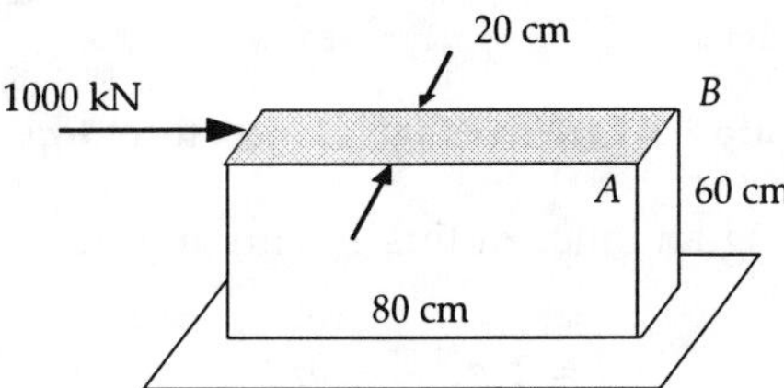

Solution. From Eq. 4.1.7

$$\tau = \frac{F}{A} = \frac{1000}{(0.8)(0.2)} = 6250 \text{ kPa}.$$

From Eq. 4.1.9

$$\gamma_{xy} = \frac{\tau_{xy}}{G} = \frac{6250}{83 \times 10^6} = 7.53 \times 10^{-5} \text{ rad}.$$

From Eq. 4.1.8 the horizontal displacement is

$$\Delta x = h\gamma_{xy} = 600\,(7.53 \times 10^{-5}) = 0.0452 \text{ mm}.$$

4.2 Torsion

A straight member of constant circular cross section subjected to a twisting couple at each end is said to be in *torsion*, and such a member is called a *shaft*. For an elastic shaft of length L and radius c subjected to a *torque* T (pair of equilibrium couples), as shown in Fig. 4.3a, the angular displacement of one end relative to the other is given by the angle ϕ (in radians) as

$$\phi = \frac{TL}{JG} \tag{4.2.1}$$

where $J = \pi a^4/2$ is the *polar moment of inertia* of the circular cross section. Also, the *torsional shear stress* at the radial distance r from the axis of the shaft in a given cross section will be

$$\tau = \frac{Tr}{J} \tag{4.2.2}$$

which increases linearly as shown in Fig. 4.3b. Thus, the maximum shear stress occurs at $r = a$,

$$\tau_{\max} = \frac{Ta}{J}. \tag{4.2.3}$$

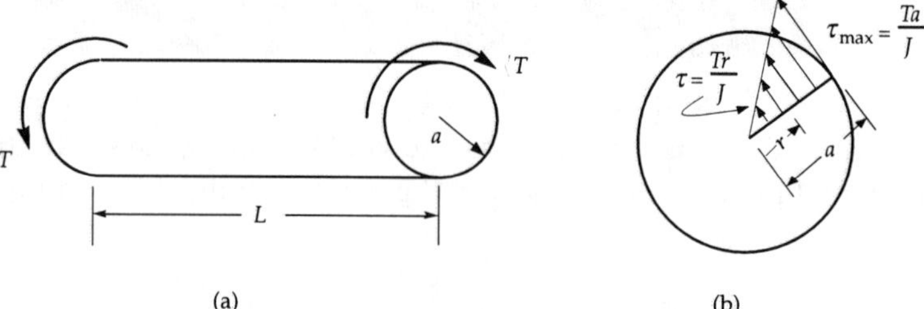

(a) (b)

Figure 4.3 Circular shaft subject to a torque.

For a hollow shaft having an inner radius a_i and an outer a_o the above formulas are all valid, but with

$$J = \pi(a_o^4 - a_i^4)/2. \tag{4.2.4}$$

Example 4.4

A 6 cm diameter, 2 m long magnesium ($G = 17\times10^9$ Pa) shaft is welded to a hollow ($c_o = 3$ cm and $c_i = 1.5$ cm) aluminum ($G = 27\times10^9$ Pa) shaft 1.2 m long. A moment of 2000 m·N is applied at end A. Determine the maximum torsional stress in each material and the angle of twist of end A relative to fixed end B.

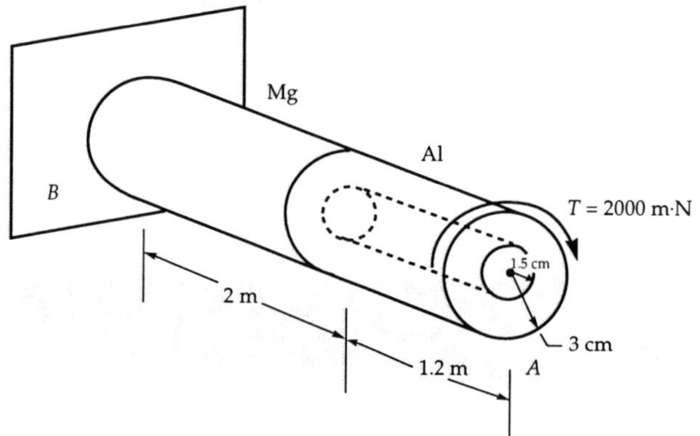

Solution. The polar moments of inertia are

$$J_{mg} = \pi a^4/2 = \pi(0.03)^4/2 = 1.272\times10^{-6}\ \text{m}^4.$$

$$J_{al} = \pi(a_o^4 - a_i^4)/2 = \pi(0.03^4 - 0.015^4)/2 = 1.193\times10^{-6}\ \text{m}^4.$$

From Eq. 4.2.3

$$\tau_{mg} = \frac{Ta}{J} = \frac{2000\times0.03}{1.272\times10^{-6}} = 47.17\times10^6\ \text{Pa}.$$

$$\tau_{al} = \frac{2000\times0.03}{1.193\times10^{-6}} = 50.29\times10^6\ \text{Pa}.$$

From Eq. 4.2.1 the angle of twist is

$$\phi = \phi_{mg} + \phi_{al} = \left(\frac{TL}{JG}\right)_{mg} + \left(\frac{TL}{JG}\right)_{al}$$

$$= \frac{2000 \times 2}{1.272 \times 10^{-6} \times 17 \times 10^{9}} + \frac{2000 \times 1.2}{1.193 \times 10^{-6} \times 27 \times 10^{9}}$$

$$= 0.1850 + 0.0745 = 0.2595 \text{ rad.}$$

4.3 Beam Theory

The usual geometry of a beam is that of a member having the length much larger than the depth with the forces applied perpendicular to this long dimension. The beams considered here have a longitudinal plane of symmetry in which the forces act and in which beam deflections occur. To illustrate we consider a T shaped beam having the cross section shown in Fig. 4.4b, supported either as a cantilever beam, Fig. 4.4a, or as a simply-supported beam, Fig. 4.4c. The longitudinal axis of the beam (x-axis here) passes through the centroidal points of all cross sections. The xy-plane is the plane of symmetry. Any combination of concentrated and distributed loads may act on the beam.

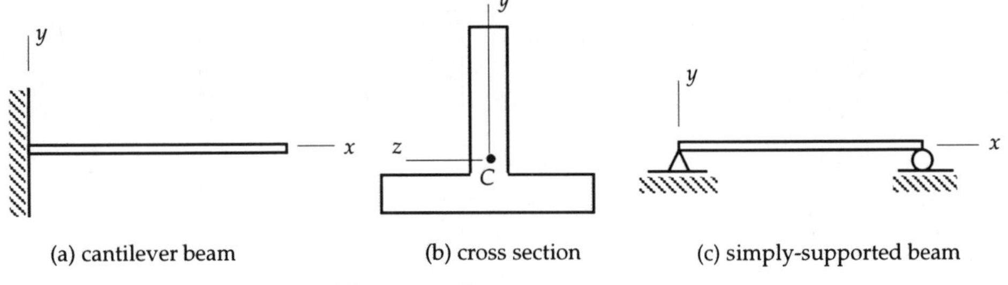

Figure 4.4 Beam geometry.

At the typical cross section of the loaded beam there is an internal force V called the *shear force,* and an internal moment M called the *bending moment.* Both V and M may be determined by a free-body diagram of the left hand portion of the beam, and are, in general, functions of x as shown in Fig. 4.5, where positive values of V and M are displayed. A plot of $V(x)$ is called a *shear diagram* and a plot of $M(x)$ is a *moment diagram.* At a given cross section where the moment has the value M, the (longitudinal) bending stress acting normal to the cross section is

$$\sigma_x = -\frac{My}{I} \tag{4.3.1}$$

where I is the plane moment of inertia of the cross sectional area relative to the centroidal axis. The minus sign is needed to assure a compressive stress for positive y values when the moment M is positive. The stress is a linear function of y as is shown in Fig. 4.6b, with the maximum compression occurring at the top of the beam, and the maximum tension at the bottom for the positive M. The bending stress is zero at $y = 0$, the so-called *neutral axis.*

The stress due to the shear force V is a vertical shear stress

$$\tau_{xy} = \frac{VQ}{Ib} \qquad \textbf{(4.3.2)}$$

where b is breadth, or thickness of the beam at the position (y coordinate) at which the shear stress is calculated. The symbol Q stands for the *statical moment* about the neutral axis of the area between the position and the top of the beam. For a rectangular beam τ_{xy} is parabolic. For the T beam it has the shape shown in Fig. 4.6c. In both cases the maximum shear stress occurs at the neutral axis (centroidal position).

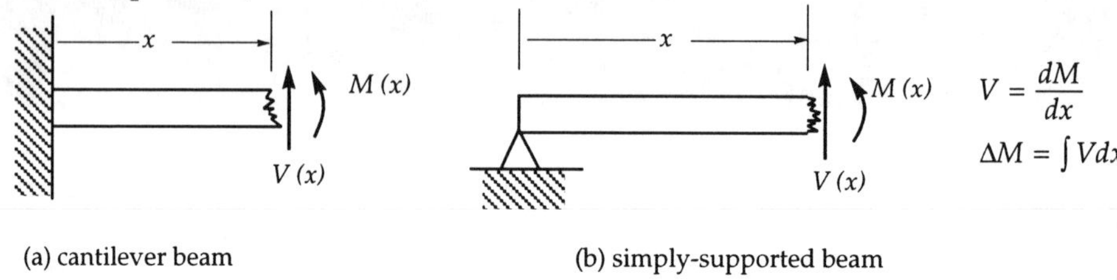

Figure 4.5 Internal shear force and bending moment.

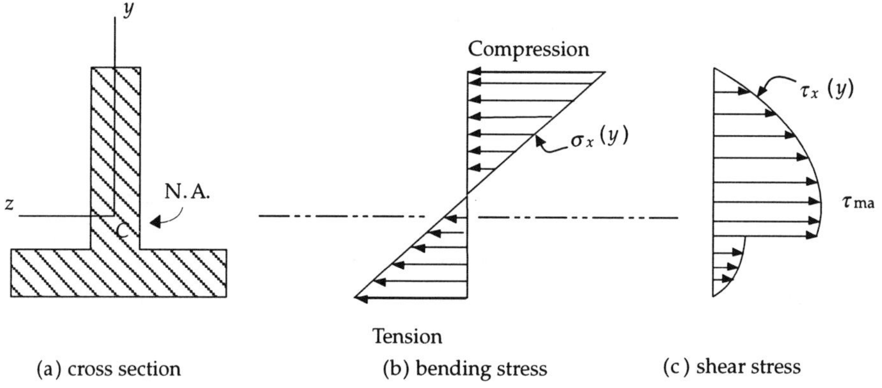

Figure 4.6 Beam stresses for positive *M*.

The vertical displacement of the x-axis of a loaded beam measures the beam deflection. The curve of this deflection $v = v(x)$ is called the *equation of the elastic line*, shown in Figure 4.7. Also, the slope $\theta = \theta(x) = dv / dx$ of the deflection curve is an important quantity in beam theory. Table 4.2 lists some useful formulas for basic beams.

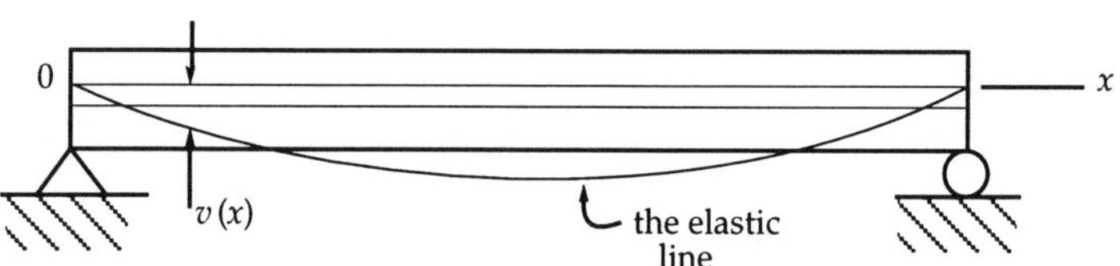

Figure 4.7 Beam deflection.

TABLE 4.2 Beam Formulas.

Number	Max Shear	Max Moment	Max Deflection	Max Slope
1	P	PL	$PL^3/3EI$	$PL^2/2EI$
2	wL	$wL^2/2$	$wL^4/8EI$	$wL^3/6EI$
3	$5wL/8$	$wL^2/8$	$wL^4/185EI$	
4	$P/2$	$PL/4$	$PL^3/48EI$	$PL^2/16EI$
5	$wL/2$	$wL^2/8$	$5wL^4/384EI$	$wL^3/24EI$
6		M_o	$M_oL^2/2EI$	M_oL/EI

Example 4.5

Sketch the shear and moment diagrams for the beam shown.

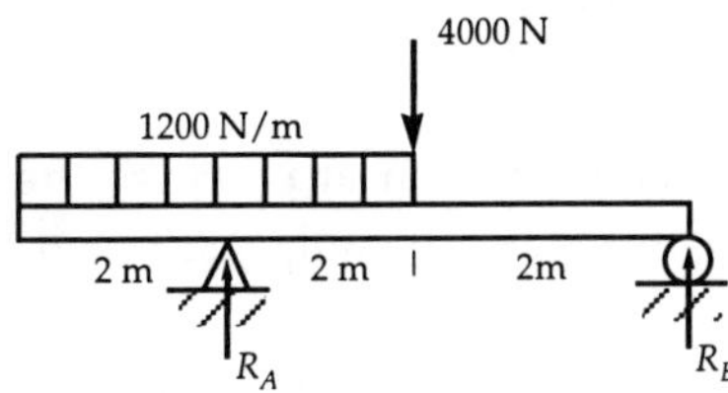

Solution.

First, determine the support reactions R_A and R_B:

$$\sum M_B = 4R_A - 4000(2) - 4(1200)4 = 0$$

$$\therefore\ R_A = 6800\text{ N}$$

$$\sum M_A = 4R_B - 4000(2) = 0$$

$$\therefore\ R_B = 2000\text{ N}$$

The values of V and M as functions of x are shown in the sketches; V is the resultant of all forces acting on the portion of the beam to the left of the x-location. The change in moment is the area under the shear diagram, $\Delta M = \int V dx$. Note from the diagrams that the maximum positive and negative moments occur at locations where the shear plot crosses the x-axis; the values of –2400 and 4000 are simply the appropriate areas under the shear diagram. Note that $M = 0$ at both ends and $V = 0$ at the left end.

Example 4.6

If the beam of Example 4.5 has the cross section shown below, determine the maximum tensile and compressive stresses, and the maximum shear stress.

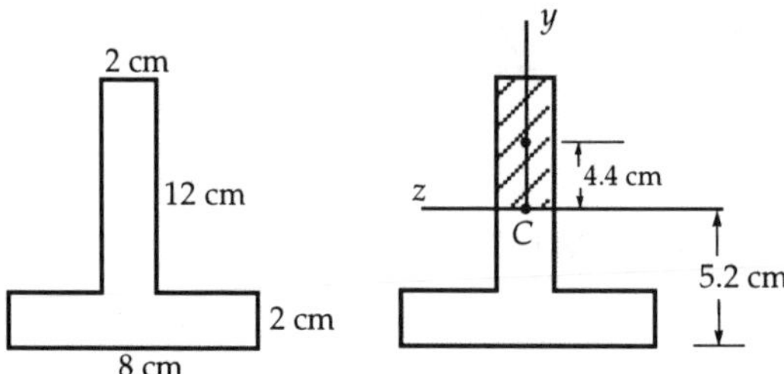

Solution. First, we locate the neutral axis by determining C relative to the bottom of the beam, as shown above:

$$\bar{y} = \frac{\sum y_i A_i}{\sum A_i} = \frac{16(1)+24(8)}{16+24} = 5.2 \text{ cm}.$$

The moment of inertia is (use $I = \bar{I} + Ad^2$)

$$I = \frac{8(2^3)}{12} + 16(4.2)^2 + \frac{2(12)^3}{12} + 24(2.8)^2$$
$$= 763.7 \text{ cm}^4.$$

The maximum positive M is 4000 N·m at $x = 4$, so the maximum compressive stress is (intuitively, we can visualize compression in the top fibers under the 4000 N force)

$$\left(\sigma_c\right)_{\max} = \frac{My}{I} = \frac{4000(0.088)}{764 \times 10^{-8}} = 46.1 \times 10^6 \text{ Pa}.$$

The maximum negative M is 2400 N·m at $x = 2$, so the maximum tensile stress is (intuitively, we know that tension occurs in the top fibers to the left of R_A)

$$\left(\sigma_t\right)_{\max} = \frac{My}{I} = \frac{2400(0.088)}{764 \times 10^{-8}} = 27.6 \times 10^6 \text{ Pa}.$$

The maximum V is 4400 N. The moment of the shaded area with respect to the neutral axis is Q. Therefore,

$$Q = \bar{y}A = 0.044 \times (0.02 \times 0.088) = 7.744 \times 10^{-5} \text{ m}^3.$$

Since $b = 0.02$ at $y = 0$, the maximum shear stress is

$$\tau_{\max} = \frac{VQ}{Ib} = \frac{4400 \times 7.744 \times 10^{-5}}{764 \times 10^{-8} \times 0.02} = 2.23 \times 10^6 \text{ Pa}.$$

Example 4.7

Determine the maximum deflection of a 3 cm × 24 cm rectangular aluminum beam, 5 m long, if a concentrated load of 800 N acts downward at its mid-point.

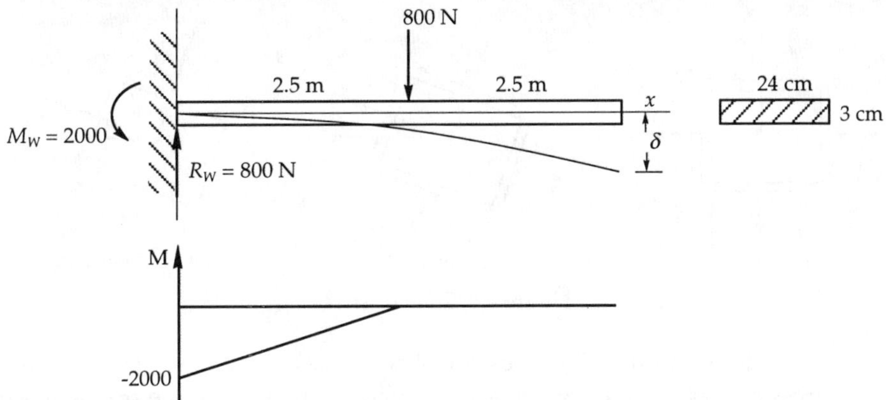

Solution. As the moment diagram shows, M is zero for the right-hand half of the beam. The right-hand half remains straight, but is inclined at the slope of the beam at mid-point. The left hand half is a simple end-loaded cantilever. From Table 4.2 with $L = 2.5$ m,

$$\delta = \delta_{\text{middle}} + \theta L = \frac{PL^3}{3EI} + \frac{PL^2}{2EI}(L) = \frac{5PL^3}{6EI}$$

$$= \frac{5(800)(2.5)^3}{6 \times (70 \times 10^9)(0.24)0.03^3 / 12} = 0.276 \text{ m}$$

4.4 Combined Stress

It often happens that structural members are simultaneously subjected to some combination of axial, torsional and bending loads. In such cases the state of stress at points on the surface of the member consists of both normal and shear components, and is called *combined stress*. At any given point of interest on the surface we introduce a local set of coordinate axes and focus attention on the stresses acting on a very small rectangular element of material at the same point P as shown by Fig. 4.8a. For an element aligned with a rotated set of $x'y'$-axes at the same point, located by the c.c.w. angle θ, shown in Fig. 4.8b, the primed stresses will differ from the original unprimed stresses, the relationship being a function of θ. At a certain angle θ_p, with which we associate the axes x^* and y^*, Fig. 4.8c, the normal stresses will reach their maximum (x^*-direction) and minimum (y^*-direction) values, while the shear stresses vanish. These axes are called *principal axes of stress*, and the values σ_1 and σ_2 are called the *principal stresses*. It turns out that

$$\sigma_1 = \sigma_{\max} = \frac{\sigma_x + \sigma_y}{2} + \sqrt{\left(\frac{\sigma_x - \sigma_y}{2}\right)^2 + \tau_{xy}^2} \quad \textbf{(4.4.1a)}$$

$$\sigma_2 = \sigma_{\min} = \frac{\sigma_x + \sigma_y}{2} - \sqrt{\left(\frac{\sigma_x - \sigma_y}{2}\right)^2 + \tau_{xy}^2} \quad \textbf{(4.4.1b)}$$

and the angle θ_p at which they occur is calculated from

$$\tan 2\theta_p = \frac{2\tau_{xy}}{\sigma_x - \sigma_y}. \quad \textbf{(4.4.2)}$$

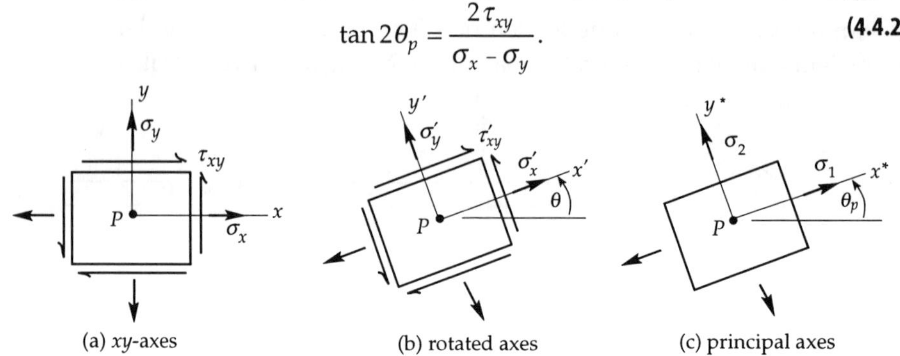

Figure 4.8 State of stress.

As stated, relative to the x^*y^*-axes the shear stresses are zero. The maximum shear stress occurs with respect to axes rotated 45° relative to the principal axes; its value is

$$\tau_{\max} = \frac{\sigma_1 - \sigma_2}{2} = \sqrt{\left(\frac{\sigma_x - \sigma_y}{2}\right)^2 + \tau_{xy}^2} \quad \textbf{(4.4.3)}$$

A graphical method is often used when obtaining stresses on a particular plane; it utilizes *Mohr's circle*. Mohr's circle is sketched by locating both ends of a diameter, whose center is always on the horizontal axis. For the stress state of Fig. 4.8a, which shows positive stresses, we plot the diameter ends as $(\sigma_x, -\tau_{xy})$ and (σ_y, τ_{xy}). The stresses on any plane, oriented at an angle θ with respect to the stresses on any known plane, are then the coordinates of a point on Mohr's circle located an angle 2θ from the known point. Once Mohr's circle is sketched, it is relatively obvious that the maximum shear stress is the circle's

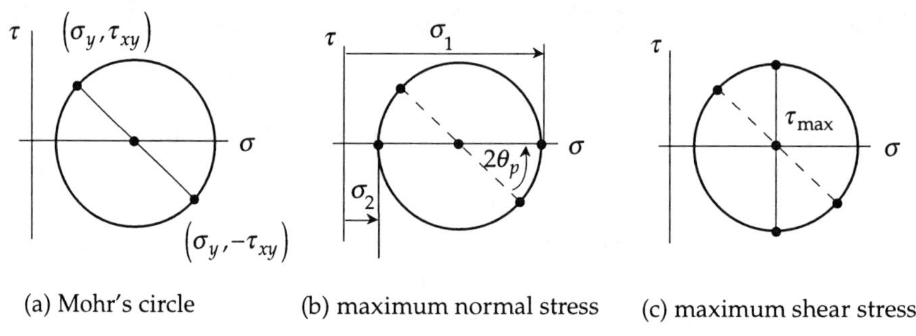

Figure. 4.9 Mohr's circle and maximum stresses.

radius, and the maximum normal stress is the circle's radius plus $(\sigma_x + \sigma_y)/2$, as observed in Fig. 4.9a. These are, in fact, equivalent to the formulas in the above equations. Rather than refer to the formulas, we can simply sketch Mohr's circle and easily find $\tau_{\max}$ and σ_1, the quantities often of interest, since they may lead to failure.

Example 4.8

A solid circular shaft of radius 5 cm and length 3 m has a 2 m rigid bar welded to end A, and is "built in" to the vertical wall at B. A load of 8 kN acts at end C, and an axial force of 80 kN compresses the shaft as shown. Determine the maximum normal, and maximum shear stress at point P on the top of the shaft, midway between A and B.

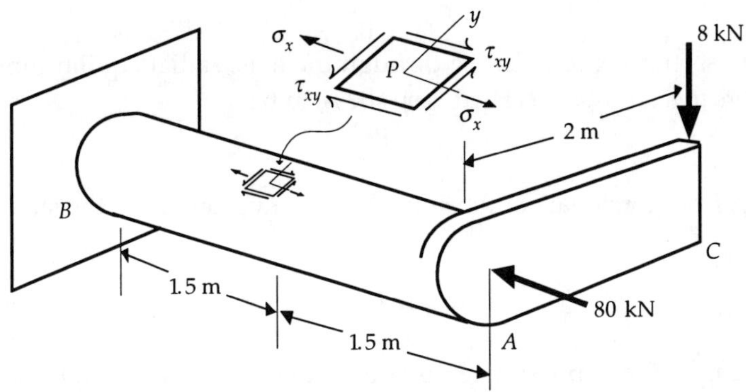

Solution. The torque on the shaft is T = (8000)(2) = 16 000 N·m. The torsional shear stress on the element at P is

$$\tau_{xy} = \frac{Tr}{J} = \frac{16\,000(0.05)}{\pi(0.05)^4/2} = 81.5\times10^6 \text{ Pa} \quad \text{or} \quad 81.5 \text{ MPa}.$$

The axial compressive stress is

$$\sigma_x = \frac{F}{A} = \frac{-80\,000}{\pi(0.05)^2} = -10.2\times10^6 \text{ Pa} \quad \text{or} \quad -10.2 \text{ MPa}.$$

The tensile bending stress on the element at P (also a σ_x stress) is

$$\sigma_x = \frac{My}{I} = \frac{8000(1.5)(0.05)}{\pi(0.05)^4/4} = 122.2\times10^6 \text{ Pa} \quad \text{or} \quad 122.2 \text{ MPa}.$$

Mohr's circle

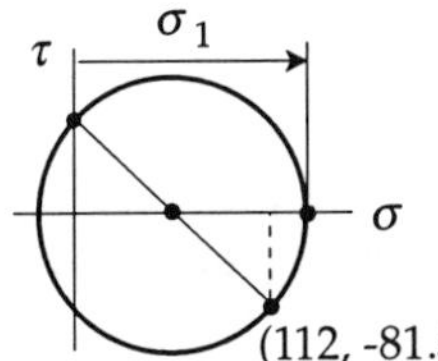

$$\text{radius} = \sqrt{(112/2)^2 + 81.5^2} = 98.9 = \tau_{max}$$

$$\sigma_1 = \text{radius} + 112/2 = 154.9$$

Thus, the stress components on the element at P (in MPa) are

$$\sigma_x = 122.2 - 10.2 = 112, \quad \sigma_y = 0, \quad \tau_{xy} = 81.5$$

so that from Eq. 4.4.1a the maximum normal stress is

$$\sigma_{max} = \frac{112+0}{2} + \sqrt{\left(\frac{112-0}{2}\right)^2 + (81.5)^2} = 155 \text{ MPa}.$$

From Eq. 4.4.3 the maximum shear stress is

$$\tau_{max} = \sqrt{\left(\frac{112-0}{2}\right)^2 + (81.5)^2} = 98.9 \text{ MPa}.$$

As another case illustrating the ideas of combined stress, let us consider a cylindrical vessel of inside diameter D and wall thickness t (with $t/D << 0.1$) containing a fluid under a pressure p, and subjected to a torque T as shown in Fig. 4.9. We consider the stresses acting upon a small element of the wall having sides parallel and perpendicular, respectively, to the axis of the cylinder. By sectioning the cylinder perpendicular to its axis at the element, we find from axial equilibrium that the *longitudinal stress*, also called *axial stress*, in the wall is

$$\sigma_a = \frac{pD}{4t}. \tag{4.4.4}$$

Similarly by sectioning lengthwise through the axis, radial equilibrium requires the *circumferential stress*, also called *hoop stress*, to be

$$\sigma_t = \frac{pD}{2t}. \tag{4.4.5}$$

And finally, from torsional equilibrium about the vessel's axis, the shear stress in the wall is

$$\tau = \frac{Tr}{J} \tag{4.4.6}$$

where $J = 2\pi r^3 t$, the approximate polar moment of inertia. From these formulas we may calculate the maximum normal and shear stresses as a problem in combined stress.

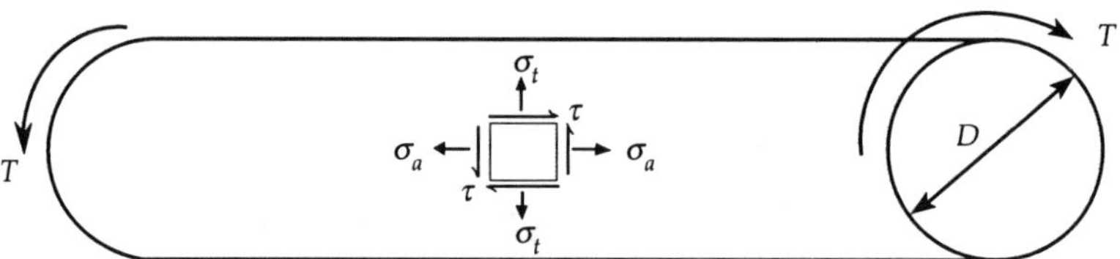

Figure 4.10 Pressurized cylinder under torque.

Finally, we note that in the absence of the torque T in Fig. 4.10, the maximum tensile stress is σ_t, the minimum tensile stress is σ_a, and the maximum shear stress is

$$\tau_{\max} = \frac{\sigma_t - \sigma_a}{2}. \tag{4.4.7}$$

Also, for a thin-walled spherical container under pressure p the normal stress in the wall is

$$\sigma = \frac{pD}{4t} \tag{4.4.8}$$

in every direction, and the shear stress in the wall is zero everywhere.

Example 4.9

A cylindrical tank of radius 40 cm and wall thickness 3 mm is subjected to an internal pressure of 2 MPa, and a torque of 0.5 $\mathrm{MN \cdot m}$. Determine the maximum normal and shear stresses in the cylinder wall.

Solution. From Eq. 4.4.4 the longitudinal stress is

$$\sigma_\ell = \frac{pr}{2t} = \frac{2(0.4)}{2(0.003)} = 133 \text{ MPa}.$$

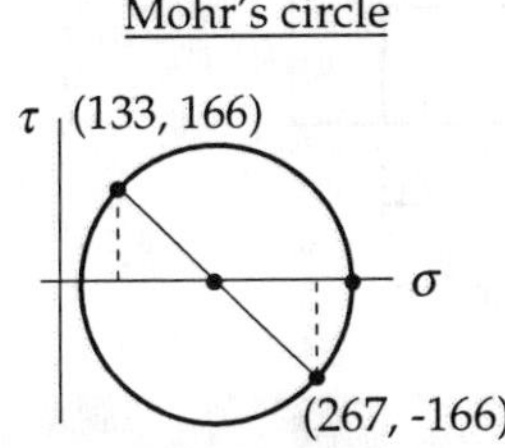

$$\text{radius} = \left[\left(\frac{267-133}{2}\right)^2 + 166^2\right]^{1/2}$$

$$= 179 \text{ MPa} = \tau_{max}$$

$$\sigma_1 = \text{radius} + (267+133)/2$$

$$= 379 \text{ MPa}$$

From Eq. 4.4.5 the circumferential stress is

$$\sigma_c = \frac{pr}{t} = \frac{(2)(0.4)}{(0.003)} = 267 \text{ MPa}.$$

From Eq. 4.4.6 the shear stress is

$$\tau_{xy} = \frac{Tr}{J} = \frac{(0.5)(0.4)}{2\pi(0.4)^3(0.003)} = 166 \text{ MPa}.$$

Thus, from Eq. 4.4.1a

$$\sigma_{max} = \frac{133+267}{2} + \sqrt{\left(\frac{133-267}{2}\right)^2 + (166)^2} = 379 \text{ MPa}$$

and from Eq. 4.4.3

$$\tau_{max} = \sqrt{\left(\frac{133-267}{2}\right)^2 + (166)^2} = 179 \text{ MPa}.$$

4.5 Composite Bars and Beams

Consider a member composed of several parallel portions, each of a particular material, securely bonded together and loaded axially. As an example, we show in Fig. 4.11 a composite bar of three materials subjected through rigid and parallel end plates to an axial force P. Let the portion of the bar have cross-sectional areas A_1, A_2 and A_3, as well as moduli of elasticity E_1, E_2, and E_3, respectively. Furthermore, let $E_1 \le E_2 \le E_3$ and form the ratios

$$m = E_2/E_1 \quad \text{and} \quad n = E_3/E_1. \tag{4.5.1}$$

Since the axial deformation is the same for each material, Hooke's law requires

$$\sigma_2 = m\sigma_1 \quad \text{and} \quad \sigma_3 = n\sigma_1 \tag{4.5.2}$$

and also that $\sigma_1 = P/A_T$ where A_T is the "transformed area" such that

$$\sigma_1 = \frac{P}{A_T} = \frac{P}{A_1 + mA_2 + nA_3}. \tag{4.5.3}$$

The generalization to a bar of any number of materials is obvious.

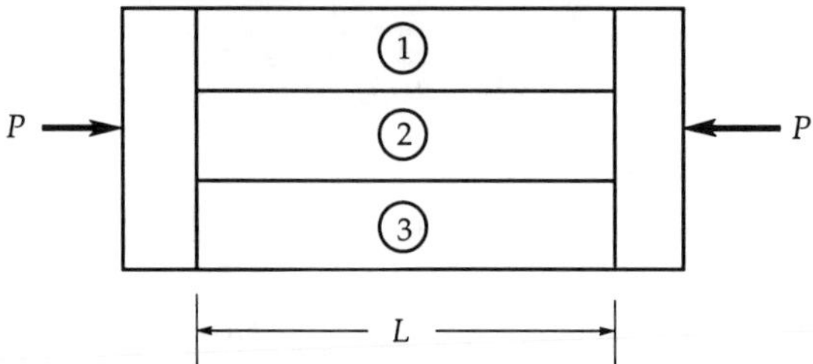

Figure 4.11 Composite bar.

Example 4.10

Let a composite bar made of aluminum, steel, and brass be subjected to an axial load of 500 kN. Determine the stress in each material if $E_{al} = 70$ GPa, $E_{st} = 210$ GPa, and $E_{br} = 105$ GPa, together with $A_{al} = 0.04\ \text{m}^2$, $A_{st} = 0.006\ \text{m}^2$, and $A_{br} = 0.08\ \text{m}^2$.

Solution. From Eq. 4.5.1, with reference to Fig. 4.10,

$$m = \frac{E_{br}}{E_{al}} = \frac{105}{70} = 1.5.$$

$$n = \frac{E_{st}}{E_{al}} = \frac{210}{70} = 3.0.$$

From Eq. 4.5.3 we have

$$\sigma_{al} = \frac{P}{A_1 + mA_2 + nA_3} = \frac{500 \times 10^3}{0.04 + 1.5(0.08) + 3(0.006)} = 2.8\ \text{MPa}.$$

From Eq. 4.5.2 there results

$$\sigma_{br} = m\sigma_1 = 1.5(2.8) = 4.2\ \text{MPa}.$$

$$\sigma_{st} = n\sigma_1 = 3.0(2.8) = 8.4\ \text{MPa}.$$

For a beam having a composite section and subjected to a bending moment M we again determine the ratios $E_2/E_1 = m$, $E_3/E_1 = n$, etc., and from them construct a "transformed cross section" by multiplying the width of each material by the corresponding ratio. We then determine the centroid of the transformed cross section, and calculate the moment of inertia I_T about the neutral axis of the transformed section. Thus,

$$\sigma_1 = -\frac{My}{I_T} \tag{4.5.4}$$

with $\sigma_2 = m\sigma_1$ and $\sigma_3 = n\sigma_1$, etc. An example illustrates the method.

Example 4.11

Let a composite steel and aluminum beam having the cross section shown be subjected to a positive bending moment of 90 N·m. Determine the maximum bending stress in each material.

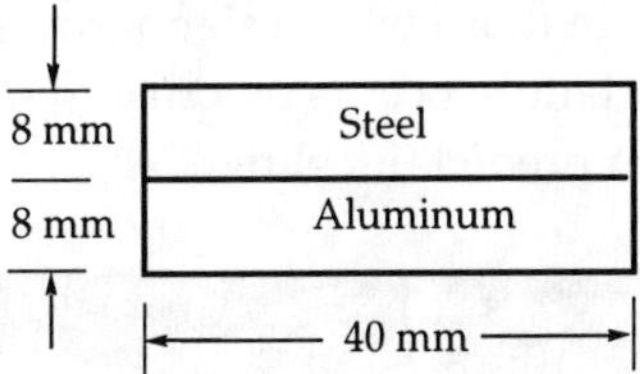

Solution. Here $m = E_s/E_a = 210/70 = 3$ so that the area has the geometry shown below with the neutral axis (N.A.) calculated to be 10 mm from the bottom.

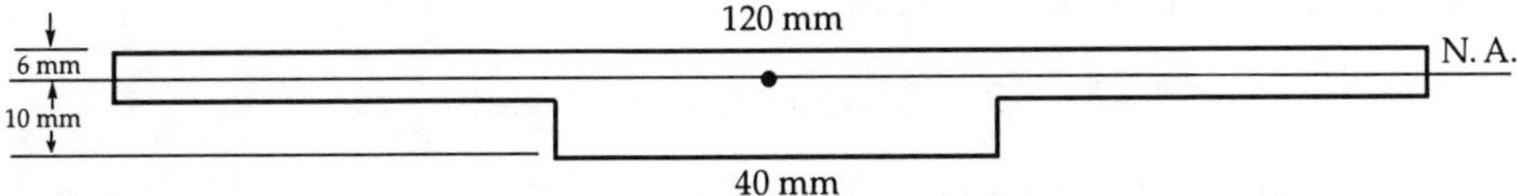

The moment of inertia about the N.A. is

$$I_T = \frac{(0.12)(0.006)^3}{3} + \frac{(0.04)(0.01)^3}{3} + \frac{(0.08)(0.002)^3}{3}$$

$$= 22 \times 10^{-9} \text{ m}^4.$$

Thus, the maximum bending stress in the aluminum is

$$(\sigma_{al})_{\max} = -\frac{My}{I_T} = -\frac{90(-0.01)}{22 \times 10^{-9}} = 41 \text{ MPa}$$

and the maximum bending stress in the steel is

$$(\sigma_{st})_{\max} = -\frac{My}{I_t} = -3\left(\frac{90(0.006)}{22 \times 10^{-9}}\right) = -74 \text{ MPa}$$

4.6 Columns

Long slender members loaded axially in compression are referred to as *columns*. Such members frequently fail by *buckling* (excessive lateral deflection) rather than by crushing. Buckling onset depends not only on the material properties but also the geometry and type of end supports of the column. The axial load at the onset of buckling is called the *critical load*.

If the *slenderness ratio* of the column, defined as *L/r* (length divided by least radius of gyration r where $r = \sqrt{I/A}$), is greater than 120, the critical load for a column is the Euler load

$$P_{cr} = \pi^2 EI/k^2L^2. \tag{4.6.1}$$

Values of k, with end supports shown in parentheses, are given as:

$$\begin{aligned} k &= 1 && \text{(pinned - pinned)} \\ k &= 0.5 && \text{(fixed - fixed)} \\ k &= 0.7 && \text{(pinned - fixed)} \\ k &= 2 && \text{(free - fixed)} \end{aligned} \qquad \textbf{(4.6.2)}$$

Intermediate columns are those whose slenderness ratios are less than 120 but greater than that at which failure occurs by crushing. For these, empirical formulas have been developed to predict buckling.

Example 4.12

Determine the critical load for a square steel (E = 210 GPa) strut 8 cm × 8 cm if its length is 6 m under (a) pinned ends, (b) fixed ends.

Solution. The moment of inertia is

$$I = bh^3/12 = (0.08)(0.08)^3/12 = 3.4 \times 10^{-6} \text{ m}^4.$$

a) The critical load for pinned ends is

$$P_{cr} = \frac{\pi^2 EI}{L^2} = \frac{\pi^2 (210 \times 10^9)(3.4 \times 10^{-6})}{6^2} = 195\ 000 \text{ N}.$$

The normal stress, which must not exceed the yield stress, is

$$\sigma = \frac{F}{A} = \frac{195 \times 10^3}{0.0064} = 30.5 \times 10^6 \text{ Pa}.$$

This is substantially less than the yield stress for all steels.

b) The critical load for fixed ends is

$$P_{cr} = \frac{\pi^2 (210 \times 10^9)(3.4 \times 10^{-6})}{0.5^2 \times 6^2} = 780\ 000 \text{ N}.$$

The normal stress for this case is

$$\sigma = \frac{780 \times 10^3}{0.0064} = 122 \times 10^6 \text{ Pa}.$$

Practice Problems (Metric Units)

(If you choose to work only a few problems, select those with a star.)

STRESS AND STRAIN

*4.1 A structural member with the same material properties in all directions at any particular point is
a) homogeneous
b) isotropic
c) isentropic
d) holomorphic
e) orthotropic

*4.2 The amount of lateral strain in a tension member can be calculated using
a) the bulk modulus.
b) the moment of inertia.
c) the yield stress.
d) Hooke's law.
e) Poisson's ratio.

4.3 Wood has grain resulting in material properties quite different normal to the grain compared with properties parallel to the grain. Such a material is
a) nonhomogeneous
b) nonholomorphic
c) nonorthotropic
d) nonisotropic
e) nonisentropic

*4.4 Find the allowable load, in kN, on a 2-cm-dia, 1-m-long, steel rod if its maximum elongation cannot exceed 0.1 cm.
a) 35 b) 45 c) 55 d) 66 e) 76

4.5 An elevator is suspended by a 2-cm-dia, 30-m-long steel cable. Twenty people, with a total weight of 14 000 N, enter. How far, in millimeters, does the elevator drop?
a) 3.5 b) 4.5 c) 5.5 d) 6.4 e) 7.4

4.6 A hole, one meter from the end of a structural steel member fixed at one end, is 0.8 mm shy of matching another hole for possible connection. What force, in kN, is necessary to stretch it for connection? The cross section is 25 mm × 3 mm.
a) 12.6 b) 13.6 c) 14.7 d) 15.8 e) 17.2

4.7 As the load is applied, edge *AB* moves 0.03 mm to the right. Determine the shear modulus, in MPa.
a) 50 300
b) 41 700
c) 38 600
d) 32 500
e) 26 300

4.8 A 5-cm-dia steel shaft is subjected to an axial tensile force of 600 kN. What is the diameter, in cm, after the force is applied? Use $\nu = 0.28$.
a) 4.998 b) 4.996 c) 4.994 d) 4.992 e) 4.990

*4.9 An aluminum cylinder carries an axial compressive load of 1500 kN. Its diameter measures exactly 12.015 cm and its height 19.311 cm. What was its original diameter, in cm?
a) 12.010 b) 12.008 c) 12.006 d) 12.004 e) 12.002

THERMAL STRESS

4.10 A tensile stress of 100 MPa exists in a 2-cm-dia steel rod that is fastened securely between two rigid walls. If the temperature increases by 30° C, determine the final stress, in MPa in the rod.

a) 46.7 b) 41.2 c) 36.9 d) 31.2 e) 26.2

*4.11 A steel bridge span is normally 300 m long. What is the difference in length, in cm, between January (–35° C) and August (40° C)?

a) 26 b) 28 c) 30 d) 32 e) 34

4.12 An aluminum bar at 30° C is inserted between two rigid stationary walls by inducing a compressive stress of 70MPa. At what temperature, in ° C, will the bar drop out?

a) 10 b) 0 c) –8 d) –14 e) –20

*4.13 Brass could not be used to reinforce concrete because

a) its density is too large.
b) its density is too low.
c) it is too expensive.
d) its coefficient of thermal expansion is not right.
e) it does not adhere well to concrete.

TORSION

*4.14 The maximum shearing stress, in MPa, that exists in a 6-cm-dia shaft subjected to a 200 N·m torque is

a) 4.72 b) 5.83 c) 7.29 d) 8.91 e) 9.97

4.15 The shaft of Prob. 4.14 is replaced with a 6-cm-outside diameter, 5-cm inside diameter hollow shaft. What is the maximum shearing stress, in MPa?

a) 5.5 b) 6.4 c) 7.3 d) 8.2 e) 9.1

4.16 The maximum allowable shear stress in a 10-cm-dia shaft is 140 MPa. What maximum torque, in N·m, can be applied?

a) 27 500 b) 21 400 c) 19 300 d) 17 100 e) 15 300

4.17 A builder uses a 50-cm-long, 1-cm-dia steel drill. If two opposite forces of 200 N are applied normal to the shaft, each with a moment arm of 15 cm, what angle of twist, in degrees, occurs in the drill?

a) 29.3 b) 24.6 c) 22.8 d) 21.1 e) 19.2

4.18 A solid circular shaft, 8 cm in diameter, transmits a torque of 1200 N·m. Calculate the maximum normal stress in the shaft.

a) 24 MPa b) 18 MPa c) 14 MPa d) 12 MPa e) 10 MPa

BENDING MOMENTS IN BEAMS

4.19 The maximum bending stress at a given cross section of an I-beam occurs

a) where the shearing stress is maximum.
b) at the outermost fiber.
c) at the joint of the web and the flange.
d) at the neutral axis.
e) just below the joint of the web and the flange

4.20 The moment diagram for a simply-supported beam with a load at the midpoint is

a) a triangle. b) a parabola. c) a trapezoid. d) a rectangle. e) a semicircle.

4.21 Find the bending moment, in N·m, at *A*.

a) 12 000 b) 14 000 c) 16 000
d) 18 000 e) 20 000

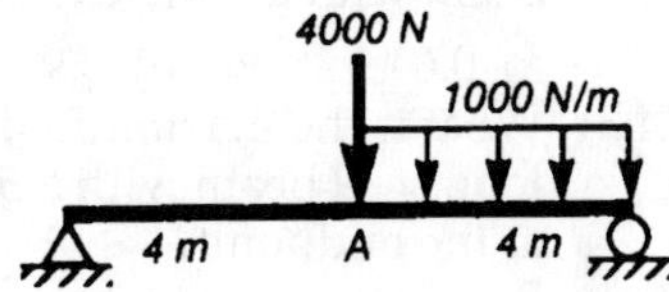

*4.22 What is the bending moment, in N·m, at *A*?

a) 26 000 b) 24 000 c) 22 000
d) 20 000 e) 18 000

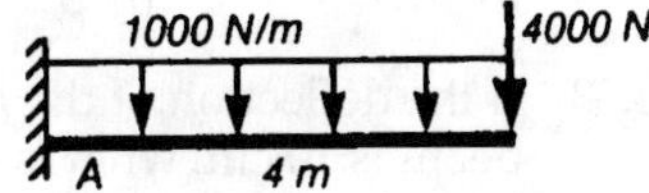

STRESSES IN BEAMS

*4.23 Find the maximum tensile stress, in MPa.

a) 94
b) 86
c) 82
d) 76
e) 72

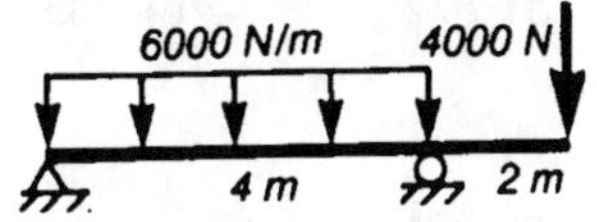

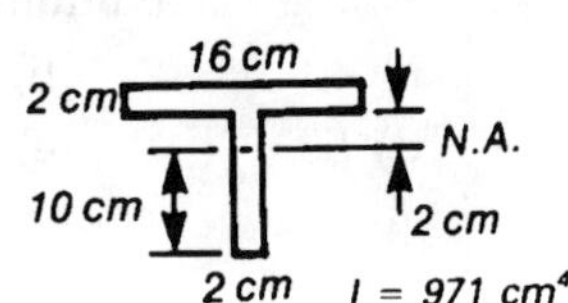

*4.24 What is the maximum compressive stress, in MPa, in the beam of Prob. 4.23?

a) 96 b) 90 c) 82 d) 76 e) 72

*4.25 What is the maximum shearing stress, in MPa, in the beam of Prob. 4.23?

a) 13.8 b) 11.3 c) 9.6 d) 8.2 e) 7.2

*4.26 The shearing stress distribution $\tau = VQ/Ib$ on the cross section of the *T*-beam in Prob. 4.23 most resembles which sketch?

a) 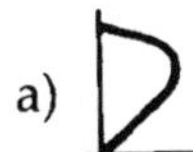 b) c) d)  e)

4.27 If the allowable bending stress is 140 MPa in the beam of Prob. 4.22, calculate the *section modulus* defined by I/y, in cm^3.

a) 196 b) 184 c) 171 d) 162 e) 153

4.28 Find the maximum bending stress, in MPa, if the 10-cm-wide beam is 5 cm deep.

a) 200 b) 180 c) 160
d) 140 e) 120

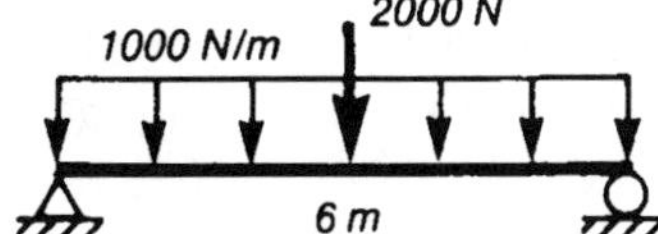

4.29 If the beam of Prob. 4.28 were 5 cm wide and 10 cm deep, find the maximum bending stress, in MPa.

a) 130 b) 120 c) 110 d) 100 e) 90

4.30 Find the maximum shearing stress, in MPa, of a simply supported, 6-m-long beam with a 5 cm × 5 cm cross section if it has a 2000 N load at the mid point.

a) 0.6 b) 0.9 c) 1.2 d) 1.6 e) 2.4

*4.31 What is the maximum deflection, in cm, of a simply supported, 6-m-long steel beam with a 5 cm x 5 cm cross-section if it has a 2000 N load at the midpoint?

DEFLECTION OF BEAMS

a) 6.35 b) 7.02 c) 7.63 d) 8.23 e) 8.92

4.32 Find the maximum deflection, in cm, for the steel beam of Prob. 4.28.

a) 39.7 b) 32.4 c) 28.3 d) 19.2 e) 11.8

4.33 If the deflection of the right end of the 5-cm-dia steel beam is 10 cm, what is the load P, in N?

a) 403 b) 523 c) 768

d) 872 e) 935

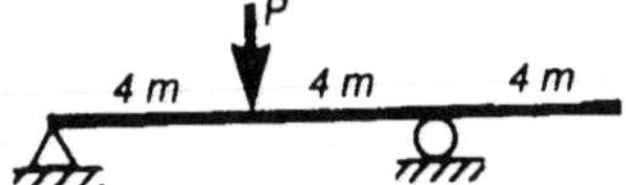

COMBINED STRESSES

*4.34 Find the maximum shearing stress, in MPa.

a) 80 b) 70 c) 60

d) 50 e) 40

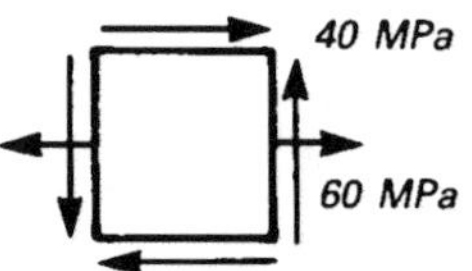

4.35 What is the maximum tensile stress, in MPa?

a) 40 b) 30 c) 20

d) 10 e) 0

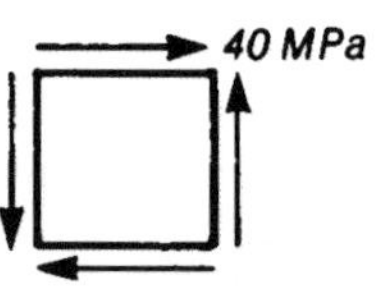

4.36 Determine the maximum shearing stress, in MPa.

a) 80 b) 60 c) 50

d) 40 e) 30

50 MPa

30 MPa

30 MPa

*4.37 Find the maximum shearing stress, in MPa, in the shaft.

a) 29.5 b) 28.5 c) 27.5

d) 26.5 e) 25.5

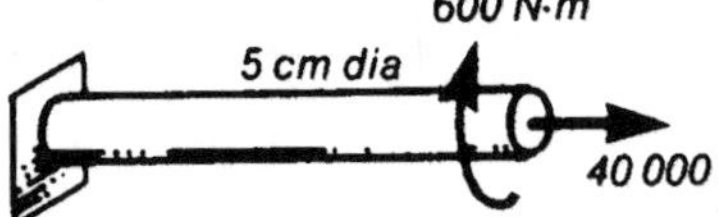

4.38 The maximum normal stress, in MPa, in the shaft of Prob. 4.37 is

a) 52.8 b) 41.7 c) 36.7 d) 30.1 e) 25.3

4.39 The normal stress, in MPa, at pt. A is

a) 263 b) 241 c) 228

d) 213 e) 201

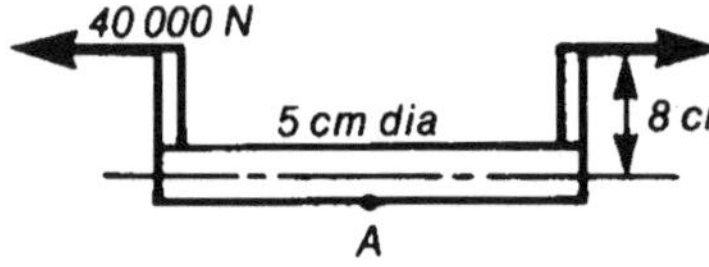

4.40 The maximum shearing stress, in MPa, at pt. A in Prob. 4.39 is

a) 140 b) 130 c) 120 d) 110 e) 100

4.41 The maximum shearing stress, in MPa, in the circular shaft is

a) 171 b) 167 c) 154
d) 142 e) 133

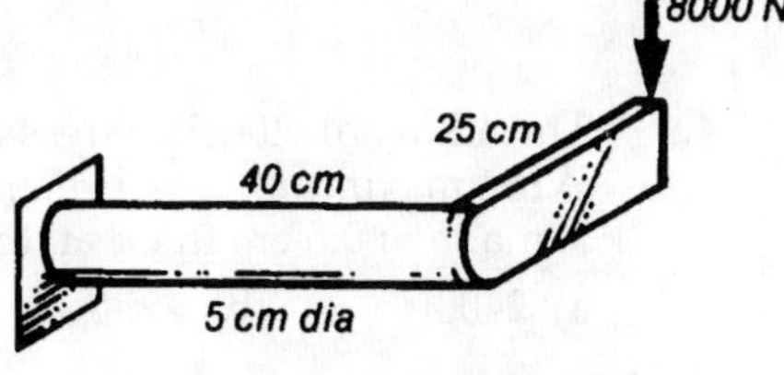

4.42 The maximum tensile stress, in MPa, in the circular shaft of Prob. 4.41 is

a) 284 b) 248 c) 223 d) 212 e) 197

Questions 4.43-4.47:

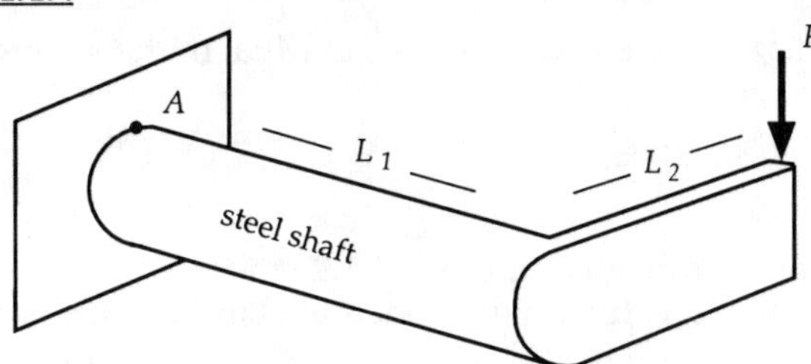

rectanglular beam — 2 cm x 5 cm
shaft — 5 cm diameter

$F = 2000$ N
$L_1 = 2$ m
$L_2 = 1$ m
$E = 210 \times 10^9$ Pa

4.43 The maximum normal stress in the rectangular beam is, in MPa,

a) 120 b) 180 c) 240 d) 360 e) 480

4.44 The maximum vertical shearing stress in the rectangular beam is, in kPa,

a) 1200 b) 1800 c) 2400 d) 3000 e) 3600

4.45 The stress situation at pt. *A* is described by which element? (Point *A* is at the top.)

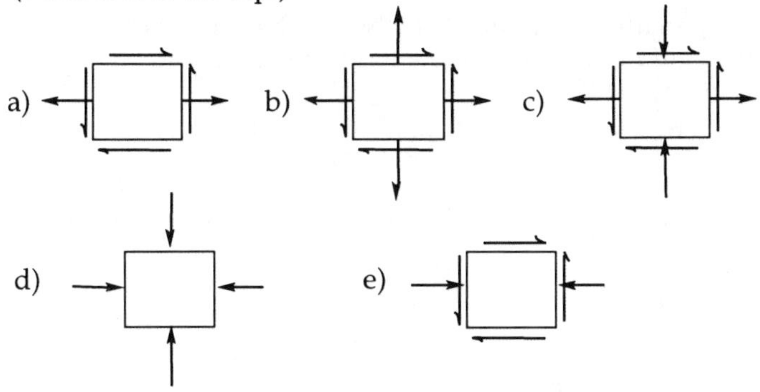

4.46 The maximum normal stress in the shaft is, in MPa,

a) 163 b) 182 c) 247 d) 326 e) 345

4.47 The maximum deflection of the shaft is, in cm,

a) 3.9 b) 5.1 c) 6.6 d) 8.3 e) 10.5

THIN-WALLED PRESSURE VESSELS

4.48 The allowable tensile stress for a pressurized cylinder is 180 MPa. What maximum pressure, in kPa, is allowed if the 80-cm-dia cylinder is made of 0.5 cm thick material?

a) 2400 b) 2250 c) 2150 d) 2050 e) 1950

4.49 The maximum normal stress that can occur in a 120-cm-dia steel sphere is 200 MPa. If it is to contain a pressure of 8000 kPa, what must be the minimum thickness, in cm?

a) 1.6 b) 1.4 c) 1.2 d) 1.0 e) 0.8

4.50 What is the maximum shearing stress, in MPa, in the sphere of Prob. 4.49?

a) 200 b) 150 c) 100 d) 50 e) 0

COMPOSITE SECTIONS

4.51 A compression member, composed of 1.2-cm-thick steel pipe with 25-cm-inside diameter, is filled with concrete. Find the stress, in MPa, in the steel if the load is 2000 kN.

a) 137 b) 145 c) 155 d) 165 e) 175

4.52 If the flanges are aluminum and the rib is steel, find the maximum tensile stress, in MPa, in the beam.

a) 7.89 b) 6.31 c) 5.72
d) 4.91 e) 3.88

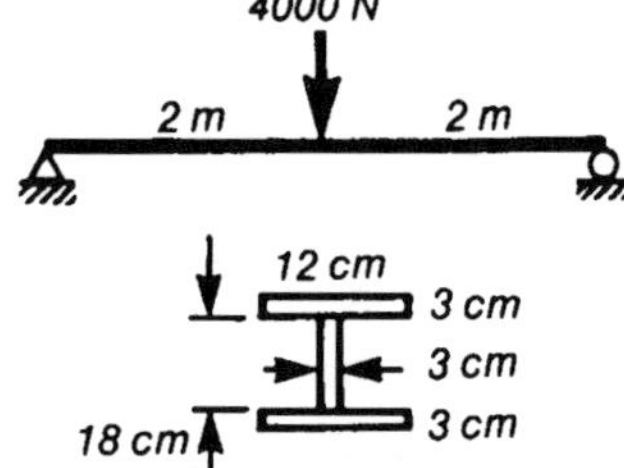

4.53 If the flanges of the I-beam of Prob. 4.52 are steel and the rib is aluminum, what is the maximum tensile stress, in MPa, in the beam?

a) 7.89 b) 6.31 c) 5.67 d) 4.91 e) 3.88

COLUMNS

4.54 What is the minimum length, in meters, for which a 10 cm × 10 cm wooden post can be considered a long column? Assume a maximum slenderness ratio of 60.

a) 4.03 b) 3.12 c) 2.24 d) 1.73 e) 1.12

*4.55 A free-standing platform, holding 2000 N, is to be supported by a 10-cm-dia vertical aluminum strut. How long, in meters, can it be if a safety factor of 2 is used?

a) 18.3 b) 16.6 c) 14.6 d) 12.2 e) 9.32

4.56 What increase in temperature, in °C, is necessary to cause a 2-cm-dia, 4-m-long, steel rod with fixed ends to buckle? There is no initial stress.

a) 9.38 b) 8.03 c) 7.12 d) 6.34 e) 5.27

4.57 A column with both ends fixed buckles when subjected to a force of 30 000 N. One end is then allowed to be free. At what force, in Newtons, will it buckle?

a) 2025 b) 1875 c) 1725 d) 1650 e) 1575

Practice Problems (English Units)

(If you choose to work only a few problems, select those with a star.)

STRESS AND STRAIN

*4.1 A structural member with the same material properties in all directions at any particular point is

a) homogeneous b) isotropic c) isentropic d) holomorphic e) orthotropic

*4.2 The amount of lateral strain in a tension member can be calculated using

a) the bulk modulus.
b) the moment of inertia.
c) the yield stress.
d) Hooke's law.
e) Poisson's ratio.

4.3 Wood has grain resulting in material properties quite different normal to the grain compared with properties parallel to the grain. Such a material is

a) nonhomogeneous
b) nonholomorphic
c) nonorthotropic
d) nonisotropic
e) nonisentropic

*4.4 Find the allowable load, in pounds, on a 1/2" dia, 4-ft-long, steel rod if its maximum elongation cannot exceed 0.04 inches.

a) 9290 b) 6990 c) 5630 d) 4910 e) 3220

4.5 An elevator is suspended by a 1/2" dia, 100-ft-long steel cable. Twenty people, with a total weight of 3500 lbs, enter. How far, in inches, does the elevator drop?

a) 1.3 b) 1.1 c) 0.9 d) 0.7 e) 0.5

4.6 A hole in a 3-ft-long structural steel member fixed at one end is 1/32" shy of matching another hole for possible connection. What force, in pounds, is necessary to stretch it for connection? The cross section is 1/8" × 1".

a) 3300 b) 3000 c) 2700 d) 2400 e) 2100

4.7 As the load is applied, edge AB moves 0.0012" to the right. Determine the shear modulus, in psi.

a) 7.5×10^6
b) 6.2×10^6
c) 5.7×10^6
d) 5.2×10^6
e) 4.5×10^6

4.8 A 2" dia steel shaft is subjected to an axial tensile force of 150,000 lbs. What is the diameter, in inches, after the force is applied? Use ν = 0.28.

a) 1.999 b) 1.998 c) 1.997 d) 1.996 e) 1.995

*4.9 An aluminum cylinder carries an axial compressive load of 400, 000 lbs. Its diameter measures exactly 5.923″ and its height 8.314″. What was its original diameter, in inches?

a) 5.922 b) 5.920 c) 5.918 d) 5.916 e) 5.914

THERMAL STRESS

4.10 A tensile stress of 16,000 psi exists in a 1″ dia steel rod that is fastened securely between two rigid walls. If the temperature increases by 50° F, determine the final stress, in psi, in the rod.

a) 12,200 b) 9400 c) 8600 d) 7400 e) 6200

*4.11 A steel bridge span is normally 1000 ft long. What is the difference in length, in inches, between January (–30° F) and August (100° F)?

a) 10 b) 9 c) 8 d) 7 e) 6

4.12 An aluminum bar at 80° F is inserted between two rigid stationary walls by inducing a compressive stress of 10,000 psi. At what temperature, in ° F, will the bar drop out?

a) 36 b) 24 c) 10 d) 2 e) –10

*4.13 Brass could not be used to reinforce concrete because

a) it is not sufficiently strong.
b) its density is too large.
c) it is too expensive.
d) its coefficient of thermal expansion is not right.
e) it does not adhere well to concrete.

TORSION

*4.14 The maximum shearing stress, in psi, that exists in a 2″ dia shaft subjected to a 2000 in-lb torque is

a) 1270 b) 1630 c) 1950 d) 2610 e) 3080

4.15 The shaft of Prob. 4.13 is replaced with a 2″ outside diameter, 1.75″ inside diameter hollow shaft. What is the maximum shearing stress, in psi?

a) 1270 b) 1630 c) 1950 d) 2610 e) 3080

4.16 The maximum allowable shear stress in a 4″ dia shaft is 20,000 psi. What maximum torque, in ft-lb, can be applied?

a) 20,900 b) 15,600 c) 11,200 d) 8,600 e) 4,210

4.17 A builder uses an 18″ long, 7/8″ dia steel drill. If two opposite forces of 160 lbs are applied normal to the shaft, each with a moment arm of 12", what angle of twist, in degrees, occurs in the drill?

a) 10.3 b) 8.29 c) 6.95 d) 5.73 e) 4.68

4.18 A solid circular shaft, 4 inches in diameter, transmits a torque of 1200 ft-lb. Calculate the maximum normal stress in the shaft.

a) 2300 psi b) 1800 psi c) 1380 psi d) 1150 psi e) 940 psi

BENDING MOMENTS IN BEAMS

4.19 The maximum bending stress at a given cross section of an I-beam occurs

a) where the shearing stress is maximum.
b) at the outermost fiber.
c) just below the joint of the web and the flange.
d) at the neutral axis.
e) at the joint of the web and the flange.

4.20 The moment diagram for a simply supported beam with a load at the midpoint is a
a) triangle. b) parabola. c) trapezoid. d) rectangle. e) semicircle.

4.21 Find the bending moment, in ft-lb, at A.
a) 7500 b) 7000 c) 6500
d) 6000 e) 5000

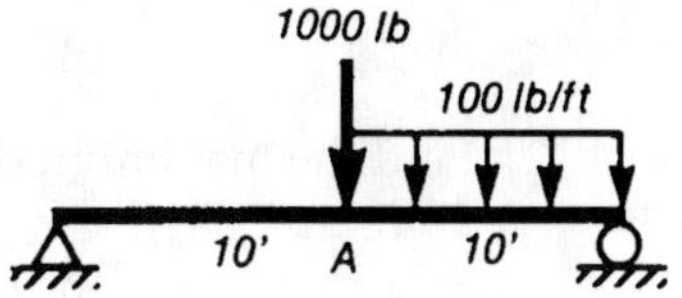

1000 lb
100 lb/ft
10' A 10'

*4.22 What is the bending moment, in ft-lb, at A?
a) 18,000 b) 15,000 c) 12,000
d) 10,000 e) 8000

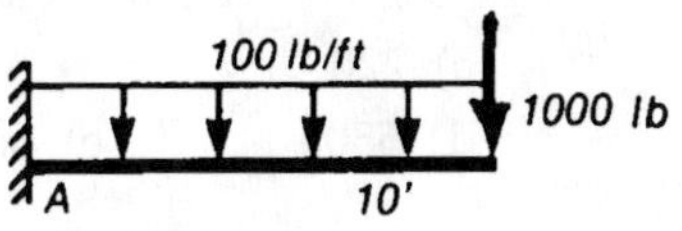

100 lb/ft
1000 lb
A 10'

*4.23 Find the maximum tensile stress, in psi.
a) 4360 b) 3960 c) 3240
d) 2860 e) 2110

500 lb/ft
1000 lb
10' 4'

8"
1"
N.A.
5"
1"
1" $I = 60.67\ in^4$

STRESSES IN BEAMS

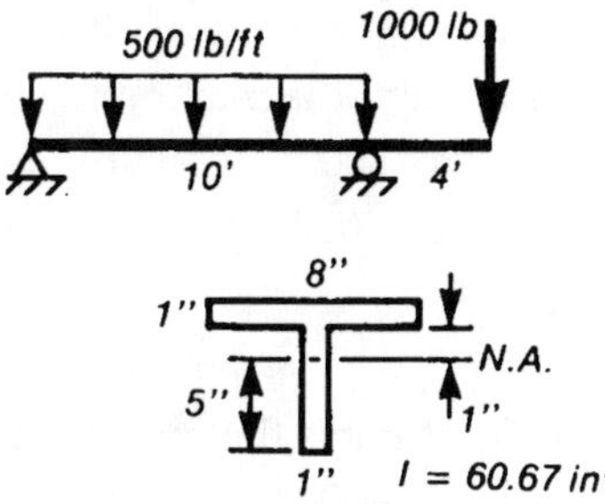

*4.24 What is the maximum compressive stress, in psi, in the beam of Prob. 4.23?
a) 4360 b) 3960 c) 3240 d) 2860 e) 2110

*4.25 What is the maximum shearing stress, in psi, in the beam of Prob. 4.23?
a) 1000 b) 900 c) 800 d) 700 e) 600

4.26 The shearing stress distribution $\tau = VQ/Ib$ on the cross section of the T-beam in Prob. 4.21 most resembles which sketch?

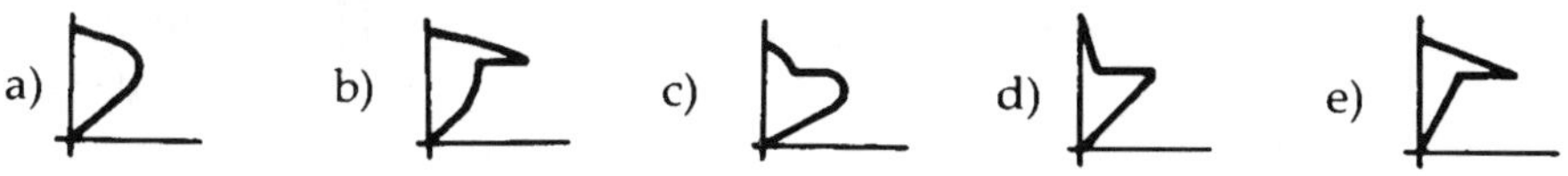

a) b) c) d) e)

4.27 If the allowable bending stress is 20,000 psi in the beam of Prob. 4.20, calculate the *section modulus* defined by I/y, in in^3.
a) 11 b) 10 c) 9 d) 8 e) 7

4.28 Find the maximum bending stress, in psi, if the 4" wide beam is 2" deep.
a) 29,200 b) 23,400 c) 18,600
d) 15,600 e) 11,700

100 lb/ft
500 lb
16'

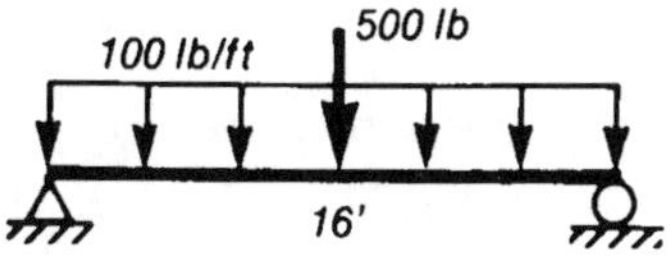

4.29 If the beam of Prob. 4.28 were 2" wide and 4" deep, find the maximum bending stress, in psi.
a) 29,200 b) 23,400 c) 18,600 d) 15,600 e) 11,700

4.30 Find the maximum shearing stress, in psi, of a simply supported, 20-ft-long beam with a 2" × 2" cross section if it has a 500-lb load at the mid point.
a) 188 b) 152 c) 131 d) 109 e) 94

DEFLECTION OF BEAMS

*4.31 What is the maximum deflection, in inches, of a simply supported, 20-ft-long steel beam with a 2″ × 2″ cross section if it has a 500-lb load at the midpoint?

a) 7.2 b) 6.0 c) 4.8 d) 3.6 e) 2.4

4.32 Find the maximum deflection, in inches, for the steel beam of Prob. 4.28.

a) 4.22 b) 4.86 c) 3.52 d) 2.98 e) 2.76

4.33 If the deflection of the right end of the 2″ dia steel beam is 4″, what is the load P, in pounds?

a) 220 b) 330 c) 440
d) 550 e) 660

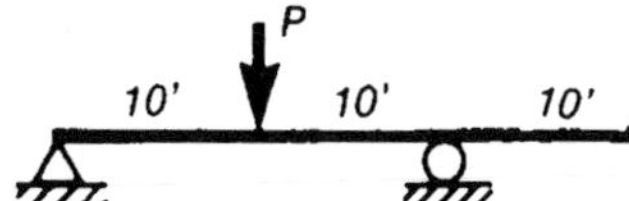

COMBINED STRESSES

*4.34 Find the maximum shearing stress, in psi.

a) 8000 b) 7000 c) 6000
d) 5000 e) 4000

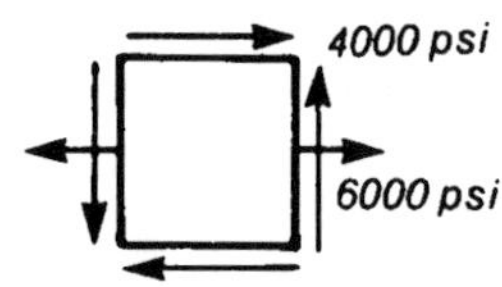

4.35 What is the maximum tensile stress, in psi?

a) 4000 b) 3000 c) 2000
d) 1000 e) 0

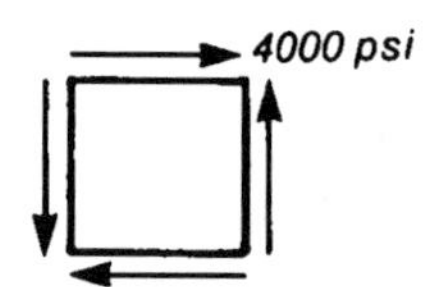

4.36 Determine the maximum shearing stress, in psi.

a) 8000 b) 6000 c) 5000
d) 4000 e) 3000

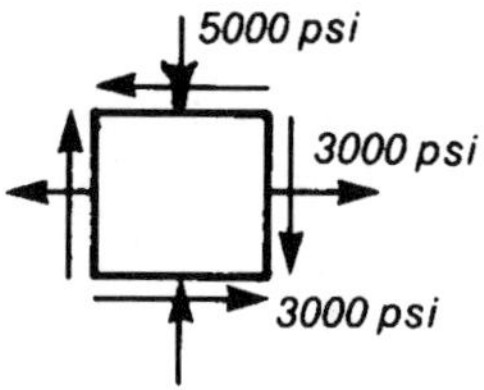

*4.37 Find the maximum shearing stress, in psi, in the shaft.

a) 7340 b) 6520 c) 5730
d) 4140 e) 3160

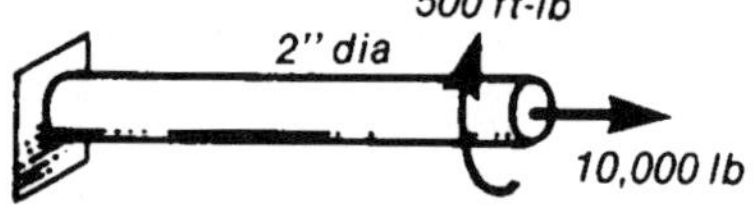

4.38 The maximum normal stress, in psi, in the shaft of Prob. 4.37 is

a) 7340 b) 6520 c) 5730 d) 4140 e) 3160

4.39 The normal stress, in psi, at pt. A is

a) 25,000 b) 35,000 c) 41,000
d) 46,000 e) 55,000

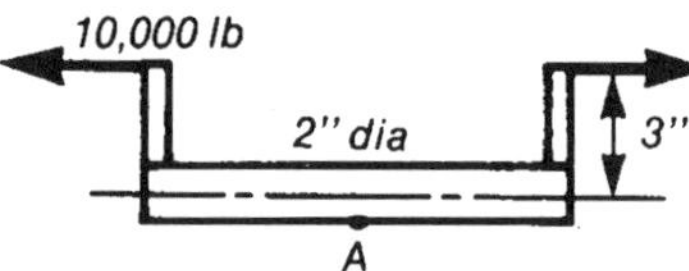

4.40 The maximum shearing stress, in psi, at pt. A in Prob. 4.39 is

a) 12,500 b) 15,000 c) 17,500 d) 20,500 e) 23,000

4.41 The maximum shearing stress, in psi, in the circular shaft is

a) 12,000 b) 18,000 c) 24,000
d) 30,000 e) 36,000

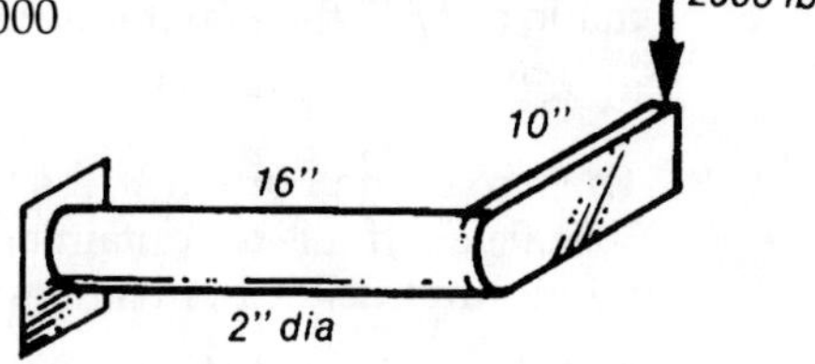

4.42 The maximum tensile stress, in psi, in the circular shaft of Prob. 4.39 is

a) 11,100 b) 22,200 c) 33,300 d) 44,400 e) 55,500

Questions 4.43-4.47:

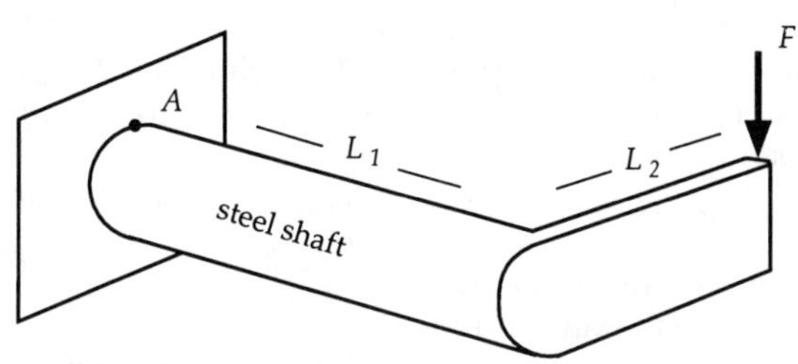

rectanglular beam — 0.787" x 1.97"
shaft — 1.97" diameter

$F =$ 450 lb
$L_1 =$ 6.56 ft
$L_2 =$ 3.28 ft
$E = 30 \times 10^6$ psi

4.43 The maximum normal stress in the rectangular beam is, in psi,

a) 17,400 b) 26,000 c) 34,800 d) 52,200 e) 69,000

4.44 The maximum vertical shearing stress in the rectangular beam is, in psi,

a) 174 b) 261 c) 348 d) 435 e) 522

4.45 The stress situation at pt. A is described by which element? (Point A is at the top.)

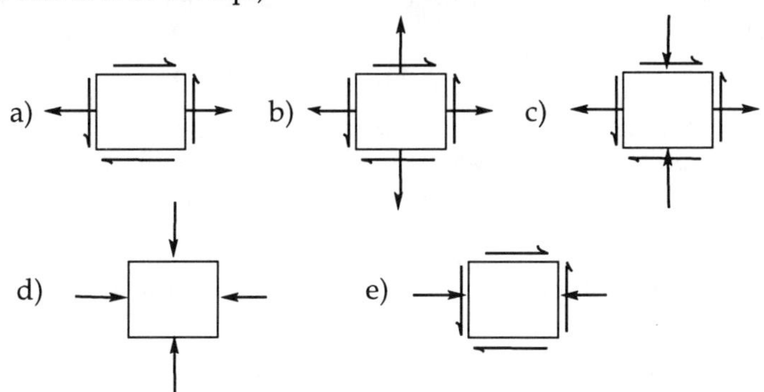

4.46 The maximum normal stress in the shaft is, in psi,

a) 23,640 b) 26,400 c) 35,800 d) 47,300 e) 50,000

4.47 The maximum deflection of the shaft is, in in.,

a) 1.54 b) 2 c) 2.6 d) 2.85 e) 4.13

*4.48 The allowable tensile stress for a pressurized cylinder is 24,000 psi. What maximum pressure, in psi, is allowed if the 2-ft-dia cylinder is made of 1/4" thick material?

a) 250 b) 500 c) 1000 d) 1500 e) 2000

4.49 The maximum normal stress that can occur in a 4-ft-dia steel sphere is 30,000 psi. If it is to contain a pressure of 2000 psi, what must be the minimum thickness, in inches?

a) 1.0 b) 0.9 c) 0.8 d) 0.7 e) 0.6

4.50 What is the maximum shearing stress, in psi, in the sphere of Prob. 4.49?

a) 30,000 b) 20,000 c) 15,000 d) 10,000 e) 0

COMPOSITE SECTIONS

*4.51 A compression member, composed of 1/2" thick steel pipe with 10" inside diameter is filled with concrete. Find the stress, in psi, in the steel if the load is 400,000 lbs.

a) 16,400 b) 14,300 c) 12,600 d) 10,100 e) 8,200

4.52 If the flanges are aluminum and the rib is steel, find the maximum tensile stress, in psi, in the beam.

a) 2170 b) 1650 c) 1320
d) 1150 e) 1110

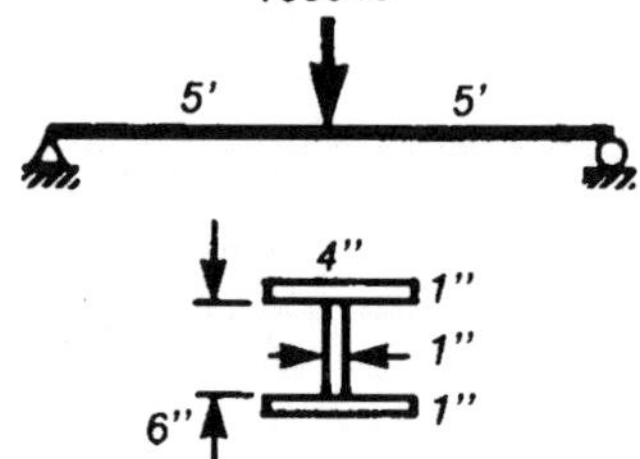

4.53 If the flanges of the I-beam of Prob. 4.52 are steel and the rib is aluminum, what is the maximum tensile stress, in psi?

a) 2170 b) 1650 c) 1320 d) 1150 e) 1110

COLUMNS

4.54 What is the minimum length, in ft, for which a 4" × 4" wooden post can be considered a long column? Assume a maximum slenderness ratio of 60.

a) 9.2 b) 7.6 c) 6.8 d) 5.8 e) 4.2

*4.55 A free-standing platform, holding 500 lb, is to be supported by a 4" dia vertical aluminum strut. How long, in ft, can it be if a safety factor of 2 is used?

a) 66 b) 56 c) 46 d) 36 e) 26

4.56 What increase in temperature, in °F, is necessary to cause a 1" dia, 10 ft long, steel rod with fixed ends to buckle? There is no initial stress.

a) 66 b) 56 c) 46 d) 36 e) 26

4.57 A column with both ends fixed buckles when subjected to a force of 8000 lbs. One end is then allowed to be free. At what force, in pounds, will it buckle?

a) 100 b) 500 c) 2000 d) 8000 e) 32,000

Solutions to Practice Problems

(Metric Units)

4.1 b) Isotropic

4.2 e) Poisson's Ratio

4.3 d)

4.4 d) $\sigma = E\delta/L = P/\pi r^2$

$\therefore P = E\delta\pi r^2/L = 210\times10^9 \times 0.001\times\pi\times0.01^2/1 = 66\,000$ N

4.5 d) $\frac{P}{\pi r^2} = E\frac{\delta}{L}. \quad \therefore \delta = \frac{14\,000\times30}{\pi\times0.01^2\times210\times10^9} = 0.0064$ m

4.6 a) $\frac{P}{A} = E\frac{\delta}{L}. \quad \therefore P = \frac{AE\delta}{L} = 0.025\times0.003\times\left(210\times10^9\right)\times\frac{0.0008}{1} = 12\,600$ N

4.7 b) $\frac{P}{A} = G\frac{\delta}{L}. \quad \therefore G = \frac{PL}{A\delta} = \frac{20\,000\times0.15}{0.012\times0.2\times0.00003} = 41.7\times10^9$ Pa

4.8 a) $P/A = E\varepsilon. \quad \varepsilon = \frac{600\,000}{\pi\times0.025^2\times210\times10^9} = 0.001455$

$\Delta d = \nu\varepsilon d = 0.28\times0.001455\times5 = 0.00204$ cm

$d = d - \Delta d = 5 - 0.00204 = 4.9980$ cm

4.9 b) $P/A = E\varepsilon. \quad \varepsilon = \frac{1500\,000}{\pi\times0.06^2\times70\times10^9} = 0.00190$

$\Delta d = \nu\varepsilon d = 0.33\times0.0019\times12.015 = 0.0075$ cm

$d = d - \Delta d = 12.015 - 0.0075 = 12.008$ cm

4.10 e) $\sigma = E\delta/L. \quad 100\times10^6 = 210\times10^9(\delta/L). \quad \therefore \quad \delta = 4.76\times10^{-4}L$

$\delta_T = \alpha L\Delta T = 11.7\times10^{-6}\times30L = 3.51\times10^{-4}L. \quad \delta_{final} = \delta - \delta_T = 1.25\times10^{-4}L$

$\therefore \sigma = \left(210\times10^9\right)\times\left(1.25\times10^{-4}\right) = 26.2\times10^6$ Pa

4.11 a) $\delta = \alpha L\Delta T = \left(11.7\times10^{-6}\right)\times300\times75 = 0.263$ m

4.12 d) $\frac{\delta}{L} = \frac{\sigma}{E} = \alpha\Delta T. \quad \frac{70\times10^6}{70\times10^9} = 23\times10^{-6}(30 - T). \quad \therefore T = -13.5°C$

4.13 d) It expands at a different rate.

4.14 a) $\tau = \frac{Tr}{J} = \frac{200\times0.03}{\pi\times0.06^4/32} = 4.72\times10^6$ Pa

4.15 e) $J = \pi\left(a_1^4 - a_2^4\right)/2 = \pi\left(0.03^4 - .025^4\right)/2 = 65.9 \times 10^{-8}$

$$\tau = \frac{Tr}{J} = \frac{200 \times 0.03}{65.9 \times 10^{-8}} = 9.10 \times 10^6 \text{ Pa}$$

4.16 a) $T = \dfrac{\tau J}{r} = \dfrac{\left(140 \times 10^6\right) \times \pi \times .05^4/2}{0.05} = 27\,500 \text{ N} \cdot \text{m}$

4.17 d) $\theta = \dfrac{TL}{JG} = \dfrac{200 \times 0.3 \times 0.5}{\left(83 \times 10^9\right) \times \pi \times .05^4/2} = 0.368 \text{ rad or } 21.1°$

4.18 d) $\tau_{max} = \dfrac{Tr}{J} = \dfrac{1200 \times 0.04}{\pi \times 0.04^4/2} = 11.9 \times 10^6 \text{ Pa.}$

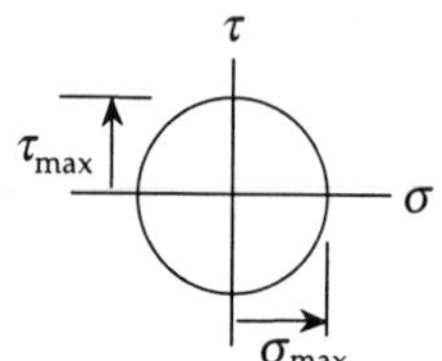

$\therefore \sigma_{max} = 11.9 \times 10^6 \text{ Pa.}$

4.19 b) $\sigma = My/I$. σ_{max} occurs at $y = y_{max}$

4.20 a) A triangle.

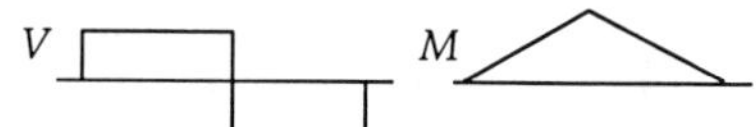

4.21 a) $\sum M_{right} = 0. \quad \therefore 8F_{left} = 4\,000 \times 4 + 4\,000 \times 2. \quad \therefore F = 3\,000.$

$M_A = 3\,000 \times 4 = 12\,000 \text{ N} \cdot \text{m}$

4.22 b) $M_A = 4\,000 \times 4 + 4\,000 \times 2 = 24\,000 \text{ N} \cdot \text{m}$

4.23 b) $4F_{right} = 24\,000 \times 2 + 4\,000 \times 6. \quad \therefore F_{right} = 18\,000 \text{ N}. \quad 4F_{left} = 24\,000 \times 2 - 4\,000 \times 2.$

$\therefore F_{left} = 10\,000 \text{ N}. \quad M_{max} = \text{area under diagram} = 10\,000 \times 1.667/2 = 8\,330 \text{ N} \cdot \text{m}.$

$$\sigma = \frac{My}{I} = \frac{8\,330 \times 0.1}{971 \times 10^{-8}} = 85.8 \times 10^6 \text{ Pa}$$

V
10 000
1.667 m
-14 000
4000

4.24 c) Compression occurs in bottom fibers over right support.

There, $M = 4\,000 \times 2 = 8\,000 \text{ N} \cdot \text{m}$. $\sigma = \dfrac{My}{I} = \dfrac{8\,000 \times 0.1}{971 \times 10^{-8}} = 82.4 \times 10^6 \text{ Pa}$

4.25 e) $\tau_{max} = \dfrac{VQ}{Ib} = \dfrac{14\,000(0.002 \times 0.05)}{\left(971 \times 10^{-8}\right) \times 0.02} = 7.21 \times 10^6 \text{ Pa}$

4.26 c) τ_{max} occurs on the N.A. with a sudden decrease when b goes from 2 to 16 cm. Also, it is a parabolic distribution.

4.27 c) $\sigma = \dfrac{My}{I}. \quad \dfrac{I}{y} = \dfrac{M}{\sigma} = \dfrac{24\,000}{140 \times 10^6} = 171 \times 10^{-6} \text{ m}^3$

4.28 b) Using the area under the curve: $M_{max} = 1000 \times 3 + 3000 \times 3/2 = 7500\ \text{N} \cdot \text{m}$

$$\sigma_{max} = \frac{My}{I} = \frac{7500 \times 0.025}{0.1 \times 0.05^3/12} = 180 \times 10^6\ \text{Pa}$$

V
1000
-1000
4000

4.29 e) $\sigma_{max} = \dfrac{My}{I} = \dfrac{7500 \times 0.05}{0.05 \times 0.1^3/12} = 90 \times 10^6\ \text{Pa}$

4.30 a) $V_{max} = 1000.\quad \tau_{max} = \dfrac{VQ}{Ib} = \dfrac{1000(0.025 \times 0.05 \times 0.0125)}{\left(0.05 \times 0.05^3/12\right) \times 0.05} = 600 \times 10^3\ \text{Pa}$

4.31 d) $\delta = \dfrac{PL^3}{48EI} = \dfrac{2000 \times 6^3}{48 \times \left(210 \times 10^9\right) \times 0.05^4/12} = 0.0823\ \text{m}$

4.32 e) $\delta = \dfrac{PL^3}{48EI} + \dfrac{5wL^4}{384EI}.\qquad I = \dfrac{bh^3}{12} = \dfrac{0.1 \times 0.05^3}{12} = 1.04 \times 10^{-6}$

$$= \frac{2000 \times 6^3}{48 \times \left(210 \times 10^9\right) \times 1.04 \times 10^{-6}} + \frac{1000 \times 5 \times 6^4}{384 \times \left(210 \times 10^9\right) \times 1.04 \times 10^{-6}} = 0.118\ \text{m}$$

4.33 a) $\delta = \theta L_2 = \dfrac{PL^2}{16EI} \times 4.\quad 0.1 = \dfrac{P \times 8^2 \times 4}{16 \times \left(210 \times 10^9\right) \pi \times .025^4/4}.\quad \therefore P = 403\ \text{N}$

θ δ

4.34 d) $\tau_{max} = \dfrac{1}{2}\sqrt{\left(\sigma_x - \sigma_y\right)^2 + 4\tau^2} = \dfrac{1}{2}\sqrt{60^2 + 4 \times 40^2} = 50\ \text{MPa}$

4.35 a) $\tau_{max} = \dfrac{1}{2}\left(\sigma_x + \sigma_y\right) + \tau_{max} = 0 + 40 = 40\ \text{MPa}$

4.36 c) $\tau_{max} = \dfrac{1}{2}\sqrt{(30 + 50)^2 + 4 \times 30^2} = 50\ \text{MPa}$

4.37 d) $\tau = Tr/J = 600 \times 0.025 \Big/ \dfrac{\pi \times .025^4}{2} = 24.45 \times 10^6\ \text{Pa}$

$\sigma = P/A = 40\,000/\pi \times 0.025^2 = 20.37 \times 10^6\ \text{Pa}$

$\therefore \tau_{max} = \dfrac{1}{2}\sqrt{20.37^2 + 4 \times 24.45^2} = 26.5 \times 10^6\ \text{Pa}$

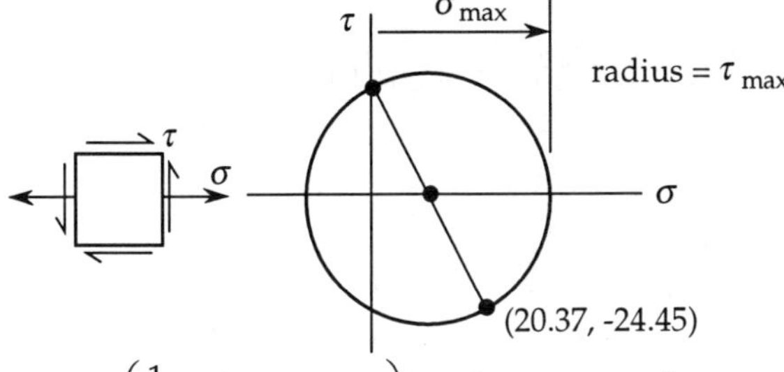

4.38 c) $\sigma_{max} = \left(\dfrac{1}{2} \times 20.37 + 26.5\right) \times 10^6 = 36.7 \times 10^6\ \text{Pa}$

4.39 b) $My/I = 3200 \times 0.025 \Big/ \frac{\pi \times .025^4}{4} = 261 \times 10^6$ comp.

$P/A = 40\,000/\pi \times 0.025^2 = 20.4 \times 10^6$ Pa tension.

$\sigma_A = (261 - 20.4) \times 10^6 = 241 \times 10^6$ Pa

4.40 c) $\tau_{max} = \sigma/2 = 120 \times 10^6$ Pa. $VQ/Ib = 0$ on outer fibers.

4.41 c) $\tau = \frac{Tr}{J} = \frac{8\,000 \times 0.25 \times 0.025}{\pi \times .025^4/2} = 81.5 \times 10^6$ Pa

$\sigma = \frac{My}{I} = \frac{8\,000 \times 0.4 \times 0.025}{\pi \times .025^4/4} = 261 \times 10^6$ Pa

$\tau_{max} = \frac{1}{2}\sqrt{261^2 + 4 \times 81.5^2} \times 10^6 = 154 \times 10^6$ Pa

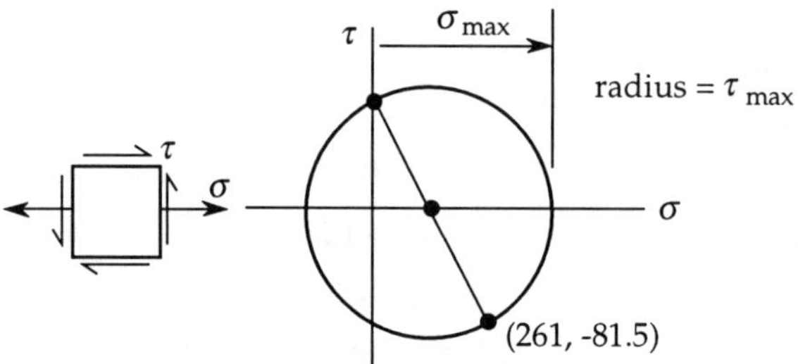

4.42 a) $\sigma_{max} = \left(\frac{1}{2} \times 261 + 154\right) \times 10^6 = 284 \times 10^6$ Pa

4.43 c) $M_{max} = FL_2 = 2000 \times 1 = 2000$ N·m.

$\sigma = \frac{My}{I} = \frac{2000 \times 0.025}{0.02 \times 0.05^3/12} = 240 \times 10^6$ Pa.

4.44 d) $\frac{VQ}{Ib} = \frac{2000 \times (0.02 \times 0.025) \times 0.0125}{\left(0.02 \times 0.05^3/12\right) \times 0.02} = 3 \times 10^6$ Pa.

4.45 a) The force F provides a torque of 2000 N·m and a force of 2000 N acting on the end of the steel shaft. This is best described by (a) since the top fibers experience tension. There is no normal stress in the circumferential direction, ruling out (b), (c) and (d).

4.46 e) $\sigma = \frac{My}{I} = \frac{(2000 \times 2) \times 0.025}{\pi \times 0.025^4/4} = 326 \times 10^6$ Pa.

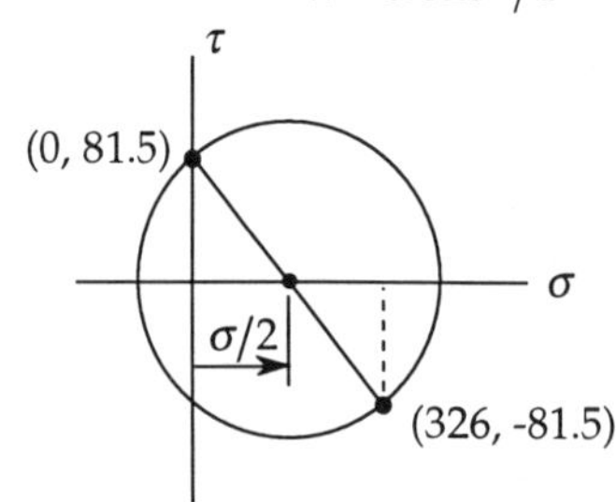

$$\tau = \frac{Tr}{J} = \frac{(2000 \times 1) \times 0.025}{\pi \times 0.025^4/2} = 81.5 \times 10^6 \text{ Pa.}$$

$$\therefore \sigma_{max} = 163 + \left[163^2 + 81.5^2\right]^{1/2}$$

$$= 345 \text{ MPa.}$$

4.47 d) $\delta = \frac{PL^3}{3EI} = \frac{2000 \times 2^3}{3 \times 210 \times 10^9 \times \pi \times 0.025^4/4} = 0.0828$ m.

4.48 b) $\sigma_t = pD/2t. \quad \therefore p = 180 \times 10^6 \times 2 \times 0.005/0.8 = 2250 \times 10^3$ Pa

4.49 c) $\sigma_a = pD/4t. \quad \therefore t = \frac{8000 \times 10^3 \times 1.2}{4 \times 200 \times 10^6} = 0.012$ m

4.50 e) $\tau_{max} = \frac{1}{2}\sqrt{(200 - 200) + 0 \times 4} = 0$

4.51 a) $\left(\frac{\Delta L}{L}\right)_s = \left(\frac{\Delta L}{L}\right)_c. \quad \therefore \varepsilon_s = \varepsilon_c. \quad \therefore \sigma_s = \frac{E_s}{E_c}\sigma_c = 10.5\,\sigma_c.$

$F_s + F_c = 2000000$ or $A_s\sigma_s + A_c\sigma_c = 2000000.$

$\sigma_s\left[\pi\left(0.137^2 - 0.125^2\right) + \pi \times \frac{0.125^2}{10.5}\right] = 2 \times 10^6. \quad \therefore \sigma_s = 137 \times 10^6$ Pa.

4.52 e) $n = E/E_{min} = 3.$ The area is transformed:

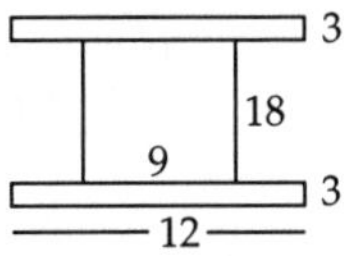

$$I_t = \frac{0.12 \times 0.24^3}{12} - \frac{0.03 \times 0.18^3}{12} = 1.237 \times 10^{-4} \text{ m}^4.$$

$$\therefore \sigma_{al} = \frac{My}{I} = \frac{2000 \times 2 \times 0.12}{1.237 \times 10^{-4}} = 3.88 \times 10^6 \text{ Pa}$$

4.53 c) $n = E/E_{min} = 3.$ The area is transformed:

$$I_t = \frac{0.36 \times 0.24^3}{12} - \frac{0.33 \times 0.18^3}{12} = 2.54 \times 10^{-4} \text{ m}^4.$$

$$\therefore \sigma_s = \frac{nMy}{I} = \frac{3 \times 4000 \times 0.12}{2.54 \times 10^{-4}} = 5.67 \times 10^6 \text{ Pa}$$

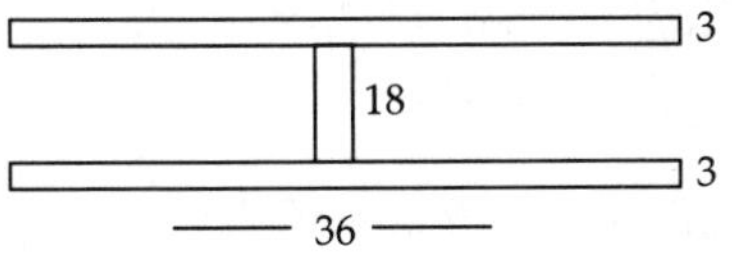

4.54 d) $60 = \frac{L}{r} = \frac{L}{\sqrt{I/A}} = \frac{L}{\sqrt{0.1 \times \left(0.1^3/12\right)/0.01}}. \quad \therefore L = 1.73$ m

4.55 c) Assume $P = 4000$ N using a factor of safety of 2.

$$4000 = \frac{\pi^2 \times 70 \times 10^9 \times \pi \times \left(0.1^4/64\right)}{4L^2}. \quad \therefore L = 14.55 \text{ m}$$

4.56 e) $P_{cr} = 4\pi^2 EI/L^2 = \alpha\Delta TEA.\quad \Delta T = \frac{4\pi^2 I}{\alpha AL^2} = \frac{4\pi^2 \times \pi \times (0.02^4/64)}{11.7\times10^{-6}\times\pi\times0.01^2\times4^2} = 5.27\ °C$

4.57 b) $P_{cr} = 4\pi^2 EI/L^2 = 30\,000.\quad \therefore \pi^2 EI/L^2 = 7500.\quad \therefore P_{cr} = \pi^2 EI/4L^2 = 7500/4 = 1875$ N

Solutions to Practice Problems

(English Units)

4.1 b) Isotropic

4.2 e) Poisson's Ratio

4.3 d)

4.4 d) $\sigma = E\delta/L = P/\pi r^2.\quad \therefore P = E\delta\pi r^2/L = 30\times10^6\times0.04\times\pi(1/4)^2/48 = 4909\text{ lb}.$

4.5 d) $\frac{P}{\pi r^2} = E\frac{\delta}{L}.\quad \therefore \delta = \frac{PL}{\pi r^2 E} = \frac{3500\times100\times12}{\pi\times(1/4^2)\times30\times10^6} = 0.71\text{ in}$

4.6 a) $\frac{P}{A} = E\frac{\delta}{L}.\quad \therefore P = \frac{AE\delta}{L} = \frac{1}{8}\times1\times30\times10^6\times\frac{1/32}{36} = 3255\text{ lb}.$

4.7 b) $\frac{P}{A} = G\frac{\delta}{L}.\quad \therefore G = \frac{PL}{A\delta} = \frac{5000\times6}{8\times(1/2)\times0.0012} = 6.25\times10^6$

4.8 a) $\frac{P}{A} = E\varepsilon.\quad \therefore \varepsilon = \frac{150{,}000}{\pi\times1^2\times30\times10^6} = 0.00159$

$\therefore \Delta d = \nu\varepsilon d = 0.28\times0.00159\times2 = 0.00089.\quad \therefore d_{after} = 2 - 0.00095 = 1.9991\text{ in}$

4.9 b) $\frac{P}{A} = E\varepsilon.\quad \therefore \varepsilon = \frac{400{,}000}{\pi\times(5.923/2)^2\times10\times10^6} = 0.00145$

$\Delta d = \nu\varepsilon d = 0.33\times0.00145\times5.923 = 0.00283.$

$\therefore d_f = d - \Delta d = 5.923 - 0.00283 = 5.920\text{ in}$

4.10 e) $\sigma = E\delta/L.\quad 16{,}000 = 30\times10^6(\delta/L).\quad \therefore \delta = 5.33\times10^{-4}L.$

$\delta_T = \alpha L\Delta T = 6.5\times10^{-6}\times50L = 3.25\times10^{-4}L.$

$\therefore \delta_{final} = 2.08\times10^{-4}L.\quad \therefore \sigma = 30\times10^6\times2.08\times10^{-4} = 6{,}240\text{ psi}$

4.11 a) $\delta = \alpha L\Delta T = 6.5\times10^{-6}\times1000\times12\times130 = 10.1''$

4.12 d) $\frac{\delta}{L} = \frac{\sigma}{E} = \alpha\Delta T.\quad \therefore \frac{10{,}000}{10\times10^6} = 12.8\times10^{-6}(80 - T).\quad \therefore T = 1.88°\text{F}$

4.13 d) It expands at a different rate.

4.14 a) $\tau = \frac{Tr}{J} = \frac{2000 \times 1}{\pi \times 1^4/2} = 1,273$ psi

4.15 e) $J = \pi\left(a_1^4 - a_2^4\right)/2 = \pi\left(1^4 - .875^4\right)/2 = 0.650 \text{ in}^4$

$\tau = \frac{Tr}{J} = 2,000 \times 1/0.650 = 3,077$ psi

4.16 a) $T = \frac{\tau J}{r} = 20,000 \times \left(\pi 2^4/2\right)/2 = 251,300 \text{ in} \cdot \text{lb or } 20,940 \text{ ft} \cdot \text{lb}$

4.17 d) $\theta = \frac{TL}{JG} = \frac{160 \times 24 \times 18}{\left[\pi(7/16)^4/2\right] \times 12 \times 10^6} = 0.1001 \text{ rad } (5.73°)$

4.18 d) $\tau_{max} = \frac{Tr}{J} = \frac{(1200 \times 12) \times 2}{\pi \times 2^4/2} = 1146 \text{ psi.} \quad \therefore \sigma_{max} = 1146 \text{ psi.}$

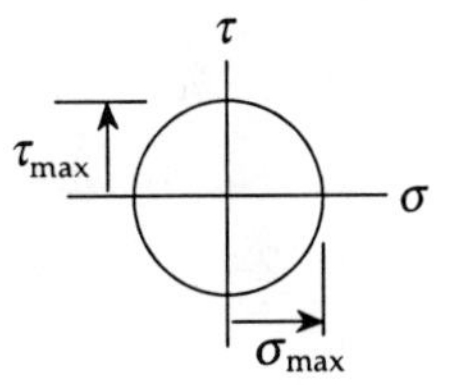

4.19 b) $\sigma = My/I. \quad \therefore \sigma = \sigma_{max}$ at $y = y_{max}$

4.20 a) A triangle.

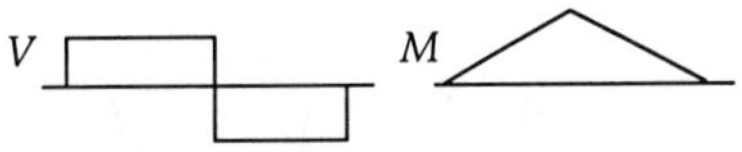

4.21 a) $\sum M_{right\,end} = 0. \quad \therefore 20F_{left} = 1,000 \times 10 + 1,000 \times 5. \quad \therefore F_{left} = 750.$

$M_A = 750 \times 10 = 7,500 \text{ lb} \cdot \text{ft}$

4.22 b) $M_A = 1,000 \times 10 + 1,000 \times 5 = 15,000 \text{ ft} \cdot \text{lb}$

4.23 a) $10F_{right} = 5,000 \times 5 + 1,000 \times 14. \quad \therefore F_{right} = 3,900 \text{ lb.}$

$10F_{left} = 5,000 \times 5 - 1,000 \times 4. \quad \therefore F_{left} = 2,100 \text{ lb.}$

$M_{max} = 2,100 \times (4.2/2) = 4,410 \text{ ft} \cdot \text{lb} = \text{area under V - diagram.}$

$\sigma = \frac{My}{I} = \frac{4,410 \times 12 \times 5}{60.67} = 4,361 \text{ psi}$

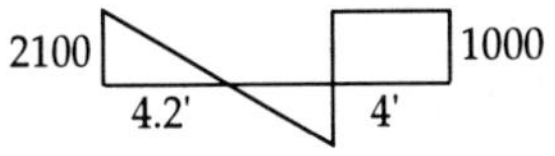

4.24 b) Compression occurs in bottom fibers over right support.

$\sigma = \frac{My}{I} = \frac{4,000 \times 12 \times 5}{60.67} = 3,956 \text{ psi}$

4.25 e) $\tau_{max} = \frac{VQ}{Ib} = \frac{2,900 \times (5 \times 2.5)}{60.67 \times 1} = 597 \text{ psi}$

4.26 c) τ_{max} occurs on the N.A. with a sudden decrease when b goes from 1" to 8".

Also, it is parabolic.

4.27 c) $\sigma = \dfrac{My}{I}$. $\therefore \dfrac{I}{y} = \dfrac{M}{\sigma} = \dfrac{15{,}000 \times 12}{2{,}000} = 9 \text{ in}^3$

4.28 b) Area under curve: $M_{max} = 250 \times 8 + 800 \times (8/2) = 5{,}200$

$\sigma = \dfrac{My}{I} = \dfrac{5{,}200 \times 12 \times 1}{4 \times 2^3/12} = 23{,}4000 \text{ psi}$

V 250 -250 1050

4.29 e) $\sigma_{max} = \dfrac{5{,}200 \times 12 \times 2}{2 \times 4^3/12} = 11{,}700 \text{ psi}$

4.30 e) $V_{max} = 250.$ $\tau_{max} = \dfrac{VQ}{Ib} = \dfrac{250 \times 2 \times (1/2)}{2 \times (2^3/12) \times 2} = 93.8$

4.31 d) $\delta = PL^3/48EI = 500 \times 240^3 / (48 \times 30 \times 10^6 \times 2^4/12) = 3.6''$

4.32 e) $\delta = \dfrac{PL^3}{48EI} + \dfrac{5wL^4}{384EI}$ $\qquad I = \dfrac{4 \times 2^3}{12} = 2.67 \text{ in}^4$

$= \dfrac{500 \times (16 \times 12)^3}{48 \times 30 \times 10^6 \times 2.67} + \dfrac{5 \times (100/12) \times (16 \times 12)^4}{384 \times 30 \times 10^6 \times 2.67} = 2.76''$

4.33 a) $\delta = \theta L_2 = \dfrac{PL^2}{16EI} \times L_2.$ $\therefore 4 = \dfrac{P \times 240^2 \times 120}{16 \times 30 \times 10^6 \times \pi \times 1^4/4}.$ $\therefore P = 218 \text{ lb}$

θ δ

4.34 d) $\tau_{max} = \dfrac{1}{2}\sqrt{(\sigma_x - \sigma_y)^2 + 4\tau^2} = \dfrac{1}{2}\sqrt{6{,}000^2 + 4 \times 4{,}000^2} = 5{,}000 \text{ psi}$

4.35 a) $\sigma_{max} = \dfrac{1}{2}(\sigma_x + \sigma_y) + \tau_{max} = 0 + 4{,}000 = 4{,}000 \text{ psi}$

4.36 c) $\tau_{max} = \dfrac{1}{2}\sqrt{(3{,}000 + 5{,}000)^2 + 4 \times 3{,}000^2} = 5{,}000 \text{ psi}$

4.37 d) $\tau = Tr/J = 500 \times 12 \times 1/(\pi \times 1^4/2) = 3{,}820 \text{ psi.}$

$\sigma = P/A = 10{,}000/\pi \times 1^2 = 3{,}180 \text{psi.}$

$\tau_{max} = \dfrac{1}{2}\sqrt{3{,}180^2 + 4 \times 3{,}820^2} = 4{,}137 \text{ psi}$

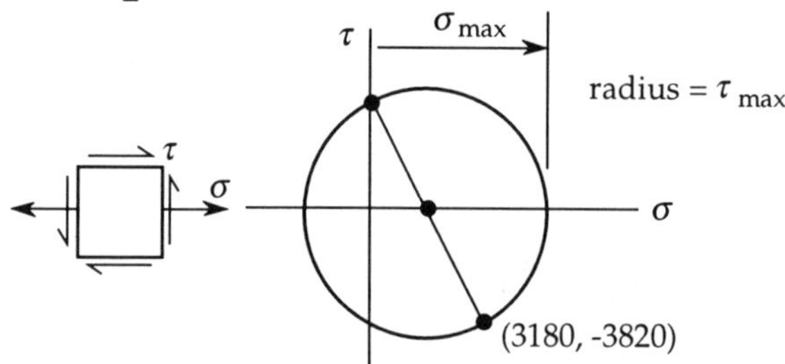

4.38 c) $\sigma_{max} = \dfrac{1}{2} \times 3{,}180 + 4{,}137 = 5{,}727 \text{ psi}$

4.39 b) $M = 30{,}000 \text{ in} \cdot \text{lb.} \quad \dfrac{My}{I} = \dfrac{30{,}000 \times 1}{\pi \times \left(1^4/4\right)} = 38{,}200 \text{ comp.}$

$\dfrac{P}{A} = \dfrac{10{,}000}{\pi \times 1^2} = 3{,}183 \text{ tens.}$

$\sigma_A = 38{,}200 - 3{,}183 = 35{,}000 \text{ comp.}$

4.40 c) $\tau_{max} = \sigma/2 = 17{,}500 \text{ psi.} \quad VQ/Ib = 0$ on outer fibers

4.41 c) $\tau = \dfrac{Tr}{J} = \dfrac{2{,}000 \times 10 \times 1}{\pi \times \left(1^4/2\right)} = 12{,}732 \text{ psi}$

$\sigma = \dfrac{My}{I} = \dfrac{2{,}000 \times 16 \times 1}{\pi \times \left(1^4/2\right)} = 40{,}744 \text{ psi}$

$\tau_{max} = \dfrac{1}{2}\sqrt{40{,}744^2 + 4 \times 12{,}732^2} = 24{,}023 \text{ psi}$

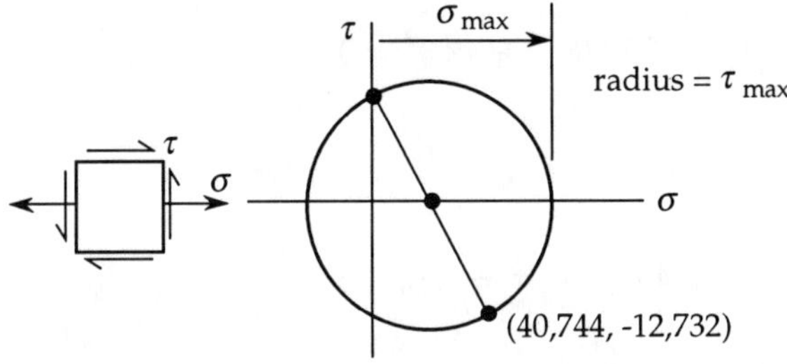

4.42 d) $\sigma_{max} = \dfrac{1}{2} \times 40{,}744 + 24{,}023 = 44{,}395 \text{ psi}$

4.43 c) $M_{max} = FL_2 = 450 \times 3.28 = 1476 \text{ ft} \cdot \text{lb.}$

$\sigma = \dfrac{My}{I} = \dfrac{(1476 \times 12) \times 1.97/2}{0.787 \times 1.97^3/12} = 34{,}790 \text{ psi.}$

4.44 d) $\dfrac{VQ}{Ib} = \dfrac{450 \times (0.787 \times 0.985) \times 0.4925}{\left(0.787 \times 1.97^3/12\right) \times 0.787} = 435 \text{ psi.}$

4.45 a) The force F provides a torque and a force acting on the end of the steel shaft. This is best described by (a) since the top fibers experience tension. There is no normal stress in the circumferential direction, ruling out (b), (c) and (d).

4.46 e) $\sigma = \dfrac{My}{I} = \dfrac{(450 \times 6.56 \times 12) \times 0.985}{\pi \times 0.985^4/4} = 47{,}200 \text{ psi.}$

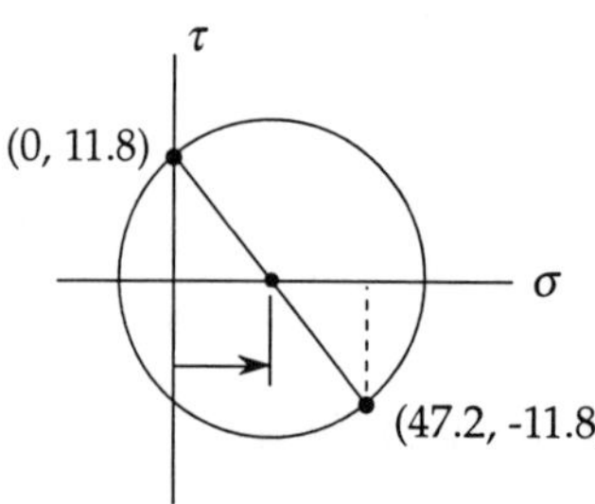

$$\tau = \frac{Tr}{J} = \frac{(450 \times 3.28 \times 12) \times 0.985}{\pi \times 0.985^4/2} = 11,800 \text{ psi.}$$

$$\therefore \sigma_{\max} = 23,600 + \left[23,600^2 + 11,800^2\right]^{1/2}$$

$$= 50,000 \text{ psi.}$$

4.47 d) $\delta = \frac{PL^3}{3EI} = \frac{450 \times (6.25 \times 12)^3}{3 \times 30 \times 10^6 \times \pi \times 0.985^4/4} = 2.85$ in.

4.48 b) $\sigma_t = pD/2t. \quad \therefore p = 24,000 \times 2 \times \frac{1}{4}/24 = 500$ psi

4.49 c) $\sigma_a = pD/4t. \quad \therefore t = \frac{2000 \times 48}{4 \times 30,000} = 0.800$ in

4.50 e) $\tau_{\max} = \frac{1}{2}\sqrt{(30,000 - 30,000)^2 + 4 \times 0^2} = 0$

4.51 a) $\left(\frac{\Delta L}{L}\right)_s = \left(\frac{\Delta L}{L}\right)_c. \quad \therefore \varepsilon_s = \varepsilon_c. \quad \therefore \sigma_s = \frac{E_s}{E_c}\sigma_c = 10\,\sigma_c.$

$F_s + F_c = 400,000$ or $A_s\sigma_s + A_c\sigma_c = 400,000.$

$$\sigma_s\left[\pi\left(5.5^2 - 5^2\right) + \pi \times \frac{5^2}{10}\right] = 400,000. \quad \therefore \sigma_s = 16,430 \text{ psi}$$

4.52 e) $n = E/E_{\min} = 3.$ The area is transformed:

1"
6"
3"
1"
4"

$$I_t = \frac{4 \times 8^3}{12} - \frac{1 \times 6^3}{12} = 152.7 \text{ in}^4.$$

$$\therefore \sigma_{al} = \frac{My}{I} = \frac{2,500 \times 12 \times 4}{152.7} = 786 \text{ psi}$$

4.53 d) $n = 3.$ The area is transformed:

1"
1"
6"
1"
12"

$$I_t = \frac{12 \times 8^3}{12} - \frac{11 \times 6^3}{12} = 314 \text{ in}^4.$$

$$\therefore \sigma_s = \frac{nMy}{I} = \frac{3 \times 2,500 \times 12 \times 4}{314} = 1,146 \text{ psi.}$$

4.54 d) $60 = \frac{L}{r} = \frac{L}{\sqrt{I/A}} = \frac{L}{\sqrt{4 \times \left(4^3/12\right)/16}}. \quad \therefore L = 69.3\text{"}$ or $5.77'$

Assume $P = 1,000$ lb using a factor of safety of 2.

4.55 c) $1,000 = \frac{\pi^2 \times 10 \times 10^6 \times \pi \times 4^4/64}{4L^2}. \quad \therefore L = 557\text{"}$ or $46.4'$

4.56 e) $P_{cr} = 4\pi^2 EI/L^2 = \alpha\Delta TEA.$ $\Delta T = \dfrac{4\pi^2 I}{\alpha A L^2} = \dfrac{4\pi^2 \times \pi \times 1^4/64}{6.5\times 10^{-6} \times \pi \times 0.5^2 \times 120^2} = 26.4\ °\text{F}$

4.57 b) $P_{cr} = 4\pi^2 EI/L^2 = 8,000.$ $\therefore \pi^2 EI/L^2 = 2,000.$

$\therefore P_{cr} = \pi^2 EI/4L^2 = 2,000/4 = 500$ lb.

Thermodynamics

by Merle C. Potter

Overview, Definitions and Laws

Thermodynamics involves the storage, transformation and transfer of energy. It is stored as internal energy, kinetic energy and potential energy; it is transformed between these various forms; and, it is transferred as work or heat transfer.

The *macroscopic* approach is used in this presentation, that is, we assume matter occupies all points in a region of interest. This is acceptable providing the density is sufficiently large, which it is in most engineering situations.

Both a *system*, a fixed quantity of matter, and a *control volume*, a volume into which and/or from which a substance flows, can be used in thermodynamics. (A control volume may also be referred to as an *open system*.) A system and its surroundings make up the *universe*. Some useful definitions follow:

phase — matter that has the same composition throughout; it is homogeneous

mixture — a quantity of matter that has more than one phase

property — a quantity which serves to describe a system

simple system — a system composed of a single phase, free of magnetic, electrical, and surface effects. Only two properties are needed to fix a simple system

state — the condition of a system described by giving values to its properties at the given instant

intensive property — a property that does not depend on the mass

extensive property — a property that depends on the mass of the system

specific property — an extensive property divided by the mass

thermodynamic equilibrium — when the properties do not vary from point to point in a system and there is no tendency for additional change

process — the path of successive states through which a system passes

quasi-equilibrium — if, in passing from one state to the next, the deviation from equilibrium is infinitesimal. It is also called a quasistatic process

reversible process — a process which, when reversed, leaves no change in either the system or surroundings

isothermal — temperature is constant

isobaric — pressure is constant

isometric — volume is constant

isentropic — entropy is constant

adiabatic — no heat transfer

Experimental observations are organized into mathematical statements or *laws*. Some of those used in thermodynamics follow:

zeroith law of thermodynamics— If two bodies are equal in temperature to a third, they are equal in temperature to each other.

first law of thermodynamics — During a given process, the net heat transfer minus the net work output equals the change in energy.

second law of thermodynamics — A device cannot operate in a cycle and produce work output while exchanging heat at a single temperature.

Boyle's law — The volume varies inversely with pressure for an ideal gas.

Charles' law — The volume varies directly with temperature for an ideal gas.

Avogadro's law — Equal volumes of different ideal gases with the same temperature and pressure contain the equal number of molecules.

5.1 Density, Pressure, and Temperature

The density ρ is the mass divided by the volume,

$$\rho = \frac{m}{V}. \tag{5.1.1}$$

The specific volume is the reciprocal of the density,

$$v = \frac{1}{\rho} = \frac{V}{m}. \tag{5.1.2}$$

The pressure P is the normal force divided by the area upon which it acts. In thermodynamics, it is important to use *absolute pressure,* defined by

$$P_{abs} = P_{gauge} + P_{atmospheric} \tag{5.1.3}$$

where the atmospheric pressure is taken as 100 kPa (14.7 psi), unless otherwise stated. If the gauge pressure is negative, it is a *vacuum.*

The temperature scale is established by choosing a specified number of divisions, called degrees, between the ice point and the steam point, each at 101 kPaolute. In the Celsius scale, the ice point is set at 0°C and the steam point at 100°C. **In thermodynamics, pressures are always assumed to be given absolute pressures.** Expressions for theolute temperature in kelvins, and degrees Rankine are, respectively,

$$T = T_{celsius} + 273; \quad T = T_{fahrenheit} + 460. \tag{5.1.4}$$

The temperature, pressure and specific volume for an ideal (perfect) gas are related by the *ideal gas law*

$$Pv = RT, \quad P = \rho RT, \quad \text{or } PV = mRT \qquad R = \overline{R}/M \tag{5.1.5}$$

where the *universal gas constant* $\overline{R} = 8.134 \text{ kJ/kmol}\cdot\text{K}$ $(1545 \text{ ft-lbf/lbmol-°R})$, M is the molar mass, and R is the gas constant; for air it is 0.287 kJ/kg.K (53.3 ft-lbf/lbm-°R). (Note: the NCEES Handbook does not list R_{air}. It is a good idea to memorize R_{air} since it is used so often.) Below moderate pressure this equation can be used if the temperature exceeds twice the critical temperature, the temperature above which the liquid and vapor phases do not coexist. Thus, if the temperature of the steam is greater than 800°C (1500°F), Equation 5.1.5 can be used with $R = 0.462$ kJ/kg.K (85.7 ft-lbf/lbm-°R) if the pressure is not excessive; for high pressure ($P > 10$ MPa) the temperature should be higher than 1000°C if Equation 5.1.5 is to be used.

If the temperature is below twice the critical temperature, tables relating the three variables P, v and T must be used. For water such tables are called the "steam tables" and are presented in Table 5.3.1. Data is presented in Table 5.3.1 and 5.3.2 that is used to relate P, v and T when water exists in the liquid phase, as a liquid/vapor mixture, or in the saturated vapor phase. These situations are best described referring to a T-v diagram, shown in Figure 5.1. Suppose we start with a constant pressure container (a cylinder with a floating piston) containing water

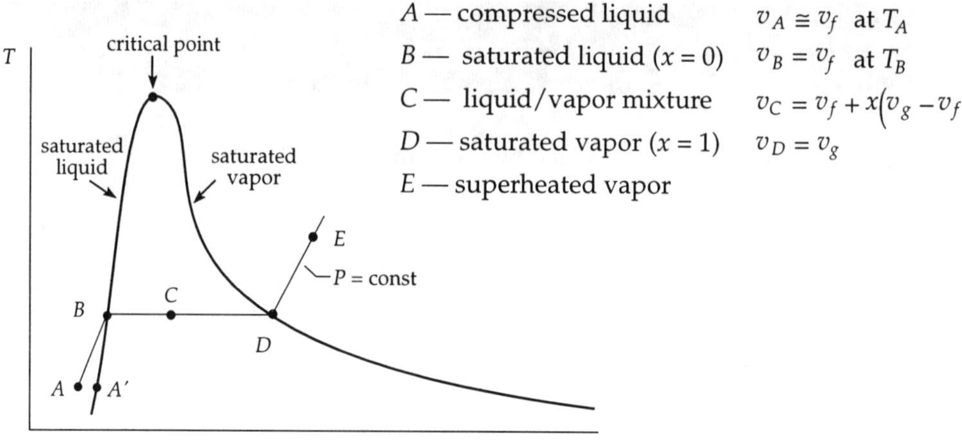

Figure 5.1 A *T*-*v* diagram.

at room temperature; it would undergo the following changes if heat is transferred to the volume:

- The temperature would rise above T_A in the *compressed liquid* and $v_A \cong v_{A'}$.
- At state *B* *saturated liquid* phase results and vaporization (boiling) begins; $v_B = v_f$. A subscript *f* will always denote a saturated liquid.
- At state C the liquid phase and the vapor phase are in equilibrium; a *liquid/vapor mixture* occurs and

$$v = v_f + x(v_g - v_f) \quad \text{or} \quad v = xv_g + (1-x)v_f \tag{5.1.6}$$

  where *x* is the *quality*. Sometimes we let $v_{fg} = v_g - v_f$.
- At state *D* a *saturated vapor* exists and vaporization is complete; a subscript *g* will always denote a saturated vapor: $v_D = v_g$.
- At state *E* the vapor is *superheated* and *v* is found in Table 5.3.3.

Equation 5.1.6 above results from the definition of *quality*, the ratio of the vapor mass to the total mass:

$$x = \frac{m_{vapor}}{m_{total}} = \frac{m_g}{m_g + m_f}. \tag{5.1.7}$$

Note that the entries in the tables at the end of this chapter are inolute pressure.

An alternative to the tabulated properties is the *P*-*h* diagram which includes the properties *P*, *h*, *T*, *v* and *s*. If internal energy is required we use the relation $u = h - Pv$ (see Eq. 5.2.10). We will illustrate its use in Article 5.3.

Example 5.1

What mass of air is contained in a room 20 m × 40 m × 3 m at standard conditions?

Solution. Standard conditions are T = 25°C and P = 100 kPa. Using M = 28.97,

$$\begin{aligned}\rho &= \frac{P}{RT}\\ &= \frac{100}{0.287\times 298} = 1.17\ \text{kg/m}^3, \quad \text{where } R = \frac{8.314}{28.97} = 0.287.\\ \therefore m &= \rho V\\ &= 1.17\times 20\times 40\times 3 = 2810\ \text{kg}.\end{aligned}$$

Example 5.2

What mass of water contained in 2m^3 at 2000 kPa and 200°C?

Solution. The specific volume of compressed water is insensitive to pressure so we use v_f at 200°C. It is, using Table 5.3.1,

$$v_f = 0.001156\ \text{m}^3/\text{kg}.$$

The mass is then

$$\begin{aligned}m &= V/v\\ &= 2/0.001156 = 1730\ \text{kg}.\end{aligned}$$

Note: Liquid water has mass density of about 1000 kg/m^3 or $v = 0.001\ \text{m}^3/\text{kg}$. This is good if the temperature is relatively low, say below 100°C.

Example 5.3

The volume occupied by 20 kg of a water-vapor mixture at 200°C is 2m^3. Calculate the quality.

Solution. The specific volume is

$$\begin{aligned}v &= V/m\\ &= 2/20 = 0.1\ \text{m}^3/\text{kg}.\end{aligned}$$

Using Equation 5.1.6, and Table 5.3.1, we have

$$\begin{aligned}x &= \frac{v - v_f}{v_g - v_f}\\ &= \frac{0.1 - 0.001156}{0.1274 - 0.001156} = 0.783 \text{ or } 78.3\%.\end{aligned}$$

Example 5.4

Find the volume occupied by 20 kg of steam at 4 MPa and 400°C.

Solution. The specific volume is found in Table 5.3.3 to be

$$v = 0.07341\ \text{m}^3/\text{kg}.$$

The volume is then

$$\begin{aligned}V &= mv\\ &= 20\times 0.07341 = 1.468\ \text{m}^3.\end{aligned}$$

5.2 The First Law of Thermodynamics for a System

The first law of thermodynamics, referred to as the "first law" or the "energy equation," is expressed for a cycle as

$$Q_{net} = W_{net} \tag{5.2.1}$$

and for a process as

$$Q - W = \Delta E \tag{5.2.2}$$

where Q is the heat transfer, W is the work, and E represents the energy (kinetic, potential, and internal) of the system*. In thermodynamics attention is focused on internal energy with kinetic and potential energy changes neglected (unless otherwise stated) so that we have

$$Q - W = \Delta U \quad \text{or} \quad q - w = \Delta u \tag{5.2.3}$$

where the specific internal energy is

$$q = \frac{Q}{m}, \quad w = \frac{W}{m}, \quad u = \frac{U}{m}. \tag{5.2.4}$$

Heat transfer may occur during any of the three following modes :

• ***Conduction***—heat transfer due to molecular activity. For steady-state heat transfer through a constant wall area *Fourier's law* states†

$$\dot{Q} = kA\Delta T/L = A\Delta T/R \tag{5.2.5}$$

where k is the *conductivity* (dependent on the material), R is the *resistance factor,* and the length, L, is normal to the heat flow.

• ***Convection***—heat transfer due to fluid motion. The mathematical expression used is

$$\dot{Q} = hA\Delta T = A\Delta T/R \tag{5.2.6}$$

where h is the *convective heat transfer coefficient* (dependent on the surface geometry, the fluid velocity, the fluid viscosity and density, and the temperature difference). The *Nusselt number* ($Nu = hd/k$) is the dimensionless convective heat transfer coefficient.

• ***Radiation***—heat transfer due to the transmission of waves. The heat transfer from body 1 is

$$\dot{Q} = \sigma\varepsilon A\left(T_1^4 - T_2^4\right)F_{1-2} \tag{5.2.7}$$

---

*In this book heat transferred to the system is positive and work done by the system is positive. It is also conventional to define work done on the system as positive so that $Q + W = \Delta U$.

†A dot signifies a rate, so that $\dot{Q}$ has units of J/s.

in which the *Stefan-Boltzmann constant* is $\sigma = 5.67 \times 10^{-11} \text{kJ/s} \cdot \text{m}^2 \cdot \text{K}^4$ (1.714 × 10^{-9} Btu / hr - ft^2-°R^4), ε is the *emissivity* (ε = 1 for a black body), and $F_{1\text{-}2}$ is the *shape factor* ($F_{1-2} = 1$ if body 2 encloses body 1).

For a two layer composite wall with an inner and outer convection layer we use the resistance factors and obtain

$$\dot{Q} = A\Delta T/(R_i + R_1 + R_2 + R_0) = UA\Delta T \tag{5.2.8}$$

where U is the *overall heat transfer coefficient*. It is not to be confused with internal energy of Eq. 5.2.3.

In thermodynamics, the heat transfer is usually specified or calculated using Equation 5.2.3; it is usually not calculated with the previous four equations. The equations in the one dimensional form above and in two-dimensions are the focus of attention in a course in Heat Transfer. Thermal properties are presented in Tables 5.5 and 5.6.

Work can be accomplished mechanically by moving a boundary, resulting in a quasi-equilibrium work mode

$$W = \int PdV \tag{5.2.9}$$

It can also be accomplished in non-quasi-equilibrium modes such as with paddle wheel or by electrical resistance. But then Equation 5.2.9 cannot be used.

We introduce *enthalpy* for convenience, and define it to be

$$\begin{aligned} H &= U + PV \\ h &= u + Pv. \end{aligned} \tag{5.2.10}$$

For substances such as steam, the specific internal energy and specific enthalpy are either found in the steam tables or from a *P-h* diagram. For a compressed liquid, u and h are insensitive to pressure and are found in the Table 5.3.1 with the "*f*" subscripts under the specified temperature. For the liquid/vapor mixture we use

$$\begin{aligned} u &= u_f + x(u_g - u_f) \\ h &= h_f + xh_{fg} \end{aligned} \tag{5.2.11}$$

where $h_{fg} = h_g - h_f$.

For ideal gases we assume constant specific heats and use

$$\Delta u = c_v \Delta T \tag{5.2.12}$$

$$\Delta h = c_p \Delta T \tag{5.2.13}$$

where c_v is the *constant volume specific heat*, and c_v is the *constant pressure specific heat*. From the differential forms of the above we can find

$$c_p = c_v + R. \tag{5.2.14}$$

We also define the *ratio of specific heats* k to be

$$k = c_p / c_v. \tag{5.2.15}$$

For air $c_v = 0.716$ kJ/kg · K (0.171 BTU/lbm - °R), $c_p = 1.00$ kJ/kg · K (0.24 BTU/lbm - °R), $k = 1.4$. For most solids and liquids we can find the heat transfer using

$$Q = mc_p \Delta T. \tag{5.2.16}$$

For water $c_p = 4.18$ kJ/kg·°K (1.00 BTU/lbm-°R), and for ice $c_p \cong 2.1$ kJ/kg·°K (0.50 BTU/lbm-°R).

When a substance changes phase, *latent heat* is involved. The energy necessary to melt a unit mass of a solid is the *heat of fusion;* the energy necessary to vaporize a unit mass of liquid is the *heat of vaporization*, equal to $(h_g - h_f)$; the energy necessary to vaporize a unit mass of solid is the *heat of sublimation*. For ice, the heat of fusion is approximately 320 kJ/kg (140 BTU/lbm) and the heat of sublimation is about 2040 kJ/kg (877 BTU/lbm); the heat of vaporization varies from 2050 kJ/kg at 0°C (1075 BTU/lbm at 32°F) to zero at the critical point.

For specific processes, we consider the preceding paragraphs and summarize as follows:

Isothermal Process

$$\text{1st law:} \quad Q - W = m\Delta u \quad \text{or} \quad q - w = \Delta u \tag{5.2.17}$$

$$\text{ideal gas:} \quad Q = W = mRT \ln \frac{v_2}{v_1} = mRT \ln \frac{P_1}{P_2} \tag{5.2.18}$$

$$P_2 = P_1 v_1 / v_2 \tag{5.2.19}$$

Constant Pressure (Isobaric)

$$\text{1st law:} \quad Q = m\Delta h \quad \text{or} \quad q = \Delta h \tag{5.2.20}$$

$$W = mP\Delta v \tag{5.2.21}$$

$$\text{ideal gas:} \quad Q = mc_p \Delta T \tag{5.2.22}$$

$$T_2 = T_1 v_2 / v_1 \tag{5.2.23}$$

Constant Volume

$$\text{1st law:} \quad Q = m\Delta u \quad \text{or} \quad q = \Delta u \tag{5.2.24}$$

$$W = 0 \tag{5.2.25}$$

$$\text{ideal gas:} \quad Q = mc_v \Delta T \tag{5.2.26}$$

$$T_2 = T_1 P_2 / P_1 \tag{5.2.27}$$

Adiabatic Process (Isentropic)

$$\text{1st law:} \quad -W = m\Delta u \quad \text{or} \quad -w = \Delta u \tag{5.2.28}$$

$$Q = 0 \tag{5.2.29}$$

$$\text{ideal gas:} \quad -W = mc_v \Delta T \tag{5.2.30}$$

$$T_2 = T_1 (v_1/v_2)^{k-1} = T_1 (P_2/P_1)^{(k-1)/k} \tag{5.2.31}$$

$$P_2 = P_1 (v_1/v_2)^k \tag{5.2.32}$$

A *polytropic process* results if k in Equations 5.2.31 and 5.2.32 is replaced with n. Then n must be specified. Note that the adiabatic, quasi-equilibrium process is often referred to as an *isentropic process*.

Example 5.5

How much heat must be added to 2 kg of steam contained in a rigid volume, if the initial pressure of 2 MPa is increased to 5 MPa? $T_1 = 300°C$.

Solution. The first law with $\Delta KE = \Delta PE = 0$ is

$$Q - W = \Delta U.$$

For a rigid container $W = 0$ so that

$$Q = m(u_2 - u_1).$$

From the steam Table 5.3.3, we find $u_1 = 2772.6$ kJ/kg and $v_1 = 0.1255\ \text{m}^3/\text{kg}$. We can locate state 2 because the container is rigid, so that

$$v_2 = v_1 \cong 0.126\ \text{m}^3/\text{kg}.$$

The temperature T_2 that has $P_2 = 5$ MPa and $v_2 = 0.126\ \text{m}^3/\text{kg}$ is 1100°C. At that state $u_2 = 4246$ kJ/kg. Thus,

$$\begin{aligned} Q &= m(u_2 - u_1) \\ &= 2(4246 - 2773) = 2946 \text{ kJ}. \end{aligned}$$

Example 5.6

Calculate the heat transfer necessary to raise the temperature of 2 kg of saturated water vapor to 600°C if the pressure is maintained constant at 2000 kPa?

Solution. The first law, for a constant pressure process, is

$$Q = m(h_2 - h_1).$$

Using Tables 5.3.2 and 5.3.3, we find $h_1 = 2799.5$ and $h_2 = 3690.1$ kJ/kg. Hence, we have

$$Q = 2(3690.1 - 2799.5) = 1781 \text{ kJ}.$$

Example 5.7

How much heat is needed to completely vaporize 100 kg of ice at $T_1 = -10°C$ if the pressure is held constant at 200 kPa?

Solution. The heat transfer is related to the enthalpy by

$$\begin{aligned} Q &= m\Delta h \\ &= m(c\Delta T_{ice} + \text{heat of fusion} + c\Delta T_{water} + \text{heat of vaporization}). \end{aligned}$$

Using the values given in Article 5.2, Table 5.2, and Table 5.3.2,

$$\begin{aligned} Q &= 100(2.1 \times 10 + 320 + 4.18 \times 120.2 + 2201.9) \\ &= 304\ 500\ \text{kJ or } 304.5\ \text{MJ}. \end{aligned}$$

Example 5.8

Estimate the heat transfer necessary to increase the pressure of 50% quality steam from 200 kPa to 800 kPa if the volume is kept constant.

Solution. The first law, with $W = 0$ for a constant volume process, is

$$q = u_2 - u_1.$$

To find state 2 we must use $v_1 = v_2$. At state 1 we have

$$\begin{aligned} v_1 &= v_f + x(v_g - v_f) \\ &= 0.00106 + 0.5(0.8857 - 0.00106) = 0.4434 \ \text{m}^3/\text{kg}, \end{aligned}$$

$$\begin{aligned} u_1 &= u_f + x(u_g - u_f) \\ &= 504.5 + 0.5(2529.5 - 504.5) = 1517 \ \text{kJ/kg}. \end{aligned}$$

At state 2, $P_2 = 0.8$ MPa and $v_2 = 0.4434$ m^3/kg (note: state 2 is superheat) so that at this superheated state $u_2 = 3126$ kJ/kg. Hence,

$$q = 3126 - 1517 = 1609 \text{ kJ/kg}.$$

Note that we have used q rather than Q since mass was not specified.

Example 5.9

Calculate the work done by a piston if the 2 m^3 volume of air is tripled while the temperature is maintained at 40°C. The initial pressure is 400 kPa.

Solution. The mass is needed in order to use Equation 5.2.18 to find the work; it is, using $R = \overline{R}/M = 8.314/28.97 = 0.287 \text{ kJ/kg}\cdot\text{K}$,

$$\begin{aligned} m &= \frac{PV}{RT} \\ &= \frac{400 \times 2}{0.287 \times 313} = 8.91 \text{ kg}, \qquad R = \frac{8.314}{28.97} = 0.287. \end{aligned}$$

The work is then found to be

$$\begin{aligned} W &= mRT \ln v_2/v_1 \\ &= 8.91 \times 0.287 \times 313 \ln 3 = 879 \text{ kJ}. \end{aligned}$$

Note: The temperature is expressed as 40 + 273 = 313 K.

Example 5.10

How much work is necessary to compress air in an insulated cylinder from 0.2 m^3 to 0.01 m^3? Use $T_1 = 20°\text{C}$ and $P_1 = 100$ kPa.

Solution. For an adiabatic process $Q = 0$ so that the first law is

$$\begin{aligned} -W &= m(u_2 - u_1) \\ &= mc_v(T_2 - T_1). \end{aligned}$$

To find the mass m we use the ideal gas equation with

$$R = \overline{R}/M = 8.314/28.97 = 0.287\,\text{kJ/kg}\cdot\text{K},:$$

$$m = \frac{PV}{RT}$$

$$= \frac{100 \times 0.2}{0.287 \times 293} = 0.2378\ \text{kg}.$$

The temperature T_2 is found to be

$$T_2 = T_1(v_1 / v_2)^{k-1}$$

$$= 293(0.2 / 0.01)^{0.4} = 971.1\ \text{K}.$$

The work is then

$$W = 0.2378 \times 0.716(971.1 - 293) = 115.5\ \text{kJ}.$$

Example 5.11

A 10-cm-thick wall made of pine wood is 3 m high and 10 m long. Calculate the heat transfer rate if the temperature is 25°C on the inside and –20°C on the outside. Neglect convection.

Solution. The heat transfer occurs due to conduction. Using k from Table 5.5, Eq. 5.2.5 provides

$$\dot{Q} = KA\Delta T/L$$

$$= 0.15 \times (3 \times 10) \times [25 - (-20)]/0.1 = 2025\ \text{J/s}.$$

Example 5.12

The surface of the glass in a 1.2 m × 0.8 m skylight is maintained at 20°C. If the air temperature is –20°C, estimate the rate of heat loss from the window. Use $h = 12\ \text{J/s}\cdot\text{m}^2\cdot{}^\circ\text{C}$.

Solution. The convective heat transfer coefficient depends on several parameters so, as usual, it is specified. Using Equation 5.2.6 the rate of heat loss is

$$\dot{Q} = hA\Delta T$$

$$= 12 \times (1.2 \times 0.8) \times [20 - (-20)] = 461\ \text{J/s}.$$

Example 5.13

A 2-cm-diameter heating oven is maintained at 1000°C and the oven walls are at 500°C. If the emissivity of the element is 0.85, estimate the rate of heat loss from the 2-m-long element.

Solution. The heat loss will be primarily due to radiation. Neglecting any convection loss, using $F_{1-2} = 1$ since the oven encloses the element, and Eq. 5.2.7 provides us with

$$\dot{Q} = \sigma\varepsilon A(T_1^{\,4} - T_2^{\,4})$$

$$= 5.67 \times 10^{-11} \times 0.85 \times (\pi \times 0.02 \times 2)(1273^4 - 773^4) = 13.7 \text{ kJ/s}.$$

Note that absolute temperature must be used. Also, the area A is the surface area of the cylinder.

Example 5.14

Estimate the rate of heat loss from a $3 \text{ m} \times 10 \text{ m}$ wall if it is composed of 2.5 cm of pine wood, 8 cm of wool insulation, and 1.5 cm of plaster. The room is at 22°C. and the outside air is at –20°C. Also, calculate the overall heat transfer coefficient.

Solution. The resistance coefficient is L/k for a material layer and $1/h$ for an air layer. Using Equation 5.2.8, and values from Table 5.5, we have

$$\dot{Q} = A\Delta T/(R_i + R_1 + R_2 + R_3 + R_o)$$

$$= A\Delta T\Big/\left(\frac{1}{h_i} + \frac{L_1}{k_1} + \frac{L_2}{k_2} + \frac{L_3}{k_3} + \frac{1}{h_o}\right)$$

$$= \frac{(3 \times 10)[22 - (-20)]}{\frac{1}{12} + \frac{0.025}{0.15} + \frac{0.08}{0.038} + \frac{0.015}{0.81} + \frac{1}{12}} = 513 \text{ J/s}.$$

The overall heat transfer coefficient is found as follows:

$$\dot{Q} = UA\Delta T = U \times 30 \times 42 = 513.$$

$$\therefore U = 0.407 \text{ J/s} \cdot \text{m}^2 \cdot {}^\circ\text{C}.$$

5.3 The First Law of Thermodynamics for a Control Volume

The continuity equation, which accounts for the conservation of mass, may be used in certain situations involving control volumes. It is stated as

$$\dot{m} = \rho_1 A_1 V_1 = \rho_2 A_2 V_2 \tag{5.3.1}$$

where, in control volume formulations, V is the velocity; $\dot{m}$ is called the *mass flux*. In the above continuity equation, we assume *steady flow*, that is, the variables are independent of time. For such steady state flow situations, the first law takes the form:

$$\frac{\dot{Q} - \dot{W}_S}{\dot{m}} = \frac{V_2^2 - V_1^2}{2} + h_2 - h_1 + g(z_2 - z_1) \tag{5.3.2}$$

where the dot signifies a rate so that $\dot{Q}$ and $\dot{W}_S$ have the units of kJ/s. In most devices the potential energy is negligible. Also, the kinetic energy change can often be ignored (but if sufficient information is given, it should be included) so that the first law is most often used in the simplified form

$$\dot{Q} - \dot{W}_S = \dot{m}(h_2 - h_1) \qquad \text{or} \qquad q - w_S = h_2 - h_1. \tag{5.3.3}$$

Particular devices are of special interest. The energy equation for a *valve* or a *throttle plate* is simply

$$h_2 = h_1 \tag{5.3.4}$$

providing kinetic energy can be neglected.

For a *turbine* expanding a gas, the heat transfer is negligible so that

$$\dot{W}_T = \dot{m}(h_1 - h_2) \qquad \text{or} \qquad w_T = h_1 - h_2. \tag{5.3.5}$$

The work input to a *gas compressor* with negligible heat transfer is

$$\dot{W}_C = \dot{m}(h_2 - h_1) \qquad \text{or} \qquad w_C = h_2 - h_1. \tag{5.3.6}$$

A *boiler* and a *condenser* are simply heat transfer devises. The first law then simplifies to

$$\dot{Q} = \dot{m}(h_2 - h_1) \qquad \text{or} \qquad q = h_2 - h_1. \tag{5.3.7}$$

For a *nozzle* or a *diffuser* there is no work or heat transfer; we must include, however, the kinetic energy change resulting in

$$0 = \frac{V_2^2 - V_1^2}{2} + h_2 - h_1. \tag{5.3.8}$$

For a *pump* or a *hydroturbine* we take a slightly different approach. We return to Equation 5.3.2 and write it using $v = 1/\rho$, and with Equation 5.2.10, as

$$\frac{\dot{Q} - \dot{W}_S}{\dot{m}} = \frac{V_2^2 - V_1^2}{2} + u_2 - u_1 + \frac{P_2 - P_1}{\rho} + g(z_2 - z_1). \tag{5.3.9}$$

For an ideal situation we do not transfer heat. Neglecting kinetic and potential energy changes we find that

$$-\dot{W}_S = \dot{m}\frac{P_2 - P_1}{\rho}. \tag{5.3.10}$$

This would provide the minimum pump power requirement or the maximum turbine power output. The inclusion of an efficiency would increase the pump power requirement or decrease the turbine output.

A gas turbine or compressor efficiency is based on an isentropic process ($s_2 = s_1$) as the ideal process. For a gas turbine or a compressor we have

$$\eta_T = \frac{\dot{W}_a}{\dot{W}_s} \qquad \eta_C = \frac{\dot{W}_s}{\dot{W}_a} \tag{5.3.11}$$

where $\dot{W}_a$ is the actual power and $\dot{W}_s$ is the power assuming an isentropic process.

Example 5.15

Refrigerant-134a expands through a valve from a state of saturated liquid at 800 kPa to a pressure of 100 kPa. What is the final quality?

Solution. The first law states that

$$h_1 = h_2.$$

Using Fig. 5.6 (at the end of the chapter) for Refrigerant-134a we find, using $h_1 = h_f$,

$$h_1 = 240 \text{ kJ/kg}.$$

There follows at $P_2 = 100$ kPa, using $h_2 = 240$,

$$x = 0.35 \quad \text{or} \quad 35\%.$$

Example 5.16

Steam expands through a turbine from 6 MPa, at 600°C to 2 kPa with $x_2 = 1.0$. Find the work output.

Solution. The first law gives

$$\begin{aligned}\dot{w}_T &= h_1 - h_2 \\ &= 3658.4 - 2533.5 = 1125 \text{ kJ/kg},\end{aligned}$$

where h_2 is h_g at $P_2 = 0.002$ MPa, as given in Table 5.3.2.

Example 5.17

What is the turbine efficiency in Example 5.16?

Solution. The turbine efficiency is based on an isentropic process. Let state 2′ be at 2 kPa with

$$s_{2'} = s_1 = 7.1685 \text{ kJ/kg}\cdot\text{K}.$$

At 0.002 MPa we find, from Table 5.3.2

$$s_{2'} = s_f + x_{2'} s_{fg}.$$

$$7.1685 = 0.2606 + x_{2'}(8.4639).$$

$$\therefore x_{2'} = 0.816.$$

At this ideal state we find

$$h_{2'} = h_f + x_{2'} h_{fg}$$

$$= 73.5 + 0.816 \times 2460 = 2080 \text{ kJ / kg}.$$

Finally, we use $\dot{W}_a$ as the answer in Example 5.16 and obtain

$$\eta_T = \frac{\dot{W}_a}{\dot{W}_s} = \frac{\dot{W}_a}{\dot{m}(h_1 - h_{2'})}$$

$$= \frac{11\,250}{10(3658.4 - 2080)} = 0.713 \text{ or } 71.3\%.$$

Example 5.18

What is the minimum power requirement of a pump that is to increase the pressure from 2 kPa to 6 MPa for a mass flux of 10 kg/s of water?

Solution. With a liquid we let $h_2 - h_1 = u_2 - u_1 + (P_2 - P_1)v$ since $v = \text{const.}$ We let $u_2 - u_1 = 0$ so that the first law simplifies to

$$\dot{W}_P = \dot{m}\frac{P_2 - P_1}{\rho}$$

$$= 10\frac{6000 - 2}{1000} = 59.98 \text{ kW}.$$

Note how small this is relative to the power output of the turbine of example 5.16 operating between the same pressures. Because this is less than 1% of the turbine output, the pump work may usually be neglected in the analysis of a cycle involving a turbine and a pump.

Example 5.19

A nozzle accelerates air from 100 m/s, 400°C and 400 kPa to a receiver where $P = 20$ kPa. Assuming an isentropic process, find V_2.

Solution. The energy equation takes the form

$$0 = \frac{V_2^2 - V_1^2}{2} + h_2 - h_1.$$

Assuming air to be an ideal gas with constant c_p we have

$$c_p(T_1 - T_2) = \frac{V_2^2 - V_1^2}{2}.$$

We can find T_2 from Equation 5.2.31 to be

$$T_2 = T_1(P_2 / P_1)^{k-1/k}$$

$$= 673(20 / 400)^{0.4/1.4} = 286 \text{ K}.$$

The exiting velocity is found as follows:

$$1000(673 - 286) = \frac{V_2^{\,2} - 100^2}{2}.$$

$$\therefore \; V_2 = 874 \text{ m/s} .$$

Note: c_p must be used as $1000 \text{ J / kg} \cdot \text{K}$ so that the units are consistent.

5.4 The Second Law of Thermodynamics

The two scientific statements of the second law of thermodynamics can be shown to be equivalent. They are stated and shown schematically in Figure 5.2.

Clausius Statement—A device which operates in a cycle cannot transfer heat from a cooler body to a hotter body without a work input.

Kelvin-Planck Statement—A device which operates in a cycle cannot produce work while exchanging heat at a single temperature.

Figure 5.3 shows an engine (it produces work) and a refrigerator (it transfers heat from a body at low temperature) that satisfy the second law. The devices of Figure 5.2 do not violate the first law—energy is conserved—however, they represent impossibilities, violations of the second law.

To write a mathematical statement of the second law, we use entropy, defined by $dS = \delta Q/T$. The net entropy change during any process is given by

$$\Delta S_{universe} = \Delta S_{system} + \Delta S_{surroundings} \geq 0 \tag{5.4.1}$$

The equal sign applies to a reversible process, the greater-than sign applies to an irreversible (or real) process. Losses, friction, unrestrained expansion, and heat transfer across a finite temperature difference all lead to irreversibilities. Entropy changes can be found using tables, or for an ideal gas with constant specific heats we can use

$$\Delta s = c_p \ln \frac{T_2}{T_1} - R \ln \frac{P_2}{P_1} \tag{5.4.2}$$

For constant temperature processes, such as heat transfer to a reservoir, we use

$$\Delta S = \frac{Q}{T} \tag{5.4.3}$$

For a solid or a liquid we use

$$\Delta S = mc \ln \frac{T_2}{T_1} \tag{5.4.4}$$

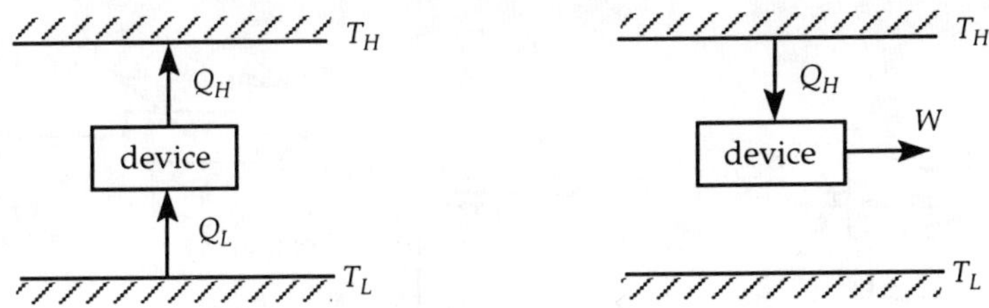

Figure 5.2 Violations of the second law.

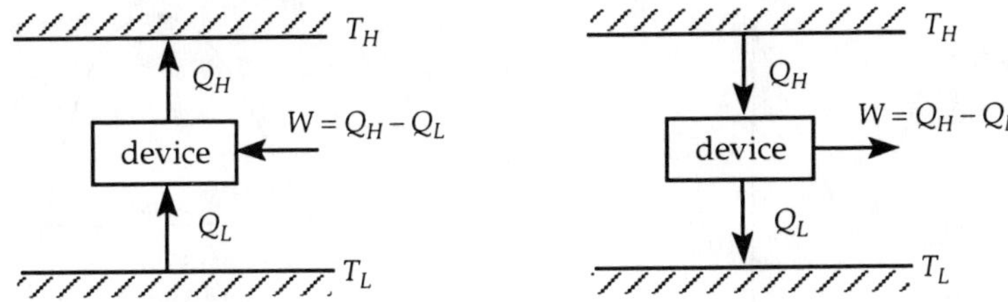

Figure 5.3 Devices that satisfy the second law.

where $c = c_p$ found in Table 5.2.

In general, for a reversible process, we have, from the definition of entropy,

$$Q = \int TdS \tag{5.4.5}$$

which is analogous to $W = \int PdV$ for such a process.

A Carnot engine or a refrigerator is a fictitious device that operates with reversible processes. It provides us with the maximum possible efficiency of an engine in terms of temperatures:

$$\eta = \frac{W_{out}}{Q_{in}}. \qquad \eta_{\max} = 1 - \frac{T_L}{T_H} \tag{5.4.6}$$

It provides the maximum possible *coefficient of performance*, COP, of a refrigerator,

$$\text{COP} = \frac{Q_L}{W_{in}}. \qquad \text{COP}_{\max} = \frac{1}{T_H/T_L - 1} \tag{5.4.7}$$

or, the upper limit for the COP of a heat pump,

$$\text{COP} = \frac{Q_H}{W_{in}}. \qquad \text{COP}_{\max} = \frac{1}{1 - T_L/T_H} \tag{5.4.8}$$

T_H is the temperature of the high temperature reservoir and T_L is the temperature of the low temperature reservoir.

The vapor power cycle is the basic cycle for most power plants; it is sketched in Figure 5.4a. The vapor refrigeration cycle is shown in Figure 5.4b.

The Carnot cycle is sketched in Figure 5.5 along with some other common gas cycles. The efficiency of each of the other cycles is less than that of Carnot cycle, primarily due to the transfer of heat across a finite temperature difference.

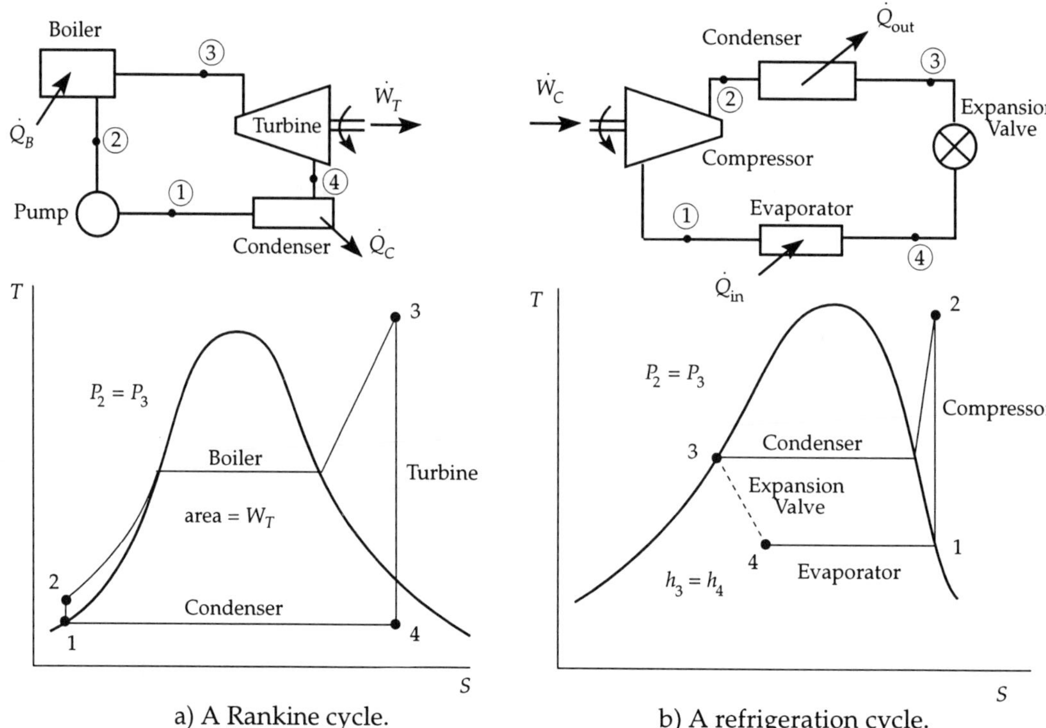

Figure 5.4 Vapor cycles.

Note that all of the above cycles are ideal cycles; the entropy is assumed constant in two processes in each cycle (except for the refrigeration cycle). Actual processes deviate from these ideal processes resulting in lower cycle efficiencies than predicted by the above.

To determine an expression for the maximum work output of a steady flow device, use the first law, neglecting kinetic and potential energy changes, in the form:

$$\dot{Q} - \dot{W}_s = \dot{m}(h_2 - h_1) \tag{5.4.9}$$

If we assume heat is transferred to the surroundings at atmospheric temperature T_a we can relate

$$\dot{m}(s_2 - s_1) = \dot{Q}/T_o. \tag{5.4.10}$$

Substituting into the Equation 5.4.9 we have

$$\begin{aligned} \dot{W}_{max} &= \dot{m}(h_1 - T_o s_1) - \dot{m}(h_2 - T_o s_2) \\ &= \dot{m}(\phi_1 - \phi_2) \end{aligned} \tag{5.4.11}$$

where ϕ is the *availability*. Hence, the maximum work output is the change in the availability.

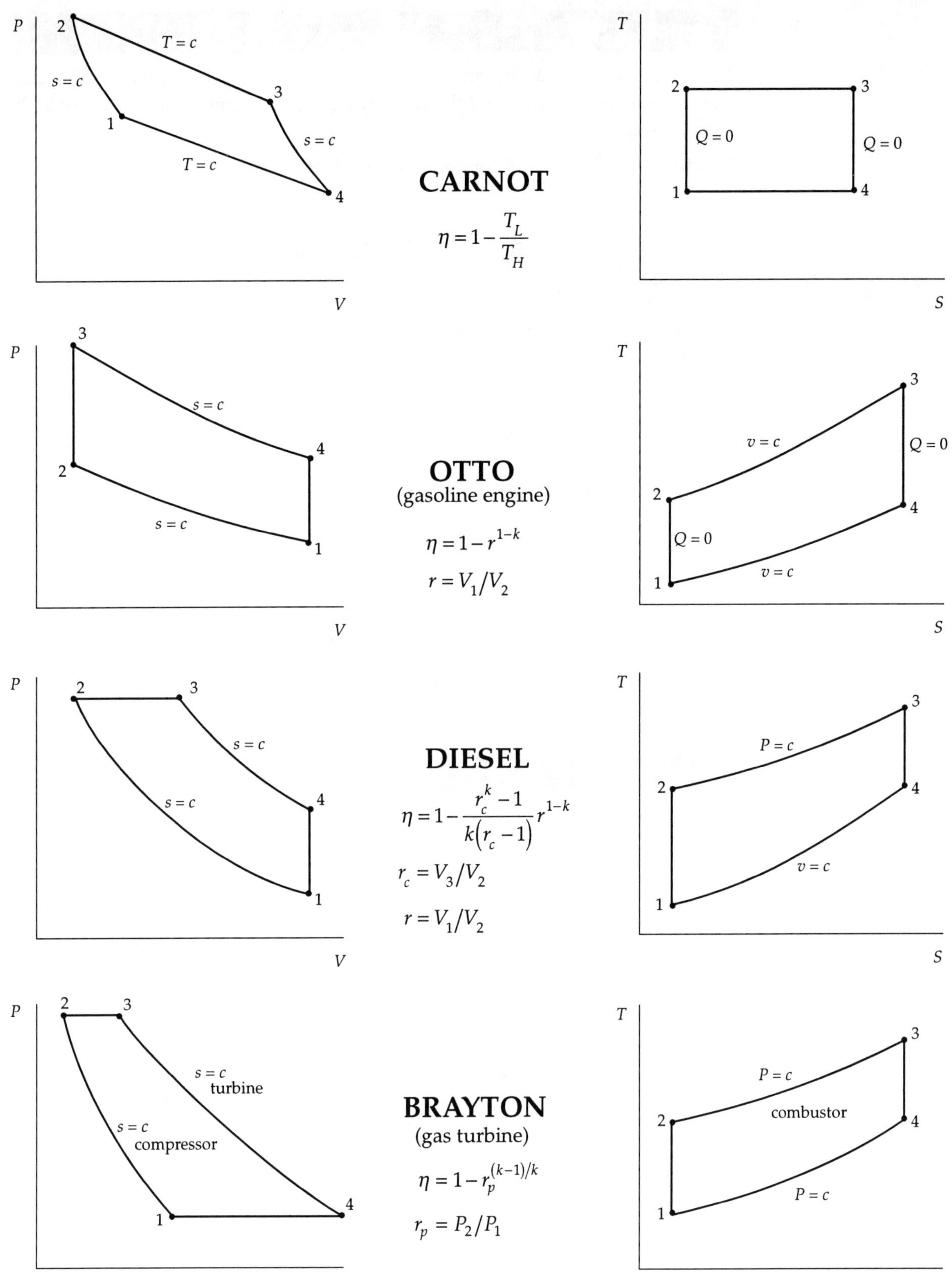

Figure 5.5 Common thermodynamic gas cycles.

Example 5.20

Ten kilograms of ice at 0°C are melted in 100 kg of water initially at 25°C. Calculate the final temperature and the entropy change. Assume no heat transfer to the surroundings.

Solution. The first law is applied to the ice-water system:

$$Q_{gain} = Q_{lost}$$

$$10\times 320 + 10(T-0)\times 4.18 = 100(25-T)\times 4.18.$$

$$\therefore T = 15.8^\circ\text{C}.$$

The entropy change is found as follows:

$$\text{ice}: \quad \Delta S = \frac{Q}{T_1} + mc\ln\frac{T_2}{T_1}$$

$$= \frac{10\times 320}{273} + 10\times 4.18\ \ln\frac{288.8}{273} = 14.073\ \text{kJ/K}.$$

$$\text{water}: \quad \Delta S = mc\ln\frac{T_2}{T_1}$$

$$= 100\times 4.18\ \ln\frac{288.8}{298} = -13.107\ \text{kJ/K}.$$

The net entropy change is

$$\Delta S_{net} = 14.073 - 13.107 = 0.966\ \text{kJ / K}.$$

This is positive, as required by the second law.

Example 5.21

An inventor claims to have invented an engine, using a 160°C geothermal heat source, which operates with an efficiency of 30%. If it exhausts to the 20°C atmosphere, is the invention possible?

Solution. The maximum possible efficiency, as limited by the second law, is given by

$$\eta_{\max} = 1 - \frac{T_L}{T_H}$$

$$= 1 - \frac{293}{433} = 0.323 \ \text{ or } \ 32.3\%.$$

The proposal is a possibility. However, the proposed efficiency is quite close to the maximum efficiency. It would be extremely difficult to obtain the 30% because of the losses due to heat transfer across a finite temperature difference and friction.

Example 5.22

A heat pump delivers 20 000 kJ /hr of heat with a 1.39 kW input. Calculate the COP.

Solution. Using the definition of the COP we find

$$\text{COP} = \frac{\dot{Q}_H}{\dot{W}_{in}}$$

$$= \frac{20\ 000 / 3600}{1.39} = 4.00.$$

Note, the factor 3600 converts hours into seconds.

Example 5.23

Compare the efficiency of an Otto cycle operating on an 8 to 1 compression ratio ($r = 8$) with a Diesel cycle that has a 20 to 1 compression ratio and a cut-off ratio of 2 to 1 ($r_c = 2$). Use air.

Solution. The efficiency of an Otto cycle (see Figure 5.5) is

$$\eta = 1 - r^{1-k}$$

$$= 1 - 8^{-0.4} = 0.565 \quad \text{or} \quad 56.5\%$$

where air is assumed to be the working fluid.

The efficiency of the Diesel cycle is

$$\eta = 1 - \frac{r_c^{\ k} - 1}{k(r_c - 1)} r^{1-k}$$

$$= 1 - \frac{2^{1.4} - 1}{1.4(2-1)} 20^{-0.4} = 0.647 \quad \text{or} \quad 64.7\%.$$

The efficiency of the Diesel cycle is higher than that of the Otto cycle because it operates at a higher compression ratio. If the Otto cycle could operate at $r = 20$, its efficiency would be greater than that of the Diesel.

Example 5.24

If a power plant operates on simple Rankine cycle using water between 600°C, 6 MPa and a low pressure of 10 kPa, calculate $\eta_{\max}$.

Solution. Referring to Figure 5.4a, we define the efficiency to be

$$\eta_{\max} = \frac{\dot{W}_T}{\dot{Q}_B}.$$

The pump work is neglected (see Example 5.18). To find $\dot{W}_T$ we must find h_4. This is accomplished using $P_3 = 6$ MPa and $T_3 = 600\,°\text{C}$, as follows:

$$s_4 = s_3 = 7.168 \text{ kJ/kg}\cdot\text{K}.$$

$$\text{At } P = 10 \text{ kPa}: \quad 7.168 = 0.6491 + x_4(7.5019).$$

$$\therefore\ x_4 = 0.869.$$

$$\therefore\ h_4 = 191.8 + 0.869 \times 2392.8 = 2271 \text{ kJ/kg}.$$

The turbine output is then (assuming $\dot{m} = 1$ kg / s since it is not given; alter-na-tively we could le t $w_T = h_3 - h_4$)

$$\dot{W}_T = \dot{m}(h_3 - h_4)$$

$$= 1 \times (3658.4 - 2271) = 1387 \text{ kW}.$$

The energy input occurs in the boiler. It is

$$\dot{Q}_B = \dot{m}(h_3 - h_2)$$

$$= 1 \times (3658.4 - 191.8) = 3467 \text{ kW}.$$

Note: Be careful finding h_2. We ignore the energy of the pump since it is always quite small; then $h_2 \cong h_1 = h_f$ at 10 kPa.

Finally, the efficiency of this idealized cycle is

$$\eta_{\max} = \frac{1387}{3467} = 0.400 \text{ or } 40.0\%.$$

Example 5.25

A refrigeration system, using Refrigerant-134a, operates between –20°C and 40°C. What is the maximum possible COP?

Solution. The refrigeration effect takes place in the evaporator. Hence, referring to Figure 5.4, the COP is defined as

$$\text{COP} = \frac{\dot{Q}_{in}}{\dot{W}_C}.$$

To find $\dot{W}_C$ we must locate state 2. This is done by following the constant en-tropy line $(s_1 = s_2 \cong 1.74)$ in Fig. 5.4b at the end of this chapter. Locating states 1 and 2 we find

$$h_1 \cong 385 \text{ kJ/kg}$$

$$h_2 \cong 425 \text{ kJ/kg}$$

The compressor work is, assuming $\dot{m} = 1$ kg / s,

$$\dot{W}_C = \dot{m}(h_2 - h_1)$$

$$= 1 \times (425 - 385) = 40 \text{ kW}.$$

To find $\dot{Q}_{in}$ we recognize that $h_4 = h_3 = 255$ kJ/kg, using Fig. 5.4b. Thus we find,

$$\dot{Q}_{in} = \dot{m}(h_1 - h_4)$$

$$= 1 \times (385 - 255) = 130 \text{ kW}.$$

The maximum COP for this idealized cycle is

$$\text{COP} = \frac{130}{40} = 3.25.$$

Example 5.26

The maximum pressure in an Otto cycle is 8 MPa. Air is compressed from 85 kPa and 22°C with a compression ratio of 8. What is the required heat addition?

Solution. Referring to the diagrams of the Otto cycle

$$P_1 = 85 \text{ kPa} \ , \quad T_1 = 22°\text{C}. \qquad \therefore v_1 = \frac{RT_1}{P_1} = \frac{0.287 \times 295}{85} = 0.9961 \ \text{m}^3/\text{kg}.$$

Since the compression ratio is a volume ratio we have

$$v_3 = v_2 = v_1/8 = 0.9961/8 = 0.1245 \ \text{m}^3/\text{kg} .$$

We know that $P_3 = 8$ MPa . Thus

$$T_3 = \frac{P_3 v_3}{R} = \frac{8000 \times 0.1245}{0.287} = 3471 \text{ K}$$

The isentropic process from 1 to 2 allows us to find T_2:

$$T_2 = T_1 \left(\frac{v_1}{v_2} \right)^{k-1} = 295 \times 8^{0.4} = 677.7 \text{ K}.$$

Finally, since the process from 2 to 3 is at constant volume with a system,

$$q_{2-3} = c_v (T_3 - T_2) = 0.717(3471 - 677.7) = 2003 \text{ kJ/kg}.$$

Example 5.27

Steam at 200°C and 200 kPa is available to produce work by expanding it to the atmosphere at 20°C and 100 kPa. What is $\dot{W}_{max}$ if $\dot{m} = 2$ kg / s?

Solution. We will use the equation

$$\dot{W}_{max} = \dot{m}(\phi_1 - \phi_2).$$

The availabilities are found to be (using Table 5.3)

$$\begin{aligned} \phi_1 &= h_1 - T_0 s_1 \\ &= 2870.5 - 293 \times 7.5074 = 670.8 \quad \text{kJ/kg}. \\ \phi_2 &= h_2 - T_0 s_2 \\ &= 83.9 - 293 \times 0.2965 = -3.0 \quad \text{kJ/kg}. \end{aligned}$$

Note: State 2 is at 20°C and 100 kPa. This is the liquid state so we use $h_2 = h_f$ and $s_2 = s_f$ at 20°C (see Table 5.3.1), and simply ignore the pressure. Finally,

$$\dot{W}_{max} = 2[670.8 - (-3.0)] = 1348 \ \text{kW}.$$

5.5 Psychrometrics

Air is primarily a mixture of nitrogen, oxygen, argon and water vapor. When water vapor isent, it is *dry air*. In most problems water vapor does not influence calculations, e.g., the problems considered earlier in this chapter. It must, however, be accounted for in problems involving combustion and air conditioning. Here we will present the primary quantities when considering *atmospheric air*, air that contains water vapor. We will make our calculations using the *psychrometric chart* , found in the NCEES Handbook.

The atmospheric pressure P is the sum of the partial pressure P_a of dry air and the partial pressure P_v of the water vapor:

$$P = P_a + P_v \,. \tag{5.5.1}$$

Relative humidity ϕ is

$$\phi = \frac{m_v}{m_g} = \frac{P_v}{P_g} \tag{5.5.2}$$

where P_g is the saturation pressure at temperature of the air (found in Table 5.3.1, and m_g is the maximum mass of water vapor that the air can hold at air temperature, usually called the *dry-bulb temperature* T_{db}. The *specific humidity* (or *humidity ratio*) ω is

$$\omega = \frac{m_v}{m_a} = 0.622\frac{P_v}{P_a} \tag{5.5.3}$$

The enthalpy of the atmospheric air is

$$h = h_a + \omega h_v \tag{5.5.4}$$

The *dew-point temperature* T_{dp} is the temperature at which condensation begins. The *wet-bulb temperature* T_{wb} is the temperature that a thermometer would read if a saturated wick were attached to its bulb and whirled around through the air. On the *T-s* diagram shown here, $T_1 = T_2 = T_{db}$, $T_3 = T_{dp}$, and T_{wb} would be between T_2 and T_3.

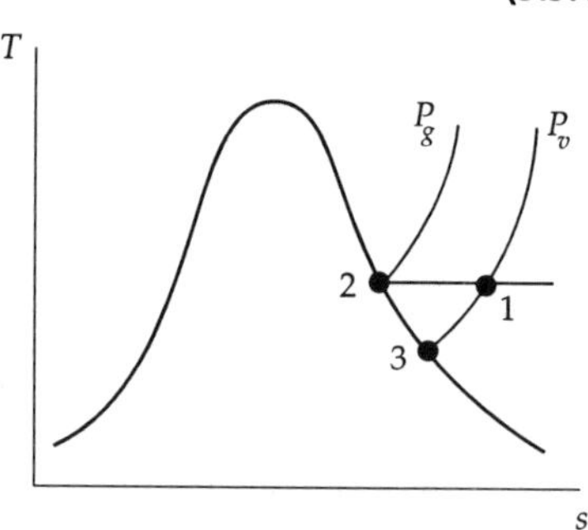

Figure 5.6 Water vapor.

The quantities T_{db}, T_{dp}, T_{wb}, ω, ϕ, and h can be read directly from the psychrometric chart so that the above equations are not needed in most air conditioning problems. Consider a state A, on the psychrometric chart shown, located by specifying the dry-bulb temperature T_{db} and the relative humidity ϕ. We would read T_{wb} at 1, T_{dp} at 2, h at 3, and ω at 4. (A detailed psychrometric chart is in the NCEES handbook.)

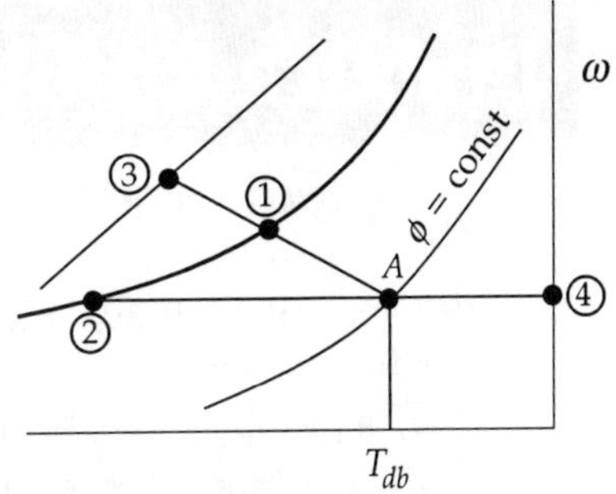

Figure 5.7 Psychrometric chart.

Example 5.28

A thermometer with a wet cloth attached to its bulb reads 20°C when air is blown around it. If the atmospheric air is 33°C, what is the relative humidity and dew-point temperature? How much water could be condensed out of a 100-m³ volume?

Solution. From the psychrometric chart we read

$$\phi = 30\% \quad \text{and} \quad T_{dp} = 13.2°\text{C}. \quad \text{Also,} \quad \omega = 0.0093 \text{ kg water/kg air}.$$

The amount of water that could be condensed out is

$$\omega \rho_{\text{air}} V_{\text{air}} = 0.0093 \times \frac{100}{0.287 \times 3.6} \times 100 = 1.06 \text{ kg} \quad \text{or} \quad 1.06 \text{ Liters}.$$

Example 5.29

It is desired to condition 35°C, 80% humidity air to 24°C and 50% humidity. If 100 m³/min of air is to be conditioned, how much energy is required in the cooling process, and how much in the heating process?

Solution. To decrease the moisture content in the air we first cool the air from 1 to 2 at constant ω and then remove water by cooling it further to 3 (along the 100% humidity line). In this process

$$\dot{Q}_{\text{cool}} = \dot{m}\Delta h = \frac{100}{60}\rho_a(h_1 - h_3) + h_f(\omega_1 - \omega_3) \quad \text{(neglect)}$$

$$= \frac{100}{60}\frac{1}{0.913}(105 - 37) = 124 \text{ kJ/s}.$$

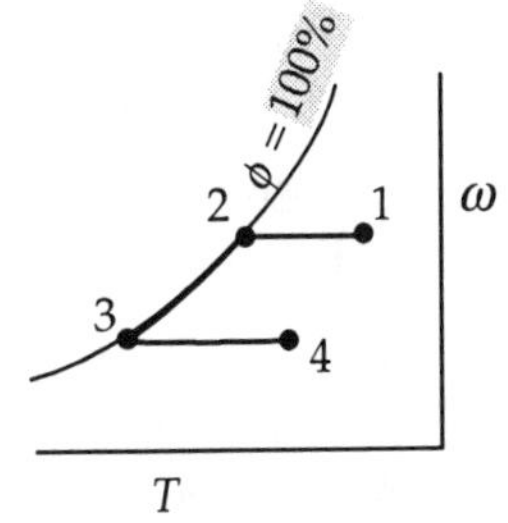

Heating takes place from 3 to 4 at constant ω:

$$\dot{Q}_{\text{heat}} = \dot{m}(h_4 - h_3)$$

$$= \frac{100}{60}\frac{1}{0.823}(48 - 37) = 18.3 \text{ kJ/s}.$$

We obtained $v = 1/\rho$ and the above properties from the psychrometric chart found in the NCEES Handbook.

Practice Problems

(If you choose to work only a few problems, select those with a star.)

GENERAL

*5.1 Which of the following would be considered a system rather than a control volume?

a) a pump b) a tire c) a pressure cooker
d) a turbine e) a jet nozzle

*5.2 Which of the following is an extensive property?

a) Temperature b) Velocity c) Pressure d) Mass e) Stress

5.3 An automobile heats up while sitting in a parking lot on a sunny day. The process can be assumed to be:

a) isothermal b) isobaric c) isometric d) isentropic e) adiabatic

*5.4 If a quasi-equilibrium process exists, we have assumed

a) the pressure at any instant to be everywhere constant.
b) an isothermal process.
c) the heat transfer to be small.
d) the boundary motion to be infinitesimally small.
e) that no friction exists.

*5.5 A scientific law is a statement that

a) we postulate to be true.
b) is generally observed to be true.
c) is derived from a mathematical theorem.
d) is agreed upon by scientific community.
e) is a summary of experimental observation.

5.6 Which of the following is not an extensive property?

a) Momentum b) Kinetic energy c) Enthalpy
d) Density e) Mass

5.7 Which of the following is not an acceptable SI unit?

a) Distance measured in centimeters.
b) Temperature measured in kelvins.
c) Pressure measured in newtons per square meter.
d) Volume measured in cubic centimeters.
e) Density measured in grams per cubic centimeter.

5.8 A long metal rod can be used as a very stiff spring. The deformation of a rod is given by $\delta = PL/AE$, as defined in Mechanics of Materials. What is the "spring constant" of the metal rod?

a) AE/L b) $1/AE$ c) L/AE d) AE e) $AE\delta/L$

DENSITY, PRESSURE, AND TEMPERATURE

*5.9 The density of air at vacuum of 40 kPa and –40°C is, in kg/m^3,

a) 0.598 b) 0.638 c) 0.697 d) 0.753 e) 0.897

5.10 The specific volume of water is 0.5 m^3/kg at a pressure of 200 kPa. Find the quality.

a) 0.623 b) 0.564 c) 0.478 d) 0.423 e) 0.356

5.11 The specific volume of steam at 4 MPa and 1200°C, in m^3/kg, is

a) 0.20 b) 0.19 c) 0.18 d) 0.17 e) 0.16

5.12 Water with quality $x = 50\%$ is contained in a rigid vessel. If more heat is added

a) x increases.
b) x decreases.
c) x remains relatively constant.
d) x may increase or decrease.
e) superheating occurs.

5.13 There are 20 kg of steam contained in 2 m^3 at 4 MPa. What is the temperature in °C?

a) 600 b) 610 c) 620 d) 630 e) 640

5.14 The pressure in a cylinder containing water is 200 kPa and the temperature is 115°C. What state is it in?

a) compressed liquid
b) saturated liquid
c) liquid/vapor mixture
d) saturated vapor
e) superheated vapor

*5.15 A cold tire has a volume of 0.03 m^3 at –10°C and 180 kPa gage. If the pressure and temperature increase to 210 kPa gage and 30°C, find the final volume in m^3.

a) 0.0304 b) 0.0308 c) 0.0312 d) 0.0316 e) 0.0320

THE FIRST LAW FOR A SYSTEM

*5.16 A 300-watt light bulb provides energy in a 10-m-dia spherical space. If the outside temperature is 20°C, find the inside steady-state temperature, in °C, if $R = 1.5\ \text{hr}\cdot\text{m}^2\cdot°\text{C}/\text{kJ}$ for the wall.

a) 40.5 b) 35.5 c) 32.6 d) 29.4 e) 25.2

5.17 A 10 kg mass, which is attached to a pulley and a paddle wheel submerged in water, drops 3 m. Find the subsequent heat transfer, in joules, needed to return the temperature of the water to its original value.

a) –30 b) –95 c) –126 d) –195 e) –294

*5.18 A cycle undergoes the following processes. All units are kJ. Find E_{after} for the process $1 \to 2$.

	Q	W	ΔE	E_{before}	E_{after}
$1 \to 2$	20	5		10	
$2 \to 3$		–5	5		
$3 \to 1$	30			30	

a) 10 b) 15 c) 20 d) 25 e) 30

5.19 For the cycle of Problem 5.18 find W_{3-1}.

a) 50 b) 60 c) 70 d) 80 e) 90

5.20 A 2000 kg automobile traveling at 25 m/s strikes a plunger in 10 000 cm^3 of water, bringing the auto to a stop. What is the maximum temperature rise, in °C, in the water?

a) 5 b) 10 c) 15 d) 20 e) 25

5.21 Ten kilograms of –10°C ice is added to 100 kg of 20°C water. What is the eventual temperature, in °C, of the water? Assume an insulated container.

a) 9.2 b) 10.8 c) 11.4 d) 12.6 e) 13.9

*5.22 Of the following first law statements, choose the one that is wrong:

a) the net heat transfer equals the net work for a cycle.

b) the heat transfer cannot exceed the work done.

c) the heat transfer equals the work plus the energy change.

d) the heat transfer equals the energy change if no work is done.

e) the energy of an isolated system remains constant.

5.23 Select the incorrect statement. ($Q - W = \Delta U$)

a) Work and heat transfer represent energy crossing a boundary.

b) The differentials of work and heat transfer are exact.

c) Work and heat transfer are path integrals.

d) Net work and net heat transfer are equal for a cycle.

e) On a molecular level, heat transfer results from work being done.

ISOTHERMAL PROCESS

*5.24 Determine the work, in kJ, necessary to compress 2 kg of air from 100 kPa to 4000 kPa if the temperature is held constant 300°C.

a) –1210 b) –1105 c) –932 d) –812 e) –733

*5.25 Steam is compressed from 100 kPa to 4000 kPa holding the temperature constant at 300°C. What is the internal energy change, in kJ, for m = 2 kg?

a) –180 b) –170 c) –160 d) –150 e) –140

5.26 How much heat transfer, in kJ, is needed to convert 2 kg of saturated liquid water to saturated vapor if the temperature is held constant at 200°C?

a) 2380 b) 2980 c) 3210 d) 3520 e) 3880

CONSTANT PRESSURE PROCESS

5.27 There are 200 people in a 2000 m^2 room, lighted with 30 W/m^2. Estimate the maximum temperature increase, in °C, if the ventilation system fails for 20 min. Each person generates 400 kJ/h. The room is 3 m high.

a) 5.6 b) 6.8 c) 8.6 d) 11.4 e) 13.4

*5.28 Estimate the average c_p value, in kJ/kg·K, of a gas if 522 kJ of heat are necessary to raise the temperature from 300 K to 800 K holding the pressure constant.

a) 1.000 b) 1.026 c) 1.038 d) 1.044 e) 1.052

*5.29 How much heat, in kJ, must be transferred to 10 kg of air to increase the temperature from 10°C to 230°C if the pressure is maintained constant?

a) 1780 b) 1620 c) 1890 d) 2090 e) 2200

5.30 Ten kilograms of water, initially at 10°C, are heated until $T = 300$°C. Estimate the heat transfer, in MJ, if the pressure is held constant at 200 kPa.

a) 29.2 b) 29.7 c) 30.3 d) 30.9 e) 31.3

5.31 Calculate the work done, in MJ, in Problem 5.30.

a) 2.63 b) 4.72 c) 8.96 d) 11.4 e) 15.6

CONSTANT VOLUME PROCESS

5.32 How much heat, in kJ, must be added to a rigid volume, containing 2 kg of water/vapor mixture with $x = 0.5$, to increase the temperature from 200°C to 500°C?

a) 2730 b) 2620 c) 2510 d) 2390 e) 2250

*5.33 A tire is pressurized to 100 kPa gauge in Michigan where $T = 0$°C. In Arizona the tire is at 70°C. Assuming a rigid tire, estimate the pressure in kPa gauge.

a) 120 b) 130 c) 140 d) 150 e) 160

5.34 A sealed, rigid, 10 m^3 air tank is heated by the sun from 20°C to 80°C. How much energy, in kJ, is transferred to the tank? Assume $P_1 = 100$ kPa.

a) 720 b) 680 c) 620 d) 560 e) 510

5.35 Ten kilograms of water are heated in a rigid container from 10°C to 200°C. What is the final quality if $Q = 9000$ kJ?

a) 0.258 b) 0.162 c) 0.093 d) 0.078 e) 0.052

ISENTROPIC PROCESS

*5.36 Air expands from an insulated cylinder from 200°C and 400 kPa to 20 kPa. Find T_2 in °C.

a) –24 b) –28 c) –51 d) –72 e) –93

*5.37 During an isentropic expansion of air, the volume triples. If the initial temperature is 200°C, find T_2 in °C.

a) 32 b) 28 c) 16 d) 8 e) –12

5.38 Superheated steam expands isentropically from 600°C and 6 MPa to 10 kPa. Find the final quality.

a) 0.79 b) 0.83 c) 0.87 d) 0.91 e) 0.95

5.39 Superheated steam expands isentropically from 600°C and 6 MPa to 400 kPa. Find T_2 in °C.

a) 220 b) 200 c) 190 d) 160 e) 140

5.40 Find the work, in kJ/kg, needed to compress air isentropically from 20°C and 100 kPa to 6 MPa.

a) –523 b) –466 c) –423 d) –392 e) –376

*5.41 During an isentropic process, which one of the following is true?

a) The temperature increases as the pressure decreases.
b) The temperature increases as the volume increases.
c) The heat transfer equals the enthalpy change.
d) The heat transfer is zero.
e) The volume decreases as the pressure decreases.

POLYTROPIC PROCESS

*5.42 Find T_2, in °C, if the pressure triples and $T_1 = 10$°C. Let $n = 1.2$.

a) 179 b) 113 c) 93 d) 79 e) 67

THE FIRST LAW FOR A CONTROL VOLUME

5.43 Steam enters a turbine in a 20-cm-dia pipe at 600°C and 6 MPa. It exits from a 5-cm-dia pipe at 20 kPa with $x = 1$. What is the velocity ratio V_{out}/V_{in}?

a) 1880 b) 1640 c) 1210 d) 820 e) 420

*5.44 Water enters a boiler at 60°C and 4 MPa. How much energy, in kJ/kg, must be added to obtain 600°C at the exit if the pressure remains constant?

a) 2340 b) 2630 c) 2970 d) 3280 e) 3420

5.45 A condenser is cooled by heating water from 20°C to 30°C. If the condenser inlets 10 kg/s of saturated water vapor at 20 kPa and exits saturated liquid, what is the mass flux, in kg/s, of the cooling water?

a) 640 b) 560 c) 500 d) 410 e) 350

5.46 Steam at 400°C and 4 MPa expands isentropically through a turbine to 10 kPa. Estimate the maximum work output in kJ/kg.

a) 1030 b) 1050 c) 1070 d) 1090 e) 1110

*5.47 What is the energy requirement, in kW, for a pump that is 75% efficient if it increases the pressure of 10 kg/s of water from 10 kPa to 6 MPa?

a) 40 b) 50 c) 60 d) 70 e) 80

5.48 A river 60 m wide and 2 m deep flows at 2 m/s. A hydro plant develops a pressure of 300 kPa gage just before the turbine. What maximum power, in MW, is possible?

a) 40 b) 48 c) 56 d) 64 e) 72

*5.49 A nozzle expands air isentropically from 400°C and 2 MPa to the atmosphere at 80 kPa. If the inlet velocity is small, what exit velocity, in m/s, can be expected?

a) 500 b) 600 c) 700 d) 800 e) 900

5.50 If the efficiency of a turbine that expands steam at 400°C and 6 MPa to 20 kPa is 87%, find the work output in kJ/kg.

a) 723 b) 891 c) 933 d) 996 e) 1123

THE SECOND LAW

*5.51 An inventor proposes to take 10 kg/s of geothermal water at 120°C and generate 4180 kW of energy by exhausting very near to the ambient temperature of 20°C. This proposal should not be supported because

a) it violates the first law.
b) it violates the second law.
c) it would be too expensive.
d) friction must be accounted for.
e) the geothermal water is not hot enough.

*5.52 The net entropy change in the universe during any real process

a) is equal to zero.
b) is positive.
c) is negative.
d) must be calculated to determine its sense.
e) is positive if $T_{system} > T_{surroundings}$.

*5.53 A Carnot engine

a) provides a fictitious model which is of little use.
b) can be experimentally modeled.
c) supplies us with the lower limit for engine efficiency.
d) operates between two constant temperature reservoirs.
e) has two reversible and two irreversible processes.

5.54 Ninety kilograms of ice at 0°C are completely melted. Find the entropy change, in kJ/K, if $T_2 = 0$°C.

a) 0 b) 45 c) 85 d) 105 e) 145

5.55 Forty kilograms of ice at 0°C are mixed with 100 kg of water at 20°C. What is the entropy change, in kJ/K?

a) 2.36 b) 2.15 c) 1.96 d) 1.53 e) 1.04

*5.56 An inventor proposes to have developed a small power plant that operates at 70% efficiency. It operates between temperature extremes of 600°C and 50°C. Your analysis shows that the maximum possible efficiency is

a) 56 b) 63 c) 67 d) 72 e) 81

POWER CYCLES

*5.57 A Carnot cycle operates on air such that $P_4 = 160$ kPa and $v_4 = 0.5\ \text{m}^3/\text{kg}$ (see Fig. 5.5). If 30 kJ/kg of heat are added at $T_H = 200$°C, find the work produced in kJ/kg.

a) 20.3 b) 18.2 c) 16.8 d) 14.2 e) 12.3

5.58 An Otto cycle operates with volumes of 40 cm^3 and 400 cm^3 at top dead center (TDC) and bottom dead center (BDC), respectively. If the power output is 100 kW, what is the heat input in kJ/s? Assume $k = 1.4$.

a) 166 b) 145 c) 110 d) 93 e) 60

5.59 The volumes of states 1, 2, and 3 in Fig. 5.5 of the Diesel cycle are 450 cm^3, 25 cm^3 and 45 cm^3, respectively. If the power produced is 120 kW, what is the required heat input in kJ/s? Assume that $k = 1.4$.

a) 187 b) 172 c) 157 d) 146 e) 132

5.60 A simple Rankine cycle operates between superheated steam at 600°C and 6 MPa entering the turbine, and 10 kPa entering the pump. What is the maximum possible efficiency, as a percent?

a) 30 b) 35 c) 40 d) 45 e) 50

5.61 The water vapor that expands from 600°C and 6 MPa in a turbine of a Rankine cycle is intercepted at 200 kPa and reheated at constant pressure to 600°C in the boiler, after which it is re-injected in the turbine and expanded to 10 kPa. (This is the *reheat cycle*.) Calculate the maximum possible efficiency, as a percent?

a) 32 b) 37 c) 41 d) 45 e) 49

5.62 The primary effect of reheating, as illustrated in Problem 5.61 (compare with the result in Problem 5.60), is to

a) increase the efficiency.
b) decrease the pump work.
c) decrease the heat requirement.
d) eliminate unnecessary piping.
e) decrease or eliminate moisture condensation in the turbine.

REFRIGERATION CYCLES

5.63 A refrigeration cycle operates with Refrigerant-134a between 100 kPa and 1000 kPa. What is the maximum possible COP?

a) 1.5 b) 2 c) 2.5 d) 3 e) 3.5

5.64 The quality of the refrigerant immediately after the expansion valve of the cycle of Problem 5.63 is

a) 0.58 b) 0.54 c) 0.50 d) 0.46 e) 0.42

AVAILABILITY

5.65 Determine the maximum possible power output, in kW, if 10 kg/s of air is expanded in a turbine from 100°C and 6 MPa to the surroundings at 20°C and 80 kPa.

a) 3720 b) 2610 c) 2030 d) 1890 e) 1620

5.66 The change in thermodynamic availability is equal to

a) the quantity $T\Delta S$.
b) the change in enthalpy.
c) the actual work output.
d) the maximum possible work output.
e) the reversible work output minus actual work output.

HEAT TRANSFER (See Article 5.2)

*5.67 Heat transfer decreases with increased

a) conductivity b) *R*-factor c) emissivity d) *U*-factor e) molecular activity

*5.68 Heat transfer due to convection can be determined if we know the temperature difference, fluid properties, the geometry, and the

a) Reynolds No. b) Nusselt No. c) Prandtl No. d) Froude No. e) Weber No.

*5.69 Frozen soil at –15°C and water at 10°C are separated by a 50 cm-thick concrete wall. Estimate the rate of heat transfer, in J/s, if the wall is 2 m × 8 m.

a) 180 b) 140 c) 100 d) 50 e) 20

5.70 The rate of heat loss, due to 150 °C air blowing over a 4-cm-dia cylinder, is measured to be 1500 kJ over a 10 minute time period. If the cylinder is 1.0-m-long and is maintained at 40°C, estimate the convection heat transfer coefficient in $\text{J/s}\cdot\text{m}^2\cdot{}^\circ\text{C}$.

a) 180 b) 140 c) 100 d) 50 e) 20

5.71 A filament with a surface area of 0.8 cm^2 is positioned in a spherical space whose surface is at 80°C. Estimate the temperature of the filament, in °C, if 100 W of power is dissipated. Assume a black body filament.

a) 3000 b) 2600 c) 2300 d) 1900 e) 1600

5.72 Calculate the overall *R*-factor, in $\text{m}^2\cdot{}^\circ\text{C}\cdot\text{s/J}$, for a wall composed of 2.5 cm of polystyrene, 8 cm of wool insulation, and 2 cm of pine wood. Assume $h = 10\ \text{J/s}\cdot\text{m}^2\cdot{}^\circ\text{C}$.

a) 3.2 b) 3.0 c) 2.8 d) 2.6 e) 2.4

PSYCHROMETRICS

5.73 How much liquid water, in liters, is contained in a $3\text{m} \times 10\text{m} \times 20\text{m}$ room if $T = 25°\text{C}$ and the humidity is 60%?

a) 5.1 b) 6.6 c) 7.2 d) 7.8 e) 8.2

5.74 A "swamp" cooler operates by blowing relatively dry air past a network of wet wicks. What maximum exiting temperature can be realized if the air is at 35°C with a humidity of 20%?

a) 28°C b) 25°C c) 22°C d) 19°C e) 16°C

Solutions to Practice Problems

5.1 b) A tire. All other devices have fluid entering and/or leaving.

5.2 d) Mass. All other quantities do not depend on the mass.

5.3 b) Isobaric. The pressure will remain constant due to the inlets and outlets for air.

5.4 a) All properties are uniform throughout the volume.

5.5 e) A scientific law results from experimental observations.

5.6 d) Density does not depend on the mass, i.e., when the mass is increased the density does not increase simultaneously.

5.7 e) Distance, area, or volume may be measured in cm, cm^2, or cm^3, respectively, but grams are not acceptable in SI and cm cannot be used in combination with other units.

5.8 a) $F = Kx. \qquad P = \frac{AE}{L}\delta. \qquad \therefore K = \frac{AE}{L}.$

5.9 e) $\rho = 1/v = P/RT = (-40+100)/.287 \times (-40+273) = 0.897 \text{ kg/m}^3.$
$R = 8.314/28.97 = .287.$

5.10 b) $v = v_f + x(v_g - v_f)$ $\qquad P = 0.2$ MPa

$0.5 = .001 + x(.8857 - .001)$ $\qquad$ TABLE 5.3.2

$\therefore x = 0.564$

5.11 d) If the state is beyond Table 5.3.3 use the equation of state with

$R = \frac{8.314}{18} = 0.462.$

$$v = \frac{RT}{P} = \frac{0.462 \times 1473}{4000} = 0.17 \text{ m}^3/\text{kg}.$$

Use $T = 1200 + 273 = 1473$ K, $P = 4000$ kPa

5.12 a) In a rigid vessel, the volume remains constant. If we move straight up on the T-v diagram, x increases. (Temperature increases as heat is added.) Note: it is important to know that the diagram is skewed quite markedly to the left.

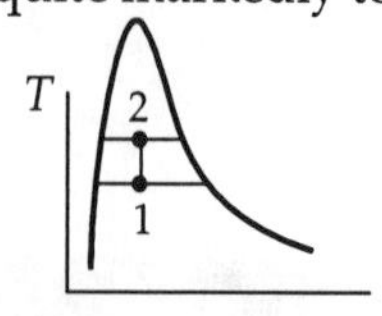

5.13 b) $v = \frac{V}{m} = \frac{2}{20} = 0.1, \quad P = 4 \text{ MPa}.$

From Table 5.3.3 interpolation gives

$$T = \frac{0.1 - 0.09885}{0.1109 - 0.09885} \times 100 + 600 = 610° \text{ C}.$$

5.14 a) Using Table 5.3.2 we see that 115°C < 120.2°C. ∴ compressed.

5.15 c) $m = \frac{P_1 V_1}{RT_1} = \frac{280 \times .03}{.287 \times 263} = 0.1113 \text{ kg}, \text{ where } R = \frac{8.3.4}{28.97} = 0.287.$

$$\therefore V_2 = \frac{mRT_2}{P_2} = \frac{.1113 \times .287 \times 303}{310} = 0.0312 \text{ m}^3$$

5.16 e) $\dot{Q} = \frac{1}{R} A \Delta T \qquad A = 4\pi r^2$

$$\frac{300}{1000} = \frac{1}{1.5} 4\pi \times 5^2 (T - 20) \frac{1}{3600}. \quad \therefore T = 25.2° \text{C}.$$

The 1000 converts J/s to kJ/s.

The 3600 converts hr to seconds.

5.17 e) $W = F \times d = (10 \times 9.8) \times 3 = 294 \text{ N} \cdot \text{m input}. \quad \therefore W = -294 \text{ N} \cdot \text{m}.$

$\therefore Q = W = -294 \text{ J}.$ (Heat output is negative.)

5.18 d) $Q - W = E_a - E_b.$

$20 - 5 = E_a - 10. \quad \therefore E_a = 25 \text{ kJ}.$

5.19 a) For $2 \to 3 \quad Q - (-5) = 5. \quad \therefore Q_{2-3} = 0$

$\therefore Q_{net} = 20 + 0 + 30 = 50$

$Q_{net} = W_{net} = 5 - 5 + W_{3-1} = 50.$

$\therefore W_{3-1} = 50 \text{ kJ}.$

5.20 c) $Q = mc\Delta T.$ Here $Q = \frac{1}{2} mV^2, \quad m = ¥\rho$

$$\frac{1}{2} \times 2000 \times 25^2 = 10\,000 \times 10^{-6} \times 1000 \times 4180 \; \Delta T$$

$\therefore \Delta T = 14.95° \text{C}$

5.21 b) $(mc\Delta T)_{ice} + (mc\Delta T)_{\substack{melted \\ ice}} + mh_{fusion} = (mc\Delta T)_{water}$

$$10 \times 2.1 \times 10 + 10 \times 4.18(T - 0) + 10 \times 320 = 100 \times 4.18(20 - T). \quad \therefore T = 10.77° \text{C}.$$

5.22 b) The heat transfer exceeds the work done if $\Delta E > 0$.

5.23 b) δW and δQ are inexact differentials. That's why we don't write dW and dQ.

5.24 a) $W = mRT \ln P_1/P_2$

$$= 2 \times .287 \times 573 \ln 100/4000 = -1213 \text{ kJ}$$

5.25 b) $\Delta U = m(u_2 - u_1)$ Table 7.3.3

$$= 2(2725.3 - 2810.4) = -170.2 \text{ kJ}$$

5.26 e) $Q = m\Delta h = mh_{fg} = 2 \times 1940.8 = 3881.6 \text{ kJ}$

5.27 e) Assume P = const since doors would be opened. Cracks abound.

$$Q = mC_p\Delta T .$$

$$\frac{400 \times 200}{3600} + \frac{30 \times 2000}{1000} = \frac{6000 \times 1.23 \times 1.00\Delta T}{20 \times 60} . \qquad \therefore \Delta T = 13.37° \text{ C}$$

5.28 d) $Q = m\Delta h = mc_p\Delta T$

$$522 = 1 \times c_p \times (800 - 300). \quad \therefore c_p = 1.044 \frac{\text{kJ}}{\text{kg} \cdot \text{K}}$$

5.29 e) $Q = m\Delta h = mc_p\Delta T = 10 \times 1.00 \times (230 - 10) = 2200 \text{ kJ}$

5.30 c) $Q = m\Delta h = 10[3072 - 42] = 30\,300 \text{ kJ}$

5.31 a) $W = P\Delta V = mP\Delta v = 10(1.316 - .001) \times 200 = 2630 \text{ kJ}$

5.32 a) $Q = m\Delta u$ $\qquad u_1 = 851 + .5(2595 - 851) = 1723 \text{ kJ/kg}.$

$T_2 = 500°\text{C}$

$v_2 = .0642$ $\qquad v_1 = .001 + .5(.1274 - .001) = .0642 \text{ m}^3/\text{kg}.$

Use Table 5.3.3

$$u_2 = \frac{.0642 - .0566}{.0686 - .0566}(3091 - 3082) + 3082 = 3088$$

$$\therefore Q = 2(3088 - 1723) = 2730 \text{ kJ}$$

5.33 d) $\dfrac{P_1}{T_1} = \dfrac{P_2}{T_2}$. $\quad (p_2 + 100)273 = (100 + 100)343$

$$\therefore P_2 = 151 \text{ kPa gage}$$

5.34 e) $Q = m\Delta u = mc_v\Delta T$ $\qquad m = \dfrac{PV}{RT} = \dfrac{100 \times 10}{.287 \times 293} = 11.89 \text{ kg}$

$$= 11.89 \times .716 \times 60 = 511 \text{ kJ}$$

5.35 e) $Q = m\Delta u.\ \ 9000 = 10(u_2 - 42)$ $\qquad \therefore u_2 = 942 \text{ kJ/kg}$

$942 = 851 + x_2(2595 - 851)$ $\qquad \therefore x_2 = .0522$

5.36 d) $T_2 = T_1\left(\frac{P_2}{P_1}\right)^{k-1/k} = 473\left(\frac{20}{400}\right)^{0.4/1.4} = 201\text{K} \text{ or } -72°\text{C}$

5.37 a) $T_2 = T_1\left(\frac{v_1}{v_2}\right)^{k-1} = 473\left(\frac{1}{3}\right)^{0.4} = 305\text{K} \text{ or } 31.8°\text{C}$

5.38 c) $s_2 = s_1 = 7.1685.$ from Table 7.3.2, $7.1685 = .649 + 7.502x_2$. $\therefore$ $x_2 = .869$

5.39 b) $s_2 = s_1 = 7.1685$ from Table 7.3.3. This is slightly less than $s = 7.171$ at $T = 200°\text{C}$ and $P = 0.4\text{MPa}$. $\therefore T_2 = 199°\text{C}$.

5.40 b) $T_2 = T_1\left(\frac{P_2}{P_1}\right)^{k-1/k} = 293\left(\frac{6000}{100}\right)^{.4/1.4} = 944 \text{ K}.$

$W = -m\Delta u = -mc_v\Delta T$ (let $m = 1$)

$= -.716(944 - 293) = -466 \text{ kJ/kg}$.

5.41 d) The heat transfer must be zero.

5.42 e) $T_2 = T_1\left(\frac{P_1}{P_2}\right)^{n-1/n} = 283 \times 3^{.2/1.2} = 340 \text{ K}$ or $66.9°\text{C}$

5.43 a) $\rho_1 = \frac{1}{v_1} = \frac{1}{.06525} = 15.33$ Table 7.3.3

$\rho_2 = \frac{1}{v_2} = \frac{1}{7.649} = 0.1307$ Table 7.3.2

$\rho_1 A_1 V_1 = \rho_2 A_2 V_2$. $15.33\pi \times \frac{.2^2}{4} V_1 = .1307\pi \times \frac{.05^2}{4} V_2$. $\therefore$ $\frac{V_2}{V_1} = 1877$

5.44 e) $Q = h_2 - h_1$. $h_1 = 251$ from Table 7.3.1.

$= 3674 - 251 = 3423 \text{ kJ/kg}$

5.45 b) $\dot{Q} = \dot{m}_w c\Delta T = \dot{m}_s \Delta h$. Table 7.3.1, $h_{fg} = 2358 \text{ kJ/kg}$.

$\dot{m} \times 4.18(30 - 20) = 10 \times 2358$. $\therefore \dot{m} = 564 \text{ kg/s}$

5.46 c) $s_2 = s_1 = 6.7698 = .649 + 7.502x_2$. $\therefore x_2 = .816$.

$\therefore h_2 = 192 + .816 \times 2393 = 2144$. $W = h_1 - h_2 = 3213.5 - 2144 = 1069 \text{ kJ/kg}$

5.47 e) $\dot{W}_P = \dot{m}\frac{\Delta P}{\rho\eta} = 10\frac{6000 - 10}{1000 \times .75} = 79.9 \text{ kW}$

5.48 e) $\dot{W}_T = \dot{m}\frac{\Delta P}{\rho} = 60 \times 2 \times 2 \times 1000\frac{300}{1000} = 72\ 000 \text{ kW}$

5.49 e) $T_2 = T_1\left(\frac{P_2}{P_1}\right)^{k-1/k} = 673\left(\frac{80}{2000}\right)^{.4/1.4} = 268 \text{ K}.$

$V_2^2/2 = h_1 - h_2 = c_p(T_1 - T_2) = 1000(673 - 268)$. $\therefore V_2 = 900 \text{ m/s}$.

5.50 b) $h_1 = 3177 \quad s_{2'} = s_1 = 6.5415 = .832 + x_{2'}7.0774. \quad \therefore x_{2'} = .807.$
$\therefore h_{2'} = 251 + .807 \times 2358 = 2153. \quad w = 0.87 \times (3177 - 2153) = 891 \text{ kJ/kg}.$

5.51 b) $\eta_{\max} = 1 - \frac{T_L}{T_H} = 1 - \frac{293}{393} = 0.254$

$\eta = \frac{4180}{10 \times 4.18(120-20)} = 1.00. \quad \therefore \text{impossible}$

5.52 b) $\Delta s_{\text{net}} > 0$ for all processes.

5.53 d) Operates between const. temp. reservoirs.

5.54 d) $\Delta S > Q/T = 90 \times 320/273 = 105.5 \text{ kJ/K}$

5.55 e) $40(320 + 4.18T_2) = 100 \times 4.18(20 - T_2). \quad \therefore T_2 = -7.6°\text{C}.$
Impossible. $\therefore T_2 = 0°\text{C}. \quad 320\ m = 100 \times 4.8 \times 20. \quad \therefore m = 26.1 \text{ kg}.$
$\Delta s = 26.1 \times 320/273 + 100 \times 4.18 \ln 273/293 = 1.04$

5.56 b) $\eta_{\max} = 1 - \frac{T_L}{T_H} = 1 - 323/873 = 0.63$

5.57 e) $T_4 = P_4 v_4 / R = 160 \times .5/.287 = 279 \text{ K}.$
$w = q\eta = 30(1 - 279/473) = 12.3 \text{ kJ/kg}$

5.58 a) $\eta = 1 - (400/40)^{-.4} = 0.602. \quad \therefore \dot{Q} = \dot{W}/\eta = 100/.602 = 166 \text{ kW}.$

5.59 a) $r_C = 45/25 = 1.8. \qquad r = 450/25 = 18.$

$\dot{Q} = \dot{W}/\eta = 120 \Big/ \left[1 - \frac{1.8^{1.4} - 1}{1.4(1.8-1)} 18^{-.4}\right] = 187 \text{ kJ/s}.$

5.60 c) $s_4 = s_3 = 7.1685 = .649 + 7.502x_4. \quad \therefore x_4 = .869.$
$h_4 = 192 + .869 \times 2393 = 2272. \quad h_3 = 3658, \quad h_1 = 192.$

$\eta = \frac{w}{q} = (3658 - 2272)/(3658 - 192) = 0.40$

5.61 c) $s_4 = s_3 = 7.168. \quad \therefore$ ④ superheat $\therefore h_4 = 2722. \quad h_3 = 3658.$
After reheat $h_5 = 3704. \quad s_6 = s_5 = 8.778.$

$\therefore h_6 = \frac{8.778 - 8.689}{8.9046 - 8.689}(2879.5 - 2783) + 2783 = 2823.$

$\eta = \frac{w_{3-4} + w_{5-6}}{q_{2-3} + q_{4-5}} = \frac{3658 - 2722 + 3704 - 2823}{3658 - 192 + 3704 - 2722} = 0.408$

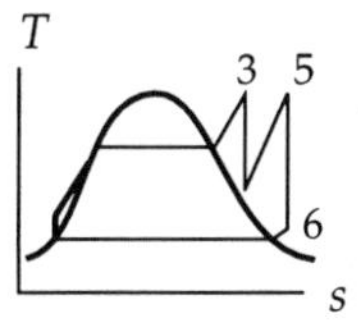

5.62 e) Decrease moisture content in the turbine.

5.63 b) $s_2 = s_1 = 1.75$. Refer to Fig. 5.4a: $h_1 = 380, \ h_2 = 430, \ h_3 = 255 = h_4.$

$COP = \frac{q_{4-1}}{w_{1-2}} = \frac{380 - 255}{430 - 380} = 2.5.$

5.64 e) $h_4 = h_3 = 255 = h_f + x_4(h_g - h_f)$

$$= 165 + x_4(380 - 165). \quad \therefore x_4 = 0.42 .$$

5.65 a) $\dot{W}_{max} = \dot{m}(h_1 - T_o s_1) - \dot{m}(h_2 - T_o s_2)$

$$= \dot{m}\left[(h_1 - h_2) + T_o(s_2 - s_1)\right] = \dot{m}\left[c_p \Delta T + T_o\left(c_p \ln\frac{T_2}{T_1} - R\ln\frac{P_2}{P_1}\right)\right]$$

$$= 10\left[1.00(100 - 20) + 293\left(1.00\ln\frac{293}{373} - .287\ln\frac{80}{6000}\right)\right] = 3723 \text{ kW}$$

5.66 d) $\phi_2 - \phi_1 = w_{max}$.

5.67 b)

5.68 b)

5.69 c) $\dot{Q} = .13 \times 2 \times 8 \times \frac{25}{.5} = 104 \text{ J/s}$

5.70 a) $\frac{1\,500\,000}{10 \times 60} = h \times \pi \times .04 \times (150 - 40). \quad \therefore h = 181 \text{ J/s} \cdot \text{m}^2 \cdot °\text{C}$

5.71 d) $100 = 5.67 \times 10^{-8} \times (.8 \times 10^{-4})(T^4 - 353^4)$.

$\therefore T = 2167$ K or 1894° C

5.72 b) $R = \frac{1}{h} + \frac{L_1}{k_1} + \frac{L_2}{k_2} + \frac{L_3}{k_3} + \frac{1}{h}$

$$= \frac{1}{10} + \frac{.025}{.04} + \frac{.08}{.038} + \frac{.02}{.15} + \frac{1}{10} = 3.06 \ \frac{\text{s} \cdot \text{m}^2 \cdot °\text{C}}{\text{J}}$$

5.73 e) $\omega = 0.0118. \quad \therefore \ m_v = \omega m_a = \omega V/v$

$$= .0118 \times 600/.861$$

$$= 8.2 \text{ kg} \quad \text{or } 8.2 \text{ L}$$

5.74 d) The enthalpy remains constant since $\dot{W} = \dot{Q} = 0$. States 1 and 2 lie on a constant enthalpy line on the psychrometric chart with $\phi_2 = 100\%$. $\therefore T_2 = 19°\text{C}$.

Tables for Thermodynamics

TABLE 5.1 Properties of Ideal Gases — Metric Units

Gas	Chemical Formula	Molar Mass	$R\,\frac{kJ}{kg\cdot K}$	$c_p\,\frac{kJ}{kg\cdot K}$	$c_v\,\frac{kJ}{kg\cdot K}$	k
Air	—	28.97	0.287 00	1.0035	0.7165	1.400
Argon	Ar	39.948	0.208 13	0.5203	0.3122	1.667
Butane	C_4H_{10}	58.124	0.143 04	1.7164	1.5734	1.091
Carbon Dioxide	CO_2	44.01	0.188 92	0.8418	0.6529	1.289
Carbon Monoxide	CO	28.01	0.296 83	1.0413	0.7445	1.400
Ethane	C_2H_6	30.07	0.276 50	1.7662	1.4897	1.186
Ethylene	C_2H_4	28.054	0.296 37	1.5482	1.2518	1.237
Helium	He	4.003	2.077 03	5.1926	3.1156	1.667
Hydrogen	H_2	2.016	4.124 18	14.2091	10.0849	1.409
Methane	CH_4	16.04	0.518 35	2.2537	1.7354	1.299
Neon	Ne	20.183	0.411 95	1.0299	0.6179	1.667
Nitrogen	N_2	28.013	0.296 80	1.0416	0.7448	1.400
Octane	C_8H_{18}	114.23	0.072 79	1.7113	1.6385	1.044
Oxygen	O_2	31.999	0.259 83	0.9216	0.6618	1.393
Propane	C_3H_8	44.097	0.188 55	1.6794	1.4909	1.126
Steam	H_2O	18.015	0.461 52	1.8723	1.4108	1.327

TABLE 5.2 Specific Heats of Liquids and Solids — Metric Units

c_p kJ/(kg·°C)

A. LIQUIDS

Substance	State	c_p	Substance	State	c_p
Water	1 atm, 25°C	4.177	Glycerin	1 atm, 10°C	2.32
Ammonia	sat., —20°C	4.52	Bismuth	1 atm, 425°C	0.144
	sat., 50°C	5.10	Mercury	1 atm, 10°C	0.138
Refrigerant 12	sat., —20°C	0.908	Sodium	1 atm, 95°C	1.38
	sat., 50°C	1.02	Propane	1 atm, 0°C	2.41
Benzene	1 atm, 15°C	1.80	Ethyl Alcohol	1 atm, 25°C	2.43

B. SOLIDS

Substance	T, °C	c_p	Substance	T, °C	c_p
Ice	–11	2.033	Lead	–100	0.118
	–2.2	2.10		0	0.124
Aluminum	–100	0.699		100	0.134
	0	0.870	Copper	–100	0.328
	100	0.941		0	0.381
Iron	20	0.448		100	0.393
Silver	20	0.233			

TABLE 5.3 Thermodynamic Properties of Water (Steam Tables) — Metric Units

5.3.1 Saturated H_2O — Temperature Table

		Volume, m³/kg		Energy, kJ/kg		Enthalpy, kJ/kg			Entropy, kJ/(kg·K)		
T, °C	*P*, MPa	v_f	v_g	u_f	u_g	h_f	h_{fg}	h_g	s_f	s_{fg}	s_g
0.010	0.0006113	0.001000	206.1	0.0	2375.3	0.0	2501.3	2501.3	0.0000	9.1571	9.1571
5	0.0008721	0.001000	147.1	21.0	2382.2	21.0	2489.5	2510.5	0.0761	8.9505	9.0266
10	0.001228	0.001000	106.4	42.0	2389.2	42.0	2477.7	2519.7	0.1510	8.7506	8.9016
20	0.002338	0.001002	57.79	83.9	2402.9	83.9	2454.2	2538.1	0.2965	8.3715	8.6680
30	0.004246	0.001004	32.90	125.8	2416.6	125.8	2430.4	2556.2	0.4367	8.0174	8.4541
40	0.007383	0.001008	19.52	167.5	2430.1	167.5	2406.8	2574.3	0.5723	7.6855	8.2578
50	0.01235	0.001012	12.03	209.3	2443.5	209.3	2382.8	2592.1	0.7036	7.3735	8.0771
60	0.01994	0.001017	7.671	251.1	2456.6	251.1	2358.5	2609.6	0.8310	7.0794	7.9104
70	0.03119	0.001023	5.042	292.9	2469.5	293.0	2333.8	2626.8	0.9549	6.8012	7.7561
80	0.04739	0.001029	3.407	334.8	2482.2	334.9	2308.8	2643.7	1.0754	6.5376	7.6130
90	0.07013	0.001036	2.361	376.8	2494.5	376.9	2283.2	2660.1	1.1927	6.2872	7.4799
100(atm)	0.1013	0.001044	1.673	418.9	2506.5	419.0	2257.0	2676.0	1.3071	6.0486	7.3557
120	0.1985	0.001060	0.8919	503.5	2529.2	503.7	2202.6	2706.3	1.5280	5.6024	7.1304
140	0.3613	0.001080	0.5089	588.7	2550.0	589.1	2144.8	2733.9	1.7395	5.1912	6.9307
160	0.6178	0.001102	0.3071	674.9	2568.4	675.5	2082.6	2758.1	1.9431	4.8079	6.7510
180	1.002	0.001127	0.1941	762.1	2583.7	763.2	2015.0	2778.2	2.1400	4.4466	6.5866
200	1.554	0.001156	0.1274	850.6	2595.3	852.4	1940.8	2793.2	2.3313	4.1018	6.4331
220	2.318	0.001190	0.08620	940.9	2602.4	943.6	1858.5	2802.1	2.5183	3.7686	6.2869
240	3.344	0.001229	0.5977	1033.2	2604.0	1037.3	1766.5	2803.8	2.7021	3.4425	6.1446
260	4.688	0.001276	0.04221	1128.4	2599.0	1134.4	1662.5	2796.9	2.8844	3.1184	6.0028
280	6.411	0.001332	0.03017	1227.4	2586.1	1236.0	1543.6	2779.6	3.0674	2.7905	5.8579
300	8.580	0.001404	0.02168	1332.0	2563.0	1344.0	1405.0	2749.0	3.2540	2.4513	5.7053
320	11.27	0.001499	0.01549	1444.6	2525.5	1461.4	1238.7	2700.1	3.4487	2.0883	5.5370
340	14.59	0.001638	0.01080	1570.3	2464.6	1594.2	1027.9	2622.1	3.6601	1.6765	5.3366
360	18.65	0.001892	0.006947	1725.2	2351.6	1760.5	720.7	2481.2	3.9154	1.1382	5.0536
374.136	22.088	0.003155	0.003155	2029.6	2029.6	2099.3	0.0	2099.3	4.4305	0.0000	4.4305

5.3.2 Saturated H_2O — Pressure Table

		Volume, m³/kg		Energy, kJ/kg		Enthalpy, kJ/kg			Entropy, kJ/(kg·K)		
P, MPa	*T*, °C	v_f	v_g	u_f	u_g	h_f	h_{fg}	h_g	s_f	s_{fg}	s_g
0.001	7.0	0.001000	129.2	29.3	2385.0	29.3	2484.9	2514.2	0.1059	8.8706	8.9765
0.002	17.5	0.001001	67.00	73.5	2399.5	73.5	2460.0	2533.5	0.2606	8.4639	8.7245
0.01	45.8	0.001010	14.67	191.8	2437.9	191.8	2392.8	2584.6	0.6491	7.5019	8.1510
0.02	60.1	0.001017	7.649	251.4	2456.7	251.4	2358.3	2609.7	0.8319	7.0774	7.9093
0.04	75.9	0.001026	3.993	317.5	2477.0	317.6	2319 1	2636.7	1.0260	6.6449	7.6709
0.06	85.9	0.001033	2.732	359.8	2489.6	359.8	2293.7	2653.5	1.1455	6.3873	7.5328
0.08	93.5	0.001039	2.087	391.6	2498.8	391.6	2274.1	2665.7	1.2331	6.2023	7.4354
0.1	99.6	0.001043	1.694	417.3	2506.1	417.4	2258.1	2675.5	1.3029	6.0573	7.3602
0.12	104.8	0.001047	1.428	439.2	2512.1	439.3	2244.2	2683.5	1.3611	5.9378	7.2989
0.16	113.3	0.001054	1.091	475.2	2521.8	475.3	2221.2	2696.5	1.4553	5.7472	7.2025
0.2	120.2	0.001061	0.8857	504.5	2529.5	504.7	2201.9	2706.6	1.5305	5.5975	7.1280
0.4	143.6	0.001084	0.4625	604.3	2553.6	604.7	2133.8	2738.5	1.7770	5.1197	6.8967
0.6	158.9	0.001101	0.3157	669.9	2567.4	670.6	2086.2	2756.8	1.9316	4.8293	6.7609
0.8	170.4	0.001115	0.2404	720.2	2576.8	721.1	2048.0	2769.1	2.0466	4.6170	6.6636
1	179.9	0.001127	0.1944	761.7	2583.6	762.8	2015.3	2778.1	2.1391	4.4482	6.5873
1.2	188.0	0.001139	0.1633	797.3	2588.8	798.6	1986.2	2784.8	2.2170	4.3072	6.5242
1.6	201.4	0.001159	0.1238	856.9	2596.0	858.8	1935.2	2794.0	2.3446	4.0780	6.4226
2	212.4	0.001177	0.09963	906.4	2600.3	908.8	1890.7	2799.5	2.4478	3.8939	6.3417
4	250.4	0.001252	0.04978	1082.3	2602.3	1087.3	1714.1	2801.4	2.7970	3.2739	6.0709
6	275.6	0.001319	0.03244	1205.4	2589.7	1213.3	1571.0	2784.3	3.0273	2.8627	5.8900
8	295.1	0.001384	0.02352	1305.6	2569.8	1316.6	1441.4	2758.0	3.2075	2.5365	5.7440
12	324.8	0.001527	0.01426	1472.9	2513.7	1491.3	1193.6	2684.9	3.4970	1.9963	5.4933
16	347.4	0.001711	0.009307	1622.7	2431.8	1650.0	930.7	2580.7	3.7468	1.4996	5.2464
20	365.8	0.002036	0.005836	1785.6	2293.2	1826.3	583.7	2410.0	4.0146	0.9135	4.9281
22.088	374.136	0.003155	0.003155	2029.6	2029.6	2099.3	0.0	2099.3	4.4305	0.0000	4.4305

5.3.3 Superheated Steam

°C	v	u	h	s	v	u	h	s	v	u	h	s
	P = .010 MPa				P = .050 MPa				P = .10 MPa			
100	17.196	2515.5	2687.5	8.4479	3.418	2511.6	2682.5	7.6947	1.6958	2506.7	2676.2	7.3614
200	21.825	2661.3	2879.5	8.9038	4.356	2659.9	2877.7	8.1580	2.172	2658.1	2875.3	7.8343
300	26.445	2812.1	3076.5	9.2813	5.284	2811.3	3075.5	8.5373	2.639	2810.4	3074.3	8.2158
400	31.063	2968.9	3279.6	9.6077	6.209	2968.5	3278.9	8.8642	3.103	2967.9	3278.2	8.5435
500	35.679	3132.3	3489.1	9.8978	7.134	3132.0	3488.7	9.1546	3.565	3131.6	3488.1	8.8342
600	40.295	3302.5	3705.4	10.1608	8.057	3302.2	3705.1	9.4178	4.028	3301.9	3704.7	9.0976
700	44.911	3479.6	3928.7	10.4028	8.981	3479.4	3928.5	9.6599	4.490	3479.2	3928.2	9.3398
800	49.526	3663.8	4159.0	10.6281	9.904	3663.6	4158.9	9.8852	4.952	3663.5	4158.6	9.5652
900	54.141	3855.0	4396.4	10.8396	10.828	3854.9	4396.3	10.0967	5.414	3854.8	4396.1	9.7767
1000	58.757	4053.0	4640.6	11.0393	11.751	4052.9	4640.5	10.2964	5.875	4052.8	4640.3	9.9764
	P = .20 MPa				P = .40 MPa				P = .60 MPa			
200	1.080	2654.4	2870.5	7.5066	.5342	2646.8	2860.5	7.1706	.3520	2638.9	2850.1	6.9665
300	1.316	2808.6	3071.8	7.8926	.6548	2804.8	3066.8	7.5662	.4344	2801.0	3061.6	7.3724
400	1.549	2966.7	3276.6	8.2218	.7726	2964.4	3273.4	7.8985	.5137	2962.1	3270.3	7.7079
500	1.781	3130.8	3487.1	8.5133	.8893	3129.2	3484.9	8.1913	.5920	3127.6	3482.8	8.0021
600	2.013	3301.4	3704.0	8.7770	1.0055	3300.2	3702.4	8.4558	.6697	3299.1	3700.9	8.2674
700	2.244	3478.8	3927.6	9.0194	1.1215	3477.9	3926.5	8.6987	.7472	3477.0	3925.3	8.5107
800	2.475	3663.1	4158.2	9.2449	1.2372	3662.4	4157.3	8.9244	.8245	3661.8	4156.5	8.7367
900	2.706	3854.5	4395.8	9.4566	1.3529	3853.9	4395.1	9.1362	.9017	3853.4	4394.4	8.9486
1000	2.937	4052.5	4640.0	9.6563	1.4685	4052.0	4639.4	9.3360	.9788	4051.5	4638.8	9.1485
1100	3.168	4257.0	4890.7	9.8458	1.5840	4256.5	4890.2	9.5256	1.0559	4256.1	4889.6	9.3381
	P = .80 MPa				P = 1.00 MPa				P = 2.00 MPa			
200	.2608	2630.6	2839.3	6.8158	.2060	2621.9	2827.9	6.6940				
300	.3241	2797.2	3056.5	7.2328	.2579	2793.2	3051.2	7.1229	.1255	2772.6	3023.5	6.7664
400	.3843	2959.7	3267.1	7.5716	.3066	2957.3	3263.9	7.4651	.1512	2945.2	3247.6	7.1271
500	.4433	3126.0	3480.6	7.8673	.3541	3124.4	3478.5	7.7622	.1757	3116.2	3467.6	7.4317
600	.5018	3297.9	3699.4	8.1333	.4011	3296.8	3697.9	8.0290	.1996	3290.9	3690.1	7.7024
700	.5601	3476.2	3924.2	8.3770	.4478	3475.3	3923.1	8.2731	.2232	3470.9	3917.4	7.9487
800	.6181	3661.1	4155.6	8.6033	.4943	3660.4	4154.7	8.4996	.2467	3657.0	4150.3	8.1765
900	.6761	3852.8	4393.7	8.8153	.5407	3852.2	4392.9	8.7118	.2700	3849.3	4389.4	8.3895
1000	.7340	4051.0	4638.2	9.0153	.5871	4050.5	4637.6	8.9119	.2933	4048.0	4634.6	8.5901
1100	.7919	4255.6	4889.1	9.2050	.6335	4255.1	4888.6	9.1017	.3166	4252.7	4885.9	8.7800
	P = 3.0 MPa				P = 4.0 MPa				P = 5.0 MPa			
300	.08114	2750.1	2993.5	6.5390	.05884	2725.3	2960.7	6.3615	.04532	2698.0	2924.5	6.2084
400	.09936	2932.8	3230.9	6.9212	.07341	2919.9	3213.6	6.7690	.05781	2906.6	3195.7	6.6459
500	.11619	3108.0	3456.5	7.2338	.08643	3099.5	3445.3	7.0901	.06857	3091.0	3433.8	6.9759
600	.13243	3285.0	3682.3	7.5085	.09885	3279.1	3674.4	7.3688	.07869	3273.0	3666.5	7.2589
700	.14838	3466.5	3911.7	7.7571	.11095	3462.1	3905.9	7.6198	.08849	3457.6	3900.1	7.5122
800	.16414	3653.5	4145.9	7.9862	.12287	3650.0	4141.5	7.8502	.09811	3646.6	4137.1	7.7440
900	.17980	3846.5	4385.9	8.1999	.13469	3843.6	4382.3	8.0647	.10762	3840.7	4378.8	7.9593
1000	.19541	4045.4	4631.6	8.4009	.14645	4042.9	4628.7	8.2662	.11707	4040.4	4625.7	8.1612
1100	.21098	4250.3	4883.3	8.5912	.15817	4248.0	4880.6	8.4567	.12648	4245.6	4878.0	8.3520
1200	.22652	4460.9	5140.5	8.7720	.16987	4458.6	5138.1	8.6376	.13587	4456.3	5135.7	8.5331
	P = 6.0 MPa				P = 8.0 MPa				P = 10.0 MPa			
300	.03616	2667.2	2884.2	6.0674	.02426	2590.9	2785.0	5.7906				
400	.04739	2892.9	3177.2	6.5408	.03432	2863.8	3138.3	6.3634	.02641	2832.4	3096.5	6.2120
500	.05665	3082.2	3422.2	6.8803	.04175	3064.3	3398.3	6.7240	.03279	3045.8	3373.7	6.5966
600	.06525	3266.9	3658.4	7.1677	.04845	3254.4	3642.0	7.0206	.03837	3241.7	3625.3	6.9029
700	.07352	3453.1	3894.2	7.4234	.05481	3443.9	3882.4	7.2812	.04358	3434.7	3870.5	7.1687
800	.08160	3643.1	4132.7	7.6566	.06097	3636.0	4123.8	7.5173	.04859	3628.9	4114.8	7.4077
900	.08958	3837.8	4375.3	7.8727	.06702	3832.1	4368.3	7.7351	.05349	3826.3	4361.2	7.6272
1000	.09749	4037.8	4622.7	8.0751	.07301	4032.8	4616.9	7.9384	.05832	4027.8	4611.0	7.8315
1100	.10536	4243.3	4875.4	8.2661	.07896	4238.6	4870.3	8.1300	.06312	4234.0	4865.1	8.0237
1200	.11321	4454.0	5133.3	8.4474	.08489	4449.5	5128.5	8.3115	.06789	4444.9	5123.8	8.2055
	P = 15.0 MPa				P = 20.0 MPa				P = 40.0 MPa			
400	.01564	2740.7	2975.5	5.8811	.00994	2619.3	2818.1	5.5540	.001908	1854.6	1930.9	4.1135
500	.02080	2996.6	3308.6	6.3443	.01477	2942.9	3238.2	6.1401	.005622	2678.4	2903.3	5.4700
600	.02491	3208.6	3582.3	6.6776	.01818	3174.0	3537.6	6.5048	.008094	3022.6	3346.4	6.0114
700	.02861	3410.9	3840.1	6.9572	.02113	3386.4	3809.0	6.7993	.009941	3283.6	3681.2	6.3750
800	.03210	3610.9	4092.4	7.2040	.()2385	3592.7	4069.7	7.0544	.011523	3517.8	3978.7	6.6662
900	.03546	3811.9	4343.8	7.4279	.02645	3797.5	4326.4	7.2830	.012962	3739.4	4257.9	6.9150
1000	.03875	4015.4	4596.6	7.6348	.02897	4003.1	4582.5	7.4925	.041324	3954.6	4527.6	7.1356
1100	.04200	4222.6	4852.6	7.8283	.03145	4211.3	4840.2	7.6874	.015642	4167.4	4793.1	7.3364
1200	.04523	4433.8	5112.3	8.0108	.03391	4422.8	5101.0	7.8707	.016940	4380.1	5057.7	7.5224

Figure 5.6 *P-h* Diagram for Refrigerant 134a — Metric Units

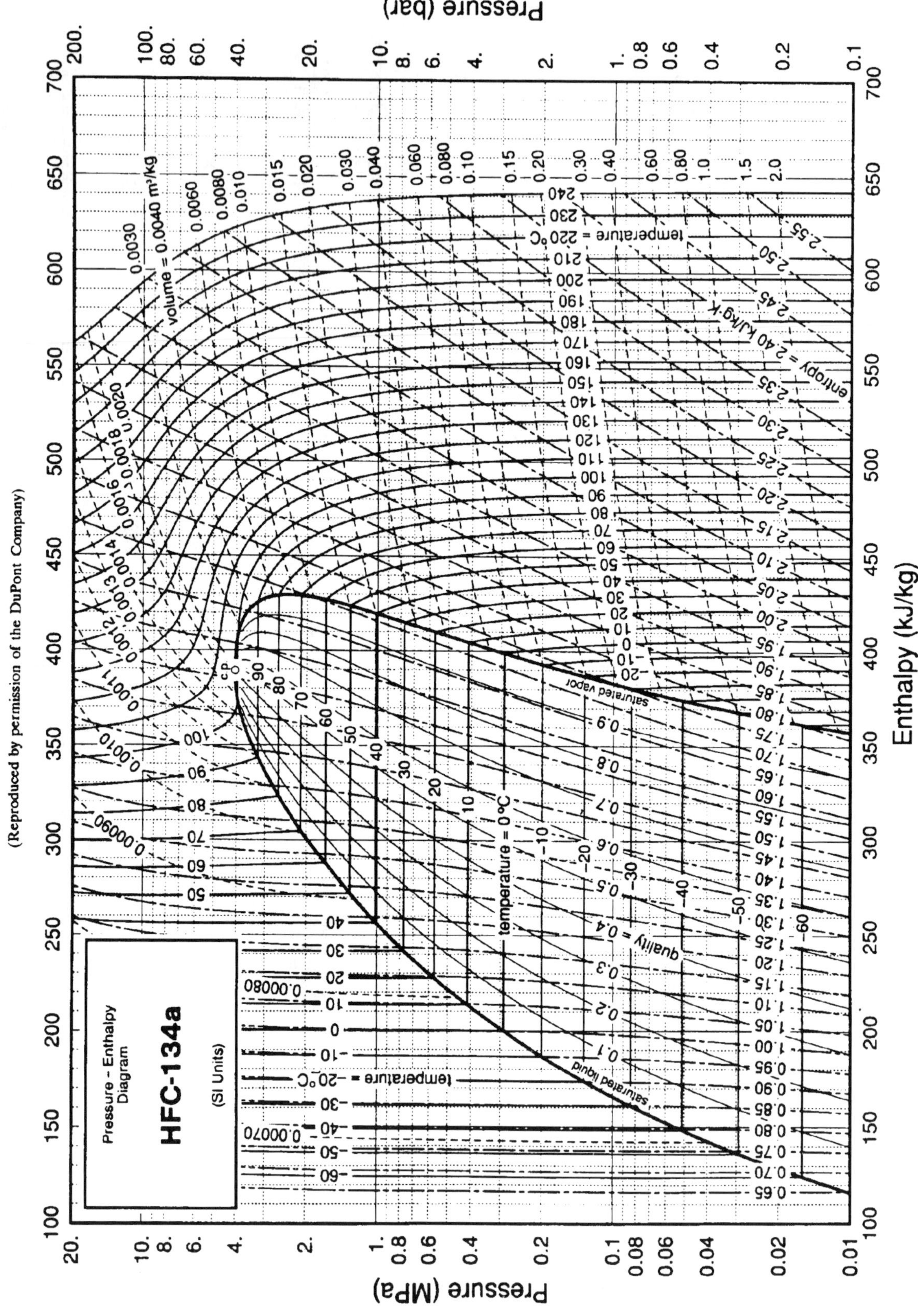

(Reproduced by permission of the DuPont Company)

TABLE 5.5 Thermal Properties of Various Materials at 20°C or 68°F.

Material	Density – ρ		Specific Heat – c_p		Conductivity – k	
	kg/m³	lb/ft³	J/kg·K	BTU/lbm-°R	J/s·m·°C	BTU/hr-ft-°F
Aluminum	2700	167	896	0.214	237	136
Bronze	8670	541	343	0.0819	26	15
Concrete	500	30	840	0.2	0.13	0.075
Copper	8930	557	383	0.0915	400	230
Glass	2800	175	800	0.2	0.81	0.47
Ice	910	57	1830	0.44	2.2	1.3
Plaster	1800	112			0.81	0.47
Polystyrene	1210	75			.04	.023
Steel (1% C)	7800	487	473	0.113	43	25
Wood (pine)	420	26	2700	0.64	0.15	0.087
Wool (insulation)	200	12			0.038	0.022

TABLE 5.6 Typical Convective Coefficients h.

Fluid	J/s · m² · °C	BTU/hr-ft²-°F
Air, free convection	6 to 30	1 to 5
Air, forced convection	30 to 300	5 to 50
Superheated steam, forced convection	30 to 300	5 to 50
Water, forced convection	300 to 12 000	50 to 2000
Water, boiling	3000 to 60 000	500 to 20,000
Steam, condensing	6000 to 120 000	1000 to 20,000

TABLE 5.1 Properties of Ideal Gases—English Units

Gas	Chemical Formula	Molecular Weight	$R \frac{\text{ft}-\text{lb}}{\text{lbm}\cdot°\text{R}}$	$c_p \frac{\text{BTU}}{\text{lbm}\cdot°\text{R}}$	$c_v \frac{\text{BTU}}{\text{lbm}\cdot°\text{R}}$	k
Air	. . .	28.97	53.34	0.240	0.171	1.400
Argon	Ar	39.94	38.68	0.1253	0.0756	1.667
Butane	C_4H_{10}	58.124	26.58	0.415	0.381	1.09
Carbon Dioxide	CO_2	44.01	35.10	0.203	0.158	1.285
Carbon Monoxide	CO	28.01	55.16	0.249	0.178	1.399
Ethane	C_2H_6	30.07	51.38	0.427	0.361	1.183
Ethylene	C_2H_4	28.054	55.07	0.411	0.340	1.208
Helium	He	4.003	386.0	1.25	0.753	1.667
Hydrogen	H_2	2.016	766.4	3.43	2.44	1.404
Methane	CH_4	16.04	96.35	0.532	0.403	1.32
Neon	Ne	20.183	76.55	0.246	0.1477	1.667
Nitrogen	N_2	28.016	55.15	0.248	0.177	1.400
Octane	C_8H_{18}	114.22	13.53	0.409	0.392	1.044
Oxygen	O_2	32.000	48.28	0.219	0.157	1.395
Propane	C_3H_8,	44.097	35.04	0.407	0.362	1.124
Steam	H_2O	18.016	85.76	0.445	0.335	1.329

TABLE 5.2 Specific Heats of Liquids and Solids—English Units
c_p BTU/(lbm-°F)

A. LIQUIDS

Substance	State	c_p	Substance	State	c_p
Water	1 atm, 77°F	1.00	Glycerin	1 atm, 50°F	0.555
Ammonia	sat., –4°F	1.08	Bismuth	1 atm, 800°F	0.0344
	sat., 120°F	1.22	Mercury	1 atm, 50°F	0.0330
Refrigerant-12	sat.,–4°F	0.217	Sodium	1 atm, 200°F	0.330
	sat., 120°F	0.244	Propane	1 atm, 32°F	0.577
Benzene	1 atm, 60°F	0.431	Ethyl Alcohol	1 atm, 77°F	0.581

B. SOLIDS

Substance	T, °F	c_p	Substance	T, °F	c_p
Ice	–76	0.392	Silver	–4	0.0557
	12	0.486	Lead	–150	0.0282
	28	0.402		30	0.0297
Aluminum	–150	0.167		210	0.0321
	30	0.208	Copper	–150	0.0785
	210	0.225		30	0.0911
Iron	–4	0.107		210	0.0940

TABLE 5.3 Thermodynamic Properties of Water (Steam Tables)—English Units

5.3.1 Saturated H_2O—Temperature Table

		Volume, ft³/lbm		Energy, BTU/lbm		Enthalpy, BTU/lbm			Entropy, BTU/lbm-°R		
T, °F	*P*, psia	v_f	v_g	u_f	u_g	h_f	h_{fg}	h_g	s_f	s_{fg}	s_g
32.018	0.08866	0.016022	3302	0.00	1021.2	0.01	1075.4	1075.4	0.00000	2.1869	2.1869
40	0.12166	0.016020	2445	8.02	1023.9	8.02	1070.9	1078.9	0.01617	2.1430	2.1592
60	0.2563	0.016035	1206.9	28.08	1030.4	28.08	1059.6	1087.7	0.05555	2.0388	2.0943
80	0.5073	0.016073	632.8	48.08	1037.0	48.09	1048.3	1096.4	0.09332	1.9423	2.0356
100	0.9503	0.016130	350.0	68.04	1043.5	68.05	1037.0	1105.0	0.12963	1.8526	1.9822
120	1.6945	0.016205	203.0	87.99	1049.9	88.00	1025.5	1113.5	0.16465	1.7690	1.9336
140	2.892	0.016293	122.88	107.95	1056.2	107.96	1014.0	1121.9	0.19851	1.6907	1.8892
160	4.745	0.016395	77.23	127.94	1062.3	127.96	1002.2	1130.1	0.23130	1.6171	1.8484
180	7.515	0.016509	50.20	147.97	1068.3	147.99	990.2	1138.2	0.26311	1.5478	1.8109
200	11.529	0.016634	33.63	168.04	1074.2	168.07	977.9	1145.9	0.29400	1.4822	1.7762
212	14.698	0.016716	26.80	180.11	1077.6	180.16	970.3	1150.5	0.31213	1.4446	1.7567
220	17.188	0.016772	23.15	188.17	1079.8	188.22	965.3	1153.5	0.32406	1.4201	1.7441
240	24.97	0.016922	16.327	208.36	1085.3	208.44	952.3	1160.7	0.35335	1.3609	1.7143
260	35.42	0.017084	11.768	228.64	1090.5	228.76	938.8	1167.6	0.38193	1.3044	1.6864
280	49.18	0.017259	8.650	249.02	1095.4	249.18	924.9	1174.1	0.40986	1.2504	1.6602
300	66.98	0.017448	6.472	269.52	1100.0	269.73	910.4	1180.2	0.43720	1.1984	1.6356
340	117.93	0.017872	3.792	310.91	1108.0	311.30	879.5	1190.8	0.49031	1.0997	1.5901
380	195.60	0.018363	2.339	352.95	1114.3	353.62	845.4	1199.0	0.54163	1.0067	1.5483
420	308.5	0.018936	1.5024	395.81	1118.3	396.89	807.2	1204.1	0.59152	0.9175	1.5091
460	466.3	0.019614	0.9961	439.7	1119.6	441.4	764.1	1205.5	0.6404	0.8308	1.4712
500	680.0	0.02043	0.6761	485.1	1117.4	487.7	714.8	1202.5	0.6888	0.7448	1.4335
540	961.5	0.02145	0.4658	532.6	1111.0	536.4	657.5	1193.8	0.7374	0.6576	1.3950
580	1324.3	0.02278	0.3225	583.1	1098.9	588.6	589.3	1178.0	0.7872	0.5668	1.3540
620	1784.4	0.02465	0.2209	638.3	1078.5	646.4	505.0	1151.4	0.8398	0.4677	1.3075
660	2362	0.02767	0.14459	702.3	1042.3	714.4	391.1	1105.5	0.8990	0.3493	1.2483
700	3090	0.03666	0.07438	801.7	947.7	822.7	167.5	990.2	0.9902	0.1444	1.1346
705.44	3204	0.05053	0.05053	872.6	872.6	902.5	0	902.5	1.0580	0	1.0580

5.3.2 Saturated H_2O—Pressure Table

		Volume, ft³/lbm		Energy, BTU/lbm		Enthalpy, BTU/lbm			Entropy, BTU/lbm-°R		
P, psia	*T*, °F	v_f	v_g	u_f	u_g	h_f	h_{fg}	h_g	s_f	s_{fg}	s_g
1.0	101.70	0.016136	333.6	69.74	1044.0	69.74	1036.0	1105.8	0.13266	1.8453	1.9779
2.0	126.04	0.016230	173.75	94.02	1051.8	94.02	1022.1	1116.1	0.17499	1.7448	1.9198
4.0	152.93	0.016358	90.64	120.88	1060.2	120.89	1006.4	1127.3	0.21983	1.6426	1.8624
6.0	170.03	0.016451	61.98	137.98	1065.4	138.00	996.2	1134.2	0.24736	1.5819	1.8292
10	193.19	0.016590	38.42	161.20	1072.2	161.23	982.1	1143.3	0.28358	1.5041	1.7877
14.696	211.99	0.016715	26.80	180.10	1077.6	180.15	970.4	1150.5	0.31212	1.4446	1.7567
20	227.96	0.016830	20.09	196.19	1082.0	196.26	960.1	1156.4	0.33580	1.3962	1.7320
30	250.34	0.017004	13.748	218.84	1088.0	218.93	945.4	1164.3	0.36821	1.3314	1.6996
40	267.26	0.017146	10.501	236.03	1092.3	236.16	933.8	1170.0	0.39214	1.2845	1.6767
50	281.03	0.017269	8.518	250.08	1095.6	250.24	924.2	1174.4	0.41129	1.2476	1.6589
60	292.73	0.017378	7.177	262.06	1098.3	262.25	915.8	1178.0	0.42733	1.2170	1.6444
70	302.96	0.017478	6.209	272.56	1100.6	272.79	908.3	1181.0	0.44120	1.1909	1.6321
80	312.07	0.017570	5.474	281.95	1102.6	282.21	901.4	1183.6	0.45344	1.1679	1.6214
90	320.31	0.017655	4.898	290.46	1104.3	290.76	895.1	1185.9	0.46442	1.1475	1.6119
100	327.86	0.017736	4.434	298.28	1105.8	298.61	889.2	1187.8	0.47439	1.1290	1.6034
120	341.30	0.017886	3.730	312.27	1108.3	312.67	878.5	1191.1	0.49201	1.0966	1.5886
140	353.08	0.018024	3.221	324.58	1110.3	325.05	868.7	1193.8	0.50727	1.0688	1.5761
160	363.60	0.018152	2.836	335.63	1112.0	336.16	859.8	1196.0	0.52078	1.0443	1.5651
180	373.13	0.018273	2.533	345.68	1113.4	346.29	851.5	1197.8	0.53292	1.0223	1.5553
200	381.86	0.018387	2.289	354.9	1114.6	355.6	843.7	1199.3	0.5440	1.0025	1.5464
300	417.43	0.018896	1.5442	393.0	1118.2	394.1	809.8	1203.9	0.5883	0.9232	1.5115
400	444.70	0.019340	1.1620	422.8	1119.5	424.2	781.2	1205.5	0.6218	0.8638	1.4856
500	467.13	0.019748	0.9283	447.7	1119.4	449.5	755.8	1205.3	0.6490	0.8154	1.4645
600	486.33	0.02013	0.7702	469.4	1118.6	471.7	732.4	1204.1	0.6723	0.7742	1.4464
800	518.36	0.02087	0.5691	506.6	1115.0	509.7	689.6	1199.3	0.7110	0.7050	1.4160
1000	544.75	0.02159	0.4459	538.4	1109.9	542.4	650.0	1192.4	0.7432	0.6471	1.3903
1400	587.25	0.02307	0.3016	592.7	1096.0	598.6	575.5	1174.1	0.7964	0.5497	1.3461
2000	636.00	0.02565	0.18813	662.4	1066.6	671.9	464.4	1136.3	0.8623	0.4238	1.2861
3000	695.52	0.03431	0.08404	783.4	968.8	802.5	213.0	1015.5	0.9732	0.1843	1.1575
3203.6	705.44	0.05053	0.05053	872.6	872.6	902.5	0	902.5	1.0580	0	1.0580

5.3.3 SuperheatedVapor

°F	*v*	*u*	*h*	*s*	*v*	*u*	*h*	*s*	*v*	*u*	*h*	*s*
	1 psia				10 psia				14.7 psia			
200	392.5	1077.5	1150.1	2.0508	38.85	1074.7	1146.6	1.7927	...	...	...	...
300	452.3	1112.1	1195.8	2.1153	45.00	1110.6	1193.9	1.8595	30.53	1109.7	1192.8	1.8160
400	511.9	1147.0	1241.8	2.1720	51.03	1146.1	1240.5	1.9171	34.67	1145.6	1239.9	1.8741
500	571.5	1182.8	1288.5	2.2235	57.04	1182.2	1287.7	1.9690	38.77	1181.8	1287.3	1.9263
600	631.1	1219.3	1336.1	2.2706	63.03	1218.9	1335.5	2.0164	42.86	1218.6	1335.2	1.9737
700	690.7	1256.7	1384.5	2.3142	69.01	1256.3	1384.0	2.0601	46.93	1256.1	1383.8	2.0175
800	750.3	1294.9	1433.7	2.3550	74.98	1294.6	1433.3	2.1009	51.00	1294.4	1433.1	2.0584
1000	869.5	1373.9	1534.8	2.4294	86.91	1373.8	1534.6	2.1755	59.13	1373.7	1534.5	2.1330
1200	988.6	1456.7	1639.6	2.4967	98.84	1456.5	1639.4	2.2428	67.25	1456.5	1639.3	2.2003
1400	1107.7	1543.1	1748.1	2.5584	110.76	1543.0	1748.0	2.3045	75.36	1543.0	1747.9	2.2621
	20 psia				60 psia				100 psia			
300	22.36	1108.8	1191.6	1.7808	7.259	1101.0	1181.6	1.6492	...	...	...	...
400	25.43	1145.1	1239.2	1.8395	8.353	1140.8	1233.5	1.7134	4.934	1136.2	1227.5	1.6517
500	28.46	1181.5	1286.8	1.8919	9.399	1178.6	1283.0	1.7678	5.587	1175.7	1279.1	1.7085
600	31.47	1218.4	1334.8	1.9395	10.425	1216.3	1332.1	1.8165	6.216	1214.2	1329.3	1.7582
700	34.47	1255.9	1383.5	1.9834	11.440	1254.4	1381.4	1.8609	6.834	1252.8	1379.2	1.8033
800	37.46	1294.3	1432.9	2.0243	12.448	1293.0	1431.2	1.9022	7.445	1291.8	1429.6	1.8449
1000	43.44	1373.5	1534.3	2.0989	14.454	1372.7	1533.2	1.9773	8.647	1371.9	1532.1	1.9204
1200	49.41	1456.4	1639.2	2.1663	16.452	1455.8	1638.5	2.0448	9.861	1455.2	1637.7	1.9882
1400	55.37	1542.9	1747.9	2.2281	18.445	1542.5	1747.3	2.1067	11.060	1542.0	1746.7	2.0502
1600	61.33	1633.2	1860.1	2.2854	20.44	1632.8	1859.7	2.1641	12.257	1632.4	1859.3	2.1076
	120 psia				160 psia				200 psia			
400	4.079	1133.8	1224.4	1.6288	3.007	1128.8	1217.8	1.5911	2.361	1123.5	1210.8	1.5600
500	4.633	1174.2	1277.1	1.6868	3.440	1171.2	1273.0	1.6518	2.724	1168.0	1268.8	1.6239
600	5.164	1213.2	1327.8	1.7371	3.848	1211.1	1325.0	1.7034	3.058	1208.9	1322.1	1.6767
700	5.682	1252.0	1378.2	1.7825	4.243	1250.4	1376.0	1.7494	3.379	1248.8	1373.8	1.7234
800	6.195	1291.2	1428.7	1.8243	4.631	1289.9	1427.0	1.7916	3.693	1288.6	1425.3	1.7660
1000	7.208	1371.5	1531.5	1.9000	5.397	1370.6	1530.4	1.8677	4.310	1369.8	1529.3	1.8425
1200	8.213	1454.9	1637.3	1.9679	6.154	1454.3	1636.5	1.9358	4.918	1453.7	1635.7	1.9109
1400	9.214	1541.8	1746.4	2.0300	6.906	1541.4	1745.9	1.9980	5.521	1540.9	1745.3	1.9732
1600	10.212	1632.3	1859.0	2.0875	7.656	1631.9	1858.6	2.0556	6.123	1631.6	1858.2	2.0308
1800	11.209	1726.2	1975.1	2.1413	8.405	1725.9	1974.8	2.1094	6.722	1725.6	1974.4	2.0847
	250 psia				300 psia				400 psia			
500	2.150	1163.8	1263.3	1.5948	1.7662	1159.5	1257.5	1.5701	1.2843	1150.1	1245.2	1.5282
600	2.426	1206.1	1318.3	1.6494	2.004	1203.2	1314.5	1.6266	1.4760	1197.3	1306.6	1.5892
700	2.688	1246.7	1371.1	1.6970	2.227	1244.6	1368.3	1.6751	1.6503	1240.4	1362.5	1.6397
800	2.943	1287.0	1423.2	1.7401	2.442	1285.4	1421.0	1.7187	1.8163	1282.1	1416.6	1.6844
900	3.193	1327.6	1475.3	1.7799	2.653	1326.3	1473.6	1.7589	1.9776	1323.7	1470.1	1.7252
1000	3.440	1368.7	1527.9	1.8172	2.860	1367.7	1526.5	1.7964	2.136	1365.5	1523.6	1.7632
1200	3.929	1453.0	1634.8	1.8858	3.270	1452.2	1633.8	1.8653	2.446	1450.7	1631.8	1.8327
1400	4.414	1540.4	1744.6	1.9483	3.675	1539.8	1743.8	1.9279	2.752	1538.7	1742.4	1.8956
1600	4.896	1631.1	1857.6	2.0060	4.078	1630.7	1857.0	1.9857	3.055	1629.8	1855.9	1.9535
1800	5.376	1725.2	1974.0	2.0599	4.479	1724.9	1973.5	2.0396	3.357	1724.1	1972.6	2.0076
	600 psia				800 psia				1000 psia			
500	0.7947	1128.0	1216.2	1.4592	...	...	...	...	...	...	...	...
600	0.9456	1184.5	1289.5	1.5320	0.6776	1170.1	1270.4	1.4861	0.5140	1153.7	1248.8	1.4450
700	1.0727	1231.5	1350.6	1.5872	0.7829	1222.1	1338.0	1.5471	0.6080	1212.0	1324.6	1.5135
800	1.1900	1275.4	1407.6	1.6343	0.8764	1268.5	1398.2	1.5969	0.6878	1261.2	1388.5	1.5664
900	1.3021	1318.4	1462.9	1.6766	0.9640	1312.9	1455.6	1.6408	0.7610	1307.3	1448.1	1.6120
1000	1.4108	1361.2	1517.8	1.7155	1.0482	1356.7	1511.9	1.6807	0.8305	1352.2	1505.9	1.6530
1200	1.6222	1447.7	1627.8	1.7861	1.2102	1444.6	1623.8	1.7526	0.9630	1441.5	1619.7	1.7261
1400	1.8289	1536.5	1739.5	1.8497	1.3674	1534.2	1736.6	1.8167	1.0905	1531.9	1733.7	1.7909
1600	2.033	1628.0	1853.7	1.9080	1.5218	1626.2	1851.5	1.8754	1.2152	1624.4	1849.3	1.8499
1800	2.236	1722.6	1970.8	1.9622	1.6749	1721.0	1969.0	1.9298	1.3384	1719.5	1967.2	1.9046
	2000 psia				3000 psia				4000 psia			
700	0.2487	1147.7	1239.8	1.3782	0.09771	1003.9	1058.1	1.1944	0.02867	742.1	763.4	0.9345
800	0.3071	1220.1	1333.8	1.4562	0.17572	1167.6	1265.2	1.3675	0.10522	1095.0	1172.9	1.2740
900	0.3534	1276.8	1407.6	1.5126	0.2160	1241.8	1361.7	1.4414	0.14622	1201.5	1309.7	1.3789
1000	0.3945	1328.1	1474.1	1.5598	0.2485	1301.7	1439.6	1.4967	0.17520	1272.9	1402.6	1.4449
1200	0.4685	1425.2	1598.6	1.6398	0.3036	1408.0	1576.6	1.5848	0.2213	1390.1	1553.9	1.5423
1400	0.5368	1520.2	1718.8	1.7082	0.3524	1508.1	1703.7	1.6571	0.2603	1495.7	1688.4	1.6188
1600	0.6020	1615.4	1838.2	1.7692	0.3978	1606.3	1827.1	1.7201	0.2959	1597.1	1816.1	1.6841
1800	0.6656	1712.0	1958.3	1.8249	0.4416	1704.5	1949.6	1.7769	0.3296	1697.1	1941.1	1.7420
2000	0.7284	1810.6	2080.2	1.8765	0.4844	1803.9	2072.8	1.8291	0.3625	1797.3	2065.6	1.7948

Figure 5.6 *P-h* Diagram for Refrigerant 134a — English Units

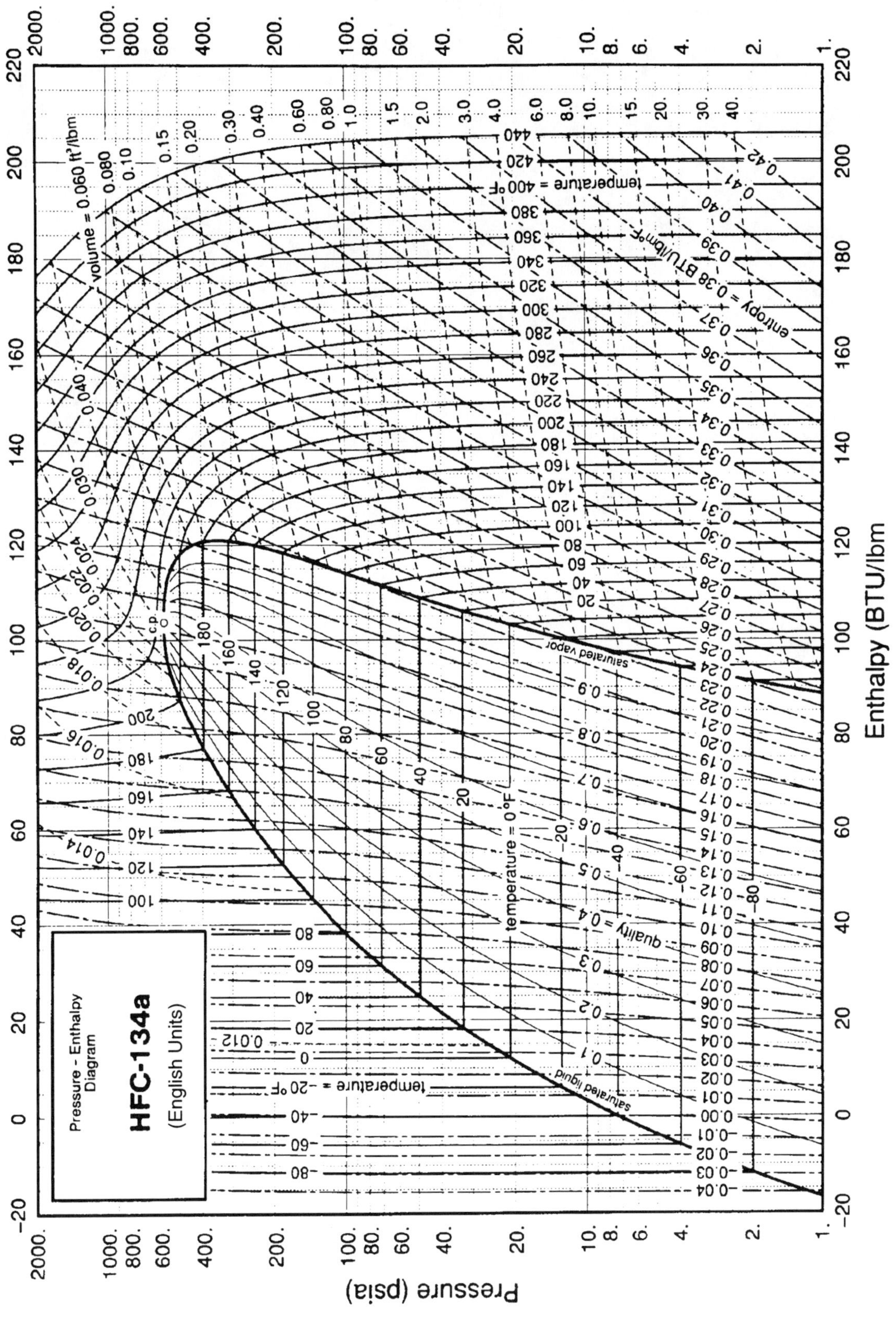
Pressure - Enthalpy Diagram
HFC-134a
(English Units)
Pressure (psia)
Enthalpy (BTU/lbm)
saturated liquid
saturated vapor
c.p.

(Reproduced by permission of the DuPont Company)

Fluid Mechanics

by Merle C. Potter

Fluid Mechanics deals with the statics, kinematics and dynamics of fluids, including both gases and liquids. Most fluid flows can be assumed to be incompressible (constant density); such flows include liquid flows as well as low speed gas flows (with velocities less than about 100 m/s). In addition, particular flows are either viscous or inviscid. Viscous effects dominate internal flows—such as flow in a pipe—and must be included near the boundaries of external flows (flow near the surface of an airfoil). Viscous flows are laminar if well-behaved, or turbulent if chaotic and highly fluctuating. Inviscid flows occur primarily as external flows outside the boundary layers that contain viscous effects. This review will focus on *Newtonian fluids*, that is, fluids which exhibit linear stress-strain-rate relationships; Newtonian fluids include air, water, oil, gasoline and tar. Please note that we have used upper case P for pressure in thermo, whereas in fluids we use lower case p; that's simply conventional.

We have used metric (SI) units only since those are the units most often used. Make sure you are familiar with SI units. We have given some quantities in both sets of units since this book is also often used as a general reference.

6.1 Fluid Properties

Some of the more common fluid properties are defined below and listed in Tables 6.1 and 6.2 for water and air at standard conditions.

density $$\rho = \frac{M}{\forall} \quad \textbf{(6.1.1)}$$

specific weight $$\gamma = \rho g = \frac{W}{\forall} \quad \textbf{(6.1.2)}$$

viscosity $$\mu = \frac{\tau}{du/dy} \quad \textbf{(6.1.3)}$$

kinematic viscosity $\nu = \dfrac{\mu}{\rho}$ **(6.1.4)**

specific gravity $SG = \dfrac{\rho_x}{\rho_{H_2O}}$ **(6.1.5)**

bulk modulus $K = -\forall \dfrac{\Delta p}{\Delta \forall}$ **(6.1.6)**

speed of sound $c_{liquid} = \sqrt{K/\rho}$ $c_{gas} = \sqrt{kRT}$ $(k_{air} = 1.4)$ **(6.1.7)**

Example 6.1

A velocity difference of 2.4 m/s is measured between radial points 2 mm apart in a pipe in which 20°C water is flowing. What is the shear stress?

Solution. Using Eq. 6.1.3 we find, with $\mu = 10^{-3}$ N·s/m² from Table 6.1,

$$\tau = \mu \frac{du}{dy} \cong \mu \frac{\Delta u}{\Delta y} = 10^{-3} \frac{2.4}{0.002} = 1.2 \text{ Pa}.$$

TABLE 6.1 Properties

Property	Symbol	Definition	Water (20°C,)	Air (STP)
density	ρ	$\dfrac{\text{mass}}{\text{volume}}$	1000 kg/m^3	1.23 kg/m^3
viscosity	μ	$\dfrac{\text{shear stress}}{\text{velocity gradient}}$	$10^{-3} \text{ N}\cdot\text{s/m}^2$	$2.0\times10^{-5} \text{ N}\cdot\text{s/m}^2$
kinematic viscosity	ν	$\dfrac{\text{viscosity}}{\text{density}}$	$10^{-6} \text{ m}^2/\text{s}$	$1.6\times10^{-5} \text{ m}^2/\text{s}$
speed of sound	c	velocity of propagation of a small wave	1480 m/s	343 m/s
specific weight	γ	$\dfrac{\text{weight}}{\text{volume}}$	9800 N/m^3	12 N/m^3
surface tension	σ	stored energy per unit area	0.073 J/m^2	
bulk modulus	K	$-\text{volume} \dfrac{\Delta \text{ pressure}}{\Delta \text{ volume}}$	220×10^4 kPa	
vapor pressure	p_v	pressure at which liquid & vapor are in equilibrium	2.45 kPa	

Notes:
- Kinematic viscosity is used because the ratio μ/ρ occurs frequently.
- Surface tension is used primarily for calculating capillary rise.
- Vapor pressure is used to predict *cavitation* which exists whenever the local pressure falls below the vapor pressure (for water at standard temperatures, the vapor pressure is close to absolute zero).

TABLE 6.2 Properties of Water and Air

Properties of Water (Metric)

Temperature °C	Density kg/m^3	Viscosity $N \cdot s/m^2$	Kinematic Viscosity m^2/s	Bulk Modulus kPa	Surface Tension N/m	Vapor Pressure kPa
0	999.9	1.792×10^{-3}	1.792×10^{-6}	204×10^4	7.62×10^{-2}	0.588
5	1000.0	1.519	1.519	206	7.54	0.882
10	999.7	1.308	1.308	211	7.48	1.176
15	999.1	1.140	1.141	214	7.41	1.666
20	998.2	1.005	1.007	220	7.36	2.447
30	995.7	0.801	0.804	223	7.18	4.297
40	992.2	0.656	0.661	227	7.01	7.400
50	988.1	0.549	0.556	230	6.82	12.220
60	983.2	0.469	0.477	228	6.68	19.600
70	977.8	0.406	0.415	225	6.50	30.700
80	971.8	0.357	0.367	221	6.30	46.400
90	965.3	0.317	0.328	216	6.12	68.200
100	958.4	0.284×10^{-3}	0.296×10^{-6}	207×10^4	5.94×10^{-2}	97.500

Properties of Air at Standard Pressure (Metric)

Temperature	Density kg/m^3	Specific Weight N/m^3	Viscosity $N \cdot s/m^2$	Kinematic Viscosity m^2/s
–20°C	1.39	13.6	1.56×10^{-5}	1.13×10^{-5}
–10°C	1.34	13.1	1.62×10^{-5}	1.21×10^{-5}
0°C	1.29	12.6	1.68×10^{-5}	1.30×10^{-5}
10°C	1.25	12.2	1.73×10^{-5}	1.39×10^{-5}
20°C	1.20	11.8	1.80×10^{-5}	1.49×10^{-5}
40°C	1.12	11.0	1.91×10^{-5}	1.70×10^{-5}
60°C	1.06	10.4	2.03×10^{-5}	1.92×10^{-5}
80°C	0.99	9.71	2.15×10^{-5}	2.17×10^{-5}
100°C	0.94	9.24	2.28×10^{-5}	2.45×10^{-5}

Properties of the Atmosphere (Metric)

Altitude m	Temperature K	p/p_0 (p_0=101 kPa)	ρ/ρ_0 (ρ_0=1.23 kg/m^3)
0	288.2	1.000	1.000
1 000	281.7	0.8870	0.9075
2 000	275.2	0.7846	0.8217
4 000	262.2	0.6085	0.6689
6 000	249.2	0.4660	0.5389
8 000	236.2	0.3519	0.4292
10 000	223.3	0.2615	0.3376
12 000	216.7	0.1915	0.2546
14 000	216.7	0.1399	0.1860
16 000	216.7	0.1022	0.1359
18 000	216.7	0.07466	0.09930
20 000	216.7	0.05457	0.07258
22 000	218.6	0.03995	0.05266
26 000	222.5	0.02160	0.02797
30 000	226.5	0.01181	0.01503
40 000	250.4	0.2834×10^{-2}	0.3262×10^{-2}
50 000	270.7	0.7874×10^{-3}	0.8383×10^{-3}
60 000	255.8	0.2217×10^{-3}	0.2497×10^{-3}
70 000	219.7	0.5448×10^{-4}	0.7146×10^{-4}
80 000	180.7	0.1023×10^{-4}	0.1632×10^{-4}
90 000	180.7	0.1622×10^{-5}	0.2588×10^{-5}

TABLE 6.2(E) Properties of Water and Air

Properties of Water (English)

Temperature °F	Density slugs/ft³	Viscosity lb-sec/ft²	Surface Tension lb/ft	Vapor Pressure lb/in²	Bulk Modulus lb/in²
32	1.94	3.75×10^{-5}	0.518×10^{-2}	0.089	293,000
40	1.94	3.23×10^{-5}	0.514×10^{-2}	0.122	294,000
50	1.94	2.74×10^{-5}	0.509×10^{-2}	0.178	305,000
60	1.94	2.36×10^{-5}	0.504×10^{-2}	0.256	311,000
70	1.94	2.05×10^{-5}	0.500×10^{-2}	0.340	320,000
80	1.93	1.80×10^{-5}	0.492×10^{-2}	0.507	322,000
90	1.93	1.60×10^{-5}	0.486×10^{-2}	0.698	323,000
100	1.93	1.42×10^{-5}	0.480×10^{-2}	0.949	327,000
120	1.92	1.17×10^{-5}	0.465×10^{-2}	1.69	333,000
140	1.91	0.98×10^{-5}	0.454×10^{-2}	2.89	330,000
160	1.90	0.84×10^{-5}	0.441×10^{-2}	4.74	326,000
180	1.88	0.73×10^{-5}	0.426×10^{-2}	7.51	318,000
200	1.87	0.64×10^{-5}	0.412×10^{-2}	11.53	308,000
212	1.86	0.59×10^{-5}	0.404×10^{-2}	14.7	300,000

Properties of Air at Standard Pressure (English)

Temperature °F	Density slugs/ft³	Viscosity lb-sec/ft²	Kinematic Viscosity ft²/sec
0	0.00268	3.28×10^{-7}	12.6×10^{-5}
20	0.00257	3.50×10^{-7}	13.6×10^{-5}
40	0.00247	3.62×10^{-7}	14.6×10^{-5}
60	0.00237	3.74×10^{-7}	15.8×10^{-5}
68	0.00233	3.81×10^{-7}	16.0×10^{-5}
80	0.00228	3.85×10^{-7}	16.9×10^{-5}
100	0.00220	3.96×10^{-7}	18.0×10^{-5}
120	0.00215	4.07×10^{-7}	18.9×10^{-5}

Properties of the Atmosphere (English)

Altitude ft	Temperature °F	Pressure lb/ft²	Density slugs/ft³	Kinematic Viscosity ft²/sec	Velocity of Sound ft/sec
0	59.0	2116	0.00237	1.56×10^{-4}	1117
1,000	55.4	2041	0.00231	1.60×10^{-4}	1113
2,000	51.9	1968	0.00224	1.64×10^{-4}	1109
5,000	41.2	1760	0.00205	1.77×10^{-4}	1098
10,000	23.4	1455	0.00176	2.00×10^{-4}	1078
15,000	5.54	1194	0.00150	2.28×10^{-4}	1058
20,000	–12.3	973	0.00127	2.61×10^{-4}	1037
25,000	–30.1	785	0.00107	3.00×10^{-4}	1016
30,000	–48.0	628	0.000890	3.47×10^{-4}	995
35,000	–65.8	498	0.000737	4.04×10^{-4}	973
36,000	–67.6	475	0.000709	4.18×10^{-4}	971
40,000	–67.6	392	0.000586	5.06×10^{-4}	971
50,000	–67.6	242	0.000362	8.18×10^{-4}	971
100,000	–67.6	22.4	3.31×10^{-5}	89.5×10^{-4}	971
110,000	–47.4	13.9	1.97×10^{-5}	1.57×10^{-6}	996
150,000	113.5	3.00	3.05×10^{-6}	13.2×10^{-6}	1174
200,000	160.0	0.665	6.20×10^{-7}	68.4×10^{-6}	1220
260,000	–28	0.0742	1.0×10^{-7}	321×10^{-6}	1019

Example 6.2

Find the speed of sound in air at an elevation of 1000 m.

Solution. From Table 6.2 we find $T = 281.7$ K. Using Eq. 6.1.7, with $R = 287$ J/kg·K, there results

$$c = \sqrt{kRT} = \sqrt{1.4 \times 287 \times 281.7} = 336.4 \text{ m/s}.$$

Note: Temperature must be absolute.

6.2 Fluid Statics

Typical problems in fluid statics involve manometers, forces on plane and curved surfaces, and buoyancy. All of these problems are solved by using the pressure distribution derived from summing forces on an infinitesimal element of fluid; in differential form with h positive downward, it is

$$dp = \gamma dh. \tag{6.2.1}$$

For constant specific weight, assuming $p = 0$ at $h = 0$, we have

$$p = \gamma h. \tag{6.2.2}$$

Equation 6.2.2 can be used to interpret manometer readings directly. By summing forces on elements of a plane surface, we would find the magnitude and location of a force acting on one side (refer to Fig. 6.1) to be

$$F = \gamma h_c A \tag{6.2.3}$$

$$y_p = y_c + \frac{I_c}{y_c A} \tag{6.2.4}$$

where y_c locates the centroid and I_c is the second moment* of the area about the centroidal axis.

To solve problems involving curved surfaces, we simply draw a free-body diagram of the liquid contained above the curved surface and, using the above formulas, solve the problem.

To solve buoyancy-related problems we use Archimedes' principle which states: the buoyant force on a submerged object is equal to the weight of displaced liquid; that is,

$$F_b = \gamma V\!\!\!\!-_{displaced}\,. \tag{6.2.5}$$

---

*The second moment I_c (often symbolized by $\bar{I}$) of three common areas:

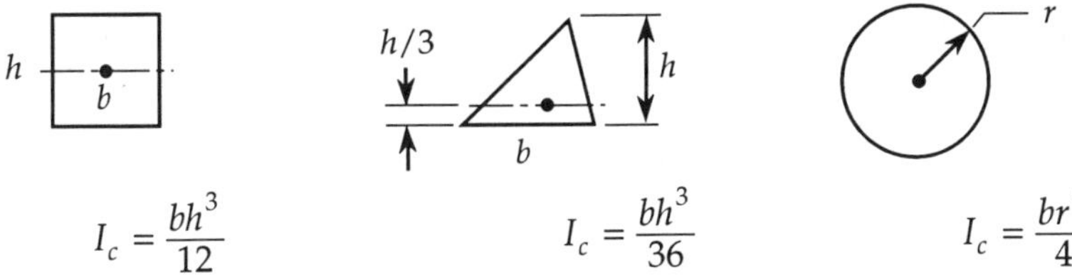

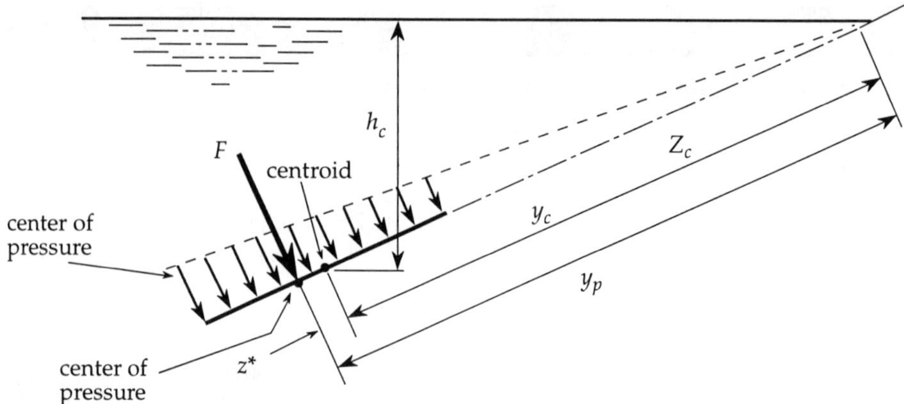

Figure 6.1 Force on a plane surface.

Example 6.3

Find the pressure difference between the air pipe and the water pipe.

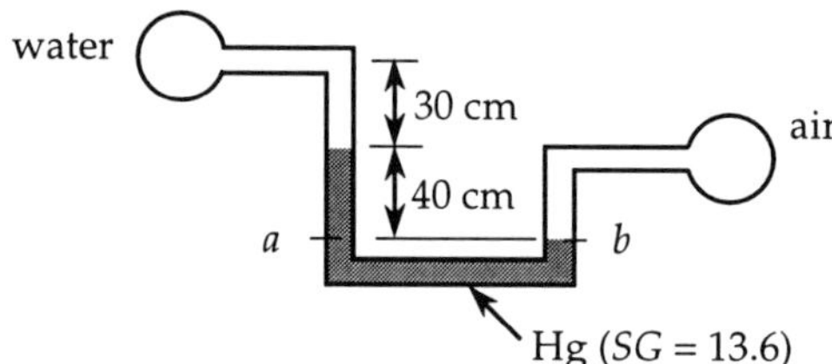

Solution. We first locate points "a" and "b" in the same fluid where $p_a = p_b$; then using Eq. 6.2.2

$$p_{water} + 9800 \times 0.3 + (9800 \times 13.6) \times 0.4 = p_{air} + \overset{\text{neglect}}{\cancel{\gamma_{air}}} \times 0.4$$

$$\therefore p_{air} - p_{water} = 56\,300\,\text{Pa} \quad \text{or} \quad 56.3\ \text{kPa}.$$

Example 6.4

What is the pressure in pipe A?

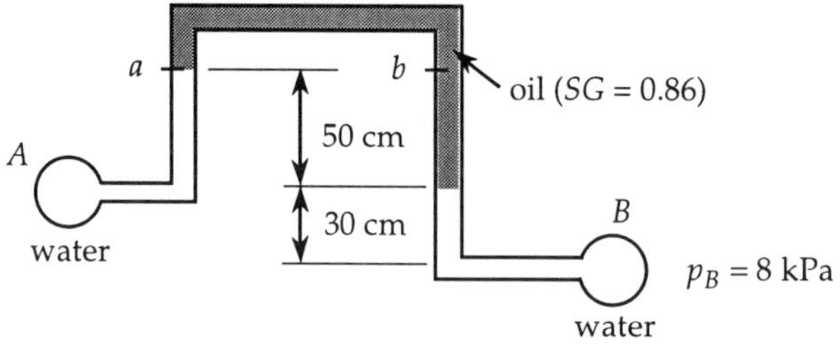

Solution. Locate points "a" and "b" so that $p_a = p_b$. Then, using Eq. 6.2.2 there results

$$p_A - 9800 \times 0.5 = 8000 - 9800 \times 0.3 - (9800 \times 0.86) \times 0.5$$

$$\therefore p_A = 5750\ \text{Pa} \quad \text{or} \quad 5.75\ \text{kPa}.$$

Example 6.5

Find the force P needed to hold the 5-m-wide gate closed.

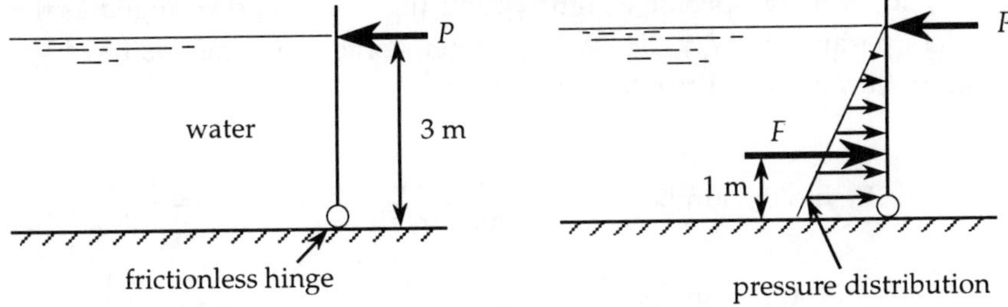

Solution. First, we note that the pressure distribution is triangular, as shown. Hence, the resultant force F acts 1/3 up from the hinge through the centroid of the triangular distribution. Summing moments about the hinge gives

$$F \times 1 = P \times 3$$

$$\therefore P = F/3 = \gamma h_c A/3$$

$$= 9800 \times \frac{3}{2} \times (5 \times 3)/3 = 73\ 500 \text{ N}.$$

Note: If the top of the gate were not at the free surface, we would find y_p using Eq. 6.2.4.

Example 6.6

A rectangular 4 m × 20 m vessel has a mass of 40 000 kg. How far will it sink in water when carrying a load of 100 000 kg?

Solution. The total weight of the loaded vessel must equal the weight of the displaced water. This is expressed as

$$W = \gamma \forall$$

$$(40\ 000 + 100\ 000) \times 9.8 = 9800 \times 4 \times 20 \times h$$

$$\therefore h = 1.75 \text{ m}.$$

6.3 Dimensionless Parameters and Similitude

Information involving phenomena encountered in fluid mechanics is often presented in terms of dimensionless parameters. For example, the lift force F_L on a streamlined body can be represented by a lift coefficient C_L, a dimensionless parameter. Rather than plotting the lift force as a function of velocity, the lift coefficient could be plotted as a function of the Reynolds number, or the Mach number—two other dimensionless parameters.

To form dimensionless parameters, we first list various quantities encountered in fluid mechanics in Table 6.3. A dimensionless parameter involving several quantities is then formed by combining the quantities so that the combination of quantities is dimensionless. If all units are present in the quantities to be combined, this usually requires four quantities. For example, the four quantities power $\dot{W}$, flow rate Q, specific weight γ, and head H can be arranged as the dimensionless parameter $\dot{W}/\gamma QH$. Many dimensionless parameters have special significance; they are identified as follows:

$$\text{Reynolds number} = \frac{\text{inertial force}}{\text{viscous force}} \qquad \text{Re} = \frac{V\ell\rho}{\mu}$$

$$\text{Froude number} = \frac{\text{inertial force}}{\text{gravity force}} \qquad \text{Fr} = \frac{V^2}{\ell g}$$

$$\text{Mach number} = \frac{\text{inertial force}}{\text{compressibility force}} \qquad \text{M} = \frac{V}{c}$$

$$\text{Weber number} = \frac{\text{inertial force}}{\text{surface tension force}} \qquad \text{We} = \frac{V^2\ell\rho}{\sigma}$$

$$\text{Pressure coefficient} = \frac{\text{pressure force}}{\text{inertial force}} \qquad C_p = \frac{\Delta p}{\frac{1}{2}\rho V^2}$$

$$\text{Drag coefficient} = \frac{\text{drag force}}{\text{inertial force}} \qquad C_D = \frac{\text{drag}}{\frac{1}{2}\rho V^2 A}$$

So, rather than writing the drag force on a cylinder as a function of length ℓ, diameter D, velocity V, viscosity μ, and density ρ, i.e.,

$$F_D = f(\ell, D, V, \mu, \rho) \tag{6.3.2}$$

we express the relationship using dimensionless parameters as

$$C_D = f\left(\frac{V\rho D}{\mu}, \frac{\ell}{D}\right). \tag{6.3.3}$$

The subject of similarity is encountered when attempting to use the results of a model study in predicting the performance of a prototype. We always assume *geometric similarity*, that is, the model is constructed to scale with the prototype; the length scale $\ell_p/\ell_m = \lambda$ is usually designated. The primary notion is simply stated: *Dimensionless quantities associated with the model are equal to corresponding dimensionless quantities associated with the prototype.* For example, if viscous effects dominate we would require

$$\text{Re}_m = \text{Re}_p. \tag{6.3.4}$$

Then if we are interested in, for example, the drag force, we would demand the dimensionless forces to be equal:

$$(F_D)_m^* = (F_D)_p^* \tag{6.3.5}$$

where the asterisk * denotes a dimensionless quantity. Since force is pressure (ρV^2) times area (ℓ^2), the above equation can be expressed in terms of dimensional quantities:

$$\frac{(F_D)_m}{\rho_m {V_m}^2 {\ell_m}^2} = \frac{(F_D)_p}{\rho_p {V_p}^2 {\ell_p}^2}. \tag{6.3.6}$$

TABLE 6.3 Symbols and Dimensions of Quantities Used in Fluid Mechanics

Quantity	Symbol	Dimensions	Quantity	Symbol	Dimensions
Length	ℓ	L	Pressure	p	M/LT^2
Time	t	T	Stress	τ	M/LT^2
Mass	m	M	Density	ρ	M/L^3
Force	F	ML/T^2	Specific Weight	γ	M/L^2T^2
Velocity	V	L/T	Viscosity	μ	M/LT
Acceleration	a	L/T^2	Kinematic Viscosity	ν	L^2/T
Frequency	ω	T^{-1}	Work	W	ML^2/T^2
Gravity	g	L/T^2	Power	$\dot{W}$	ML^2/T^3
Area	A	L^2	Heat Flux	$\dot{Q}$	ML^2/T^3
Flow Rate	Q	L^3/T	Surface Tension	σ	M/T^2
Mass Flux	$\dot{m}$	M/T	Bulk Modulus	K	M/LT^2

This would allow us to predict the drag force expected on the prototype as

$$(F_D)_p = (F_D)_m \frac{\rho_p \ell_p^{\,2} V_p^{\,2}}{\rho_m \ell_m^{\,2} V_m^{\,2}}. \tag{6.3.7}$$

The same strategy is used for other quantities of interest.

Example 6.7

Combine $\dot{W}$, Q, γ, and H as a dimensionless parameter.

Solution. First, let us note the dimensions on each variable:

$$[\dot{W}] = \frac{ML^2}{T^3} \qquad [Q] = \frac{L^3}{T} \quad [\gamma] = \frac{M}{L^2T^2} \qquad [H] = L$$

Now, by inspection we simply form the dimensionless parameter. Note that to eliminate the mass unit, $\dot{W}$ and γ must appear as the ratio, $\dot{W}/\gamma$. This puts an extra time unit in the denominator; hence, Q must appear with γ as $\dot{W}/\gamma Q$. Now, we inspect the length unit and find one length unit still in the numerator. This requires H in the denominator giving the dimensionless parameter as

$$\frac{\dot{W}}{\gamma QH}$$

Example 6.8

If a flow rate of 0.2 m^3/s is measured over a 9-to-1 scale model of a weir, what flow rate can be expected on the prototype?

Solution. First, we recognize that gravity forces dominate (as they do in all problems involving weirs, dams, ships, and open channels), and demand that

$$\text{Fr}_p = \text{Fr}_m \quad \text{or} \quad \frac{V_p^{\ 2}}{\ell_p g_p} = \frac{V_m^{\ 2}}{\ell_m g_m}.$$

$$\therefore \frac{V_p}{V_m} = \sqrt{\frac{\ell_p}{\ell_m}} = 3.$$

The dimensionless flow rates are now equated:

$$Q_p{}^* = Q_m{}^*$$

$$\frac{Q_p}{V_p \ell_p^{\ 2}} = \frac{Q_m}{V_m \ell_m^{\ 2}}$$

recognizing that velocity times area ($V \times \ell^2$) give the flow rate. We have

$$Q_p = Q_m \frac{V_p \ell_p^{\ 2}}{V_m \ell_m^{\ 2}}$$

$$= 0.2 \times 3 \times 9^2 = 48.6 \ \text{m}^3/\text{s}.$$

6.4 Control Volume Equations

When solving problems in fluid dynamics, we are most often interested in volumes into which and from which fluid flows; such volumes are called *control volumes*. The control volume equations include the conservation of mass (the continuity equation), Newton's second law (the momentum equation), and the first law of thermodynamics (the energy equation). We will not derive the equations but simply state them and then apply them to some situations of interest. We will assume *steady, incompressible flow* with *uniform velocity profiles*. The equations take the following forms:

continuity: $$A_1 V_1 = A_2 V_2 \tag{6.4.1}$$

momentum: $$\Sigma \mathbf{F} = \rho Q (\mathbf{V}_2 - \mathbf{V}_1) \tag{6.4.2}$$

energy $$-\frac{\dot{W}_S}{\gamma Q} = \frac{V_2^{\ 2} - V_1^{\ 2}}{2g} + \frac{p_2 - p_1}{\gamma} + z_2 - z_1 + h_f \tag{6.4.3}$$

where

$$Q = AV = \text{flow rate} \tag{6.4.4}$$

$\dot{W}_S$ = shaft work (positive for a turbine)

h_f = head loss.

If there is no shaft work term $\dot{W}_S$ (due to a pump or turbine) between the two sections and the losses are zero, then the energy equation reduces to the Bernoulli equation, namely,

$$\frac{V_2^{\ 2}}{2g} + \frac{p_2}{\gamma} + z_2 = \frac{V_1^{\ 2}}{2g} + \frac{p_1}{\gamma} + z_1. \tag{6.4.5}$$

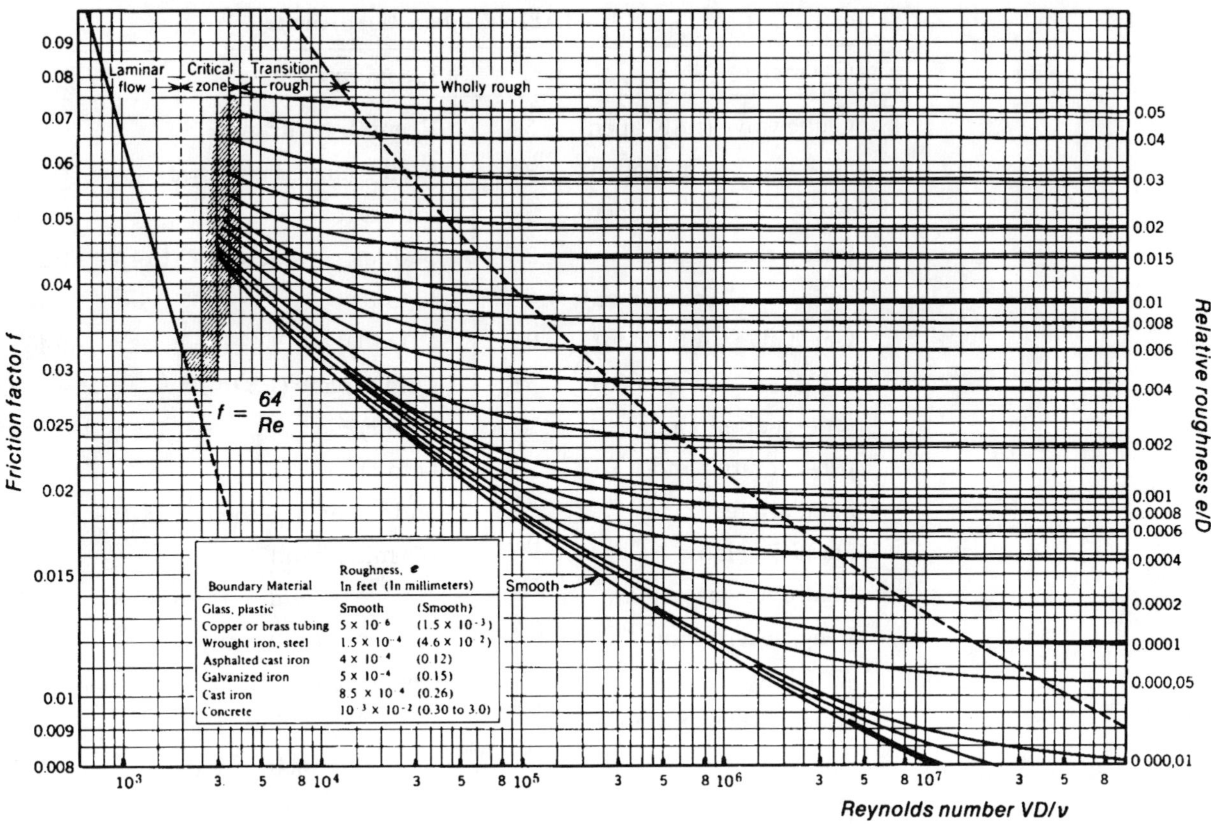

Boundary Material	Roughness, e In feet	(In millimeters)
Glass, plastic	Smooth	(Smooth)
Copper or brass tubing	5×10^{-6}	(1.5×10^{-3})
Wrought iron, steel	1.5×10^{-4}	(4.6×10^{-2})
Asphalted cast iron	4×10^{-4}	(0.12)
Galvanized iron	5×10^{-4}	(0.15)
Cast iron	8.5×10^{-4}	(0.26)
Concrete	$10^{-3} \times 10^{-2}$	(0.30 to 3.0)

Figure 6.2 The Moody Diagram.

For flow in a pipe, the head loss can be related to the friction factor by the Darcy-Weisbach equation,

$$h_f = f \frac{L}{D} \frac{V^2}{2g} \tag{6.4.6}$$

where the friction factor is related to the Reynolds number, $\text{Re} = VD/\nu$, and the relative roughness e/D by the Moody diagram, Fig. 6.2; the roughness e is given for various materials. Note that for completely turbulent flows, the friction factor is constant so that the head loss varies with the square of the velocity. For laminar flow the head loss is directly proportional to the velocity.

For sudden geometry changes, such as valves, elbows, and enlargements, the head loss (often called a minor loss) is determined by using a loss coefficient C; that is,

$$h_f = C \frac{V^2}{2g} \tag{6.4.7}$$

where V is the characteristic velocity associated with the device. Typical values are given in Table 6.4.

In engineering practice, the loss coefficient is often expressed as an *equivalent length* L_e of pipe; if that is done, the equivalent length is expressed as

$$L_e = C \frac{D}{f}. \tag{6.4.8}$$

TABLE 6.4 Loss Coefficients

Geometry	C	Geometry	C
Globe valve (fully open)	6.4	Reentrant entrance	0.8
(half open)	9.5	Well-rounded entrance	0.03++
Angle valve (fully open)	5.0	Pipe exit	1.0
Swing check valve	2.5	Sudden contraction (2 to 1)*	0.25++
(fully open)		(5 to 1)*	0.41++
Gate valve (fully open)	0.2	(10 to 1)*	0.46++
(half open)	5.6	Orifice plate (1.5 to 1)*	0.85
(one-quarter open)	24.	(2 to 1)*	3.4
Close return bend	0	(4 to 1)*	29
Standard tee	2.2	Sudden enlargement+	$(1-A_1/A_2)^2$
Standard elbow	0.9	90° miter bend	
Medium sweep elbow	0.9	(without vanes)	1.1
Long sweep elbow	0.7	(with vanes)	0.2
	5		
45° elbow		General contraction	
Square-edged entrance	0.6	(30° included angle)	0.02
	0.4	(70° included angle)	0.07
	0.5		

*Area ratio +Based on V_1 ++Based on V_2

The above analysis, using the Moody diagram and the loss coefficients, can be applied directly to only circular cross-section conduits; if the cross section is non-circular but fairly "open" (rectangular with aspect ratio less than four, oval, or triangular), a good approximation can be obtained by using the *hydraulic radius* defined by

$$R_H = A/P \tag{6.4.9}$$

where A is the cross sectional area and P is the *wetted perimeter* (that perimeter where the fluid is in contact with the solid boundary). Using this formula the diameter of a pipe is $D = 4R_H$. The Reynolds number then takes the form

$$\text{Re} = \frac{4VR_H}{\nu}. \tag{6.4.10}$$

If the shape is not "open," such as flow in an anulus, the error in using the above relationships will be quite significant.

A final note in this article defines the energy grade line (*EGL*) and the hydraulic grade line (*HGL*). The distance $(z+p/\gamma)$ above the datum (the zero elevation line) locates the *HGL*, and the distance $(z+p/\gamma+V^2/2g)$ above the datum locates the *EGL*. These are shown in Fig. 6.3. Note that the pump head H_P is given by

$$H_P = -\frac{\dot{W}_P}{\gamma Q}. \tag{6.4.11}$$

The negative sign is necessary since the pump power $\dot{W}_P$ is negative.

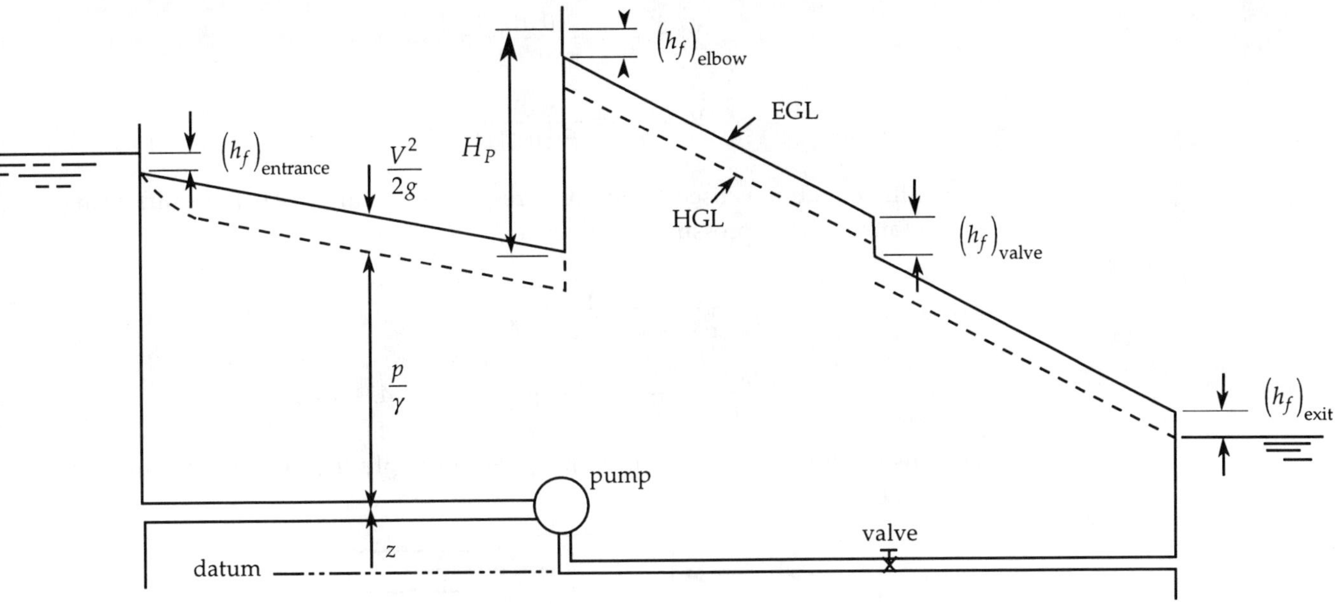

Figure 6.3 The energy grade line *(EGL)* and the hydraulic grade line *(HGL)*.

Example 6.9

The velocity in a 2-cm-dia pipe is 10 m/s. If the pipe enlarges to 4-cm-dia, find the velocity and the flow rate.

Solution. The continuity equation is used as follows:

$$A_1V_1 = A_2V_2$$

$$\frac{\pi D_1^2}{4}V_1 = \frac{\pi D_2^2}{4}V_2.$$

$$\therefore V_2 = V_1\frac{D_1^2}{D_2^2} = 10\times\frac{2^2}{4^2} = 2.5 \text{ m/s}.$$

The flow rate is

$$Q = A_1V_1$$

$$= \pi\times 0.01^2\times 10 = 0.00314 \text{ m}^3/\text{s}.$$

Example 6.10

What force is exerted on the joint if the flow rate of water is 0.01 m^3/s?

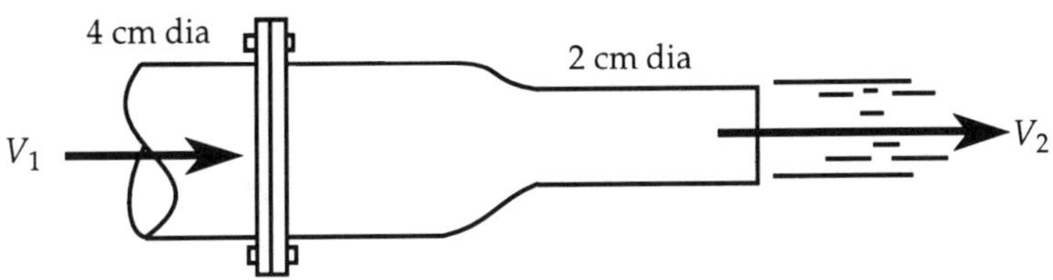

Solution. The velocities are found to be

$$V_1 = \frac{Q}{A_1} = \frac{0.01}{\pi \times 0.02^2} = 7.96 \text{ m/s}$$

$$V_2 = \frac{Q}{A_2} = \frac{0.01}{\pi \times 0.01^2} = 31.8 \text{ m/s}.$$

Bernoulli's equation is used to find the pressure at section 1. There results, using $p_2 = 0$ (atmospheric pressure is zero gage),

$$\frac{V_1^2}{2g} + \frac{p_1}{\gamma} = \frac{V_2^2}{2g} + \frac{p_2}{\gamma}$$

$$\frac{7.96^2}{2 \times 9.8} + \frac{p_1}{9800} = \frac{31.8^2}{2 \times 9.8}. \qquad \therefore p_1 = 474\ 000 \text{ Pa}.$$

Now, using the control volume shown, we can apply the momentum equation (6.4.2) in the x-direction:

$p_1 A_1$ → F_j ← x

F_j = the force of the contraction on the water

$$p_1 A_1 - F_j = \rho Q (V_2 - V_1)$$

$$474\ 000 \times \pi \times 0.02^2 - F_j = 1000 \times 0.01(31.8 - 7.96)$$

$$\therefore F_j = 357 \text{ N}.$$

Note: Remember, all forces on the control volume must be included. Never forget the pressure force.

Example 6.11

What is the force F needed to hold the plate as shown?

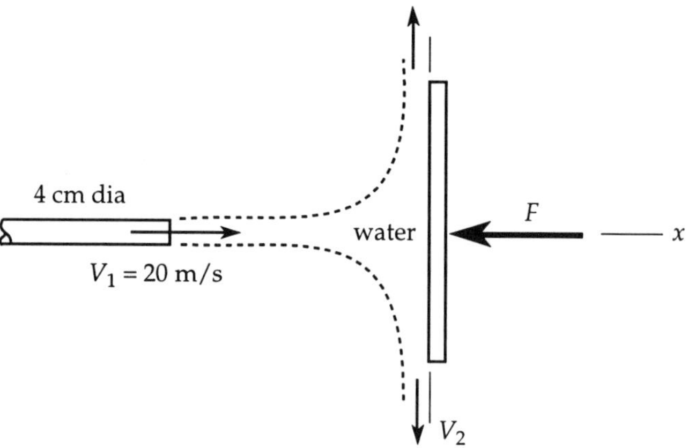

Solution. The momentum equation (6.4.2) is a vector equation; applying it in the x-direction results in

$$-F = \rho Q(\overset{0}{\cancel{V_{2x}}} - V_{1x}).$$

$$\therefore F = \rho A_1 V_1^2$$

$$= 1000 \times \pi \times 0.02^2 \times 20^2 = 503\,\text{N}.$$

Note: Since the water is open to the atmosphere, $p_2 = p_1$, and if we neglect elevation changes, Bernoulli's equation requires $V_2 = V_1$. However, here $V_{2x} = 0$ so V_2 was not necessary.

Example 6.12

What is the pump power needed to increase the pressure by 600 kPa in a 8-cm-dia pipe transporting 0.04 m^3/s of water?

Solution. The energy equation (6.4.3) is used:

$$-\frac{\dot{W}_S}{\gamma Q} = \overset{0}{\cancel{\frac{V_2^2 - V_1^2}{2g}}} + \frac{p_2 - p_1}{\gamma} + \overset{0}{\cancel{z_2 - z_1}}$$

$$-\frac{-\dot{W}_P}{9800 \times 0.04} = \frac{600\,000}{9800}. \qquad \therefore \dot{W}_P = 24\,000 \text{ W} \quad \text{or} \quad 24 \text{ kW}.$$

Example 6.13

A pitot tube is used to measure the velocity in the pipe. If $V = 15$ m/s, what is H ?

Solution. Bernoulli's equation can be used to relate the pressure at pt. 2, which is just inside the tube, to the velocity V. It gives

$$\overset{0}{\cancel{\frac{V_2^2}{2g}}} + \frac{p_2}{\gamma} + z_2 = \frac{V_1^2}{2g} + \frac{p_1}{\gamma} + z_1$$

$$\therefore p_2 = p_1 + \gamma \frac{V_1^2}{2g}.$$

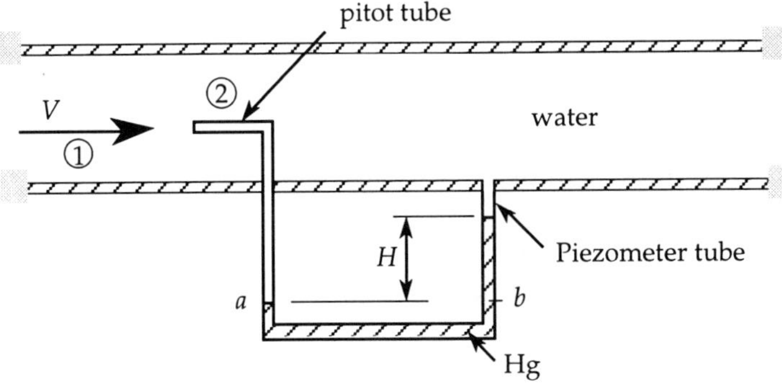

The manometer allows us to write

$$p_a = p_b$$

$$\gamma H + p_2 = \gamma_{Hg} H + p_1$$

where the distance from pt. 2 to the top of H balances on each side. Substituting for p_2 we have

$$\gamma H + p_1 + \gamma \frac{V_1^2}{2g} = \gamma_{Hg} H + p_1$$

$$\therefore H = \frac{\gamma}{\gamma_{Hg} - \gamma} \cdot \frac{V_1^2}{2g}$$

$$= \frac{9800}{13.6 \times 9800 - 9800} \cdot \frac{15^2}{2 \times 9.8} = 0.91 \text{ m}.$$

Note: The piezometer tube on the right leg measures the pressure p_1 in the pipe.

Example 6.14

For a flow rate of 0.02 m^3/s, find the turbine output if it is 80% efficient.

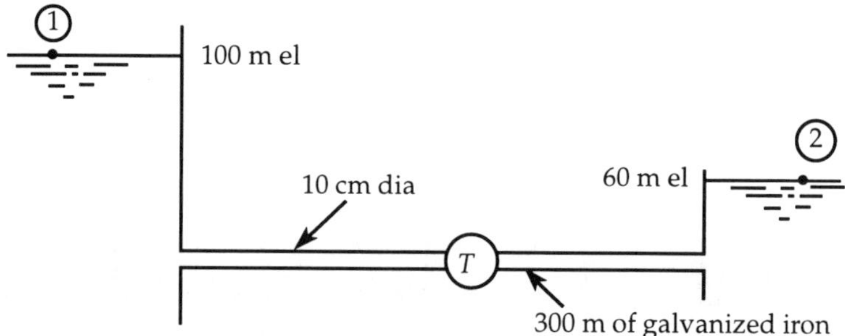

Solution. The energy equation (6.4.3) takes the form:

$$-\frac{\dot{W}_T}{\gamma Q} = \frac{V_2^2 - V_1^2}{2g} + \frac{p_2 - p_1}{\gamma} + z_2 - z_1 + \left(C_{inlet} + C_{exit} + f\frac{L}{D} \right) \frac{V^2}{2g}$$

(with $\frac{V_2^2 - V_1^2}{2g}$ and $\frac{p_2 - p_1}{\gamma}$ struck out and marked 0)

where Eqs. 6.4.6 and 6.4.7 have been used for the head loss. To find f, using the Moody diagram, we need

$$V = \frac{0.02}{\pi \times 0.05^2} = 2.55 \text{ m/s} \quad \text{Re} = \frac{2.55 \times 0.1}{10^{-6}} = 2.55 \times 10^5 \qquad \frac{e}{D} = \frac{0.15}{100} = 0.0015$$

$$\therefore f = 0.022.$$

Using the loss coefficients from Table 6.4 we have

$$-\frac{\dot{W}_T / 0.8}{9800 \times 0.02} = 60 - 100 + \left(0.5 + 1.0 + 0.022 \frac{300}{0.1} \right) \frac{2.55^2}{2 \times 9.8}$$

$$\therefore \dot{W}_T = 2760 \text{ W} \quad \text{or} \quad 2.76 \text{ kW}.$$

Example 6.15

The pressure drop over a 4-cm-dia, 300-m-long section of pipe is measured to be 120 kPa. If the elevation drops 25 m over that length of pipe and the flow rate is 0.003 m^3/s, calculate the friction factor and the power loss.

Solution. The velocity is found to be

$$V = \frac{Q}{A} = \frac{0.003}{\pi \times 0.02^2} = 2.39 \text{ m/s}.$$

The energy equation (6.4.3) with Eq. 6.4.6 then gives

$$-\frac{\dot{W}_S}{\gamma Q}^{\nearrow 0} = \frac{V_2^2 - V_1^2}{2g}^{\nearrow 0} + \frac{p_2 - p_1}{\gamma} + z_2 - z_1 + f\frac{L}{D}\frac{V^2}{2g}$$

$$0 = -\frac{120\,000}{9800} - 25 + f\frac{300}{0.04}\frac{2.39^2}{2 \times 9.8}$$

$$\therefore f = 0.0170, \qquad h_f = 37.2 \text{ m}$$

The power loss is

$$\dot{W}_{friction} = h_f \gamma Q$$
$$= 37.2 \times 9800 \times 0.003 = 1095 \text{ W}$$

where the head loss is analogous to the pump head in Eq. 6.4.11.

Example 6.16

Estimate the loss coefficient for an orifice plate if $A_1/A_0 = 2$.

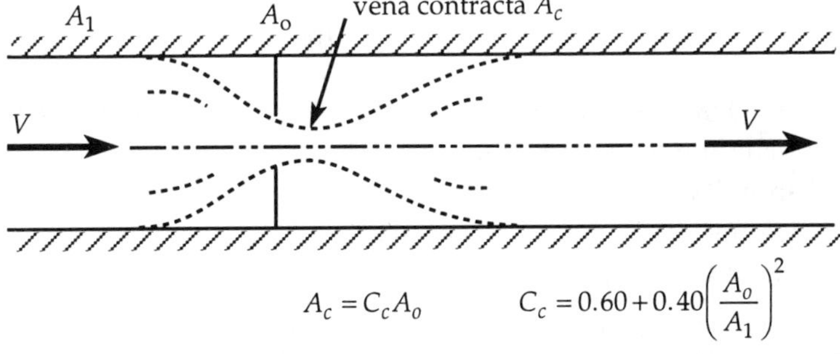

$$A_c = C_c A_o \qquad C_c = 0.60 + 0.40\left(\frac{A_o}{A_1}\right)^2$$

Solution. We approximate the flow situation shown as a gradual contraction up to A_c and a sudden enlargement from A_c back to A_1. The loss coefficient for the contraction is very small so it will be neglected. For the enlargement, we need to know A_c; it is

$$A_c = C_c A_0 = \left[0.6 + 0.4\left(\frac{A_0}{A_1}\right)^2\right] A_0 = \left[0.6 + 0.4\left(\frac{1}{2}\right)^2\right]\frac{A_1}{2} = 0.35 A_1.$$

Using the loss coefficient for an enlargement from Table 6.4, there results

$$h_f = C\frac{V_c^{\,2}}{2g}$$

$$= \left(1-\frac{A_c}{A_1}\right)^2 \frac{V_c^{\,2}}{2g} = (1-0.35)^2 \frac{1}{0.35^2}\frac{V_1^{\,2}}{2g} = 3.45\frac{V_1^{\,2}}{2g}$$

where the continuity equation $A_cV_c = A_1V_1$ has been used. The loss coefficient for the orifice plate is thus

$$C = 3.4$$

Note: Two-place accuracy is assumed since C_c is known to only two significant figures.

6.5 Open Channel Flow

If liquid flows down a slope in an open channel at a constant depth, the energy equation (6.4.3) takes the form

$$-\cancelto{0}{\frac{\dot{W}_S}{\gamma Q}} = \cancelto{0}{\frac{V_2^{\,2} - V_1^{\,2}}{2g}} + \cancelto{0}{\frac{p_2 - p_1}{\gamma}} + z_2 - z_1 + h_f \tag{6.5.1}$$

which shows that the head loss is given by

$$h_f = z_1 - z_2 = LS \tag{6.5.2}$$

where L is the length of the channel between the two sections and S is the slope. Since we normally have small angles, we can use $S = \tan\theta = \sin\theta = \theta$ where θ is the angle that the channel makes with the horizontal.

The Chezy-Manning equation is used to relate the flow rate to the slope and the cross section; it is

$$Q = \frac{1.0}{n} A R_H^{\,2/3} S^{1/2} \qquad \text{(metric-SI)} \tag{6.5.3}$$

where R_H is the hydraulic radius given by Eq. 6.4.9, A is the cross sectional area, and n is the Manning n, given in Table 6.5. The constant 1.0 must be replaced by 1.49 if English units are used. The most efficient cross section occurs when the width is twice the depth for a rectangular section, and when the sides make angles of 60° with the horizontal for a trapezoidal cross section.

TABLE 6.5 Average Values* of the Manning n

Wall Material	Manning n	Wall Material	Manning n
Planed wood	.012	Concrete pipe	.015
Unplaned wood	.013	Riveted steel	.017
Finished concrete	.012	Earth, straight	.022
Unfinished concrete	.014	Corrugated metal flumes	.025
Sewer Pipe	.013	Rubble	.03
Brick	.016	Earth with stones and weeds	.035
Cast iron, wrought iron	.015	Mountain streams	.05

*If $R_H > 3$ m, increase n by 15%.

Example 6.17

A 2-m-dia concrete pipe transports water at a depth of 0.8 m. What is the flow rate if the slope is 0.001?

Solution. Calculate the geometric properties:

$$\alpha = \sin^{-1}\frac{0.2}{1.0} = 11.54°$$

$\therefore \theta = 156.9°$.

$$A = \pi \times 1^2 \times \frac{156.9}{360} - 0.2 \times \cos 11.54° \times \frac{1}{2} \times 2 = 1.174 \text{ m}^2$$

$$P = \pi \times 2 \times \frac{156.9}{360} = 2.738 \text{ m}.$$

For concrete pipe, $n = 0.015$, so

$$Q = \frac{1.0}{n} A R_H^{\;2/3} S^{1/2} = \frac{1.0}{0.015} \times 1.174 \times \left(\frac{1.174}{2.738}\right)^{2/3} \times 0.001^{1/2} = 1.41 \text{ m}^3/\text{s}.$$

6.6 Compressible Flow

A gas flow with a Mach number below 0.3 (at standard conditions this means velocities less than about 100 m/s) can be treated as an incompressible flow, as in previous articles. If the Mach number is greater than 0.3, the density variation must be accounted for. For such problems, we use the control volume equation as

continuity: $$\rho_1 A_1 V_1 = \rho_2 A_2 V_2 \tag{6.6.1}$$

momentum (x-direction): $$\sum F = \dot{m}(V_2 - V_1) \tag{6.6.2}$$

energy: $$\frac{\dot{Q} - \dot{W}_s}{\dot{m}} = \frac{V_2^{\,2} - V_1^{\,2}}{2} + c_p(T_2 - T_1) \tag{6.6.3}$$

where

$$\dot{m} = \rho A V = \text{mass flux} \tag{6.6.4}$$

c_p = constant pressure specific heat

We often use the ideal gas relations (they become inaccurate at high pressure or low temperature)

$$p = \rho R T \qquad c_p = c_v + R \qquad k = c_p/c_v \tag{6.6.5}$$

The energy equation for an ideal gas takes the form

$$\frac{\dot{Q} - \dot{W}_S}{\dot{m}} = \frac{V_2^{\,2} - V_1^{\,2}}{2} + \frac{k}{k-1}\left(\frac{p_2}{\rho_2} - \frac{p_1}{\rho_1}\right) \tag{6.6.6}$$

We recall that the speed of sound and Mach number are given by

$$c = \sqrt{kRT} \qquad \text{M} = V/c \tag{6.6.7}$$

Subsonic flow occurs whenever M < 1 and supersonic flow whenever M > 1. In subsonic flows, losses are quite small and isentropic flows can usually be assumed; thus, we can relate the properties by the isentropic relations

$$\frac{T_2}{T_1} = \left(\frac{p_2}{p_1}\right)^{\frac{k-1}{k}} \qquad \frac{p_2}{p_1} = \left(\frac{\rho_2}{\rho_1}\right)^{k} \tag{6.6.8}$$

For air $k = 1.4$, $c_p = 1.00\ \text{kJ/kg}\cdot\text{K}$, $R = 0.287\ \text{kJ/kg}\cdot\text{K}$. The isentropic flow table (Table 6.6) can also be used.

For supersonic flows, shock waves are encountered; across a normal shock wave, Table 6.7 can be used with isentropic flow assumed before and after the shock. If the entropy change is desired, it is given by

$$\Delta s = c_p \ln\frac{T_2}{T_1} - R\ln\frac{p_2}{p_1} \tag{6.6.9}$$

Supersonic flow behaves quite differently from subsonic flow; its velocity increases with increasing area and decreases with decreasing area. Hence, to obtain a supersonic flow from a reservoir, the flow must first accelerate through a converging section to a throat where $M = 1$; then an enlarging section will allow it to reach supersonic speed with $M > 1$. Supersonic flow cannot occur in a converging section only.

Example 6.18

Air is flowing from a 20°C reservoir to the atmosphere through the converging-diverging nozzle shown. What reservoir pressure will locate a normal shock wave at the exit? Also, find V_C and the mass flux.

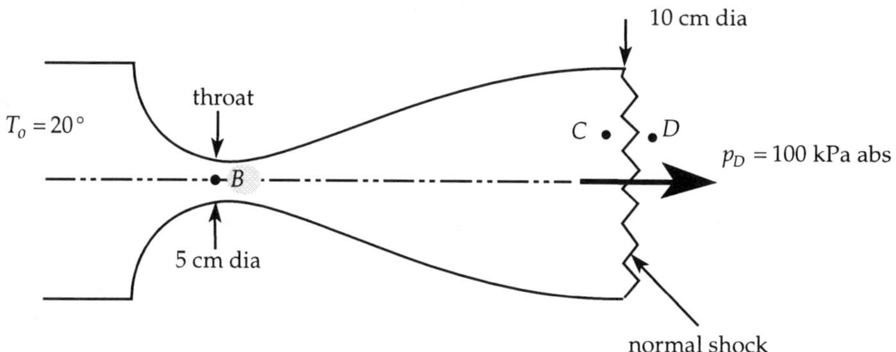

Solution. We know that at the throat the Mach number is unity. Such an area is the *critical area* and is designated A^* in the isentropic flow table. Between points B and C we have

$$\frac{A_C}{A_B} = \frac{A}{A^*} = \frac{\pi \times 5^2}{\pi \times 2.5^2} = 4.$$

Using Table 6.6 we find

$$\text{M}_C = 2.94.$$

Then from Table 6.7 we find, across the shock,

$$\frac{p_C}{p_D} = 9.92 \qquad \therefore p_C = 9.92 \times 100 = 992\ \text{kPa}$$

The isentropic flow Table 6.6 then gives, at M = 2.94,

$$\frac{p_C}{p_o} = 0.0298 \qquad \therefore p_o = \frac{992}{0.0298} = 33\,300 \text{ kPa}$$

To find V_C we must determine T_C. Using Table 6.6 at $M = 2.94$ we find

$$\frac{T_C}{T_o} = 0.3665 \qquad \therefore T_C = (273 + 20) \times 0.3665 = 107.4 \text{ K}.$$

The velocity is then found to be

$$\begin{aligned} V_C &= M_C\sqrt{kRT_C} \\ &= 2.94\sqrt{1.4 \times 287 \times 107.4} = 611 \text{ m/s}. \end{aligned}$$

The mass flux is

$$\begin{aligned} \dot{m} &= \rho_C A_C V_C \\ &= \frac{p_C}{RT_C} A_C V_C = \frac{992\,000}{287 \times 107.4} \times \pi \times 0.05^2 \times 611 = 154 \text{ kg/s}. \end{aligned}$$

TABLE 6.6 Isentropic Flow (air, k = 1.4)

M_1	A/A^*	p/p_0	ρ/ρ_0	T/T_0	V/V^*
0.00	∞	1.000	1.000	1.000	0.000
0.05	11.591	0.998	0.999	1.000	0.055
0.10	5.822	0.993	0.995	0.998	0.109
0.15	3.910	0.984	0.989	0.996	0.164
0.20	2.964	0.972	0.980	0.992	0.218
0.25	2.403	0.957	0.969	0.988	0.272
0.30	2.035	0.939	0.956	0.982	0.326
0.35	1.778	0.919	0.941	0.976	0.379
0.40	1.590	0.896	0.924	0.969	0.431
0.45	1.449	0.870	0.906	0.961	0.483
0.50	1.340	0.843	0.885	0.952	0.535
0.55	1.255	0.814	0.863	0.943	0.585
0.60	1.188	0.784	0.840	0.933	0.635
0.65	1.136	0.753	0.816	0.922	0.684
0.70	1.094	0.721	0.792	0.911	0.732
0.75	1.062	0.689	0.766	0.899	0.779
0.80	1.038	0.656	0.740	0.887	0.825
0.85	1.021	0.624	0.714	0.874	0.870
0.90	1.009	0.591	0.687	0.861	0.915
0.95	1.002	0.559	0.660	0.847	0.958
1.00	1.000	0.528	0.634	0.833	1.000
1.10	1.008	0.468	0.582	0.805	1.081
1.20	1.030	0.412	0.531	0.776	1.158
1.30	1.066	0.361	0.483	0.747	1.231
1.40	1.115	0.314	0.437	0.718	1.300
1.50	1.176	0.272	0.395	0.690	1.365
1.60	1.250	0.235	0.356	0.661	1.425
1.70	1.338	0.203	0.320	0.634	1.482
1.80	1.439	0.174	0.287	0.607	1.536
1.90	1.555	0.149	0.257	0.581	1.586
2.00	1.687	0.128	0.230	0.556	1.633
2.10	1.837	0.109	0.206	0.531	1.677
2.20	2.005	0.094	0.184	0.508	1.718
2.30	2.193	0.080	0.165	0.486	1.756
2.40	2.403	0.068	0.147	0.465	1.792
2.50	2.637	0.059	0.132	0.444	1.826
2.60	2.896	0.050	0.118	0.425	1.857
2.70	3.183	0.043	0.106	0.407	1.887
2.80	3.500	0.037	0.085	0.389	1.914
2.90	3.850	0.032	0.085	0.373	1.940
3.00	4.235	0.027	0.076	0.357	1.964
3.50	6.790	0.013	0.045	0.290	2.064
4.00	10.719	0.007	0.028	0.238	2.138
4.50	16.562	0.003	0.017	0.198	2.194
5.00	25.000	0.002	0.011	0.167	2.236
5.50	36.869	0.001	0.008	0.142	2.269
6.00	53.180	0.001	0.005	0.122	2.295
6.50	75.134	0.000	0.004	0.106	2.316
7.00	104.143	0.000	0.003	0.093	2.333
7.50	141.842	0.000	0.002	0.082	2.347
8.00	190.110	0.000	0.001	0.072	2.359
8.50	251.086	0.000	0.001	0.065	2.369
8.00	327.189	0.000	0.001	0.058	2.377
9.50	421.130	0.000	0.001	0.052	2.384
10.00	535.936	0.000	0.000	0.048	2.390
∞	∞	0.000	0.000	0.000	∞

TABLE 6.7 Normal Shock Wave (air, $k = 1.4$)

M_1	M_2	p_2/p_1	T_2/T_1	ρ_2/ρ_1	p_{02}/p_{01}
1.00	1.000	1.000	1.000	1.000	1.000
1.05	0.953	1.120	1.033	1.084	1.000
1.10	0.912	1.245	1.065	1.169	0.999
1.15	0.875	1.376	1.097	1.255	0.997
1.20	0.842	1.513	1.128	1.342	0.993
1.25	0.813	1.656	1.159	1.429	0.987
1.30	0.786	1.805	1.191	1.516	0.979
1.35	0.762	1.960	1.223	1.603	0.970
1.40	0.740	2.120	1.255	1.690	0.958
1.45	0.720	2.286	1.287	1.776	0.945
1.50	0.701	2.458	1.320	1.862	0.930
1.55	0.684	2.636	1.354	1.947	0.913
1.60	0.668	2.820	1.388	2.032	0.895
1.65	0.654	3.010	1.423	2.115	0.876
1.70	0.641	3.205	1.458	2.198	0.856
1.75	0.628	3.406	1.495	2.279	0.835
1.80	0.617	3.613	1.532	2.359	0.813
1.85	0.606	3.826	1.569	2.438	0.790
1.90	0.596	4.045	1.608	2.516	0.767
1.95	0.586	4.270	1.647	2.592	0.744
2.00	0.577	4.500	1.687	2.667	0.721
2.05	0.569	4.736	1.729	2.740	0.698
2.10	0.561	4.978	1.770	2.812	0.674
2.15	0.554	5.226	1.813	2.882	0.651
2.20	0.547	5.480	1.857	2.951	0.628
2.25	0.541	5.740	1.901	3.019	0.606
2.30	0.534	6.005	1.947	3.085	0.583
2.35	0.529	6.276	1.993	3.149	0.561
2.40	0.523	6.553	2.040	3.212	0.540
2.45	0.518	6.836	2.088	3.273	0.519
2.50	0.513	7.125	2.137	3.333	0.499
2.55	0.508	7.420	2.187	3.392	0.479
2.60	0.504	7.720	2.238	3.449	0.460
2.65	0.500	8.026	2.290	3.505	0.442
2.70	0.496	8.338	2.343	3.559	0.424
2.75	0.492	8.656	2.397	3.612	0.406
2.80	0.488	8.980	2.451	3.664	0.389
2.85	0.485	9.310	2.507	3.714	0.373
2.90	0.481	9.645	2.563	3.763	0.358
2.95	0.478	9.986	2.621	3.811	0.343
3.00	0.475	10.333	2.679	3.857	0.328
3.50	0.451	14.125	3.315	4.261	0.213
4.00	0.435	18.500	4.047	4.571	0.139
4.50	0.424	23.458	4.875	4.812	0.092
5.00	0.415	29.000	5.800	5.000	0.062
5.50	0.409	35.125	6.822	5.149	0.042
6.00	0.404	41.833	7.941	5.268	0.030
6.50	0.400	49.125	9.156	5.365	0.021
7.00	0.397	57.000	10.469	5.444	0.015
7.50	0.395	65.458	11.879	5.510	0.011
8.00	0.393	74.500	13.387	5.565	0.008
8.50	0.391	84.125	14.991	5.612	0.006
9.00	0.390	94.333	16.693	5.651	0.005
9.50	0.389	105.125	18.492	5.685	0.004
10.00	0.388	116.500	20.387	5.714	0.003
∞	0.378	∞	∞	6.000	0.000

Practice Problems

(If you choose to work only a few problems, select those with a star.)

GENERAL

*6.1 A fluid is a substance that
  a) is essentially incompressible.
  b) always moves when subjected to a shearing stress.
  c) has a viscosity that always increases with temperature.
  d) has a viscosity that always decreases with temperature.
  e) expands until it fills its space.

*6.2 Viscosity has dimensions of
  a) FT^2/L b) F/TL^2 c) M/LT^2 d) M/LT e) ML/T

*6.3 The viscosity of a fluid varies with
  a) temperature. d) temperature and pressure.
  b) pressure. e) temperature, pressure, and density.
  c) density.

*6.4 In an isothermal atmosphere the pressure
  a) is constant with elevation.
  b) decreases linearly with elevation.
  c) cannot be related to elevation.
  d) decreases near the surface but approaches a constant value.
  e) decreases exponentially with elevation.

6.5 If a fluid "fails" in a piping system, the "failure" is related to which of the following properties?
  a) surface tension d) speed of sound
  b) bulk modulus e) vapor pressure
  c) viscosity

*6.6 A torque of 1.6 N·m is needed to rotate the cylinder at 1000 rad/s. Estimate the viscosity $(\text{N}\cdot\text{s/m}^2)$.
  a) 0.1 b) 0.2 c) 0.3
  d) 0.4 e) 0.5

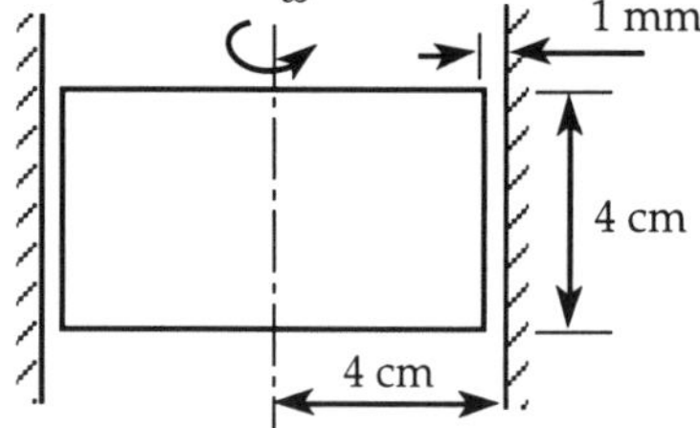

6.7 A pressure of 500 kPa applied to 2 m^3 of liquid results in a volume change of 0.004 m^3. The bulk modulus, in MPa, is
  a) 2.5 b) 25 c) 250 d) 2500 e) 2.5×10^6

*6.8 Water at 20°C will rise, in a clean 1-mm-dia glass tube, a distance, in cm, of
  a) 1 b) 2 c) 3 d) 4 e) 5

6.9 Water at 20°C flows in a piping system at a low velocity. At what pressure, in kPa abs, will cavitation result?
a) 35.6 b) 20.1 c) 10.6 d) 5.67 e) 2.45

6.10 A man is observed to strike an object and 1.2 s later the sound is heard. How far away, in meters, is the man?
a) 220 b) 370 c) 410 d) 520 e) 640

6.11 The viscosity of a fluid with specific gravity 1.3 is measured to be 0.0034 $N \cdot s/m^2$. Its kinematic viscosity, in m^2/s, is
a) 2.6×10^{-6} b) 4.4×10^{-6} c) 5.8×10^{-6} d) 7.2×10^{-6} e) 9.6×10^{-6}

FLUID STATICS

6.12 Fresh water 2 m deep flows over the top of 4 m of salt water $(SG = 1.04)$. The pressure at the bottom, in kPa, is
a) 60.4 b) 58.8 c) 55.2 d) 51.3 e) 47.9

*6.13 What pressure, in kPa, is equivalent to 600 mm of Hg?
a) 100 b) 95.2 c) 80.0 d) 55.2 e) 51.3

6.14 What pressure, in MPa, must be maintained in a diving bell, at a depth of 1200 m, to keep out the ocean water $(SG = 1.03)$?
a) 1.24 b) 5.16 c) 9.32 d) 12.1 e) 14.3

6.15 Predict the pressure, in kPa, at an elevation of 2000 m in an isothermal atmosphere assuming $T = 20°\text{C}$. Assume $p_{atm} = 100$ kPa.
a) 87 b) 82 c) 79 d) 71 e) 63

*6.16 The force F, in Newtons, is
a) 25
b) 8.9
c) 2.5
d) 1.5
e) 0.36

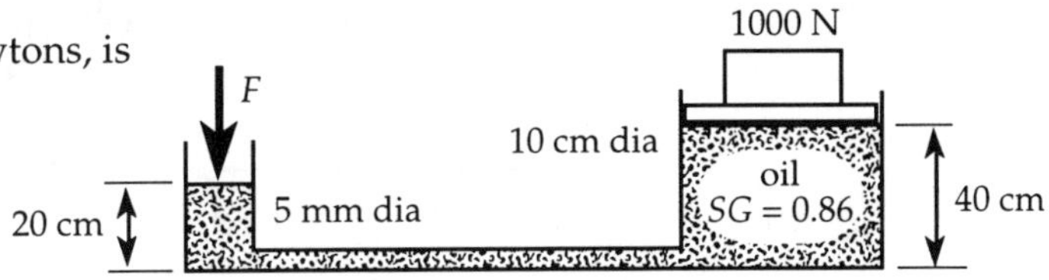

6.17 A U-tube manometer, attached to an air pipe, measures 20 cm of mercury. The pressure, in kPa, in the air pipe is
a) 26.7 b) 32.4 c) 38.6 d) 42.5 e) 51.3

*6.18 The pressure p, in kPa, is
a) 51.3
b) 48.0
c) 45.2
d) 40.0
e) 37.0

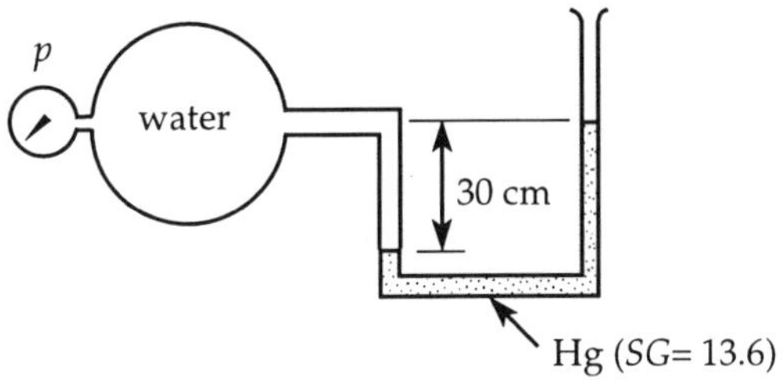

6.19 A 2-m-dia, 3-m-high, cylindrical water tank is pressurized such that the pressure at the top is 20 kPa. The force, in kN, acting on the bottom is
a) 195 b) 176 c) 155 d) 132 e) 106

6.20 The force, in kN, acting on one of the 1.5-m sides of an open cubical water tank (which is full) is
a) 18.2 b) 16.5 c) 15.3 d) 12.1 e) 10.2

*6.21 The force P, in kN, to just open the 3-m-wide gate is
a) 55
b) 60
c) 65
d) 70
e) 75

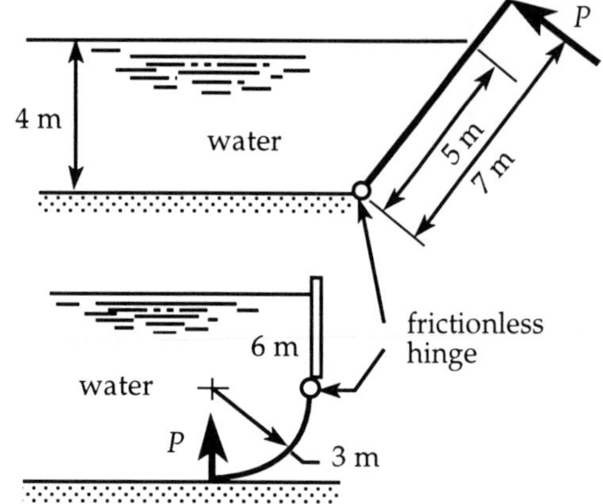

6.22 The force P, in kN, to just open the 4-m-wide gate is
a) 710
b) 762
c) 831 d) 983
e) 1220

*6.23 The force P, in kN, on the 5-m-wide gate is

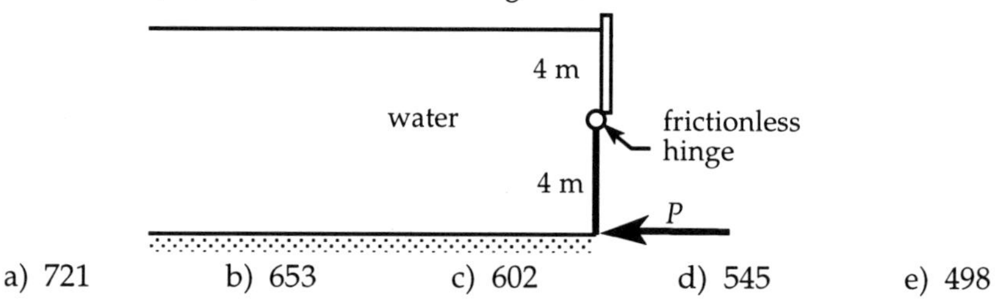

a) 721 b) 653 c) 602 d) 545 e) 498

6.24 Four cars, with a mass of 1500 kg each, are loaded on a 6-m-wide, 12-m-long small-car ferry. How far, in cm, will it sink in the water?
a) 15.2 b) 11.5 c) 10.2 d) 9.6 e) 8.3

*6.25 An object weighs 100 N in air and 25 N when submerged in water. Its specific gravity is
a) 1.11 b) 1.22 c) 1.33 d) 1.44 e) 1.55

6.26 What pressure differential, in Pascals, exists at the bottom of a 3 m vertical wall if the temperature inside is 20°C and outside it is –20°C? Assume equal pressures at the top.
a) 15 b) 12 c) 9 d) 6 e) 3

DIMENSIONLESS PARAMETERS

6.27 Arrange pressure p, flow rate Q, diameter D, and density ρ into a dimensionless group.
a) $pQ^2/\rho D^4$ b) $p/\rho Q^2 D^4$ c) $pD^4\rho/Q^2$ d) $pD^4/\rho Q^2$ e) $p/\rho Q^2$

*6.28 Combine surface tension σ, density ρ, diameter D, and velocity V into a dimensionless para-meter.
a) $\sigma/\rho V^2 D$ b) $\sigma D/\rho V$ c) $\sigma\rho/VD$ d) $\sigma V/\rho D$ e) $\sigma D^2/\rho V$

*6.29 The Reynolds number is a ratio of
a) velocity effects to viscous effects.
b) inertial forces to viscous forces.
c) mass flux to viscosity.
d) flow rate to kinematic viscosity.
e) mass flux to kinematic viscosity.

6.30 The Froude number is a ratio of
a) inertial forces to viscous forces.
b) body forces to viscous forces.
c) body forces to pressure forces.
d) inertial forces to body forces.
e) inertial forces to pressure forces.

SIMILITUDE

*6.31 What flow rate, in m^3/s, is needed using a 20:1 scale model of a dam over which 4 m^3/s of water flows?
a) 0.010 b) 0.0068 c) 0.0047 d) 0.0022 e) 0.0015

6.32 It is proposed to model a submarine moving at 10 m/s by testing a 10:1 scale model. What velocity, in m/s, would be needed in the model study?
a) 1 b) 10 c) 40 d) 80 e) 100

6.33 The drag force on a 40:1 scale model of a ship is measured to be 10 N. What force, in kN, is expected on the ship?
a) 640 b) 520 c) 320 d) 160 e) 80

6.34 The power output of a 10:1 scale model of a water wheel is measured to be 20 W. The power output in kW expected from the prototype is
a) 200 b) 150 c) 100 d) 63 e) 2

CONTINUITY

6.35 The velocity in a 2-cm-dia pipe is 20 m/s. If the pipe enlarges to 5-cm-dia, the velocity, in m/s, will be
a) 8.0 b) 6.4 c) 5.2 d) 4.8 e) 3.2

*6.36 A 2-cm-dia pipe transports water at 20 m/s. If it exits out 100 small 2-mm-dia holes, the exiting velocity, in m/s, will be
a) 120 b) 80 c) 40 d) 20 e) 10

6.37 Water flows through a 2-cm-dia pipe at 20 m/s. It then flows radially outward between two discs, 2 mm apart. When it reaches a radius of 40 cm, its velocity, in m/s, will be
a) 5.0 b) 2.5 c) 2.25 d) 1.85 e) 1.25

BERNOULLI'S EQUATION

6.38 Select the false statement for the Bernoulli equation.
a) It can be used for an unsteady flow.
b) It can be used along a streamline.
c) It can be used in an inertial coordinate system.
d) It can be used in an inviscid flow.
e) It can be used in a low speed air flow.

6.39 The pressure force, in Newtons, on the 15-cm-dia headlight of an automobile traveling at 25 m/s is

a) 10.4 b) 6.8 c) 5.6 d) 4.8 e) 3.2

*6.40 The pressure inside a 4-cm-dia hose is 700 kPa. If the water exits through a 2-cm-dia nozzle, what velocity, in m/s, can be expected inside the hose?

a) 20.4 b) 16.3 c) 12.4 d) 10.5 e) 9.7

6.41 Calculate V, in m/s.

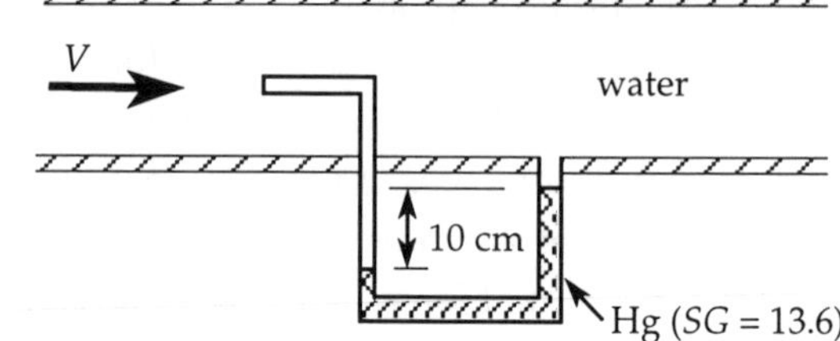

a) 8
b) 7
c) 6
d) 5
e) 4

6.42 Water enters a turbine at 900 kPa with negligible velocity. What maximum speed, in m/s, can it reach before it enters the turbine rotor?

a) 52 b) 47 c) 45 d) 42 e) 38

MOMENTUM

*6.43 If the density of the air is 1.2 kg/m^3, find F, in Newtons.

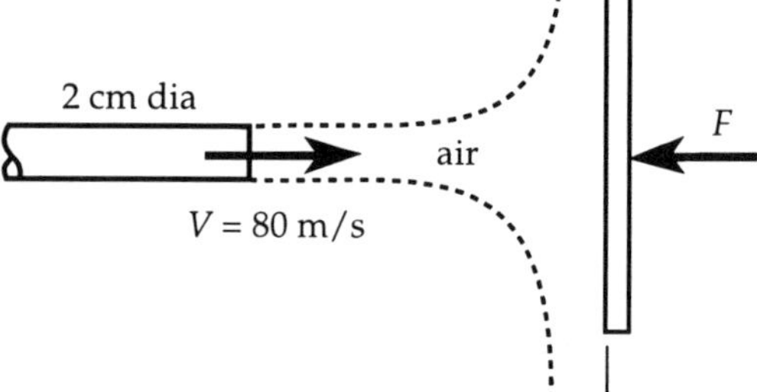

a) 2.4
b) 3.6
c) 4.8
d) 7.6
e) 9.6

6.44 If the single plate is to move at 10 m/s to the right the force F, in Newtons, is

a) 1000 b) 2000 c) 3000 d) 4000 e) 5000

6.45 A rocket exits exhaust gases with $\rho = 0.5\ kg/m^3$ out a 50-cm-dia nozzle at a velocity of 1200 m/s. Estimate the thrust, in kN.

a) 420 b) 280 c) 140 d) 90 e) 40

6.46 A high-speed vehicle, traveling at 50 m/s, dips an 80-cm-wide scoop into water and deflects the water 180°. If it dips 5 cm deep, what force, in kN, is exerted on the scoop?

a) 200 b) 150 c) 100 d) 50 e) 25

6.47 What force, in Newtons, acts on the nozzle?

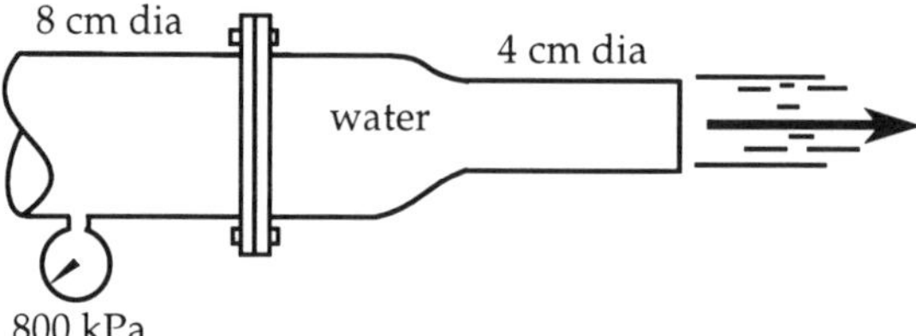

a) 4020 b) 3230 c) 2420 d) 1830 e) 1610

ENERGY

*6.48 The locus of elevations that water will rise in a series of pitot tubes is called
a) the hydraulic grade line.
b) the energy grade line.
c) the velocity head.
d) the pressure head.
e) the head loss.

6.49 The velocity head in a section of a pipe is given by
a) the difference between the *EGL* and the *HGL* at the section.
b) the elevation of the *EGL*.
c) the elevation of the *HGL*.
d) Bernoulli's equation applied from the inlet to the section.
e) a momentum balance on the fluid between the inlet and the section.

*6.50 A pressure rise of 500 kPa is needed across a pump in a pipe transporting 0.2 m^3/s of water. If the pump is 85% efficient , the power needed, in kW, would be
a) 118 b) 100 c) 85 d) 65 e) 60

6.51 An 85% efficient turbine accepts 0.8 m^3/s of water at a pressure of 600 kPa. What is the max-imum power output, in kW?
a) 820 b) 640 c) 560 d) 480 e) 410

6.52 If the turbine is 88% efficient, the power output, in kW, is

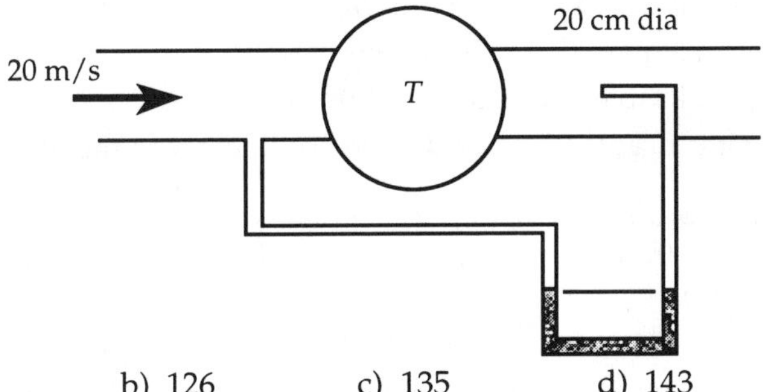

a) 111 b) 126 c) 135 d) 143 e) 176

6.53 To determine the flow rate using a square-edged orifice, the pressure must be measured upstream of the orifice and
a) at the orifice.
b) just upstream of the orifice.
c) just downstream of the orifice at the vena contracta.
d) at least 10 diameters downstream of the orifice.
e) at least 20 diameters downstream of the orifice.

6.54 Freon flows through an expansion valve. Select the best statement.
a) The temperature increases.
b) The internal energy remains constant.
c) The pressure remains constant.
d) The enthalpy remains constant.
e) The entropy decreases.

LOSSES

6.55 A laminar flow exists in a pipe flow. We know that
a) the velocity profile is linear.
b) the Reynolds number is less than 2000.
c) the shear stress distribution is linear.
d) the pipe is smooth.
e) inertial effects dominate the viscous effects.

*6.56 In a completely turbulent flow the head loss
a) increases with the velocity.
b) increases with the velocity squared.
c) decreases with wall roughness.
d) increases with diameter.
e) increases with flow rate.

*6.57 The shear stress in a turbulent pipe flow
a) varies parabolically with the radius.
b) is constant over the pipe radius.
c) varies according to the 1/7th power law.
d) is zero at the center and increases linearly to the wall.
e) is zero at the wall and increases linearly to the center.

*6.58 The velocity distribution in a turbulent flow in a pipe is often assumed to
a) vary parabolically.
b) be zero at the wall and increase linearly to the center.
c) vary according to the 1/7th power law.
d) be unpredictable and is thus not used.
e) be maximum at the wall and decrease linearly to the center.

6.59 The velocity profile between parallel plates is calculated to be Vy/δ, where y is measured from the bottom plate and δ is the distance between plates. We know that:
I. The flow is laminar. II. The flow is turbulent. III. The flow is steady.
IV. One plate is stationary and the other is moving with velocity V.
a) I, III, and IV only
b) II, III and IV only
c) I and III only
d) II and IV only
e) I and IV only

*6.60 The Moody diagram is sketched. The friction factor for turbulent flow in a smooth pipe is given by curve
a) *A*
b) *B*
c) *C*
d) *D*
e) *E*

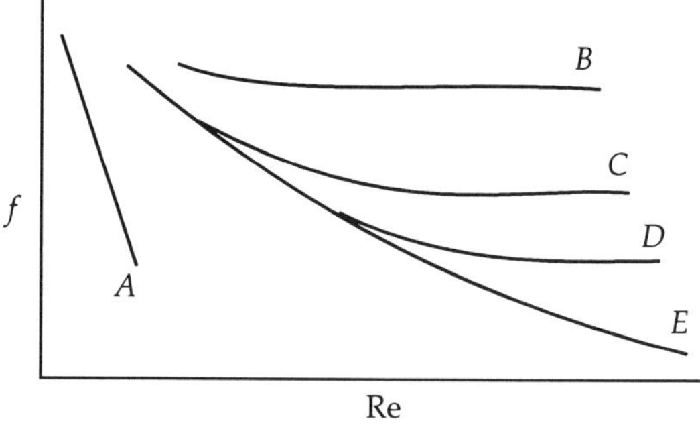

*6.61 For the Moody diagram given in Problem 6.60, the completely turbulent flow is best represented by curve
a) *A* b) *B* c) *C* d) *D* e) *E*

*6.62 The pressure gradient $(\Delta p/\Delta x)$ in a developed turbulent flow in a horizontal constant diameter pipe

a) is constant.
b) varies linearly with axial distance.
c) is zero.
d) decrease exponentially.
e) varies directly with the average velocity.

*6.63 The head loss in a pipe flow can be calculated using

a) the Bernoulli equation.
b) Darcy's law.
c) the Chezy-Manning equation.
d) the Momentum equation.
e) the Darcy-Weisbach equation.

*6.64 Minor losses in a piping system are

a) less than the friction factor losses, $f\dfrac{L}{D}\dfrac{V^2}{2g}$.
b) due to the viscous stresses.
c) assumed to vary linearly with the velocity.
d) found by using loss coefficients.
e) independent of the flow rate.

*6.65 In a turbulent flow in a pipe we know the

a) Reynolds number is greater than 10 000.
b) fluid particles move in straight lines.
c) head loss varies linearly with the flow rate.
d) shear stress varies linearly with radius.
e) viscous stresses dominate.

*6.66 Water flows through a 10-cm-dia, 100-m-long pipe connecting two reservoirs with an elevation difference of 40 m. The average velocity is 6 m/s. Neglecting minor losses, the friction factor is

a) 0.020 b) 0.022 c) 0.024 d) 0.026 e) 0.028

*6.67 Find the energy required, in kW, by the 85% efficient pump if $Q = 0.02\,\text{m}^3/\text{s}$.

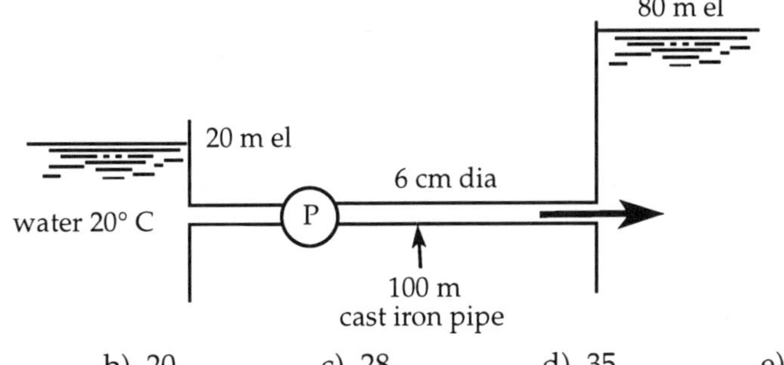

a) 14 b) 20 c) 28 d) 35 e) 44

6.68 The pressure at section A in a 4-cm-dia, wrought-iron, horizontal pipe is 510 kPa. A fully open globe valve, two elbows, and 50 meters of pipe connect section B. If $Q = 0.006\ \text{m}^3/\text{s}$ of water, the pressure p_B, in kPa, is

a) 300 b) 250 c) 200 d) 150 e) 100

6.69 Air at 20°C and 100 kPa abs is transported through 500 m of smooth, horizontal, 15 cm × 40 cm rectangular duct with a flow rate of 0.3 m^3/s. The pressure drop, in Pascals, is

a) 300 b) 400 c) 500 d) 600 e) 700

6.70 Estimate the loss coefficient C in a sudden contraction $A_1/A_2 = 2$ by neglecting the losses up to the vena contracta A_c. Assume that $A_c/A_2 = 0.62 + 0.38(A_2/A_1)^3$ and $h_f = CV_2^2/2g$.

a) 0.40 b) 0.35 c) 0.30 d) 0.25 e) 0.20

6.71 An elbow exists in a 6-cm-dia galvanized iron pipe transporting 0.02 m^3/s of water. Find the equivalent length of the elbow, in meters.

a) 6.3 b) 4.5 c) 3.6 d) 2.8 e) 2.2

OPEN CHANNEL FLOW

*6.72 The depth of water in a 3-m-wide, rectangular, finished concrete channel is 2 m. If the slope is 0.001, estimate the flow rate, in m^3/s.

a) 14 b) 13 c) 12 d) 11 e) 10

6.73 At what depth, in meters, will 10 m^3/s of water flow in a 4-m-wide, rectangular, brick channel if the slope is 0.001?

a) 1.3 b) 1.4 c) 1.5 d) 1.6 e) 1.7

6.74 A 2-m-dia, brick storm sewer transports 10 m^3/s when it's nearly full. Estimate the slope of the sewer.

a) 0.0070 b) 0.0065 c) 0.0060 d) 0.0055 e) 0.0050

COMPRESSIBLE FLOW

*6.75 The pressure and temperature in a 10-cm-dia pipe are 500 kPa abs and 40°C, respectively. What is the mass flux, in kg/s, if the velocity is 100 m/s?

a) 6.63 b) 5.81 c) 5.36 d) 4.92 e) 4.37

*6.76 Air in a reservoir at 20°C and 500 kPa abs exits a hole with a velocity, in m/s, of

a) 353 b) 333 c) 313 d) 293 e) 273

6.77 A farmer uses 20°C nitrogen pressurized to 800 kPa abs. Estimate the temperature, in °C, in the nitrogen as it exits a short hose fitted to the tank.

a) –110 b) –90 c) –70 d) –50 e) –30

6.78 Select the correct statement concerning a supersonic flow.

a) It occurs after a converging nozzle for sufficiently low receiver pressures.
b) In a converging nozzle the velocity increases.
c) It may occur in a converging-diverging nozzle attached to a reservoir.
d) In a diverging nozzle the temperature decreases.
e) In a converging nozzle the enthalpy decreases.

6.79 Estimate p_1 in kPa.

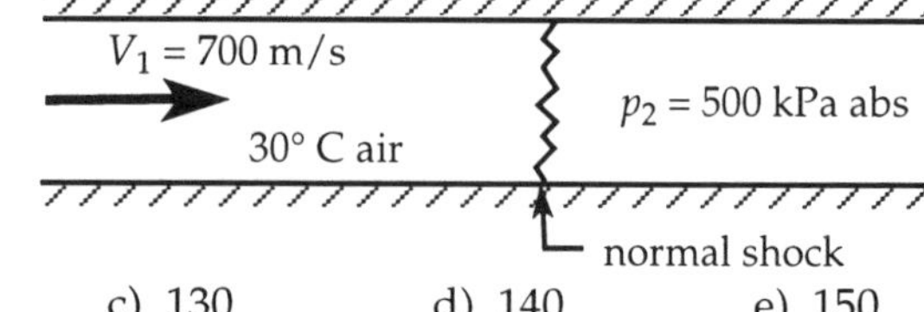

a) 110 b) 120 c) 130 d) 140 e) 150

6.80 Air leaves a reservoir and accelerates until a shock wave is encountered at a diameter of 10 cm. If the throat diameter is 6 cm, what is the Mach number before the shock wave?
a) 2.03 b) 2.19 c) 2.25 d) 2.40 e) 2.56

6.81 A supersonic aircraft flies at M=2 at an elevation of 1000 m. How long, in seconds, after it passes overhead is the aircraft heard?
a) 2.3 b) 2.5 c) 2.7 d) 2.9 e) 3.1

6.82 Air at 20°C is to exit a nozzle from a reservoir. What maximum pressure, in kPa abs, can the reservoir have if compressibility effects can be neglected? $p_{atm} = 100$ kPa.
a) 115 b) 111 c) 109 d) 106 e) 104

Solutions to Practice Problems

6.1 b) a) is true for a liquid and low speed gas flows. c) and e) are true of gases. d) is true for a liquid.

6.2 d) $\tau = \mu\, du/dy. \quad \therefore \mu = \tau/du/dy. \quad [\mu] = \dfrac{F/L^2}{\frac{L}{T}/L} = \dfrac{FT}{L^2} = \dfrac{(ML/T^2)T}{L^2} = \dfrac{M}{LT}$

6.3 a) Viscosity μ varies with temperature only.

6.4 e) $dp = -\gamma dz = -\rho g dz. \quad p = \rho RT$ (ideal gas)

$\therefore\ dp = -\dfrac{p}{RT} g dz \quad \text{or} \quad \dfrac{dp}{p} = -\dfrac{g}{RT} dz.$

$\int \dfrac{dp}{p} = -\dfrac{g}{RT}\int dz. \quad \therefore\ \ln p = -Cz. \quad \therefore\ p = e^{-Cz}.$

6.5 e)

6.6 a) $T = \tau A r = \mu \dfrac{du}{dy} A r = \mu \dfrac{r\omega}{t} 2\pi r L r.$

$1.6 = \mu \dfrac{0.04 \times 1000}{0.001} 2\pi \times 0.04 \times 0.04 \times 0.04. \quad \therefore\ \mu = 0.1\ \text{N}\cdot\text{s/m}^2$

6.7 c) $K = -V \dfrac{\Delta p}{\Delta V} = -2 \dfrac{500}{-0.004} = 250\ 000$ or 250 MPa.

6.8 c) $\sigma 2\pi D = \gamma \pi r^2 L$

$0.0736 \times \pi \times 0.001 = 9800\pi \times 0.0005^2 L$

$\therefore L = 0.03$ m or 3 cm.

$\sigma\pi D$

$W = \gamma V$

L

6.9 e) Cavitation occurs when the pressure reaches the vapor pressure = 2.45 kPa. (See Table 6.2.)

6.10 c) $L = V\Delta t = \sqrt{kRT}\Delta t = \sqrt{1.4 \times 287 \times 288} \times 1.2 = 408$ m.
Assume $T = 15°C$. T must be in absolute. $\therefore T = 288$.

6.11 a) $\nu = \frac{\mu}{\rho} = \frac{0.0034}{1.3 \times 1000} = 2.6 \times 10^{-6}$.

6.12 a) $p = \gamma_1 \Delta h_1 + \gamma_2 \Delta h_2$ where $\gamma_2 = SG \times \gamma_{H_2O}$
$= 9800 \times 2 + 1.04 \times 9800 \times 4 = 60\ 400$ Pa or 60.4 kPa.

6.13 c) $p = \gamma h$ where $\gamma_{Hg} = 13.6\ \gamma_{H_2O}$
$= 13.6 \times 9800 \times 0.6 = 80\ 000$ Pa.

6.14 d) $p = \gamma h = 1.03 \times 9800 \times 1200 = 12.1 \times 10^6$ Pa or 12.1 MPa.

6.15 c) $dp = -\gamma dz = -\rho g dz = -p \frac{g}{RT} dz$ using $p = \rho RT$
$\therefore \frac{dp}{p} = -\frac{g}{RT} dz.$ $\int_{100}^{p} dp/p = -\frac{g}{RT} \int_0^{2000} dz.$
$\therefore \ln \frac{p}{100} = -\frac{9.8}{287 \times 293} \times 2000.$ $\therefore p = 79.2$ kPa.

6.16 c) $F = pA.$ $p = \frac{1000}{\pi \times 0.05^2} + 0.86 \times 9800 \times 0.2 = 129\ 000$ Pa.
$\therefore F = 129\ 000 \times \pi \times 0.0025^2 = 2.53$ N.

6.17 a) $p = \gamma h = (13.6 \times 9800) \times 0.2 = 26\ 700$ Pa or 26.7 kPa.

6.18 e) $p + 9800 \times 0.3 = 13.6 \times 9800 \times 0.3.$ $\therefore p = 37\ 000$ Pa.

6.19 c) $F = pA = (20\ 000 \times 9800 \times 3)\pi \times 1^2 = 155\ 000$ N.

6.20 b) $F = p_c A = 9800 \times \frac{1.5}{2} \times 1.5^2 = 16\ 500$ N.

6.21 d) $7P = \frac{5}{3} F = \frac{5}{3} \gamma h_c A.$ $\therefore P = \frac{5}{21} \times 9800 \times 2 \times 15 = 70\ 000$ N.

6.22 d) All pressures on the curved section pass through the center. Moments about the hinge give
$P = F_v = \gamma \times \text{Volume} = 9800 \times (9\pi/4) \times 4 + 9800 \times 6 \times 3 \times 4 = 983\ 000$ N.

6.23 b) $y_p = y_c + \frac{I_c}{y_c} = 6 + \frac{5 \times 4^3/12}{6 \times 20} = 6.22$ m.
$4P = F \times 2.22 = \gamma h_c A \times 2.22.$ $\therefore P = 9800 \times 6 \times 20 \times 2.22 / 4 = 653\ 000$ N.

6.24 e) $W = \gamma \mathcal{V}.$ $4 \times 1500 \times 9.8 = 9800 \times 6 \times 12 \times h.$ $\therefore h = 0.0833$ m.

6.25 c) $25 = 100 - 9800 \mathcal{V}.$ $\therefore \mathcal{V} = 0.00765$ m^3.
$100 = 9800(SG) \times 0.00765.$ $\therefore SG = 1.33.$

6.26 d) $\Delta p = \Delta\gamma \times h = \left(\frac{1}{253} - \frac{1}{293}\right) \times \frac{100}{0.287} \times 3 \times 9.8 = 5.53 \text{ Pa}.$

6.27 d) $[p] = \frac{M}{LT^2} \quad [Q] = \frac{L^3}{T} \quad [D] = L \quad [\rho] = \frac{M}{L^3}.$

First, eliminate M, then T, then L:

$$\frac{M}{LT^2} \cdot \frac{L^3}{M} \cdot \frac{T^2}{L^6} \cdot L^4 = p \cdot \frac{1}{\rho} \cdot \frac{1}{Q^2} \cdot D^4 = \frac{pD^4}{\rho Q^2}.$$

6.28 a) $[\sigma] = \frac{M}{T^2} \quad [\rho] = \frac{M}{L^3} \quad [D] = L \quad [V] = \frac{L}{T}.$ Combine: $\frac{\sigma}{\rho} \frac{1}{V^2} \frac{1}{D}$

6.29 b) Inertial force to viscous forces.

6.30 d)

6.31 d) $(\text{Fr})_m = (\text{Fr})_p. \quad \therefore \frac{V_m^2}{l_m g} = \frac{V_p^2}{l_p g}. \quad \therefore \frac{V_m^2}{V_p^2} = \frac{1}{20}.$

$Q_m^* = Q_p^*$ or $\frac{Q_m}{V_m l_m^2} = \frac{Q_p}{V_p l_p^2}. \quad \therefore Q_m = 4 \times \frac{1}{20^2} \times \frac{1}{\sqrt{20}} = 0.0022.$

6.32 a) $\text{Re}_m = \text{Re}_p. \quad \left(\frac{Vl}{\nu}\right)_m = \left(\frac{Vl}{\nu}\right)_p. \quad \therefore \frac{V_m}{V_p} = \frac{l_p}{l_m} = 10.$

$\therefore V_m = 10\, V_p = 10 \times 10 = 100 \text{ m/s}.$

6.33 a) $\text{Fr}_m = \text{Fr}_p. \quad \therefore \left(\frac{V^2}{lg}\right)_m = \left(\frac{V^2}{lg}\right)_p. \quad \therefore \frac{V_p^2}{V_m^2} = \frac{\ell_p}{\ell_m}.$

$(F_D)_m^* = (F_D)_p^*$ or $\frac{(F_D)_m}{\rho_m V_m^2 l_m^2} = \frac{(F_D)_p}{\rho_p V_p^2 l_p^2}. \quad \therefore (F_D)_p = 10 \frac{\rho_p}{\rho_m} \frac{V_p^2}{V_m^2} \frac{l_p^2}{l_m^2} = 10 \frac{l_p^3}{l_m^3}.$

$\therefore (F_D)_p = 10 \times 40^3 = 640\ 000 \text{ N}.$

6.34 d) $\text{Fr}_m = \text{Fr}_p. \quad \left(\frac{V^2}{lg}\right)_m = \left(\frac{V^2}{lg}\right)_p. \quad \therefore \frac{V_p^2}{V_m^2} = 10.$

$\dot{W}_p^* = \dot{W}_m^*. \quad \frac{\dot{W}_m}{\rho_m V_m^3 l_m^2} = \frac{\dot{W}_p}{\rho_p V_p^3 l_p^2}.$

$$\therefore \dot{W}_p = \frac{V_p^3}{V_m^3} \frac{l_p^2}{l_m^2} \dot{W}_m = 10^3 \sqrt{10} \times 20 = 63\ 250 \text{ W}.$$

6.35 e) $V_2 = 20\pi \times 2^2 / \pi \times 5^2 = 3.2 \text{ m/s}.$

6.36 d) $V_2 = 20\pi \times 1^2 / 100 \times \pi \times 0.1^2 = 20 \text{ m/s}.$

6.37 e) $V_2 \times 2\pi \times 40 \times 0.2 = 20 \times \pi \times 1^2$. $\therefore V_2 = 1.25$ m/s.

6.38 a) Bernoulli's equation is only applicable in a steady, viscous, incompressible, flow along a streamline in an inertial coordinate system. It must be modified to accommodate other effects, such as compressible viscous flow.

6.39 b) $p = \rho V^2/2 = 1.23 \times 25^2/2 = 384$ Pa. $F = pA = 384 \times \pi \times 0.075^2 = 6.79$ N.

6.40 e) $V_2 A_2 = V_1 A_1$. $\therefore V_2 = V_1 \times 4^2/2^2 = 4V_1$.

$$\frac{p_1}{\rho} + \frac{V_1^2}{2} = \frac{\cancelto{0}{p_2}}{\rho} + \frac{V_2^2}{2}. \quad \frac{700\ 000}{1000} + \frac{V_1^2}{2} = \frac{16V_1^2}{2}. \quad \therefore V_1 = 9.66 \text{ m/s}$$

6.41 d) $p + \rho\dfrac{V^2}{2} + 9800 \times 0.1 = p + 13.6 \times 9800 \times 0.1$.

$\therefore V^2 = 12.6 \times 9800 \times 0.1 \times 2/1000$. $\therefore V = 4.97$ m/s.

6.42 c) Cavitation results if $p_2 = -100$ kPa.

$$\frac{p_1}{\rho} + \frac{\cancelto{0}{V_1^2}}{2} = \frac{p_2}{\rho} + \frac{V_2^2}{2}. \quad 900\ 000/1000 = -100\ 000/1000 + V_2^2/2. \quad \therefore V_2 = 44.7 \text{ m}$$

6.43 a) $F = \rho A V^2 = 1.2 \times \pi \times 0.01^2 \times 80^2 = 2.41$.

6.44 b) Momentum: $-F = \rho A v_r (v_{r2} - v_{r1})_x$. $F = 1000 \times 0.02 \times 10(10 - 0) = 2000$ N.

(v_r is relative speed)

Note: With a single blade we must also use relative speed for $\dot{m} = \rho AV$ since not all of the fluid leaving the jet has its momentum changed.

6.45 c) $F = \rho A V^2 = .5 \times \pi \times 0.25^2 \times 1200^2 = 141\ 000$ N.

6.46 a) $-F = \rho AV(-V - V)$. $\therefore F = 2\rho AV^2$.

$\therefore F = 2 \times 1000 \times 0.05 \times 0.8 \times 50^2 = 200\ 000$ N.

6.47 c) $p_1 A_1 - F = \rho A_1 V_1 (V_2 - V_1)$. $V_2 = 4V_1$. $\dfrac{p_1}{\rho} + \dfrac{V_1^2}{2} = \dfrac{\cancelto{0}{p_2}}{\rho} + \dfrac{V_2^2}{2} = \dfrac{16V_1^2}{2}$.

$\therefore V_1 = \sqrt{\dfrac{800\ 000}{1000}} \times \dfrac{2}{15} = 10.3$. $V_2 = 41.2$.

$\therefore F = 800\ 000 \times \pi \times 0.04^2 - 1000 \times \pi \times 0.04^2 \times 10.3 \times (41.2 - 10.3) = 2420$ N.

6.48 b) The energy grade line.

6.49 a) EGL has elevation $\dfrac{V^2}{2g} + \dfrac{p}{\gamma} + z$. The HGL has elevation $\dfrac{p}{\gamma} + z$.

6.50 a) $\dot{W}_P = \gamma Q \dfrac{\Delta p}{\gamma} \Big/ 0.85 = 0.2 \times 500 / 0.85 = 117.6$ kW.

6.51 e) $\dot{W}_T = \gamma Q \dfrac{\Delta p}{\gamma} \times 0.85 = 0.8 \times 600 \times 0.85 = 408 \text{ kW}$.

6.52 a) manometer: $p_1 = p_2 + \rho V_2^2/2$. $-\dfrac{\dot{W}_T}{\gamma Q} = \dfrac{V_2^2}{2g} + \dfrac{p_2}{\gamma} - \dfrac{V_1^2}{2g} - \dfrac{p_1}{\gamma}$ (100% efficient)

$\therefore \dot{W}_T = Q\dfrac{V_2^2}{2}\rho\eta = \left(20 \times \pi \times 0.1^2\right)\dfrac{20^2}{2} \times 1000 \times .88 = 111\ 000 \text{ W}.$

6.53 c) In an orifice, the minimum pressure (maximum velocity) occurs at the vena contracta. The loss coefficient is based on this minimum pressure.

6.54 d) The energy equation for a control volume, with no work or heat transfer demands that $\Delta h = 0$.

6.55 c) Laminar flow can occur in a pipe upwards of 40 000 Reynolds number. At low Reynolds number a laminar flow occurs in even a rough pipe. Viscous effects dominate in a laminar flow.

6.56 b) Increases with the velocity squared.

6.57 d) Increases linearly to the wall.

6.58 c) Vary as the 1/7th power law.

6.59 a) The flow is laminar and steady, and $v = V$ at $y = \delta$ and $v = 0$ at $y = 0$.

6.60 e) By curve E.

6.61 b) By curve B.

6.62 a) Pressure varies linearly. $\therefore \dfrac{\Delta p}{\Delta x} = \text{Const.}$

6.63 e) The Darcy-Weisbach equation.

6.64 d) Found by using loss coefficients.

6.65 d) Shear stress varies linearly with radius.

6.66 b) $h_f = f\dfrac{L}{D}\dfrac{V^2}{2g}$. $\therefore f = 40\dfrac{0.1}{100}\dfrac{2 \times 9.8}{6^2} = 0.0218$.

6.67 e) $V = Q/A = \dfrac{0.02}{\pi \times 0.03^2} = 7.07 \text{ m/s}$. $\text{Re} = \dfrac{VD}{\nu} = 7.07 \times 0.06/10^{-6} = 4.2 \times 10^5$.

$\dfrac{e}{D} = \dfrac{.26}{60} = .0043$. From Fig. 11.2 $f = 0.03$.

$$\dot{W}_p = \frac{\gamma Q}{\eta}\left(\cancelto{0}{\frac{V_2^2}{2g}} + \cancelto{0}{\frac{p_2}{\gamma}} + z_2 - \cancelto{0}{\frac{V_1^2}{2g}} - \cancelto{0}{\frac{p_1}{\gamma}} - z_1 + f\frac{L}{D}\frac{V^2}{2g} + C\frac{V^2}{2g}\right)$$

$$= \frac{9800 \times 0.02}{0.85}\left[80 - 20 + \left(0.03\frac{100}{.06} + 1 + .5\right)\frac{7.07^2}{2 \times 9.8}\right] = 44\ 000 \text{ W}.$$

6.68 e) $V = Q/A = \dfrac{0.006}{\pi \times 0.03^2} = 4.77 \text{ m/s}$. $\text{Re} = \dfrac{VD}{\nu} = 4.77 \times .04/10^{-6} = 1.9 \times 10^5$.

$\dfrac{e}{D} = \dfrac{0.046}{40} = 0.0011$. From Fig. 4.2 $f = 0.022$. $0 = \dfrac{p_B - p_A}{\gamma} + f\dfrac{L}{D}\dfrac{V^2}{2g} + C\dfrac{V^2}{2g}$

$\therefore p_B = 510\,000 - \left(.022\dfrac{50}{.04} + 6.4 + 2 \times .9\right)\dfrac{4.77^2}{2 \times 9.8} \times 9800 = 104\,000 \text{ kPa}$.

6.69 e) $V = Q/A = 0.3/(.15 \times .4) = 5 \text{ m/s}$. $R_H = \dfrac{40 \times 15}{110} = 5.45 \text{ cm}$.

$\text{Re} = \dfrac{5 \times 4 \times .0545}{1.6 \times 10^{-5}} = 6.8 \times 10^4$. With $\dfrac{e}{D} = 0$, $f = 0.02$.

$\Delta p = f\dfrac{L}{4R_H}\dfrac{V^2}{2g}\gamma = .02\dfrac{500}{4 \times .0545}\dfrac{5^2}{2} \times 1.23 = 705 \text{ Pa}$.

6.70 d) $A_c/A_2 = .62 + .38(.5)^3 = 0.668$. $C_1 = (1 - .668)^2 = 0.11$.

$0.11\, V_c^2/2g = C\, V_2^2/2g$. $\therefore C = 0.11\left(\dfrac{A_2}{A_c}\right)^2 = .11 \times \dfrac{1}{.668^2} = 0.25$.

6.71 e) $V = Q/A = 0.02/\pi \times .03^2 = 7.07 \text{ m/s}$. $e/D = .15/60 = .0025$

$\text{Re} = \dfrac{7.07 \times .06}{10^{-6}} = 4.2 \times 10^5$. $\therefore f = 0.024$.

$L_e = CD/f = 0.9 \times 0.06/0.025 = 2.16 \text{ m}$.

6.72 a) $Q = \dfrac{1}{n}AR_H^{2/3}S^{1/2} = \dfrac{1}{.012} \times 6 \times .86^{2/3} \times .001^{1/2} = 14.3$

where $R_H = 6/(3+4) = 0.86 \text{ m}$.

6.73 b) $Q = \dfrac{1}{n}AR_H^{2/3}S^{1/2} = \dfrac{1}{.016}4h\left(\dfrac{4h}{4+2h}\right)^{2/3} \times .001^{1/2} = 10$.

Trial-and-error: $h = 1.4 \text{ m}$.

6.74 b) $Q = \dfrac{1}{n}AR_H^{2/3}S^{1/2} = \dfrac{1}{.016}\pi \times 1^2 \times .5^{2/3}S^{1/2} = 10$, where $R_H = \dfrac{A}{P} = \dfrac{\pi \times 1^2}{2\pi} = .5$.

$\therefore S = 0.00654$.

6.75 e) $\rho = p/RT = 500/.287 \times 313 = 5.57 \text{ kg/m}^3$

$\dot{m} = \rho AV = 5.57 \times \pi \times .05^2 \times 100 = 4.37 \text{ kg/s}$

6.76 e) $T_e = T_o(p_e/p_o)^{k-1/k} = 293\left(\dfrac{100}{500}\right)^{.286} = 185\text{K}$.

$V = \text{M}c = 1\sqrt{1.4 \times 287 \times 185} = 273 \text{ m/s}$.

6.77 a) $T_e = T_o(p_e/p_o)^{k-1/k} = 293\left(\dfrac{100}{800}\right)^{.286} = 162\text{K}$. $\therefore T_e = 162 - 273 = -111°\text{C}$.

6.78 c) To obtain supersonic flow from a reservoir the subsonic velocity must increase in a converging section until $m = 1$ at the throat and then continue to increase in the diverging section. As velocity increases temperature (and enthalpy) decrease. Velocity decreases in a converging section if $M > 1$.

6.79 a) $M_1 = \frac{V_1}{c_1} = \frac{700}{\sqrt{1.4 \times 287 \times 303}} = 2.01 \quad \therefore \frac{p_2}{p_1} = 4.54$ (from Normal Shock Table)

$\therefore p_1 = p_2/4.54 = 500/4.54 = 110$ kPa

6.80 e) $\frac{A}{A^*} = \frac{10^2}{6^2} = 2.78. \quad \therefore 2.5 < M_1 < 2.6$ (Isentropic flow Table)

6.81 b) $\sin\phi = \frac{1}{M} = \frac{1}{2}. \quad \therefore \phi = 30°$

$\tan 30° = \frac{1000}{L}. \quad \therefore L = 1732$ m

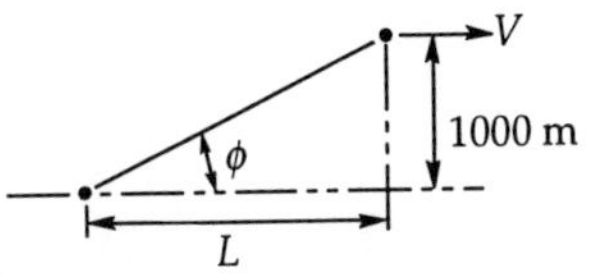

$\Delta t = \frac{L}{V} = \frac{1732}{2\sqrt{1.4 \times 287 \times 293}} = 2.52$ sec.

V
1000 m
ϕ
L

6.82 d) Assume $M_e = 0.3$, the maximum if the density is assumed constant (i.e., $\rho_e = 0.97\rho_o$). $V_e = M_e c_e = 0.3\sqrt{1.4 \times 287\, T_e}$.

$\therefore V_e^2 = 36.2\, T_e$. energy: $0 = \frac{V_e^2 - V_o^2}{2} + c_p(T_e - T_o)$.

$36.2 T_e = 2 \times 1000(293 - T_e). \quad \therefore T_e = 287.8$

$p_o = p_e (T_o/T_e)^{\frac{k}{k-1}} = 100\left(\frac{293}{287.8}\right)^{\frac{1.4}{.4}} = 106.$

Kinematics and Dynamics of Machinery

by Charles W. Radcliffe and Clark J. Radcliffe

Introduction

Mechanical machinery is designed as an assemblage of structures, linkages, cams, gears, controllers and power sources made to satisfy a functional need. Some examples of common machinery would include clocks, automobiles, robots, computer peripherals and numerically controlled machine tools. In each case the machine allows a source of power to *do something useful mechanically.* Increasingly, the design of modern machines involves both mechanical and electrical components, and, must also include an analysis of the forces, energy, stresses, and control systems that would be involved in the operation of the machine.

7.1 Kinematics of Machinery

Preliminary to an investigation of the forces involved in the operation of a machine an analysis of the position, velocity, and acceleration of each member is required. The design may also require analytical or graphical geometrical synthesis of the dimensions of the machine elements as necessary to create desired motions.

Fundamental Concepts

A *mechanism*, as defined here, is an element of a *machine* that is used to transform mechanical motion. The motion of an input member is converted to that of an output member in a consistent pattern dictated by the dimensions, connections and arrangement of rigid bodies that form the mechanism. The design of a mechanism is primarily *kinematic design* since it is the geometry of the device that determines the mechanical input-output relationship.

The Rigid Body

The concept of a rigid body is important in the kinematic design and/or analysis of mechanisms. Kinematic design typically is carried out before the actual shape of the parts has been established. Each member is assumed to be made of an imaginary non-deformable weightless material in which all points of the body maintain the same relative position at all times. An exception would be where a functional elastic member such as a spring is included in the design. Once the geometric arrangements of links, pivot points, sliders, slots, cam surfaces, and gear sizes are known, the shape and mass of the members can be estimated. With the shape defined, a preliminary dynamic analysis would result in masses, forces and torques necessary for calculations of stresses, deflections, and fatigue analysis.

In kinematics the representation of a rigid body includes all points that move with the body. Each point in a rigid body is assumed to maintain a constant distance relative to any other point in the body. In order to define the *motion of a rigid body* the motion of at least one vector fixed in the body must be known.

If the motion is described as *plane motion,* all points of the body move in parallel planes. The motion of any point in a rigid body can be found if the motions of any two points fixed in the body are specified. Another method for defining plane motion of a rigid body is to specify the motion of one point on the rigid body and the angular orientation of a vector passing through the point. The motion of the rigid body is assumed to include all points that would move with a large sheet of paper (lamina) attached to the body.

In the analysis of *spatial mechanisms* the motions are in three dimensional space and the motion of at least three arbitrary points fixed in the body must be specified to completely define the rigid body motion.

The connections between rigid bodies in a planar mechanism involve a pair of joint surfaces which slide, roll, or slip relative to each other. This *pairing* of the joint surfaces leads to the definition of joints as *kinematic pairs.* A *lower pair* is a joint pair with one degree of freedom, e.g., specification of one geometric parameter is required to describe the motion at the joint. A *higher pair* requires more than one geometric parameter to define the relative motion. Figure 7.1 shows some common lower pair connections found in planar mechanisms.

The *revolute pair* is formed by mating cylindrical surfaces and is typically referred to as a *pin joint.* The revolute pair allows only relative rotation about the axis of the cylindrical surface. In spatial mechanisms the *cylindrical* pair allows both rotation about the axis and sliding along the axis.

Lower Kinematic Pairs

The *prismatic pair* is formed by mating parallel surfaces and is sometimes referred to as a *slider joint.* The prismatic pair allows only a *relative translation* along the open direction of the parallel surfaces.

The *rolling pair* is formed by two surfaces which *roll without sliding* relative to each other. Examples would include the hypothetical rolling of the pitch circles of two spur gears or the rolling of a roller cam follower on the surface of a cam.

All of the pairs shown in Figure 7.1 have one degree of freedom. In each case a single kinematic motion parameter will describe the relative motion at the pair. Lower pairs with surface contact are often described as *form closed* since the kinematic constraint is independent of the force transmitted at the joint.

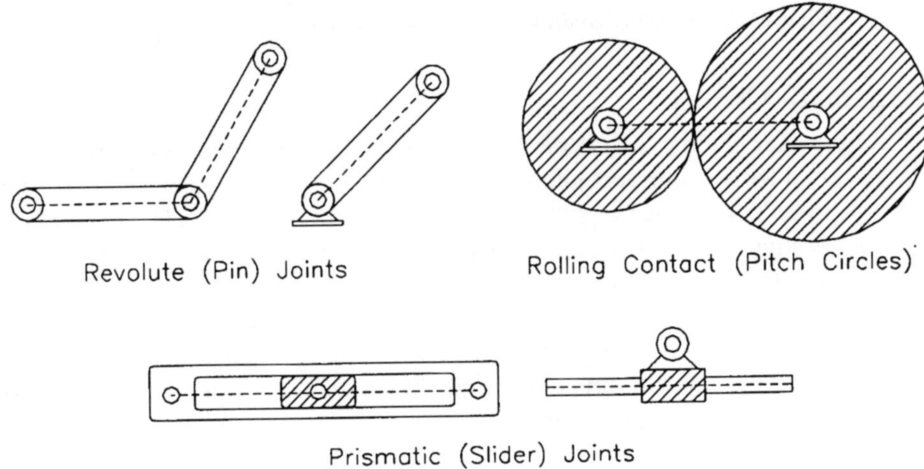

Figure 7.1 Kinematic pairs with one degree of freedom

Higher Kinematic Pairs

A *higher pair* is a connection or joint that requires more than one motion parameter to describe the relative motion at the pair, that is, it has *more than one degree of freedom* at the joint. A higher pair is distinguished by *point or line contact* or its kinematic equivalent. An example for a planar mechanism would be the roll-slide pair at the point of contact of a non-roller cam follower.

In spatial (3d) mechanisms the spherical motion at a *ball and socket* joint is an example of a higher pair. Note that the position of the point at the center of the ball is the geometric position parameter. Three angular motion parameters are required to completely describe the relative motion. In order to proceed with the position analysis some analytical methods require that the *spherical* joint be replaced by a series of intersecting revolute joints.

Many higher pairs and rolling pairs must be *force-closed*. In the absence of a force to hold the pairs together the surfaces could separate and lose the *kinematic constraint* necessary for proper function of the mechanism.

In the kinematic analysis of mechanisms involving *higher* pairs the multi degree of freedom joint is usually replaced by an equivalent series of *lower* pairs.

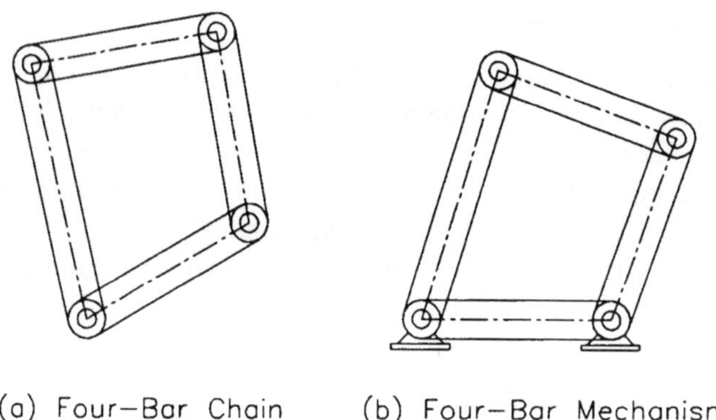

Figure 7.2 Kinematic chain vs. linkage mechanism

Kinematic Chains

A kinematic chain is an assembly of link and pair combinations to form one or more closed loops. Four bars joined together by four pin joints is a *four-bar chain.* Imagine four links pinned together and lying on a table as shown in Figure 7.2(a). The connectivity is defined but the possible relative motion between the links is not referred to a known reference.

If any one of the four bars is fixed to the ground or *reference member,* the four-bar chain becomes a *four-bar mechanism.* Once one of the bars is grounded the motion of the mechanism can be described as *constrained motion.* The relative position of any point on a given member now can be described relative to any other member or point.

Schematic Diagrams

For purposes of kinematic design it is usually sufficient to represent the actual mechanism using a schematic diagram. A skeleton diagram of this type displays the necessary geometric parameters such as point locations, link lengths, pitch circles for gears, slots, and sliders. For graphical analysis such diagrams must be drawn with great precision since the accuracy of the results are determined by the scale and precision of the layout. When used as a guide for setting up an analytical solution, the diagram need not be drawn to scale. A carefully drawn and annotated freehand sketch often will be sufficient to define the coordinate system and prevent errors in sign or other mistakes in calculated values.

The kinematic or skeleton diagram of the mechanism is sufficient to define the basic geometry. The simple diagram has the advantage of displaying the required dimensions and points required without the sometimes confusing shape of the actual parts. In the analysis of an existing machine the identification of the basic kinematic diagram is often a major challenge. Some examples of common mechanisms found in machines are shown in Figure 7.3. In order to proceed with the analysis an appropriate coordinate system must be defined.

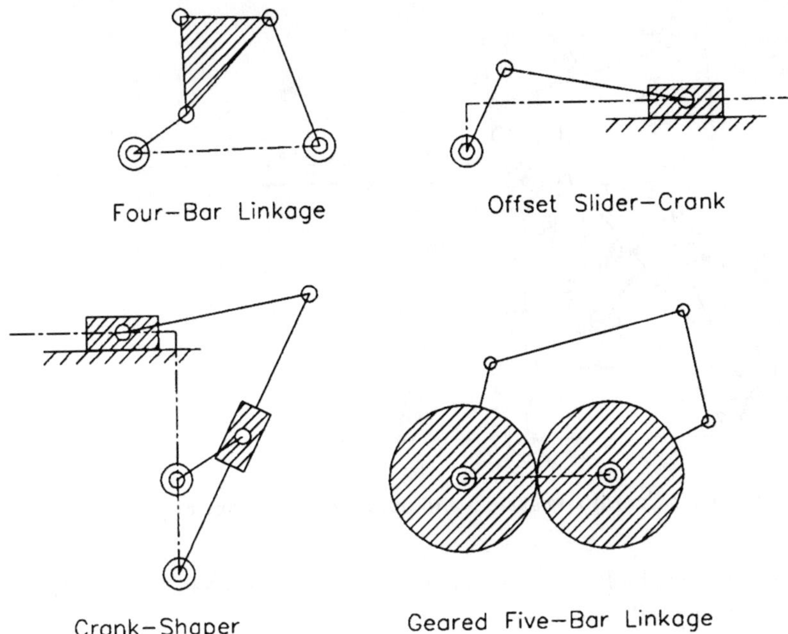

Figure 7.3 Kinematic (skeleton) diagrams of common planar mechanisms

Coordinate Systems

For either graphical or analytical kinematic analysis a coordinate system and notation similar to Figure 7.4 can be used. Members are identified by circled numbers and points by upper case letters. Angles may be either absolute, referred to the x-axis of a set of orthogonal x-y axes, or a relative angle between any two vectors. Positive angles are counterclockwise. Note that fixed points are designated by the letter "O" with a subscript corresponding to the member connected at that point. The center of gravity of any member is identified by the letter "G" with the member number as a subscript.

The vectors forming the basic kinematic loops are labeled with the letter "**r**" or "**R**" and a subscript corresponding to the member number. The vector which locates any point relative to the **R**-vector is identified by an **S**-vector at an angle ϕ.

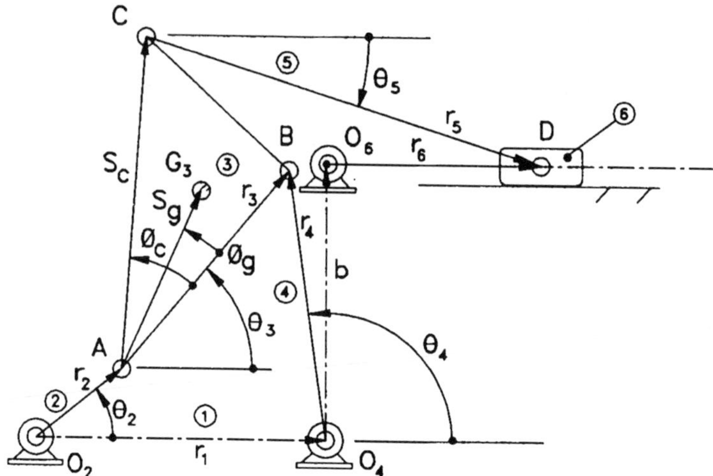

Figure 7.4 Coordinate system and notation for linkage mechanisms

Graphical Methods

If the tools are available, a graphical analysis may be the fastest method of analysis for a *single position* of the mechanism. However, with desktop computers available to most engineers, analytical methods can accomplish the same task for *many positions*, and the program is then available to solve similar problems. A sketch of the graphical layout often will help prevent errors in sign or blunders in the computer program.

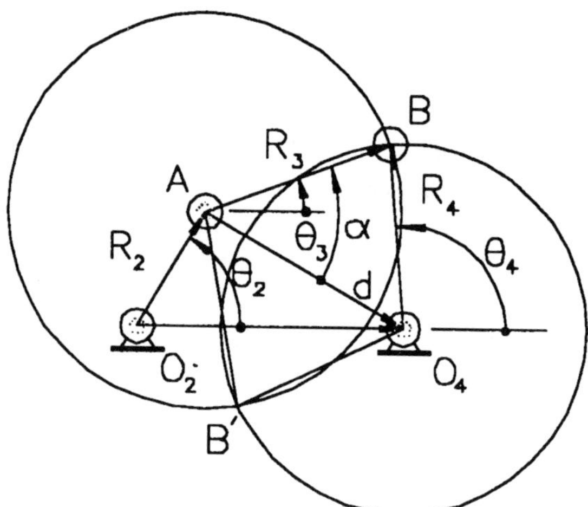

Figure 7.5 Graphical position analysis - four bar linkage

Graphical Position Analysis

Fig. 7.5 shows a graphical layout for the position analysis of a four-bar linkage. The diagram was drawn using a computer-aided-design (CAD) system. The graphical accuracy of computer graphics is of considerable advantage here. Points B and B′ correspond to the two *branches* or *modes of assembly* of the linkage. The position of the points B and B′, common to members 3 and 4, are found as the intersection of two circles of radius R_3 and R_4 using the *circle* command. Once the intersections have been found the *line* command is used to draw vectors from A to B (or B′) and from 0_4 to the point B. Once point B has been located, the *dimension, angle* command will determine the magnitude of the unknown position angles. After the position analysis is complete, the circles used to find the intersection can be erased.

Graphical Velocity Analysis

Graphical velocity analysis continues using the results from the position analysis. Sufficient space around the position layout should be reserved for the velocity and acceleration vector diagrams. The instant center for member AB is a convenient choice for the origin O_V of the velocity diagram as shown in Figure 7.6. The velocity of both points O_2 and O_4 is zero and is represented by point O_V in the velocity diagram. All velocity vectors with an origin at point O_V are absolute velocity vectors referred to the fixed reference member. The vector from A to B in the velocity diagram is a relative velocity vector.

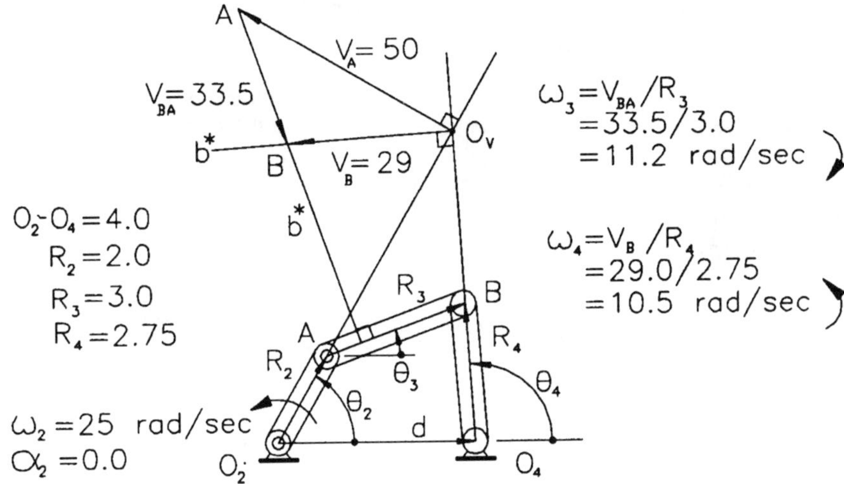

Figure 7.6 Graphical velocity analysis - four bar linkage

Begin the graphical velocity analysis by writing the velocity vector equation in the form $\mathbf{V}_B = \mathbf{V}_A + \mathbf{V}_{BA}$.

This equation reads "The velocity of point B equals the velocity of point A plus the velocity of point B relative to point A". The magnitude of $\mathbf{V}_A$ equals (25.0)(2.0)= 50 ips. The direction for the vector $\mathbf{V}_A$ is determined as follows. Rotate the vector $\mathbf{R}_2$ (O_2A) 90 degrees in the direction of its angular velocity and the rotated vector will point in the direction of point A relative to O_V in the velocity diagram.

Based on the magnitude of $\mathbf{V}_A$ determine a scale factor for the velocity diagram which will keep the diagram in the space available. Draw the vector $\mathbf{V}_A$ from the origin O_V perpendicular to an extension of the line O_2A. Note that the *directed line segment* O_VA represents the vector $\mathbf{V}_A$. The vector equation indicates that the unknown vector $\mathbf{V}_{BA}$ is added to $\mathbf{V}_A$. A construction line is drawn from point A in the known direction of B relative to A, that is, perpendicular to line AB. The point O_V represents both O_2 and O_4 in the velocity diagram since both have zero velocity. Therefore to complete the diagram a second construction line is drawn from point O_V perpendicular to O_4B. A convenient bookkeeping procedure is to label these two construction lines with the symbol **b***. The intersection of the two *"b-star"* lines locates the point B in the velocity diagram.

Once the point B has been located the magnitude of the two directed line segments AB and O_VB can be scaled from the diagram to give $\mathbf{V}_{BA}$ and $\mathbf{V}_B$.

Example 7.1 Acceleration analysis of a four-bar linkage

Perform an acceleration analysis of the four-bar linkage shown.

Solution. The analysis is shown in the following figure:

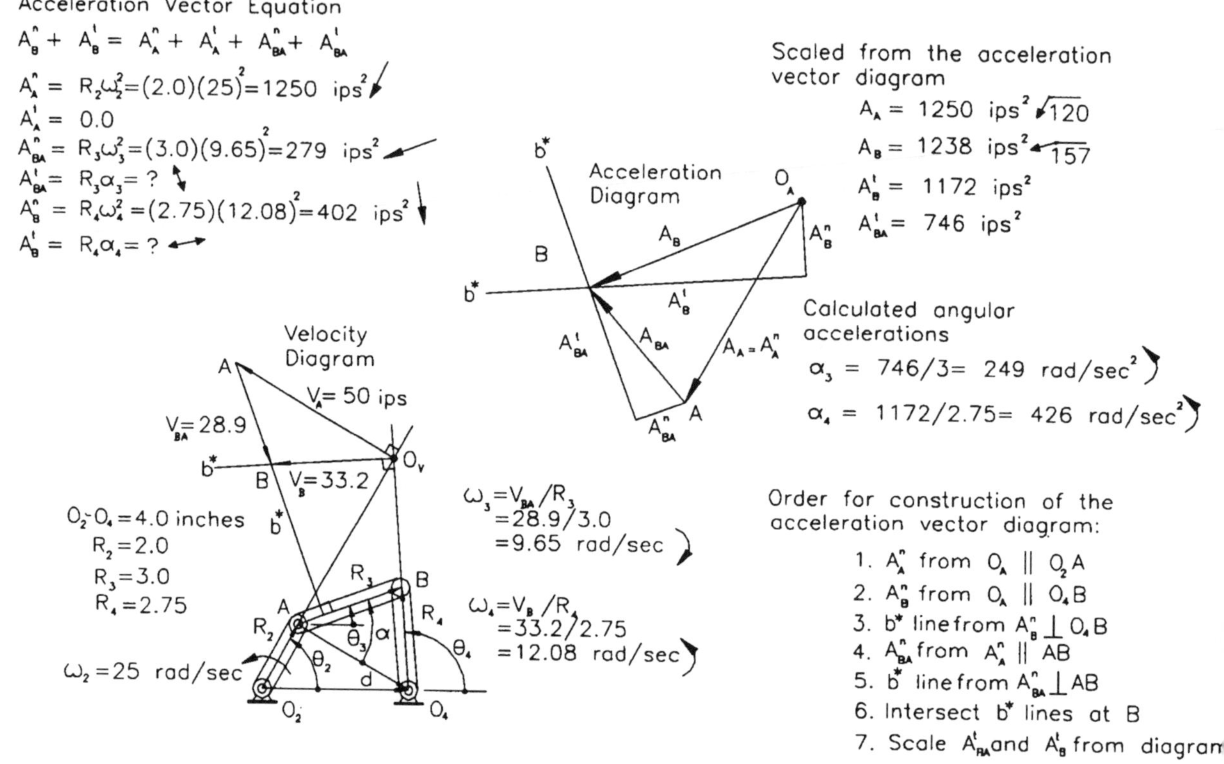

Figure 7.7 Graphical acceleration analysis - four-Bar Linkage

Figure 7.7 shows the results for the graphical acceleration analysis of a typical four-bar linkage. Where space is available it is good practice to show both the velocity and acceleration diagrams on the same sheet since many direction vectors are common to both. The graphical analysis should begin by writing the acceleration vector equation and identifying all of the normal and tangential components.

Before starting the graphical construction of the acceleration diagram note that all normal components of acceleration can always be calculated from the results of the velocity analysis. These values are summarized below the acceleration vector equation. Also note that the acceleration of the point A is completely specified by the motion of the input crank.

All normal components are directed toward the center of curvature of the relative path. The magnitude of the normal component of the acceleration of point B relative to A can be calculated using either of the two forms:

$$\mathbf{A}^n_{BA} = R_3\omega_3^2$$

$$\mathbf{A}^n_{BA} = \mathbf{V}^2_{BA} / R_3$$

In this case the direction of the normal component of $\mathbf{A}_{BA}$ is directed from point B towards point A, along the $\mathbf{R}_3$ vector. The tangential component is always perpendicular to the corresponding normal component, i.e., perpendicular to the $\mathbf{R}_3$ vector . The two primary unknowns for the analysis are the angular accelerations of members α_3 and α_4 which can be calculated from the relative tangential accelerations using the relations:

$$\alpha_3 = \mathbf{A}_{BA}^t / R_3$$

$$\alpha_4 = \mathbf{A}_B^t / R_4$$

once the results from the acceleration vector diagram are available.

The actual procedure for completing the acceleration vector diagram is outlined step-by-step in the lower-right of Figure 7.7. The two construction lines which locate the unknown point B are labeled **b*** (read *contains B).*

Instant Centers

Given two planar bodies rotating relative to each other there will always be one point in space common to the two bodies which has the same velocity on each of the two bodies. At that point the *relative velocity* is zero and the point is called *the instantaneous center of zero relative velocity*. This rather long description is usually shortened to the term *instant center,* but the concept of zero relative velocity is fundamental to its application to the velocity analysis of multi-link mechanisms.

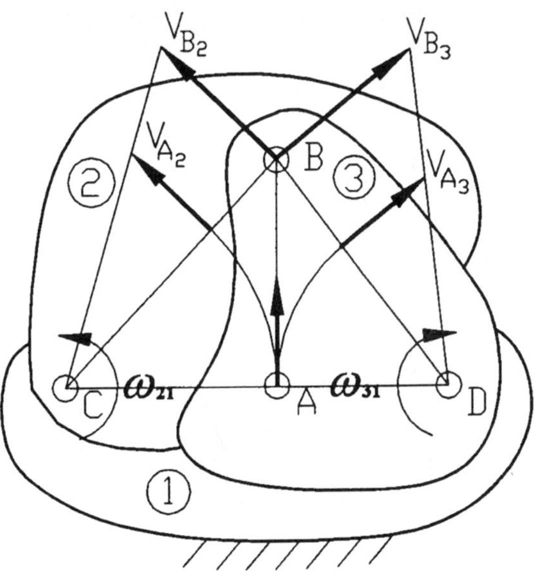

Figure 7.8 The theorem of three centers

The Kennedy-Arnhold Theorem of Three Centers

For any three bodies in planar motion the there will be exactly three instant centers and these three instant centers must lie on the same straight line. This, the *Kennedy-Arnhold Theorem,* is a very powerful tool for relative velocity analysis of a wide variety of planar mechanisms. Figure 7.8 shows three bodies with relative motion. For purposes of proof of the theorem, body number 1 is shown fixed to a reference member. Body 2 is joined to body 1 by a pin joint at point C. Body 3 is joined to body 1 at point D. The angular velocity of body 2 relative to body 1, ω_{21}, is counterclockwise as shown. The angular velocity of body 3 relative to body 1,

ω_{31}, is clockwise. Point A has been selected such that it has the same velocity on both bodies 2 and 3. This point is therefore the instant center for bodies 2 and 3, since for each of the two bodies the velocity vector has the same magnitude and direction. If a point such as point B is selected it is clear that it would be impossible for the velocity of point B on links 2 and 3 to have the same velocity vector since their directions could never be parallel.

We now introduce a notation which will be useful for *instant center analysis.* The point C in Figure 7.8 will be relabeled as the *relative instant center* **12** , the point D becomes **13**, and point A would be **23**. Note that all three instant centers lie on the same straight line.

Instant Centers for a Four-Bar Linkage

Another useful tool will be the *circle diagram*. Note in Figure 7.9 that a circle has been drawn with four equally spaced points on its circumference, one point for each of the four bodies. Whenever a relative instant center is located a solid line is drawn between the two points on the circle corresponding to the two digits of the instant center. A triangle can be drawn for any three points on the circle. When attempting to locate an unknown center two sides of a triangle will be known, and the third side , drawn first as a dashed line, represents the unknown instant center. The strategy is to try to find two triangles with the same unknown. Since each triangle represents a different straight line in the layout, the two triangles represent two intersecting lines, which locate the unknown instant center.

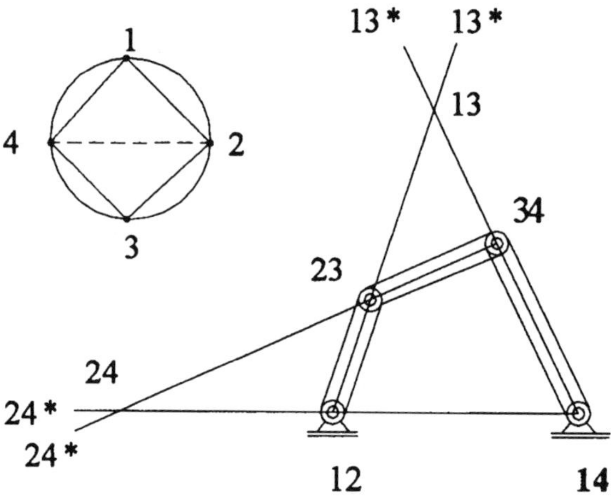

Figure 7.9 Instant centers for a four-bar linkage

Begin by locating *all points where the instant centers are obvious*, such as points where two members are pinned together, the contact point for two surfaces that roll without sliding, or a prismatic pair with relative instant center at infinity perpendicular to the sliding velocity. For the four-bar linkage of Figure 7.9 this defines obvious instant centers **12**, **14**, **23**, and **34**. Lines are then drawn between corresponding points in the circle diagram to form a square in the circle diagram.

Instant center **24** is located as follows. First draw a dashed line between points 2 and 4 in the circle diagram. Now there are two triangles **124** and **234.** Note that **24** is common when we are looking for instant center **24**. Now referring to triangle **124** extend the line from **12** to **14** in the layout of the four-bar and label the line **24***. Now referring to triangle **234** extend the line from **23** to **34** and again label it **24***. The intersection of these two *24-star* lines locates the instant center **24.** Once the center **24** is located, the line between points 2 and 4 is redrawn as a solid line, indicating that its location is known.

Instant center **13** is now found using a similar procedure, beginning with a dashed line between points 1 and 3 in the circle diagram. The triangles **123** and **134**, (note that **13** is common), show that lines **(12-23)** and **(14-34)** will locate instant center **13.**

Counting the instant centers for the four-bar linkage, we see that they total six in number. This could have been predicted from the equation for four items taken two at a time. The next possible combination of rigid bodies for constrained motion would be a six-bar linkage, with fifteen possible instant centers.

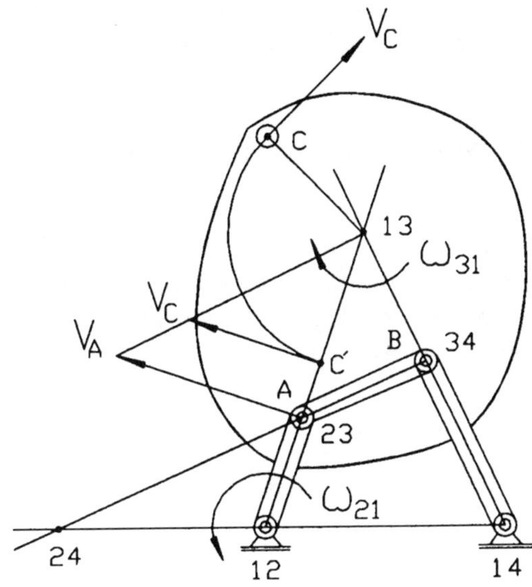

Figure 7.10 Finding the velocity of an arbitrary point C on the coupler 3

Instant Centers for Velocity Analysis of a Four-Bar Linkage

After finding all of the instant centers for the four-bar linkage of Figure 7.9, use the instant centers to find the linear velocity of an arbitrary point C on the coupler link 3, as shown in Figure 7.10. Assume the angular velocity ω_{21} is specified as the input motion parameter. First, calculate the velocity of point A and draw it to an appropriate scale as the vector $\mathbf{V}_A = \mathbf{V}_{23}$ on the diagram as shown. The motion transfer from member 2 to member 3 is through the instant center **23.** The instant center **23** becomes the *transfer point*, a point on both members with the same velocity vector. Point C rotates instantaneously about the instant center **31=13**. The angular velocity of member 3, ω_{31}, can be found using a graphical rotation of the point C about **13** to a temporary position C′. Next, using the graphical construction shown in Figure 7.10, determine the magnitude of the velocity of point C′. Once the magnitude is known, the vector $\mathbf{V}_C$ can be rotated about **13** back to its proper position at point C. With the velocity of point C known the angular velocity ω_{31} is calculated from the relation

$$\omega_{31} = V_C / (13\text{-}C) \qquad \textbf{(7.1.1)}$$

The angular velocity ω_{31} can also be found in a more direct manner using the transfer point **23** and scaling distances from the instant center diagram. First calculate the velocity of the transfer point using the expression

$$V_{23} = (12\text{-}23)\ \omega_{21} \qquad \textbf{(7.1.2)}$$

Once the velocity of the transfer point **23** is known the angular velocity of member 3 relative to the reference link 1 is determined using the relation

$$\omega_{31} = V_{23} / (13\text{-}23) \qquad \textbf{(7.1.3)}$$

where the line **(13-23)** is considered a *directed line segment* with positive direction. Combining the two expressions we have

$$\omega_{31} = \omega_{21}\ (12\text{-}23) / (13\text{-}23) \qquad \textbf{(7.1.4)}$$

In this example, the line **(13-23)** would have a negative sign relative to **(12-23)** hence the angular velocity ω_{31} would be negative (in the clockwise direction).

Instant Center Analysis for a Four-Bar Function Generator

Referring to Figure 7.9, if the four-bar linkage is to be used in an application where the angular velocity of member 2, ω_{21}, is the input, and the angular velocity of member 4, ω_{41}, is the output, the linkage is described as a *function generator.* Instant centers provide an efficient method of finding the velocity ratio $\omega_{41} / \omega_{21}$.

Using the instant center **24** as the transfer point its velocity can be calculated from the expression

$$V_{24} = (12\text{-}24)\ \omega_{21} \qquad \textbf{(7.1.5)}$$

It could also be calculated from

$$V_{24} = (14\text{-}24)\ \omega_{41} \qquad \textbf{(7.1.6)}$$

Equating the two gives

$$\omega_{41} = \omega_{21}\ (12\text{-}24) / (14\text{-}24) \qquad \textbf{(7.1.7)}$$

Note that the directed line segments **12-24** and **14-24** have the same sign. ω_{41} would have the sense as ω_{21} in the position shown.

Example 7.2 Instant center analysis - epicyclic gears

Assume pitch circle radii and angular velocities for the gears as follows:

R_2 (Sun Gear radius) = 4.0 inches $\quad \omega_2 = 100$ rad/sec
R_3 (Arm length) = 6.0 $\quad \omega_3 = 200$
R_4 (Planet Gear radius) = 2.0
R_5 (Ring Gear internal radius) = 8.0

Determine ω_4 and ω_5.

Solution. The solution is included in the following figures:

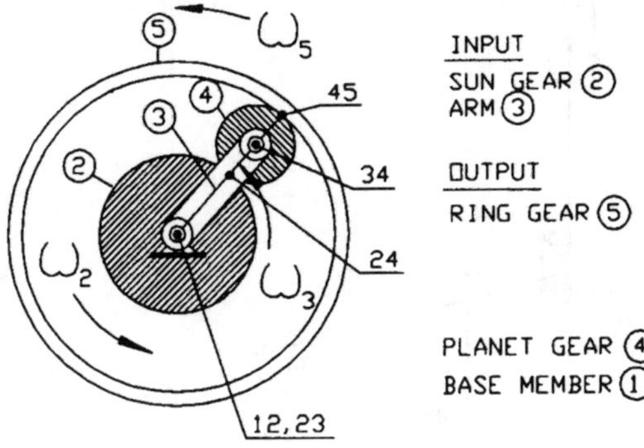

Figure 7.11 Geometry - epicyclic gear train

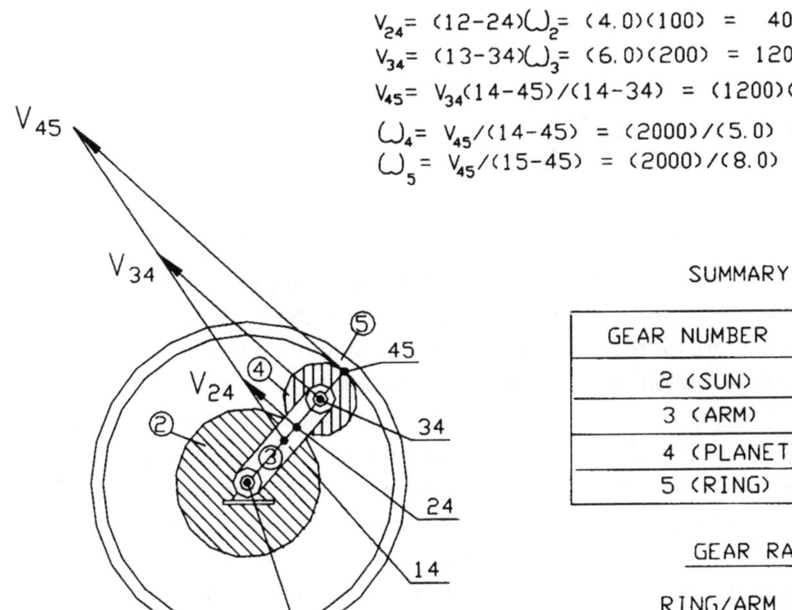

SUMMARY OF RESULTS

GEAR NUMBER	ANGULAR VELOCITY
2 (SUN)	100.0
3 (ARM)	200.0
4 (PLANET)	400.0
5 (RING)	250.0

GEAR RATIOS

RING/ARM = (250)/(200) = 1.25

RING/SUN = (250)/(100) = 2.50

Figure 7.12 Velocity analysis of an epicyclic gear train using instant centers

In the analysis of the epicyclic gear train shown in Figures 7.11 and 7.12, the obvious instant centers are located at **12, 13, 15, 23, 24, 34,** and **45**. To complete the analysis only instant center **14** is required. The location of instant center **14** lies on the line of centers containing all instant centers in this example. The circle diagram is of little use in this case, since all lines would be parallel. Center **14** can be located graphically using the construction shown, or, using Figure 7.12 as a guide, from the ratio of $\mathbf{V}_{34}$ to $\mathbf{V}_{24}$ and the radius of the planet gear 4.

Instant Center Analysis for a Crank-Shaper Mechanism

The crank-shaper mechanism provides an excellent example of the power of the instant center method for relative velocity analysis. Although there are six rigid bodies in the mechanism, including two sliders, it is not necessary to include all members in this example.

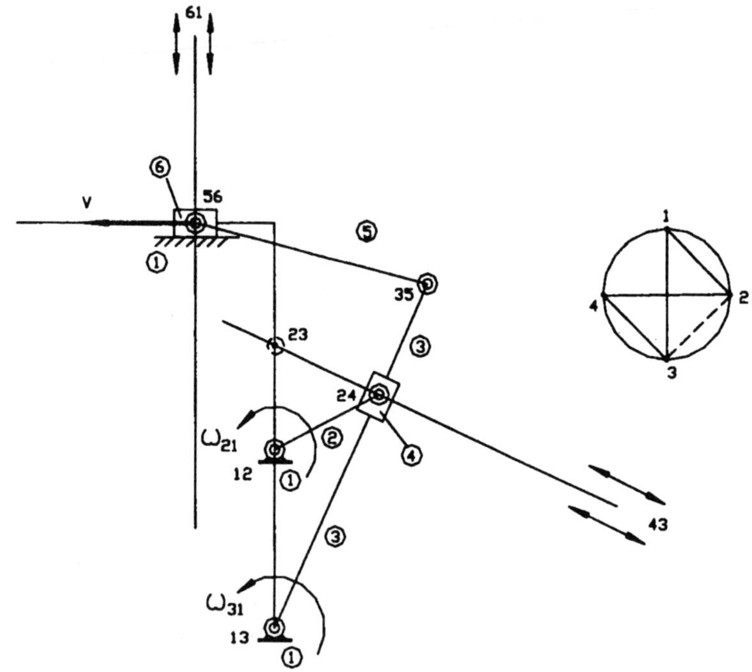

Figure 7.13 Instant centers for a crank-shaper mechanism

As the first problem assume that member 2 is the input member and the velocity ratio $\omega_{31} / \omega_{21}$ is to be determined. In this case the analysis would only require instant centers for members 1,2 and 3. The member 4 is included only to aid in the physical visualization of the relative motions.

Begin by drawing a 4-member circle diagram as shown in Figure 7.13. The obvious centers, determined by inspection, are centers **12, 13, 24,** and **43.** The instant center **43,** is best visualized by imagining member 3 temporarily fixed in space, and member 4 sliding relative to 3. Clearly the instant center **43,** for a straight line relative motion, must be located at infinity at any point on a *family of straight lines,* all of which have a direction *perpendicular to the relative sliding direction.* This is shown by the set of double-ended vectors representing the direction in which to seek the instant center. Instant center **61** also is located on a family, although it is not required to solve the present problem.

With the obvious centers noted as solid lines in the circle diagram, a dashed line is drawn from 2 to 3, indicating that this is the next center to be located. Using the circle diagram as a guide, the center **23** is located at the intersection of lines **12-13** (extended) and **24-43** (the single member of the family **43** which passes through **24**).

To complete the analysis write two expressions for the velocity of the transfer point **23** and equate. This leads to

$$(12\text{-}23)\ \omega_{21} = (13\text{-}23)\ \omega_{31} \tag{7.1.8}$$

The velocity ratio is then

$$\omega_{31} / \omega_{21} = (12\text{-}23) / (12\text{-}23) \tag{7.1.9}$$

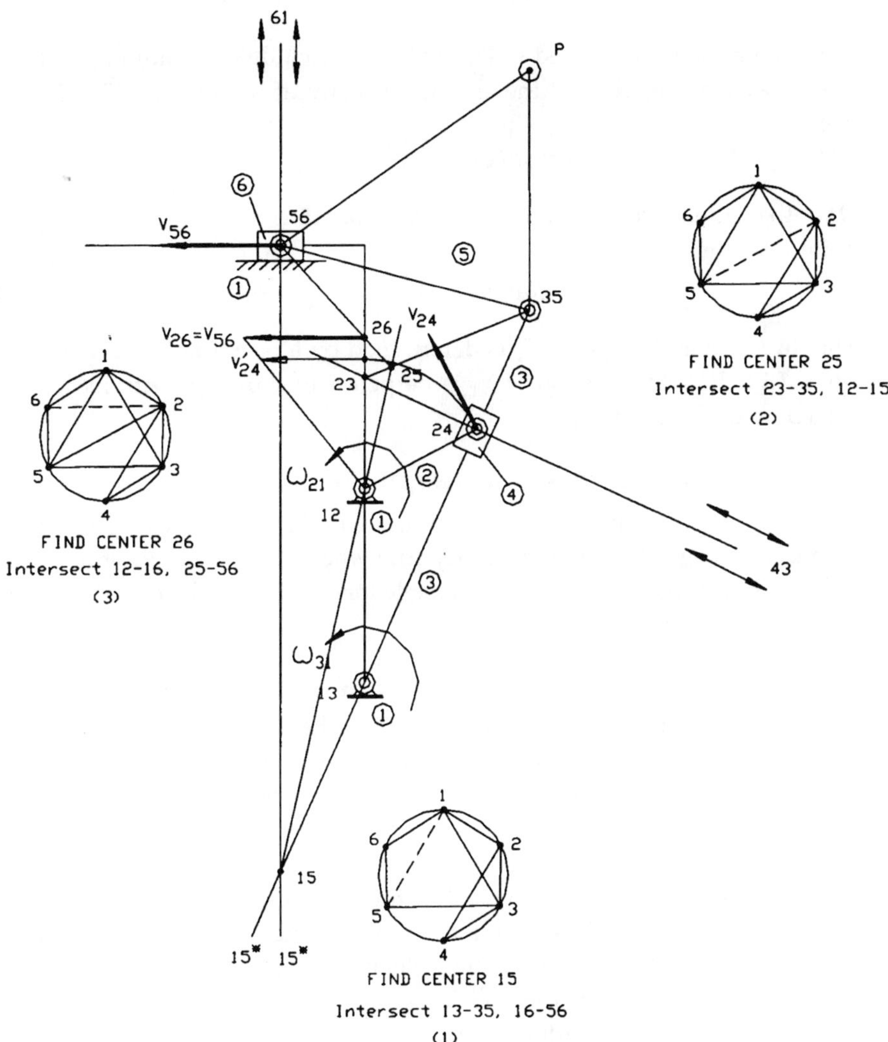

Figure 7.14 Use of the instant centers to find the velocity of slider member 6

As a second problem assume that the velocity of the slider **6** with respect to the base member **1** is the unknown. Examination of the diagram shows that the absolute velocity of the instant center **56** would be equal to the desired slider velocity. In this case all of the members are involved in the analysis and the circle diagram must have six equally spaced points.

Begin by planning the strategy to use the fewest possible construction lines. The instant center **26** would have the same velocity on either member **2** or **6**. The velocity of point **26** on member **2** is easily calculated from the equation

$$V_{26} = (12\text{-}26)\ \omega_{21} \tag{7.1.10}$$

And, since all points on the slider must have the same linear velocity, the velocity of the pivot at **56** must be equal to the velocity of point **26**.

The location of instant center **26** will require centers **25** and **15**, therefore, three constructions will be required. In a typical problem all constructions would be guided by a single circle diagram, upgraded at the end of each stage. In this example, a separate circle diagram is shown for each stage. Diagram (1) is used when seeking **15.** Diagrams (2) and (3) are used when seeking **25** and **26** respectively.

Link 5 has been extended in Figure 7.14 to include the point P. To determine the velocity of the point P the angular velocity of the member 5, ω_{51}, is found from

$$\omega_{51} = \mathbf{V_{56} / (15\text{-}56)} \qquad \textbf{(7.1.11)}$$

Once the angular velocity about the *fixed point* **15** is known, the linear velocity of point P is found from

$$\mathbf{V_P = (15\text{-}P)}\ \omega_{51} \qquad \textbf{(7.1.12)}$$

The direction and sense of $\mathbf{V_P}$ is determined by a construction line drawn from point **15** to **P**. The direction of $\mathbf{V_P}$ is perpendicular to **15-P** with its sense determined by the sense of ω_{51}.

Use of Instant Centers as a Design Tool

Instant centers can often be useful in the preliminary phase of kinematic design of a linkage. This is particularly true when a computer program, based on the location of instant centers, is available for rapid re-analysis of the mechanism and can be used interactively.

Use of Instant Centers for the Design of a Four-Bar Function Generator

There are many design situations where the angular velocity of an output shaft must be coordinated with the angular velocity of an input shaft. When a constant angular velocity ratio must be maintained over one or more revolutions of the input shaft, a gear or toothed belt driven mechanism would be the obvious choice.

If the angular velocity ratio need be maintained for only a partial revolution, perhaps 30 degrees or less, then a four-bar linkage might be considered. This is particularly true if a variation in the velocity ratio can be tolerated as the linkage moves in either direction from the design position.

To design a linkage with a given angular velocity ratio $\mathbf{R} = \omega_{41} / \omega_{21}$ in the design position, the procedure would be as follows:

1. Derive an expression to locate the instant center **24** relative to **14**

Let **a** be the unknown distance from **14** to the unknown instant center **24**. Let **R** be the desired angular velocity ratio. Assume the base link distance **(12-14)** = 1.0. After the dimensions of the linkage have been determined they can be scaled uniformly without changing the velocity ratio. The instant center **24** is assumed in the positive x-direction relative to **12** .

$$R = \omega_{41} / \omega_{21} = ((12\text{-}14) + a) / a = (1 + a) / a \tag{7.1.13}$$

Solving for **a:**

$$a = 1 / (R-1) \tag{7.1.14}$$

The sign of **a** is determined by the sign of the directed line segment **(12-14)** and the ratio R.

2. Calculate the dimension **a** for a function generator with a base link length **(12-14)** = 1.0 and a velocity ratio R = 2.5 .

$$a = 1 / (2.5-1) = 1 / 1.5 = 0.667$$

3. Locate point **24** at a distance of 0.667 units from **14** in the same sense as **(12-14)**.

4. Draw a line at an arbitrary angle from point **24** similar to Figure 7.10.

5. Select any two moving pivots **23** and **34** on the line from step 4. Test the linkage by analysis over the desired range of motion. If the instant center changes its position more than the allowable error, the design is unsatisfactory. Select either a new line in step 4 or new choices for moving pivots in step 5 then iterate until the design requirements are met. Once the design is satisfactory, the dimensions can be scaled to the desired size.

Design of a Prosthetic Knee Linkage for a Trans-Femoral Amputee

A four-bar linkage can be used in the design of a prosthetic knee joint for a trans-femoral lower extremity amputee. Consider the arrangement shown in Figure 7.15. Member **3** includes the *knee block,* the upper knee part, and the *socket,* (the receptacle for the amputation stump). The socket rotates relative to member **1**, which includes the lower knee part and the attached shank structure. The instant center **13** is located at the intersection of the centerlines of the two guiding links **2** and **4**. Instant center **13** , with the knee in full extension, is initially located well above and behind the line from the point of the heel to the hip joint. As a result, when the heel strikes the ground to begin the weight-bearing phase in walking, the knee is forced into full extension against a mechanical stop and is able to bear weight without collapsing.

At the end of the weight-bearing phase, as the amputee pushes off, the knee must flex freely to begin the swing phase of walking. As the knee flexes, the instant center **13** changes its position relative to member **1,** typically moving forward and downward rapidly toward the anatomical or *cosmetic* knee center. In the design of such devices, the shape of the path of the instant center, the *centrode,* plays an important role in the manner in which the amputee controls the flexing of the knee near the end of the weight-bearing phase of the walking cycle. The designer must weigh the *increased stability advantage* of an initially elevated and posterior position of the instant center with many other factors; including cosmesis in the seated position, range of allowable flexion for kneeling, and *con-*

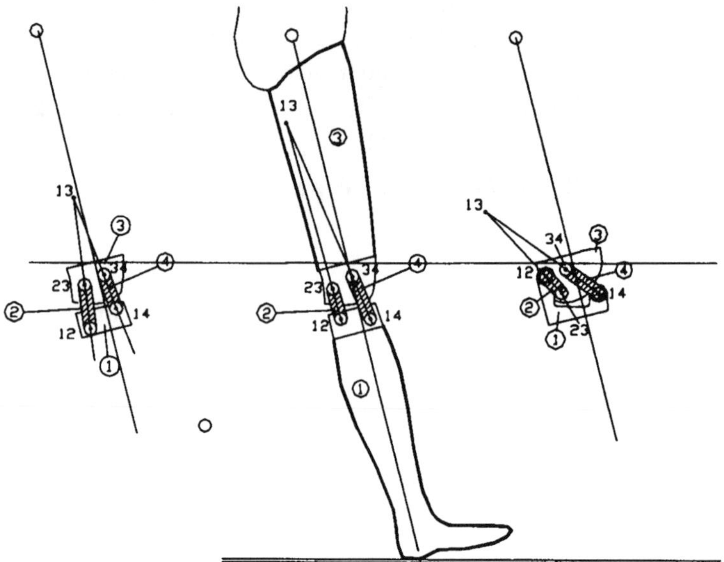

Figure 7.15 Use of instant centers in the design of a four-bar linkage prosthetic knee

trollability of the knee. The design is not based on duplication of anatomical knee motion. The kinematic design enhances the ability of the residual musculature acting at the hip on the amputated side to control knee stability .

Figure 7.15 shows several possible kinematic arrangements for four-bar prosthetic knees. The location of the instant center , with the knee in full extension at the instant of heel contact, is the primary design variable. Each arrangement has advantages, depending upon the functional needs of the individual amputee. In addition to the kinematic arrangement, which is primarily for stability against knee buckling during the initiation of weight-bearing on the prosthesis, the device can include mechanical, hydraulic, or pneumatic devices to substitute for lost muscle action which control knee motions in the swing phase of walking.

7.2 Analytical Methods for Kinematic Analysis of Mechanisms

The position, velocity, and acceleration analysis of planar mechanisms involves the time rate of change of *vector* quantities. In section 7.1.2 graphical methods for solving the kinematics of mechanisms were shown. The methods of *analytical mechanics*, using well known concepts of vector algebra, have been taught to engineering students for many years. However, before the use of personal computers, graphical methods were often emphasized in *mechanisms* or *machine design* classes, with the justification that graphical methods gave the engineer a better physical picture of the motions. Now that computer graphics have become available to most, if not all, engineers and engineering students, analytical methods are equally effective in displaying graphical results from analytical solutions.

Analytical methods, particularly when combined with interactive computer programs and graphic output, allow the engineer to try a variety of designs without building a mechanical prototype. This ability to *simulate* and *iterate* give computer-aided methods a definite advantage over graphical methods in present day engineering practice. For those situations where graphical solutions are used, the use of a *computer-aided-design* (CAD) drawing program, with its ability to extend the drawing surface at will and draw accurate intersections, can make graphical solutions much less of a chore than in the past.

Use of Cartesian Vectors for Kinematic Analysis

There are several concepts which must be reviewed before proceeding. In this presentation a *vector* quantity will be represented by a lower case bold face letter, for example, **r** or **a**. All *scalar* variables, such as the magnitude of a vector, will be non bold face.

All Cartesian vectors will be referred to a set of right-handed unit vectors **i, j** and **k** in the directions of a set of fixed right-handed orthogonal reference axes *X,Y* and *Z*. A typical vector quantity would have components in each of the three unit vector directions.

Velocity Components

The dot notation will be used to denote differentiation with respect to time, for example,

$$\dot{\mathbf{a}} = \frac{d}{dt}\mathbf{a} = \frac{d}{dt}(a_x\mathbf{i} + a_y\mathbf{j} + a_z\mathbf{k}) \qquad \textbf{(7.2.1)}$$

$$\dot{\mathbf{a}} = \dot{a}_x\mathbf{i} + \dot{a}_y\mathbf{j} + \dot{a}_z\mathbf{k} \qquad \textbf{(7.2.2)}$$

Another useful notation would define a vector quantity in the form

$$\mathbf{r} = r\hat{\mathbf{r}} \qquad \textbf{(7.2.3)}$$

where r is the scalar length of the vector and $\hat{\mathbf{r}}$ is a unit vector which defines the direction.

Differentiating Eq. 7.2.3 we obtain

$$\dot{\mathbf{r}} = \dot{r}\hat{\mathbf{r}} + r\frac{d}{dt}(\hat{\mathbf{r}}) \qquad \textbf{(7.2.4)}$$

The second term in Eq. 7.2.4 accounts for the rate of change of the *direction* of the vector $\mathbf{r}$:

$$\frac{d}{dt}\hat{\mathbf{r}} = \lim_{\Delta t \to 0} \frac{\Delta\hat{\mathbf{r}}}{\Delta t} \qquad \textbf{(7.2.5)}$$

$$= \lim_{\Delta t \to 0} \frac{|\hat{\mathbf{r}}|\,\Delta\theta}{\Delta t} \qquad \textbf{(7.2.6)}$$

$$= \dot{\theta}\hat{\mathbf{u}} = \omega$$

The two components of $\dot{\mathbf{r}}$ in Eq. 7.2.4 are orthogonal to each other. Equation 7.2.6 defines the rate of change of direction of the $\hat{\mathbf{r}}$ vector. The vector $\mathbf{u}$ is the spin axis for the rotating vector and is perpendicular to the plane defined by $\hat{\mathbf{r}}$ and $\Delta\hat{\mathbf{r}}$. In the limit, the direction for $\Delta\hat{\mathbf{r}}$ is given by $\mathbf{u} \times \hat{\mathbf{r}}$. This defines $\dot{\mathbf{r}}$ as

$$\dot{\mathbf{r}} = \dot{r}\hat{\mathbf{r}} + r\dot{\theta}(\hat{\mathbf{u}} \times \hat{\mathbf{r}}) \qquad \textbf{(7.2.7)}$$

or

$$\dot{\mathbf{r}} = \dot{r}\hat{\mathbf{r}} + \omega \times \mathbf{r} \tag{7.2.8}$$

where ω is the angular velocity vector.

Acceleration Components

Differentiating Equation 7.2.22, and noting that differentiation must include both magnitude and direction for each vector,

$$\ddot{\mathbf{r}} = \frac{d}{dt}(\dot{r}\hat{\mathbf{r}} + r\dot{\theta}(\hat{\mathbf{u}} \times \hat{\mathbf{r}})) \tag{7.2.9}$$

which leads to

$$\ddot{\mathbf{r}} = \ddot{r}\hat{\mathbf{r}} + 2(\omega \times \dot{\mathbf{r}}) + (\omega \times (\omega \times \mathbf{r})) + \dot{\omega} \times \mathbf{r} \tag{7.2.10}$$

where $\mathbf{r} = r\hat{\mathbf{r}}$.

Note that $\dot{\omega} = \ddot{\theta}\hat{\mathbf{u}} + \dot{\theta}\dot{\hat{\mathbf{u}}}$ for spatial (three-dimensional) kinematics. In planar problems the axis $\hat{\mathbf{u}}$ has a constant direction perpendicular to the plane of motion, hence, the second term is zero. In the case of *plane motion only*, the $\ddot{\theta}\hat{\mathbf{u}}$ term becomes the *angular acceleration vector*, and is written in the form

$$\alpha = \ddot{\theta}\mathbf{k} \tag{7.2.11}$$

Position Analysis of a Four-Bar Linkage

In this example the two fixed pivots have been labeled $\mathbf{A}_0$ and $\mathbf{B}_0$. The position of the point **A** on the input crank is determined by the position angle θ_2. The primary unknown, for the position analysis, is the angle α. Once α has been determined, the location of points $\mathbf{B}^+$ and $\mathbf{B}^-$, corresponding to the positive or negative *mode of assembly*, is fixed by the two possible values of θ_3.

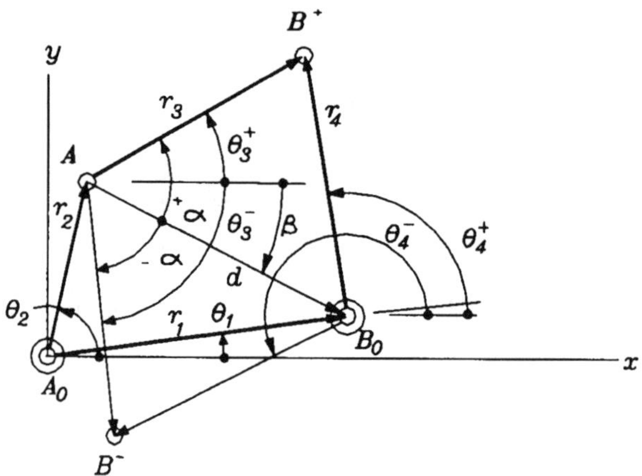

Figure 7. 16 Position analysis

With the position of points $\mathbf{A}(a_x, a_y)$ and $\mathbf{B}_0(b_{0x}, b_{0y})$ known, the next step is to determine the components of the diagonal vector **d**. First calculate

$$\begin{aligned} d_x &= b_{0x} - a_x \\ d_y &= b_{0y} - a_y \end{aligned} \tag{7.2.12}$$

from which $|\mathbf{d}| = \text{sqrt}(d_x^2 + d_y^2)$ and $\beta = \tan^{-1}(d_y/d_x)$

Before proceeding, the possibility that the lengths of $\mathbf{r}_3$ and $\mathbf{r}_4$ may not allow assembly of the linkage must be checked.

If $$d > (r_3 + r_4) \tag{7.2.13}$$

or $$d < |(r_3 - r_4)| \tag{7.2.14}$$

then the *two-link dyad* cannot be assembled.

Solve for the angle α using the law of cosines

$$r_4^2 = r_3^2 + d^2 - 2r_3 d \cos\alpha \tag{7.2.15}$$

from which

$$\alpha = \pm\cos^{-1}\frac{r_3^2 + d^2 - r_4^2}{2r_3 d} \tag{7.2.16}$$

Since α is double-valued, the two values for θ_3 are given by

$$\theta_3 = \beta \pm \alpha \tag{7.2.17}$$

Finally, calculating the coordinates of point **B** from

$$\begin{aligned} b_x &= a_x + r_3 \cos\theta_3 \\ b_y &= a_y + r_3 \sin\theta_3 \end{aligned} \tag{7.2.18}$$

then the angle θ_4 is found from

$$\theta_4 = \tan^{-1}\frac{b_y - b_{0y}}{b_x - b_{0x}} \tag{7.2.19}$$

Example 7.3

Refer to Figure 7.16. In this case assume the following coordinates and dimensions.

$\mathbf{A}_0 = (0.0, 0.0)$, $\mathbf{B}_0 = (4.0, 0.0)$, $\mathbf{r}_2 = 1.0$, $\mathbf{r}_3 = 2.0$, $\mathbf{r}_4 = 1.75$ inches, $\theta_2 = 45.0$ deg

Perform a position analysis of the linkage.

Solution. The point **A** is located at

$$\begin{aligned} a_x &= a_{0x} + r_2 \cos\theta_2 \\ a_y &= a_{0y} + r_2 \sin\theta_2 \end{aligned}$$

$$\begin{aligned} a_x &= (0) + (1.0)\cos(45) = 0.707 \\ a_y &= (0) + (1.0)\sin(45) = 0.707 \end{aligned}$$

Compute the components, length, and angle of the *diagonal vector*:

$$\begin{aligned} d_x &= b_{0x} - a_x = 4.0 - 0.707 = 3.293 \\ d_y &= b_{0y} - a_y = 0.0 - 0.707 = -0.707 \\ |d| &= sqrt(3.293^2 + 0.707^2) = 3.368 \\ \beta &= \tan^{-1}\frac{-0.707}{3.293} = -12.117 \text{ deg} \end{aligned}$$

Check the assembly requirements:

$$\begin{aligned} &r_3 + r_4 = 2.0 + 1.75 = 3.75 \\ &|(r_3 - r_4)| = 0.25 \end{aligned}$$

Hence the conditions from Equations 7.2.13 and 7.2.14 for assembly of the dyad are satisfied.

Compute the angle α:

$$\alpha = \cos^{-1}\frac{2.0^2 + 3.368^2 - 1.75^2}{2(2.0)(3.368)} = 24.274\,\text{deg}$$

from which

$$\theta_3 = -12.117 + 24.274 = 12.157\,\text{deg}$$

Compute coordinates of point $\mathbf{B}(b_x, b_y)$ using Equation 7.2.18:

$$b_x = 0.707 + 2\cos 12.157 = 0.707 + 2(.978) = 2.662$$

$$b_y = 0.707 + 2\sin 12.157 = 0.707 + 2(.211) = 1.128$$

Compute the angle θ_4 using Equation 7.2.19:

$$\theta_4 = \tan^{-1}\frac{1.128 - 0.0}{2.662 - 4.0} = 139.1\,\text{deg}$$

Velocity Analysis of a Four-Bar Linkage

Consider the four-bar linkage of Figure 7.16. Note that if the angular velocity of member 2 is specified, the velocity of point A can be calculated from the relationship

$$\dot{\mathbf{a}} = \dot{\mathbf{a}}_0 + \omega_2 \times \mathbf{r}_2 \tag{7.2.20}$$

Since $\dot{\mathbf{a}}_0 = (0,0)$, this leads to

$$\dot{\mathbf{a}} = \begin{vmatrix} \mathbf{i} & \mathbf{j} & \mathbf{k} \\ 0 & 0 & \omega_2 \\ r_{2x} & r_{2y} & 0 \end{vmatrix} = \mathbf{i}(-r_{2y}\omega_2) + \mathbf{j}(r_{2x}\omega_2) \tag{7.2.21}$$

Since the $\mathbf{i}$ direction is in the positive x-direction,

$$\begin{aligned} \dot{a}_x &= -r_{2y}\omega_2 \\ \dot{a}_y &= r_{2x}\omega_2 \end{aligned} \tag{7.2.22}$$

To calculate the angular velocity of members 3 and 4 note that two separate equations for the velocity of point B can be written in terms of ω_3 and equated. This leads to a single vector equation with two scalar unknowns, ω_3 and ω_4:

$$\dot{\mathbf{b}} = \dot{\mathbf{a}} + \omega_3 \times \mathbf{r}_3$$

$$\dot{\mathbf{b}} = \dot{\mathbf{b}}_0 + \omega_4 \times \mathbf{r}_4$$

$$\dot{\mathbf{a}} + \omega_3 \times \mathbf{r}_3 = \dot{\mathbf{b}}_0 + \omega_4 \times \mathbf{r}_4 \tag{7.2.23}$$

Expanding the cross products and collecting terms, Equation 7.2.23 leads to two linear equations

$$\begin{aligned} \omega_3(-r_{3y}) + \omega_4(r_{4y}) &= (0.0) - \dot{a}_x \\ \omega_3(r_{3x}) + \omega_4(-r_{4x}) &= (0.0) - \dot{a}_y \end{aligned} \tag{7.2.24}$$

Solve Equations 7.2.24 using Cramer's rule:

$$\omega_3 = \frac{\begin{vmatrix} -\dot{a}_x & r_{4y} \\ -\dot{a}_y & -r_{4x} \end{vmatrix}}{\det} = \frac{\dot{a}_x r_{4x} + \dot{a}_y r_{4y}}{\det} \tag{7.2.25}$$

$$\omega_4 = \frac{\begin{vmatrix} -r_{3y} & -\dot{a}_x \\ r_{3x} & -\dot{a}_y \end{vmatrix}}{\det} = \frac{\dot{a}_y r_{3y} + \dot{a}_x r_{3x}}{\det} \tag{7.2.26}$$

where $\dot{a}_x$ and $\dot{a}_y$ are computed using Equations 7.2.22 and

$$\det = r_{3y} r_{4x} - r_{3x} r_{4y} = (r_3 \sin\theta_3)(r_4 \cos\theta_4) - (r_3 \cos\theta_3)(r_4 \sin\theta_4)$$

Once ω_3 and ω_4 have been determined, $\dot{\mathbf{b}}$, the velocity of point B, can be calculated using either of the two equations which precede Equation 7.2.23.

The velocity of any arbitrary point on any member can be determined by first defining its location relative to the reference **r** vector for that member. Let any point **p** be located by a relative position vector **s** at an angle ϕ with respect to the **r** vector for that member. In this example assume a point **p**, on member 3 is located by the vector equation

$$\mathbf{p}_3 = \mathbf{a} + \mathbf{s}_3 \tag{7.2.27}$$

where the angle for the vector $\mathbf{s}_3$ with respect to the x-axis is given by $(\theta_3 + \phi_3)$. The position coordinates for point $\mathbf{p}_3$ are calculated from equation 7.2.27 as

$$\begin{aligned} p_{3x} &= a_x + s_3 \cos(\theta_3 + \phi_3) \\ p_{3y} &= a_y + s_3 \sin(\theta_3 + \phi_3) \end{aligned} \tag{7.2.28}$$

The linear velocity of point $\mathbf{p}_3$ can be calculated from the velocity vector equation

$$\dot{\mathbf{p}}_3 = \dot{\mathbf{a}}_3 + \omega_3 \times \mathbf{s}_3 \tag{7.2.29}$$

Example 7.4

Continuing with the analysis of the mechanism of Fig.7.16, assume that the input crank, member 2, has an angular velocity $\omega_2 = 500$ rpm = 52.36 rad/sec and an angular acceleration $\alpha_2 = 0.0$ rad/sec^2. Perform a velocity analysis.

Solution. First, calculate the velocity components for the point A on the input crank, then calculate the angular velocity of members 3 and 4. Once the angular velocities are known the linear velocity components of any point on members 2, 3, or 4 can be computed.

The velocity components for point A on the input crank are calculated using Equations 7.2.22:

$$\dot{a}_x = (-.707)(52.36) = -37.02 \ rad/sec$$

$$\dot{a}_y = (\ .707)(52.36) = \ 37.02 \ rad/sec$$

$$\det = (2\sin 12.16)(1.75\cos 139.9) - (2\cos 12.16)(1.75\sin 139.9) = -2.768$$

From Equation 7.2.25

$$\omega_3 = \frac{(-37.02)(1.75\cos 139.9) + (37.02)(1.75\sin 139.9)}{\det} = \frac{91.285}{-2.768} = -32.98 \ rad/sec$$

and from Equation 7.2.27

$$\omega_4 = \frac{(37.02)(2.0\sin 12.16) + (-37.02)(2.0\cos 12.16)}{\det} = \frac{-56.783}{-2.768} = 20.51 \ rad/sec$$

The velocity components for point B are calculated from the either of the relations given before Equation 7.2.37. In this example

$$\dot{b}_x = \dot{b}_{0x} - r_{4y}\omega_4 = (0.0) - (1.75\sin 139.9)(20.51) = -23.12\ ips$$

$$\dot{b}_y = \dot{b}_{0y} + r_{4x}\omega_4 = (0.0) + (1.75\cos 139.9)(20.51) = -27.45\ ips$$

Acceleration Analysis of a Four-Bar Linkage

Acceleration analysis of a four-bar linkage continues from the velocity vector Equation 7.2.20. Differentiating Equation 7.2.20 leads to an expression for the acceleration of point A on the input crank, member 2:

$$\ddot{\mathbf{a}} = \ddot{\mathbf{a}}_0 + \alpha_2 \times \mathbf{r}_2 + \omega_2 \times (\omega_2 \times \mathbf{r}_2) \tag{7.2.30}$$

Noting that $\ddot{\mathbf{a}}_0 = (0,0)$ and expanding the cross products leads to

$$\begin{aligned} \ddot{a}_x &= -r_{2y}\alpha_2 - r_{2x}\omega_2^2 \\ \ddot{a}_y &= r_{2x}\alpha_2 - r_{2y}\omega_2^2 \end{aligned} \tag{7.2.31}$$

As with the velocity analysis, the angular motion parameters are the primary unknowns. Writing two vector equations for the acceleration of point B and equating leads to

$$\ddot{\mathbf{a}} + \alpha_3 \times \mathbf{r}_3 + \omega_3 \times (\omega_3 \times \mathbf{r}_3) = \ddot{\mathbf{b}}_0 + \alpha_4 \times \mathbf{r}_4 + \omega_4 \times (\omega_4 \times \mathbf{r}_4) \tag{7.2.32}$$

Expanding the cross products and rearranging gives two linear equations with unknowns α_3 and α_4;

$$\begin{aligned} \alpha_3(-r_{3y}) + \alpha_4(r_{4y}) &= (0.0) - \ddot{a}_x + r_{3x}\omega_3^2 - r_{4x}\omega_4^2 \\ \alpha_3(r_{3x}) + \alpha_4(-r_{4x}) &= (0.0) - \ddot{a}_y + r_{3y}\omega_3^2 - r_{4y}\omega_4^2 \end{aligned} \tag{7.2.33}$$

where $\ddot{a}_x$ and $\ddot{a}_y$ are computed from Equations 7.2.31.

Solving Equations 7.2.33 using Cramer's rule leads to

$$\begin{aligned} \alpha_3 &= \frac{(\ddot{a}_x - r_{3x}\omega_3^2 + r_{4x}\omega_4^2)(r_{4x}) + (\ddot{a}_y - r_{3y}\omega_3^2 + r_{4y}\omega_4^2)(r_{4y})}{\det} \\ \alpha_4 &= \frac{(\ddot{a}_y - r_{3y}\omega_3^2 + r_{4y}\omega_4^2)(r_{3y}) + (\ddot{a}_x - r_{3x}\omega_3^2 + r_{4x}\omega_4^2)(r_{3x})}{\det} \end{aligned} \tag{7.2.34}$$

where

$$\det = \begin{vmatrix} -r_{3y} & r_{4y} \\ r_{3x} & -r_{4x} \end{vmatrix} = r_{3y}r_{4x} - r_{3x}r_{4y} \tag{7.2.35}$$

The acceleration of any point $\mathbf{p}_3$ on member 3 is given by

$$\ddot{\mathbf{p}}_3 = \ddot{\mathbf{a}}_3 + \alpha_3 \times \mathbf{s}_3 + \omega_3 \times (\omega_3 \times \mathbf{s}_3) \tag{7.2.36}$$

where the vector $\mathbf{s}_3$ has the same origin as $\mathbf{r}_3$ and is rotated a constant angle ϕ_3 relative to $\mathbf{r}_3$.

Example 7.5

Continue the analysis of the mechanism of Fig.7.16 and use the velocity results from Example 7.3 to calculate the accelerations of points A and B.

Solution. The x and y components of $\ddot{\mathbf{a}}$ are found using Equations 7.2.31 with $\omega_2 = 52.36$ rad/sec and $\alpha_2 = 0.0$:

$$\ddot{a}_x = -(.707)(52.36^2) = -1938 \ ips^2$$

$$\ddot{a}_y = -(.707)(52.36^2) = -1938 \ ips^2$$

det = −2.768 (Same as value used in velocity analysis)

Substitute into Equations 7.2.34 using

$$r_{3x} = 2\cos 12.16 = 1.955$$

$$r_{3y} = 2\sin 12.16 = 0.421$$

$$r_{4x} = 1.75\cos 139.9 = -1.339$$

$$r_{4y} = 1.75\sin 139.9 = 1.127$$

$$\alpha_3 = \frac{(-1938-(1.955)(32.98^2)+(-1.339)(20.51^2))(-1.339)+(-1938-(0.421)(32.98^2)+(1.127)(20.51^2))(1.127)}{-2.768}$$

$$= \frac{6212-2166}{-2.768} = \frac{4046}{-2.768} = -1462 \text{ rad/sec}^2$$

$$\alpha_4 = \frac{(-1938-(0.421)(32.98^2)+(1.127)(20.51^2))(0.421)+(-1938-(1.955)(32.98^2)+(-1.339)(20.51^2))(1.955)}{-2.768}$$

$$= \frac{-809-9047}{-2.768} = \frac{-9856}{-2.768} = 3561 \text{ rad/sec}^2$$

The linear acceleration components for point B are found using

$$\ddot{b}_x = \ddot{a}_x - r_{3y}\alpha_3 - r_{3x}\omega_3^2 = -1938-(0.421)(-1462)-(1.955)(32.98^2) = -3449 \text{ ips}^2$$

$$\ddot{b}_y = \ddot{a}_y + r_{3x}\alpha_3 - r_{3y}\omega_3^2 = -1938+(1.955)(-1462)-(0.421)(32.98^2) = -5313 \text{ ips}^2$$

The Coriolis Component of Acceleration

In some types of mechanisms it will be necessary to consider the motion of a point which is moving relative to a rotating reference member. This will lead to a supplemental term in the acceleration vector equation called the *Coriolis component.* The shape of the relative path can be either along a straight line or a curve.

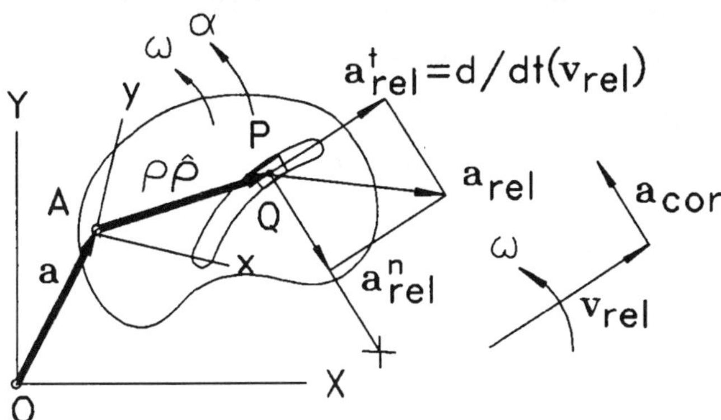

Figure 7.17 Acceleration relative to a moving reference - The Coriolis component

In many cases for planar mechanisms it will be necessary to account for the relative motion for a point that is moving relative to a rotating reference member. Consider the situation illustrated in Figure 7.17. The vector $\mathbf{a}$ locates a reference point A, with specified motion, attached to a member moving with respect to a fixed X-Y coordinate system. The vector $\rho\hat{\rho}$ locates the point P relative to point A in the x-y coordinate system attached to the moving member. The unit vectors and $\mathbf{j}$ define the directions of the x and y axes.

The position of point P in the fixed X-Y system is given by the position vector equation

$$\begin{aligned}\mathbf{p} &= \mathbf{a} + \rho\hat{\rho} \\ &= \mathbf{a} + x\mathbf{i} + y\mathbf{j}\end{aligned} \tag{7.2.37}$$

The moving member is assumed to have an angular velocity $\omega = \dot{\theta}\mathbf{k}$ and an angular acceleration $\alpha = \ddot{\theta}\mathbf{k}$ relative to the fixed X-Y system. The unit vectors $\mathbf{i}$ and $\mathbf{j}$ are of constant length and are attached to the moving member. The scalar variables x and y are of *variable length* and define the motion of the point P *relative to the moving member.* From Equation 7.2.7 it is noted that

$$\frac{d}{dt}\hat{\mathbf{r}} = \dot{\theta}(\hat{\mathbf{u}} \times \hat{\mathbf{r}})$$

hence

$$\frac{d}{dt}\mathbf{i} = \dot{\theta}(\mathbf{k} \times \mathbf{i}) = \omega \times \mathbf{i} \tag{7.2.38}$$

Differentiate the position vector Equation 7.2.37:

$$\begin{aligned}\dot{\mathbf{p}} &= \dot{\mathbf{a}} + \dot{x}\mathbf{i} + x\dot{\theta}(\mathbf{k} \times \mathbf{i}) + \dot{y}\mathbf{j} + y\dot{\theta}(\mathbf{k} \times \mathbf{j}) \\ &= \dot{\mathbf{a}} + (\dot{x}\mathbf{i} + \dot{y}\mathbf{j}) + \dot{\theta}\mathbf{k} \times (x\mathbf{i} + y\mathbf{j}) \\ &= \dot{\mathbf{a}} + (\dot{x}\mathbf{i} + \dot{y}\mathbf{j}) + \dot{\theta}\mathbf{k} \times \rho\hat{\rho} \\ &= \dot{\mathbf{a}} + \mathbf{V}_{rel} + \omega \times \rho\hat{\rho}\end{aligned} \tag{7.2.39}$$

Examine Equation 7.2.39 carefully. Note that the first and third terms would give the velocity of a point Q fixed in the moving member which is *instantaneously* coincident with the point of interest P. The second term describes the *relative velocity* of point P with respect to the moving member. The vector $\rho\hat{\rho}$ is not a physical member and is not fixed in any member. Note that $\dot{x}\mathbf{i} + \dot{y}\mathbf{j} \neq \dot{\rho}\hat{\rho}$!

Differentiating a second time, being careful to account for *all* products, we have:

$$\begin{aligned}\ddot{\mathbf{p}} &= \ddot{\mathbf{a}} + \ddot{x}\mathbf{i} + \dot{x}\dot{\theta}\mathbf{k} \times \mathbf{i} + \ddot{y}\mathbf{j} + \dot{y}\dot{\theta}\mathbf{k} \times \mathbf{j} \\ &\quad + \ddot{\theta}\mathbf{k} \times (x\mathbf{i} + y\mathbf{j}) \\ &\quad + \dot{\theta}\mathbf{k} \times (\dot{x}\mathbf{i} + x(\dot{\theta}\mathbf{k} \times \mathbf{i}) + \dot{y}\mathbf{j} + y(\dot{\theta}\mathbf{k} \times \mathbf{j})) \\ &= \ddot{\mathbf{a}} + (\ddot{x}\mathbf{i} + \ddot{y}\mathbf{j}) + \ddot{\theta}(\mathbf{k} \times \rho\hat{\rho}) + \dot{\theta}^2(\mathbf{k} \times (\mathbf{k} \times \rho\hat{\rho})) + 2(\dot{\theta}\mathbf{k} \times (\dot{x}\mathbf{i} + \dot{y}\mathbf{j})) \\ &= \ddot{\mathbf{q}} + \mathbf{A}_{rel} + \mathbf{A}_{cor}\end{aligned} \tag{7.2.40}$$

where

$$\ddot{\mathbf{q}} = \ddot{\mathbf{a}} + \ddot{\theta}\mathbf{k} \times \rho\hat{p} + \dot{\theta}\mathbf{k} \times (\dot{\theta}\mathbf{k} \times \rho\hat{p})$$
$$= \ddot{\mathbf{a}} + \alpha \times \rho\hat{p} + \omega \times (\omega \times \rho\hat{p})$$
$$\mathbf{A}_{rel} = \ddot{x}\mathbf{i} + \ddot{y}\mathbf{j}$$
$$\mathbf{A}_{cor} = 2\dot{\theta}\mathbf{k} \times (\dot{x}\mathbf{i} + \dot{y}\mathbf{j})$$
$$= 2\omega \times \mathbf{V}_{rel}$$

In Equation 7.2.40 $\ddot{\mathbf{q}}$ is the acceleration of a point Q on the moving member that is instantaneously coincident with the point of interest P. $\mathbf{A}_{rel}$ is the acceleration of point P *relative to a temporarily fixed (non-rotating) member containing point Q*. It will sometimes be useful to use normal and tangential components in describing the *relative* acceleration along and perpendicular to the relative path. $\mathbf{A}_{cor}$ is the *supplemental* Coriolis component which accounts for the angular velocity of the *x-y* system. The origin of the factor 2 can be found in terms involving $\dot{\theta}\dot{x}$ and $\dot{\theta}\dot{y}$ in Equation 7.2.40.

Kinematic Analysis of the Crank-Shaper Mechanism

The crank-shaper mechanism is an example of a *quick-return* mechanism. If the input crank, member 2, rotates at a constant angular velocity, the time for the return rotation of member 4 takes less time than the angle of advance. The ratio θ_A / θ_R is the *advance to return ratio.* The advance to return ratio can be used to create a relative quick return after a slower working stroke in a shaper or a wide variety of feed mechanisms in machinery.

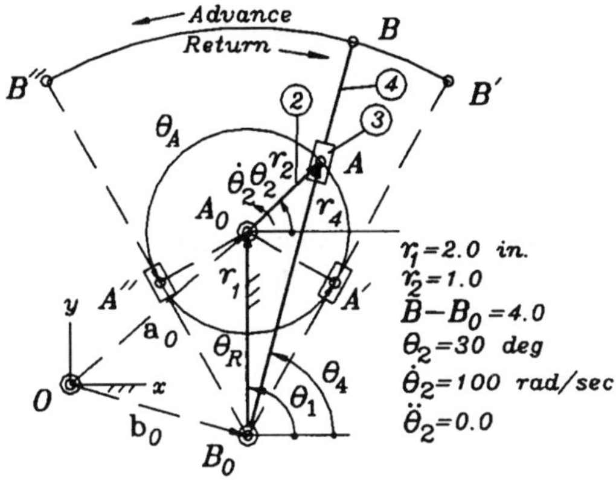

Figure 7.18 The crank-shaper mechanism - An example involving the Coriolis component

Position Analysis

The *position vector equation* is formed by equating two vector sums which describe two independent paths to point B:

$$\mathbf{a}_0 + \mathbf{r}_2 = \mathbf{b}_0 + \mathbf{r}_4 \tag{7.2.41}$$

With $\mathbf{r}_2$ specified as a function of the input angle θ_2 the unknowns for the position analysis are the length r_4 and angle θ_4 for the output member 4. To solve Equation 7.2.41 first resolve it into x and y components then rearrange with both unknowns on the left side:

$$r_4 \cos\theta_4 = a_{0x} - b_{0x} + r_2 \cos\theta_2$$
$$r_4 \sin\theta_4 = a_{0y} - b_{0y} + r_2 \sin\theta_2$$

Let $r_{1x} = (a_{0x} - b_{0x})$ and $r_{1y} = (a_{0y} - b_{0y})$. Square both sides and add to eliminate θ_4 on the left side. This leads to an expression for the length r_4 of vector:

$$r_4 = \sqrt{r_1^2 + r_2^2 + 2r_{1x}r_2\cos\theta_2 + 2r_{1y}r_2\sin\theta_2} \quad \textbf{(7.2.42)}$$

With r_4 known solve for θ_4 from

$$\theta_4 = \tan^{-1}\frac{r_{1y} + r_2\sin\theta_2}{r_{1x} + r_2\cos\theta_2} \quad \textbf{(7.2.43)}$$

Velocity Analysis

Differentiate Equation 7.2.41 to obtain the *velocity vector equation.*

$$\omega_2 \times \mathbf{r}_2 = \dot{r}_4\hat{\mathbf{r}}_4 + \omega_4 \times \mathbf{r}_4 \quad \textbf{(7.2.44)}$$

Resolving Equation 7.2.58 into x and y components

$$\begin{aligned} -r_2\omega_2\sin\theta_2 &= \dot{r}_4\cos\theta_4 - r_4\omega_4\sin\theta_4 \\ r_2\omega_2\cos\theta_2 &= \dot{r}_4\sin\theta_4 + r_4\omega_4\cos\theta_4 \end{aligned} \quad \textbf{(7.2.45)}$$

Solve Equations 7.2.59 for $\dot{r}_4$ and ω_4 by Cramer's rule:

$$\dot{r}_4 = \frac{\begin{vmatrix} -r_2\omega_2\sin\theta_2 & -r_4\sin\theta_4 \\ r_2\omega_2\cos\theta_2 & r_4\cos\theta_4 \end{vmatrix}}{\begin{vmatrix} \cos\theta_4 & -r_4\sin\theta_4 \\ \sin\theta_4 & r_4\cos\theta_4 \end{vmatrix}} = -r_2\omega_2\sin(\theta_2 - \theta_4)$$

$$\omega_4 = \frac{\begin{vmatrix} \cos\theta_4 & -r_2\omega_2\sin\theta_2 \\ \sin\theta_4 & r_2\omega_2\cos\theta_2 \end{vmatrix}}{r_4} = \frac{r_2\omega_2\cos(\theta_2 - \theta_4)}{r_4} \quad \textbf{(7.2.46)}$$

Acceleration Analysis

Since the point A has a velocity relative to the rotating member 4, a Coriolis component must be included in the acceleration analysis. Using Equation 7.2.40 as a guide, the *acceleration vector equation* is written in the form

$$\ddot{\mathbf{a}} = \alpha_4 \times \mathbf{r}_4 + \omega_4 \times (\omega_4 \times \mathbf{r}_4) + \ddot{r}_4\hat{\mathbf{r}}_4 + 2\omega_4 \times \dot{r}_4\hat{\mathbf{r}}_4 \quad \textbf{(7.2.47)}$$

with $\ddot{\mathbf{a}}$ specified by the input crank motion and $\ddot{r}_4$ and α_4 as unknowns. Rearrange with terms involving $\ddot{r}_4$ and α_4 on the left side.

$$\alpha_4 \times \mathbf{r}_4 + \ddot{r}_4\hat{\mathbf{r}}_4 = \ddot{\mathbf{a}} - \omega_4 \times (\omega_4 \times \mathbf{r}_4) - 2\omega_4 \times \dot{r}_4\hat{\mathbf{r}}_4 \quad \textbf{(7.2.48)}$$

In the solution of Equation 7.2.48 a new method will be introduced which takes advantage of the properties of the dot and cross products of unit vectors. Assume two unit vectors $\hat{\mathbf{u}}$ and $\hat{\mathbf{v}}$. The following identities will be found useful:

$$\hat{\mathbf{u}} \cdot \hat{\mathbf{v}} = \mathbf{cos}(\hat{\mathbf{u}}, \hat{\mathbf{v}})$$

$$\hat{\mathbf{u}} \times \hat{\mathbf{v}} = \mathbf{sin}(\hat{\mathbf{u}}, \hat{\mathbf{v}})$$

where $(\hat{\mathbf{u}}, \hat{\mathbf{v}})$ is the angle from $\hat{\mathbf{u}}$ to $\hat{\mathbf{v}}$. This implies that when two unit vectors are parallel the dot product =1 and when two unit vectors are perpendicular the cross product =1.

Take the dot product of each term in Equation 7.2.48 with the unit direction vector $\hat{\mathbf{r}}_4$. Since, for example, the vector $\alpha_4 \times \mathbf{r}_4$ is perpendicular to $\hat{\mathbf{r}}_4$, this leads to

$$\ddot{r}_4 = \ddot{\mathbf{a}} \cdot \hat{\mathbf{r}}_4 - (\omega_4 \times (\omega_4 \times \mathbf{r}_4) \cdot \hat{\mathbf{r}}_4 \quad \textbf{(7.2.49)}$$

where $\ddot{\mathbf{a}}$ is specified by the motion of the input member 2. This gives

$$\ddot{r}_4 = \ddot{a}_x r_4 \cos\theta_4 + \ddot{a}_y r_4 \sin\theta_4 + r_4\omega_4^2 \quad \textbf{(7.2.50)}$$

Similarly take the cross product of each term in Equation 7.2.48 with $\hat{\mathbf{r}}_4$ leads to.

$$r_4\alpha_4(\mathbf{k}\times\hat{\mathbf{r}}_4)\times\hat{\mathbf{r}}_4 = \ddot{\mathbf{a}}\times\hat{\mathbf{r}}_4 - 2\omega_4\dot{r}_4 \quad \textbf{(7.2.51)}$$

Solving for α_4

$$\alpha_4 = \frac{|\ddot{\mathbf{a}}|\sin(\theta_2-\theta_4) - 2\omega_4\dot{r}_4}{r_4} \quad \textbf{(7.2.52)}$$

Example 7.6

Use the numerical values specified in Fig.7.18 and find α_4.

Solution. From Equations 7.2.42

$$r_4 = \sqrt{2^2+1^2+0+2(2)(1)(\sin 30)} = 2.646$$

$$\theta_4 = \tan^{-1}\frac{2+(1)(\sin 30)}{0+1(\cos 30)} = 70.89\,\text{deg}$$

From Equations 7.2.46

$$\dot{r}_4 = -(1)(100)(\sin(30-70.89)) = 65.46 \text{ ips}$$

$$\omega_4 = \dot{\theta}_4 = \frac{(1)(100)\cos(30-70.89)}{2.646} = 28.57 \text{ rad/sec}$$

From Equations 7.2.50 and 7.2.52

$$\ddot{a}_x = -(1\cos 30)(100^2) = -8660 \text{ ips}^2$$

$$\ddot{a}_y = -(1\sin 30)(100^2) = -5000 \text{ ips}^2$$

$$|\ddot{\mathbf{a}}| = (-\mathbf{8660})^2 + (-5000)^2 = -10{,}000 \text{ ips}^2 (\text{In negative } r_2 \text{ direction})$$

$$\ddot{r}_4 = (-8660)(2.646)\cos 70.89 + (-5000)(2.646)\sin 70.89 + (2.646)(30.86^2)$$

$$= -5399 \text{ ips}^2$$

$$\alpha_4 = \ddot{\theta}_4 = \frac{-10{,}000\sin(30-70.89) - 2(28.57)(65.46)}{2.646}$$

$$= 1060 \text{ rad/sec}^2$$

Solution of Equation 7.2.48 by Cramer's rule:

Equation 7.2.48 could also be solved by resolving into two linear equations with the two unknowns α_4.and $\ddot{r}_4$.

The x and y components become

$$-r_{4y}\alpha_4 + \ddot{r}_4\cos\theta_4 = \ddot{a}_x - r_4\omega_4^2\cos\theta_4 - 2\omega_4\dot{r}_4\cos\theta_4$$

$$r_{4x}\alpha_4 + \ddot{r}_4\sin\theta_4 = \ddot{a}_y - r_4\omega_4^2\sin\theta_4 - 2\omega_4\dot{r}_4\sin\theta_4$$

Solving gives an alternate expression for α_4:

$$\alpha_4 = \frac{-\ddot{a}_x\sin\theta_4 + \ddot{a}_y\cos\theta_4 - 2\omega_4\dot{r}_4}{r_4}$$

$$= \frac{-(-8660)\sin 70.89 + (-5000)\cos 70.89 - 2(28.57)(65.47)}{2.646}$$

$$= 1060 \text{ rad/sec}^2$$

The above equations could also be solved for $\ddot{r}_4$.

Acceleration Analysis of Rolling Contact Mechanisms

Rolling contact mechanisms require special methods to account for the relative motion at the point of contact. Consider the cam mechanism with an oscillating roller follower shown in Figure 7.19.

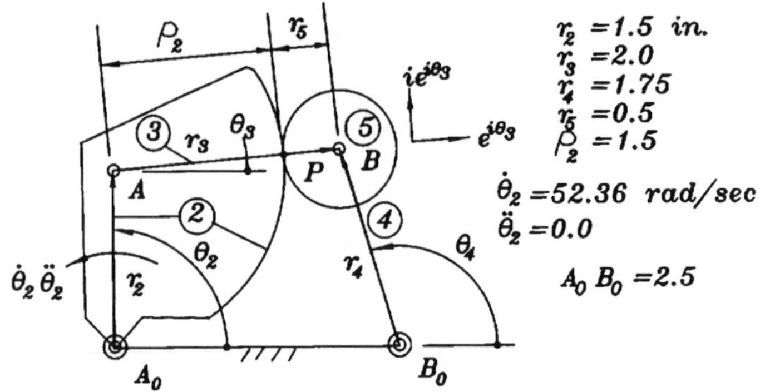

Figure 7.19 A cam mechanism with oscillating roller follower

The point of rolling contact between the cam, member 2, and the follower, member 5, is at P. The point P_2 and P_5 have the same acceleration component in the tangential direction at the instant shown. The relative acceleration at point P in the tangential direction is zero. This condition will be important in the acceleration analysis. The normal component of the acceleration of point P_5 relative to point B_5 is directed from the point of contact toward the center of curvature of the roller surface. In visualizing the relative motion components imagine the cam temporarily fixed in space with the roller follower rolling on a fixed circle of radius equal to the radius of curvature of the cam at the contact point.

Begin the analysis by temporaily bypassing the relative motion at the point P. Note that for small motions about the current position the distance between points A and B would change a very small amount and can be considered a constant. Introduce a temporary link $\mathbf{r}_3$ connecting points A and B to form an *equivalent four-bar linkage.* The analysis of the four-bar linkage would proceed using exactly the same method shown in Examples 7.3, 7.4, and 7.5. Before proceeding it is important to remember that, in general, for a cam with a variable radius of curvature, a new four-bar linkage must be defined for each new position of the input crank with point A at the center of curvature of the cam surface.

Angular Velocity of the Roller Follower

At the point of rolling contact $\mathbf{p}_2 = \mathbf{p}_5$, $\theta_5 = \theta_3$, but $\dot{\theta}_5 \neq \dot{\theta}_3$. Begin by writing two *independent position equations* for point P.

$$\mathbf{a}_0 + \mathbf{r}_2 + \rho_2\hat{\rho}_2 = \mathbf{b}_0 + \mathbf{r}_4 - \mathbf{r}_5 \tag{7.2.53}$$

Differentiate Equation 7.2.53, noting that ρ_2 is imbedded in member 2:

$$\omega_2 \times \mathbf{r}_2 + \omega_2 \times \rho_2\hat{\rho}_2 = \omega_4 \times \mathbf{r}_4 - \omega_5 \times \mathbf{r}_5 \tag{7.2.54}$$

Since $\dot{\theta}_4\mathbf{k} = \omega_4$ was determined in the velocity analysis of the equivalent four-bar linkage, $\dot{\theta}_5\mathbf{k} = \omega_5$ is the single unknown in Equation 7.2.54.

To solve for ω_5 take the dot product of each term with the unit direction $\hat{\rho}_2$. This leads to

$$\omega_2 r_2 \hat{\mathbf{r}}_2 \cdot \hat{\rho}_2 + \omega_2 \rho_2 = \omega_4 r_4 \hat{\mathbf{r}}_4 \cdot \hat{\rho}_2 - \omega_5 r_5 \qquad (7.2.55)$$

Finally

$$\omega_5 = \frac{\omega_4 r_4 \cos(\theta_4 - \theta_3) - \omega_2 \rho_2 - \omega_2 r_2 \cos(\theta_2 - \theta_3)}{r_5} \qquad (7.2.56)$$

Angular Acceleration of the Roller Follower

Differentiate Equation 7.2.54:

$$\begin{aligned} &\alpha_2 \times \mathbf{r}_2 + \omega_2 \times (\omega_2 \times \mathbf{r}_2) + \alpha_2 \times \rho_2 \hat{\rho}_2 + \omega_2 \times (\omega_2 \times \rho_2 \hat{\rho}_2) \\ &= \alpha_4 \times \mathbf{r}_4 + \omega_4 \times (\omega_4 \times \mathbf{r}_4) - \alpha_5 \times \mathbf{r}_5 - \omega_5 \times (\omega_5 \times \mathbf{r}_5) \end{aligned} \qquad (7.2.57)$$

Again, take the dot product of each term with the unit direction ρ_2. This leads to

$$\alpha_5 = \frac{\begin{aligned} &r_4 \alpha_4 \cos(\theta_4 - \theta_3) - r_4 \omega_4^2 \sin(\theta_4 - \theta_3) \\ &- r_2 \alpha_2 \cos(\theta_2 - \theta_3) + r_2 \omega_2^2 \sin(\theta_2 - \theta_3) - \rho_2 \alpha_2 \end{aligned}}{r_5} \qquad (7.2.58)$$

Example 7.7

Using the numerical values listed in Fig.7.19, analyze the linkage of Fig.7.19.

Solution. For the geometry of Figure 7.19, analysis of the equivalent four-bar linkage gave the following numerical results:

$$\begin{array}{lll} \theta_2 = 90 \text{ deg} & \omega_2 = 52.36 \text{ rad/sec} & \alpha_2 = 0.0 \text{ rad/sec}^2 \\ \theta_3 = 5.01 & \omega_3 = -11.64 & \alpha_3 = 116.4 \\ \theta_4 = 106.86 & \omega_4 = 45.68 & \alpha_4 = 805.9 \end{array}$$

The relative position angles are:

$$(\theta_2 - \theta_3) = 90 - 5.01 = 84.99 \text{ deg}$$

$$(\theta_4 - \theta_3) = 106.86 - 5.01 = 101.85 \text{ deg}$$

Substituting into equation 7.58 gives

$$\alpha_5 = \frac{1.75(805.9)\cos 101.85 - 1.75(45.68^2)\sin 101.85 - 1.5(0)\cos 84.99 + 1.5(52.36^2)\sin 84.99 - 1.5(0)}{0.5}$$

$$= 466.4 \text{ rad/sec}^2$$

7.3 Dynamics of Mechanisms

The dynamic analysis of a mechanism as part of the dynamic design of a complete machine is necessarily an iterative process, as is typical of any design procedure. If the members are to resist dynamic loads due to the acceleration of the members, the forces required to produce the accelerations must be added to the forces required to do the external work of the machine. The input forces and torques provided by a source of external energy must account for both working forces and torques and the inertia effects due to acceleration of the moving masses.

Equations of Motion

The fundamental relations between the forces acting on a rigid body and the acceleration of its center of gravity, *Newton's Laws of Motion*, can be stated in the simple form

$$\sum \mathbf{F} = m\ddot{\mathbf{g}} \qquad \sum \mathbf{T}_G = I_G\alpha \tag{7.3.1}$$

These equations can be interpreted as:

- The acceleration of the center of gravity of any mass is proportional to the sum of the external forces acting on the mass and in the direction of the resultant of the forces.
- The angular acceleration of a mass acted on by a system of external forces is proportional to the sum of the torques of all forces and pure couples acting about the center of gravity of the mass.

Consider the single mass of Figure 7.20 with a system of external forces and torques applied as shown

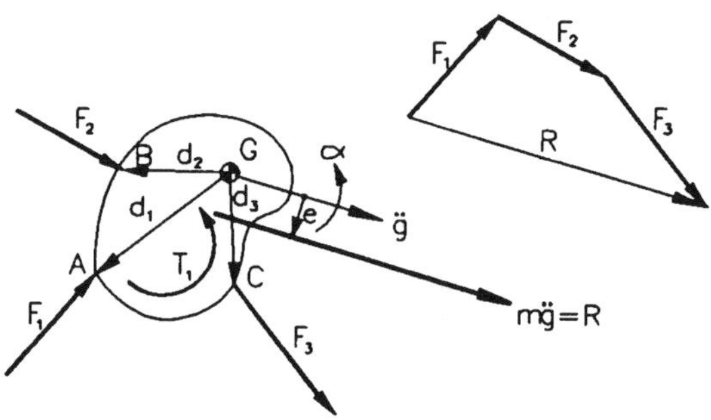

Figure 7.20 Motion of the mass center

Three forces $\mathbf{F}_1, \mathbf{F}_2, \mathbf{F}_3$ and one pure couple $\mathbf{T}_1$ are shown acting on a single rigid body of mass m and moment of inertia I_G about the center of gravity G. The sum of the three forces, the resultant force R, is found by vector addition of the forces as shown. The acceleration of the center of gravity G, shown as $\ddot{\mathbf{g}}$ in figure 7.20, is parallel to the direction found for the vector **R**.

Applying Equations 7.3.1, the equations of motion become

$$\mathbf{F}_1 + \mathbf{F}_2 + \mathbf{F}_3 = m\ddot{\mathbf{g}} \qquad \mathbf{d}_1 \times \mathbf{F}_1 + \mathbf{d}_2 \times \mathbf{F}_2 + \mathbf{d}_3 \times \mathbf{F}_3 + \mathbf{T}_1 = I_G\alpha \tag{7.3.2}$$

The **d** vectors are measured from the center of gravity G to any point on the line of action of a force **F**. The vector cross products give the moment of the individual force about the center of gravity G. Assuming the mass m and moment of inertia I_G are known, and all of the applied forces and torques are specified, Equations 7.3.2 can be solved for the three scalar unknowns $\ddot{g}_x, \ddot{g}_y$ and α.

In most mechanisms problems with constrained motion, the accelerations of the center of gravity of all moving members can be determined from the *kinematic* analysis. The dynamic analysis is then the problem of finding the forces at the joints that are consistent with the known accelerations. This class of dynamics problem is known as a *kinetostatic* problem. The descriptive term derives from the fact, that methods used to solve static force analysis problems in mechanics can be used to solve such problems.

Graphical Methods for Kinetostatic Analysis

Refer to Figure 7.20. Note that the resultant force **R** is always numerically equal to $m\ddot{\mathbf{g}}$, the product of the mass times the linear acceleration of the mass center. Therefore, if the acceleration of the mass center is known as the result of kinematic analysis, the sum of the external forces acting on the mass is known. The problem then is to separate the unknown joint forces from the known external loads acting on the body.

In Figure 7.20 three forces and one pure couple are shown acting on the body. The couple $\mathbf{T}_1$ can be moved such that it acts about the point G. The line of action of the resultant force is determined by noting that the moment of the resultant force **R** about the center of gravity G requires that the line of action of the resultant force be offset from the center of gravity by an offset distance **e**. The offset is calculated by first rewriting Equation 7.3.2 in the form

$$\mathbf{e} \times \mathbf{R} + \mathbf{T}_1 = I_G \alpha \tag{7.3.3}$$

If e is the perpendicular distance from point G to the line of action of the resultant force, Eq.7.3.3 can be solved for the *eccentricity e*:

$$e = \frac{I_G \alpha - \mathbf{T}_1}{m|\ddot{\mathbf{g}}|} \tag{7.3.4}$$

In the majority of problems the *sum of the pure couples* $\mathbf{T}_1 = 0$ and Equation 7.3.4 takes the more familiar form

$$e = \frac{I_G \alpha}{m|\ddot{\mathbf{g}}|} \tag{7.3.5}$$

Note that the line of action of the *effective force* $m|\ddot{\mathbf{g}}|$ has been established such that it accounts for both the linear acceleration of the mass center and the angular acceleration about the mass center.

Graphical Force-Torque Analysis for a Four-Bar Linkage

Graphical force-torque analysis requires that the effects of both the linear acceleration of the mass center and the angular acceleration about the mass center be accounted for in a single graphical construction. This can be done if the concept of *inertia force* is introduced. Begin by rewriting Equations 7.3.1 with all terms moved to the left side in the form

$$\sum \mathbf{F} - m\ddot{\mathbf{g}} = 0 \tag{7.3.6}$$

$$\sum \mathbf{T} - I_G \alpha = 0 \tag{7.3.7}$$

The equations of motion can now be considered *equations of static equilibrium* and solved using the methods of engineering statics. When written in this form the $-m\ddot{g}$ term can be described as a *reversed effective force* since it has the same magnitude and line of action as the effective force but with its direction reversed. This *inertia force* is only a hypothetical concept introduced to allow the use of graphical or analytical methods used to solve statics problems. Inertia forces do not exist as external physical forces acting on an accelerating mass. The $-I_G\alpha$ term is an *inertia torque or moment.*

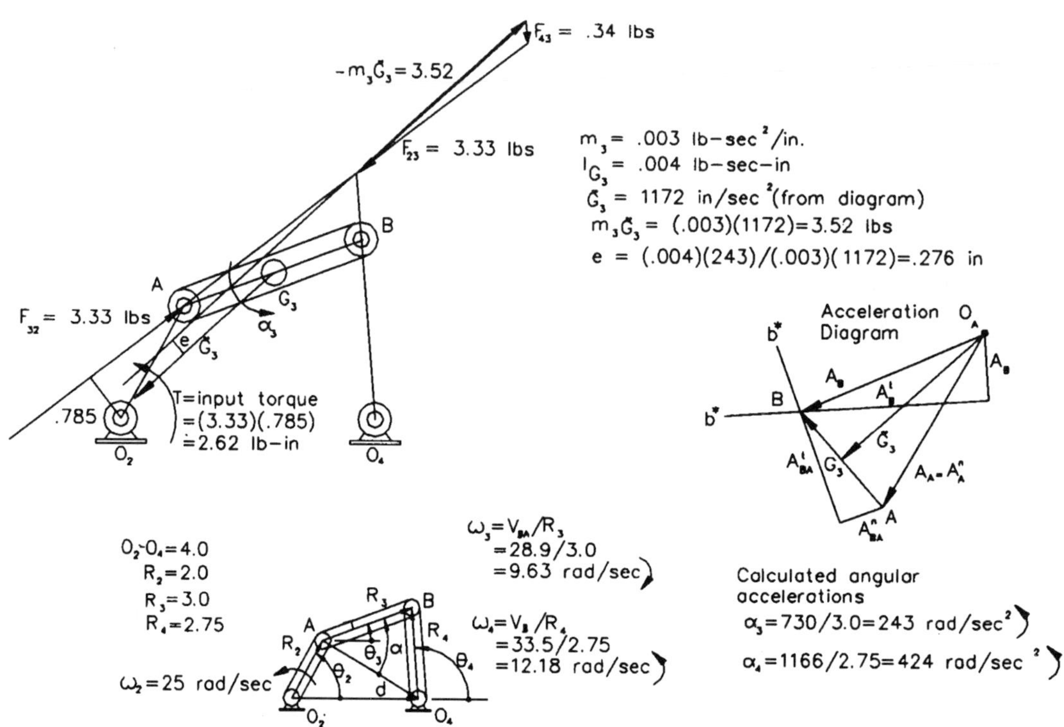

Figure 7.21 Graphical force-torque analysis of a four-bar linkage

In graphical solutions the method of superposition is often applied. The effects of the acceleration of each member are solved as a separate subproblem. After all members have been accounted for, the joint force and input torque contributions from all subproblems are then added vectorially to give the final solution. In the following example only the acceleration of the coupler in Fig.7.21, member 3, is accounted for.

Example 7.8

Analyze the linkage of Fig.7.21 using the numerical values listed in that figure.

Solution. Proceed with the following steps:

- From the results of an acceleration analysis, the acceleration of the center of mass for member 3 is equal to 1172 ips^2 in the direction from point O_v to G_3 in the acceleration diagram. The point G_3 is located at the midpoint of the line from A to B in the acceleration diagram.

- Next consider the line of action of the effective force $m_3\ddot{\mathbf{g}}_3$ acting on member 3. This force must have the same direction as the acceleration of the center of mass. To account for the angular acceleration the effective force must be offset a distance **e**, calculated using Equation 7.3.5, then drawn on the force diagram above the center of mass such that the parallel line of action of the new effective force would result in a positive angular acceleration for this example. Note that, up to this point, no mention has been made of inertia force.
- Now with the proper line of action for the effective force established, the reversed effective force, i.e., the inertia force, will act along the same line but with the opposite sense.
- Examination of the force system acting on member 3 shows that, *for the conditions of this subproblem*, there are three forces acting (including the inertia force). When three forces act on a body in static equilibrium the system of three forces must form a concurrent system, that is., intersect at a single point.
- Note that when the mass of members 2 and 4 is neglected, as in this example, member 4 becomes a two-force member and the direction of the force $\mathbf{F}_{14}$ or $\mathbf{F}_{34}$ must be along the centerline of member 4. The point of intersection of the three-force system acting on member 3 is found at the intersection of the inertia force and the force along the centerline of member 4. The line of action of the force $\mathbf{F}_{23}$ can then be drawn from the intersection to point A.
- Now with all directions known a force vector polygon is drawn. Begin with the inertia force vector , and set its origin at the point of intersection. Next draw a line in the direction of $\mathbf{F}_{43}$ from the head of the vector and a line in the direction of $\mathbf{F}_{23}$ from the tail. The intersection of these construction lines determines the magnitude of the unknown joint forces.

  The force at joint A, applied by member 3 on member 2, is $\mathbf{F}_{32} = -\mathbf{F}_{23}$. Next measure the perpendicular distance from the fixed pivot O_2 to the line of action of $\mathbf{F}_{32}$. This gives a lever arm = 0.785 inches. The component of the driving torque that accounts for the acceleration of the mass of member 3 becomes

$$\mathbf{T}'_{12} = 3.33 \times 0.785 = 2.62 \text{ in - lb.}$$

- To complete the problem, the mass and moment of inertia of members 2 and 4 would be accounted for individually and the results of all subproblems superimposed to give the total force at each joint and the total driving torque.

An Alternative Form for the Equations of Motion

Consider the force system of Figure 7.22 with the effective force $m\ddot{\mathbf{g}}$ and the effective torque $I_G\alpha$ shown in addition to the system of external loads. In many mechanisms problems, to eliminate unknowns from the equations of motion, it will advantageous to use a reference point such as point A in the torque equation. It is obvious that the force $\mathbf{F}_1$ would not appear in the torque equation, since it does not contribute to the summation of moments about point A. Writing an expression for the sum of the moments about point A

$$\mathbf{d}_2 \times \mathbf{F}_2 + \mathbf{d}_3 \times \mathbf{F}_3 + \mathbf{T}_1 = I_G\alpha + \mathbf{s} \times m\ddot{\mathbf{g}}$$

which can be generalized in the form

$$\sum \mathbf{d} \times \mathbf{F} + \sum \mathbf{T} = I_G\alpha + \mathbf{s} \times m\ddot{\mathbf{g}} \qquad \textbf{(7.3.8)}$$

Note that the moment of inertia in Equation 7.3.8 is referred to the center of gravity, point G, not the point A. The $\mathbf{s} \times m\ddot{\mathbf{g}}$ term accounts for the use of point A as a reference.

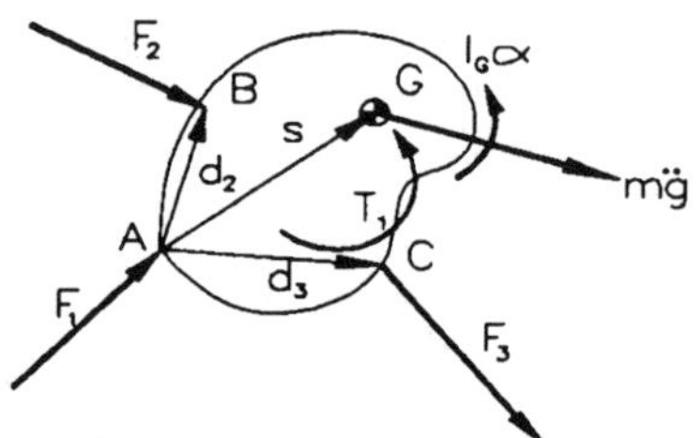

Figure 7.22 The equation of motion referred to point A

Example 7.9 Analytical force-torque solution for a four-bar linkage

The mechanism of Figure 7.23 has the following dimensions and physical properties:

$\mathbf{r}_2 = O_2 - A = 1.0$ inch	$m_2 = .002$ lb − sec²/in	$I_2 = .0003$ lb − sec² − in
$\mathbf{r}_3 = A - B = 2.828$	$m_3 = .005$	$I_3 = .0006$
$\mathbf{r}_4 = O_4 - B = 2.0$	$m_4 = .004$	$I_4 = .0005$

Perform a force-torque analysis for this linkage.

Solution. The results of an acceleration analysis are:

$\theta_2 = 0$ deg	$\dot{\theta}_2 = 100$ rad/sec	$\ddot{\theta}_2 = 0$ rad/sec²
$\theta_3 = 45$	$\dot{\theta}_3 = -50$	$\ddot{\theta}_3 = 0$
$\theta_4 = 90$	$\dot{\theta}_4 = -50$	$\ddot{\theta}_4 = 7500$

$\mathbf{a} = (1,0)$ inches	$\dot{\mathbf{a}} = (0,100)$ ips	$\ddot{\mathbf{a}} = (-10000,0)$ ips²
$\mathbf{b} = (3,2)$	$\dot{\mathbf{b}} = (100,0)$	$\ddot{\mathbf{b}} = (-15000,-5000)$
$\mathbf{g}_2 = (0.5,0)$	$\dot{\mathbf{g}}_2 = (0,50)$	$\ddot{\mathbf{g}}_2 = (-5000,0)$
$\mathbf{g}_3 = (1.388,1.449)$	$\dot{\mathbf{g}}_3 = (72.45,80.6)$	$\ddot{\mathbf{g}}_3 = (-10970,-3623)$
$\mathbf{g}_4 = (3,1)$	$\dot{\mathbf{g}}_4 = (50,0)$	$\ddot{\mathbf{g}}_4 = (-7500,-2500)$

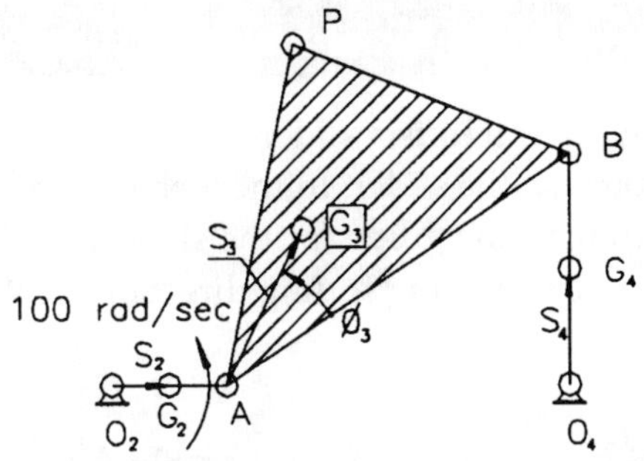

P
B
S_3
G_3
G_4
100 rad/sec
ϕ_3
S_4
S_2
O_2
G_2
A
O_4

Figure 7.23 Acceleration analysis of a four-bar linkage

The joint force $\mathbf{F}_{23}$ is the force at point A applied by member 2 on member 3. The force that member 3 applies on member 2 could be either $\mathbf{F}_{32}$ or $-\mathbf{F}_{23}$.

When solving for the joint force components the analysis always begins with the output member and proceeds toward the input. Begin with member 4 and write the torque equations for members 3 and 4 using Equation 7.3.8 as a reference.

$$\mathbf{r}_4 \times \mathbf{F}_{34} = I_4\alpha_4 + \mathbf{s}_4 \times m_4\ddot{\mathbf{g}}_4$$
$$\mathbf{r}_3 \times (-\mathbf{F}_{34}) = I_3\alpha_3 + \mathbf{s}_3 \times m_3\ddot{\mathbf{g}}_3$$

Expanding the cross products, two linear equations with unknowns F_{34x} and F_{34y} can be written in the form

$$r_{4x}F_{34y} - r_{4y}F_{34x} = I_4\alpha_4 + m_4(s_{4x}\ddot{g}_{4y} - s_{4y}\ddot{g}_{4x})$$
$$r_{3x}(-F_{34y}) - r_{3y}(-F_{34y}) = I_3\alpha_3 + m_3(s_{3x}\ddot{g}_{3y} - s_{3y}\ddot{g}_{3x})$$

Substituting numerical values

$$(0)(F_{34y}) - (2)(F_{34x}) = (.0005)(7500) + .004((0)(-2500) - (1)(-7500)) = 33.75$$
$$(2)(-F_{34y}) - (2)(-F_{34x}) = (.0006)(0) + .005((0.388)(-3623) - (1.449)(-10970)) = 72.44$$

which gives $F_{34x} = -16.9$ lbs and $F_{34y} = -53.1$ lbs. With the joint force $\mathbf{F}_{34}$ at joint B known, substitute into equation 7.80 for member 3 to complete the force analysis for $\mathbf{F}_{23}$ at joint A:

$$F_{23x} + F_{43x} = m_3\ddot{g}_{3x}$$
$$F_{23y} + F_{43y} = m_3\ddot{g}_{3y}$$

Substituting numerical values

$$F_{23x} = (.005)(-10970) - (16.9) = -71.8 \text{ lbs}$$
$$F_{23y} = (.005)(-3623) - (53.1) = -71.2$$

The required input torque $\mathbf{T}_{12}$ can be found by writing a torque equation for member 2 about the fixed point O_2. Note that $\mathbf{F}_{32} = -\mathbf{F}_{23}$ and that α_2=0 in this example:

$$\mathbf{r}_2 \times \mathbf{F}_{32} + \mathbf{T}_{12} = I_2\alpha_2 + \mathbf{s}_2 \times m_2\ddot{\mathbf{g}}_2$$

Substituting numerical values

$$T_{12} = I_2\alpha_2 + m_2(s_{2x}\ddot{g}_{2y} - s_{2y}\ddot{g}_{2x}) - (r_{2x}F_{32y} - r_{2y}F_{32x})$$
$$= (.0003)(0) + .002((0.5)(0) - (0)(-5000)) - ((1)(71.2) - (0)(71.8)) = -71.2 \text{ lb} - \text{in}$$

The null terms are included to show where they did not contribute in this example.

Example 7.10 Dynamic analysis of an offset slider-crank mechanism

The dynamic analysis of an offset slider-crank takes advantage of the known direction of the side force on the slider. In the mechanism shown in Figure 7.24 it is assumed that the straight line guide for the slider is well lubricated hence the contact force is perpendicular to the sliding direction. Perform a dynamic analysis of this mechanism.

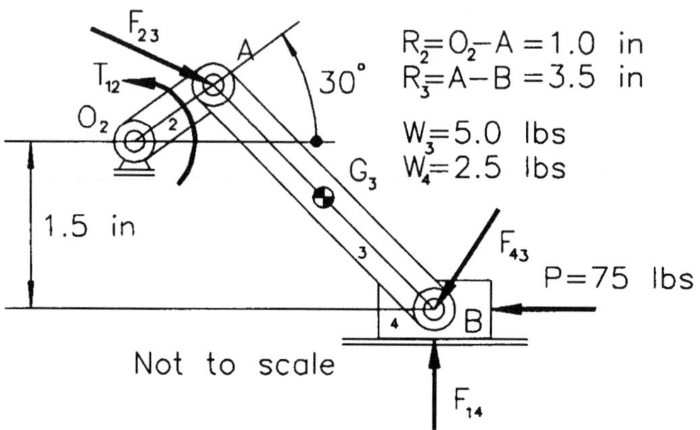

Figure 7.24 An offset slider-crank mechanism

Solution. Assuming a constant speed input angular velocity $\dot{\theta}_2 = 500\,\text{rpm}$ the results of the kinematic analysis can be summarized as

$\theta_2 = 30$ deg	$\theta_3 = -34.85$ deg	$\mathbf{a} = (0.866, 0.500$ in	$\mathbf{b} = (3.738, -1.500)$ in
$\dot{\theta}_2 = 52.36$ rad/sec	$\dot{\theta}_3 = -15.79$ rad/sec	$\dot{\mathbf{a}} = (-26.180, 45.345)$ ips	$\dot{\mathbf{b}} = (-57.754, 0.0)$ ips
$\ddot{\theta}_2 = 0$	$\ddot{\theta}_3 = 303.7$ rad/sec^2	$\ddot{\mathbf{a}} = (-2374, -1371)$ ips^2	$\ddot{\mathbf{b}} = (-2483, 0.0)$ ips^2

Ignoring the mass of member 2, the results for the mass centers of member 3 and 4 are

$$\mathbf{g}_3 = (2.302, -.500) \qquad m_3 = 5.0/386 = .013\ \text{lb}-\text{sec}^2/\text{in}$$

$$\dot{\mathbf{g}}_3 = (-41.967, 22.673) \qquad I_3 = \frac{1}{12} m_3 r_3^2 = .0133\ \text{lb}-\text{sec}^2-\text{in}$$

$$\ddot{\mathbf{g}}_3 = (-2429, -685.4) \qquad m_4 = 2.5/386 = .0065\ \text{lb}-\text{sec}^2/\text{in}$$

To solve for the joint force $\mathbf{F}_{43}$, first write the torque equation for member 3 with point A as the reference

$$\mathbf{r}_3 \times \mathbf{F}_{43} = I_3\alpha_3 + \mathbf{s}_3 \times m_3\ddot{\mathbf{g}}_3$$

The equation of motion for the slider, member 4, is

$$\mathbf{F}_{14} + (-\mathbf{F}_{43}) + \mathbf{P} = m_4\ddot{\mathbf{g}}_4$$

where in the absence of friction, $F_{14x} = 0$. Taking the x-components:

$$-F_{43x} + P = m_4\ddot{g}_{4x}$$

Expanding :

$$r_{3x}F_{43y} - r_{3y}F_{43x} = I_3\alpha_3 + m_3(s_{3x}\ddot{g}_{3y} - s_{3y}\ddot{g}_{3x})$$

Substituting in numerical values:

$$-F_{43x} - 75.0 = 0.0065(-2483)$$
$$F_{43x} = 16.14 - 75.0 = -58.86 \text{ lbs}$$

Then

$$(3.738 - 0.866)(F_{43y}) - (-1.500 - 0.500)(-58.86)$$
$$= (.0133)(303.7) + (.013)((2.302 - 0.866)(-685.4) - (-0.500 - 0.500)(-2429))$$
$$= 4.039 - 12.795 - 31.577$$
$$F_{43y} = \frac{4.039 - 12.785 - 31.577 + 117.72}{2.872} = 26.95 \text{ lbs}$$

Apply Eq.7.3.6 to member 3:

$$F_{23x} + F_{43x} = m_3\ddot{g}_{3x} \qquad F_{23x} = (.013)(-2429) - (-58.86) = 27.28 \text{ lbs}$$
$$F_{23y} + F_{43y} = m_3\ddot{g}_{3y} \qquad F_{23y} = (.013)(-685.4) - (26.95) = -35.87$$

Determine the required input torque:

$$r_{2x}(-F_{23y}) - r_{2y}(-F_{23x}) + T_{12} = 0$$
$$T_{12} = -((0.866)(35.87) - (0.500)(-27.28)) = -44.7 \text{ lb} - \text{in}$$

The Power Equation

The work done by a force acting on a rigid body is equal to the vector dot product of the force and the displacement of the point of application of the force on the body. Take the dot product of each term in the equations of motion given in Equation 7.3.2 with the instantaneous velocity of the point of application of the force. Note that each velocity vector can be considered as a displacement $\Delta\mathbf{s}$ which occurs in the time Δt . The product of force times velocity has the units of rate of change of doing work.

A new equation is formed, which when summed over the total number of moving bodies, is known as the *power equation*:

$$\sum_{k=2}^{n} \mathbf{F_k} \cdot \mathbf{v_k} + \sum_{k=2}^{n} \mathbf{T_k} \cdot \omega_\mathbf{k} = \sum_{k=2}^{n} m_k \ddot{\mathbf{g}}_\mathbf{k} \cdot \dot{\mathbf{g}}_\mathbf{k} + \sum_{k=2}^{n} I_k \alpha_\mathbf{k} \omega_\mathbf{k} \qquad \textbf{(7.3.9)}$$

where the $\mathbf{F_k}$ and $\mathbf{T_k}$ terms include only external loads or driving forces acting on the system.

The power equation is useful when only the input torque is required as the result of a dynamic analysis. Note that the joint forces, which are *internal forces*, do no work on the complete system, hence are not included in the power equation. Since the dot product of two vectors is a scalar quantity, Equation 7.3.9 is the sum of the power for all members in a multi-link mechanism. The sum begins with member 2 since member 1 is always the fixed or reference member.

Example 7.11 Input torque analysis using the power equation

Using the kinematic data and physical properties from Example 7.10, use the power equation to solve for the torque.

Solution. The external force and torque terms are

$$\mathbf{P} \cdot \dot{\mathbf{b}} = P_x \dot{b}_x + P_y \dot{b}_y = (-75)(-57.754) + (0)(0) = 4331$$
$$T_{12}\dot{\theta}_2 = T_{12}(52.36)$$

The right hand side terms of Eq.7.3.9, noting that $\mathbf{g_4} \equiv \mathbf{b}$, are

$$m_3\ddot{\mathbf{g}}_3 \cdot \dot{\mathbf{g}}_3 = m_3(\ddot{g}_{3x}\dot{g}_{3x} + \ddot{g}_{3y}\dot{g}_{3y}) = .013((-2429)(-41.967) + (-685.4)(22.673)) = 1123$$

$$m_4\ddot{\mathbf{g}}_4 \cdot \dot{\mathbf{g}}_4 = m_4(\ddot{g}_{4x}\dot{g}_{4x} + \ddot{g}_{4y}\dot{g}_{4y}) = .0065((-2483)(-57.754) + (0)(0)) = 932$$

$$I_3\ddot{\theta}_3\dot{\theta}_3 = (.0133)(303.7)(-15.79) = -63.78$$

Substituting into the power Equation 7.3.9, the input or driving torque for the mechanism is

$$T_{12} = \frac{1123 + 932 - 63.78 - 4331}{52.36} = -44.7 \text{ lb} - \text{in}$$

which is exactly the same result obtained in Example 7.10!

7.4 Gear Train Kinematics

This section will provide a brief introduction to the kinematics of spur gears and spur gear trains. A complete discussion of gear trains would include other types of gears and the stress analysis and design for strength of the gears.

The *fundamental law of gearing* states that the angular velocity ratio between the input and output gears of a gear set must remain constant throughout the mesh. The simplest model for two gears in mesh would be a pair of rolling cylinders as shown in Figure 7.25.

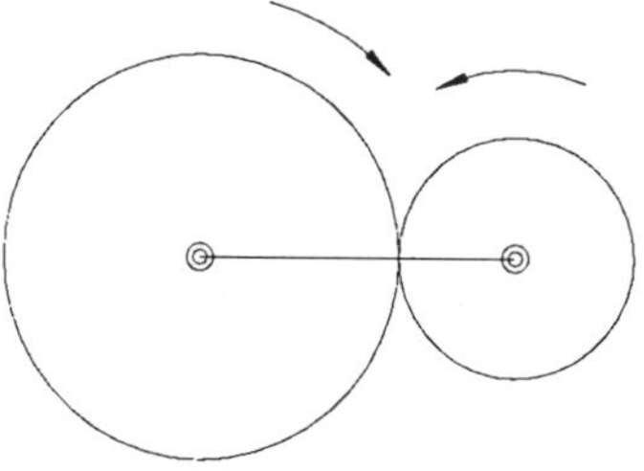

Figure 7.25 Rolling cylinders

The *angular velocity ratio* is inversely proportional to the radii of the rolling cylinders:

$$\frac{\omega_{out}}{\omega_{in}} = \mp\frac{R_{in}}{R_{out}} \qquad \textbf{(7.4.1)}$$

The *torque ratio* or *mechanical advantage* of the gearset is the reciprocal of the velocity ratio:

$$\frac{T_{out}}{T_{in}} = \mp\frac{R_{out}}{R_{in}} \qquad \textbf{(7.4.2)}$$

In Equations 7.4.1 and 7.4.2 the sign is negative for an external set and positive for an internal set of rolling cylinders.

To increase the torque capacity, meshing teeth are added to the rolling cylinders. In describing the kinematics of spur gears, the equivalent rolling cylinders become the *pitch circles* of the gears. The contact point of the rolling pitch circles is the *pitch point* of the gears.

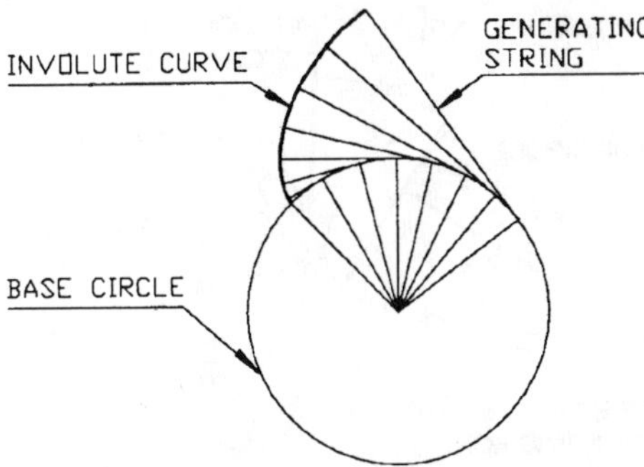

Figure 7.26 The involute curve

In order to maintain a constant velocity for mating spur gears the pitch point must remain at a constant distance from the centers of the pitch circles. This is possible if the shape of the mating teeth are *conjugates* of each other. There are several possible conjugate tooth profiles but for most gears the *involute* shape is used.

The Involute Tooth Form

An involute curve is the curve generated by unwrapping a string from a cylinder as shown in Figure 7.26. This curve has several properties which are important to the conjugate tooth action.

- The string is always tangent to the cylinder.
- The center of curvature of the involute is always at the point of tangency on the cylinder.
- The tangent to the involute is always perpendicular to the string.

The cylinders from which the string is unwrapped are called the *base circles.*

Now consider the geometry of meshing gears, as shown in Figure 7.27. For each of the pair of mating teeth, note that the profile was generated by the same tracing point on the string. A pair of conjugate profiles would be drawn, one on each lamina moving with a rolling cylinder. Note that the tooth contact point is at the pitch point in only one position. However, since the common normal at the generating point always passes through the pitch point, the angular velocity of the gearset would be constant as long as the conjugate tooth profiles are in contact.

Changing Center Distance

If the tooth form is other than involute, a change in the center distance of the pitch circles will result in a change in the velocity ratio. If the tooth profiles are involute , changing the center distance will not change the velocity ratio. In Figure 7.27 imagine the string is elongated and a change is made of the center distance. The unwrapping of the string would continue to generate conjugate involutes with their shape determined by the base circle radii. The only difference in the tooth action would be a change in the pressure angle.

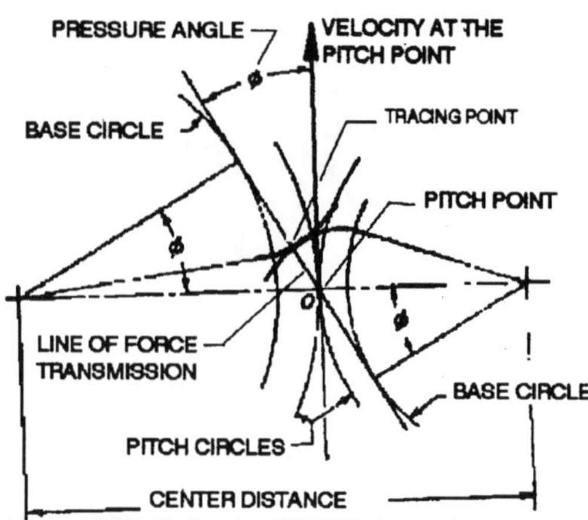

Figure 7.27 Generation of the involute gear tooth shape

Pressure Angle

The common normal at the point of contact of mating involute gears always passes through the pitch point. The contact force, neglecting friction, is also directed along the same line. The angle between the contact force and the velocity of the pitch point is called the *pressure angle*. Only the component of the contact force in the direction of the pitch point velocity contributes to the transmitted torque. Therefore, it is clear that the smaller the pressure angle the more efficient the gear set in its ability to transmit torque with minimal tooth loads.

The pressure angles for involute gear sets have been standardized at 14.5, 20, and 25 degrees, with 20 degrees the most common. Meshing gears must be cut to the same *nominal* pressure angle.

Figure 7.28 shows the geometry of a typical involute gear tooth. The length of the tooth is defined by the *addendum,* the height above the pitch circle, and the *dedendum,* the portion below the pitch circle. The amount of the dedendum is determined by the tip clearance necessary for the addendum of the mating tooth. The tooth profile is a true involute only for points outside the base circle. If the dedendum extends below the base circle the portion of the tooth below the base circle will be *undercut.* Undercutting will reduce the beam strength of the tooth and promote early tooth failure. Undercutting is caused by an attempt to use too small a number of teeth on the gear. For a 20 degree pressure angle the minimum number of teeth for no undercut is 18 teeth. The minimum with an allowable change in tooth strength is 14 teeth for a 20 degree pressure angle, 20 for a 14.5 degree pressure angle, and 10 for a 25 degree pressure angle.

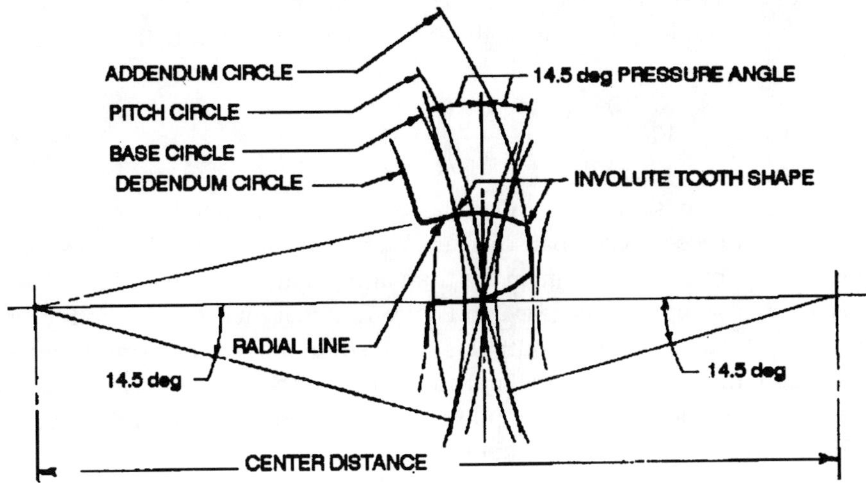

Figure 7.28 Gear tooth geometry and nomenclature

Gear Tooth Geometry and Nomenclature

The *circular pitch* p_c for a gear tooth is the arc length along the pitch circle between a point on one tooth and a similar point on the next tooth:

$$p_c = \frac{\pi d}{N} \tag{7.4.3}$$

The *diametral pitch* p_d is a more convenient description of pitch in terms of the gear diameter rather than its circumference:

$$p_d = \frac{N}{d} \tag{7.4.4}$$

The *module* m is the reciprocal of the diametral pitch:

$$m = \frac{d}{N} \tag{7.4.5}$$

Diametral pitch is used to describe US gears. In the SI or metric system the module is used.

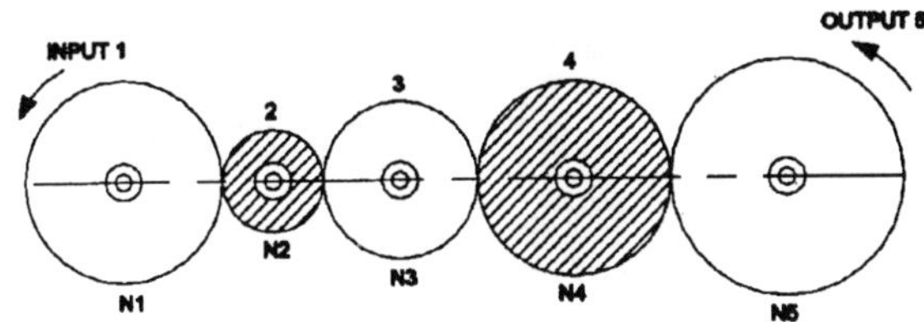

Figure 7.29 A simple gear train

Simple Gear Trains

The velocity ratio for a pair of meshing gears can also be expressed in terms of a *gear tooth ratio*. For example , Equation 7.4.1 could have been written in the form

$$VR = \frac{\omega_{out}}{\omega_{in}} = \mp \frac{N_{in}}{N_{out}} \tag{7.4.6}$$

Consider the gear train of Fig. 7.27. The velocity ratio $VR = \omega_{out} / \omega_{in}$ can be calculated by combining the gear tooth ratios for each gear pair in the train as follows:

$$VR = \left(-\frac{N_1}{N_2}\right)\left(-\frac{N_2}{N_3}\right)\left(-\frac{N_3}{N_4}\right)\left(-\frac{N_4}{N_5}\right) \tag{7.4.7}$$

The velocity ratio, sometimes called the *train ratio*, is seen to be the product of the individual ratios for each pair of gears. Note that the ratio will be positive if there is an even number of gears in the train and negative for an odd number. If there is an idler gear, such as gear 2 in Figure 7.29 that meshes with both gear 1 and gear 3 , only the sign of the output is changed. In the case of a simple gearset all of the tooth numbers except the first and last cancel and all of the intermediate gears are seen as idlers because no power is taken from them. Therefore, in Figure 7.29, the overall gear ratio would be

$$VR = -\frac{N_1}{N_5} \tag{7.4.8}$$

A simple gearset is usually limited to train ratios less than 10:1 because of the minimum number of teeth allowed to avoid undercutting of the smaller gear. Above 10:1 the gears would be large and expensive. To get a larger velocity ratio the gearset must be designed as a *compound or planetary gear train.*

Compound Gear Trains

To get velocity ratios greater than 10:1 it is necessary to compound the train. In a compound train at least one of the shafts carries more than one gear, as shown in Figure 7.30. The train ratio for this two-stage compound gearset becomes

$$VR = \left(-\frac{N_1}{N_2}\right)\left(-\frac{N_3}{N_4}\right) \tag{7.4.9}$$

This equation can be generalized in the form

$$VR = \mp\frac{\text{The product of the number of teeth on driver gears}}{\text{The product of the number of teeth on driven gears}} \tag{7.4.10}$$

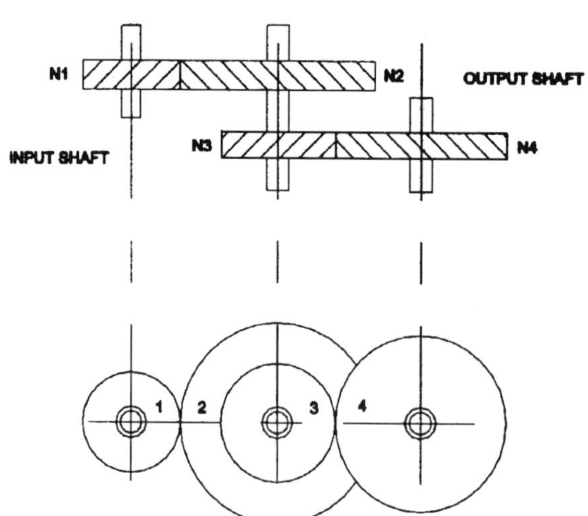

Figure 7.30 A compound gear train

Example 7.12 Kinematic design of a compound gear train

Design a compound gear train with an overall train ratio of 180:1.

Solution. First consider a two-stage set. If both stages have a tooth ratio of 10:1, the design limit, the overall ratio will be less than the design objective. Next consider a three-stage set. Let the first two stages be identical compounded sets with each set having a ratio of 6:1. This would give an overall ratio of 36:1 after the first two stages. The ratio for the third stage is then found by dividing the required overall ratio of 180 by 36 to give the required ratio for the third stage as 5:1.

Once the ratios have been determined the number of teeth on each gear can be selected using the minimum of 14 teeth on the pinion (the smaller driving gear) for each set. The results are as follows.

Gearset Number	Gearset Ratio	Driving Gear Teeth	Driven Gear Teeth
1	6:1	14	84
2	6:1	14	84
3	5:1	14	70

Overall gear train ratio = (84)(84)(70)/(14)(14)(14) = 493,920/2,744 = 180

Reverted Compound Gear Trains

In a *reverted compound gear train* the output shaft is concentric with the input shaft. This reverting of the train is particularly important in automotive transmissions. The intermediate gears are mounted on a countershaft as shown in Figure 7.31.

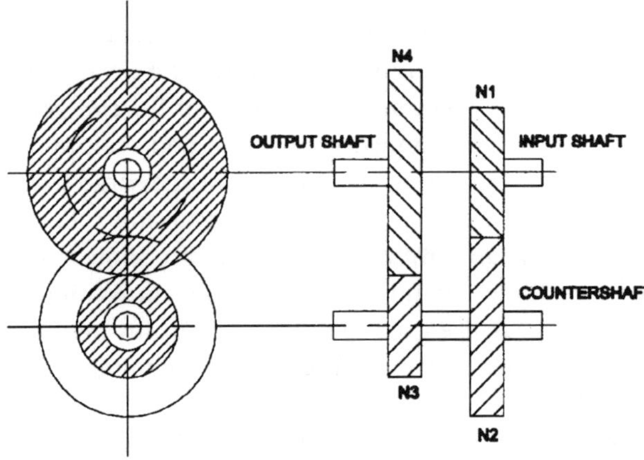

Figure 7.31 A reverted compound gear train

Example 7.13 Design of a reverted gear train

Design a reverted compound gear train to have an exact gear train ratio equal to 18. Assume that both gearset ratios and tooth numbers must be integers.

Solution. Figure 7.31 shows an arrangement for a two-stage set. The center distance for both stages must be equal. This gives the relationship

$$d_1 + d_2 = d_3 + d_4$$

Substituting Equation 7.4.4 and canceling the diametral pitch leads to

$$N_1 + N_2 = N_3 + N_4 = \mathrm{K}$$

Assume a velocity ratio of 6 for the first stage and a velocity ratio of 3 for the second stage. Substituting into Equation 7.4.10 results in

$$N_2 = 6N_1$$
$$N_4 = 3N_3$$

Substitution gives

$$N_1 + 6N_1 = K$$
$$N_3 + 3N_3 = K$$

To make equations solvable for integer values of N_1 and N_3 the value of the constant K must be set to the smallest common multiple of 7 and 4 . First, assume a value of K = 28. This gives a solution for N_1 = 4. Earlier it was shown that the minimum number of teeth to avoid undercutting is equal to 14 for a 20 degree pressure angle. This requires that the value of K be increased to 4 times 28·. This would give values of N_1 = 16 and N_2 = 96 teeth. Using a 25 degree pressure angle will reduce the number of teeth on the gears. Assume 25 degree teeth and a K value = 3 times 28. This gives

$$N_1 = \frac{84}{7} = 12 \qquad N_2 = 6N_1 = 72$$
$$N_3 = \frac{84}{4} = 21 \qquad N_4 = 3N_3 = 63$$

Epicyclic or Planetary Gear Trains

A planetary gear train is a two degree of freedom device. Two inputs are required to produce the output. Figure 7.32 shows the arrangement of a typical planetary gear system. The basic system consists of the *sun gear* and a meshing *planet gear* with an *arm* which connects the gears and maintains the center distance. Both the sun gear and arm can rotate about the same fixed center of rotation. A *ring gear* that is in mesh with the planet gear and rotates about the fixed center can also be used as an output device. There are some situations where the ring gear is fixed by adding a band brake to the system. With the ring gear fixed either the arm or sun gear are possible input or output members. The relative motions between the three possible input-output members is very difficult to predict by observation and several specialized methods for the kinematic analysis of planetary gears have been developed. The use of relative instant centers has been discussed earlier.

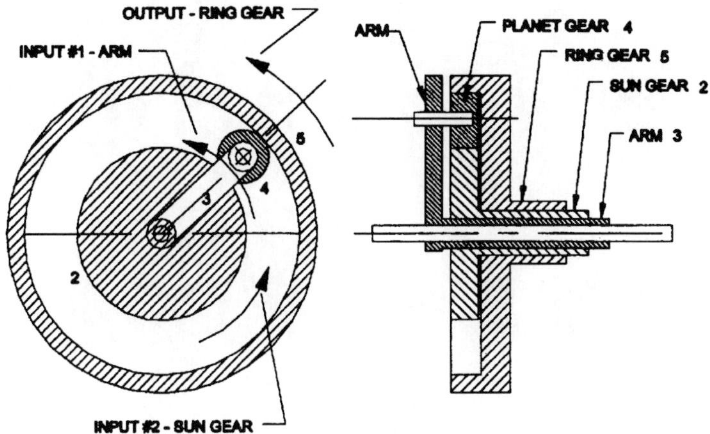

Figure 7.32 An epicyclic planetary gear train

The Formula Method

The relative motion between either of the gears connected to the arm can be expressed in the form

$$\omega_{gear} = \omega_{arm} + \omega_{gear/arm} \tag{7.4.11}$$

This method will work for input at either end of the system. Let ω_F represent the angular velocity of the first gear, ω_L the last gear, and ω_{arm} the arm. For the first gear $\omega_{F/arm} = \omega_F - \omega_{arm}$. For the last gear $\omega_{L/arm} = \omega_L - \omega_{arm}$. This leads to

$$VR = \mp \frac{\text{product of number of teeth on driver gears}}{\text{product of number of teeth on driven gears}} = \frac{\omega_L - \omega_{arm}}{\omega_F - \omega_{arm}} \tag{7.4.12}$$

Example 7.14 Analysis of a planetary gear train

The gear train of Figure 7.11 will be used again as an example. In this case the driver gear is member 2 with 40 teeth and an angular velocity ω_2 equal to 100 rad/sec. The arm, member 3, has an angular velocity equal to 200 rad/sec. The planet gear 4 has 40 teeth and the ring gear 5 has 80 teeth. Determine the output angular velocities of the driven gears 4 and 5.

Solution. Substitute into Eq.7.4.12 being careful to note the sign of the velocity ratio:

$$(-1)\left(\frac{40}{20}\right) = \frac{\omega_4 - 200}{100 - 200}$$

Solving, the output angular velocity of the planet $\omega_4 = 400$ rad/sec.

To solve for the ring gear velocity Equation 7.4.12 becomes

$$(-1)\left(\frac{40}{80}\right) = \frac{\omega_5 - 200}{100 - 200}$$

which gives $\omega_5 = 250$ rad/sec.

Stress Analysis & Structures

by Clark J. Radcliffe and Charles W. Radcliffe

8.1 Stress Analysis

Uniformly Distributed Stresses

Uniform Stresses are often assumed in design resulting in pure tension, pure compression or pure shear depending on the orientation of the load on the body under design. For a simple tension or compression load F applied to both ends of a bar, the assumption of simple (or pure) tension or compression implies that 2-3 diameters from the ends of the bar, the load is taken by a uniform stress σ across the bar cross sectional area A:

$$\sigma = \pm F/A \tag{8.1.1}$$

where F is positive for tension and negative for compression. This is also the correct expression for *average axial stress*, however, it is often an incorrect expression for the *maximum axial stress* in the cross-section. The correction to find maximum axial stress is found in the stress concentration discussion in Section 8.2.

Simple or direct shear is often assumed where nearly collinear, opposing forces are applied across a section with minimal bending of the section where the load is taken by a uniform shear stress τ across the bar cross sectional area A:

$$\tau = F/A \tag{8.1.2}$$

Examples of these two cases of uniform stresses are shown in Figure 8.1.

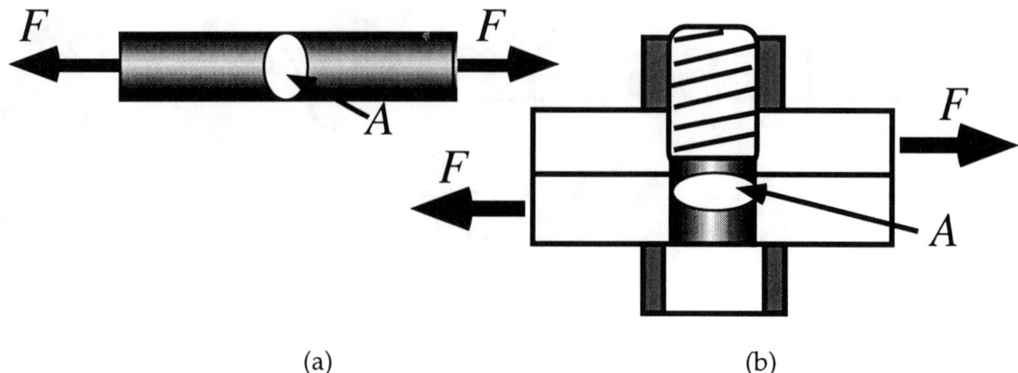

Figure 8.1. Examples Of Uniformly Distributed Stresses. (a) A Bar In Tension With Uniform Tensile Stresses Across The Bar And (b) A Bolt In Shear With Uniform Stress Across The Bolt Cross Section.

Bending

Figure 8.2 shows a bar loaded only in bending due to a pure moment. The resulting stresses are given by

$$\sigma = My/I \tag{8.1.3}$$

where M is the applied moment at the cross section, I is the moment of inertia of the cross section with respect to the neutral axis and y is the distance from the neutral axis. Bending stresses are normal stresses, the same as axial stresses. For a bending moment applied as shown, the beam "smiles" and the bending stresses are tensile below the neutral axis and compressive above.

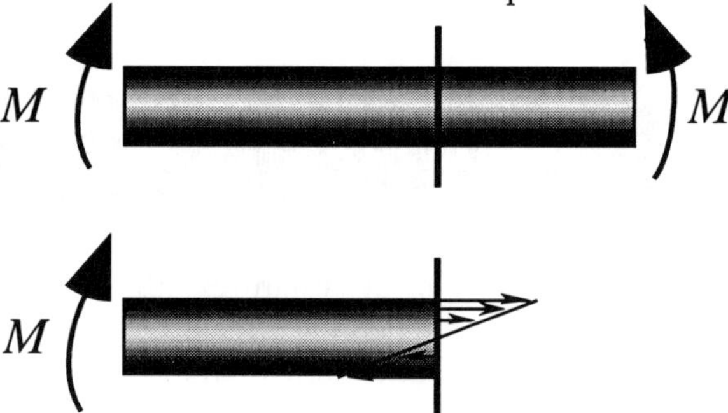

Figure 8.2 Bending Stresses Are Tensile Below The Neutral Axis And Compressive Above With An Applied Moment M.

Torsion

For a bar in torsion, the shear stresses vary linearly from zero along the central axis to a maximum at the surface. The shear stress at any radius r is

$$\tau = Tr/J \tag{8.1.4}$$

where T is the applied torsional moment, r is the radial position and J is the polar moment of inertia of the cross section. Of particular interest is the *maximum stress in round shafts* with outer diameter d and inner diameter d_i subjected to torsional moment T:

$$\tau = \frac{16dT}{\pi\left(d^4 - d_i^4\right)} \tag{8.1.5}$$

Properties of Simple Sections

Table 8.1: Common Section Properties.

Section	Properties
h, y, b	Area, $A = bh$ Centroid, $y = h/2$ Moment of Inertia, $I = \frac{bh^3}{12}$
d, d_i	Area, $A = \frac{\pi}{4}\left(d^2 - d_i^2\right)$ Centroid, $y = d/2$ Moment of Inertia, $I = \frac{\pi}{64}\left(d^4 - d_i^4\right)$ Polar Moment, $J = \frac{\pi}{32}\left(d^4 - d_i^4\right)$

Example 8.1

A bar of Length $L = 20\,\text{cm}$ and diameter $d = 2\,\text{cm}$ is loaded by a bending force $F_B = 100\,\text{N}$, an axial tension force $F_A = 50\,\text{N}$ and a torsion $T = 10\,\text{N}\cdot\text{m}$ as shown. What is the stress in the bar at the wall?

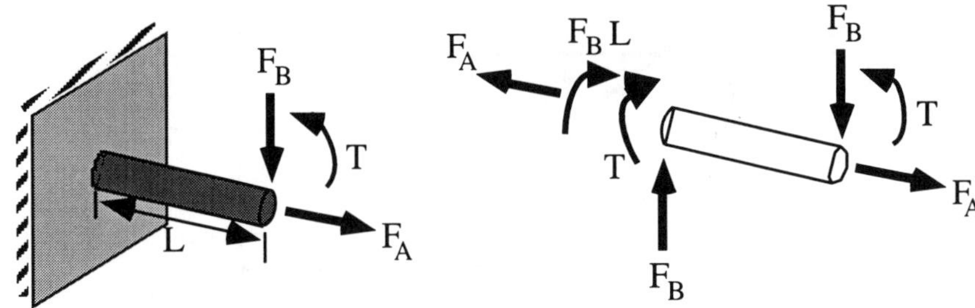

Solution. The free-body diagram indicates the reaction moments and forces at the bar attachment point. The reactions are a vertical force $F_B = 100\,\text{N}$, a horizontal force $F_A = 50\,\text{N}$, a torsional moment $T = 10\,\text{N}\cdot\text{m}$ and bending moment $M_B = F_B L = (100\,\text{N})(0.2\,\text{m}) = 20\,\text{N}\cdot\text{m}$.

Stress On Top Of Bar At Wall: Arises from tension due to bending force F_B combined with axial force F_A:

$$\text{Normal Stress: } \sigma = \frac{My}{I} + \frac{F}{A} = \frac{(F_B L)(d/2)}{\left(\pi d^4/64\right)} + \frac{F_A}{\left(\pi d^2/4\right)} = \frac{32 F_B L}{\pi d^3} + \frac{4 F_A}{\pi d^2}$$

$$= \frac{32(20)}{\pi(0.02)^3} + \frac{4(50)}{\pi(0.02)^2} = 25.46 \times 10^6 + 0.159 \times 10^6\ \text{N/m}^2 = 25.62\,\text{MPa}$$

The first component arises from the bending moment at the wall while the second arises from the axial force on the bar. Both normal stresses are tensile at the top of the bar and sum. At the bottom of the bar, the bending stress is compressive and the axial stress is tensile and the net stress would be their difference,

$$\sigma = -25.46 + 0.159 = -25.3\,\text{MPa}$$

Shear stress: $\tau = \frac{Tr}{J} = \frac{T(d/2)}{32\pi d^4} = \frac{16T}{\pi d^3} = \frac{16(10\ )}{\pi(0.02\ )^3}$

$$= 6.37 \times 10^6\ \text{N}/\text{m}^2 = 6.37\ \text{MPa}$$

8.2 Stress Concentration

Stress concentrations K_t give the local maximum stress using the average stress computed above. The stress concentration factors in Figures 8.3 to 8.11 below [1] multiply the average stress ($\sigma_{max} = K_t\sigma$ and $\tau_{max} = K_t\tau$) to obtain the actual local stress at the position of the stress concentration under the specific loading condition indicated in each figure.

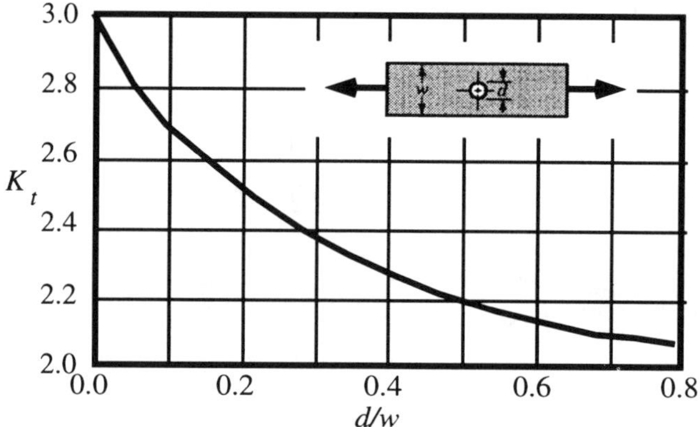

Figure 8.3 Bar In Tension Or Simple Compression With A Transverse Hole, $\sigma_o = F/A$ where $A = (w - d)t$ and t is the Thickness

[1] R.E. Peterson,"Design Factors for Stress Concentration", *Machine Design*, vol.23.

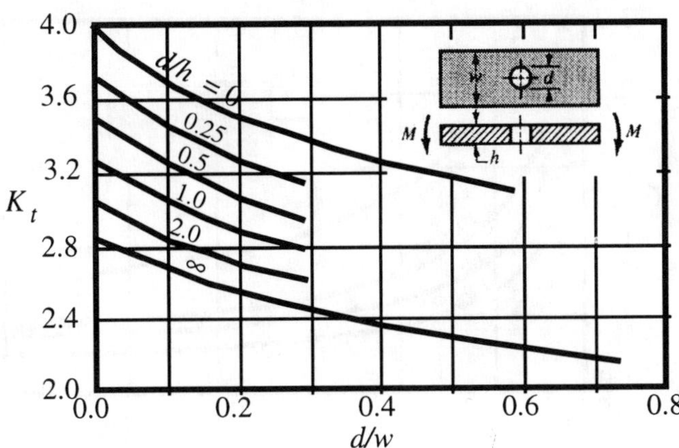

Figure 8.4 **Rectangular Bar With A Transverse Hole In Bending,** $\sigma_o = Mc/I$ **where** $I = (w-d)h^3/12$

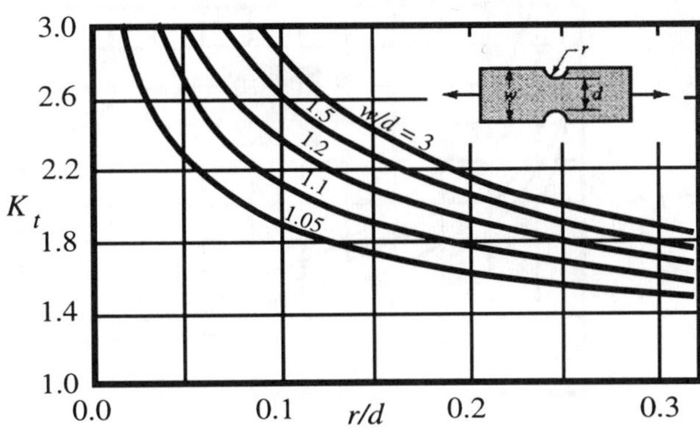

Figure 8.5 **Notched Rectangular Bar in Tension or Simple Compression,** $\sigma_o = F/A$ **where** $A = td$ **and** t **is the Thickness**

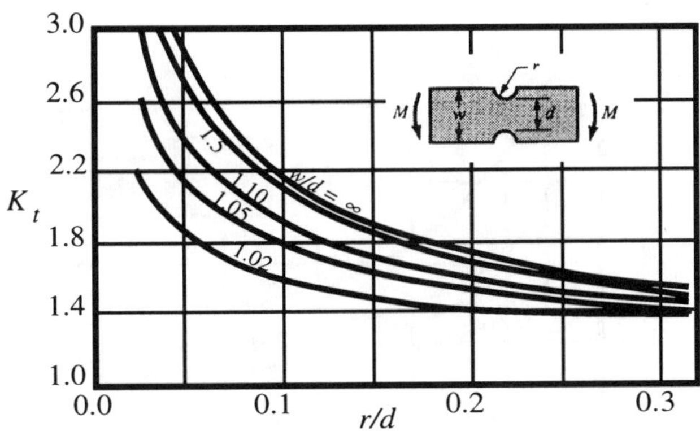

Figure 8.6 **Notched Rectangular Bar in Bending,** $\sigma_o = Mc/I$ **where** $c = d/2$ **,** $I = td^3/12$ **and** t **is the Thickness**

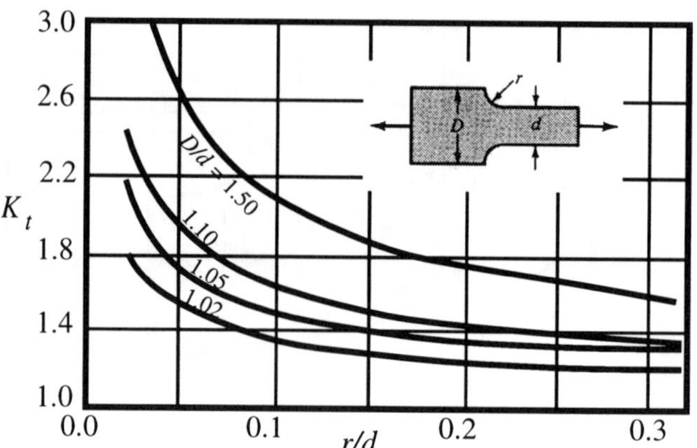

Figure 8.7 Rectangular Filleted Bar in Tension or Simple Compression, $\sigma_o = F/A$ **where** $A = td$ **and** t **is the Thickness**

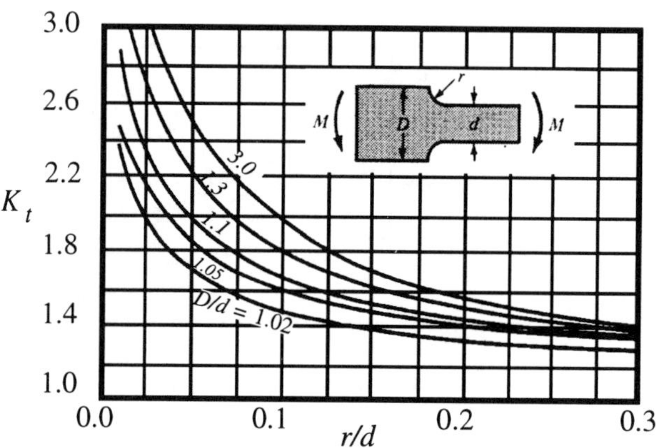

Figure 8.8 Rectangular Filleted in Bending, $\sigma_o = Mc/I$ **where** $c = d/2$ **,** $I = td^3/12$ **and** t **is the Thickness**

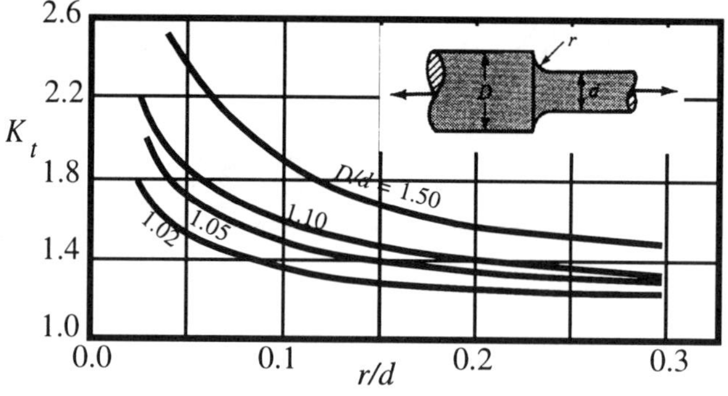

Figure 8.9 Round Shaft Bar with Shoulder in Tension, $\sigma_o = F/A$ **where** $A = \pi d^2/4$

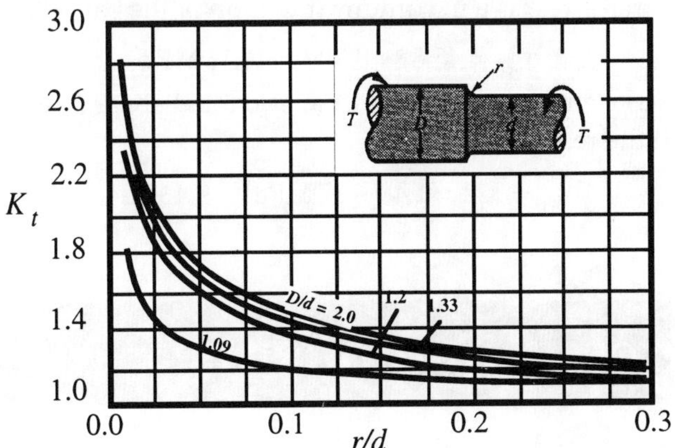

Figure 8.10 Round Shaft Bar with Shoulder in Torsion, $\tau_o = Tc/J$ where $c = d/2$, $J = \pi d^4/32$

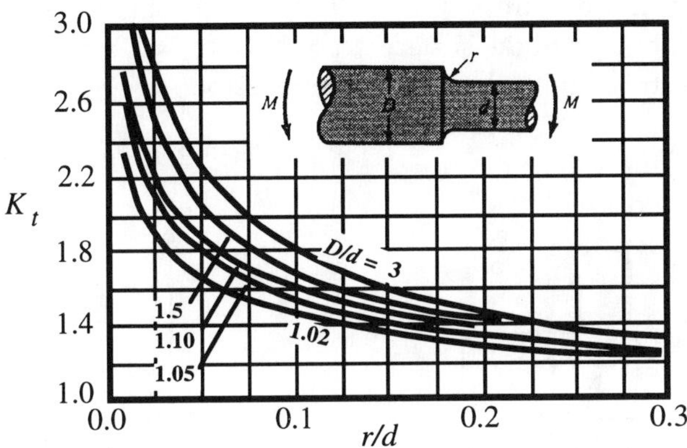

Figure 8.11 Round Shaft Bar with Shoulder in Bending, $\sigma_o = Mc/I$ where $c = d/2$, $I = \pi d^4/64$

Example 8.2

In the bar of Example 8.1, find the stress at the wall due to the stress concentration arising from bar attachment to the wall with a fillet radius $r = 2\,\text{mm}$.

Solution. For this fillet radius, the fillet radius to bar diameter ratio $r/d = 0.002/0.02 = 0.1$.

In the bar stress Figures 8.9 - 8.11, the bar diameter ratio $D/d = \infty/0.02 = \infty$ so use the largest D/d available in each figure.

The stress concentration factor for bar bending stress $K_t = 1.8$ from Fig. 8.11 and

$$\sigma_{B_{act}} = K_t \sigma_B = 1.80(25.46\,\text{MPa}) = 45.83\,\text{MPa}$$

The stress concentration factor for bar bending stress $K_t = 1.88$ from Fig. 8.9 so

$$\sigma_{A_{act}} = K_t \sigma_A = 1.88(0.159\,\text{MPa}) = 0.299\,\text{MPa}$$

The net normal stress is a maximum at the top of the bar where

$$\sigma_{act} = 45.83 + 0.299 = 46.13\,\text{MPa}$$

The stress concentration factor for bar torsional shear stress $K_t = 1.48$ from Figure 8.10 and the resulting shear

$$\tau_{Act} = K_t \tau = 1.48(6.37\,\text{MPa}) = 9.43\,\text{MPa}$$

8.3 Stress Components & Mohr's Circle

The three normal stresses and six shear stresses applied on a three-dimensional stress element are shown in Figure 8.12(a). The most common stress analysis for engineering design is the analysis of stresses at the surface of an object where the external surface normal stress $\sigma_z = 0$ along with all shear stresses related to the shear stresses applied on the surface $\tau_{zx} = \tau_{xz} = \tau_{zy} = \tau_{yz} = 0$. This second condition is plane stress and is shown in Figure 8.12(b).

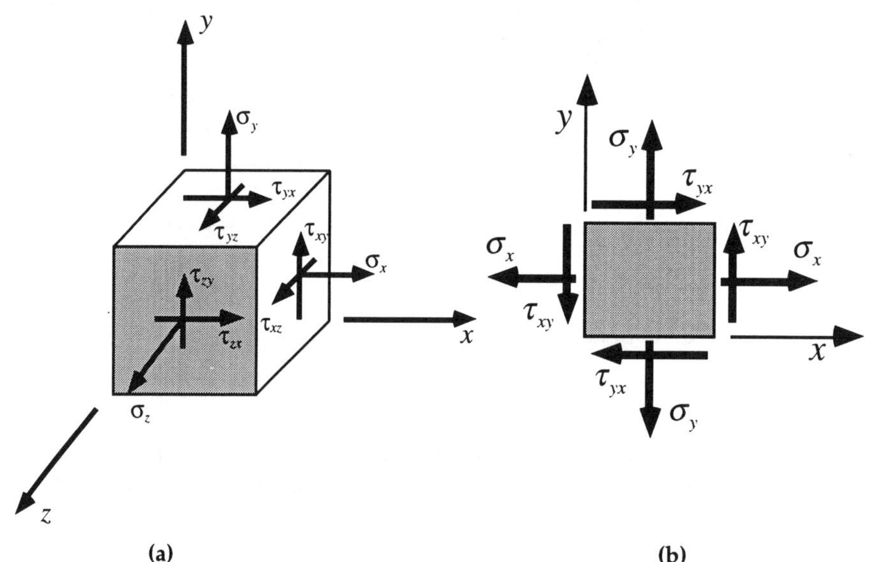

Figure 8.12 (a) Three Dimensional Stresses and (b) Plane Stresses

The *normal and shear stresses at an angle* to the *x-y* reference frame resulting from a state of plane stress is indicated in Figure 8.13. Static equilibrium requires $\tau_{xy} = \tau_{yx}$. The normal stress at angle ϕ is

$$\sigma = \frac{\sigma_x + \sigma_y}{2} + \frac{\sigma_x - \sigma_y}{2}\cos(2\phi) + \tau_{xy}\sin(2\phi) \tag{8.3.1}$$

while the shear stress at angle ϕ is

$$\tau = \frac{\sigma_x - \sigma_y}{2}\sin(2\phi) + \tau_{xy}\cos(2\phi) \tag{8.3.2}$$

At two perpendicular values of angle ϕ, the normal stresses take on maximum σ_1 and minimum σ_2 values. These maximum and minimum stresses are the *principal stresses.*

The two principal directions of these principal stresses is given by the two solutions to

$$\tan(2\phi) = \frac{2\tau_{xy}}{\sigma_x - \sigma_y} \tag{8.3.3}$$

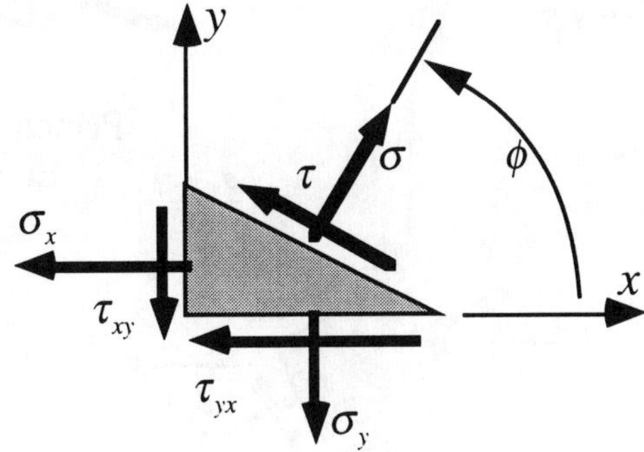

Figure 8.13 Plane Stress State at Angle ϕ from body x-y axes

The graphical *Mohr's circle* is an effective means of solving the above equations to find principal stresses. In Figure 8.14, a coordinate system is created. Normal stresses plotted along the x-axis with tensile stresses are plotted as positive. Shear stresses plotted along the y-axis with clockwise shear plotted as positive. The three important plane states of stress for different stress element rotations are given.

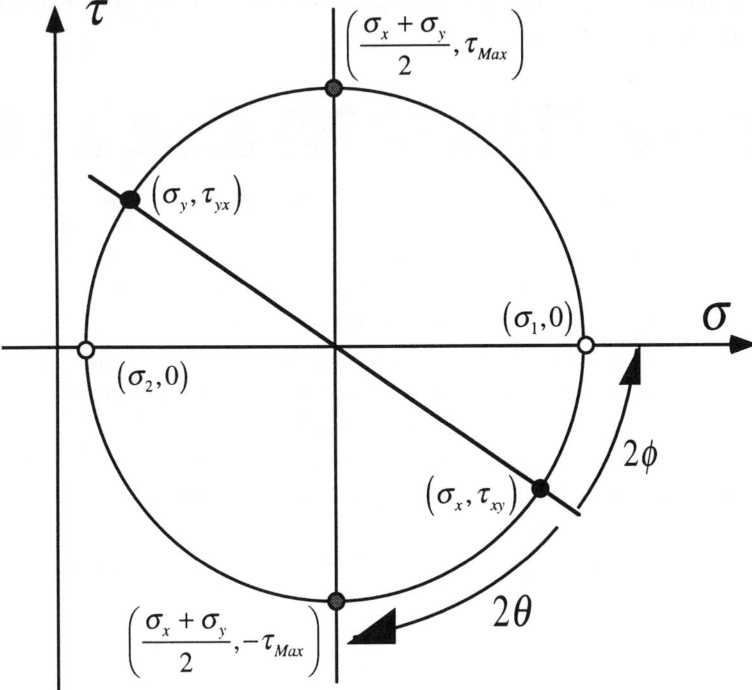

Figure 8.14 Mohr's Circle for Plane Stress With Three Element Rotations in Fig.8.15 ● (Normal, shear) stress in original x-y coordinate element, ○ Principal stresses on element rotated ϕ counterclockwise, and ● maximum shear stresses on element rotated θ clockwise

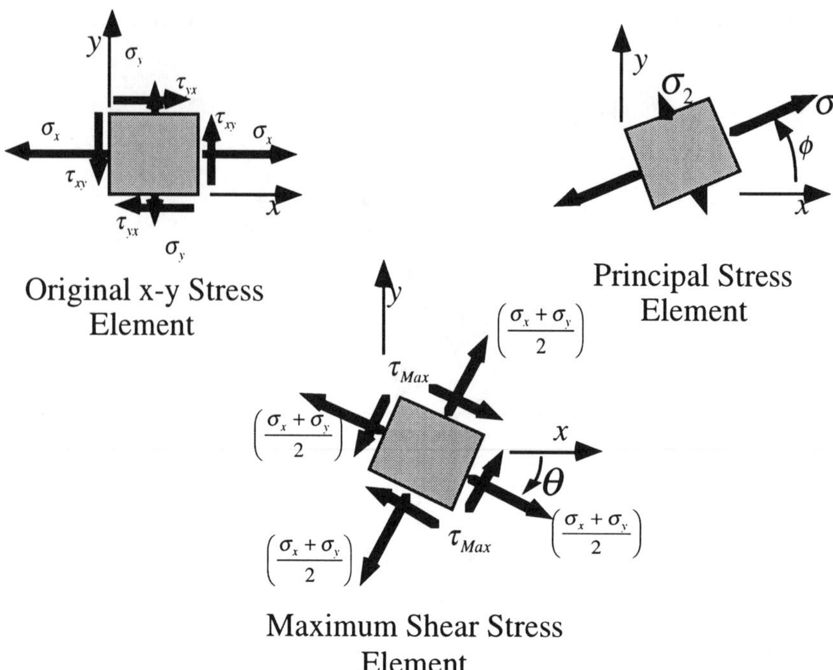

Figure 8.15 Plane Stress Elements For The Three Element Rotations in Fig.8.14. ● (Normal, shear) stress in original x-y coordinate element, ○ Principal stresses on element rotated ϕ counterclockwise, and ● Maximum Shear stresses on element rotated θ clockwise

Example 8.3

For the bar of Example 8.1 and 8.2, what is the maximum shear stress at the wall?

Solution: In the bar of Example 8.1 and 8.2, the stress at the wall on the top of the bar due to the loads and stress concentrations was tensile, $\sigma_x = \sigma_{act} = 45.83 + 0.299 = 46.13\,\text{MPa}$ combined with a shear stress $\tau_{xy} = \tau_{act} = K_t\tau = 1.48(6.37\,\text{MPa}) = 9.43\,\text{MPa}$ clockwise looking down on the beam. The normal stress in the y-(circumferential) direction is zero. The state of stress is shown in the stress element and Mohr's circle is sketched below.

The x-face stresses are plotted at coordinate $(\sigma,\tau) = (46.13,-9.43)$ while the y-face stresses are plotted at $(\sigma,\tau) = (0, 9.43)$ and the circle diameter sketched between them.

Examining the resulting Mohr's circle, the center is located at the average normal stress

$$\left(\sigma_x + \sigma_y\right)/2 = (46.13 + 0.00)/2 = 23.07 \text{ MPa}$$

From simple geometry and further inspection of the circle, the radius of the circle is

$$r = \sqrt{23.07^2 + 9.43^2} = 24.92 \text{ MPa}$$

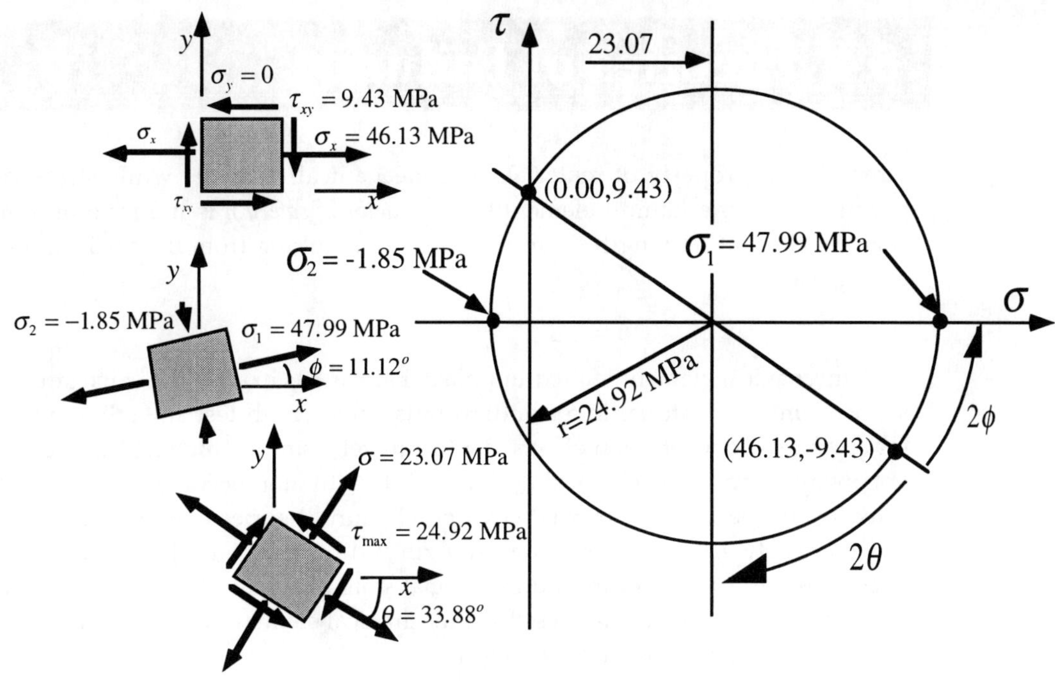

The *principle stresses* are

$$\sigma_1 = 23.07 + 24.92 = 47.99 \text{ MPa}, \quad \sigma_2 = 23.07 - 24.92 = -1.85 \text{ MPa}$$

at the element double rotation

$$2\phi = \tan^{-1}\left(\frac{9.43}{23.07}\right) = 22.23^o \text{ ccw} \Rightarrow \phi = 11.12^o \text{ ccw}$$

The *maximum shear stress* $\tau_{max} = -24.92$ MPa on the original x-face at an element double rotation

$$2\theta = \tan^{-1}(23.07/9.43) = 67.76^o \text{ cw} \Rightarrow \theta = 33.88^o \text{ cw}$$

Note that the geometry of Mohr's circle was used in this example ***not*** the fundamental equations. In general, the solution of the fundamental equations requires significantly greater effort and is more error prone.

8.4 Theories of Failure

Strength is a property of materials and mechanical elements while stress is a product of the mechanical element's use. *Factor of safety n* is the ratio of a mechanical element's strength S to the stress σ resulting from the loads applied during use,

$$n = S/\sigma \tag{8.4.1}$$

A mechanical element loaded in uniaxial stress has exactly the same strength as the element's material in a standard uniaxial strength test suitably modified for any stress concentration effects. Unfortunately, simple uniaxial loading does not always occur in mechanical design and the situation becomes more complicated for biaxial and/or triaxial stresses. Failure theories compensate for the difference between "as tested" axial strength and "as used" mechanical element's stress concentrations and multidimensional state of stress. These theories compute an equivalent uniaxial stress for a mechanical element to be compared with the element material's uniaxial strength.

Static Failure

Ductile materials (steel, aluminum) fail after significant elongation (> 3%) while brittle materials (cast iron, ceramics) fail without significant elongation (< 3%). The appropriate failure theory to apply for static failure is dependent on the material's failure type: brittle or ductile. In all cases below, the predicted material strength S is a function of the biaxial principal stresses σ_1 and σ_2 at the surface of a loaded mechanical element.

Static Failure of Brittle Materials

Brittle materials often have different ultimate strengths in compression S_{uc} and tension S_{ut}. The predicted biaxial states of stress at failure for a recommended failure criteria for a combination of Modified-Mohr and Maximum-Normal Theories is shown in Fig.8.16 and quantified in Table 8.1. In both cases, the failure stress prediction is a function of the relative signs of the principal stresses σ_1 and σ_2.

Table 8.1 Brittle Failure Stress Prediction (Modified-Mohr)

Quadrant	(σ_1, σ_2) Condition	Failure Stress Prediction
I	$\sigma_1 \geq 0, \sigma_2 \geq 0$	$\max(\sigma_1, \sigma_2) = S_{ut}$
II	$\sigma_1 < 0, \sigma_2 > 0, \lVert\sigma_2\rVert \geq \lVert\sigma_1\rVert$	$\sigma_1 = -S_{ut}$
IIa	$\sigma_1 < 0, \sigma_2 > 0, \lVert\sigma_2\rVert < \lVert\sigma_1\rVert$	$\sigma_1 - \frac{S_{ut}\sigma_2}{S_{uc} - S_{ut}} = \frac{S_{uc}S_{ut}}{S_{uc} - S_{ut}}$
III	$\sigma_1 \leq 0, \sigma_2 \leq 0$	$\min(\sigma_1, \sigma_2) = S_{uc}$
IV	$\sigma_1 > 0, \sigma_2 < 0, \lVert\sigma_2\rVert < \lVert\sigma_1\rVert$	$\max(\lVert\sigma_1\rVert, \lVert\sigma_2\rVert) = S_{ut}$
IVa	$\sigma_1 > 0, \sigma_2 < 0, \lVert\sigma_2\rVert > \lVert\sigma_1\rVert$	$\sigma_2 - \frac{S_{ut}\sigma_1}{S_{uc} - S_{ut}} = \frac{S_{uc}S_{ut}}{S_{uc} - S_{ut}}$

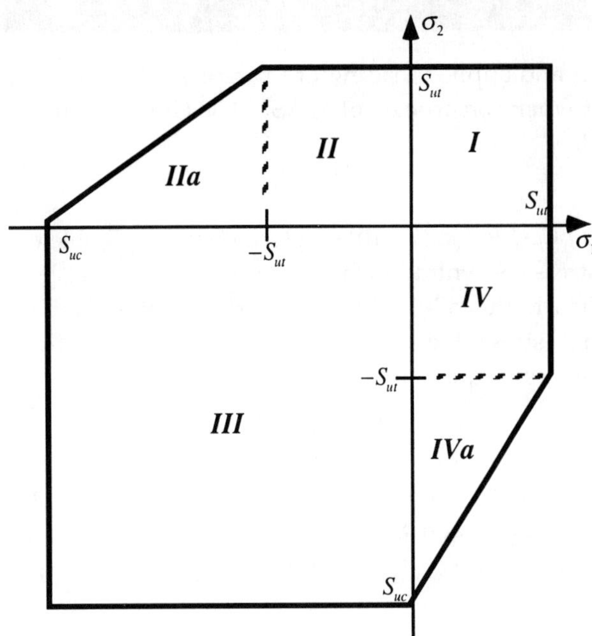

Figure 8.16 Polygon Indicates the Modified-Mohr Predicted Brittle Static Failure Strength S with Changing Biaxial (Plane) Principal Stresses σ_1 and σ_2

Static Failure of Ductile Materials

Ductile materials have the same ultimate strength in compression and tension S_u. The recommended theory for predicted biaxial states of stress at failure is the *Distortion Energy Theory*. The Maximum Shear Stress Theory is also a reasonable predictor of failure but will not be covered here. The Distortion Energy Theory states that failure occurs whenever the material ultimate strength S_u equals the *Von Mises Stress* σ':

$$\sigma' = \sqrt{\left[\frac{(\sigma_1-\sigma_2)^2+(\sigma_2-\sigma_3)^2+(\sigma_3-\sigma_1)^2}{2}\right]} \qquad \textbf{(8.4.2)}$$

Ductile materials have large elongation, distortion and reduction of area at failure so *stress concentration factors are not used to compute applied stress.*

Table 8.2 Mechanical Properties of Example Problem Metals

Metal	Tensile Strength S_{ut} MPa (kpsi)	Compressive Strength S_{uc} MPa (kpsi)	Yield Strength S_y MPa (kpsi)	Elongation (%)	Shear Strength S_{ut} MPa (kpsi)
ASTM 20 Gray Cast Iron	150 (22)	575 (83)	N/A	negligible	180 (26)
AISI 1035 Hot Rolled	500 (72)	500 (72)	270 (39.5)	18	N/A
AISI 1035 Cold Drawn	550 (80)	550 (80)	460 (67)	12	N/A

Example 8.4

For the shaft and applied loading of Examples 8.1 - 8.3, what is the factor of safety of the bar when constructed of a) ASTM 30 Gray Cast Iron and b)AISI 1035 Hot Rolled steel.

Solution:

a) *ASTM 30 Gray Cast Iron* is a brittle material with negligible elongation at rupture therefore stress concentration factors apply and failure should be evaluated with the Modified-Mohr failure theory for brittle materials. From Example 8.3, the plane (biaxial) stress due to the applied loads and the stress concentrations gives principal stresses equal to

$$\sigma_1 = 47.99 \text{ MPa and } \sigma_2 = -1.85 \text{ MPa.}$$

This loading places the stresses in the fourth quadrant (Table 8.1) where $\sigma_1 > 0$, $\sigma_2 < 0$, $\|\sigma_2\| < \|\sigma_1\|$ and the failure theory states failure occurs whenever $\max(\|\sigma_1\|, \|\sigma_2\|) = S_{ut} =$ 150 MPa (Table 8.4-2). For *ASTM 30 Gray Cast Iron* , the factor of safety

$$n = \frac{\text{Strength}}{\text{Load}} = \frac{150 \text{ MPa}}{47.99 \text{ MPa}} = 3.1$$

b) *AISI 1035 Hot Rolled steel* is a ductile material because it has more than 3% elongation at failure (Table 8.2). The material has the same applied loading as a), however, for static failure of ductile materials, *stress concentrations do not apply* and the results of Example 8.3 cannot be used directly. Using the results of Example 8.1 without stress concentrations gives the state of stress on the top of the bar as

$$\text{tensile stress, } \sigma = 25.62 \text{ MPa and shear stress, } \tau = 6.37 \text{ MPa}$$

Sketching Mohr's circle (see Section 8.3 for method) gives

$$\sigma_1 = \left(\frac{25.62}{2}\right) + \sqrt{\left(\frac{25.62}{2}\right)^2 + 6.37^2} = 27.12 \text{ MPa and}$$

$$\sigma_2 = \left(\frac{25.62}{2}\right) - \sqrt{\left(\frac{25.62}{2}\right)^2 + 6.37^2} = 1.50 \text{ MPa.}$$

Using Eq. 8.4.2 to evaluate Von Mises Stress

$$\sigma' = \sqrt{\left[\frac{(\sigma_1 - \sigma_2)^2 + (\sigma_2 - \sigma_3)^2 + (\sigma_3 - \sigma_1)^2}{2}\right]} = 27.9 \text{ MPa}$$

Using the material strength to compute the AISI 1035 hot rolled steel's factor of safety

$$n = \frac{\text{Strength}}{\text{Load}} = \frac{500 \text{ MPa}}{27.9 \text{ MPa}} = 17.9$$

Material Fatigue Strength

Fatigue Failure

Fatigue of mechanical elements can be predicted using laboratory tests of standard material specimens with the R.R. Moore rotating beam machine which applies completely reversed axial stresses to the specimen surface. All other forms of cyclic loading are then modeled by appropriate weighting applied to the Moore test results. Typical Moore test results (Shigley and Mischke) yield the Fatigue Strength S_f versus Stress Cycles N results shown in Figure 8.17. This figure shows the recommended approximation for material fatigue strength S_f The graph shows that beyond 106 cycles, the material strength is constant referred to as the *material endurance strength*:

$$S_e' = \begin{cases} 0.504S_{ut} & S_{ut} \le 200 \text{ kpsi (1400 MPa)} \\ 100 \text{ kpsi} & S_{ut} > 200 \text{ kpsi} \\ 700 \text{ MPa} & S_{ut} > 1400 \text{ MPa} \end{cases} \tag{8.4.3}$$

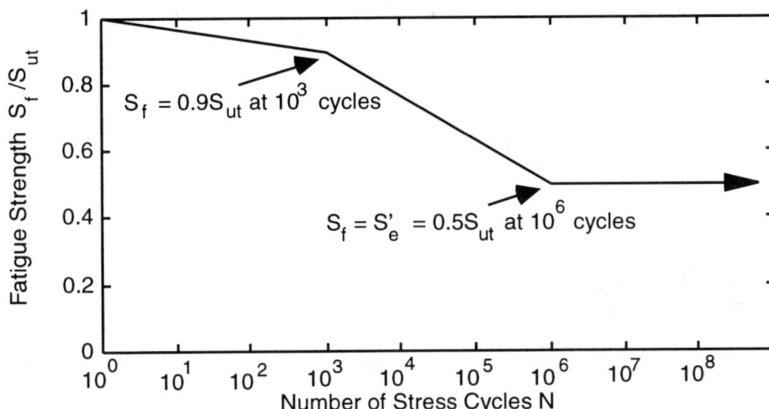

Figure 8.17 Material Fatigue Strength S_f versus Number of Stress Cycles N for Completely Reversed Cyclic Loads.

The analytical representation of this graph includes a recommendation for material fatigue strength

$$S_f = aN^b \tag{8.4.4a}$$

where N is between 10^3 and 10^6 cycles and

$$a = \frac{(0.9S_{ut})^2}{S_e} \qquad b = -\frac{1}{3}\log\frac{0.9S_{ut}}{S_e} \tag{8.4.4b}$$

The relationship between material fatigue strength S_f and Stress Cycles N can be inverted at a particular value of the fatigue strength $S_f = \sigma_a$ to find the corresponding number of stress cycles:

$$N = \left(\frac{\sigma_a}{a}\right)^{1/b} \tag{8.4.5}$$

Example 8.5

An AISI 1035 HR Steel has a minimum tensile strength of 500 MPa (Table 8.2) and a minimum yield strength of 270 MPa.

a) Estimate the rotating-beam endurance limit.

b) Estimate the fatigue strength corresponding to 10^4 cycles of life.

c) Estimate the expected life corresponding to a completely reversed stress of 200 MPa.

Solution:

a) From Eq.8.4.3, $S'_e = 0.504 S_{ut} = 0.504(500\text{ MPa}) = 252\text{ MPa}$

b) Using equation pair (8.4.4),

$$a = \frac{(0.9 S_{ut})^2}{S_e} = \frac{(0.9[500\text{ MPa}])^2}{252\text{ MPa}} = 804\text{ MPa}$$

$$b = -\frac{1}{3}\log\frac{0.9 S_{ut}}{S_e} = -\frac{1}{3}\log\frac{0.9[500\text{ MPa}]}{252\text{ MPa}} = -0.0839$$

Then, from Eq.8.4.4,

$$S_f = aN^b = (804\text{ MPa})(10^4)^{-0.0839} = 371\text{ MPa}$$

c) Using equation Eq.8.4.5, for $\sigma_a = 200$ MPa, gives

$$N = \left(\frac{\sigma_a}{a}\right)^{1/b} = \left(\frac{200}{804}\right)^{1/-0.0839} = 1.58\times 10^7 \text{ cycles}$$

Endurance Strength Modifying Factors

To account for the effects which modify the material fatigue strength when that material is used in a mechanical element, modifying factors on the material endurance limit (Shigley and Mischke) are used to define the *mechanical element endurance limit*

$$S_e = k_a k_b k_c k_d k_e S'_e \qquad \textbf{(8.4.6)}$$

where k_a = surface factor

k_b = size factor

k_c = load factor

k_d = temperature factor

k_e = miscellaneous effects factor

Surface Factor, k_a

The surface of the rotating beam material specimen is polished smooth and the material endurance strength is modified to account for the as-used surface finish. Appropriate modifying factors k_a are shown in Table 8.3.

Table 8.3 Surface Finish Factor $k_a = aS_{ut}^b$ Coefficients

Surface Finish	Factor *a*		Exponent
	kpsi	MPa	b
Ground	1.34	1.58	-0.085
Machined or Cold-Drawn	2.70	4.51	-0.265
Hot-Rolled	14.4	57.7	-0.718
As Forged	39.9	272.	-0.995

Size Factor, k_b

The size factor modifies the endurance strength for the effects of size and is dependent on load application. For *axial loading*, there is no size effect and

$$k_b = 1 \tag{8.4.7}$$

For *elements in rotating bending and torsion* (note difference for dimension in inches and millimeters):

$$k_b = \begin{cases} \left(\dfrac{d}{0.3}\right)^{-0.1133} & 0.11 \le d \le 2 \text{ in} \\ \left(\dfrac{d}{7.62}\right)^{-0.1133} & 2.79 \le d \le 51 \text{ mm} \end{cases} \tag{8.4.8a}$$

For non-rotating round sections, the effective diameter is

$$d_e = 0.370d \tag{8.4.8b}$$

For *rectangular* $(h \times b)$ *sections*, the effective diameter is

$$d_e = 0.808\sqrt{hb} \tag{8.4.8c}$$

Load Factor, k_c

$$k_c = \begin{cases} 0.923 & \text{axial loading} \quad S_{ut} \le 220 \text{ kpsi (1520 MPa)} \\ 1.0 & \text{axial loading} \quad S_{ut} \le 220 \text{ kpsi (1520 MPa)} \\ 1.0 & \text{bending} \\ 0.577 & \text{torsion and shear} \end{cases} \tag{8.4.9}$$

Temperature Factor, k_d

Results are available for carbon steel. The recommendation is to use them for all metals. Using the ratio of steel tensile strength at operating temperature to the tensile strength at room temperature S_T/S_{RT} shown in Table 8.4, the temperature factor is

$$k_d = S_T/S_{RT} \tag{8.4.10}$$

Table 8.4 Effect of Temperature on the Tensile Strength of Steel

Temperature °C	S_T/S_{RT}	Temperature °F	S_T/S_{RT}
20	1.000	70	1.000
50	1.010	100	1.008
100	1.020	200	1.020
150	1.025	300	1.024
200	1.020	400	1.018
250	1.000	500	0.995
300	0.975	600	0.963
350	0.943	700	0.927
400	0.900	800	0.872
450	0.840	900	0.797
500	0.766	1000	0.698
550	0.670	1100	0.567
600	0.546		

Miscellaneous Effects Factor, k_e

There are a number of environmental and use effects which modify endurance strengths. Most of these effects are not easily quantifiable, however, an important, quantifiable, effect is that of stress concentration K_f. The most widely accepted approach for ductile materials - *leave the low cycle fatigue strength* $0.9S_{ut}$ *unmodified and use stress concentration* K_f *to lower the endurance strength*

$$k_e = 1/K_f \tag{8.4.11}$$

Example 8.6

An AISI 1035 hot rolled steel bar has a diameter of 3 cm, a minimum tensile strength of 500 MPa (Table 8.2) and a minimum yield strength of 270 MPa.

a) Estimate the endurance limit.

b) Estimate the endurance limit for reversed bending without rotation.

c) Estimate the fatigue strength at $N = 70(103)$ cycles at an operating temperature of 300 °C.

Solution:

a) From Eq.8.4.3, $S'_e = 0.504S_{ut} = 0.504(500\text{ MPa}) = 252\text{ MPa}$. Use Table 8.3, $a = 57.7$ and $b = -0.718$ and compute the surface factor $k_a = aS_{ut}^b = 1.09$.

From Eq.8.4.8a, the size factor $k_b = \left(\frac{d}{7.62}\right)^{-0.1133} = \left(\frac{30}{7.62}\right)^{-0.1133} = 0.856$,

Also, $k_c = k_d = k_e = 1.0$, therefore from Eq.8.4.5

$$S_e = k_a k_b k_c k_d k_e S'_e = (1.09)(0.856)(1.0)(1.0)(1.0)252 = 235\text{ MPa}$$

b) From Eq.8.4.8b, find $d_e = 0.370d = (0.370)(30) = 11.1\text{ mm}$ then Eq.8.4.7a gives

$$k_b = \left(\frac{d}{7.62}\right)^{-0.1133} = \left(\frac{11.1}{7.62}\right)^{-0.1133} = 0.958$$

so the endurance strength for non-rotating bending is

$$S_e = k_a k_b k_c k_d k_e S'_e = (1.09)(0.958)(1.0)(1.0)(1.0)252 = 263\text{ MPa}$$

Note that non-rotating, reversed bending means the bending force changes direction (+ or -). It is not the same as one-way or push-push bending.

c) Assume a rotating specimen. From Table 8.4, find $S_T/S_{RT} = 1.024$ at 300 °C. Then $k_d = S_T/S_{RT} = 1.024$ and

$$S_e = k_a k_b k_c k_d k_e S'_e = (1.09)(0.856)(1.0)(1.024)(1.0)252 = 241\text{ MPa}$$

Applying the temperature effect, tensile strength $S_{ut} = 1.024(500) = 512\text{ MPa}$. Using Eq.8.4.3, the fatigue strength

$$S_f = aN^b = 881\left[70(10)^3\right]^{-0.0938} = 309\text{ MPa}$$

where

$$a = \frac{(0.9S_{ut})^2}{S_e} = \frac{(0.9[512])^2}{241} = 881 \text{ and}$$

$$b = -\frac{1}{3}\log\left[\frac{0.9S_{ut}}{S_e}\right] = -\frac{1}{3}\log\left[\frac{0.9(512)}{241}\right] = -0.0938$$

This method handles the temperature effect at both ends of the S-N diagram (Figure 8.17).

Fluctuating Stresses

Frequently, the strengths of machine elements are subjected to fluctuating stresses which are not completely reversed. A sinusoidally applied stress with maximum value σ_{max} and minimum value σ_{min} can be visualized as a mean stress σ_m summed with a completely reversed alternating stress σ_a:

$$\sigma_m = \frac{\sigma_{max} + \sigma_{min}}{2} \tag{8.4.12}$$

$$\sigma_a = \frac{\sigma_{max} - \sigma_{min}}{2} \tag{8.4.13}$$

In these cases, the material failure is commonly predicted by the Goodman Diagram (Figure 8.18). The fluctuating stress material strength is given by the *Goodman Line* (point A),

$$\frac{S_a}{S_e} + \frac{S_m}{S_{ut}} = 1 \tag{8.4.14}$$

while the ductile, static, yield strength failure line is

$$S_a + S_m = S_{yt} \tag{8.4.15}$$

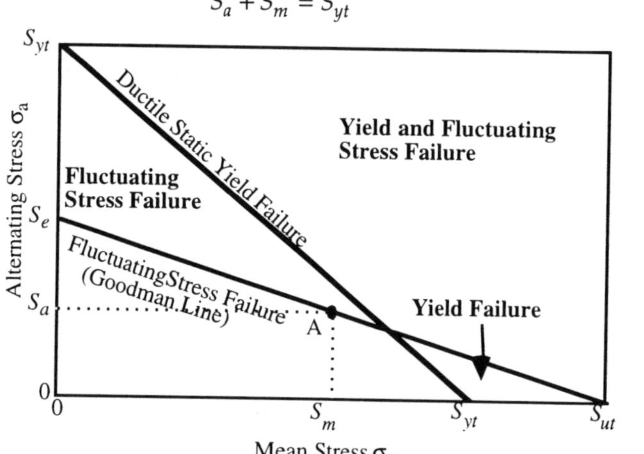

Figure 8.18 Fluctuating Stress Material Failure Showing Alternating Stress Strength S_a vs Mean Stress Strength S_m along the Goodman Line.

Example 8.7

An AISI 1035 hot rolled round steel bar has a diameter of 3 cm, a minimum tensile strength of S_{ut} = 500 MPa (Table 8.2) and a minimum yield strength of S_{yt} = 270 MPa. The part is to withstand fluctuating tensile load varying from 0 to 57 kN. Because of the design of the ends and fillet radius, a fatigue stress concentration factor K_f =1.85 exists. The remaining fatigue factors have been evaluated and are k_a = 0.797, k_b = 1.0, k_c = 0.861, and k_d = 1.0.

a) Use K_f as a fatigue strength reduction factor with the Goodman Line and determine the factor of safety n under the assumption that σ_m remains fixed.

b) The same as part (a) except σ_a remains fixed.

c) The same as part (a) except σ_a / σ_m remains fixed.

d) Find the factor of safety n guarding against yielding.

Solution. The material endurance strength

$$S_e' = 0.504(500\text{ MPa}) = 252\text{ MPa}$$

Using the stress concentration as a strength reduction factor $k_e = 1/K_f = 1/1.85 = 0.541$ and with the other modifying factors, the corrected endurance strength

$$S_e = k_a k_b k_c k_d k_e S_e' = (0.797)(1.0)(0.861)(1.0)(0.541)252 = 93.5\text{ MPa}$$

This data is plotted as two solid lines in Fig. 8.19 showing both Yield Strength failure and Fluctuating Stress Failure (Goodman Line). To show the load situations for cases (a) through (c), compute the operating point mean and fluctuating stresses using Eqs.8.4.12 and 8.4.13 with

$$\sigma_{\max} = F/A = 57kN/\pi\left[(1.5)^2\right] = 80.6\text{ MPa}$$

and $\sigma_{\min} = 0.0$ MPa we have

$$\sigma_m = (\sigma_{\max} + \sigma_{\min})/2 = (80.6 + 0.0)/2 = 40.3\text{ MPa}$$

and

$$\sigma_a = (\sigma_{\max} - \sigma_{\min})/2 = (80.6 - 0.0)/2 = 40.3\text{ MPa}$$

a) For constant mean stress $\sigma_m =$ constant, the "load line" meets the Fluctuating Stress (Goodman) Failure first so calculate the amplitude of the alternating strength S_a at that intersection for applied mean stress $\sigma_m = S_m$ using Eq.8.4.14:

$$\frac{S_a}{S_e} + \frac{S_m}{S_{ut}} = 1 \Rightarrow \qquad \frac{S_a}{93.5} + \frac{40.3}{500} = 1$$

$$\Rightarrow \qquad S_a = 93.5\left[1 - \frac{40.3}{500}\right] = 86.0\text{ MPa}$$

For constant mean stress, the factor of safety

$$n = \frac{Strength}{Stress} = \frac{S_a}{\sigma_a} = \frac{86.0}{40.3} = 2.13$$

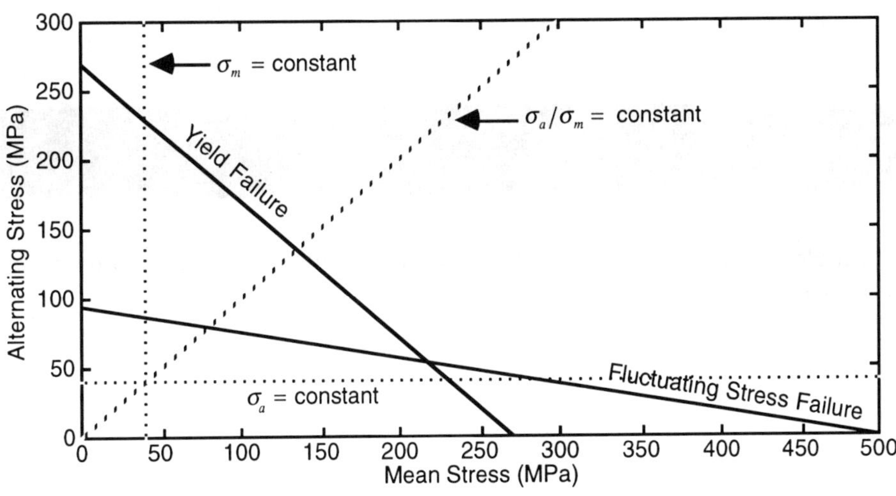

Figure 8.19 Failure Criteria for cases (a) through (c), Example 8.7

b) For constant alternating stress $\sigma_a = \text{constant}$, the "load line" meets the Yield Stress Failure first. Calculate the amplitude of the mean strength S_m at that intersection for applied alternating stress $\sigma_a = S_a$ using Eq.8.4.15:

$$S_a + S_m = S_{yt} \Rightarrow 40.3 + S_m = 270 \text{ MPa}$$

$$\Rightarrow \qquad S_m = 270 - 40.3 = 230 \text{ MPa}$$

For constant alternating stress, the factor of safety

$$n = \frac{Strength}{Stress} = \frac{S_m}{\sigma_m} = \frac{230}{40.3} = 5.7$$

c) For constant ratio of alternating to mean stress $\sigma_a/\sigma_m = \text{constant}$, the "load line" meets the Fluctuating Stress failure line first. Calculate the amplitude of the strength $S = S_a = S_m$ at that intersection for applied mean stress $\sigma_m = \sigma_a = S$ using Eq.8.4.14:

$$\frac{S_a}{S_e} + \frac{S_m}{S_{ut}} = 1 \Rightarrow \qquad \frac{S}{93.5} + \frac{S}{500} = 1$$

$$\Rightarrow \qquad S_a = \left[\frac{93.5(500)}{93.5 + 500}\right] \text{MPa} = 78.8 \text{ MPa}$$

For constant mean stress, the factor of safety

$$n = \frac{Strength}{Stress} = \frac{S}{\sigma} = \frac{78.8 \text{ MPa}}{40.3 \text{ MPa}} = 1.96$$

d) The factor of safety guarding against static yield failure is simply the ratio of yield strength to maximum stress:

$$n = \frac{Strength}{Stress} = \frac{S}{\sigma} = \frac{270}{80.6} = 3.35$$

As expected, this factor of safety is larger than the factor of safety for Fluctuating Stress failure because for this load condition ($\sigma_{max} = \sigma_a + \sigma_m$, $\sigma_a = \sigma_m$) the yield failure line is beyond the fluctuating stress line.

Reference

J.E. Shigley and C.R. Mischke, *Mechanical Engineering Design*, 5th edition, McGraw Hill, 1989

Practice Problems

Essay Problems

8.1 An automobile shift linkage which is bent to avoid an obstruction is shown below.

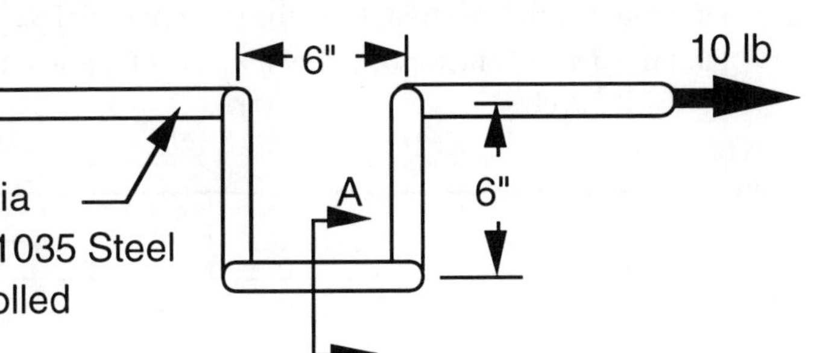

a) What is the maximum von Mises stress in plane A-A?

b) What is the factor of safety of the linkage for 500 cycles of completely reversed loading, e.g., zero to 10 lb to zero to -10 lb to zero?

c) What is the mean factor of safety of the linkage for 100,000 cycles of completely reversed loading?

8.2 The axle below is subjected to loads such that at the smaller diameter of the shoulder when under constant load, the axle has a bending stress, $\sigma_x = 8$ kpsi and a torsional stress, $\tau_{xy} = 3$ kpsi (cw).

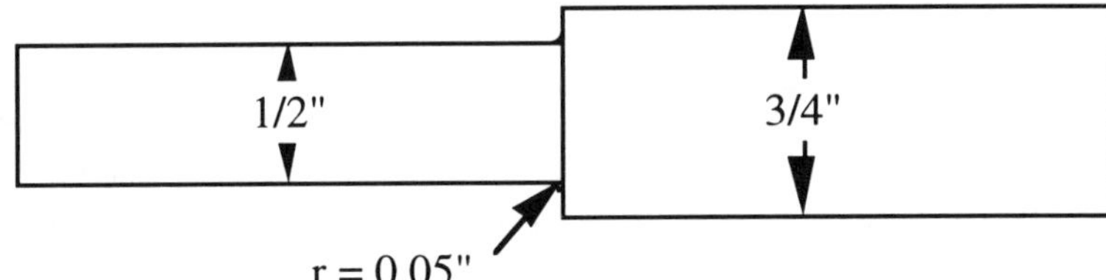

When constructed, the mild steel axle has a yield strength, $S_y = 40$ kpsi and an ultimate strength, $S_u = 50$ kpsi. with 15% elongation.

a) State the failure theory to be used to analyze the axle for static failure.

b) Find the factor of safety for static failure of the shaft.

8.3 The axle of Problem 8.2 is constructed of a material with an ultimate tensile strength, $S_{ut} = 50$ kpsi and ultimate compressive strength, $S_{uc} = 100$ kpsi with negligible elongation in both cases. At the smaller diameter of the shoulder when under constant load, the axle has a bending stress, $\sigma_x = 8$ kpsi and a torsional stress, $\tau_{xy} = 3$ kpsi (cw).

a) State the failure theory to be used to analyze the axle for static failure.

b) Find the factor of safety for static failure of the shaft using the all applicable failure theories.

8.4 The axle of Problem 8.2 is constructed of a mild steel with a machined finish which has a yield strength, S_y = 40 kpsi and an ultimate strength, S_u = 50 kpsi with 15% elongation. At the smaller diameter of the shoulder under constant load, the axle has only bending stress, $\sigma_x = 8$ kpsi when rotated at 1750 rpm.

a) Find the endurance strength of the shaft material S_e.
b) Find the design's factor of safety n against fatigue failure of the shaft at $N = 10^5$ cycles.

Multiple Choice Problems

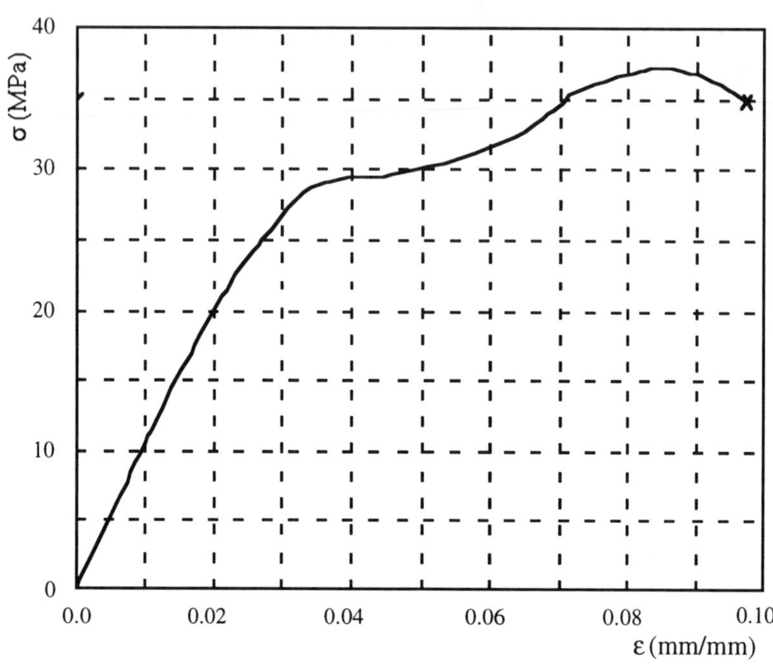

The graph above shows the result of an engineering tensile test. Answer Questions 8.5 – 8.10.

8.5 The material's Young's Modulus, E = ______

a) 25 MPa b) 37 MPa c) 435 MPa d) 1000 MPa

8.6 The material's 0.2% Yield Strength, S_{yt} = ________

a) 25 MPa b) 37 MPa c) 435 MPa d) 1000 MPa

8.7 The material's ultimate strength, S_{ut} = ______________________ MPa

a) 25 MPa b) 37 MPa c) 435 MPa d) 1000 MPa

8.8 The material's permanent elongation (%) at failure is approximately

a) 0.1% b) 2% c) 6% d) 10%

8.9 This specimen failed as a _________ material.

a) ductile b) brittle c) fatigued d) low cycled fatigued

8.10 The principle stresses on the specimen at failure are

a) $\sigma_1 = 35$ MPa, $\sigma_2 = 0$ MPa, $\sigma_3 = 0$ MPa,

b) $\sigma_1 = 35$ MPa, $\sigma_2 = 35$ MPa, $\sigma_3 = 0$ MPa,

c) $\sigma_1 = 35$ MPa, $\sigma_2 = 35$ MPa, $\sigma_3 = 35$ MPa,

d) $\sigma_1 = 35$ MPa, $\sigma_2 = 0$ MPa, $\sigma_3 = 35$ MPa,

The shaft shown below is to be designed for a factor of safety, $n = 1.8$. The shaft is driven through the pulley at "D" by a belt with top and bottom tensions of 100 and 200 lb, respectively, applied to drive a machine through the pulley at "B" with top tension F and zero bottom tension. The shaft is supported by two bearings at "A" and "C". All shoulders are constructed with fillet radii, $r = d/10$ and the larger diameter of the shaft is 1.5 times the smaller diameter ($D = 1.5\,d$). The shaft is to be constructed of UNS G10100 HR steel with a tensile strength, $S_{ut} = 47$ kpsi and yield strength, $S_y = 26$ kpsi .

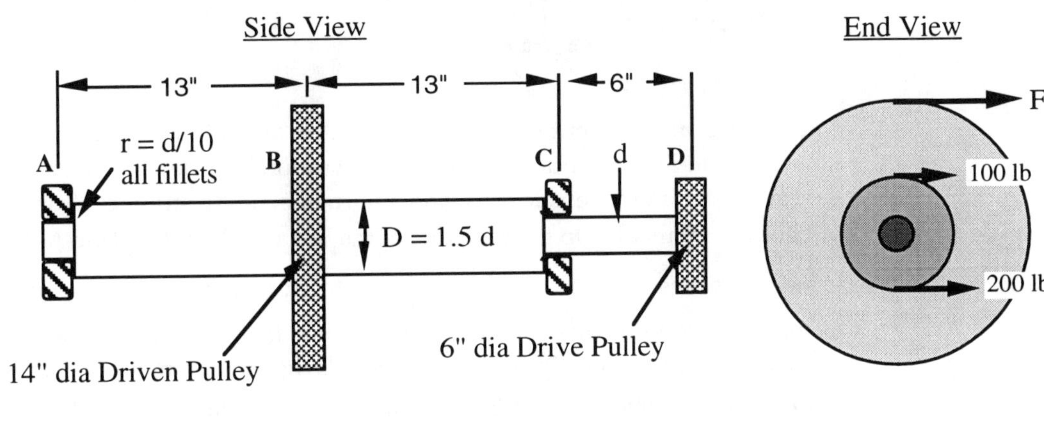

8.11 What is the driving torque T_D at pulley D?
a) 100 in lb b) 200 in lb c) 300 in lb d) 400 in lb

8.12 What is the top tension force F at pulley B?
a) 0.0 lb b) 42.9 lb c) 100 lb d) 147.8 lb

8.13 What is the transmitted torque T_B at pulley "B"?
a) 100 in lb b) 200 in lb c) 300 in lb d) 400 in lb

8.14 What is the transverse reaction force at bearing A ?
a) 47.8 lb b) 100 lb c) 148.3 lb d) 200 lb

8.15 What is the transverse reaction force at bearing C ?
a) 47.8 lb b) 100 lb c) 148.3 lb d) 390.7 lb

8.16 What is the bending moment on the shaft at the shoulder?
a) 300 in lb b) 1300 in lb c) 1800 in lb d) 2600 in lb

8.17 To prevent static failure, what minimum appropriate shaft size d should be specified for the shaft design?
a) 1.0 in b) 1.25 in c) 1.50 in d) 2.00 in

8.18 To prevent fatigue failure, what minimum appropriate shaft size d should be specified for the shaft design?

a) 1.0 in b) 1.25 in c) 1.50 in d) 2.00 in

Solutions to Practice Problems

Essay Problem Solutions

8.1 a) Draw a free-body diagram of half the linkage cut through plane A-A.

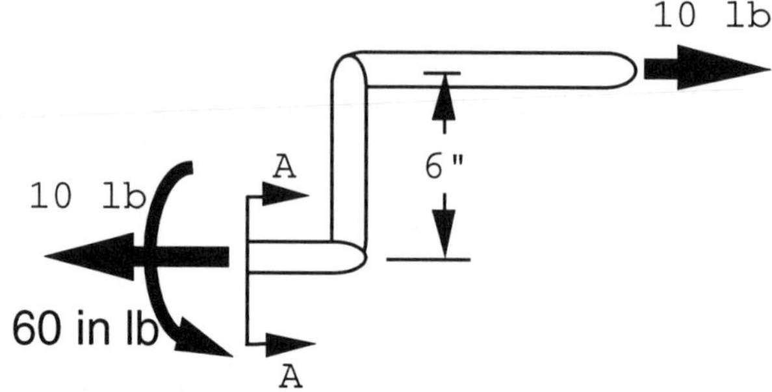

The force and moment reactions at plane A-A are a 10 lb tension force and a 60 in-lb moment. The stress in the section includes an axial normal stress σ_t from the tensile force and a normal bending stress σ_b from the bending moment. On the top edge of the linkage, the total plane A-A normal stress $\sigma = \sigma_t + \sigma_b = \frac{F}{A} + \frac{My}{I}$ is the sum of the tensile and bending stress components. Because there are no other stress components at this section, the von-Mises stress σ' equals the maximum normal stress at this point where $y = d/2$, $A = \pi d^2/4$, $I = \pi d^4/64$ and d = 0.5 in:

$$\sigma' = \sigma_t + \sigma_b = \frac{F}{A} + \frac{My}{I} = \frac{4F}{\pi d^2} + \frac{32M}{\pi d^3} = 4.94 \times 10^3 \text{ lb/in}^2$$

b) For 500 cycles, both "low cycle fatigue" and "yield" failures are analyzed, where $S_{yt} = 39.5$ kpsi and $S_{ut} = 72$ kpsi from Table 8.2. The material's low cycle fatigue strength at 500 cycles is determined using interpolation and Figure 8.17 as

$$S_f = \left[1 - 0.1\left(\frac{\log(500)}{\log(10^3)}\right)\right] S_{ut} = [0.91]72 \text{ kpsi} = 65.5 \text{ kpsi}$$

which is larger than the materials' yield strength. Using the lower material yield strength, the factor of safety for completely reversed loading is $n = S_{yt}/\sigma' = 39.5 \text{ kpsi}/4.94 \text{ kpsi} = 8.0$. Typically, ductile materials fail in yield at loads smaller than their low cycle fatigue strengths.

c) For 100,000 cycles, "high cycle" fatigue strength will be evaluated. Using interpolation and Figure 8.17,

$$S_f = \left[0.9 - 0.4\left(\frac{\log(10^5) - \log(10^3)}{\log(10^6) - \log(10^3)}\right)\right] S_{ut} = [0.633]72 \text{ kpsi} = 45.6 \text{ kpsi}$$

and the factor of safety at 100,000 cycles remains $n = 8.0$ since the material has a lower yield strength.

8.2 a) This axle's material has more than 3% elongation at failure so is considered ductile and *the axle will fail in material yield.*

b) Because the material yields, stress at stress concentrations remain approximately constant and do not affect the failure strength of the axle. The von Mises stress is evaluated by using Mohr's circle to find the principal stresses and substituting into Eq.8.4.2:

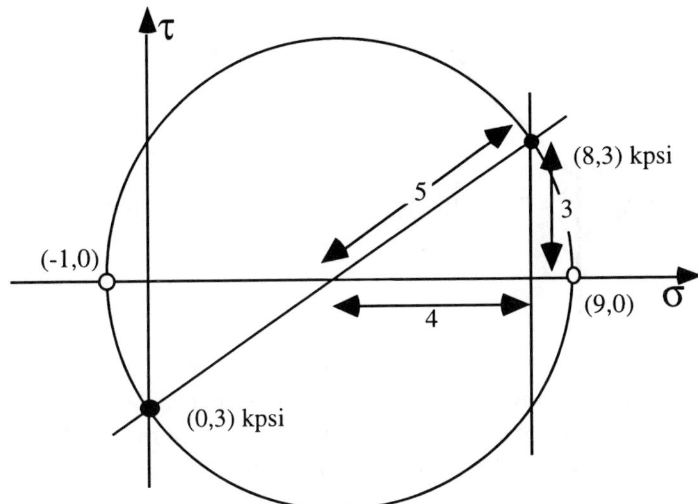

$$\sigma' = \sqrt{\frac{(9-(-1))^2 + ((-1)-0)^2 + (0-9)^2}{2}} = \sqrt{\frac{100+1+81}{2}} = \sqrt{91} = 9.54 \text{ kpsi}$$

The factor of safety for static ductile failure is $n = S_{yt}/\sigma' = 40/9.54 = 4.19$

8.3 a) This axle has the same geometry and loading as in Problem 8.2, however, the material has different compressive and tensile stresses indicating a brittle material. For this reason, *the material failure is evaluated using the Modified-Mohr failure theory.* Both load stresses must be increased by appropriate stress concentration factors for $r/d = 0.1$ and $D/d = 1.5$.

b) The torsional stress concentration $K_t = 1.5$ from Figure 8.10 gives a torsional stress $\tau = K_t\tau_0 = 1.5(3) = 4.5$ kpsi(cw). The bending stress concentration $K_t = 1.7$ from Figure 8.11 gives a bending stress $\sigma_b = K_t\sigma_0 =$ 1.7(8)= 13.6 kpsi. The principal stresses are found from a Mohr's circle analysis (graphic not reproduced here) to be

$$\sigma_1, \sigma_2 = \left(\frac{13.6}{2}\right) \pm \sqrt{\left(\frac{13.6}{2}\right)^2 + (4.5)^2} = 15.0, -1.35$$

Reproducing the fourth quadrant of Figure 8.16 the load line is determined by the combination of a $\sigma_1 = 15.0$ kpsi and $\sigma_2 = -1.35$ kpsi and crosses the Modified-Mohr failure criteria at $S_{ut} = 50$ kpsi with a slope of -1.35/15 (see the figure that follows). Because the load line slope is less than 1, the factor of safety $n = S_{ut}/\sigma_1 = 50/15 = 3.33$ is determined by the maximum normal stress.

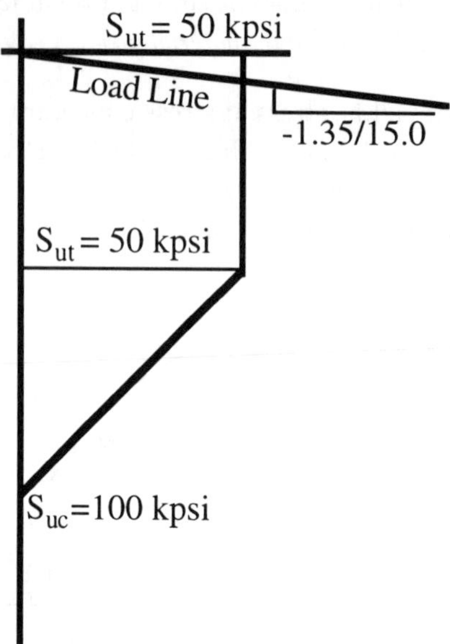

8.4 a) The same axle geometry as Problem 8.2 but constructed of mild steel with a machined finish and subjected to a rotating bending stress has a life governed by fatigue failure. The endurance limit Eq.8.4.6:

$$S_e = k_a k_a k_a k_a k_a S_e' = (0.958)(0.944)(1)(1)(.588)(0.5 \times 50) \text{ kpsi} = 13.3 \text{ kpsi}$$

where $k_a = 2.7(50)^{-0.265} = 0.958$ (from Table 8.3)

$k_b = \left(\frac{0.5}{0.3}\right)^{-0.1133} = 0.944$ (from Eq.8.4.8a)

$k_c = 1.0$ (from Eq.8.4.9)

$k_d = 1.0$ (from Eq.8.4.10 because no temperature specified)

$k_e = 1/K_f = 1/1.7 = 0.588$ (from Eq.8.4.11)

b) At N=10^5 cycles, use the unmodified load stress $\sigma_b = 8$ kpsi since stress concentration K_f was included in the endurance limit calculation. Compute the "low cycle" strength $0.9S_t$ without stress concentration K_f because the material is ductile. Reproducing Fig. 8.17, and interpolating for the modified endurance limit S_e:

$$S_f = 0.9S_{ut} - \left(\frac{(5-3)}{(6-3)}\right)(0.9S_{ut} - S_e) = 0.9(50) - \left(\frac{2}{3}\right)(0.9(50) - 13.3) = 23.9 \text{ kpsi}$$

The factor of safety

$$n = S/\sigma = 23.9/8 = 2.99$$

Multiple Choice Problem Solutions

8.5 **d)** $E = 10\text{ MPa}/0.01 = 1000\text{ MPa}$

8.6 **a)** Examine graph at $\varepsilon = 0.002$ parallel to slope 1000 MPa-mm/mm yields 25 MPa

8.7 **b)** $S = \sigma = 37$ MPa, the peak of the stress strain curve, then the load supported falls.

8.8 **c)** 6%, extend a line parallel to the initial slope of the curve to the point of failure.

8.9 **a)** ductile, more than 3% permanent elongation.

8.10 **a)** $\sigma_1 = 35$ MPa, $\sigma_2 = \sigma_3 = 0$ for pure tension.

8.11 **c)** (6 in/2) × (200 lb - 100 lb) = 300 iN-lb

8.12 **b)** F(7 in) = 300 in-lb gives $F = 300$ in-lb/7 in = 42.9 lb

8.13 **c)** 300 in-lb neglecting bearing drag.

8.14 **a)** Draw a plan-view free-body-diagram:

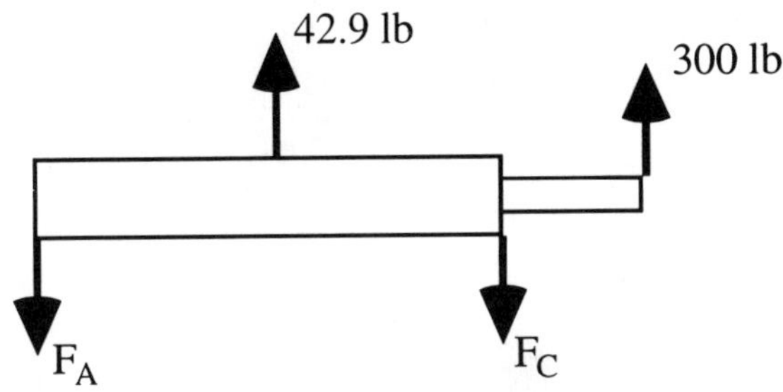

$$F_A(26\text{ in}) + 300\text{ lb}(6\text{ in}) = 42.9(13\text{ in})$$
$$\Rightarrow F_A = [42.9(13\text{ in}) - 300\text{ lb}(6\text{ in})]/(26\text{ in})$$
$$\Rightarrow F_A = -47.8\text{ lb}$$

8.15 **d)** From the free-body-diagram

$$F_C(26\text{ in}) = 300\text{ lb}(19\text{ in}) + 42.9(13\text{ in})$$
$$\Rightarrow F_C = [300\text{ lb}(32\text{ in}) + 42.9(13\text{ in})]/(26\text{ in})$$
$$\Rightarrow F_C = 390.7\text{ lb}$$

8.16 **c)** $M = (300\text{ lb})(6\text{ in}) = 1800$ in-lb

8.17 **a)** $S = \sigma = My/I \;\Rightarrow\; 26\text{ kpsi} = [1800\text{ in lb}(d/2)]/[\pi d^4/64]$ (ductile failure)

$$\Rightarrow\; d^3 = [1800\text{ in lb}(32)]/[\pi(26\text{ kpsi})] = 0.71\text{ in}^3$$
$$\Rightarrow\; d = 0.89\text{ in} \;\Rightarrow\; d = 1.0\text{ in with increase to next largest standard size}$$

8.18 **b)** $S_e = k_a k_b k_c k_d k_e S_e' = (0.91)(.81)(1)(1)(0.59)[0.5(47\text{ kpsi})] = 10.2$ kpsi using Eq.8.4.6 where $k_a = 0.91$(HR finish), $k_b = 0.81$ @ d= 2 in (0.87 @1 in), $k_c =$ 1.0 (bending), $k_d = 1.0$ (temperature unspecified), and $k_e = 1/1.7 = 0.59$ (Fig 8.11) then $S = \sigma = My/I \Rightarrow 10.2\text{ kpsi} = [1800(d/2)]/[\pi d^4/64]$

$$\Rightarrow d = 1.22\text{ in} \Rightarrow 1.25\text{ in}$$

Vibration

by Clark J. Radcliffe and Charles W. Radcliffe

9.1 One Degree of Freedom Systems

Vibration in many engineering design problems is accurately idealized with one degree of freedom models. For these systems, single coordinate motion of a dominant mass is restrained by a localized stiffness. The power flow between kinetic energy stored in the mass and potential energy stored in the stiffness results in oscillatory structural response. The examples shown in Figure 9.1 include structures, machinery mounts, and suspension systems.

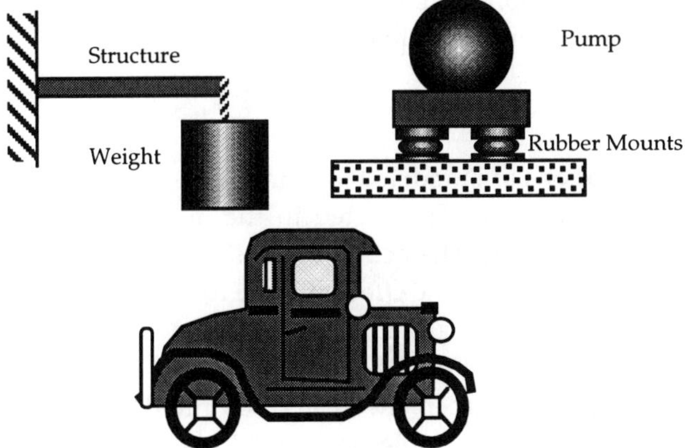

Figure 9.1 Examples Of One Degree Of Freedom Systems Include Structural Supports, Machinery Mounts, And Suspension Systems

Spring-Mass Model

The idealized spring-mass models shown in Figure 9.2 show horizontal, vertical and rotational schematics. The systems modeled here do not include any energy dissipation. These models are used to analyze systems when the energy dissipation forces are small relative to those forces associated with mass acceleration and spring compression. In all cases the significant displacement of the system from equilibrium is described by a single coordinate: x, y or θ. The differential equation model derived from all three schematics takes the same form.

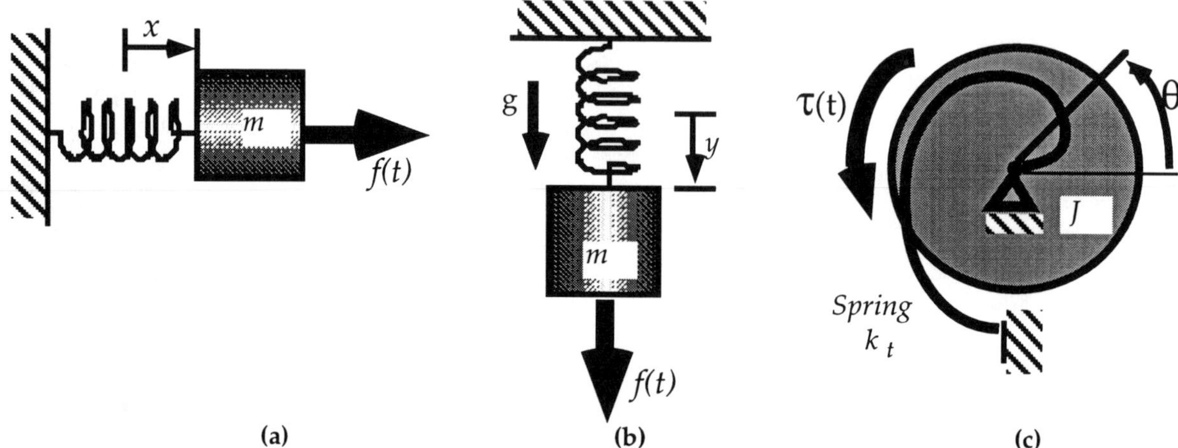

Figure 9.2 Spring-Mass System Idealized Models. (a) Horizontal and (b) Vertical and (c) Rotational

For horizontal translation $x(t)$ of the mass m from equilibrium due to the applied force $f(t)$ resisted by the spring with stiffness k, the differential equation is

$$m\ddot{x} + kx = f(t) \tag{9.1.1}$$

For vertical translation $y(t)$ of the mass m from equilibrium due to the applied force $f(t)$ resisted by the spring with stiffness k, the differential equation is

$$m\ddot{y} + ky = f(t) \tag{9.1.2}$$

For angular rotation $\theta(t)$ of the mass with rotating moment of inertia I from equilibrium due to the applied moment $\tau(t)$ resisted by the spring with rotational stiffness k_t, the differential equation is

$$I\ddot{\theta} + k_t\theta = \tau(t) \tag{9.1.3}$$

All three types of motions are modeled and analyzed identically. The actions of all constant external forces are neglected in the mathematical model for displacement from equilibrium. The action of constant external forces is only important for vibration analysis in the effect of these forces on equilibrium displacement. Specifically, the action of gravity is never included in vibration analysis.

If there is no applied force, Eq.9.1.1 can be solved for time history response of the mass given an initial displacement x_0 and/or initial velocity v_0 of the mass from equilibrium. Using the variables in Eq.9.1.1, the solution for the *undamped free vibration response* is

$$x(t) = \frac{v_0}{\omega_n}\sin(\omega_n t) + x_0\cos(\omega_n t)$$
$$= \frac{\sqrt{\omega_n^2 x_0^2 + v_0^2}}{\omega_n}\sin\left[\omega_n t + \tan^{-1}(\omega_n x_0 / v_0)\right] \quad \textbf{(9.1.4)}$$

where we used the identity of Eq. 1.2.6 and let the natural frequency of the spring-mass system

$$\omega_n = \sqrt{k/m} \quad \textbf{(9.1.5)}$$

Further discussion of vibration systems will often use the notation of Eq.9.1.1 with the understanding that the results are applicable to Eq.9.1.2 and Eq.9.1.3 with a simple change of variables.

Example 9.1

A 1000 kg automobile suspension compresses 20 cm at equilibrium. If the car is raised 10 cm from its equilibrium height and released, what is the vertical motion of vehicle assuming a one degree of freedom spring-mass model?

Solution. The mass of the automobile $m = 1000$ kg while the stiffness k can be determined from the static suspension deflection:

$$k = Force/Deflection = mg/\Delta h = [1000(9.8)]/0.20 = 49\times10^3 \text{ N / m}$$

The automobile vertical natural frequency is $\omega_n = \sqrt{k/m} = \sqrt{49\times10^3/1000}$ $= 7.0$ rad/sec with initial position $h_0 = 0.1$ m and initial velocity $v_0 = 0.0$ m/sec. Using Eq.9.1.4, the automobile motion is

$$h(t) = \frac{\sqrt{\omega_n^2 h_0^2 + v_0^2}}{\omega_n^2}\sin\left[\omega_n t + \tan^{-1}(\omega_n h_0 / v_0)\right] = \frac{\sqrt{(7)^2(0.1)^2 + (0)}}{(7)^2}\sin\left[7t + \tan^{-1}(7(0.1)/0.0)\right]$$
$$= 0.143\sin[7t + \pi/2] = 0.143\cos[7t]$$

where displacement $h(t)$ is measured from the equilibrium height of the automobile. Note that the motion is at the natural frequency of

$$\omega_n = 7.0 \text{ rad/sec} = 7.0(1 \text{ cycle}/2\pi \text{ radians}) = 1.1 \text{ cycle/sec} = 1.1 \text{ Hz}$$

Spring-Mass-Damper Model

This vibration response (9.1.4) continues indefinitely which is unrealistic for most engineering analysis so a model for vibration energy dissipation is required. The most common model for energy dissipation is a *linear viscous damping force*

$$f_d = -cv = -c\dot{x} \qquad \text{or} \quad \textbf{(9.1.6)}$$

$$\tau_d = -c_t\theta = -c_t\dot{\theta} \quad \textbf{(9.1.7)}$$

opposing the velocity of the mass. This form of energy dissipation is characterized in a physical schematic with the dashpot or rotational drag as shown in Figure 9.3. This viscous dissipation model is sometimes used as an approximation for friction. Substituting Eqs.9.1.6-7 into Eqs.9.1.1-3 yields *mathematical models for the spring-mass-damper systems*:

$$m\ddot{x} + c\dot{x} + kx = f(t) \tag{9.1.8}$$

$$m\ddot{y} + c\dot{y} + Ky = f(t) \tag{9.1.9}$$

$$I\ddot{\theta} + c_t\dot{\theta} + k_t\theta = \tau(t) \tag{9.1.10}$$

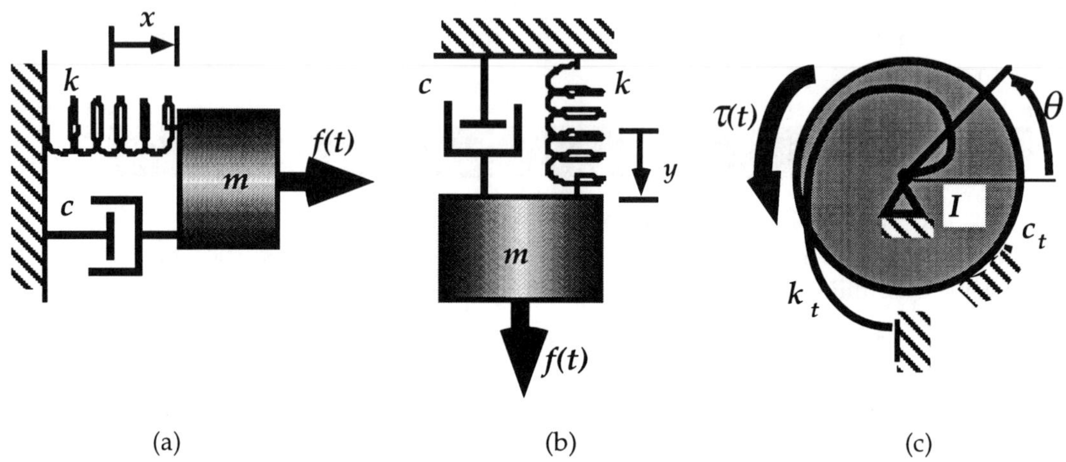

Figure 9.3 Spring-Mass-Damper System Idealized Models. (a) Horizontal, (b) Vertical and (c) Rotational showing Linear Viscous Dissipation Represented as Dampers c and Drag c_t

9.2 Free Vibrations

The free vibrations of spring-mass-damper systems are important because they reveals the characteristic frequency and persistence of vibrations excited by non-zero initial displacement and velocity. The mathematical model for the mass displacement $x(t)$ for such vibrations is

$$m\ddot{x} + c\dot{x} + kx = 0 \text{ with } x(0) = x_0 \text{ and } \dot{x}(0) = v(0) = v_0 \tag{9.2.1}$$

The solution of Eq.9.2.1 is dependent on the relative magnitude of the damping coefficient c. It is convenient to define the damping ratio ζ as

$$\zeta = \frac{c}{2\sqrt{km}} = \frac{c}{2m\omega_n} \tag{9.2.2}$$

The value of a system's damping ratio determines whether the motion will be "underdamped", "critically damped" or "overdamped".

Underdamped Free Vibration ($0 < \zeta < 1$)

This is the case of most interest because free response oscillations occur only in the underdamped cases. When $\zeta < 1$, the free damped vibration is

$$x(t) = e^{-\zeta\omega_n t}\left[A\cos(\omega_d t) + B\sin(\omega_d t)\right] = Ce^{-\zeta\omega_n t}\sin(\omega_d t + \phi) \quad \textbf{(9.2.3)}$$

where: $\omega_d = \omega_n\sqrt{1-\zeta^2}$, $A = x_0$, $B = (v_0 + \zeta\omega_n x_0)/\omega_d$,

$$C = \frac{\sqrt{(v_0+\zeta\omega_n x_0)^2 + (x_0\omega_d)^2}}{\omega_d} \text{ and } \phi = \tan^{-1}\left[\frac{x_0\omega_d}{(v_0+\zeta\omega_n x_0)}\right]$$

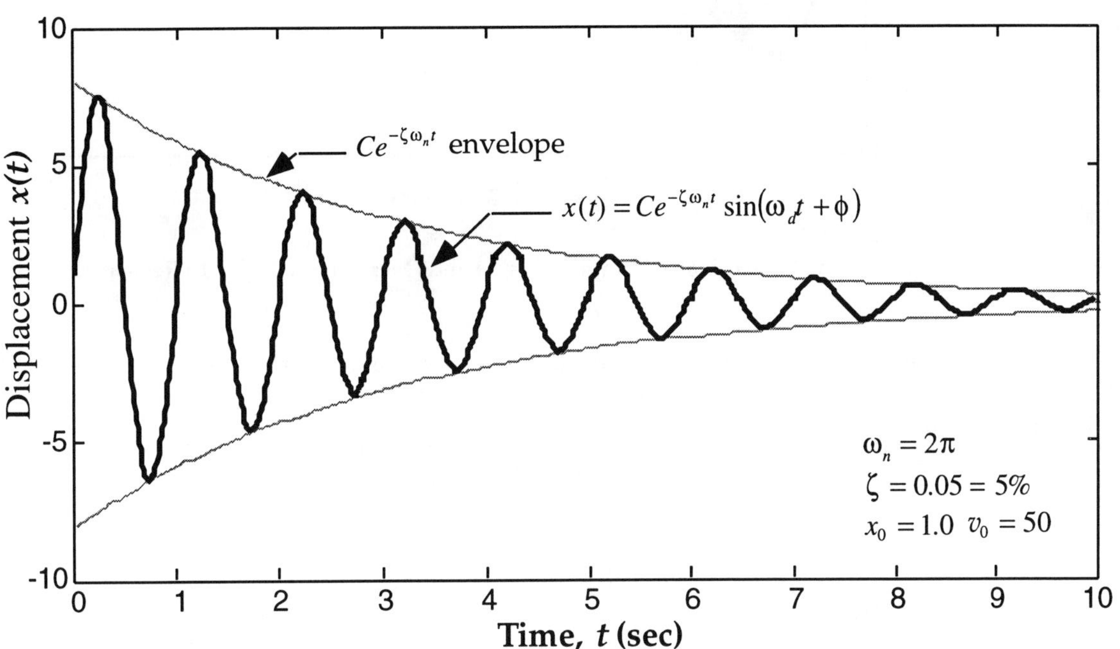

Figure 9.4 Underdamped Response

Underdamped response has the form shown in Fig.9.4 which shows the oscillatory character of the response with damped natural frequency ω_d. The underdamped response decays at a rate given by it's *time constant*

$$\tau_d = \frac{1}{\zeta\omega_n} \quad \textbf{(9.2.4)}$$

to an amplitude of approximately 37% of initial amplitude after one time constant and 2% at time $t = 4\tau_d$.

Critically Damped Free Vibration ($\zeta = 1$)

For the critical damping ratio $\zeta = \zeta_c = 1.0$, the free vibration solution for model Eq.9.2.1 becomes

$$x(t) = e^{-\zeta\omega_n t}\left[(v_0 + \omega_n x_0)t + x_0\right] \quad \textbf{(9.2.5)}$$

This critically damped is shown in Figure 9.5 and demonstrates the absence of oscillation. Shown here for completeness, the absence of oscillation in critically damped response makes it irrelevant to most vibration modeling problems.

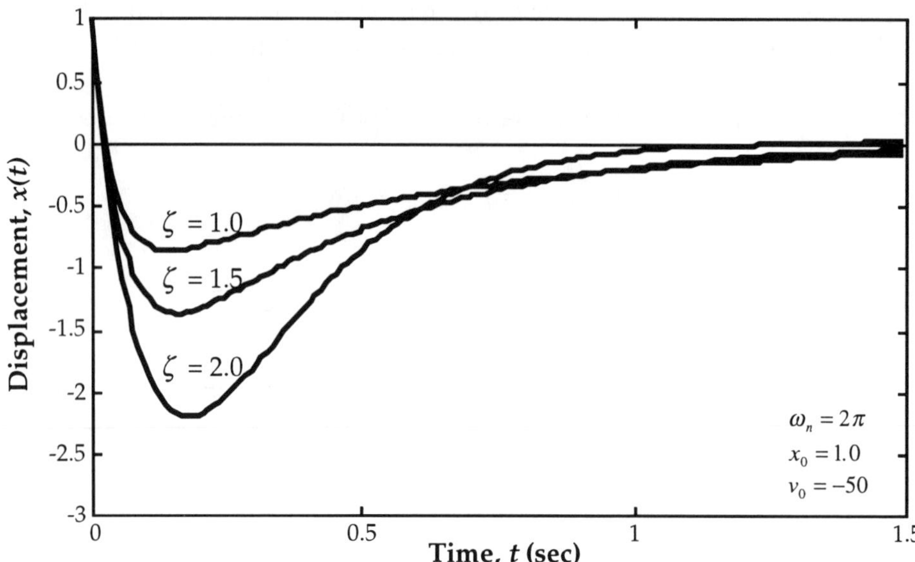

Figure 9.5 Critically Damped ($\zeta = 1$) and Over Damped ($\zeta > 1$) Responses

Overdamped Free Vibration ($\zeta > 1$)

For damping ratio larger than the critical value $\zeta > 1$, the free vibration solution for model Eq.9.2.1 becomes

$$x(t) = Ae^{\left[-\zeta+\sqrt{\zeta^2-1}\right]\omega_n t} + Be^{\left[-\zeta-\sqrt{\zeta^2-1}\right]\omega_n t} \qquad \textbf{(9.2.6)}$$

where $A = \dfrac{v_0 + \left[\zeta + \sqrt{\zeta^2 - 1}\right]\omega_n x_0}{2\omega_n\sqrt{\zeta^2 - 1}}$ and $B = \dfrac{v_0 + \left[\zeta - \sqrt{\zeta^2 - 1}\right]\omega_n x_0}{2\omega_n\sqrt{\zeta^2 - 1}}$

This overdamped response is shown in Figure 9.5 and demonstrates the absence of oscillation. Included again for completeness, the absence of oscillation in overdamped response makes it irrelevant to most vibration modeling problems.

Example 9.2

The system below has a piston of mass m traveling in a "frictionless" tube with velocity v_0 which engages a spring k and dashpot c at time $t = 0$. determine the maximum displacement of the spring and how long it takes to reach that maximum displacement.

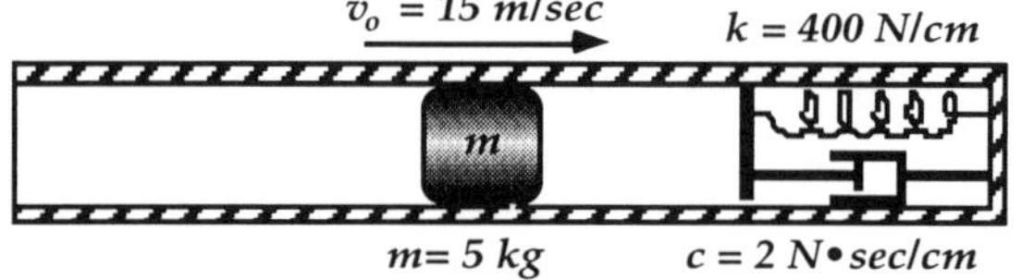

Solution. In this case, static equilibrium is the undeflected spring length and all displacements will be computed from that position. Using the one degree of freedom, spring-mass-damper Eq.9.2.1, first use Eq.9.2.2 to determine the system's damping ratio

$$\zeta = \frac{c}{2\sqrt{km}} = \frac{[(2\ \mathrm{N}\cdot\mathrm{sec}/\mathrm{cm})(100\ \mathrm{cm}/\mathrm{m})]}{2\sqrt{(400\ \mathrm{N}/\mathrm{cm})(100\ \mathrm{cm}/\mathrm{m})(5\ \mathrm{kg})}} = \frac{(200\ \mathrm{N}\cdot\mathrm{sec}/\mathrm{m})}{2\sqrt{(4\times 10^4\ \mathrm{N}/\mathrm{m})(5\ \mathrm{kg})}} = 0.2236$$

This system is underdamped ($\zeta < 1$) and its free vibration solution Eq.9.2.3 is

$x(t) = Ce^{-\zeta\omega_n t}\sin(\omega_d t + \phi)$ where initial displacement $x_0 = 0$, initial velocity $v_o = 15\ \mathrm{m}/\mathrm{s}$,

$$\omega_n = \sqrt{k/m} = \sqrt{(4\times 10^4\ \mathrm{N}/\mathrm{m})/(5\ \mathrm{kg})} = \sqrt{8\times 10^3\ /\mathrm{sec}^2} = 89.4427\ \mathrm{rad}/\mathrm{sec}$$

$$\omega_d = \omega_n\sqrt{1-\zeta^2} = 89.4427\sqrt{1-(0.2236)^2} = 87.1781\ \mathrm{rad/sec},$$

$$C = \frac{\sqrt{(v_0 + \zeta\omega_n x_0)^2 + (x_0\omega_d)^2}}{\omega_d} = \frac{\sqrt{(15\ \mathrm{m/sec} + 0)^2 + (0)^2}}{87.1728\ \mathrm{rad/sec}} = \frac{15\ \mathrm{m/sec}}{87.1728\ \mathrm{rad/sec}} = 0.1721\ \mathrm{m}$$

and $\phi = \tan^{-1}\left[\frac{x_0\omega_d}{(v_0 + \zeta\omega_n x_0)}\right] = \tan^{-1}\left[\frac{0\ \mathrm{m/sec}}{(15\ \mathrm{m/sec} + 0)}\right] = \tan^{-1}[0] = 0$ radians.

The displacement is

$$x(t) = Ce^{-\zeta\omega_n t}\sin(\omega_d t + \phi) = 0.17e^{-(0.2236)(89.44)t}\sin(87.17t + 0) = 0.17e^{-20t}\sin(87t)$$

To find the time of maximum spring compression, solve $\dot{x}(t) = 0$ for time t, then substitute back into the above solution for $x(t)$:

$\dot{x}(t) = 0$ at $t = 0, \pi, 2\pi, \cdots$ giving

$$t = \pi/87.1728 = 36\times 10^{-3}\ \mathrm{sec}$$

when the displacement

$$x_{\max} = 0.1721e^{-20(0.0360)} = 0.0838\ \mathrm{m} = 8.4\ \mathrm{cm}$$

9.3 Forced Vibrations

The preceding section demonstrated the oscillatory response of underdamped systems to non-zero initial displacement and velocity. Each system had a characteristic oscillation natural frequency ω_n dependent on the system's mass and stiffness properties. This section will demonstrate the importance of natural frequency to the system's response when subjected to harmonic excitation.

For the spring-mass-damper system forced with harmonic excitation

$$f(t) = F_0 \sin(\omega t) \tag{9.3.1}$$

the mathematical model for harmonically forced response is

$$m\ddot{x} + c\dot{x} + kx = F_0 \sin(\omega t) \tag{9.3.2}$$

The response of interest for vibration problems is the component of the solution which persists beyond the affect of the initial conditions. This persistent, steady-state, component is termed the *particular solution* (see Chapter 1):

$$x(t) = X \sin(\omega t + \phi) \tag{9.3.3}$$

where the harmonic response's *amplitude*

$$X = F_0 \Big/ \sqrt{\left(k - m\omega^2\right)^2 + (c\omega)^2} \tag{9.3.4}$$

and phase angle

$$\phi = -\tan^{-1}\left[(c\omega) / \left(k - m\omega^2\right)\right] \tag{9.3.5}$$

The forms of the harmonic response amplitude and phase angles can be simplified with the substitution of the natural frequency definition $\omega_n = \sqrt{k/m}$ from Eq.9.1.4 and the damping ratio definition $\zeta = c/2\sqrt{km}$ from Eq.9.2.2:

$$X = \frac{(F_0/k)}{\sqrt{\left(1 - \left(\frac{\omega}{\omega_n}\right)^2\right)^2 + \left(2\zeta\left(\frac{\omega}{\omega_n}\right)\right)^2}} \tag{9.3.6}$$

$$\phi = -\tan^{-1}\left[\frac{2\zeta\left(\frac{\omega}{\omega_n}\right)}{1 - \left(\frac{\omega}{\omega_n}\right)^2}\right] \tag{9.3.7}$$

The harmonic forced response in shown in Fig.9.6. The phase angle ϕ is always negative showing the response always lags behind the harmonic force excitation, as expected. The phase angle is small at low frequencies and increases with increasing excitation frequency. The response has important frequency ranges for phase angle and magnitude information.

At low frequency $\omega/\omega_n \le 1/\sqrt{2}$, the response magnitude is dominated by the effect of stiffness with little phase lag. At frequency $\omega = 0$, the magnitude of the response is simply the static deflection of the system. At other low frequencies, the deflection response follows the force with a small phase lag that increases with excitation frequency. Magnitude ratio changes little in this frequency range, especially for small damping.

At middle frequencies $1/\sqrt{2} \leq \omega/\omega_n \leq 2/\sqrt{2}$ the response magnitude is dominated by the effect of damping. At the natural frequency $\omega/\omega_n = 1$, the magnitude of the response $Xk/F_0 = 1/2\zeta$ and is governed strictly by the system's non-dimensional damping ratio. The phase angle at the natural frequency $\omega/\omega_n = 1$ always equals $\phi = -90^o = -\pi/2$ radians. The *peak amplitude* of the frequency response occurs at

$$\omega_{peak} = \omega_n \sqrt{1-2\zeta^2} \tag{9.3.8}$$

and $\omega_n \approx \omega_d \approx \omega_{peak}$ for $\zeta << 1$. The error is less than 2% for $\zeta < 0.1$ which is typical for metallic system vibration problems.

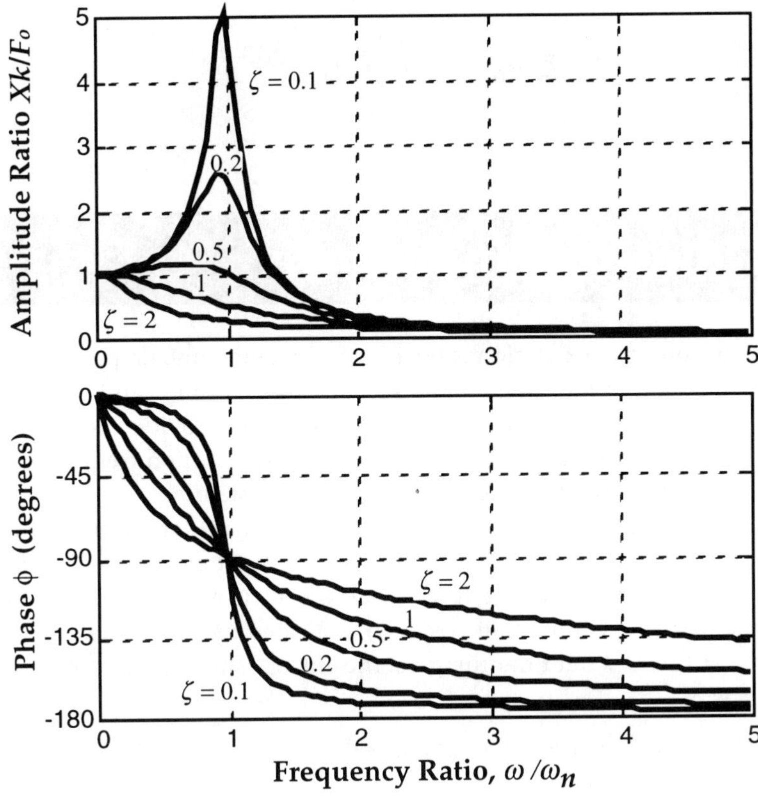

Figure 9.6 Harmonically Forced Response Amplitude Ratio and Phase Angle

At high frequencies $\omega/\omega_n \geq 2/\sqrt{2}$, the response magnitude is dominated by the effect of mass with increasing phase lag as frequencies increase. The phase lag asymptotically approaches $\phi = -180^o = -\pi$ radians as $\omega/\omega_n \Rightarrow \infty$. At these higher frequencies, the magnitude ratio Xk/F_0 decreases to $Xk/F_0 = 0$ as $\omega/\omega_n \Rightarrow \infty$.

Example 9.3

A machine part weighing 10 N vibrates with a resonant amplitude of 4 cm with a frequency of 5 cycles per second when excited by a 25 N harmonic force. Assuming the damping is viscous, what is the damping coefficient?

Solution. The resonant frequency $\omega_{res} = 5\times 2\pi = 31.4$ rad / s and $\omega_{res} \approx \omega_n$ for $\zeta << 1$. At $\omega = \omega_n$, use $X_{resonance} = X(\omega_n) = F_0/c\omega_n$ so that

$$c = \frac{F_0}{\omega_n X_{resonance}} = \frac{10}{[(5)(2\pi)]\left(4\times 10^{-2}\right)} = 8.0 \text{ N}\cdot\text{s/rad}$$

Now check the magnitude of the damping ratio using Eq.9.2.2:

$$m = W/g = (10 \text{ N})/\left(9.8 \text{ m/sec}^2\right) = 1.02 \text{ kg}$$

$$\zeta = \frac{c}{2m\omega_n} = \frac{8.0}{2(1.0)(31.4)} = 0.13 << 1$$

9.4 Rotating Unbalance

An important source of excitation is the effect of rotating imbalance. Such unbalance is common in mechanical systems such as engine pumps and other rotating machinery. The inertial forces in imbalanced systems provide excitation forces which can lead to large motions, especially when the rotation frequency equals, or is close to, the natural frequency of the vibrating system.

A one degree of freedom model of a machine with a rotating unbalance force is shown in Fig. 9.7. The model is constrained to move in the x-direction only and the source of the unbalance force is an eccentric mass m. The total mass M has a nonrotating component $(M-m)$ indicated by the shaded block. The support has stiffness k and damping coefficient c. The equation of motion for this system is

$$M\ddot{x} + c\dot{x} + kx = \left(me\omega^2\right)\sin(\omega t) \tag{9.4.1}$$

This model has frequency response magnitude X and phase angle ϕ given by

$$X = \left(me\omega^2\right)\Big/\sqrt{\left(k - M\omega^2\right)^2 + (c\omega)^2} \tag{9.4.2}$$

and phase angle

$$\phi = -\tan^{-1}\left[(c\omega)/\left(k - M\omega^2\right)\right] \tag{9.4.3}$$

The forms of the harmonic response amplitude and phase angles can be simplified with the substitution of the natural frequency definition $\omega_n = \sqrt{k/M}$ from Eq.9.1.4 and the damping ratio definition $\zeta = c/2\sqrt{kM}$ from Eq.9.2.2 where M now represents the total mass of the spring-mass-damper system:

$$\left(\frac{M}{m}\right)\frac{X}{e} = \frac{\left(\frac{\omega}{\omega_n}\right)}{\sqrt{\left(1-\left(\frac{\omega}{\omega_n}\right)^2\right)^2 + \left(2\zeta\left(\frac{\omega}{\omega_n}\right)\right)^2}} \tag{9.4.4}$$

$$\phi = -\tan^{-1}\left[\frac{2\zeta\left(\frac{\omega}{\omega_n}\right)}{1-\left(\frac{\omega}{\omega_n}\right)^2}\right] \tag{9.4.5}$$

These equations are very similar in form to Eqs.9.3.6 and 9.3.7 and have the frequency response shown in Figure 9.8.

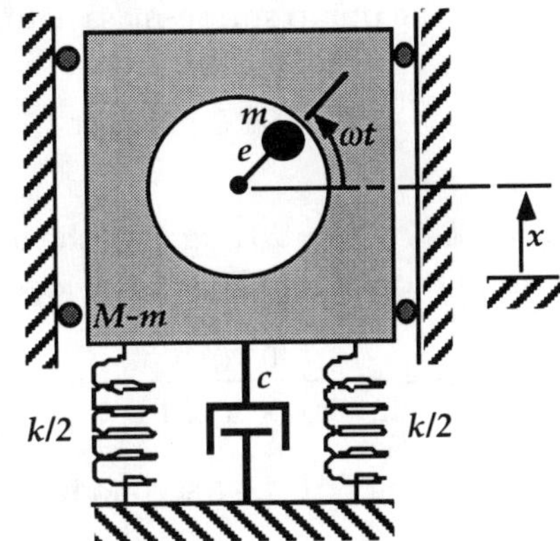

Figure 9.7 Rotating Unbalance Schematic Model

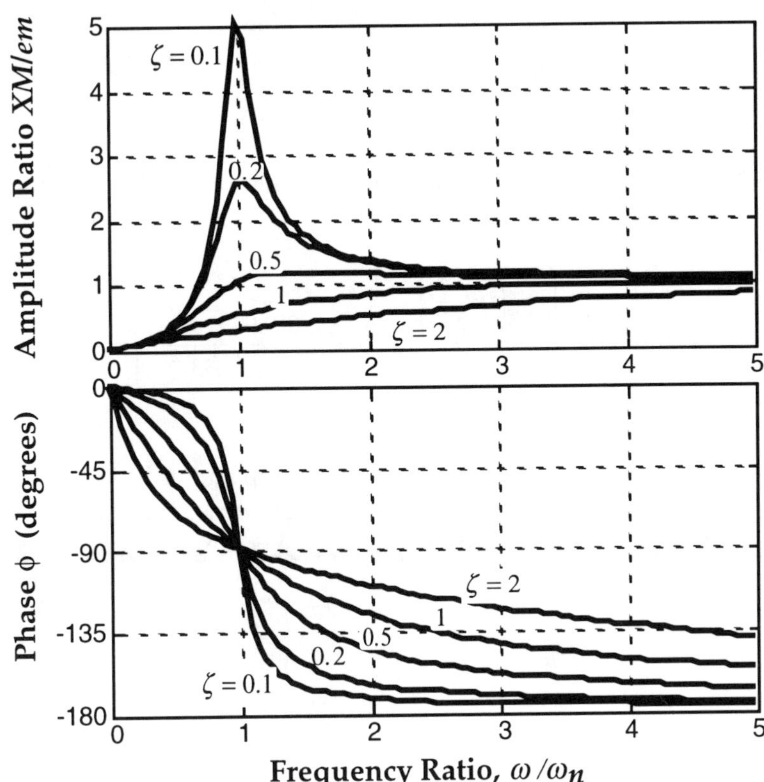

Figure 9.8 Forced Vibration Response with Rotating Unbalance

Example 9.4

A centrifugal pump and electric motor system is supported in a manner similar to Figure 9.7 and has a total weight of 2000 N. The circumferential location of the unbalance is marked on the pump rotor. At a speed of 900 rpm, a stroboscope shows the unbalance to be at the top of the rotor at the instant the structure is moving upward through its static equilibrium position and the double amplitude of the motion is 40 mm peak-to-peak. If the unbalance $me = 0.20 \text{ kg} \cdot \text{m}$, find the expected amplitude and rotor position at 1200 rpm.

Solution. The description says that the pump displacement lags the rotor position by a quarter rotation or 90° at 900 rpm. This phase lag corresponds to the phase lag at $\omega/\omega_n = 1$ in Figure 9.8. The pump spring-mass natural frequency is

$$\omega_n = \frac{900}{60} 2\pi = 94.2 \text{ rad / sec}$$

The total mass $M = W/g = 2000 / 9.81 = 204$ kg and the amplitude $X = 20$ mm. (half of the peak-to-peak amplitude). From Eq.9.4.4 at $\omega/\omega_n = 1$,

$$\left(\frac{M}{m}\right)\frac{X}{e} = \frac{1}{2\zeta} \text{ so that } \zeta = \frac{me}{2MX} = \frac{0.20}{2 \times 204 \times 20 \text{ x } 10^{-3}} = 0.025 = 2.5\%$$

At 1200 rpm $\omega/\omega_n = 1.33$ and using Eq.9.4.4 a second time, the *expected amplitude*

$$X = \frac{\left(\frac{me}{M}\right)\left(\frac{\omega}{\omega_n}\right)}{\sqrt{\left(1-\left(\frac{\omega}{\omega_n}\right)^2\right)^2 + \left(2\zeta\left(\frac{\omega}{\omega_n}\right)\right)^2}} = \frac{\left(\frac{0.2}{204}\right)(1.33)}{\sqrt{\left(1-(1.33)^2\right)^2 + \left(2(0.025)(1.33)\right)^2}} = 2.2 \times 10^{-3} \text{ m}$$

which gives an *expected double amplitude* of 4.4 mm. The *phase angle* at this speed is

$$\phi = -\tan^{-1}\left[\frac{2\zeta\left(\frac{\omega}{\omega_n}\right)}{1-\left(\frac{\omega}{\omega_n}\right)^2}\right] = -\tan^{-1}\left[\frac{2(0.025)(1.33)}{1-(1.33)^2}\right] = -175 \text{ deg}$$

The expected position of the mark is past the top and nearly back to a horizontal position on the rotor as the pump passes the equilibrium position.

9.5 Support Motion

A common source of excitation is motion of the support. Such motion is typical of suspension systems on vehicles. This support motion can cause large motion of the supported mass, especially when the frequency of the motion equals the natural frequency of a spring-mass damper system.

A one degree of freedom model of a machine with *support motion*

$$y(t) = Y\sin(\omega t) \tag{9.5.1}$$

is shown in Figure 9.9. The model is constrained to move in the x-direction only. The equation of motion for this system is

$$m\ddot{x} + c\dot{x} + kx = c\dot{y} + ky \tag{9.5.2}$$

With the definitions $\omega_n = \sqrt{k/m}$ from Eq.9.1.4 and $\zeta = c/2\sqrt{km}$, the *mass displacement* frequency response ratio

$$\frac{X}{Y} = \sqrt{\frac{k^2 + (c\omega)^2}{\left(k - m\omega^2\right)^2 + (c\omega)^2}} = \sqrt{\frac{1 + \left(2\zeta\left(\frac{\omega}{\omega_n}\right)\right)^2}{\left(1 - \left(\frac{\omega}{\omega_n}\right)^2\right)^2 + \left(2\zeta\left(\frac{\omega}{\omega_n}\right)\right)^2}} \tag{9.5.3}$$

and phase angle

$$\psi = \tan^{-1}\left[\frac{c\omega}{k}\right] - \tan^{-1}\left[\frac{c\omega}{k - m\omega^2}\right] \tag{9.5.4}$$

The mass displacement has the frequency response shown in Figure 9.10. It should be noted that the form of this result is very different from the forms of the previous frequency responses and, for all values of damping ratio ζ, the frequency response $X/Y = 1$ at $\omega/\omega_n = \sqrt{2}$.

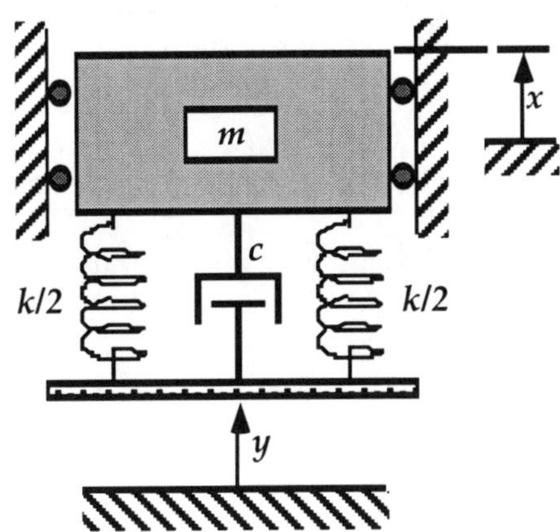

Figure 9.9 Support Motion Schematic Model

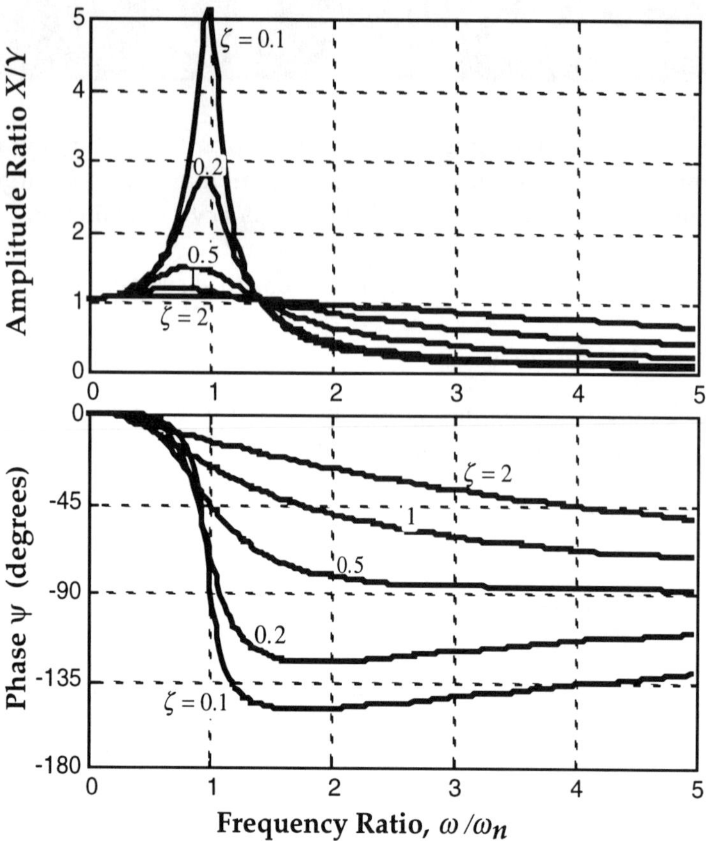

Figure 9.10 Mass Displacement Response with Support Motion

The support motion model Eq.9.5.2 also has a simple solution for the *suspension extension*

$$z = x - y \tag{9.5.5}$$

which gives the support extension model

$$m\ddot{z} + c\dot{z} + kz = -m\ddot{y} \tag{9.5.6}$$

With the definitions $\omega_n = \sqrt{k/m}$ from Eq.9.1.4 and $\zeta = c/2\sqrt{km}$, the model has the *suspension extension* frequency response ratio

$$\frac{Z}{Y} = \frac{\left(m\omega^2\right)}{\sqrt{\left(k - m\omega^2\right)^2 + \left(c\omega\right)^2}} = \frac{\left(\dfrac{\omega}{\omega_n}\right)^2}{\sqrt{\left(1 - \left(\dfrac{\omega}{\omega_n}\right)^2\right)^2 + \left(2\zeta\left(\dfrac{\omega}{\omega_n}\right)\right)^2}} \tag{9.5.7}$$

and phase angle

$$\phi = -\tan^{-1}\left(\frac{(c\omega)}{\left(k - m\omega^2\right)}\right) = -\tan^{-1}\left[2\zeta\left(\frac{\omega}{\omega_n}\right)\Big/\left(1 - \left(\frac{\omega}{\omega_n}\right)^2\right)\right] \tag{9.5.8}$$

The forms of the amplitude ratio and phase angle for *suspension extension* shown in Figure 9.11 have forms similar to those of the previous one degree of freedom systems; in fact, the normalized result is identical in form to the result for mass imbalance (Figure 9.8).

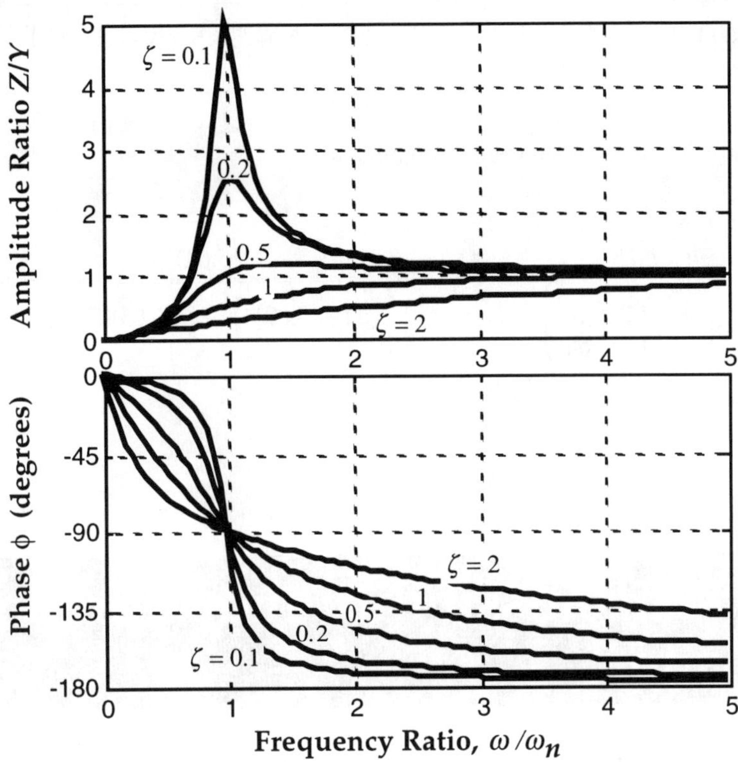

Figure 9.11 Suspension Extension Response with Support Motion

Example 9.5

A sensitive instrument of mass $m = 2$ kg is mounted on a structure vibrating at 10 Hz with an amplitude of 1 mm. If the mount stiffness $k = 0.25$ N/mm and damping ratio $\zeta = 5\%$, what is the amplitude of displacement of the instrument and the maximum mount deflection?

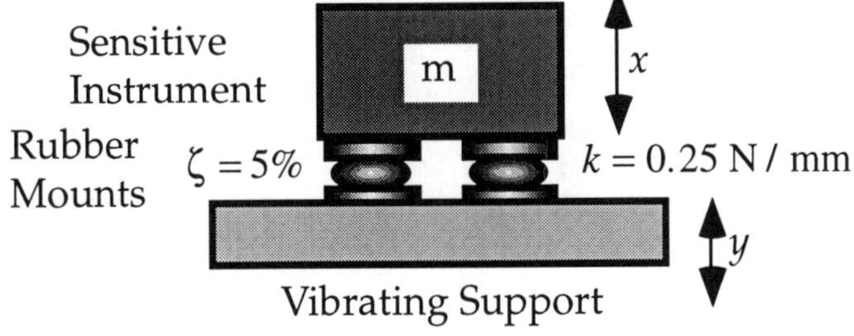

Solution. The supported mass natural frequency is

$$\omega_n = \sqrt{(250\text{ N/m})/(2\text{ kg})} = \sqrt{125}\text{ rad/sec} = 11.2\text{ rad/sec} = 1.78\text{ Hz}$$

The frequency ratio $\omega/\omega_n = 10\text{ Hz}/1.78\text{ Hz} = 5.62$.

Using Eq.9.5.3, the instrument displacement amplitude

$$X = Y\sqrt{\frac{1+\left(2\zeta\left(\frac{\omega}{\omega_n}\right)\right)^2}{\left(1-\left(\frac{\omega}{\omega_n}\right)^2\right)^2+\left(2\zeta\left(\frac{\omega}{\omega_n}\right)\right)^2}} = \left(10^{-3}\ \text{m}\right)\sqrt{\frac{1+\left(2(0.05)(5.62)\right)^2}{\left(1-(5.62)^2\right)^2+\left(2(0.05)(5.62)\right)^2}} = 0.038\ \text{mm}$$

Using Eq.9.5.7, the mount deflection amplitude

$$Z = Y\frac{\left(\frac{\omega}{\omega_n}\right)^2}{\sqrt{\left(1-\left(\frac{\omega}{\omega_n}\right)^2\right)^2+\left(2\zeta\left(\frac{\omega}{\omega_n}\right)\right)^2}} = \left(10^{-3}\ \text{m}\right)\frac{(5.62)^2}{\sqrt{\left(1-(5.62)^2\right)^2+\left(2(0.05)(5.62)\right)^2}} = 1.032\ \text{mm}$$

9.6 Transmissibility and Vibration Isolation

The problem of vibration isolation mounts and reducing the forces transmitted through these mounts receives a great deal of attention in vibration related design. The conceptual physical models shown in Fig.9.12 remain the same as models seen previously, however, we are now interested in two forms of *Transmissibilty;*

a) $T = F_t/F_0$, the ratio of force transmitted through the mount to force applied to the mass.

b) $T = |\ddot{x}|/|\ddot{y}| = X/Y$ (or $T = |\dot{x}|/|\dot{y}| = X/Y$), the ratio of mass to support acceleration (or velocity).

All forms have identical frequency responses

$$T = \frac{F_t}{F_0} = \frac{|\ddot{x}|}{|\ddot{y}|} = \frac{|\dot{x}|}{|\dot{y}|} = \frac{X}{Y} = \sqrt{\frac{k^2+(c\omega)^2}{\left(k-m\omega^2\right)^2+(c\omega)^2}} = \sqrt{\frac{1+\left(2\zeta\left(\frac{\omega}{\omega_n}\right)\right)^2}{\left(1-\left(\frac{\omega}{\omega_n}\right)^2\right)^2+\left(2\zeta\left(\frac{\omega}{\omega_n}\right)\right)^2}} \qquad \textbf{(9.6.1)}$$

For small damping $\zeta < 0.2$ and frequency $(\omega/\omega_n) > \sqrt{2}$, *Transmissibility* takes a simple form

$$T \approx \frac{1}{(\omega/\omega_n)-1} = \frac{1}{\left[\left((2\pi f)^2\Delta\right)/g\right]-1} \qquad \textbf{(9.6.2)}$$

where f is frequency in cycles per sec (Hz) and Δ is static mount deflection in consistent displacement units. The above result demonstrates the need to isolate vibration with compliant, low damping, mounts with large static deflections.

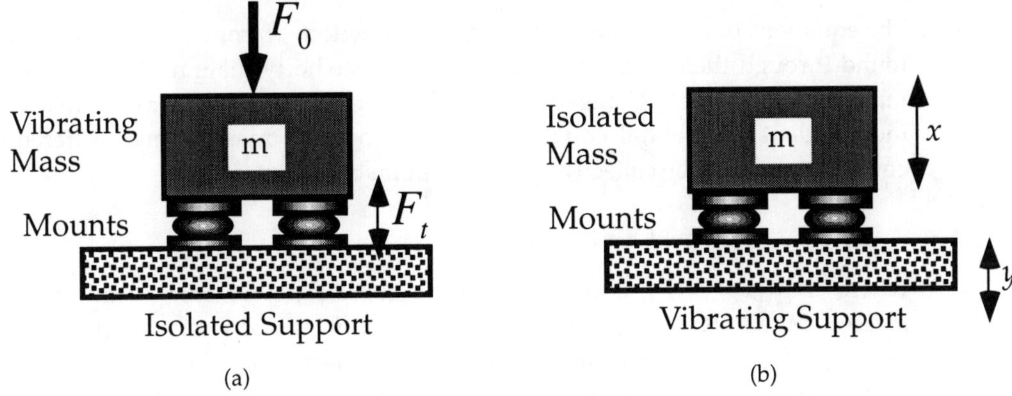

Figure 9.12 Vibration Isolation Mount Transmissibility: a) Force Transmissibility $T = |F_t/F_0|$ and b) Motion Transmissibility $T = |x/y|$

Example 9.6

A pump weighing 500 lb is mounted on a concrete floor with steel vibration isolation spring s. When placed on the spring mounts, the mounts sag 1 inch. When operated at 60 Hz with a unbalance force of 25 lb, what force is transmitted through the mounts to the floor?

Solution. Since the mounts are steel, assume the mount damping is small. Compute the mount natural frequency $\omega_n = \sqrt{g/\Delta} = \sqrt{(32.2)/(1/12)} =$ 19.7 rad/sec = 3.13 Hz and frequency ratio $(\omega/\omega_n) = (f/f_n) =$ $(60/3.13)$ $= 19.2 >> \sqrt{2}$ so the problem satisfies the simplifying assumptions.

The simplified Transmissibility Eq.9.6.2:

$$T \approx \frac{1}{(\omega/\omega_n)^2 - 1} = \frac{1}{\left[\left((2\pi f)^2 \Delta\right)/g\right] - 1} = \frac{1}{\left[\left((2\pi \times 60)^2 (1/12)\right)/32.2\right] - 1} = 2.73 \times 10^{-3}$$

so the force transmitted

$$F_t = T \times F_0 = \left(2.73 \times 10^{-3}\right) 25 \text{ lb} = 0.068 \text{ lb}$$

and the mounts are an effective vibration isolator.

9.7 Multi Degree of Freedom Systems

The equations of motion for multi-degree of freedom , lumped mass systems are found through the construction of a separate free body diagram for each of the masses in the model. Using Newton's Law $m\ddot{x}_i = f_i$ written for each mass, an independent differential equation is found for each mass. These differential equations can then be collectively written in matrix form:

$$\mathbf{M}\ddot{\mathbf{x}} + \mathbf{C}\dot{\mathbf{x}} + \mathbf{K}\mathbf{x} = \mathbf{f} \qquad (9.7.1)$$

where $\mathbf{x}^T = [x_1 \;\; x_2 \;\; \cdots \;\; x_n]$, the inertia coordinates of each of the lumped masses, $\mathbf{M}$, $\mathbf{C}$, $\mathbf{K}$ = the mass, damping and stiffness matrices, respectively, and $\mathbf{f}^T = [f_1 \;\; f_2 \;\; \cdots \;\; f_n]$, the force applied in each inertial coordinate.

The frequency response of this set of differential equations is found by a method similar to that used for the single degree of freedom forced vibration in Section 9.3. Setting

$$\mathbf{f} = \mathbf{f}_0 \sin(\omega t) \qquad (9.7.2)$$

where $\mathbf{f}_0{}^T = [F_1 \;\; F_2 \;\; \cdots \;\; F_n]$ is the vector of force amplitudes applied at frequency ω.

The response $\mathbf{x}$ is the solution of the set of linear equations

$$\left[\left(\mathbf{K} - \omega^2\mathbf{M}\right) + i\omega\mathbf{C}\right]\mathbf{x}e^{i\phi} = \mathbf{f}_0 \qquad (9.7.3)$$

where $\mathbf{x}^T = [X_1 \;\; X_2 \;\; \cdots \;\; X_n]$ is the vector of response amplitudes at frequency ω, and ϕ = is the phase angle of the response relative to the applied forces.

Example 9.7

The two-degree of freedom system shown below is a schematic of a vibration absorber system added to the oscillatory mass m_1. The vibration absorber consists of a compliantly supported mass m_2. Find an expression for the frequency response of both the left-hand mass m_1 and absorber mass m_2 for a harmonic force applied to mass m_1 with amplitude F_1 and frequency ω.

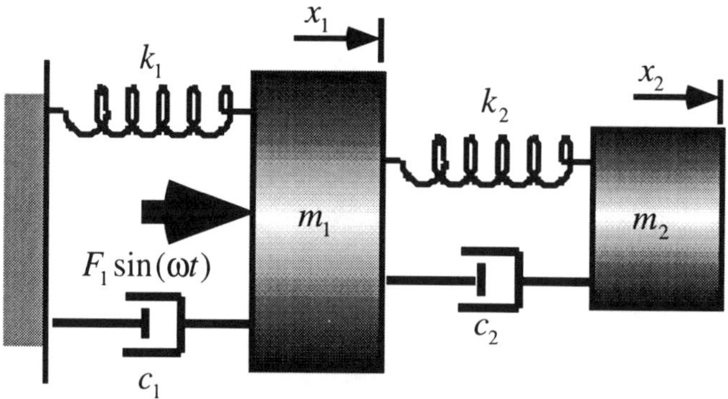

Figure 9.12 Two degree of freedom frequency response: The Vibration Absorber

Solution. The free body diagrams for the two masses follow:

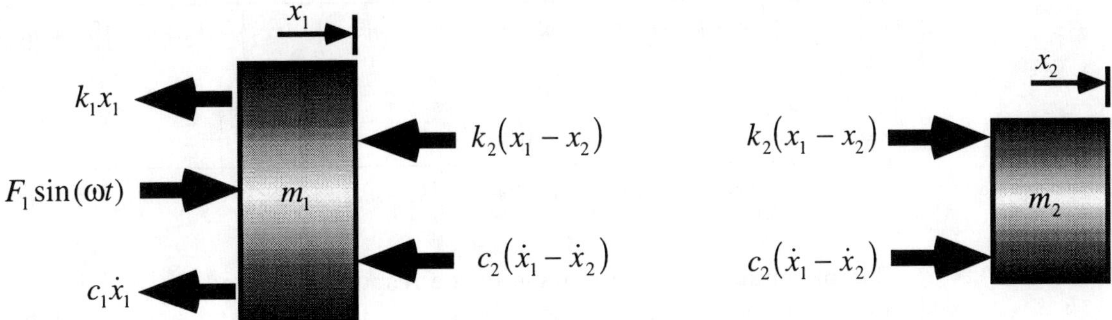

The equations of motion are

$$m_1\ddot{x}_1 = F_1\sin(\omega t) - c_1\dot{x}_1 - c_2(\dot{x}_1-\dot{x}_2) - k_1x_1 - k_2(x_1-x_2)$$

$$m_2\ddot{x}_2 = c_2(\dot{x}_1-\dot{x}_2) + k_2(x_1-x_2)$$

When written in matrix form and rearranged, the equations of motion become

$$\begin{bmatrix} m_1 & 0 \\ 0 & m_2 \end{bmatrix}\begin{bmatrix} \ddot{x}_1 \\ \ddot{x}_2 \end{bmatrix} + \begin{bmatrix} c_1+c_2 & -c_2 \\ -c_2 & c_2 \end{bmatrix}\begin{bmatrix} \dot{x}_1 \\ \dot{x}_2 \end{bmatrix} + \begin{bmatrix} k_1+k_2 & -k_2 \\ -k_2 & k_2 \end{bmatrix}\begin{bmatrix} x_1 \\ x_2 \end{bmatrix} = \begin{bmatrix} F_1 \\ 0 \end{bmatrix}\sin(\omega t)$$

Using Eq.9.7.3, the frequency response of this system of equations is

$$\begin{bmatrix} \{(k_1+k_2-m_1\omega^2)+i\omega(c_1+c_2)\} & \{(-k_2)-i\omega c_2\} \\ \{(-k_2)-i\omega c_2\} & \{(k_2-m_2\omega^2)+i\omega c_2\} \end{bmatrix}\begin{bmatrix} X_1 \\ X_2 \end{bmatrix} e^{i\phi} = \begin{bmatrix} F_1 \\ 0 \end{bmatrix}$$

Use Kramer's rule to solve these matrix algebraic equations for the responses:

$$X_1e^{i\phi} = \frac{\begin{Vmatrix} F_1 & \{(-k_2)-i\omega c_2\} \\ 0 & \{(k_2-m_2\omega^2)+i\omega c_2\} \end{Vmatrix}}{\begin{Vmatrix} \{(k_1+k_2-m_1\omega^2)+i\omega(c_1+c_2)\} & \{(-k_2)-i\omega c_2\} \\ \{(-k_2)-i\omega c_2\} & \{(k_2-m_2\omega^2)+i\omega c_2\} \end{Vmatrix}}$$

$$X_2e^{i\phi} = \frac{\begin{Vmatrix} \{(k_1+k_2-m_1\omega^2)+i\omega(c_1+c_2)\} & F_1 \\ \{(-k_2)-i\omega c_2\} & 0 \end{Vmatrix}}{\begin{Vmatrix} \{(k_1+k_2-m_1\omega^2)+i\omega(c_1+c_2)\} & \{(-k_2)-i\omega c_2\} \\ \{(-k_2)-i\omega c_2\} & \{(k_2-m_2\omega^2)+i\omega c_2\} \end{Vmatrix}}$$

Evaluating the numerator and denominator determinants gives the responses:

$$\left(\frac{X_1}{F_1}\right)e^{i\phi} = \frac{\left\{\left(k_2 - m_2\omega^2\right) + i\omega c_2\right\}}{\left[\left\{\left(k_1 + k_2 - m_1\omega^2\right) + i\omega\left(c_1 + c_2\right)\right\}\left\{\left(k_2 - m_2\omega^2\right) + i\omega c_2\right\} - \left\{k_2 + i\omega c_2\right\}\left\{k_2 + i\omega c_2\right\}\right]}$$

$$\left(\frac{X_2}{F_1}\right)e^{i\phi} = \frac{\left\{\left(k_1 + k_2 - m_1\omega^2\right) + i\omega\left(c_1 + c_2\right)\right\}}{\left[\left\{\left(k_1 + k_2 - m_1\omega^2\right) + i\omega\left(c_1 + c_2\right)\right\}\left\{\left(k_2 - m_2\omega^2\right) + i\omega c_2\right\} - \left\{k_2 + i\omega c_2\right\}\left\{k_2 + i\omega c_2\right\}\right]}$$

The magnitude and phase angle of the responses X_1 and X_2 are simply the magnitude and phase angle associated with the right hand side of each of the above expressions. The computation of magnitudes and phase angle computation was discussed in Section 9.3.

9.8 Vibration Absorbers

The addition of a compliantly supported mass to a harmonically excited vibratory system can substantially reduce the forced oscillation of the excited mass. Consider the vibration absorber system shown in the previous example (Fig. 9.12). The frequency response of the forced mass X_1 will be identically zero for any frequency where

$$\left(k_2 - m_2\omega^2\right) + i\omega c_2 = 0 \tag{9.8.1}$$

This condition is satisfied whenever the forcing frequency

$$\omega = \sqrt{k_2/m_2} \tag{9.8.2}$$

and absorber damping

$$c_2 = 0 \tag{9.8.3}$$

The first condition shows that the absorber design is independent of the parameters of the excited mass and that any effective absorber must have very small (ideally zero) damping. When the above conditions (9.8.2 and 9.8.3) are satisfied, the absorber mass has response magnitude dependent on the absorber stiffness:

$$\frac{X_2}{F_1} = \frac{-1}{k_2} \tag{9.8.4}$$

Example 9.8

A pump has a vertical shaking force of 1000 N applied at 30 cycles per second due to rotating imbalance. Design a vibration absorber that will stop any pump shaking with an absorber mass displacement limited to 1 cm.

Solution: Using Eq.9.8.4

$$\frac{X_2}{F_1} = \frac{-1}{k_2} \Rightarrow k_2 = \left\| \frac{F_1}{X_2} \right\| = \left\| \frac{1000 \text{ N}}{1 \times 10^{-3} \text{ m}} \right\| = 10^6 \text{ N/m}$$

Use Eq.9.8.2:

$$\omega = \sqrt{k_2/m_2} \Rightarrow m_2 = k_2/\omega^2 = \left(10^6\right)/\left[2\pi(30)\right]^2 = 28.1 \text{ kg}$$

Note that the above result is independent of the pump weight and only depends on the magnitude of the force to be absorbed and the force's harmonic frequency.

References

D.J. Inman, *Vibration with Control, Measurement and Stability,* Prentice Hall, 1989.

W.T Thomson, *Theory of Vibration with Applications,* Prentice Hall, 4th ed. 1993.

Practice Problems (Essay)

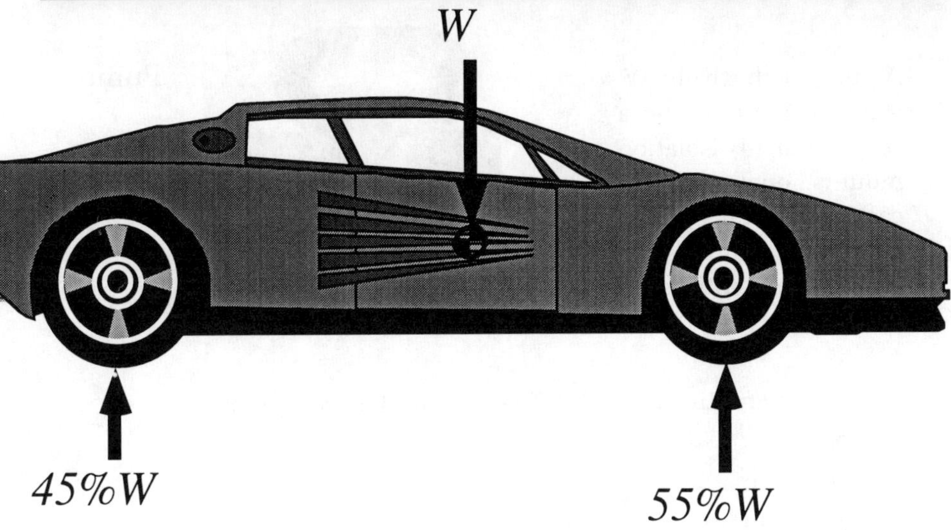

9.1 A 3200-lb car is to have a natural frequency of 1.5 Hz in heave (vertical motion). The wheelbase of the vehicle is 108 inches. The vehicle has 55% of its weight on the front wheels and 45% of its weight on the rear wheels.

a) What front and rear suspension stiffnesses are required.

b) What suspension spring preload deflections are required at rest at static equilibrium?

9.2 A vibration absorber is to be designed to absorb the vertical 100 N-shaking force from a pump which operates at 30 cycles per second. The absorber mass is constrained by the case to a 1 cm maximum deflection. Design the absorber stiffness and mass and show the analysis required to compute them.

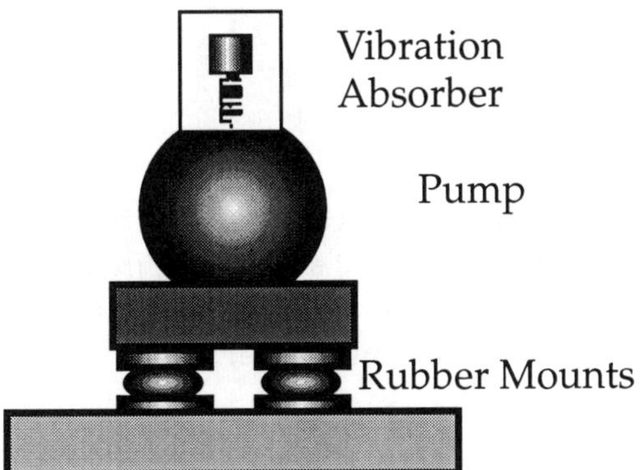

Practice Problems (Multiple Choice)

A pump with weight W = 500 lb is placed on a set of four (4) isolation mounts, one placed at each corner of the pump. Each of the four isolation mounts has a stiffness of 100 lb/in.

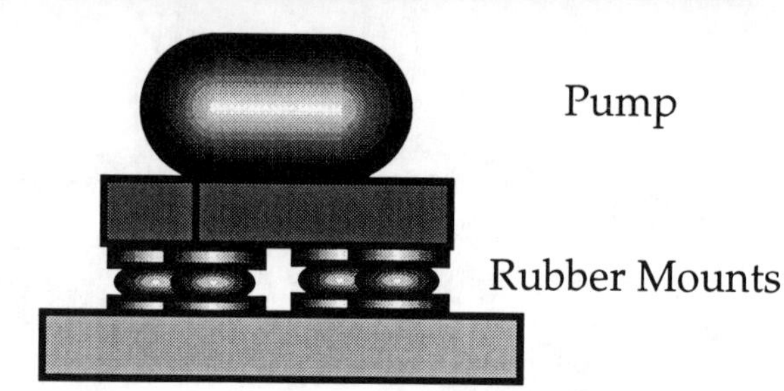

9.3 What is the effective vertical stiffness of the (4) pump mounts?

a) 20 lb/in
b) 100 lb/in
c) 125 lb/in
d) 250 lb/in
e) 400 lb/in

9.4 What is the vertical mount deflection when the pump is initially placed on the mounts?

a) 5 in
b) 2.5 in
c) 1.25 in
d) 1.0 in
e) 0.2 in

9.5 What is the vertical natural frequency ω_n of the pump on its mounts?

a) 2.80 rad/sec
b) 4.4 0 rad/sec
c) 0.70 Hz.
d) 2.80 Hz,
e) 17.6 Hz.

9.6 The pump includes an imbalance mass m with a weight of 1.0 lb at a eccentricity radius of 5 inches. At a rotational frequency much larger than the natural frequency of the pump on its mounts, what vertical motion of the pump will result from the imbalance?

a) 0.0 in
b) 0.01 in
c) 0.02 in
d) 0.04 in
e) 0.25 in

9.7 The pump has a vertical resonant amplitude $X = 1.0$ in, what is the damping ratio of the pump on its mounts?

a) 0.0 %
b) 1.0 %
c) 5.0 %
d) 10. %
e) 25. %

9.8 At a pump operating speed of 30 cycles per second, what vertical pump displacement would be expected?

a) 0.01 in
b) 0.10 in
c) 0.01 ft
d) 0.05 ft
e) 0.10 ft

9.9 At a pump operating speed of 30 cycles per second, what is the magnitude of the oscillating force due to the imbalance that is transmitted through the four mounts to the pump supporting surface?

a) 0.25 lb
b) 1.0 lb
c) 4.0 lb
d) 125 lb
e) 500 lb

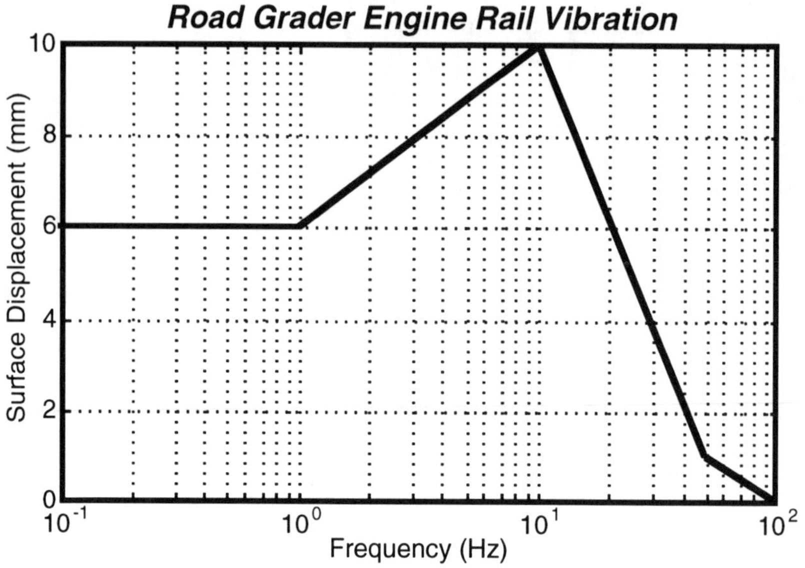

Road Grader Engine Rail Vibration

Surface Displacement (mm)

Frequency (Hz)

The engine controller for a piece of earth moving equipment is to be supported on a surface with surface vibration amplitude spectrum shown above. Answer Questions 9.10 – 9.19 below.

9.10 At what frequency does the engine rail vibration have the largest displacement amplitude?

a) 0.1 cycles/sec
b) 1.0 cycle/sec
c) 10. cycles/sec
d) 100. cycles/sec
e) none of the above

9.11 At what frequency does the largest velocity amplitude due to engine rail vibration occur?

a) 0.1 cycles/sec
b) 1.0 cycle/sec
c) 10. cycles/sec
d) 50. cycles/sec
e) 100. cycles/sec

9.12 What is the largest velocity amplitude due to engine rail vibration ?

a) 0.6 mm/sec
b) 6.0 mm/sec
c) 10. mm/sec
d) 100. mm/sec
e) 628. mm/sec

9.13 At what frequency does the largest acceleration amplitude due to engine rail vibration occur?

a) 0.1 cycles/sec
b) 1.0 cycle/sec
c) 10. cycles/sec
d) 50 cycles/sec
e) 100 cycles/sec

9.14 What is the largest acceleration amplitude due to engine rail vibration?

a) 10.0 mm/sec^2
b) 39.5 m/sec^2
c) 98.7 m/sec^2
d) 39.5×10^3 m/sec^2
e) 98.7×10^3 m/sec^2

9.15 The engine controller has a mass $m = 1$ kg, what isolator stiffness k will provide the best isolation and limit static deflection to 5 mm?

a) $k =$ 5.0 N/mm
b) $k = 1.96 \times 10^3$ N/mm
c) $k =$ 200 N/mm
d) $k = 1.96 \times 10^3$ N/m
e) $k =$ 2.00 N/m

9.16 When designed with isolator stiffness $k =$ 2 N/mm what is the natural frequency ω_n of the engine controller on the isolator ?

a) 0.707 rad/sec
b) 7.12 rad/sec
c) 44.7 rad/sec
d) 0.707 Hz
e) 44.7 Hz

9.17 When designed with isolator stiffness $k =$ 2 N/mm and a damping ratio $\zeta = 0.1$, what is the transmissibility T of the engine controller isolator at 50 Hz.?

a) 1.0
b) 0.5
c) 0.05
d) 0.036
e) 0.001

9.18 What is the maximum stiffness allowed such that the constraint $(\omega/\omega_n) > \sqrt{2}$ for good isolation occurs at $f = 50$ Hz.

a) 49.3 N/mm
b) 221. N/mm
c) 49.3 N/m
d) 221. N/m
e) None of the above

9.19 To maximize the isolator performance by minimizing Transmissibility T for $(\omega/\omega_n) > \sqrt{2}$,

a) minimize both stiffness k and damping ratio ζ
b) maximize both stiffness k and damping ratio ζ
c) minimize stiffness k and maximize damping ratio ζ
d) maximize stiffness k and minimize damping ratio ζ
e) reduce the controller mass

Practice Problem Solutions

9.1

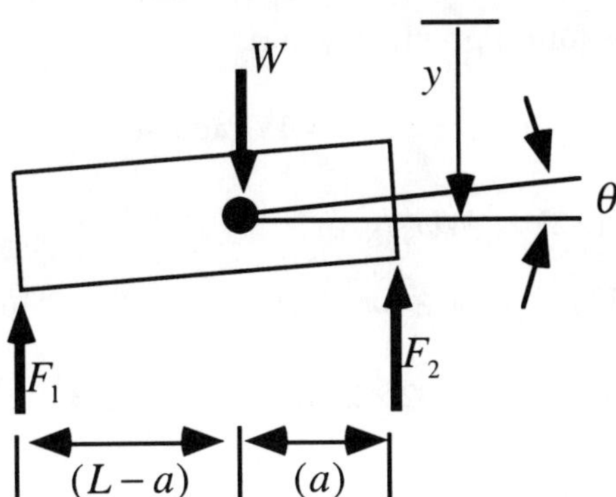

a) First sketch the free-body diagram. Here the wheelbase L determines the distance between the suspension forces while the distance a determines the position of the vehicle Center-of-Gravity (CG is traditionally measured back from the front of the vehicle). Here the vertical motion y of the CG combined with rotation θ yields spring forces $F_1 = k_1\Delta_1 = k_1(y+(L-a)\sin\theta)$ and $F_2 = k_2\Delta_2 = k_2(y - a\sin\theta)$.

Now find the vertical and rotational equations of motion:

$$M\ddot{y} = W - F_1 - F_2 = W - k_1(y+(L-a)\sin\theta) - k_2(y - a\sin\theta)$$
$$J\ddot{\theta} = F_1(L-a) - F_2 a = k_1(y+(L-a)\sin\theta)(L-a) - k_2(y-a\sin\theta)a$$

For the springs to provide no net rotation on the vehicle, we require $\ddot{\theta} = \dot{\theta} = \theta = 0$. When these conditions are substituted into the second equation of motion we derive

$$0 = k_1 y(L-a) - k_2 y a \textbf{ or } k_1 = \frac{a}{(L-a)}k_2$$

From the same rotation equation we get the condition

$$0 = F_1(L-a) - F_2 a = 0.45W(L-a) - 0.55Wa$$

which yields $a = 0.45L$, so that

$$k_1 = \frac{a}{(L-a)}k_2 = \frac{0.45L}{(L-0.45L)}k_2 = \frac{0.45L}{0.55L}k_2 = 0.82k_2$$

Substituting the constraint on rotation and the relation between stiffnesses into the vertical equation of motion yields

$$M\ddot{y} = W - k_1 y - k_2 y = W - (k_1 + k_2)y$$

and rearranging

$$M\ddot{y} + (k_1 + k_2)y = W$$

This equation of motion gives the natural frequency

$$\omega_n = \sqrt{(k_1 + k_2)/M} = \sqrt{(k_1 + k_2)g/W}$$

where $k_1 = 0.82k_2$ was found previously and

$$\omega_n = 2\pi \times 2 = 4\pi \text{ rad / sec}$$

Solving for k_2 yields $1.82k_2 = W\omega_n^2/g$ or

$$k_2 = W\omega_n^2/1.82g = \left(3200 \times 4\pi^2\right)/(1.82 \times 32.2) = 2.16 \times 10^3 \text{ lb / ft}$$

$$k_1 = 0.82k_2 = 0.82 \times 2.16 \times 10^3 = 1.76 \times 10^3 \text{ lb/ft}$$

b) Because we have constrained $\theta = 0$ in the solutions for k_1 and k_2, the deflections of both springs will be the same $\Delta = \Delta_1 = \Delta_2$. Solving for deflection Δ_2 of spring k_2 supporting 55% of the vehicle weight

$$\Delta = \Delta_1 = \Delta_2 = 0.55W/k_2 = 0.55 \times 3200/2.16 \times 10^3 = 0.81 \text{ ft} = 9.8 \text{ in.}$$

9.2
Using Section 9.8, the absorber natural frequency must match the frequency of the pump shaking force so that

$$\omega_n = 2\pi \times 30 = 188 \text{ rad / sec}$$

Using Eq.9.8.4, the absorber stiffness is found from the displacement constraint

$$k_2 = F_2/X_2 = 100/0.1 = 1000 \text{ N / m}$$

Using Eq.9.8.2, the absorber mass should be

$$m_2 = \omega_n^2/k_2 = 188^2/1000 = 35.5 \text{ kg}$$

9.3 **e)** The four isolator stiffness are in parallel and each provide 1/4 of the mounting system's effective stiffness

$$k = 4 \times 100 \text{ lb / in} = 400 \text{ lb / in}$$

9.4 **b)** The isolator each support 1/4 of the weight of the pump and

$$\Delta = (500 \text{ lb}/4)/\ (100 \text{ lb/in}) = 1.25 \text{ in}$$

9.5 **d)** Each isolator provides 1/4 of the total stiffness of the mounts applied to the total mass of the pump. The natural frequency

$$\omega_n = \sqrt{(4k)/(W/g)} = \sqrt{(4 \times 100 \times 12)/(500/32.2)} = 17.6 \text{ rad / sec} = 2.80 \text{ Hz}$$

9.6 **b)** The frequency response of the pump (9.4.2) is dominated by the inertia of the pump at high rotational frequency

$$X = \left(me\omega^2\right)\Big/\sqrt{\left(k - M\omega^2\right)^2 + (c\omega)^2} \approx me/M = \left((mg)/Mg\right)e = (1.0/500.)5 \text{ in} = 0.01 \text{ in}$$

9.7 **c)** The non-dimensional frequency response (Eq.9.4.4) of the pump at its natural frequency $(\omega/\omega_n) = 1$

$$\left(\frac{M}{m}\right)\frac{X}{e} = \frac{\left(\frac{\omega}{\omega_n}\right)}{\sqrt{\left(1-\left(\frac{\omega}{\omega_n}\right)^2\right)^2 + \left(2\zeta\left(\frac{\omega}{\omega_n}\right)\right)^2}} = \frac{1}{\sqrt{\left(1-1^2\right)^2 + (2\zeta)^2}} = \frac{1}{2\zeta} \text{ so that}$$

$$\zeta = \frac{1}{2}\left(\frac{m}{M}\right)\frac{e}{X} = \frac{1}{2}\left(\frac{1.0}{500}\right)\frac{5.0 \text{ in}}{1.0 \text{ in}} = \frac{1}{200} = 0.05 = 5\%$$

9.8 **a)** The frequency response (Eq.9.4.2) of the pump at the operating rotational frequency is

$$\omega = 2\pi f = 2\pi(30 \text{ cps}) = 188.5 \text{ rad/sec}$$

with damping coefficient (Eq.9.2.2)

$$c = 2\zeta\sqrt{km} = 2\times 0.05\sqrt{(12\times 400)\times(500/32.2)} = 1.76 \text{ lb}\cdot\text{sec}^2 / \text{ft}$$

is

$$X = \left(me\omega^2\right)\Big/\sqrt{\left(k - M\omega^2\right)^2 + (c\omega)^2}$$

$$= \left[(1/32.2)(5/12)188.5^2\right]\Big/\sqrt{\left[(4\times 100) - (500/32.2)188.5^2\right]^2 + (1.76\times 188.5)^2}$$

$$= [459.8]\Big/\left[551\times 10^3\right] = 0.834\times 10^{-3} \text{ ft} = 0.01 \text{ in}$$

Note that the effect of damping is negligible at a damping ratio of 5%.

9.9 **c)** Assuming the surface does not move, the transmitted force

$$F = kx + c\dot{x} = kX + j\omega cX = (k + j\omega c)X$$

$$= [(400)12 + j(188.5)(1.76)](0.01/12)$$

$$= [(4800) + j(332)]\left(0.834\times 10^{-3}\right)$$

$$= [4.00 + j0.277] = \sqrt{4.00^2 + 0.277^2} \text{ lb at } \theta = \tan^{-1}(0.277/4)$$

$$= 4.01 \text{ lb at } \theta = 3.96 \text{ deg}$$

9.10 **c)** From the graph, the displacement amplitude varies from 1 mm at 0.1 cyc/sec to 10 mm at 10 cyc/sec.

9.11 **c)** Velocity amplitude is proportional to the product of frequency and displacement. From the graph, the largest velocity must therefore occur at $f =$ 10 Hz = 10 cyc/sec by inspection.

9.12 **e)** The velocity magnitudes ($V = \omega X = 2\pi f X$) at the corner points in the graph are

V = 3.77, 37.7,628. and 314. mm/sec at

frequencies f = 0.1, 1.0, 10, and 50 Hz, respectively.

The largest magnitude is V = 628 mm/sec at f = 10 Hz.

9.13 **d)** Acceleration amplitude is proportional to the product of frequency-squared and displacement. From the graph, the largest velocity must therefore occur at f = 50 Hz = 50 cyc/sec by inspection.

9.14 **c)** The acceleration magnitudes ($A = \omega^2 X = 4\pi^2 f^2 X$) at the corner points in the graph are

$$2.00,\ 237,\ 39.5\times10^3,\text{and } 98.7\times10^3\,\text{mm}/\text{sec}^2\,\text{at}$$

frequencies f = 0.1, 1.0, 10, and 50 Hz, respectively.

The largest magnitude is A = 98.7 m/sec at f = 50 Hz. Note that the linear scale indicates a zero displacement amplitude at 100 Hz, however, even a very small displacement amplitude at higher frequencies can contribute significant acceleration.

9.15 **d)** The stiffness required is the smallest value consistent with the maximum displacement constraint,

$$k = F / x_{\max} = (mg) / x_{\max} = (1\times 9.80) / \left(5.0\times10^{-3}\right) = 1.96\times10^3\ \text{N}/\text{m}$$

9.16 **c)** The natural frequency (Eq.9.1.5) of the isolated engine controller is

$$\omega_n = \sqrt{k/m} = \sqrt{(2\ \text{N/mm})(1{,}000\ \text{mm/m})/(1\ \text{kg})}$$
$$= 44.7\ \text{rad/sec} = 7.12\ \text{cyc/sec}$$

9.17 **d)** The isolator transmissibility T is found using Eq.9.6.1 with ω_n = 44.7 rad/sec = 7.12 Hz. The frequency ratio in this case is

$$f/f_n = 50/7.12 = 7.02$$

9.18 **c)** The relationship, $(\omega/\omega_n) > \sqrt{2}$ yields $\omega_n^2 = \left[(50\ \text{cyc/sec})\times(2\pi\ \text{rad/cyc}\ )\right]^2/2 = 49.3\times10^3\ \text{rad}^2/\text{sec}^2$ and solving for the stiffness using (9.1.5) gives $k = \omega_n^2 m = \left(49.3\times10^3\ \text{rad}^2/\text{sec}^2\right)(1\ \text{kg}) = 49.3\times10^3\ \text{N/m} = 49.3\ \text{N/mm}$

9.19 **a)** The relationship (Eq.9.1.5)for Transmissibility T shows that transmissibility is reduced by increasing frequency ratio(ω/ω_n) and decreasing ζ for $(\omega/\omega_n) > \sqrt{2}$. The frequency ratio is increased by decreasing stiffness k to decrease ω_n. The answer is minimize both stiffness and damping ratio ζ.

Machine Design

by Clark J. Radcliffe and Charles W. Radcliffe

10.1 Bolts and Fasteners

Thread Standards and Definitions

Bolted joints are important when the ability to disassemble is required of a design. The primary design issues are joint stiffness, fastener strength, preload torque and fatigue loading.

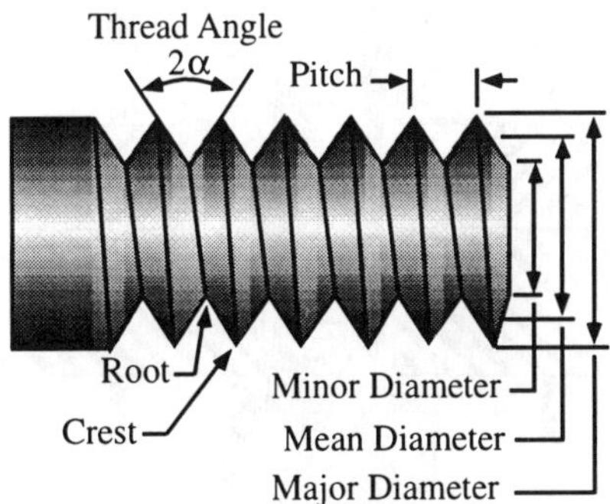

Figure 10.1 Screw thread terminology

Referring to Figure 10.1, the *pitch* p is the distance between adjacent thread forms. The pitch in U.S. units is the reciprocal of the number of thread forms per inch. The *Major diameter* d is the largest diameter of the screw thread. The *Minor diameter* d_r is the smallest diameter of the screw thread. The lead l (not shown) is the distance the nut moves along the thread axis in one rotation. Only for a single thread does the lead equal the pitch. Standardized products such as screws, bolts and nuts have single threads. A *multiple-threaded* product has two or more threads cut beside each other. All threads are cut according to the right-hand rule unless otherwise noted.

The American National (Unified) thread standard is approved for use in the United States and Great Britain and has a thread angle $2\alpha = 60^o$. The crests of the thread may be flat or rounded. The metric M and MJ profiles also use a thread angle $2\alpha = 60^o$. Standard thread dimensions are given in Table 10.1 and Table 10.2.

The Unified National and International Standards Organization (ISO) standards define tolerance ranges for both internal and external threads to control their fit: Class 1 "hardware quality" inexpensive fasteners for casual use, Class 2 "general machine" design fasteners, and Class 3 "highest precision" for closer fits. A letter designator indicates external (A) or internal (B) thread. A typical UNS thread specification is

1/4-20 UNC-2A

which defines a 0.250" diameter thread with 20 threads per inch, Class 2 fit and external thread. An example metric thread specification is

$$M8 \times 1.25$$

which defines an 8 mm diameter thread with a pitch of 1.25 mm in the ISO coarse series.

Tensile Threaded Fastener Connections

When a bolted or screw connection is made between two components which can be disassembled, the connection must be carefully designed and assembled by a competent, trained technician. The bolted joint shown in Figure 10.2 is typical of such a connection design. Note that the threads extend into the body of the connection. The *grip* of the connection is the total thickness of the clamped material and any washers. In Figure 10.2, the grip includes two washers.

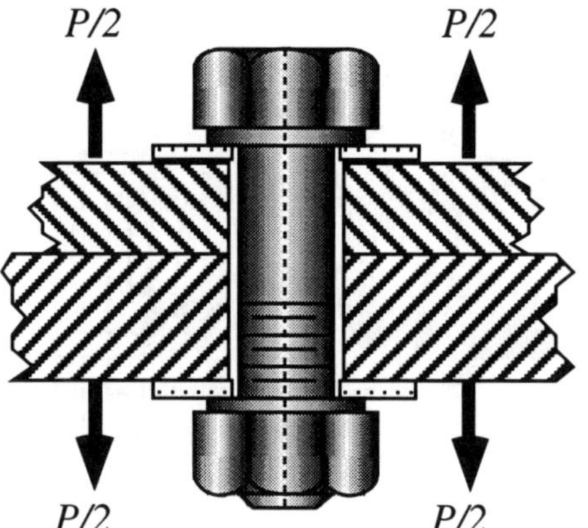

Figure 10.2 A Typical Threaded Fastener Joint

Table 10.1 American National Thread Standard

Size	Nom. Major Diameter d (in)	Coarse Series - UNC: Threads per inch N	Coarse Series - UNC: Tensile Stress Area A_t (in^2)	Coarse Series - UNC: Minor Dia. Area A_r (in^2)	Fine Series - UNF: Threads per inch N	Fine Series - UNF: Tensile Stress Area A_t ((in^2)	Fine Series - UNF: Minor Dia. Area A_r (in^2)
0	0.0600				80	0.001 80	0.001 51
1	0.0730	64	0.002 63	0.002 18	72	0.002 78	0.002 37
2	0.0860	56	0.003 70	0.003 10	64	0.003 94	0.003 39
3	0.0990	48	0.004 87	0.004 06	56	0.005 23	0.004 51
4	0.1120	40	0.006 04	0.004 96	48	0.006 61	0.005 66
5	0.1250	40	0.007 96	0.006 72	44	0.008 80	0.007 16
6	0.1380	32	0.009 09	0.007 45	40	0.010 15	0.008 74
7	0.1640	32	0.014 0	0.011 96	36	0.014 74	0.012 85
10	0.1900	24	0.017 5	0.014 50	32	0.020 0	0.017 5
12	0.2160	24	0.024 2	0.020 6	28	0.025 8	0.022 6
1/4	0.2500	20	0.031 8	0.026 9	28	0.036 4	0.032 6
5/16	0.3125	18	0.052 4	0.045 4	24	0.058 0	0.052 4
3/8	0.3750	16	0.077 5	0.067 8	24	0.087 8	0.080 9
7/16	0.4375	14	0.106 3	0.093 3	20	0.118 7	0.109 0
1/2	0.5000	13	0.141 9	0.125 7	20	0.159 9	0.148 6
9/16	0.5625	12	0.182	0.162	18	0.203	0.189
5/8	0.6250	11	0.226	0.202	18	0.256	0.240
3/4	0.7500	10	0.334	0.302	16	0.373	0.351
7/8	0.8750	9	0.462	0.419	14	0.509	0.480
1	1.0000	8	0.606	0.551	12	0.663	0.625
1 - 1/4	1.2500	7	0.969	0.890	12	1.073	1.024
1-1/2	1.5000	6	1.405	1.294	12	1.581	1.521

Note: This table was compiled from ANSI B1.1-1974 (Shigley and Mischke, 1989). The minor diameter was found from $d_r = d - 1.299038p$ and the pitch diameter from $d_m = d - 0.649519p$. The mean of the pitch and minor diameters was used to compute the tensile stress area.

Table 10.2 Metric Thread Standard

Nom. Major Diameter d (mm)	Coarse Pitch Series: Pitch p (mm)	Coarse Pitch Series: Tensile Stress Area A_t (mm^2)	Coarse Pitch Series: Minor Dia. Area A_r (mm^2)	Fine Pitch Series: Pitch p (mm)	Fine Pitch Series: Tensile Stress Area A_t (mm^2)	Fine Pitch Series: Minor Dia. Area A_r (mm^2)
1.6	0.35	1.27	1.07			
2	0.40	2.07	1.79			
2.5	0.45	3.39	2.98			
3	0.5	5.03	4.47			
3.5	0.6	6.78	6.00			
4.	0.7	8.78	7.75			
5	0.8	14.2	12.7			
6	1	20.1	17.9			
8	1.25	36.6	32.8	1	39.2	36.0
10	1.5	58.0	52.3	1.25	61.2	56.3
12	1.75	84.3	76.3	1.25	92.1	86.0
14	2	115	104	1.5	125	116
16	2	157	144	1.5	167	157
20	2.5	245	225	1.5	272	259
24	3	353	324	2	384	365
30	3.5	561	519	2	621	596
36	4	817	759	2	915	884
42	4.5	1120	1050	2	1260	1230
48	5	1470	1380	2	1670	1630
56	5.5	2030	1910	2	2300	2250
64	6	2680	2520	2	3030	2980
72	6	3460	3280	2	3860	3800
80	6	4340	4140	1.5	4850	4800
90	6	5590	5360	2	6100	6020
100	6	6990	6740	2	7560	7470
110				2	9180	9080

Note: This table was compiled from ANSI B1.1-1974 and B18.3.1-1978 (Shigley and Mischke, 1989). The minor diameter was found from $d_r = d$ -1.226 869p and the pitch diameter from $d_m = d$ -0.649 519p. The mean of the pitch and minor diameters was used to compute the tensile stress area.

The stiffness of a threaded joint k_J is the parallel combination of the stiffness of the fastener k_B and the stiffness of the material in the grip k_G:

$$k_J = k_B + k_G \tag{10.1.1}$$

The stiffness of the fastener is the series combination of the stiffness of the unthreaded shank k_u and the stiffness of the threaded shank k_t outside the face of the nut:

$$\frac{1}{k_B} = \frac{1}{k_u} + \frac{1}{k_t} = \frac{k_u k_t}{k_u + k_t} \tag{10.1.2}$$

where the unthreaded shank stiffness $k_u = A_u E / L_u$ is dependent on the nominal diameter's area $A_u = \pi d^2/4$ and unthreaded shank length L_u. The threaded shank stiffness $k_t = A_t E / L_t^*$ is dependent on the tensile stress area of the threaded shank area A_t (Tables 10.1 and 10.2) and exposed length of threaded shank L_t^*. Note this is the actual exposed length of thread not the total threaded length shown in Figure 10.3.

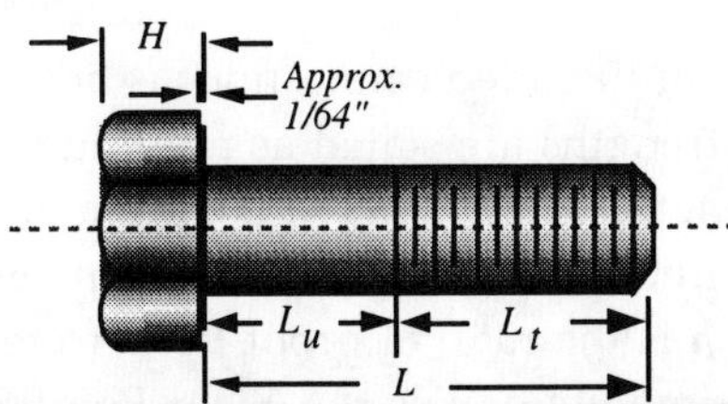

UNC and UNF	Metric
$L_u = L - L_t$	$L_u = L - L_t$
$L_t = \begin{cases} 2d + \frac{1}{4}\text{ in} & L \le 6\text{ in} \\ 2d + \frac{1}{2}\text{ in} & L > 6\text{ in} \end{cases}$	$L_t = \begin{cases} 2d + 6\text{ mm} & L \le 125\text{ mm} \\ 2d + 12\text{ mm} & 125 \le L \le 200 \\ 2d + 25\text{ mm} & L > 200\text{ mm} \end{cases}$

Figure 10.3 A Hexagon Head Bolt

The stiffness of the material in the grip is the series combination of the stiffnesses of the components in the grip. This is a difficult computation in general, however, when all components of the grip are the same material, Shigley and Mischke (1989, p. 340) recommend

$$k_m = \frac{0.577\pi Ed}{2\ln\left(5\dfrac{0.577L + 0.5d}{0.577L + 2.5d}\right)} \tag{10.1.3}$$

Note that the components of the grip may include both relatively stiff washers with relatively compliant gasket material. When the compliance of the gasket is important, the reader is referred to Shigley and Mischke (1989) or other design texts for additional details of the grip stiffness computation.

Grip stiffness is important because it reduces the deviation in the load taken by the bolt when an external load P is applied to the bolted connection (Fig. 10.2). When a bolt is preloaded with tensile force F_i the resultant bolt load is

$$F_b = \frac{k_b}{k_b + k_m}P + F_i \tag{10.1.4}$$

and the resultant force in the grip members is

$$F_m = \frac{k_m}{k_b + k_m}P - F_i \tag{10.1.5}$$

The above relationships are only valid as long as some clamping force remains in the grip, $F_m < 0$, or

$$F_i > \frac{k_m}{k_b + k_m}P \tag{10.1.6}$$

Typically values of grip member stiffness k_m are much larger than the stiffness of the bolt shank k_b and substantially reduce the stress fluctuation in the bolt. Examination of a Goodman Diagram shows reduction in fluctuating load will improve the fatigue strength of the connection. Bolt preload also provides compression of the joint so that friction can absorb a large portion of shear loads applied through the members, again increasing connection strength.

Bolt Strength

A key factor in the design of threaded connections is bolt strength. In specifications, threaded fastener strengths are stated as minimum proof strength (load), and minimum tensile strength. The *Proof Load* is the maximum load (force) that a fastener can withstand without acquiring a permanent set when the load is removed. The *Proof Strength* is the ratio of proof Load to tensile stress area. The Proof Strength S_P approximately equals the Yield Point Strength and is about 90% of the 0.2% offset Yield Strength of the fastener material. Strength specifications are given in Table 10.3 for SAE inch sizes and in Table 10.4 for Metric Sizes.

Table 10.3 SAE Bolt Strength Specifications

SAE Grade No.	Size Range (in)	Min. Proof Strength (kpsi)	Min Tensile Strength (kpsi)	Min Yield Strength (kpsi)	Material	Head Marking
1	$\frac{1}{4}-1\frac{1}{2}$	33	60	36	Low or Medium Carbon	
2	$\frac{1}{4}-\frac{3}{4}$ $\frac{7}{8}-1\frac{1}{2}$	55 33	74 60	57 36	Low or Medium Carbon	
4	$\frac{1}{4}-1\frac{1}{2}$	65	115	100	Medium Carbon, Cold Drawn	
5	$\frac{1}{4}-1$ $1\frac{1}{8}-1\frac{1}{2}$	85 74	120 105	92 81	Medium Carbon, Q & T	
5.2	$\frac{1}{4}-1$	85	120	92	Low Carbon Martensite, Q & T	
7	$\frac{1}{4}-1\frac{1}{2}$	105	133	115	Medium Carbon Alloy, Q & T	
8	$\frac{1}{4}-1\frac{1}{2}$	120	150	130	Medium Carbon Alloy, Q & T	
8.2	$\frac{1}{4}-1$	120	150	130	Low Carbon Martensite, Q & T	

Torque Requirements

Preload is very important to threaded connections and the computation of the torque required to generate an appropriate preload requires detailed analysis of threaded connection geometry and friction coefficients. This complex analysis gives a very simple result for preload torque:

$$T = KF_i d \tag{10.1.7}$$

where d is the nominal diameter of the threaded fastener, F_i is the preload force required and K is a preload torque factor (Table 10.5). When the condition and/or material of the threaded fastener is not known, K =0.2 is used.

Table 10.4 Metric Bolt Strength Specifications

Property Class	Size Range	Min. Proof Strength (MPa)	Min Tensile Strength (MPa)	Min Yield Strength (MPa)	Material	Head Marking
4.6	M5-M36	225	400	240	Low or Medium Carbon	4.6
4.8	M1.6-M16	310	420	340	Low or Medium Carbon	4.8
5.8	M5-M24	380	520	420	Low or Medium Carbon	5.8
8.8	M16-M36	600	830	660	Medium Carbon, Q & T	8.8
9.8	M1.6-M16	650	900	720	Medium Carbon, Q & T	9.8
10.9	M5-M36	830	1040	940	Low Carbon Martensite, Q & T	10.9
12.9	M1.6-M36	970	1220	1100	Alloy, Q & T	12.9

Table 10.5 Threaded Fastener (Bolt) Torque Factors for use with Eq.10.1.7

Bolt Condition	K
Nonplated, Black Finish	0.30
Zinc-Plated	0.20
Lubricated	0.18
Cadmium-Plated	0.16
with Bowman Anti-Seize	0.12
with Bowman-Grip Nuts	0.09

Example 10.1

The bolted joint shown in Figure 10.2 has a grip length $L = 2$ inches without washers, a constant design load $P = 2{,}000$ lb and all parts are steel. Determine a suitable bolt size, preload and torque for the joint. Find the factor of safety guarding against separation. Determine the optimum preload as a percentage of proof strength to maximize the minimum static design factors of safety.

Solution: A preload of 90% of the bolt's Proof Strength S_P will be applied as a first trial. For a solution, the bolt designs must be evaluated iteratively. Assume a bolt length of 2.5 inches to allow for the nut. First use Tables 10.1 and 10.3 and select a trial design of a 3/8 inch diameter UNC fastener of SAE grade 5.

Table 10.3 shows the Proof Strength to be 85 kpsi. The tensile area from Table 10.1 is 0.0524 in^2. The preload is

$$F_i = 0.9 S_P A_t = 0.9 \times 85 \times 0.0524 = 4009 \text{ lb}$$

The lengths of the unthreaded shank and exposed thread are computed using Figure 10.3. The unthreaded shank length is

$$L_u = L_{Total} - L_{thread} = 2.5 - [2(0.3125) + 0.25] = 1.625 \text{ in}$$

while the exposed thread length is

$$L_t^* = L_{Grip} - L_u = 2.0 - 1.625 = 0.325 \text{ in}$$

Using Eq. 10.1.2, the bolt stiffness is

$$k_B = \left[\frac{1}{k_u} + \frac{1}{k_t}\right]^{-1} = \left[\frac{L_u}{A_u E} + \frac{L_t^*}{A_t E}\right]^{-1} = \left[\frac{1.625}{(0.0767)(30 \times 10^6)} + \frac{0.325}{(0.0524)(30 \times 10^6)}\right]^{-1} = 1.06 \times 10^6$$

Using Eq. 10.1.3, the grip member stiffness is

$$k_m = \frac{0.577 \pi E d}{2 \ln\left(5 \dfrac{0.577L + 0.5d}{0.577L + 2.5d}\right)} = \frac{0.577 \pi (30 \times 10^6)(0.3125)}{2 \ln\left(5 \dfrac{0.577(2.0) + 0.5(0.3125)}{0.577(2.0) + 2.5(0.3125)}\right)} = 6.97 \times 10^6$$

The joint stiffness factor

$$C = \frac{k_B}{k_B + k_m} = \frac{1.06 \times 10^6}{1.06 \times 10^6 + 6.97 \times 10^6} = 0.132$$

allows the computation of the load in the bolt Eq. 10.1.4 and member Eq. 10.1.5 under the load $P = 2{,}000$ lb:

$$F_b = \frac{k_B}{k_B + k_m} P + F_i = 0.132(2{,}000) + 4{,}009 = 4{,}273 \text{ lb}$$

$$F_m = \frac{k_m}{k_b + k_m} P - F_i = (1 - 0.132)(2{,}000) - 4009 = -2273 \text{ lb.}$$

Using the UNC Tensile Stress Area in Table 10.1 and Bolt Strength Standards in Table 10.3, the maximum tensile stress on the bolt and yield factor of safety are

$$\sigma = \frac{F_b}{A_t} = \frac{4{,}273}{0.0524} = 81.5 \text{ kpsi} \quad \text{and} \quad n_{Yield} = \frac{S_y}{\sigma} = \frac{92}{81.5} = 1.13$$

the load required to separate the joint and the joint separation factor of safety are

$$P_0 = \frac{F_i}{1 - C} = \frac{4{,}009}{1 - 0.132} = 4{,}619 \text{ lb} \quad \text{and} \quad n_{sep} = \frac{P_0}{P} = \frac{4{,}619}{2{,}000} = 2.31$$

These two factors of safety are quite different. To increase the yield factor of safety, decrease the initial preload. Decreased preload will reduce the joint separation factor of safety. Repeated trials to "balance" the factors of safety indicate a preload of about 65% of Proof Strength. The preload is

$$F_i = 0.65 S_P A_t = 0.65 \times 85 \times 0.0524 = 2895 \text{ lb}$$

with $\sigma = \dfrac{F_b}{A_t} = \dfrac{3,159}{0.0524} = 60.3$ kpsi and $n_{Yield} = \dfrac{S_y}{\sigma} = \dfrac{92}{60.3} = 1.53$. The load required to separate the joint and the joint separation factor of safety are

$$P_0 = \frac{F_i}{1-C} = \frac{2,895}{1-0.132} = 3,335 \text{ lb and } n_{sep} = \frac{P_0}{P} = \frac{3,335}{2,000} = 1.67$$

Using Table 10.5 for a Cadmium plated bolt and nut, the required preload torque (Eq.10.1.7) for this bolted joint is

$$T = KF_i d = 0.16(2895)0.3125 = 145 \text{ in - lb}$$

and the appropriate SAE bolts specification for machine tolerances is a

$$5/16 - 18\ UNC - 2A, \text{ Grade 5 bolt, } 2.5 \text{ in. long, preloaded with a 145 in - lb torque}$$

Fatigue Loading

The fatigue strength of threaded fastener joints can be analyzed with the fatigue strength analysis included in Chapter 8, Stress Analysis & Structures. In most cases, infinite life is desired and a Modified Goodman diagram is drawn to compute the fatigue factor of safety for the joint. The Fatigue Strength Reduction Factors K_t shown in Table 10.6 are provided for fatigue failure due to notch effects in both rolled and cut threads as well as the fillet radius under a bolt head. Most manufactured bolts have rolled threads which fail under the nut; the fatigue strength reduction factors associated with rolled threads and machined finish are typically chosen for design. This procedure is used in the example below.

Table 10.6 Fatigue Strength Reduction Factors for Threaded Fasteners

SAE Grade	Metric Grade	Rolled Threads	Cut Threads	Bolt Head Fillet
0 to 2	3.6 to 5.8	2.2	2.8	2.1
4 to 8	6.6 to 10.9	3.0	3.8	2.3

Example 10.2

The bolted joint shown in Figure 10.2 has a grip length L= 2 inches without washers, a fluctuating design load 1000 lb ± 1000 lb, and all parts are steel. Determine a suitable bolt size, preload and torque for the joint. Find the factor of safety guarding against separation. Determine the optimum preload as a percentage of proof strength to maximize the minimum fatigue design factors of safety.

Solution: Starting with the static loading design of Example 10.1, the computed joint stiffness factor and preload are

$$C = k_b/(k_b + k_m) = 0.132 \quad \text{and} \quad F_i = 0.65 S_P A_t = 2{,}895 \text{ lb}$$

The minimum bolt tension force is

$$\min F_b = \frac{k_B}{k_B + k_m} P + F_i = 0.132(0) + 4{,}009 = 4{,}009 \text{ lb}$$

The maximum bolt tension force is

$$\max F_b = \frac{k_B}{k_B + k_m} P + F_i = 0.132(2000) + 4{,}009 = 4{,}273 \text{ lb}$$

The mean and alternating tensile forces are

$$\text{mean } F_b = (4{,}273 + 4{,}009)/2 = 4{,}141 \text{ lb}$$

$$\text{alternating } F_b = (4{,}273 - 4{,}009)/2 = 132 \text{ lb}$$

Using these forces, the mean and alternating stress components can be computed:

$$\sigma_m = \text{mean } F_b/A_t = 4{,}141 \text{ lb}/0.0524 \text{ in}^2 = 79.0 \text{ kpsi}$$

$$\sigma_a = \text{alternating } F_b/A_t = 132 \text{ lb}/0.0524 \text{ in}^2 = 2.52 \text{ kpsi}$$

Note the effectiveness of the joint stiffness in reducing the size of the alternating stress relative to the mean stress. The reduction in relative magnitude of the alternating stress will substantially improve the fatigue performance of the joint. To find the strengths required to construct a Modified Goodman Diagram, use Table 10.3 for the 5/16 - 18 UNC - 2A Grade 5 bolt designed in Example 10.1; this gives the minimum ultimate and yield strengths S_{ut} = 120 kpsi and S_y = 92 kpsi.

Compute the endurance strength of the bolt:

$$S_e = k_a k_b k_c k_d k_e S_e' = k_{surface} k_{size} k_{load} k_{temp} k_{misc} S_e' = 0.76(0.995)0.923(1)0.333(120/2) = 13.9 \text{ kpsi}$$

where $k_a = aS_{ut}^b = 2.7(120)^{-0.265} = 0.76$, $k_b = (0.3125/0.3)^{-0.1133} = 0.995$,

$k_c = 0.923$, $k_d = 1.0$ and $k_e = 1/K_f = 1/3 = 0.333$.

The Modified Goodman Diagram shown in Figure 10.4 is then drawn. The load line begins at the constant stress associated with no applied load and only the bolt preload (Point A). The load line then extends through the design load point (σ_m, σ_a) at point B. The strength of the bolt is given by the intersection of the load line with the Modified Goodman line at Point C and the fatigue factor of safety of the bolted joint

$$n = \frac{\text{Strength}}{\text{Load}} = \frac{AC}{AB} \propto \frac{S_a}{\sigma_a} = \frac{4.52}{2.52} = 1.79$$

where

$$S_a = \frac{S_{ut} - (F_i/A_t)}{1 + (S_{ut}/S_e)} = \frac{120 - (76.5)}{1 + (120/13.9)} = 4.52 \text{ kpsi}$$

This factor of safety is comparable to that derived for static loading so we will accept the original bolt selected for the design.

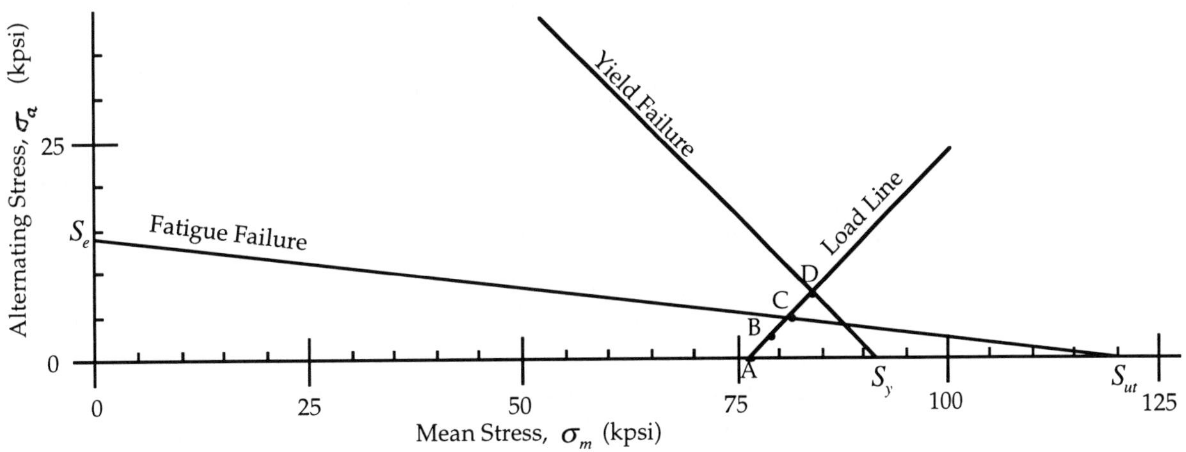

Figure 10.4 Modified Goodman Diagram for Example 10.2 bolt fatigue strength analysis

10.2 Helical Spring Design

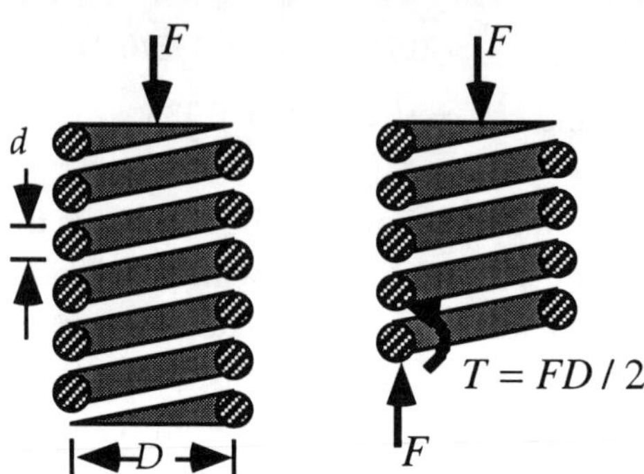

Figure 10.5 Helical compression spring schematic and free body diagram

Stresses in Helical Springs

A schematic of a helical spring in compression and the associated free body diagram are shown in Figure 10.5. The maximum shear stress in the wire is

$$\tau_{max} = \pm\frac{Tr}{J} + \frac{F}{A} \tag{10.2.1}$$

where the term Tr/J is due to the torsional load on the wire and F/A is the direct shear from the applied force on the spring. Substituting in the results from the free body diagram analysis and the spring geometry gives the actual uncorrected maximum shear stress on the inner spring coil surface:

$$\tau_{max} = \frac{8FD}{\pi d^3} + \frac{4F}{\pi d^2} = K_s \frac{8FD}{\pi d^3} \tag{10.2.2}$$

where the curvature of the spring is indicated by the *spring index*

$$C = \frac{D}{d} \tag{10.2.3}$$

and the shear stress correction factor

$$K_s = \frac{2C+1}{2C} \tag{10.2.4}$$

For most springs, the spring index C varies between 6 and 12 yielding an increase in maximum shear limited to between 8.3% and 4.2%, respectively . The important and surprising fact is that helical springs derive most of their stress (and stiffness) due to torsion and not direct shear. No bending stresses appear when the free body diagram is analyzed! This example points out the importance of drawing free body diagrams whenever design stress analysis is performed.

For fatigue stress computation only, the shear stress must be further corrected for the curvature in the spring wire. The wire curvature correction is the Bergsträsser Factor

$$K_B = \frac{4C+2}{4C-3} \tag{10.2.5}$$

The corrected maximum spring shear stress for fatigue loading is

$$\tau_{max} = K_s K_B \frac{8FD}{\pi d^3} = \left(\frac{2C+1}{2C}\right)\left(\frac{4C+2}{4C-3}\right)\frac{8FD}{\pi d^3} \tag{10.2.6}$$

This curvature stress is highly localized and is neglected for static loading ($K_B = 1$) because the curvature stress is relieved in the first load application cycle.

The deflection of a helical spring can be derived using Castigliano's Theorem and the strain energy of the torsional and direct components of the shear stresses given above (Shigley and Mischke, 1989):

$$y = \frac{8FD^3N_a}{d^4G} + \frac{4FDN_a}{d^2G} = \frac{8FD^3N_a}{d^4G}\left(1 + \frac{1}{2C^2}\right) \approx \frac{8FD^3N_a}{d^4G} \tag{10.2.7}$$

where the number of active coils N_a in the spring, the spring material shear modulus $G = E/2(1+\nu)$ and Poisson's ration ν are the additional parameters

Table 10.7 Preferred US and SI (metric) wire diameters

US (in)	SI (mm)	US (in)	SI (mm)	US (in)	SI (mm)	US (in)	SI (mm)	US (in)	SI (mm)	US (in)	SI (mm)
0.004	0.10	0.022	0.55	0.051		0.092		0.192	5.00	0.406	10.0
0.005	0.12	0.024	0.60	0.055	1.40	0.098	2.50	0.207	5.50	0.437	11.0
0.006	0.16	0.026	0.65	0.059		0.105		0.225	6.00	0.469	12.0
0.008	0.20	0.028	0.70	0.063	1.60	0.112	2.80	0.250	6.50	0.500	13.0
0.010	0.25	0.030	0.80	0.067		0.125	3.00	0.281	7.00	0.531	14.0
0.012	0.30	0.035	0.90	0.072	1.80	0.125	3.00	0.312	8.00	0.562	15.0
0.014	0.35	0.038	1.00	0.076		0.135	3.50	0.343	9.00	0.625	16.0
0.016	0.40	0.042	1.10	0.081	2.00	0.148		0.362			
0.018	0.45	0.045		0.081	2.00	0.162	4.00	0.375			
0.020	0.50	0.048	1.20	0.085	2.20	0.177	4.50	0.406	10.0		

needed for the analysis. Dividing the applied force by this deflection yields the *helical spring stiffness*

$$k = \frac{F}{y} = \frac{d^4G}{8D^3N_a} \tag{10.2.8}$$

where the number of active coils in compression springs are reduced from the total number of coils N in the spring by surface contact at the end conditions:

$$N_a = \begin{cases} N & \text{for "Plain" Ends} \\ N-1 & \text{for "Plain and Ground" Ends} \\ N-2 & \text{for "Squared" and "Squared and Ground" Ends} \end{cases} \quad \textbf{(10.2.9)}$$

The number of active coils in tension springs $N_a = N$, the total number of coils in the spring.

Compression Spring Stability

A compression spring will be unstable and buckle when the spring deflection becomes too large. There maximum, critical deflection y_{cr} when buckling occurs which, like column buckling, is dependent on the spring end conditions. The condition for compression spring stability is an initial spring length L_0 which satisfies

$$L_0 < \frac{\pi D}{\alpha}\left[\frac{2(E-G))}{2G+E}\right] = 2.63\frac{D}{\alpha} \text{ for steels} \quad \textbf{(10.2.10)}$$

where the end condition constant α is given in Table 10.8.

Table 10.8 Compression Spring Buckling End Condition Constant for Eq. 10.2.10

END CONDITION	α
Spring Supported between parallel (perpendicular) surfaces [fixed ends]	0.5
One end supported by a perpendicular flat surface (fixed), the other pivoted (hinged)	0.707
Both ends pivoted (hinged)	1.0
One end clamped, the other end free	2.0

Source: J.E. Shigley & C.R. Mischke, *Mechanical Engineering Design*, Mc-Graw Hill, NY, NY, 5thEd. 1989, pg. 420.

Spring Material Strengths

Spring materials vary in tensile strength with wire size. When plotted on a log-log graph, the relationship between tensile strength S_{ut} and wire diameter d is almost a straight line:

$$S_{ut} = b/d^m \quad \textbf{(10.2.11)}$$

where the y intercept b and the slope m are given in Table 10.9.

The torsional yield strength S_{sy} of spring steels is the maximum allowable torsional stress τ_{all} and is typically approximated from tensile strength S_{ut}:

$$S_{sy} = \tau_{\text{all}} = \begin{cases} 0.45S_{ut} & \text{Patented and Cold - Drawn Carbon Steel} \\ 0.50S_{ut} & \text{Hardened \& Tempered Carbon, Low Alloy Steel} \\ 0.35S_{ut} & \text{Nonferrous Alloys} \end{cases} \quad \textbf{(10.2.12)}$$

Table 10.9 Tensile strength constants for common spring steels using Eq. 10.2.11

MATERIAL	ASTM NO.	EXPONENT m	INTERCEPT b, kpsi	b, MPa
Cold Drawn Wire[a]	A227	0.182	141	1753
Music Wire[b]	A228	0.163	185	2154
Oil-Tempered Wire[c]	A229	0.183	147	1831
Chrome Vanadium[d]	A232	0.145	173	1910
Chrome Silicone[e]	A401	0.093	221	1059

[a]Surface is smooth and bright with no visible marks.

[b]Surface is smooth, free from defects and has a bright lustrous finish.

[c]Has a slight heat treating scale which must be removed before plating.

[d]Aircraft quality tempered wire, can also be obtained annealed.

[e]Tempered to Rockwell C49, but may be obtained untempered

Source: Associated Spring, *Design Handbook: Engineering Guide to Spring Design*, Barnes Group Inc., Bristol, Conn., 1987.

Example 10.3

A helical compression spring is made from 0.038 in diameter music wire. The outside diameter of the spring is 0.5 in. The ends are squared and there are 15 turns.

a) Estimate the torsional yield strength of the wire.

b) Find the static loading corresponding to yield strength of the wire.

c) What is the linear stiffness of the spring?

d) Compute the deflection for which the spring will be permanently deformed.

e) Compute the fully compressed "solid" length of the spring.

f) What is the maximum length of the undeflected spring so that its free length is unchanged by compression to its solid length?

g) Given the length found in part (f), is buckling a possibility?

Solution: a) Using Eq.10.2.11 and Table 10.9, we find b = 186 kpsi and $m = 0.163$ so that

$$S_{ut} = b/d^m = 186/(0.137)^{0.163} = 316 \text{ kpsi}$$

Then from Eq.10.2.12

$$S_{sy} = \tau_{\text{all}} = 0.45 S_{ut} = 0.45(316) = 143 \text{ kpsi}$$

b) The mean spring diameter $D = 0.5 - 0.038 = 0.462$ in. Using Eq.10.2.3, the spring index $C = 0.462/0.038 = 12.2$. Use Eq.10.2.4 to find the shear stress correction factor:

$$K_s = \frac{2C+1}{2C} = \frac{2(12.2)+1}{2(12.2)} = 1.04$$

Rearranging Eq.10.2.2 with $K_B = 1$ for static loading, substituting the torsional yield strength for shear stress and solving for the compressive load, we find

$$F_s = \frac{S_{sy}\pi d^3}{8K_s D} = \frac{(143 \text{ kpsi})\pi(0.038 \text{ in})^3}{8(1.04)(0.462 \text{ in})} = 6.41 \text{ lb}$$

c) From Eq.10.2.9, $N_a = N - 2 = 15 - 2 = 13$ turns.

Using $G = E/2(1+\nu) = (30 \text{ Mpsi})/2(1+0.3) = 11.5$ Mpsi and Eq.10.2.8, the stiffness of the spring is

$$k = \frac{F}{y} = \frac{d^4 G}{8D^3 N_a} = \frac{(0.038 \text{ in})^4 (11.5 \text{ Mpsi})}{8(0.462 \text{ in})^3 (13)} = 2.34 \text{ lb/in}$$

d) To find the deflection at yield load from part (b):

$$y_s = F_s/k = (6.41 \text{ lb})/(2.34 \text{ lb/in}) = 2.74 \text{ in}$$

e) The solid length of the spring L_{solid} is the wire diameter d times the number of coil heights $(N+1)$. If "Ground", this length is then reduced by the material removed to grind the end of the spring:

$$L_{solid} = d(N+1) = 0.038(15+1) = 0.608 \text{ in}$$

f) To avoid yielding, the spring's undeflected length L_0 must be no greater than the total of the solid length L_{solid} plus the deflection y_s which would cause the spring to yield:

$$L_0 = L_{solid} + y_s = 0.608 \text{ in} + 2.74 \text{ in} = 3.35 \text{ in}$$

g) For a spring with "squared" ends supported between parallel surfaces, Table 10.8 gives $\alpha = 0.5$ and from Eq.10.2.10, we find

$$2.63(D/\alpha) = 2.63(0.462/0.5) = 2.43 \text{ in}$$

which is less than L_0 = 3.35 in. The spring fails the stability condition (Eq.10.2.10) and will buckle before yielding.

Tension Springs

A typical end for a tension spring is shown in Figure 10.6. The "hook" end has bending stresses applied in contrast to the shear stresses present in the windings of the spring which makes it impossible to design the hook as strong as the body. These bending stresses are dominated by bending stresses at locations A and B so that the "hook" at the end of a tension spring is often weaker than the helical windings to which it is attached. The appropriate bending stress evaluated separately at points A and B is

$$\sigma_B = K\frac{My}{I} = \left(\frac{r_m}{r_i}\right)\frac{32Fr_m}{\pi d^3} = \frac{32Fr_m^2}{\pi d^3 r_i} \qquad \textbf{(10.2.13)}$$

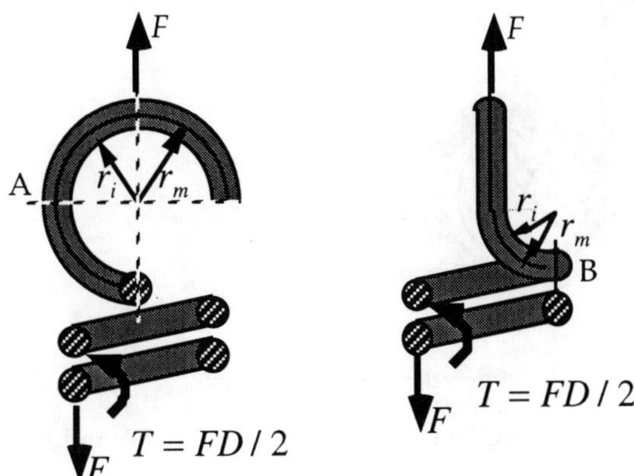

Figure 10.6 Tension spring end hook indicating bending stresses applied at points A and B

Example 10.4

A helical extension spring is made from No. 16 (0.037 in.) diameter music wire. The outside diameter of the spring is 7/16 in. The ends are to be attached through hooked ends with minimum mean radius $r_m = 0.1$ in. What is the force when the spring end hooks fail?

Solution: Using Figure 10.6, the minimum inner radius $r_i = r_m - (d/2)$ $= 0.1 - (0.037/2) = 0.0815$ in. Rearrange Eq.10.2.11 and substitute the wire ultimate strength $S_{yt} = 316$ kpsi for the bending stress σ_B. At this force level, cracks will form on the inner radius of the hooks:

$$F = \frac{\pi d^3 r_i \sigma_B}{32 r_m^2} = \frac{\pi (0.037)^3 (0.0815) 316{,}000}{32(0.1)^2} = 12.8 \text{ lb}$$

Helical Torsion Springs

Helical torsion springs are often used in torsion in components such as door hinges, and overhead garage doors. They are wound helically but have ends shaped to allow transmission of axial torsion. A torsion spring is subjected to a bending moment $M = F \times r$ which produces a normal, tensile stress in the wire (Fig. 10.7):

$$\sigma_B = K\frac{My}{I} = \left[\frac{4C^2 - C - 1}{4C(C-1)}\right]\frac{32Fr}{\pi d^3} = \left[\frac{4C^2 - C - 1}{4C(C-1)}\right]\frac{32M}{\pi d^3} \qquad \textbf{(10.2.14)}$$

where $C = D/d$ is the spring index defined in Eq.10.2.3 and M is the applied torsional load.

The helical torsion spring constant can be found using Castigliano's Theorem as

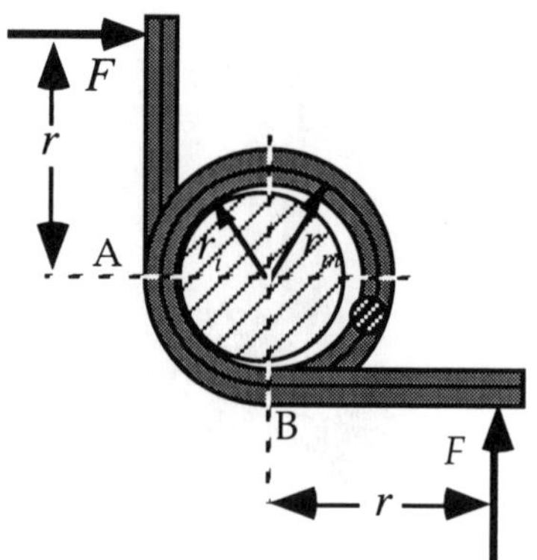

Figure 10.7 **Torsion spring indicating bending stresses at points A and B due to applied load F**

$$k = \frac{Fr}{\theta} = \frac{M}{\theta} = \frac{d^4E}{64DN} \tag{10.2.15}$$

where E is Young's Modulus of the spring material, D is the mean spring diameter (see Fig. 10.5) and N is the number of spring coils. It is often convenient to express torsional stiffness as the ratio of torsional moment to deflection turns by multiplying Eq.10.2.13 by 2π. With an empirical correction to account for spring wire curvature, the "turn stiffness" is

$$k^* = \frac{M}{n} = \frac{d^4E}{10.8DN} \tag{10.2.16}$$

Allowable tensile stresses for torsion springs can be found using distortion-energy theory combined with the empirical data given in Eq.10.2.12. Dividing the constant by 0.577 gives

$$S_y = \begin{cases} 0.78S_{ut} & \text{Patented and Cold - Drawn Carbon Steel} \\ 0.87S_{ut} & \text{Hardened \& Tempered Carbon, Low Alloy Steel} \\ 0.61S_{ut} & \text{Nonferrous Alloys} \end{cases} \tag{10.2.17}$$

Example 10.5

Design a helical torsion spring with a minimum moment of 50 lb-in and a maximum moment of 80 lb-in. after a rotation of 0.25 turn.

Solution: To get started: Assume $d = 0.2$ in, $D = 1.75$ in. so that

$$C = D/d = 1.75/0.2 = 8.75$$

The required "turn stiffness" is

$$k^* = \frac{M}{n} = \frac{(80-50)\ \text{lb}-\text{in}}{0.25\ \text{turn}} = 120\ \text{lb - in/turn}$$

Rearrange Eq.10.2.16 to compute the number of spring coils required in the spring:

$$N = \frac{M}{n} = \frac{d^4E}{10.8Dk^*} = \frac{(0.2\ \text{in})^4(30x10^6\ \text{lb/in}^2)}{10.8(1.75\ \text{in})(120\ \text{lb - in/turn})} = 21.2\ \text{turns}$$

To generate the required preload $M = 50$ lb-in, use the definition of "turn stiffness" (Eq.10.2.6) to find

$$n = \frac{M}{k^*} = \frac{50 \text{ lb} - \text{in}}{120 \text{ lb - in/turn}} = 0.417 \text{ turn}$$

The design with $d = 0.2$ in. wire wound with a mean diameter $D = 1.75$ in. requires 21.2 coil turns for the required stiffness and a preload generated by an initial "twist angle" 0.417 turn = 150 degrees. Note that the spring deflections are invariant with spring size (they are specified indirectly in the problem statement) but they can be achieved with different combinations of wire and mean coil diameters. In this case, the torsion spring length is approximately $L = (N+1)d =$ (21.2+1)0.2 = 4.44 in. This length will change with the selected wire d and mean coil D diameters.

10.3 Shaft Design

Shafts are one of the most common machine components. They transmit power by means of rotational motion and torque from one location to another. The design of shafts for strength and endurance involves several applications of basic strength of materials and dynamics. Specific design issues will be addressed in each of the sections below. Shaft design has some basic "Rules-of-Thumb".

- *Use no more than two bearings on a shaft* —The misalignment of more than two bearings on a shaft results in large bending loads and rapid fatigue failure. The use of more than two bearings is successful only in special cases when the bearings are extremely well located, such as the "line bored" bearings on an automobile crankshaft. Additionally, shafts with three or more bearings are statically indeterminate and require more complex analysis to determine bearing reaction loads.
- *Use flexible elements to connect shafts* —The use of machine elements which do not transmit bending loads reduce shaft stresses. There are many flexible couplings available which transmit torque while reducing bending loads due to shaft misalignment.
- *Use keys and keyways only to locate components and not to transmit torque* —Splines, flanges and tapered ends are able to transmit torque without local stress concentrations while keys fail rapidly due to localized shear.
- *Reduce stress concentrations though the use of generous radii* —Changes in shaft radii contribute stress concentrations which are minimized when large radii are used at these shoulders.

External Loading

Shafts are subjected to both rotational and bending loads. Free body diagrams are used to compute bearing reaction forces and moments. Normally, bearings are assumed to be "pinned" connections which do not transmit bending moments. The relationships between power, speed and transmitted torque are

$$HP = \frac{FV}{33 \times 10^3} = \frac{Tn}{63 \times 10^3} \quad \textbf{(10.3.1)}$$

where HP = power in horsepower, F = force in lb, V = velocity in ft/min, T = torque in lb-in and n = shaft speed in rpm. In SI units,

$$W = T\omega = \frac{Tn}{9.55} \quad \textbf{(10.3.2)}$$

where W = power in Watts, T = torque in N-m, ω = shaft speed in rad/s, and n = shaft speed in rpm.

The most common simplifying assumption for shaft design is to assume constant rotational torque loading due to transmitted power and completely reversed bending loads on the shaft due to its rotation. In the case of large reciprocating loads such as those in piston engines and compressors, the validity of this assumption should be carefully examined.

Example 10.6

Consider the required shaft geometry shown. The shaft is to be used to transmit 3 HP at 1000 rpm. Find

a) the bearing reactions and loads and

b) the torsional and bending moments.

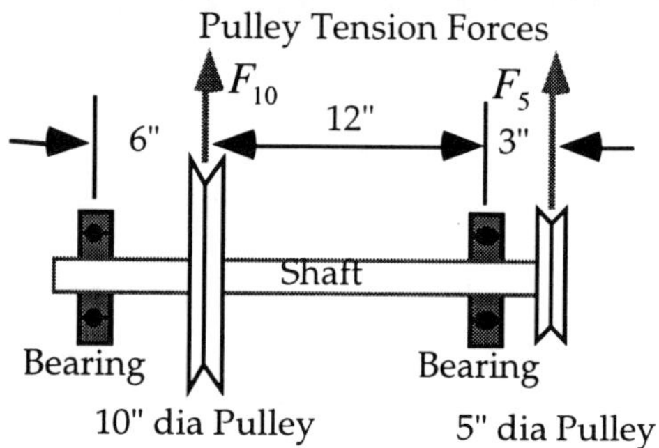

Solution: Using Eq.10.3.1, at 1000 rpm

$$T = HP(63 \times 10^3)/n = (3)(63 \times 10^3)/1000 = 189 \text{ lb-in}$$

At maximum torque, assume the belts on the pulleys are "slack" on the back side and using $T = F \times r$, the forces at the pulley outside diameters are

$$F_{10''} = T/(10 \text{ in}/2) = 189 \text{ lb-in}/5 \text{ in} = 37.8 \text{ lb on the 10'' pulley}$$

$$F_{5''} = T/(5 \text{ in}/2) = 189 \text{ lb-in}/2.5 \text{ in} = 75.6 \text{ lb on the 5'' pulley}$$

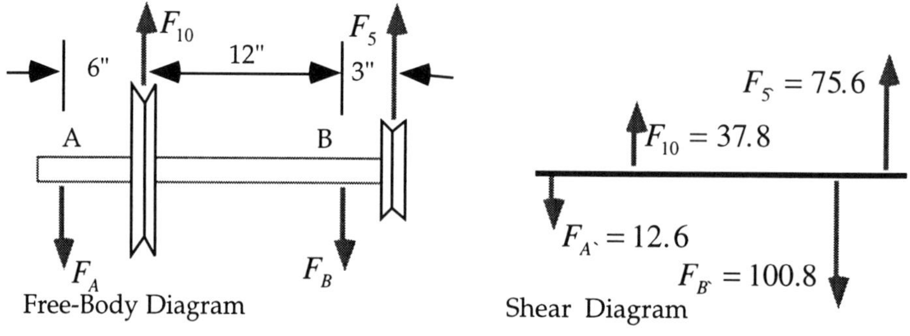

To find the bearing reactions, draw a free-body diagram and find the forces required for static equilibrium of the shaft. Summing moments about point B,

$$18F_A + 3F_5 = 12F_{10}$$
$$F_A = (12F_{10} - 3F_5)/18 = (12(37.8) - 3(75.6))/18 = 12.6 \text{ lb}$$

Summing moments about point A,

$$18F_B = 21F_5 + 6F_{10}$$
$$F_B = (21F_5 + 6F_{10})/18 = (21(75.6) + 6(37.8))/18 = 100.8 \text{ lb}$$

Note, check this result by noting static equilibrium for the (4) forces is satisfied by the moment solution

$$\sum F = F_{10} + F_{5'} - F_A - F_B = 37.8 + 75.6 - 12.6 - 100.8 = 0$$

The torsional load on the shaft is a constant torque of 189 lb-in between the two pulleys and zero elsewhere. Bending moments on the shaft are found by drawing a shear and bending moment diagram.

The Bending moment diagram shows the peak bending moment occurs at the position of bearing "B" where the moment is 226.8 in-lb. The second bending load for concern is at the position of the 10" pulley where the bending load is -75.6 in-lb.

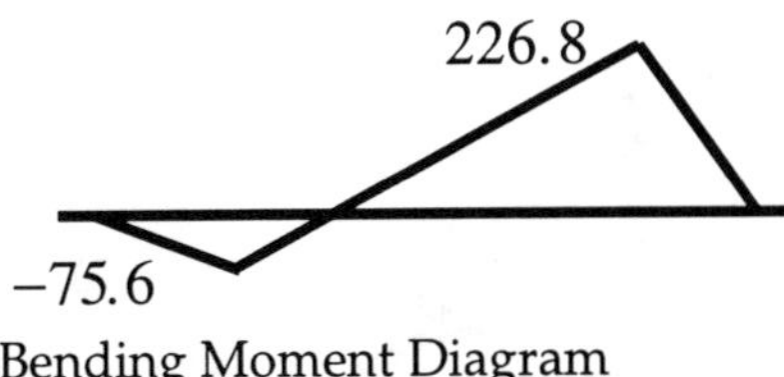

Bending Moment Diagram

Design for Fatigue Strength

The standard design assumptions of constant torsional load accompanied by completely reversed bending stresses yields a shaft design problem which is dependent on the "fluctuating stress" for the shaft (see Chapter 8). The geometry of the shaft should be chosen to satisfy

a) Part connection requirements

b) Standard geometry

c) Fatigue strength

Using the procedures outlined in Chapter 8 and the distortion energy theory, the mean stress component is computed from the constant torsional shear stresses for a solid shaft using

$$\sigma_m = \sqrt{3}\tau_m = \sqrt{3}\frac{Tr}{J} = \sqrt{3}\frac{16T}{\pi d^3} = 27.7\frac{T}{\pi d^3} \quad \textbf{(10.3.3)}$$

where d is the shaft diameter. No torsional shear stress concentration factor is applied because this stress does not fluctuate (see Section 8.4). The alternating bending stresses are computed *for a solid shaft* using

$$\sigma_a = \frac{Mr}{I} = 32\frac{M}{\pi d^3} \tag{10.3.4}$$

For alternating stresses on shafting, ASME recommends the use of the elliptic failure criteria

$$\left(\frac{n\sigma_a}{S_e}\right)^2 + \left(\frac{n\sigma_m}{S_y}\right)^2 = 1 \tag{10.3.5}$$

which uses factor of safety n, alternating stress σ_a, mean stress σ_m, endurance strength S_e, and yield strength S_y. The ASME criteria is compared to the Goodman (fluctuating stress) and Yield failure criteria (Section 8.4) in Figure 10.8. Notice that the ASME failure criteria exceeds static yield during the first half cycle of loading and this failure mechanism needs to be checked separately. Note that any the bending stress concentration factor will be applied to reduce the endurance strength S_e. This procedure is illustrated in the next example.

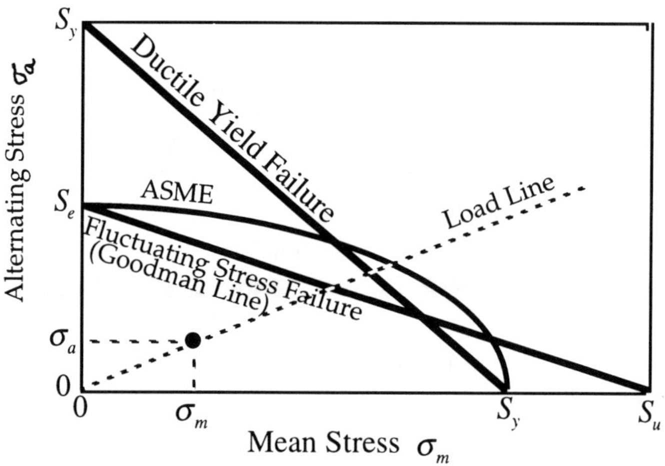

Figure 10.8 ASME shaft failure criteria Equation 10.3.5

Example 10.7

For the shaft analyzed in Example 10.6, the pulleys selected have the geometry shown at the right. The tapered clamp is tightened to hold the pulley on the shaft and a typical installation uses a shoulder to locate the pulley along the shaft.

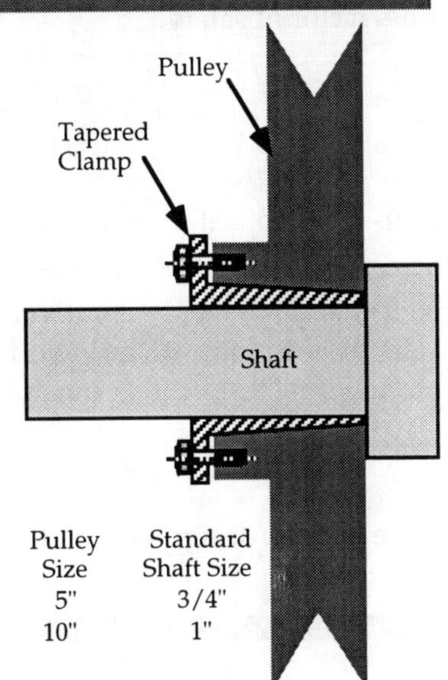

Design a shaft with infinite life which includes a means of locating standard size bearings at the centerline locations shown in Example 10.6.

Solution: The initial design shaft design will use a pair of 3/4″ bearings located by shoulders at A and B as shown below. The larger diameter portion of the shaft both accepts the torsional/bending load and serves to locate the 10″ pulley at C. The 5″ pulley is located by the additional shoulder at D.

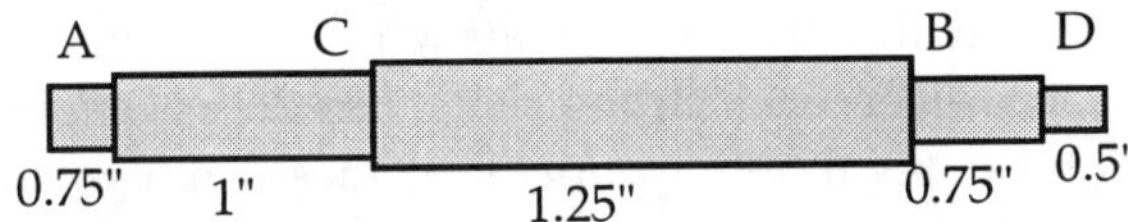

Point B is the point on the shaft most likely to fail because the torsional stresses act from point C to B and the bending stresses are highest at B. Following the stress and failure analysis procedures given in Section 8.4 and the ASME failure criteria, compute the steady torsional stress as

$$\sigma_m = 27.7\frac{T}{\pi d^3} = 27.7\frac{(189 \text{ lb - in})}{\pi(0.75 \text{ in})^3} = 3.95 \text{ kpsi}$$

and the alternating stress as

$$\sigma_a = \frac{Mr}{I} = 32\frac{M}{\pi d^3} = 32\frac{(226.8 \text{ lb - in})}{\pi(0.75 \text{ in})^3} = 5.48 \text{ kpsi}$$

Assuming a low strength G10060 HR steel with $S_y = 24$ kpsi and $S_u = 43$ kpsi machined with a shoulder radius of 1/16 inch, compute the endurance strength of the shaft material (see Section 8.4);

$$S_e = 0.5k_a k_b k_c k_d k_e S_u = 0.5(0.997)(0.901)(1.0)(1.0)(0.571)(43 \text{ kpsi}) = 11.0 \text{ kpsi}$$

where

surface factor: $k_a = aS_u^b = 2.70(43 \text{ kpsi})^{-0.265} = 0.997$,

size factor: $k_b = (d/0.3)^{-0.1133} = (0.75/0.3)^{-0.1133} = 0.901$ (Equation 8.4.8a),

load factor: $k_c = 1.0$ (Equation 8.4.9),

temperature factor: $k_d = 1.0$ (Equation 8.4.10), and

miscellaneous effects factor: $k_e = 1/K_f = 1/1.75 = 0.571$ (Equation 8.4.11).

Rearranging the ASME shaft criterion (Eq.10.3.5) gives the design alternating stress factor of safety,

$$n = 1\Bigg/\sqrt{\left(\frac{\sigma_a}{S_e}\right)^2 + \left(\frac{\sigma_m}{S_y}\right)^2} = 1\Bigg/\sqrt{\left(\frac{5.48\text{ kpsi}}{11.0\text{ kpsi}}\right)^2 + \left(\frac{3.95\text{ kpsi}}{24\text{ kpsi}}\right)^2} = 1.91$$

If this factor of safety is insufficient, either the shaft material can be changed to increase strength, the shoulder radius increased to reduce stress concentration or other changes made to iteratively improve the design's operating strength.

In this design case, the yield factor of safety

$$n_{yield} = \frac{S_y}{\sigma_{\max}} = \frac{24\text{ kpsi}}{(3.95 + 5.48)\text{ kpsi}} = \frac{24\text{ kpsi}}{9.43\text{ kpsi}} = 2.55$$

so the current design is even less likely to fail by yielding than through fluctuating stresses.

Design for Deflection

Maximum shaft stiffness is often a part of a shaft's design specifications. The previous section dealt with designing a shaft which is strong enough to support the required loads, however, even strong shafts can deflect enough to fail in use. Excessive deflection can result in improper gear mesh, excessive vibration or inability to retain adjustments in clutches and flywheels. Two forms of deflection computations will be presented here for round solid shafts based on standard beam bending and torsion deflection analysis. Note that the dynamic analysis for torsional resonance is not covered here and the material in Chapter 9 should be reviewed if vibration resonance problems are suspected of being important.

The bending deflection of a shaft is governed by the standard beam bending relationship:

$$y(x) = \frac{1}{EI}\iint M\,dx\,dx \tag{10.3.6}$$

As shown in an earlier section, bending moment M is often easily found graphically from the applied force F; further graphical integration to find $y(x)$ is an easy to implement method for design. For a shaft of constant diameter supported between two bearings separated by distance L with transverse load applied at $x = a$, the maximum deflection between the bearings is

$$y_{\max} = \frac{F}{3EI}\left\{2a^3 - \frac{a^4}{L} - La^2\right\} \tag{10.3.7}$$

and the deflection at $x = a$ is

$$y(a) = \frac{F}{6EI}\left\{\left(1 - \frac{a}{L}\right)x^3 + \frac{a}{L}\left(-a^2 + 3aL - 2L^2\right)x\right\} \tag{10.3.8}$$

where E is Young's Modulus, $I = \pi d^4/64$ is the second area moment of the shaft, and d is the shaft section diameter. The analysis will be demonstrated by Example 10.8.

Torsional deflection is computed along a shaft by adding the deflections of sections of varying diameter between the points of the applied torsional load. A section of l with constant torsional load T has torsional deflection

$$\theta = \frac{TL}{GJ} = 583.6\frac{TL}{Gd^4} = 1{,}167\frac{TL(1+\nu)}{Ed^4} \tag{10.3.9}$$

where Modulus of Rigidity $G = E/2(1+\nu)$, second polar moment of area $J = \pi d^4/32$, d is the shaft section diameter, E is Young's Modulus and ν is Poisson's Ratio for the shaft's material. The analysis will be demonstrated by Example 10.8.

Example 10.8

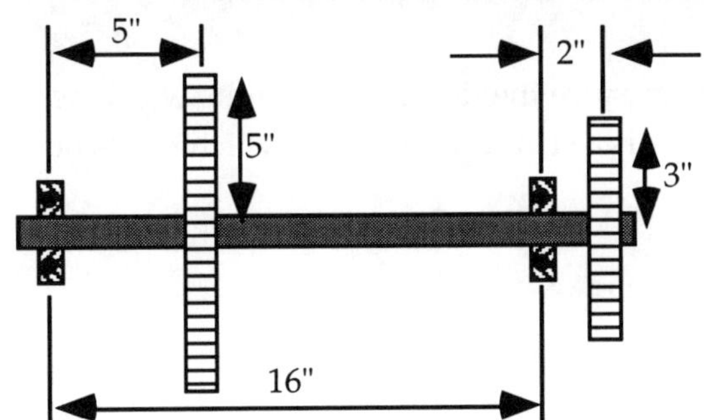

A steel shaft is to be designed to support two gears as shown above and limit maximum torsional deflection between the gears to 0.5 degrees and maximum lateral deflection to 0.002 inches. The gears are to transmit 2 HP at 3600 rpm and have the contact radii shown.

Solution: By inspection of the shaft, the maximum lateral deflection will between the two bearings. The transverse load F on the gear at $x = a = 5''$ is that required to generate the torque T needed to transmit power P at shaft rotation speed ω:

$$\omega = (3600)(2\pi)/(60) = 377 \text{ rad / sec}$$

$$T = P/\omega = 2 \times 550 / 377 = 2.92 \text{ ft - lb}$$

$$F = T/r = 2.92 \times 12 / 5 = 7.00 \text{ lb}$$

The maximum shaft lateral deflection (Eq.10.3.7) is rearranged to compute the minimum shaft diameter required for maximum bending deflection:

$$I_{\min} = \pi d_{\min}^4/64 = \frac{F}{3Ey_{\max}}\left\{2a^3 - \frac{a^4}{L} - La^2\right\} \text{ or}$$

$$d_{min}^4 = \frac{64F}{3\pi E y_{max}}\left\{2a^3 - \frac{a^4}{L} - La^2\right\}$$

$$= \left\|\frac{64(7.00 \text{ lb})}{3\pi\left(30\times10^6 \text{ psi}\right)(0.002 \text{ in})}\left\{2(5 \text{ in})^3 - \frac{(5 \text{ in})^4}{(16 \text{ in})} - (16 \text{ in})(5 \text{ in})^2\right\}\right\| = 0.4706$$

or $d_{min} = 0.828$ in for maximum bending deflection

The torsion deflection (Eq.10.3.8) is rearranged to give the minimum diameter required to meet this design constraint. In this case, $L = 13$ inches because this is the distance between the gears applying the torsional load to the shaft:

$$d_{min}^4 = 583.6\frac{TL}{G\theta_{max}} = 1167\frac{TL(1+\upsilon)}{E\theta_{max}} = 1167\frac{[(2.92)(12)](13)(1+0.3)}{\left(30\times10^6\right)[(0.5)(\pi/180)]} = 2.6397 \text{ in}^4$$

$d_{min} = 1.27$ in. to satisfy the torsional deflection design constraint

The shaft diameter chosen is $d = 1\frac{1}{4}$ in since the design constraints were specified to only 1 significant figure and this is a standard diameter which meets both criteria to that level of precision.

Practice Problems (Essay)

10.1 A bolted joint supporting 1/2 inch thick square tubing with a 1000 lb load is shown below. The bolt is an 1/2 inch UNC bolt of SAE grade 1.

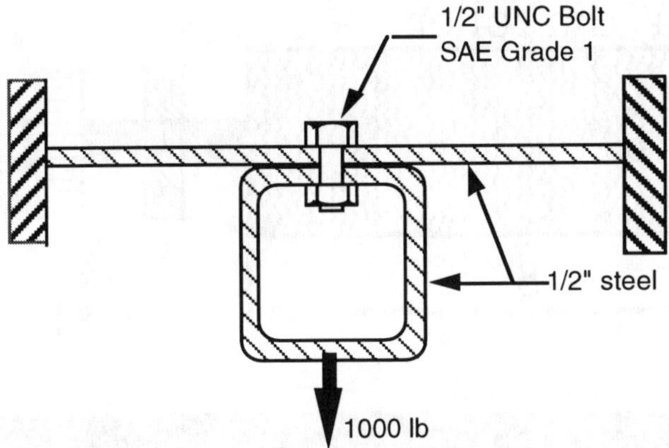

a) What is the stress in the bolt when properly preloaded for a permanent connection?
b) Under normal conditions, what wrench torque on the nut is required to get this preload?
c) What is the factor of safety guarding against joint separation when the bolt is properly preloaded?
d) Give two reasons for preloading a bolted joint.

10.2 The front end of a car weighs 1200 pounds and is to have a natural frequency of 2 Hz. Two steel helical springs are to be used to support the front end of the vehicle. The frame of the vehicle can deflect dynamically 1-1/2 inches relative to the axle before bottoming out on a rubber bumper. The maximum allowable spring shear stress is to be 95,000 psi and fatigue effects are to be neglected. The spring is to have square ends and the mean diameter is 4.5 inches.

a) Determine the required minimum wire diameter for the spring.

b) Determine the number of active spring coils.

c) Determine the minimum free length of the spring.

10.3 The geared industrial roller shown in the figure is driven at 300 rpm through the 3 in diameter gear. The roller exerts a force of 30 lb/in on material rolled through the roller with a coefficient of friction of 0.40. Assume a steel shaft material with $S_{ut} = 72$ kpsi and $S_y = 39$ kpsi. Based on static loading and a factor of safety of 3.5, find the required shaft diameter.

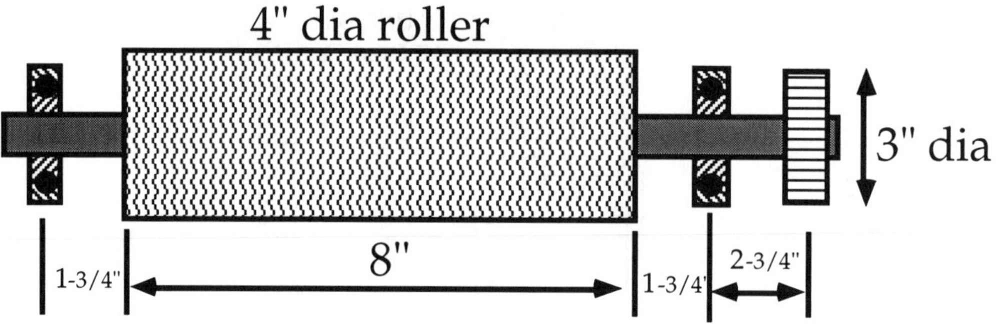

Practice Problems (Multiple Choice)

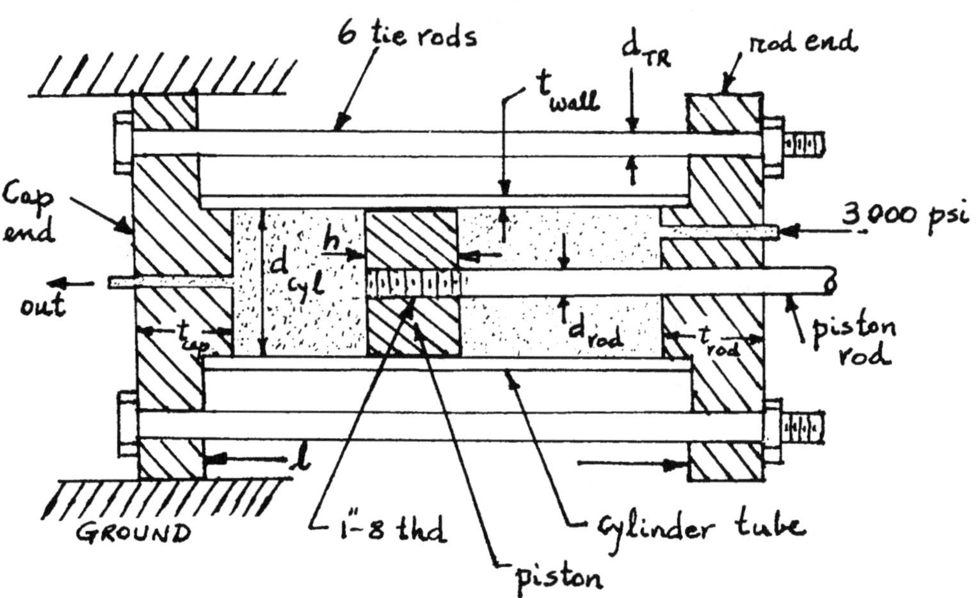

A hydraulic cylinder is made up of a tube section with two ends and is held together by 6 tie rod bolts as shown. The piston is pressurized to 3000 psi.

Data List:

a) diameter of cylinder, $d_{cyl} = 4.0$ inches
b) wall thickness of cylinder, $t_{wall} = 0.25$ inch
c) length of cylinder, $l = 14.0$ inches
d) piston thickness, $h = 1.5$ inches
e) piston rod diameter, $d_{rod} = 1.5$ inches
f) thickness of rod end, $t_{rod} = 2.5$ inch
g) tie rod diameter, $t_{TR} = 0.75$ inch
h) thickness of cap end, $t_{cap} = 2.5$ inches
i) piston-rod thread, 1"- 8 UNC

10.4 The nominal stress in the piston rod due to the rod end 3000 psi cylinder tube pressure is

a) 6200 psi tension
b) 8,600 psi tension
c) 12,000 psi tension
d) 8,300 psi tension
e) 27,700 psi tension

10.5 Which data from the data list are necessary to determine the stress condition in the cylinder tube?

a) a, b
b) a. b, c
c) b, c, e
d) a, c, e, g
e) a, b, d

10.6 The maximum hoop stress in the cylinder tube due to the rod end 3000 psi tube pressure is

a) 0 psi
b) 8,000 psi
c) 16,000 psi
d) 20,000 psi
e) 25,000 psi

10.7 The maximum longitudinal stress in the cylinder tube due to the rod end 2500 psi tube pressure is

a) 0 psi
b) 8,000 psi
c) 16,000 psi
d) 20,000 psi
e) 27,000 psi

10.8 The axial stress in a tie rod due to the rod end 3000 psi tube pressure and neglecting preload is

a) 7,650 psi
b) 12,200 psi
c) 30,600 psi
d) 38,200 psi
e) 61,200 psi

10.9 Which data from the data list are necessary to determine the shear stress in the threads at the piston rod to piston attachment?

a) a. d
b) a, c, d, i
c) c, d, e
d) d, f, h
e) a, d, e, i

10.10 The average shear stress in the threads at the piston rod to piston attachment due to the rod end 3000 psi tube pressure and neglecting preload is
a) 4,910 psi
b) 5,700 psi
c) 11,800 psi
d) 14,400 psi
e) 19,600 psi

10.11 If the cylinder is now pressurized at the cap end with 3000 psi, and opened to the atmosphere at the rod end, the force on the piston rod is
a) 4,400 lbs
b) 15,500 lbs
c) 27,000 lbs
d) 30,900 lbs
e) 37,700 lbs

10.12 Under cap end pressurization with 3000 psi as compared to rod end pressurization with 3000 psi, which item below is most accurate?
a) shear stress in the piston rod threads increases
b) nominal stress in the tie rod increases
c) hoop stress in the cylinder increases by 25%
d) longitudinal stress in the cylinder increases by 25%
e) hoop stress in the cylinder increases by 16%

Solutions to Practice Problems

10.1

A preload of 90% of the bolt's Proof Strength S_P will be applied as a first trial. For a solution, the bolt design must be evaluated iteratively. Assume a bolt length of 1.5 inches long to allow for the nut. First use Tables 10.1 and 10.3 and select a trial design of a 1/2 inch diameter UNC fastener of SAE grade 1. Table 10.3 shows the Proof Strength to be 33 kpsi. The tensile area from Table 10.1 is 0.1419 in^2. The preload is

$$F_i = 0.9 S_P A_t = 0.9 \times 33,000 \times 0.1419 = 4214 \text{ lb}$$

The lengths of the unthreaded shank and exposed thread are computed using Figure 10.3. The unthreaded shank length

$$L_u = L_{Total} - L_{thread} = 1.5 - [2(0.5) + 0.25] = 0.25 \text{ in}$$

while the exposed thread length

$$L_t^* = L_{Grip} - L_u = 1.0 - 0.25 = 0.75 \text{ in}$$

Using Eq.10.1.2, the bolt stiffness

$$k_B = \left[\frac{1}{k_u} + \frac{1}{k_t}\right]^{-1} = \left[\frac{L_u}{A_u E} + \frac{L_t^*}{A_t E}\right]^{-1} = \left[\frac{0.25}{(0.1963)(30 \times 10^6)} + \frac{0.75}{(0.1419)(30 \times 10^6)}\right]^{-1} = 4.57 \times 10^6$$

Using Eq.10.1.3, the grip member stiffness

$$k_m = \frac{0.577 \pi E d}{2 \ln\left(5 \dfrac{0.577L + 0.5d}{0.577L + 2.5d}\right)} = \frac{0.577 \pi (30 \times 10^6)(0.5)}{2 \ln\left(5 \dfrac{0.577(1.0) + 0.5(0.5)}{0.577(1.0) + 2.5(0.5)}\right)} = 16.6 \times 10^6$$

The joint stiffness factor

$$C = \frac{k_B}{k_B + k_m} = \frac{4.57 \times 10^6}{4.57 \times 10^6 + 16.6 \times 10^6} = 0.216$$

allows the computation of the load in the bolt (Eq.10.1.4) and member (Eq.10.1.5) under the load $P = 1000$ lb.

$$F_b = \frac{k_B}{k_B + k_m} P + F_i = 0.216(1,000) + 4,214 = 4,430 \text{ lb and}$$

$$F_m = \frac{k_m}{k_b + k_m} P - F_i = (1 - 0.216)(1,000) - 4214 = -3430 \text{ lb.}$$

Using the 1/2 - UNC Tensile Stress Area in Table 10.1 and Bolt Strength Standards in Table 10.3, the maximum tensile stress on the bolt and yield factor-of-safety are

$$\sigma = \frac{F_b}{A_t} = \frac{4,430}{0.1419} = 31.2 \text{ kpsi and } n_{Yield} = \frac{S_y}{\sigma} = \frac{36}{31.2} = 1.15$$

The load required to separate the joint and the joint separation factor-of-safety are

$$P_0 = \frac{F_i}{1-C} = \frac{4,214}{1-0.216} = 5,375 \text{ lb. and } n_{sep} = \frac{P_0}{P} = \frac{5,375}{1,000} = 5.38$$

These two factors-of-safety are quite different. To increase the yield factor of safety, decrease the initial preload. Decreased preload will reduce the joint separation factor of safety. Repeated trials to balance the factors of safety indicate a preload of about 40% of Proof Strength. The preload is then

$$F_i = 0.65 S_P A_t = 0.40 \times 33,000 \times 0.1419 = 1873 \text{ lb}$$

with $\sigma = \frac{F_b}{A_t} = \frac{2,089}{0.1419} = 14.7$ kpsi and $n_{Yield} = \frac{S_Y}{\sigma} = \frac{36}{14.7} = 2.45$.

The load required to separate the joint and the joint separation factor of safety are

$$P_0 = \frac{F_i}{1-C} = \frac{1,873}{1-0.216} = 2,389 \text{ lb. and } n_{sep} = \frac{P_0}{P} = \frac{2,389}{1,000} = 2.39$$

Using Table 10.5 for a Cadmium plated bolt and nut, the required preload torque (Eq.10.1.7) for this bolted joint is

$$T = KF_i d = 0.16(2,389)0.5 = 191 \text{ in - lb}$$

and the appropriate SAE bolts specification for machine tolerances is a

1/2 = 13 UNC - 2A Grade 1 bolt, 1.5 " long, preloaded with a 191 in-lb torque

10.2

Natural frequency of the car/spring

$$\omega = 2Hz = 4\pi \text{ (rad / sec)} = 12.57 \text{ (rad/sec)} = \sqrt{2K/M}$$

The factor of 2 arises from the two springs acting in parallel on the vehicle. The vehicle mass $M = 1200 / 32.2 = 37.27$ slugs so that the required stiffness is

$$K = \omega^2 M/2 = (12.57)^2 (37.27)/2 = 2.94 \times 10^3 \text{ lb / ft} = 245 \text{ lb / in}$$

a) Using Eq.10.2.2

$$\tau_{max} = K_s \frac{8FD}{\pi d^3} \text{ where } F = (1200 / 2) + (1.5)(245) = 968 \text{ lb}, \; D = 4.5 \text{ in},$$

d is unknown, and using Eq.10.2.4 $K_s = (2C+1)/2C$ where the spring index $C = D/d$. Setting $\tau_{max} = 95{,}000$ and solving iteratively (trial and error) for d yields $d = 0.498$ in.

The closest standard size is $d = 0.500$ in. which reduces the stress to 93.7 kpsi.

b) Rearranging the helical compression spring stiffness given by Eq.10.2.8 and solving for active turns,

$$N_a = \frac{(0.5 \text{ in})^4 (11.5 \text{ MPa})}{8(4.5 \text{ in})^3 (245.2 \text{ lb/in})} = 4.02 \text{ active turns}$$

For square ends, add 2 additional turns (Eq.10.2.9) yielding $N = 6$ total turns on the springs with square ends.

c) The minimum free length is the static deflection plus maximum deflection plus the summation of the wire coils,

$$L_o = \text{static} + \text{dynamic} + \text{wire} = 600/245.2 + 1.5 + 6 \times 0.5 = 6.96 \text{ in}$$

10.3

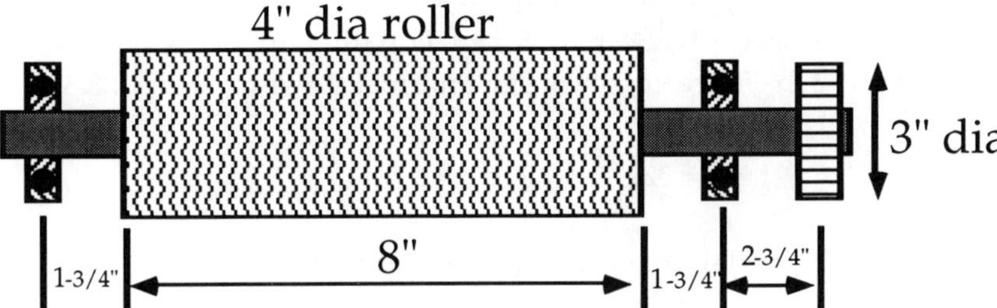

The applied load on the roller is a combination of normal load $F_n = (8 \text{ in})(30 \text{ lb/in}) = 240 \text{ lb}$ given by the roller load/unit length and tangential load $F_t = mF_n = (0.4)240 \text{ lb} = 96 \text{ lb}$ limited by the coefficient of friction. The tangential load generates a torque on the roller which is supplied by a second tangential load on the drive gear, $F_{gear} = 96 \text{ lb}(4 \text{ in}/3 \text{ in}) = 128 \text{ lb}$. The orientation of the drive gear is unspecified so assume a "worst case" and orient that force parallel to the largest roller force to draw the free-body diagram below.

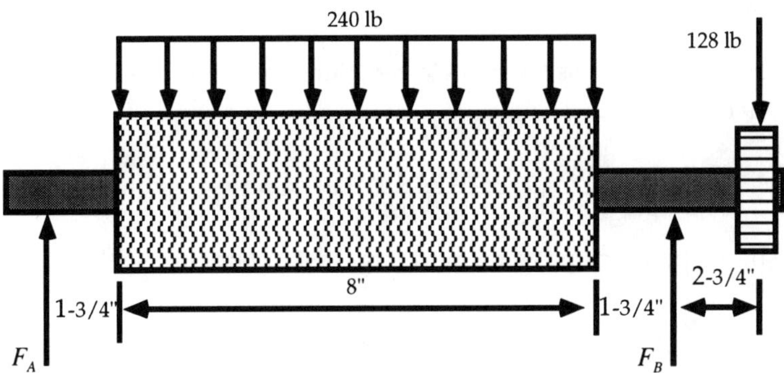

The bearing reaction forces are found by summing moments about "B" to find

$$F_A = [(240 \text{ lb})(4" + 1.75") - (128 \text{ lb})(2.75")]/(1.75" + 8" + 1.75") = 89.4 \text{ lb}$$

and by summing moments about "A" to find

$$F_B = [(240 \text{ lb})(4" + 1.75") + (128 \text{ lb})(1.75" + 8" + 1.75" + 2.75")]/(1.75" + 8" + 1.75") = 278.6 \text{ lb}$$

Shaft Bending and Shear Stresses (see Sect 8.1) are computed for the shaft in the middle of the roller using a free-body diagram for the "sectioned" roller. This location has the largest bending moment.

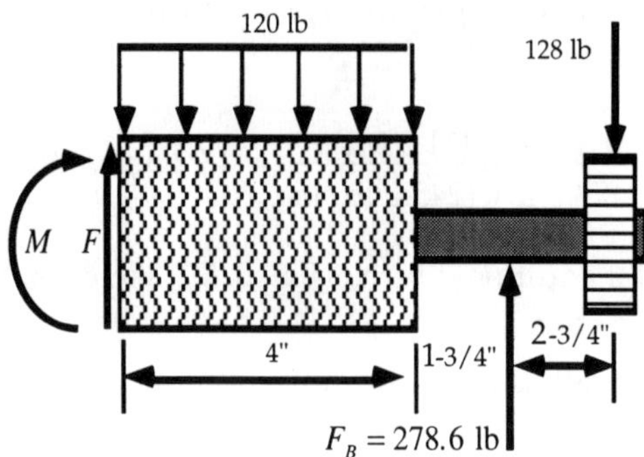

In this case, the shaft shear force $F = 128 \text{ lb} + 120 \text{ lb} - 278.6 \text{ lb} = -30.6 \text{ lb}$ and the shaft bending moment $M = (278.6 \text{ lb})(5.75") - (120 \text{ lb})(2") - (128 \text{ lb})(8.5")$ $= 274 \text{ lb-in}$

These loads result in stresses on the outside surface of the shaft. The bending moment M creates a fully reversed, normal, longitudinal stress σ_a dependent on shaft diameter (see Sect 8.1 , Sect. 10.3 and Equation 10.3.4) while the torsion $T = F_{gear}(d/2) = 192 \text{ lb-in}$ creates at constant normal stress σ_m (Eq.10.3.3) where

$$\sigma_a = 32\frac{M}{\pi d^3} \text{ and } \sigma_m = 27.7\frac{T}{\pi d^3}$$

Using the ASME elliptic failure criteria (Eq.10.3.5) with factor-of-safety $n = 3.5$ yields

$$1 = \left(\frac{n\sigma_a}{S_e}\right)^2 + \left(\frac{n\sigma_m}{S_y}\right)^2 = \left(\frac{32nM}{\pi d^3 S_e}\right)^2 + \left(\frac{27.7nT}{\pi d^3 S_y}\right)^2 \text{ or } d^3 = \left(\frac{35.7M}{S_e}\right)^2 + \left(\frac{30.9T}{S_y}\right)^2$$

The yield strength $S_y = 39$ kpsi is provided while the endurance strength S_e of the shaft is found by assuming a constant cross section with a machined finish at room temperature (see Sect 8.4):

$$S_e = 0.5k_a k_b k_c k_d k_e S_u = 0.5(0.869)(1.0)(1.0)(1.0)(1.0)(72 \text{ kpsi}) = 31.3 \text{ kpsi}$$

where the size factor k_b has been neglected. Substituting in the known strengths and loads yields

$$d^3 = \left(\frac{35.7(274 \text{ lb-in})}{31.3 \text{ kpsi}}\right)^2 + \left(\frac{30.9(192 \text{ lb-in})}{39 \text{ kpsi}}\right)^2 = 0.1208 \text{ in}^3 \text{ or } d = 0.494 \text{ in}$$

The shaft should be specified as the next largest standard size so that $d = 1/2$ in

10.4 **d)** Area, $A = \frac{\pi}{4}\left[4.0^2 - 1.5^2\right] = 10.8 \text{ in}^2$ so that

$$F_{rod} = 3000 \text{ psi}\left(10.8 \text{ in}^2\right) = 32,400 \text{ lb and}$$

$$\sigma_{rod} = (32,400 \text{ lb})\Big/\left(\frac{\pi}{4}1.5^2 \text{ in}\right) = 18,300 \text{ psi}$$

10.5 **a)** Because the internal and external pressures are specified in the problem statement, only the cylinder diameter d_{cyl} and thickness t_{wall} are required to compute the "long cylinder" ($l > d_{cyl}$) stresses in the cylinder tube where the stress state is assumed constant over the length of the cylinder.

Note: the "long cylinder" constant stress solution is used for design because it is conservative. The "short cylinder", non-constant stress problem yields smaller stresses because the end caps support the tube, cannot be solved analytically and is not typically used in design.

10.6 **e)** Using a half cylinder free-body diagram and a "thin-wall" constant stress assumption

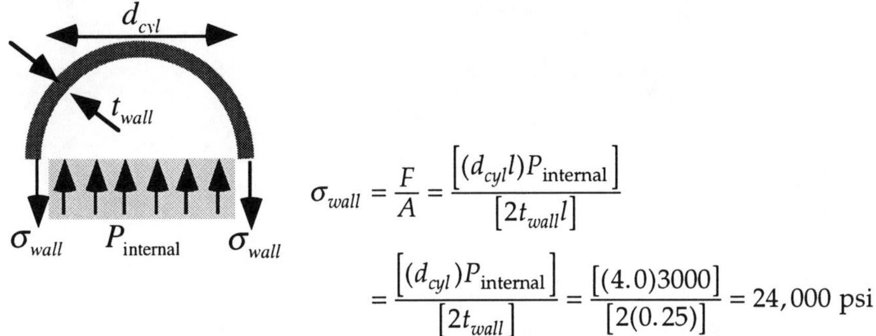

$$\sigma_{wall} = \frac{F}{A} = \frac{[(d_{cyl}l)P_{internal}]}{[2t_{wall}l]}$$

$$= \frac{[(d_{cyl})P_{internal}]}{[2t_{wall}]} = \frac{[(4.0)3000]}{[2(0.25)]} = 24,000 \text{ psi}$$

(The thick-walled, "exact" model uses Lame's equation and yields $\sigma_{wall} = 21,300$ psi)

10.7 **a)** Examining the design, the tie rods carry the longitudinal tensile stress due to pressure yielding no longitudinal stress in the cylinder due to pressure. The only stress in the tube is the compressive stress due to any preload of the tie rods BEFORE the tube is pressurized.

10.8 **b)** The axial stress on the tie rods due to the internal pressure and neglecting preload is

$$\sigma_{tie\ rod} = \frac{F}{A} = [32,400 / 6] \Big/ \left(\frac{\pi}{4} 0.75^2 \text{ in}\right) = 12,200 \text{ psi}$$

10.9 **e)** The shear stress on the piston rod threads is determined by the piston rod diameter d_{rod}, the thread size and length (from h and thread specification), and the piston rod load (from d_{cyl}, d_{rod} and $P_{internal}$). $P_{internal}$ is specified in the problem statement. You need a) d_{cyl}, d) h, e) d_{rod}, and i) the piston thread specification.

10.10 **c)** The thread shear stress acts at half the height of the thread:

$$d_{THD} = 1.0 - \tfrac{1}{8} = 0.875 \text{ in}$$

$$\tau_{THD} = \frac{F}{A} = \frac{F}{[(\pi d_{THD})h/2]} = \frac{2F}{[(\pi d_{THD})h]} = \frac{2(32,400)}{[(\pi 0.875)2.0]} = 11,800 \text{ psi}$$

10.11 **e)** The rod force is now compressive and due to the internal pressure acting over the full area of the piston,

$$F_{rod} = 3000 \text{ psi} \left(\frac{\pi}{4}(4.0^2) \text{ in}^2\right) = 37,700 \text{ lb}$$

10.12 **a)** The tie rod load goes to zero. The hoop stress and longitudinal stress in the cylinder remain unchanged. The shear stress in the threads increases because the compressive longitudinal load is larger than the past tensile load.

Control Systems

by Clark J. Radcliffe and Charles W. Radcliffe

11.1 Dynamic Systems Modeling and Analysis

Differential equations describing the dynamic response of a physical system are obtained from the physical laws which govern the system's components. The approach applies equally well to mechanical, electrical, fluid and thermal systems. A summary of the variables governing the behavior of dynamic systems is given in Table 11.1. A summary of the describing equations for lumped, linear dynamic elements is given in Table 11.2.

With the exception of thermal modeling, all energy domains are characterized by both kinetic and potential energy as well as energy interchange between these two energy storage mechanisms. Because of their common characteristics, modelers of physical system dynamics make use of the analogous behavior of these systems when compiling models. Additionally, simulation methods such as analog computers have often been used to simulate mechanical and fluid systems through the construction of analogous electrical systems.

Mechanical Systems

The most common modeling method is the *Free Body Diagram*. Only planar motion will be discussed here because three-dimensional dynamics is covered in some detail in Chapter 3. Newton's Laws for either translation or planar rotation result in a second order ordinary differential equation (ODE) for each translational or rotational degree of freedom of each lumped mass in the system.

Mechanical Translation

Newton's law for *linear translation* of a mass takes the form

$$m\ddot{x} = \sum_i f_i \tag{11.1.1}$$

where m = rigid body mass,

$\ddot{x}$ = linear acceleration of the mass, and

f_i = applied external force(s).

Mechanical Rotation

Newton's law for *planar rotation* of a mass requires that

$$J\ddot{\theta} = \sum_i \tau_i \tag{11.1.2}$$

where J = rigid body moment-of-inertia about the body center-of-gravity,

$\ddot{\theta}$ = planar angular acceleration of the body, and

τ_i = applied external moment(s).

Table 11.1 Summary of Dynamic Variables for Physical Systems

System	Variable Through	Integrated Through	Variable Across	Integrated Across
Mechanical Translation	Velocity, v (m/s)	Displacement, x (m)	Force Difference, $f_{21} = f_2 - f_1$ (N)	Linear Momentum, mv (kg·m/sec)
Mechanical Rotation	Velocity, ω (rad/s)	Displacement, θ (rad)	Net Torque, $\tau_{21} = \tau_2 - \tau_1$ (N·m)	Angular Momentum, h (kg·m^2/sec)
Electrical	Current, i (A)	Charge, q (Coulomb)	Voltage Difference, e_{21} (Volt)	Flux Linkage, λ_{21} (Volt·s)
Fluid	Flowrate, Q (m^3/s)	Flow Volume, V (m^3)	Pressure Difference, p_{12} (Pa)	Pressure Momentum, γ_{21} (N·s/m^2)

Table 11.2 Lumped, Linear Describing Differential Equations for Physical Systems

Type of Element	Physical Element	Describing Equation	Energy Stored or Power Dissipated
Inertial or Inductive Kinetic Energy Storage	Translational Inertia	$m\frac{dv}{dt} = f_{21}$	$E = \frac{1}{2}mv^2$
	Rotational Inertia	$J\frac{d\omega}{dt} = \tau_{21}$	$E = \frac{1}{2}J\omega^2$
	Electrical Inductance	$L\frac{di}{dt} = e_{21}$	$E = \frac{1}{2}Li^2$
	Fluid Inertia	$I_f\frac{dQ}{dt} = P_{21}$	$E = \frac{1}{2}I_fQ^2$
Stiffness or Capacitive Potential Energy Storage	Linear Stiffness	$\frac{1}{K}\frac{dF_{21}}{dt} = v$	$E = \frac{1}{2}Kx^2$
	Rotational Stiffness	$\frac{1}{K}\frac{d\tau_{21}}{dt} = \omega$	$E = \frac{1}{2}K\theta^2$
	Electrical Capacitance	$C\frac{de_{21}}{dt} = i$	$E = \frac{1}{2}Ce_{21}^2$
	Fluid Capacitance	$C_f\frac{dP_{21}}{dt} = Q$	$E = \frac{1}{2}C_fP_{21}^2$
	Thermal Capacitance	$C_t\frac{dT}{dt} = q$	$E = C_tT$
Energy Dissipation	Linear Damper	$Rv = F_{21}$	$P = Rv^2$
	Rotational Damper	$R\omega = \tau_{21}$	$P = R\omega^2$
	Electrical Resistance	$Ri = e_{21}$	$P = Ri^2$
	Fluid Resistance	$R_fQ = P$ (Note 1)	$P = R_fQ^2$

Note 1: Fluid resistance is often nonlinear and must be linearly approximated (see Section 11.2)

Table 11.3 Physical Properties for Physical System Lumped Elements

Element	Physical Property	Units
Mechanical Translation	Inertia, m	kg
	Viscous Friction, R	N·sec/m
	stiffness, K	N/m
Mechanical Rotation	Inertia, J	kg·m^2
	Viscous Friction, R	N·m·sec/rad
	stiffness, K	N·m/rad
Electrical	Inductance, I	H = v·s/A
	Resistance	Ohm, W
	Capacitance	F = v/(A·s)
Fluid	Inertia	kg/m^4
	Resistance	Pa·sec/m^3= N·sec/m^5
	Stiffness	Pa/m^3=N/m^5

Note 1: the Metric force unit N = kg·m/sec, W=J/sec, Pa = N/m^2

After the Ordinary Differential Equation (ODE) that models the system of interest is obtained, we take the Laplace transform of the ODE and obtain the system's transfer function. This process does not require initial conditions for the ODE because the transfer function generates the ratio of the "forced" response (particular solution) to the input. An example is shown below for a mechanical system's rotational displacement output when excited by a torque input.

Example 11.1

The torsional pendulum in the schematic diagram below is commonly used in timepieces. Find a differential equation for the torsional motion output θ to a torque input τ, where J is the torsional inertia of the rotating mass, R is the viscous drag torsional damping coefficient and K is the torsional stiffness of the support spring.

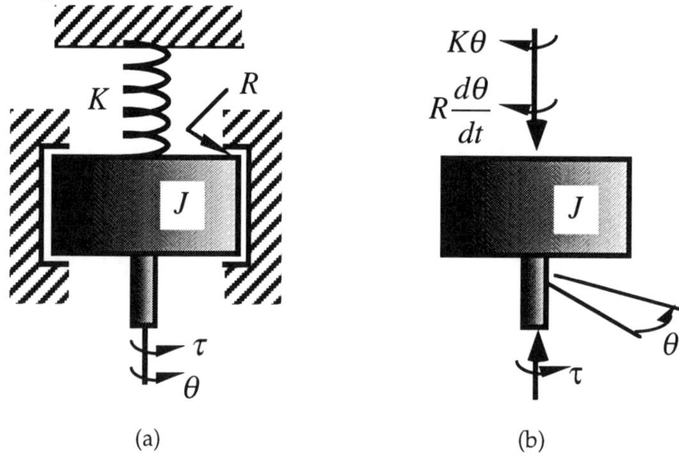

Figure 11.1 (a) Torsional pendulum and (b) free body diagram.

Solution: Apply Newton's law to the torsional pendulum's free body diagram:

$$J\ddot{\theta} = \sum_i \tau_i = \tau - R\dot{\theta} - K\theta$$

Rearrange this result to find a differential equation model in standard form:

$$J\ddot{\theta} + B\dot{\theta} + K\theta = \tau$$

The Laplace transform of this differential equation (see Chapter 1) with zero initial conditions $\theta(0) = \dot{\theta}(0) = 0$ provides

$$Js^2\Theta(s) + Bs\Theta(s) + K\Theta(s) = T(s)$$

Rearrange to yield the *system transfer function*;

$$\frac{\Theta(s)}{T(s)} = \frac{1}{Js^2 + Bs + K}$$

Electrical Circuits

Models of electrical circuits are developed from the five basic circuit elements shown in Figure 11.2 combined with Kirchoff's current and voltage laws.

Kirchoff's Voltage Law: The algebraic sum of voltages around any closed loop in an electrical circuit is zero.

Kirchoff's Current Law: The algebraic sum of currents into any junction in an electrical circuit is zero.

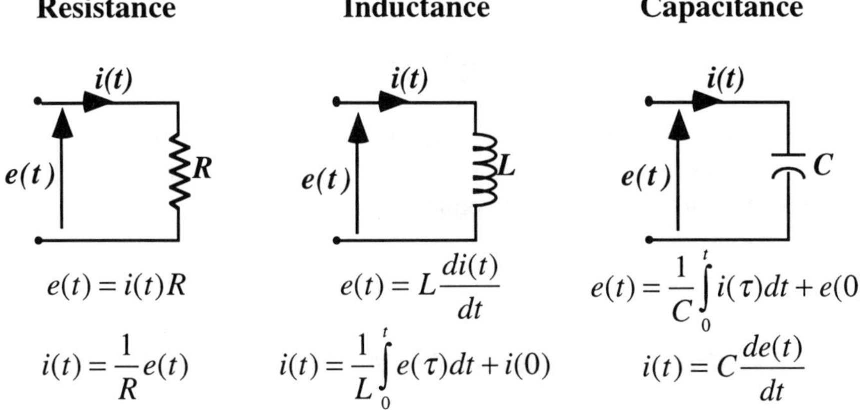

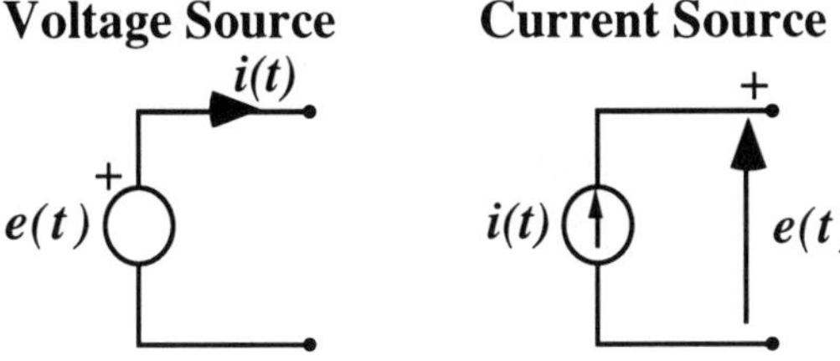

Figure 11.2 Electrical circuit elements.

Example 11.2

In the circuit below, the input is a specified voltage $e_1(t)$ while the output is the measured voltage $e_2(t)$. Find a set of equations representing the circuits dynamics.

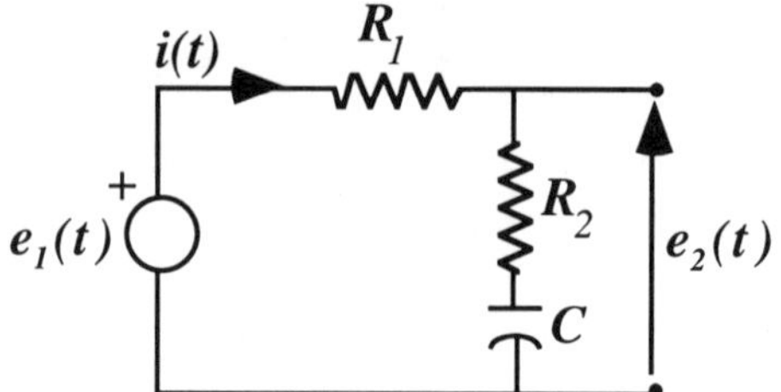

Solution: Use Kirchoff's Voltage Law on the left hand loop:

$$e_1(t) - R_1 i(t) - R_2 i(t) - \frac{1}{C}\int_0^t i(\tau)d\tau = 0$$

Kirchoff's Voltage Law on the right hand loop gives

$$e_2(t) - R_2 i(t) - \frac{1}{C}\int_0^t i(\tau)d\tau = 0$$

Note that both of the above equations assume that the initial current $i(0) = 0$. Now take the Laplace transform of the differential equations (see Chapter 1):

$$E_1(s) - R_1 I(s) - R_2 I(s) - \frac{1}{Cs} I(s) = 0$$

$$E_2(s) - R_2 I(s) - \frac{1}{Cs} I(s) = 0$$

Combine the two transformed equations to remove the variable $I(s)$. This yields the transfer function $G(s)$ that relates the transform of the output voltage $E_2(s)$ to the transform of the input voltage $E_1(s)$:

$$G(s) = \frac{E_2(s)}{E_1(s)} = \frac{R_1 Cs + 1}{(R_1 + R_2)Cs + 1}$$

Hydraulic/Pneumatic Systems

Hydraulic and pneumatic piping systems (see Chapter 6) are modeled (Fig. 11.3) using Newton's Law applied on a control volume mass (see Chapter 6). In this case, Newton's law for an accelerating fluid becomes

$$I_f \frac{dQ}{dt} = \left(\frac{\rho L}{A}\right)\frac{dQ}{dt} = P_1 - P_2 \qquad \textbf{(11.1.3)}$$

where dQ/dt $(\mathrm{m}^3/\mathrm{s}^2)$ is the time derivative of the flow rate through the pipe, ρ $(\mathrm{kg/m}^3)$ is the mass density of the fluid in the pipe and P_i $(\mathrm{N/m}^2 = \mathrm{Pa})$ are the applied pressures on the pipe control volume. The forces on the mass can arise from external pressures, fluid resistance or fluid compliance. Fluid resistance $R(Q)$ is usually nonlinear and often modeled by a pressure difference proportional to the square of flowrate:

$$P_1 - P_2 = R(Q) = K_R Q^2 \qquad \textbf{(11.1.4)}$$

Fluid compliance C_f models the time derivative of pressure with net volume flowrate into a piping segment, $\Delta Q = Q_1 - Q_2$. One example of a linear fluid compliance is the flow of a fluid of mass density ρ into a constant cross section tank of horizontal section area A where

$$\Delta Q = Q_1 - Q_2 = C_f \frac{dP}{dt} = \left(\frac{A}{\rho g}\right)\frac{dP}{dt} \qquad \textbf{(11.1.5)}$$

Fluid Inertia

$$\Delta P = P_1 - P_2 = I_f \frac{dQ}{dt}$$

Fluid Resistance

P_1 P_2 Q Q

$$\Delta P = P_1 - P_2 = R(Q)$$

Fluid Compliance

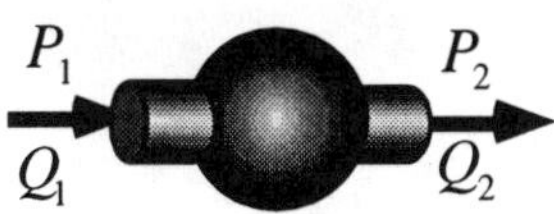

$$\Delta Q = Q_1 - Q_2 = C_f \frac{dP}{dt}$$

Fluid Junction

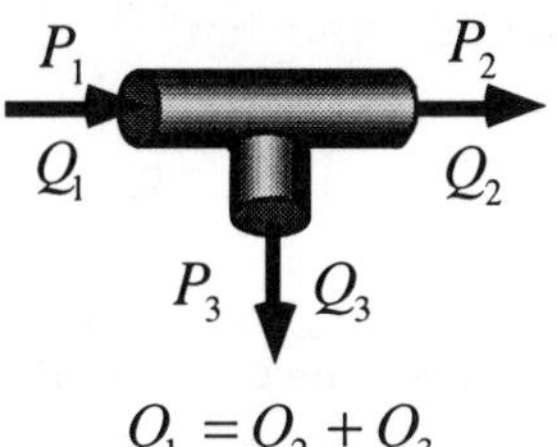

$$Q_1 = Q_2 + Q_3$$

Figure 11.3 Hydraulic and pneumatic fluid system elements

Example 11.3

For a length of pneumatic system piping, find the differential equations describing the relationship between an inlet pressure $P_1(t)$ and the resulting outlet flowrate $Q_3(t)$ for pipe segment back pressure $P_3(t)$. The piping segment model should include the effects of fluid inertia, fluid compliance and fluid resistance.

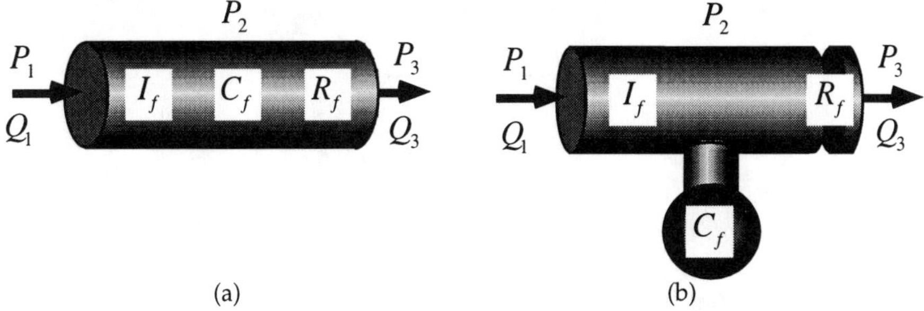

Figure 11.4 Pipe segment including the effects of fluid inertia, compliance and resistance: (a) schematic of pipe segment, (b) pipe model elements

Solution: The solution is developed using the elements of Figure 11.3 with pressures and flows numbered to correspond to Figure 11.4. Modeling of this system requires identification of the model independent and dependent variables. In this case, the model independent (or model input) variables specified are pipe inlet pressure $P_1(t)$ and outlet pressure $P_3(t)$ from which the model dependent (or model output) variable $Q_3(t)$ is derived using the fluid system model. A second model output available is the pipe inlet flowrate $Q_1(t)$ resulting from the specified inlet and outlet pressures.

The pressure across the fluid inertia element I_f is a function of the pipe length L cross sectional area A and fluid mass density ρ:

$$I_f \frac{dQ_1}{dt} = \left(\frac{\rho L}{A}\right)\frac{dQ_1}{dt} = P_1 - P_2$$

The fluid compliance C_f is modeled here to depend on thermodynamic, compressible fluid effects as contrasted with the hydraulic tank model developed earlier. The particular fluid system modeled is dependent on the particular fluid application (see Chapter 6) but fluid compliance $C_f = (Q_1 - Q_2)/(dP/dt)$ is always the ratio of net flowrate to time rate of change in pressure for the particular fluid system modeled. The linear expansion model used here assumes isentropic expansion of the volume of fluid and small changes in fluid density ρ and speed of sound c, so that

$$C_f \frac{dP_2}{dt} = \left(\frac{V}{\rho c^2}\right)\frac{dP_2}{dt} = Q_1 - Q_3$$

The fluid resistance R_f is often a nonlinear function of flow conditions. The Darcy-Weisbach pipe loss relation (see Chapter 6) provides a pressure loss relationship which is a quadratic function of flowrate $Q_3(t)$:

$$\Delta P = (P_2 - P_3) = \rho g \Delta h = \rho g\left(\frac{8fL}{g\pi^2 D^5}\right)Q^2 = \left(\frac{\rho f L}{2A^2 D}\right)Q_3^2 = R_f Q_3^2$$

where f = pipe friction factor and D = pipe diameter. Assemble the three coupled nonlinear algebraic and differential equations:

$$\frac{dQ_1}{dt} = \frac{P_1}{I_f} - \frac{P_2}{I_f}$$

$$\frac{dP_2}{dt} = \frac{Q_1}{C_f} - \frac{Q_3}{C_f}$$

$$Q_3 = \sqrt{(P_2 - P_3)/R_f}$$

The second and third equations can then be combined to form the two nonlinear differential equations

$$\frac{dQ_1}{dt} = \frac{P_1}{I_f} - \frac{P_2}{I_f}$$

$$\frac{dP_2}{dt} = \frac{Q_1}{C_f} - \frac{\sqrt{(P_2 - P_3)}}{C_f\sqrt{R_f}}$$

Given the time histories for inlet pressure $P_1(t)$ and outlet pressure $P_3(t)$, the above model can be numerically simulated for time response of the two states P_2 and Q_1. The output flowrate can then be computed from the solutions for the two states P_2 and Q_1:

$$Q_3 = \sqrt{(P_2 - P_3)\Big/R_f}$$

To derive a fluid system transfer function requires linearization of these equations which will be discussed in the next section. The need to linearize differential equations modeling engineering systems is common.

11.2 Linearization

Linearized models are important because most dynamic and control analysis is based on linear mathematics, however, mechanical engineering system components almost never have linear behavior. To perform a linearization of a nonlinear engineering system mathematical model requires three distinct steps.

1. ***Identification of the System Operating Point*** of interest and the system variable values at that point. This operating point must be a steady-state (or equilibrium) state for the system.
2. ***Application of the Taylor Expansion*** about the Operating Point for each differential equation in terms of the system variables.
3. ***A change of variables*** from the original system variables to a new set of variables representing deviations from the operating point values.

Note that the change of variables is always required except in the trivial case where the operating point corresponds to all system variables equal to zero.

As an example of linearization, consider the nonlinear set of equations describing the dynamics of the hydraulic system found in the fluid system modeling example:

$$\frac{dQ_1}{dt} = \frac{P_1}{I_f} - \frac{P_2}{I_f}$$

$$\frac{dP_2}{dt} = \frac{Q_1}{C_f} - \frac{\sqrt{(P_2 - P_3)}}{C_f\sqrt{R_f}} \qquad \textbf{(11.2.1)}$$

$$Q_3 = \sqrt{(P_2 - P_3)\Big/R_f}$$

1. ***Identification of the System Operating Point***

At an equilibrium operating point defined by the steady-state pressure inputs to the pipe segment $P_1 = \overline{P}_1$ and $P_3 = \overline{P}_3$, all state derivatives are zero yielding coupled non-linear, algebraic equations:

$$0 = \frac{\overline{P}_1}{I_f} - \frac{\overline{P}_2}{I_f}$$

$$0 = \frac{\overline{Q}_1}{C_f} - \frac{\sqrt{(\overline{P}_2 - \overline{P}_3)}}{C_f\sqrt{R_f}} \quad \textbf{(11.2.2)}$$

$$\overline{Q}_3 = \sqrt{(\overline{P}_2 - \overline{P}_3)/R_f}$$

The equilibrium operating equations can be solved for the steady-state operating point values of the system variables $\overline{P}_2$, $\overline{Q}_3$ and $\overline{Q}_1$ in terms of the specified steady-state pressure inputs $\overline{P}_1$ and $\overline{P}_3$:

$$\overline{P}_2 = \overline{P}_1$$

$$\overline{Q}_3 = \sqrt{(\overline{P}_1 - \overline{P}_3)/R_f} \quad \textbf{(11.2.3)}$$

$$\overline{Q}_1 = \overline{Q}_3$$

Because model equations are nonlinear, no single equilibrium solution is guaranteed. Multiple steady-state equilibria are possible. For each steady-state equilibrium, a separate linearized model is necessary.

2. ***Application of the Taylor Expansion***

The second differential equation and the algebraic equation in Eq.11.2.1 have nonlinear right hand sides and need to be linearized:

$$\frac{dP_2}{dt} = f_2(Q_1, P_2, P_3) = \frac{Q_1}{C_f} - \frac{\sqrt{(P_2 - P_3)}}{C_f\sqrt{R_f}}$$

$$Q_3 = f_3(P_2, P_3) = \sqrt{(P_2 - P_3)/R_f} \quad \textbf{(11.2.4)}$$

The two nonlinear functions $f_2(Q_1, P_2, P_3)$ and $f_3(P_2, P_3)$ are linearized by a Taylor expansion about the steady-state operating point values for steady-state pressure inputs $\overline{P}_1$ and $\overline{P}_3$:

$$f_2(Q_1, P_2, P_3) = f_2(\overline{Q}_1, \overline{P}_2, \overline{P}_3) + \left.\frac{\partial f_2}{\partial Q_1}\right|_{P_2, P_3}(Q_1 - \overline{Q}_1) + \left.\frac{\partial f_2}{\partial P_2}\right|_{Q_1, P_3}(P_2 - \overline{P}_2)$$

$$+ \left.\frac{\partial f_2}{\partial P_3}\right|_{Q_1, P_2}(P_3 - \overline{P}_3) + HOT \quad \textbf{(11.2.5)}$$

$$f_3(P_2, P_3) = f_3(\overline{P}_2, \overline{P}_3) + \left.\frac{\partial f_3}{\partial P_2}\right|_{P_3}(P_2 - \overline{P}_2) + \left.\frac{\partial f_3}{\partial P_3}\right|_{P_2}(P_3 - \overline{P}_3) + HOT$$

The first terms in the Taylor expansions represent the steady state values of the functions. Evaluating the partial derivative at the steady state values and neglecting the higher order terms (HOT) in the expansion yields the almost-linear functions

$$f_2(Q_1,P_2,P_3) \approx f_2(\overline{Q}_1,\overline{P}_2,\overline{P}_3)+\left(\frac{1}{C_f}\right)(Q_1-\overline{Q}_1)-\left(\frac{1}{C_fR}\right)(P_2-\overline{P}_2)+\left(\frac{1}{C_fR}\right)(P_3-\overline{P}_3)$$

$$f_3(P_2,P_3) \approx f_3(\overline{P}_2,\overline{P}_3) + \left(\frac{1}{R}\right)(P_2-\overline{P}_2) - \left(\frac{1}{R}\right)(P_3-\overline{P}_3)$$

(11.2.6)

where the constant $R = 2\sqrt{R_f(\overline{P}_2-\overline{P}_3)}$. Note that the above equations are not yet linear because they do not obey the superposition property for the six system variables P_i and Q_i required of linear equations . The next step will solve this final difficulty.

3. *A Change of Variables*

To complete linearization, a new set of system variables must be defined as deviations from the steady state operating values of the original system variables:

$$\begin{aligned}\hat{Q}_1 &= (Q_1-\overline{Q}_1), & \hat{Q}_2 &= (Q_2-\overline{Q}_2), & \hat{Q}_3 &= (Q_3-\overline{Q}_3)\\ \hat{P}_1 &= (P_1-\overline{P}_1), & \hat{P}_2 &= (P_2-\overline{P}_2), & \hat{P}_3 &= (P_3-\overline{P}_3)\end{aligned} \qquad \textbf{(11.2.7)}$$

This system model change of variables has two important properties. First, because the new system variables are defined around an equilibrium point, the system model time derivatives at the operating point vanish. Second, the derivatives of the new deviation variables are equal to the derivatives of the original variables:

$$\begin{aligned}\frac{d\hat{P}_2}{dt} &= \frac{d}{dt}(P_2-\overline{P}_2) = \frac{dP_2}{dt}-0 = \frac{dP_2}{dt}\\ \frac{d\hat{Q}_1}{dt} &= \frac{d}{dt}(Q_2-\overline{Q}_2) = \frac{dQ_2}{dt}-0 = \frac{dQ_2}{dt}\end{aligned} \qquad \textbf{(11.2.8)}$$

With these two important properties, substituting in the new deviation variables into the truncated Taylor expansion gives a fully linear set of system differential and algebraic equations:

$$\begin{aligned}\frac{d\hat{Q}_1}{dt} &= \frac{1}{I_f}\hat{P}_1 - \frac{1}{I_f}\hat{P}_2\\ \frac{d\hat{P}_2}{dt} &= \frac{1}{C_f}\hat{Q}_1-\left(\frac{1}{C_fR}\right)\hat{P}_2+\left(\frac{1}{C_fR}\right)\hat{P}_3\\ \hat{Q}_3 &= \left(\frac{1}{R}\right)\hat{P}_2-\left(\frac{1}{R}\right)\hat{P}_3\end{aligned} \qquad \textbf{(11.2.9)}$$

where again, the constant $R = 2\sqrt{R_f(\overline{P}_2-\overline{P}_3)}$. These linearized equations can now be Laplace transformed for initial conditions $\hat{P}_i(0)=0$ and $\hat{Q}_i(0)=0$:

$$s\hat{Q}_1(s) = \frac{1}{I_f}\hat{P}_1(s) - \frac{1}{I_f}\hat{P}_2(s)$$

$$s\hat{P}_2 s = \frac{1}{C_f}\hat{Q}_1(s) - \frac{1}{C_f R}\hat{P}_2(s) + \frac{1}{C_f R}\hat{P}_3(s) \qquad \textbf{(11.2.10)}$$

$$\hat{Q}_3(s) = \frac{1}{R}\hat{P}_2(s) - \frac{1}{R}\hat{P}_3(s)$$

To find the system transfer write the three simultaneous (11.2.10) equations in matrix form:

$$\begin{bmatrix} s & \frac{1}{I_f} & 0 \\ \frac{-1}{C_f} & s + \left(\frac{1}{C_f R}\right) & 0 \\ 0 & \frac{1}{R} & -1 \end{bmatrix} \begin{bmatrix} \hat{Q}_1(s) \\ \hat{P}_2(s) \\ \hat{Q}_3(s) \end{bmatrix} = \begin{bmatrix} \frac{\hat{P}_1(s)}{I_f} \\ \frac{\hat{P}_3(s)}{C_f R} \\ \frac{\hat{P}_3(s)}{R} \end{bmatrix} \qquad \textbf{(11.2.11)}$$

where $R = 2\sqrt{R_f\left(P_{2s} - P_{3s}\right)}$. This set of equations can be solved using Kramer's Rule (see Chapter 1) to find the system transfer functions between output flow deviation and the inlet pressure deviations $\hat{P}_1(s)$ and $\hat{P}_3(s)$:

$$\hat{Q}_3(s) = \frac{\frac{1}{I_f C_f R}}{s^2 + \frac{1}{RC_f}s + \frac{1}{I_f C_f}}\hat{P}_1(s) - \frac{\frac{1}{R}s^2 + \frac{1}{I_f C_f R}}{s^2 + \frac{1}{RC_f}s + \frac{1}{I_f C_f}}\hat{P}_3(s) \qquad \textbf{(11.2.12)}$$

The transfer function between inlet pressure $\hat{P}_1(s)$ and the resulting flowrate $\hat{Q}_3(s)$ is

$$G(s) = \frac{\hat{Q}_3(s)}{\hat{P}_1(s)} = \frac{\frac{1}{I_f C_f R}}{\left(s^2 + \left(\frac{1}{RC_f}\right)s + \left(\frac{1}{I_f C_f}\right)\right)} \qquad \textbf{(11.2.13)}$$

11.3 Block Diagrams

Control system analysis requires the dynamic performance analysis of control systems assembled from independent components modeled by their transfer functions. The most common method of modeling these connected components is the *block diagram*. The simplest block diagram (Fig. 11.5) is for a component

$$G(s) = \frac{C(s)}{E(s)} \tag{11.3.1}$$

The combination of two linked components requires the two *linked blocks* shown in Figure 11.6. The system transfer function is the product of the transfer functions of the two components

$$G(s) = \frac{C(s)}{E(s)} = \frac{C(s)}{A(s)}\frac{A(s)}{E(s)} = G_1(s)G_2(s) \tag{11.3.2}$$

The second block diagram element is the *summing junction* shown in Figure 11.7 which sums the input signals with the sign noted to form the single system output signal

$$E(s) = R_1(s) + R_2(s) - C(s) \tag{11.3.3}$$

The third block diagram element is the *signal node* shown in Figure 11.8 in which all node output signals are equal to the single node input signal

$$E_1(s) = E_2(s) = R(s) \tag{11.3.4}$$

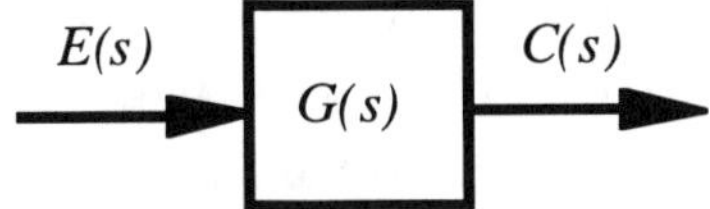

Figure 11.5 Block diagram component for single transfer function, $G(s) = C(s)/E(s)$

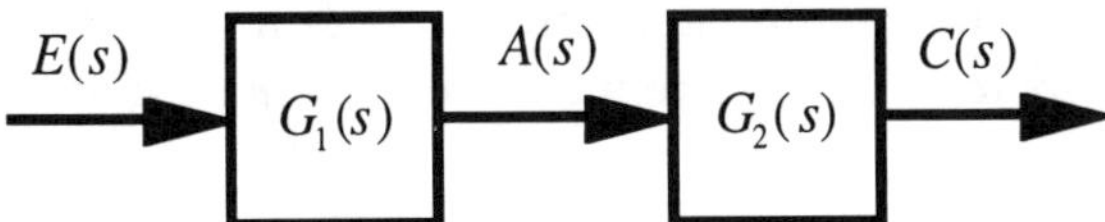

Figure 11.6 Block diagram for a system composed of two components, $G(s) = C(s)/E(s) = G_1(s)G_2(s)$

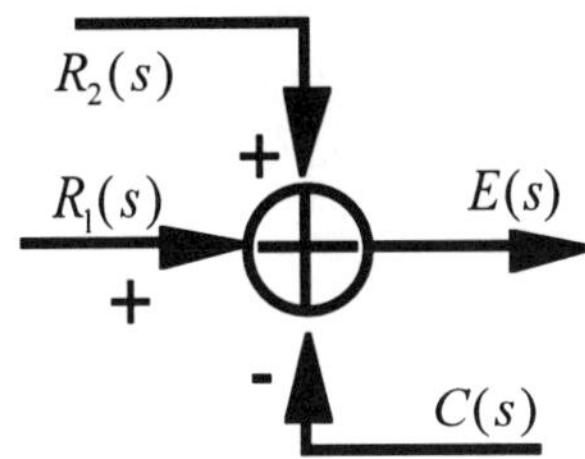

Figure 11.7 Block diagram summing junction, $E(s) = R_1(s) + R_2(s) - C(s)$

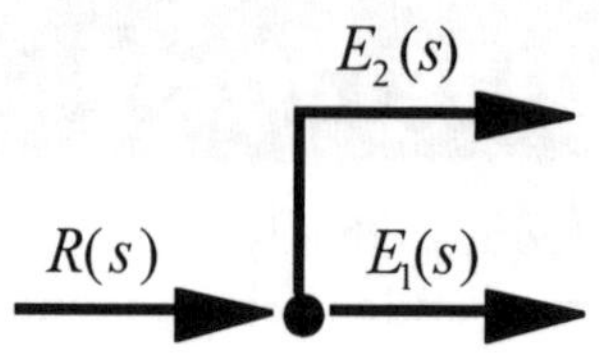

Figure 11.8 Block diagram signal node, $E_1(s) = E_2(s) = R(s)$

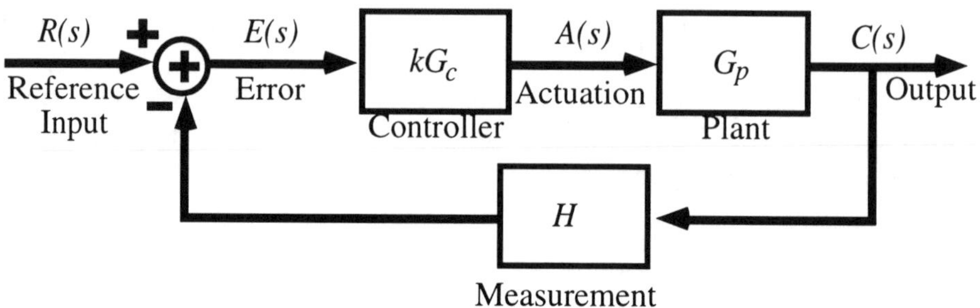

Figure 11.9 Block diagram of a standard feedback loop

The three block diagram elements: blocks, summing junctions and signal nodes can be used to solve for the system transfer function of an assembly of large numbers of system components.

The simplest and most important such assembly is the *standard feedback loop* shown in Figure 11.9. The three component transfer function blocks in this feedback system are the controller $kG_c(s)$, the plant $G_p(s)$, and the measurement system $H(s)$. The signals represented by arrows are the reference input $R(s)$, the error $E(s)$, the actuation $A(s)$, and the system output $C(s)$.

Notice that the feedback loop causes the output of the summing junction $E(s)$ to be a function of itself:

$$E(s) = R(s) - H(s)C(s) = R(s) - kG_c(s)G_p(s)H(s)E(s) \qquad \textbf{(11.3.5)}$$

The output of each summing junction in one or more feedback loops must be found algebraically to find the system transfer function. In this case,

$$E(s) = \frac{R(s)}{1 + kG_c(s)G_p(s)H(s)} \qquad \textbf{(11.3.6)}$$

The system output, $C(s) = kG_c(s)G(s)E(s)$, so that the overall transfer function of the standard feedback loop system is

$$T(s) = \frac{C(s)}{R(s)} = \frac{kG_c(s)G_p(s)}{1 + kG_c(s)G_p(s)} \qquad \textbf{(11.3.7)}$$

11.4 Control Components

Feedback controllers have many different designs, however, the two most common designs encountered in applications are Proportional-Integral-Derivative (PID) and Lead-Lag controllers. These controllers can be designed to optimize transient response, frequency response or steady-state error. The control function $kg_c(e,t)$ generates output actuation $a(t)$ in response to input measured error $e(t)$:

$$a(t) = kg_c(e,t) \tag{11.4.1}$$

The above time function relationship has a Laplace transform based transfer function,

$$kG_c(s) = \frac{A(s)}{E(s)} \tag{11.4.2}$$

Proportional-Integral-Derivative Control

The Proportional-Integral-Derivative (PID) controller has three terms

$$a(t) = k\left[e(t) + k_I \int_0^t e(\tau)d\tau + k_D \frac{de(t)}{dt}\right] \tag{11.4.3}$$

The actuation signal generated by the PID controller is the weighted sum of the current error, the error integral and the error derivative through the proportional gain $k_P = k$, the integral gain k_I, and the derivative gain k_D. Taking the Laplace transform of the above time function, the transfer function of the PID controller is

$$K(s) = k\left[1 + \frac{k_I}{s} + k_D s\right] \tag{11.4.4}$$

The form cited here is often used in industrial practice because setting proportional gain $k = k_p = 0$ turns off all actuation output independent of integral and derivative gains. In general,

- Proportional action increases system speed
- Integral action removes steady-state error but destabilizes response
- Derivative action stabilizes response while reducing system response speed

Lead-Lag Control

Producing stable approximations for both integration and differentiation can be difficult in practice. One common approximation is the lead-lag controller with the transfer function,

$$K(s) = k\frac{(s+a)}{(s+b)} \tag{11.4.5}$$

The numerator contributes a zero at $s = a$ and the lead-lag controller can be configured to approximate differentiation when $b >> a$ and $a \to 0$. In contrast, the denominator contributes a pole at $s = b$ and the lead-lag controller can be configured to approximate integration when as $a >> b$ and $b \to 0$.

11.5 Stability and The Characteristic Polynomial

The stability of a dynamic system can be evaluated through the *eigenvalues* of the differential equation model of the system. The eigenvalues of the system model are the roots of the model's characteristic polynomial. The characteristic polynomial is the denominator of the system's transfer function.

For system transfer function $G(s) = N(s)/D(s)$, the polynomial $D(s)$ is the system characteristic polynomial.

For system state equations, $\dot{\mathbf{x}} = \mathbf{A}\mathbf{x} + \mathbf{b}u$, the polynomial, $D(s) = \text{determinant}(s\mathbf{I} - \mathbf{A})$ is the system characteristic polynomial.

For linear, time-invariant system models, the characteristic polynomial has real coefficients and has polynomial order equal to the order of the differential equation model from which it was derived. The roots of the characteristic polynomial (eigenvalues) are either real values or pairs of complex conjugate values. The position of the roots (eigenvalues) gives the system stability associated with each root.

A root (eigenvalue), $\lambda_i = \alpha + \beta j$, (see Fig. 11.10)

- is stable if (and only if), $\alpha < 0$ (termed "negative real")
- is marginally stable if (and only if), $\alpha = 0$ (termed "imaginary")
- is unstable if (and only if), $\alpha > 0$ (termed "positive real")

A system is stable if and only if all its roots (eigenvalues) are negative real. Because the location of the roots, the "root locus", is so important to the analysis of system stability, special methods for plotting the location of system roots have been developed.

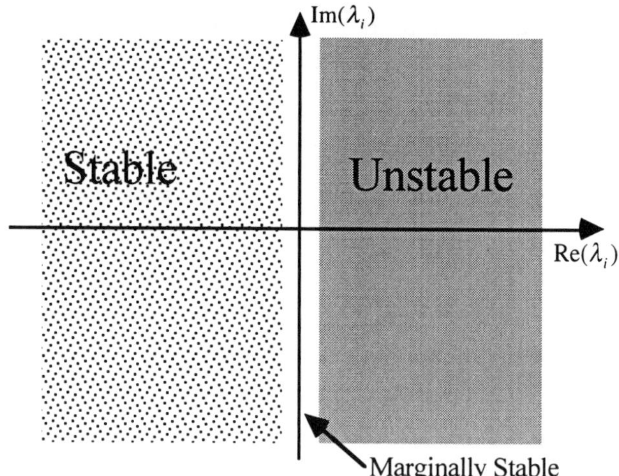

Figure 11.10 Stability and Eigenvalue Location where the Eigenvalues, $\lambda_i = \alpha + \beta j$ are the Roots of the System Characteristic Equation

11.6 Construction of a Root Locus Diagram

Control Design Problem

Analyzing differential equation models, through transfer functions and block diagram always gives a *closed-loop transfer function* in the form,

$$T(s)=\frac{C(s)}{R(s)}=\frac{kG_cG_p}{1+kG_cG_pH} \quad \textbf{(11.6.1)}$$

where G_c, G_p, H are the open-loop control, plant and measurement transfer functions and k is the open-loop gain constant (Fig. 11.9). The root locus method finds the roots of the closed-loop characteristic equation with the observation that the roots of the closed-loop characteristic equation are completely governed by the term referred to as the "open-loop transfer function", kG_cG_pH. The first step is to write the closed-loop denominator in the special form,

$$1+kG_cG_pH=1+k\frac{N(s)}{D(s)}=0 \quad \textbf{(11.6.2)}$$

and further observing that, at solutions (roots) of "$1+kG_cG_pH=0$", the condition

$$k\frac{N(s)}{D(s)}=-1 \quad \textbf{(11.6.3)}$$

For positive real gain k, the transfer function ratio must satisfy,

$$angle\left[\frac{N(s)}{D(s)}\right]=\pm 180\,\text{deg} \quad \textbf{(11.6.4)}$$

Root Locus Construction Rules ($k\geq 0$)

To sketch the root loci:

1. The $k=0$ points on the root loci are the roots of $D(s)$ (poles of $G_c(s)G_p(s)H(s)$).
2. The $k=\infty$ points on the root loci are the roots of $N(s)$ (zeros of $G_c(s)G_p(s)H(s)$).
3. The number of branches on the root loci is equal to the order of the characteristic equation, max(n, m).
4. The root loci are symmetric with respect to the real axis of the s-plane. (In general, the root loci are symmetric with respect to axes of symmetry of the roots of $D(s)$ and $N(s)$).
5. The root loci have asymptotes with angles,

   $$\theta_k=\frac{(2k+1)\pi}{\|n-m\|} \text{ for } n\neq m \text{ and } k=0,1,2,\cdots,[\|n-m\|-1]$$

6. The intersection of the asymptotes lies on the real axis at

   $$s=\sigma=\frac{\sum_n\left(\text{poles of } G_c(s)G_p(s)H(s)\right)-\sum_m\left(\text{zeros of } G_c(s)G_p(s)H(s)\right)}{(n-m)}$$

7. The root loci includes the real axis to the left of an odd number of roots of $D(s)$ and $N(s)$ (poles and zeros of $G_c(s)G_p(s)H(s)$).

8. a) The angle of departure, θ_d, from a complex root of $D(s)$ (an O.L. pole) is $\sum[\theta_d + \text{ angles from other poles}] - \sum[\text{angles from zeros}] = 180^o$

   b) The angle of arrival, θ_a, at a complex root of $N(s)$ (an O.L. zero) is $\sum[\text{angles from poles}] - \sum[\theta_a + \text{ angles from other zeros}] = 180^o$

9. The intersection of the root loci with the imaginary axis and the corresponding values of k are given by the *Routh Hurwitz criterion*.

10. a) Break-away points occur along a root locus at a relative maximum value of k.

    b) Break-in points occur along a root locus at a relative minimum value of k.

11. The magnitude of the control gain $k = \left\|\frac{D(s)}{N(s)}\right\| = \left\|\frac{1}{G_c(s)G_p(s)H(s)}\right\|$

General Parameter Design Problem

Start by factoring the system characteristic equation with a variable parameter k, into the form

$$D(s) + kN(s) = 0 \tag{11.6.5}$$

where $D(s)$ and $N(s)$ are polynomials in s and the parameter takes on the role of the control gain k. Then convert the parameter design problem to the control design problem:

$$1 + k\frac{N(s)}{D(s)} = 0$$

Note that $\frac{N(s)}{D(s)}$ is equivalent to $G_c(s)G_p(s)H(s)$ in most texts.

11.7 Bode Diagram

The log-magnitude and phase angle of transfer functions plotted against log frequency are called *Bode plots* or *Bode diagrams.* (See Fig. 11.12 and 11.13). Bode plots of transfer functions can be easily approximated as a sequence of straight lines. These straight line approximations simplify the evaluation of the magnitude and phase angle of transfer function frequency responses. The derivation of the asymptotic, straight line approximation is available in all texts on control theory and only the templates for these approximations are repeated here. The construction of a *Bode diagram* involves four steps.

1. ***Factor the numerator and denominator polynomials*** of the transfer function into a numerator constant, first order terms, second order terms, and a power of s term:

$$G(s) = \frac{k\,\Pi\left[(s/a_z)+1\right]\,\Pi\left[(s/\omega_n)^2 + 2\zeta(s/\omega_n)+1\right]_z}{s^m\,\Pi\left[(s/a_p)+1\right]\,\Pi\left[(s/\omega_n)^2 + 2\zeta(s/\omega_n)+1\right]_p} \tag{11.7.1}$$

The transfer function as expressed above has a slightly different form from that most conveniently used to express transfer functions as ratios of polynomials. All terms are normalized such that their steady-state gain is unity and included in gain k. This form has poles and/or zeros at the origin, s^m, factored from the first order, $[(s/a)+1]$, and second order, $\left[(s/\omega_n)^2 + 2\zeta(s/\omega_n)+1\right]$, terms in the transfer function. Each first and second order term is then divided by its steady-state gain so as to leave each term's steady state gain equal to one (0 decibel).

2. ***Plot each term in the factored transfer function*** separately. The transfer function log magnitude and phase angle of the complete transfer function versus log frequency is formed by adding the contributions from each term evaluated separately. The contributions from each terms are easily approximated using the straight line approximations below. All results are plotted on decibels where

$$k(dB) = 20\log_{10}(k)$$

3. ***Compute magnitude corrections*** at each term's corner frequency. Here the magnitudes associated with zeros will be given followed by the magnitude for poles in parentheses:

   First Order Zero (Pole): $1/\sqrt{2} = 0.707 = 3$ dB (-3 dB)

   Second Order Zero (Pole): 2ζ $(1/2\zeta)$

4. ***Sketch in the frequency response*** using the straight line asymptotes and the magnitude corrections at the corner frequencies.

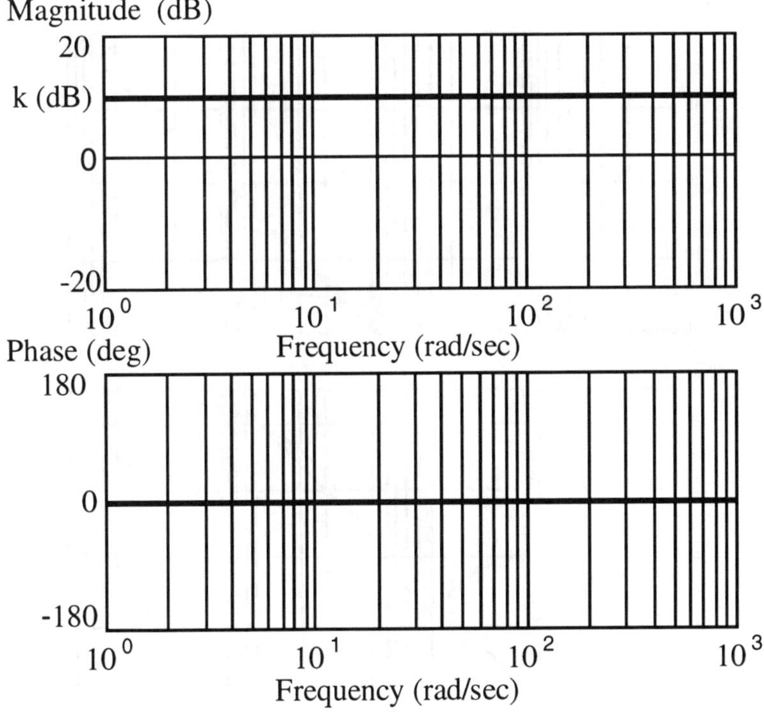

Figure 11.12 Bode Diagram for a Pure Gain, *k*

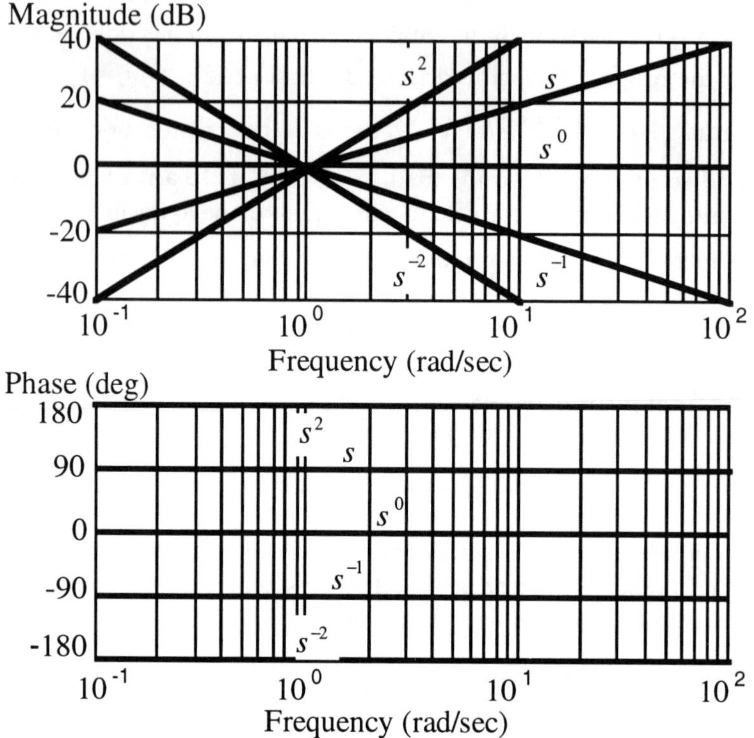

Figure 11.13 Bode diagram for a increasing powers of *s*. Note the 20 dB/ decade increase in slope and 90 degree phase accompanying each power of *s*.

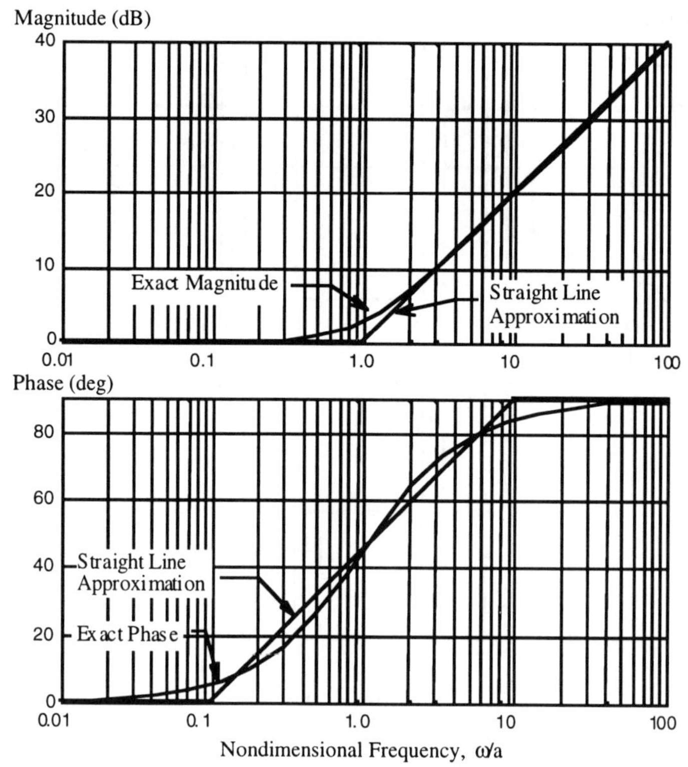

Figure 11.14 Bode diagram for a first order zero $\left[(s/a)+1\right]$

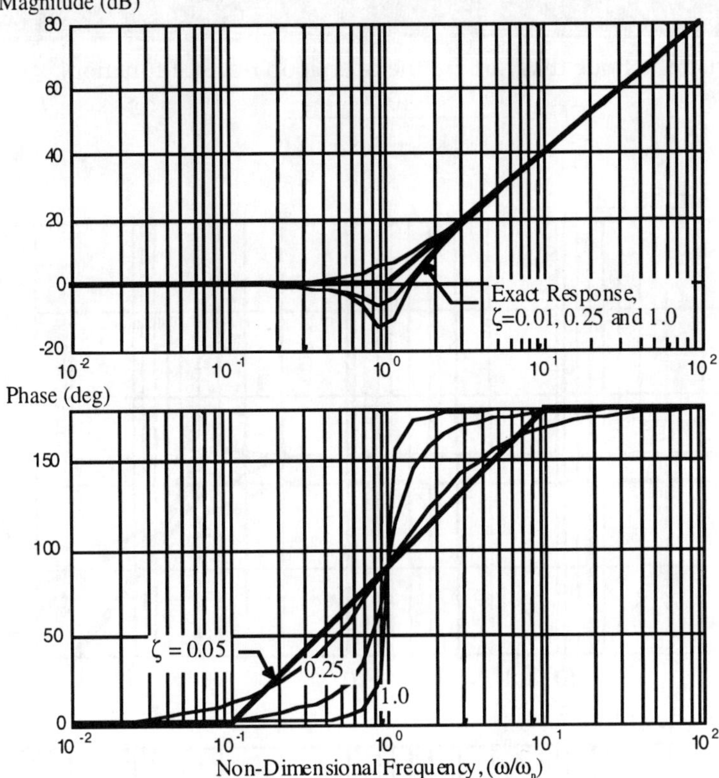

Figure 11.15 Bode diagram for a second order zero

$\left[(s/\omega_n)^2 + 2\zeta(s/\omega_n) + 1\right]$.

Example 11.6

Construct the Bode diagram for the open loop transfer function,

$$GH = \frac{10}{\left(s^3 + 4s^2 + 30s + 52\right)}$$

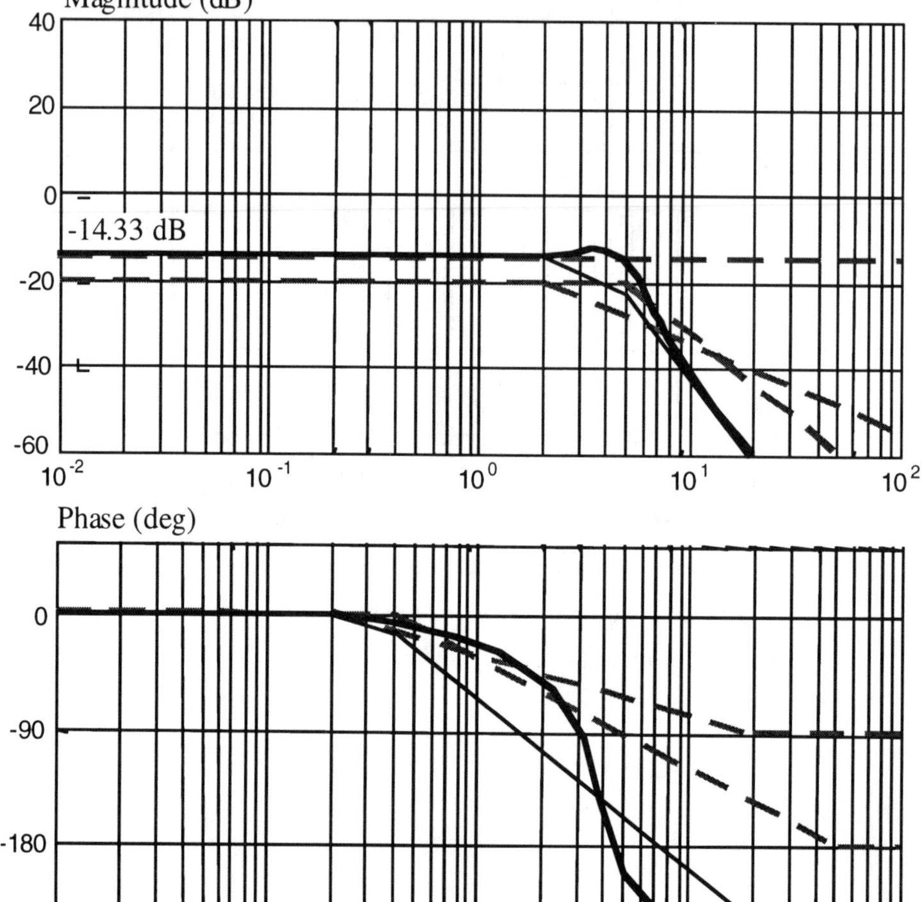

Bode Diagram using Straight Line Approximation

Solution: First factor the transfer function:

$$GH = \frac{1.92}{\left[(s/2)+1\right]\left[(s/5.10)^2 + 2(0.196)(s/5.10) + 1\right]}$$

The transfer function has a real pole at $s = -2$, and a pair of complex conjugate poles at $-1 \pm 5j$. The complex poles have natural frequency $\omega_n = \sqrt{5^2 + 1^2} = \sqrt{26} = 5.10$ rad/sec and damping ratio $\zeta = \left(1/\sqrt{26}\right) = 0.196$. The steady-state gain is 20 log(10/52)= 20 log(0.192) = -14.3 dB.

Sketching numerator gain and each of the denominator terms and adding the results yields the straight line approximation to the transfer function shown. The individual term straight line asymptotes are shown sketched in as dashed lines while the sums of the three terms are sketched as thin solid lines in both the log-magnitude and phase diagrams in the approximate Bode diagram. The solid black lines then indicate the "exact" Bode Diagram sketched with the corner frequency magnitude correction. The magnitude at the first order pole's corner frequency is -3 dB while at the magnitude at the second order pole's corner frequency is computed as

$$\Delta GH(\omega_n) = 1/2\zeta = 1/0.4 = 2.5 = 7.96\,dB$$

This approximate example indicates a stability gain margins of approximately 16 dB with and infinite phase margin. Exact Bode diagram use to determine stability is demonstrated in the next example.

Example 11.7

Find the Gain and Phase margins from the exact Bode Diagram for the previous example's open loop transfer function,

$$GH = \frac{10}{\left(s^3 + 4s^2 + 30s + 52\right)}$$

Solution: The stability gain margins is approximately 17 dB while the phase margin is infinite because the Bode diagram gain does not pass through the 0 dB level.

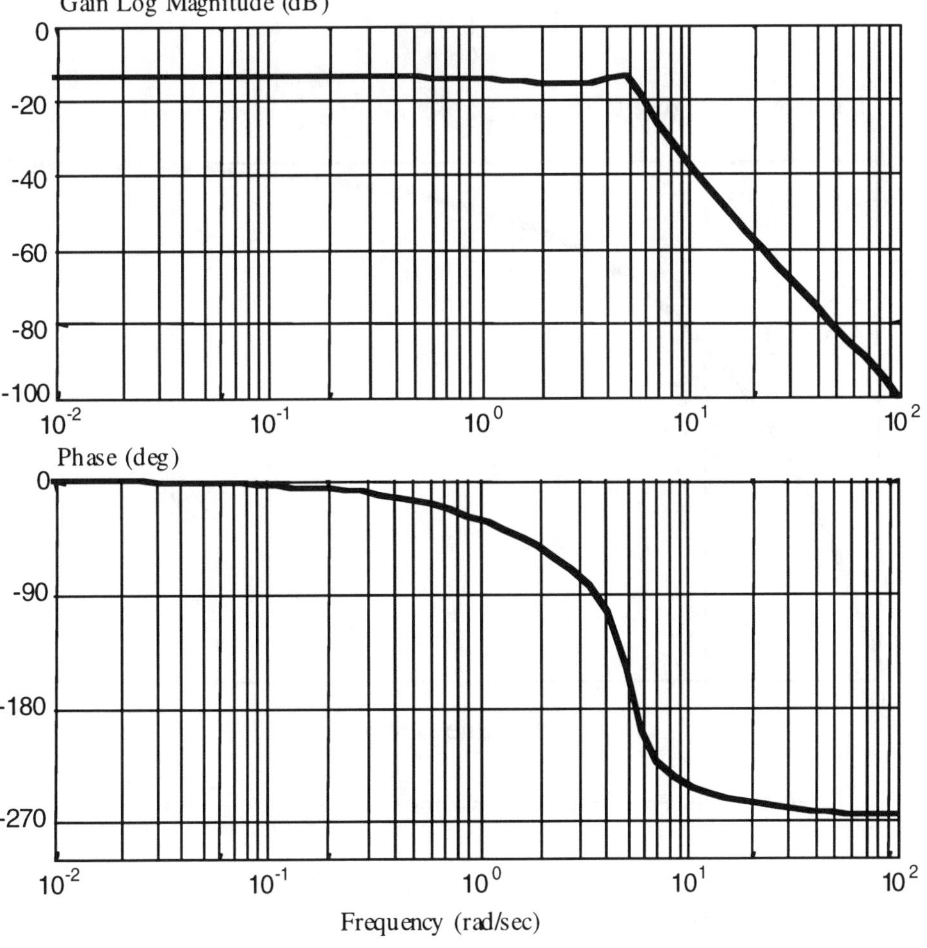

Bode Diagram for $GH = \dfrac{10}{\left(s^3 + 4s^2 + 30s + 52\right)}$

11.8 First and Second Order Systems

The dynamic response of systems is often described through the use of typical first and second order system dynamic responses. Responses similar to that of a first order system typically arise in systems whose dynamic behavior does not have the property of energy alternately passing between kinetic and potential energy. Second order system response is characterized by the interchange of energy between different energy storage mechanisms resulting in oscillatory system response.

First Order Step Response

The standard first order transfer function is

$$G(s) = \frac{C(s)}{R(s)} = \frac{K}{\tau s + 1} = \frac{b}{s + a} \tag{11.8.1}$$

where $K = b / a$ is the steady-state gain, and $\tau = 1 / a$ is the system time-constant. Transfer function (11.8.1) has the unit step response

$$c(t) = K\left(1 - e^{-\frac{t}{\tau}}\right) \tag{11.8.2}$$

It is shown in Figure 11.16.

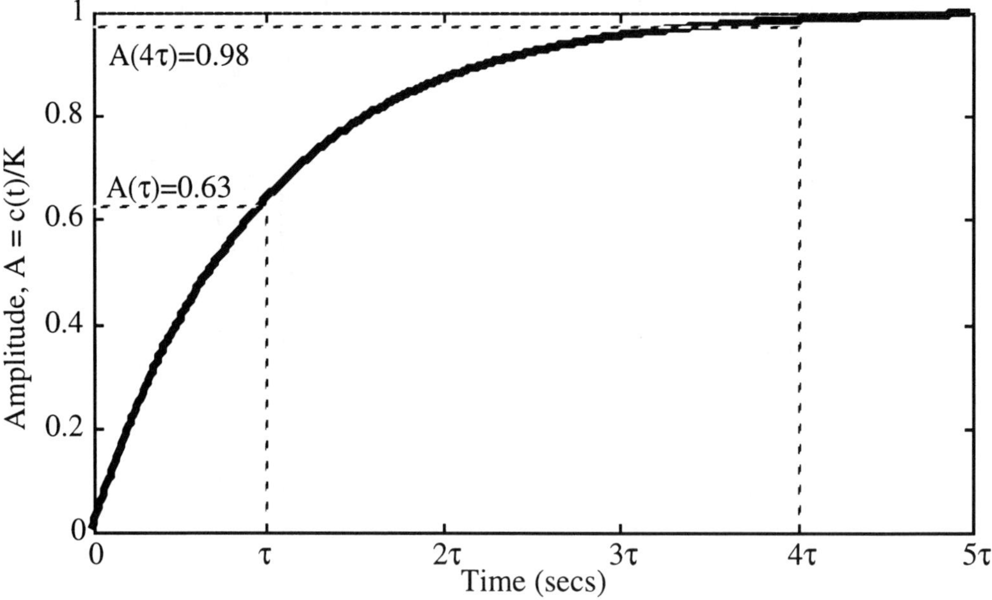

Figure 11.16 First Order Response for $G(s) = \frac{C(s)}{R(s)} = \frac{K}{\tau s + 1} = \frac{b}{s + a}$

There are three important points to notice on the first order step response. The first is the system travels 63% of the way to steady state in one time constant. The second is the system is within 2% of the steady state value after 4 time constants. The third is the output/input ratio at steady-state as $t \to \infty$, $c(\infty)/r(\infty) = K$ the steady state gain of the system.

Second Order Step Response

The second order transfer function,

$$G(s) = \frac{C(s)}{R(s)} = \frac{\omega_n^2}{s^2 + 2\zeta\omega_n + \omega_n^2} \qquad (11.8.3)$$

has the following unit step response and is shown in Figure 11.17:

$$c(t) = 1 - \frac{e^{-\zeta\omega_n t}}{\sqrt{1-\zeta^2}} \sin(\omega_n \sqrt{1-\zeta^2}\, t + \cos^{-1}\zeta) \qquad (11.8.4)$$

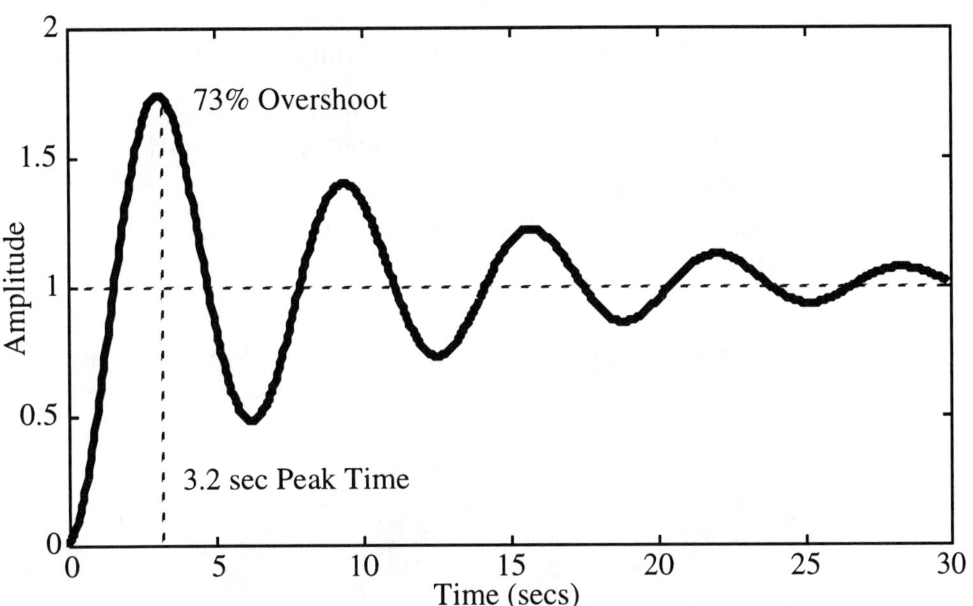

Figure 11.17 Second Order Step Response for 0.1 Damping Ratio and 1 rad/sec Natural Frequency

For the above case, the rise time

$$t_r \cong \frac{(0.8 + 2\zeta)}{\omega_n} = \frac{(0.8 + 0.2)}{1.0} = 1.0 \qquad (11.8.5)$$

and

$$\text{overshoot} \cong e^{\left[-\pi\zeta / \sqrt{1-\zeta^2}\right]} = \exp\left[-0.314 / 0.99\right] = 0.7292 = 73\% \qquad (11.8.6)$$

Practice Problems (Essay)

11.1 A Block Diagram of a controlled system is shown. The peak magnification M_p of the closed-loop frequency response is specified to be 1.155.

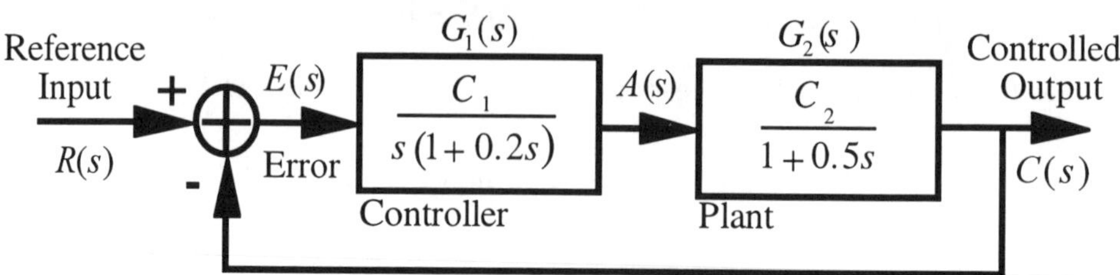

a) Find the damping ratio ζ associated with the specified peak magnification M_p.

b) Determine the corresponding open-loop gain constant K_g gain margin and phase margin.

c) Describe how you would determine the frequency ω_m at which the peak magnification occurs.

11.2 Aerodynamic drag is an inherently non-linear phenomenon that is commonly encountered in the control of vehicle speed. Examine the vehicle below traveling at velocity $v = 30\ \text{m/s}$.

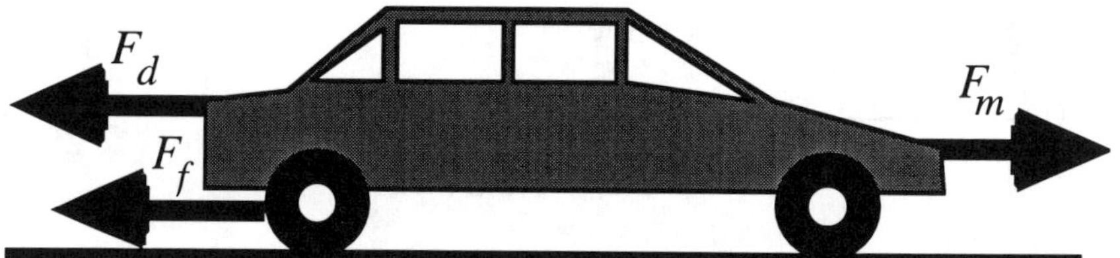

Figure 1: Vehicle with Non-Linear Drag Forces

The vehicle is pushed forward by the tractive force $F_m = F_m(t)$ from the engine, while the vehicle is slowed down by the action of linear viscous friction $F_f = K_f v$ and aerodynamic drag $F_a = K_a v^2$ proportional to vehicle velocity squared. Model the tractive force as an input to the system. Use $m = 1{,}000$ kg, $K_f = 0.1\ \text{N} \cdot \text{sec} / \text{m}$, and $K_a = 1.0\ \text{N} \cdot \text{sec}^2 / \text{m}$.

a) Using Newton's law and the above free-body diagram to find the vehicle equation of motion.

b) Linearize the model at a 100 kph operating speed, express the linearized model in state space form, and find the system transfer function.

c) Design a control system with a 10 second time constant and zero steady state error at this operating speed.

Practice Problems (Multiple Choice)

Questions 11.3-10:

A control system has a Laplace domain transfer function,

$$G(s) = \frac{Y(s)}{F(s)} = \frac{3s+1}{(s+1)(s+2)}$$

11.3 What is the system's impulse response $y(t)$ for $f(t)$ = unit impulse?

a) $-3e^{-t} + e^{-2t}$

b) $-2e^{-t} + 5e^{-2t}$

c) $\left(\frac{7}{3}\right)e^{-t} + \left(\frac{4}{3}\right)e^{-2t}$

d) $2e^{-t} - 5e^{-2t}$

e) $\left(\frac{4}{3}\right)e^{-t} + \left(\frac{7}{3}\right)e^{-2t}$

11.4 What is the system's step response $y(t)$ for $f(t)$ = unit step?

a) 0.5

b) $0.5\left(1-e^{-t}\right)\left(1-e^{-2t}\right)$

c) $2e^{-t} - 2.5e^{-2t} + 0.5$

d) 1.0

e) $0.5e^{-t}e^{-2t}$

11.5 What is the system's steady-state gain?

a) 0.5

b) 0.75

c) 1.0

d) 1.5

e) 2.0

11.6 What is the system's frequency response at $f = 3$ cycles/sec ?

a) magnitude = 0.1581, phase angle = 81.92 degree lag

b) magnitude = 0.1581, phase angle = 81.92 degree lead

c) magnitude = 6.327, phase angle = 81.92 degree lag

d) magnitude = 6.327, phase angle = 81.92 degree lead

e) magnitude = 10, phase angle = 0.00 degree

11.7 What is the system order determined from the transfer function?

a) 1

b) 2

c) 3

d) 4

e) 5

11.8 What is the system's stability indicated by the transfer function?

a) stable

b) marginally stable

11.9 What is/are the location(s) of the system pole(s)?

a) 1, 2 and 3
b) 1, 2, and 1/3
c) −1, −2
d) −1/3
e) 0.0

11.10 What is/are the location(s) of the system zero(s)?

a) 1, 2 and 3
b) 1, 2, and 1/3
c) −1, −2
d) −1/3
e) 0.0

Questions 11.11-12:

Unit negative feedback and proportional control with gain K_p is shown:

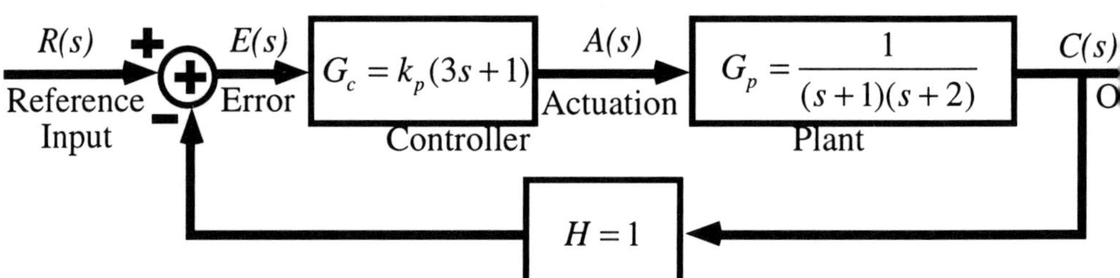

11.11 The system above is

a) unstable for $k_p > 0$
b) stable for all values of k_p
c) stable for $k_p > -1$
d) unstable for $k_p > -1$
e) unstable for all values of k_p

11.12 What value of gain k_p yields the least stable closed-loop system eigenvalue?

a) $k_p = 0$
b) $k_p = 1$
c) $k_p = 2$
d) $k_p = 10$
e) $k_p = \infty$

Questions 11.13-22:

For a plant with state equation model shown,

$$\dot{c} = -3c + 5a(t)$$

A proportional control is developed using a sensor with transfer function $H(s)$.

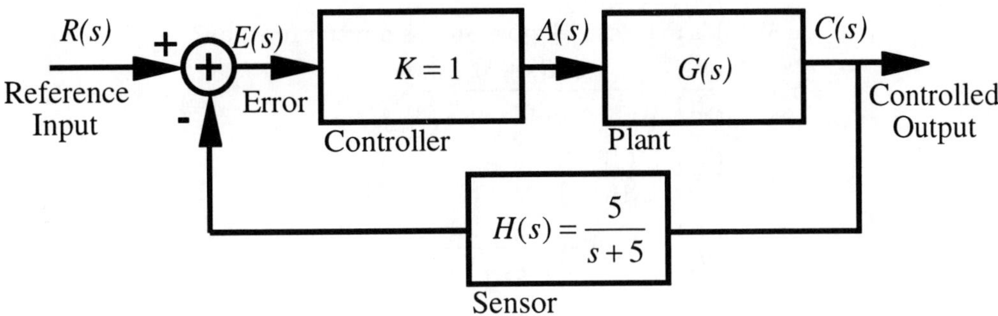

11.13 What is the plant's transfer function?

a) $C(s)/A(s) = 3/(s+5)$
b) $C(s)/A(s) = 5/(s+3)$
c) $C(s)/A(s) = 1/(3s+5)$
d) $C(s)/A(s) = 1/(5s+3)$
e) $C(s)/A(s) = 5/(s+5)$

11.14 What is the plant's steady-state gain?

a) 3/5
b) 3
c) 5
d) 5/3
e) 1

11.15 What is the magnitude of the plant's frequency response at 4 rad/sec?

a) 5.0
b) 3.0
c) 1.0
d) 0.5
e) 0.0

11.16 What is the plant's slowest time constant?

a) 0.2 sec
b) 0.33 sec
c) 3.0 sec
d) 5.0 sec
e) 10 sec

11.17 Based on the plant's model, what is the plant's stability?
a) stable
b) marginally stable
c) marginally unstable
d) unstable
e) Can't determine stability from the plant model

11.18 What is the closed-loop system's transfer function?
a) $\frac{C(s)}{R(s)} = \frac{5K(s+5)}{s^2+8s+(15+25K)}$
b) $\frac{C(s)}{R(s)} = \frac{K(s+5)}{(s+3)}$
c) $\frac{C(s)}{R(s)} = 1+K\frac{25}{s^2+8s+15}$
d) $\frac{C(s)}{R(s)} = \frac{s^2+8s+25}{s^2+8s+15}$
e) $\frac{C(s)}{R(s)} = \frac{1}{s^2+8s+15}$

11.19 What is the closed-loop system's steady-state gain?
a) $\frac{25K}{15+25K}$
b) $\frac{5K}{3}$
c) $\frac{15+25K}{15}$
d) 5/3
e) 1

11.20 What is the closed-loop system's frequency response magnitude at 4 rad/sec for $K = 1$?
a) 0
b) 0.32
c) 0.64
d) 1.28
e) 2.56

11.21 At $K = 1$, what is the closed-loop system pole contributes the slowest response component to the closed-loop system response?
a) $s \cong -3$
b) $s \cong -4$
c) $s \cong -4 \pm 5i$
d) $s \cong -5$
e) $s = 0$

11.22 Based on the closed-loop system's model, what positive values of K yield a stable closed-loop system?

a) $0 \le K \le 0.04$
b) $0.04 \le K \le \infty$
c) $0 \le K \le \infty$
d) $0 < K$
e) can't determine stability from the plant model

Solutions to Problems

11.1 First find the damping ratio ζ associated with the specified peak magnification M_p. The specified peak magnification $M_m = 1.15$ is the maximum magnitude of the closed loop transfer function, $G_{CL} = G_{OL}/(1+G_{OL})$ at some frequency to be determined.

a) For a second order system, the relationship between peak magnification M_p and damping ratio ζ is $M_m = 1/\left[2\zeta\sqrt{1-\zeta^2}\right] \Leftrightarrow \zeta = \sqrt{M_m - \sqrt{M_m^2 - 1}}$. For our case,

$$\zeta = \sqrt{1.155 - \sqrt{1.155^2 - 1}} = \sqrt{1.155 - \sqrt{0.3340}} \cong 0.76$$

b) The open-loop transfer function is $G_{OL} = K_g/[s(1+0.2s)(1+0.5s)]$ where $K_g = C_1C_2$.

Set up the closed-loop transfer function,

$$M_m = \|G_{CL}\| = \left\|\frac{G_{OL}}{1+G_{OL}}\right\| = \left\|\frac{\left[\dfrac{K_g}{s(1+0.2s)(1+0.5s)}\right]}{\left[1+\dfrac{K_g}{s(1+0.2s)(1+0.5s)}\right]}\right\| = \left\|\frac{10K_g}{s^3+7s^2+10s+10K_g}\right\|$$

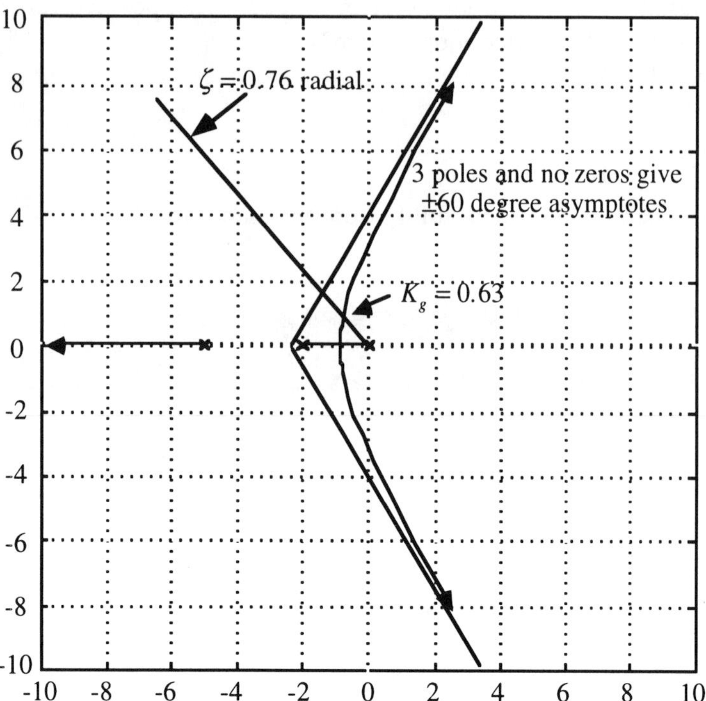

Draw the Root Locus and indicate the radial where pole positions have the required damping ratio, $\zeta = 0.76$ using the special form of the characteristic equation,

$$1 + K_g \frac{10}{s^3 + 7s^2 + 10s} = 1 + K_g \frac{10}{s(s+2)(s+5)} = 0$$

The characteristic equation is third order with one real and two complex conjugate poles.

The Root Locus displays an intersection of the $\zeta = 0.76$ radial at $s \cong 0.82 + 0.70j$. Substituting this value of s into the relationship for control gain yields,

$$K_g = \frac{-s(s+2)(s+5)}{10}$$

$$\cong \frac{-(-0.82 + 0.7i)(-0.82 + 0.7i + 2)(-0.82 + 0.7i + 5)}{10} \cong 0.63$$

Drawing a Bode Diagram yields the gain and phase margins. Using this $K_g = 0.63$ in the open-loop gain, $G_{OL} = 6.3/s(s+2)(s+5)$.

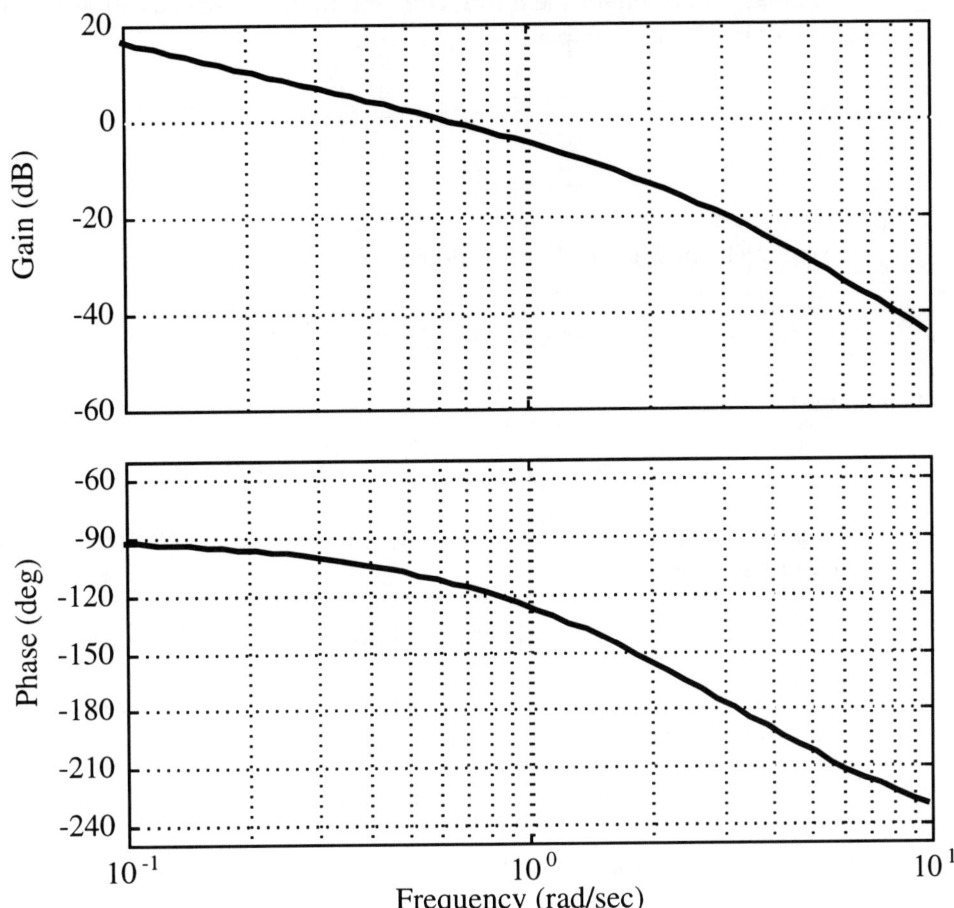

The Bode Diagram shows a gain margin of 20 dB at $\omega = 3$ rad / sec and a phase margin of approximately 70 degrees at $\omega = 0.6$ (rad / sec).

c) The frequency at which the peak closed loop response is expected is at the damped natural frequency associated with the complex conjugate poles for $K_g = 0.63$. Since the closed loop poles are located at $cs = 0.82 + 0.70j$, the damped natural frequency for these poles is at $\omega_m = 0.70$ rad / sec.

11.2

a) The first order, nonlinear, ordinary differential equation system model is

$$m\dot{v} = F_m(t) - F_f(v) - F_a(v) = F_m(t) - K_f v - K_a v^2$$

b) The aerodynamic drag term is the cause of the nonlinearity. Linearization about the "operating point" $v = v_0 =$ 30 meters/sec gives the relationship,

$$F_a(v) = K_a v^2 = F_a(v_0) + \left.\frac{\partial F_a}{\partial v}\right|_{v=v_0} (v - v_0) + HOT$$

$$= K_a v_0^2 + (2K_a v_0)(v - v_0) + HOT$$

A similar expansion for the linear, coulomb friction force has all higher order derivatives in the Taylor expansion equal to zero:

$$F_f(v) = K_f v = F_f(v_0) + \left.\frac{\partial F_a}{\partial v}\right|_{v=v_0} (v - v_0) + HOT$$

$$= K_f v_0 + (K_f)(v - v_0)$$

To complete the linearization, we define a new velocity,

$$\tilde{v} = (v - v_0), \text{ with } \dot{\tilde{v}} = \frac{d(v - v_0)}{dt} = \dot{v}$$

and the ODE becomes

$$m\dot{\tilde{v}} = F_m(t) - (K_f v_0 + K_f \tilde{v}) - (K_a v_0^2 + 2K_a v_0 \tilde{v} + HOT)$$

We now can find an linear approximation to the original ODE by neglecting second order terms.

$$m\dot{\tilde{v}} + (K_f + 2K_a v_0)\tilde{v} = F_m(t) - K_f v_0 - K_a v_0^2 = \tilde{F}_m(t)$$

where $\tilde{F}_m(t) = F_m(t) - K_f v_0 - K_a v_0^2$ is the tractive force deviation from the operating point. We have

$$m = 1000 \text{ kg}, \; K_f = 0.1 \text{ N} \cdot \text{sec} / \text{m}, \text{ and } K_a = 1.0 \text{ N} \cdot \text{sec}^2 / \text{m}^2$$

For steady-state velocity, v = 100 kph = 27.8 m/sec with $\dot{v} = 0$, the operating point tractive effort F_m=(0.1)(27.8)+(1.0)(27.8²) = 776 N so that

$$\tilde{F}_m(t) = F_m(t) - 776 \text{ N}$$

For $K_f + 2K_a v_0 = 0.1 + 2(1)(27.8) = 55.7$, the linearized system differential equation becomes

$$1000\dot{\tilde{v}} + 55.7\tilde{v} = \tilde{F}_m(t)$$

which can be expressed in state space form as

$$\dot{\tilde{v}} = -0.0557\tilde{v} + 0.001\tilde{F}_m(t)$$

The system transfer function is found from the Laplace transform of the system model:

$$G(s) = \frac{\tilde{V}(s)}{\tilde{F}_m(s)} = \frac{0.001}{s + 0.0557} = \frac{0.018}{18s + 1}$$

The vehicle model has a single real root at s = -0.0557 and a time constant of 18.0 sec with a steady state gain of 0.018.

c) The open-loop vehicle model is has a single real pole and no poles at the origin so requires integral control action for zero steady-state error. The specification requires a time constant of $\tau = 10$ sec which is faster than the 18 sec open-loop time constant requiring control action proportional to error to speed up the closed loop system response. Propose an Proportional-Integral (PI) control $G_1(s) = K_P + K_I/s$ and sketch the closed-loop block diagram below assuming unity measurement gain.

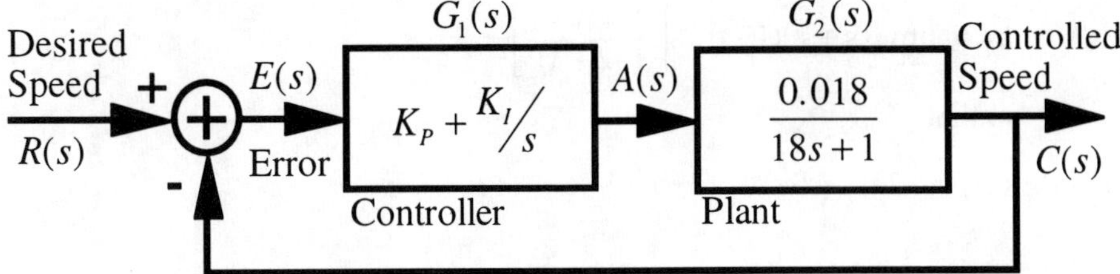

The open-loop transfer function becomes

$$G_{OL}(s) = G_1(s)G_2(s) = K_P\left[\frac{s+a}{s}\right]\left[\frac{0.018}{18s+1}\right] = 0.001K_P\left[\frac{s+a}{s}\right]\left[\frac{1}{s+0.0557}\right]$$

where $a = K_I/K_P$, the ratio of integral to proportional gains. To examine the effect of proportional and integral gains on closed-loop response, sketch the system root locus in terms of the parameter a.

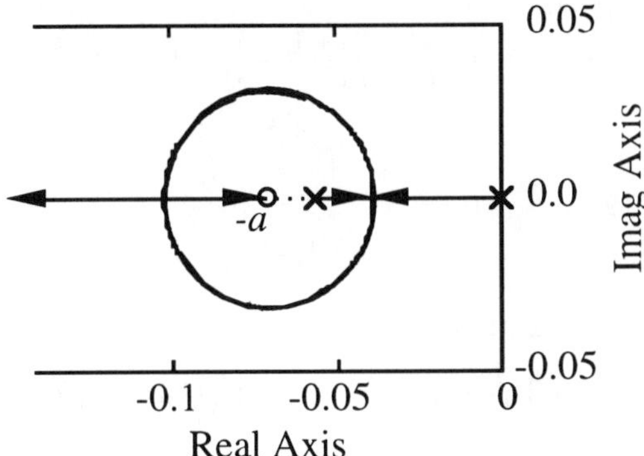

We desire a closed-loop pole at s = -0.1 for $\tau = 10$ sec so place $a = K_I/K_P$ = 0.07 then calibrate for K_P where s = -0.1 based on the Root Locus rules. Calibrating the Root Locus (rule 11), for $a = K_I/K_P$ =0.07 at s = -0.1 yields

$$K_P = \left[\frac{-s}{s+a}\right]\left[\frac{18s+1}{0.018}\right] = \left[\frac{0.1}{-0.1+0.07}\right]\left[\frac{18(-0.1)+1}{0.018}\right] = \left[-3.3\overline{31}\right]\left[-44.\overline{4}\right] = 148$$

and for $a = K_I/K_P$ =0.07

$$K_I = 0.07K_P = 10.4$$

For the system transfer function,

$$G(s) = \frac{Y(s)}{F(s)} = \frac{3s+1}{(s+1)(s+2)}$$

11.3 **b)** The impulse response is

$$g(t) = \mathcal{L}^{-1}(G(s)) = \mathcal{L}^{-1}\left(\frac{3s+1}{(s+1)(s+2)}\right) = \mathcal{L}^{-1}\left(\frac{k_1}{s+1} + \frac{k_2}{s+2}\right)$$

where $k_1 = \lim_{s\to -1}\left((s+1)G(s)\right) = \left[\frac{3(-1)+1}{(-1+2)}\right] = -2$

$$k_2 = \lim_{s\to -2}\left((s+2)G(s)\right) = \left[\frac{3(-2)+1}{(-2+1)}\right] = 5$$

so that

$$g(t) = -2e^{-t} + 5e^{-2t}$$

11.4 **c)** The step response is

$$y(t) = \mathcal{L}^{-1}(G(s)U(s)) = \mathcal{L}^{-1}\left(\frac{3s+1}{(s+1)(s+2)}\frac{1}{s}\right) = \mathcal{L}^{-1}\left(\frac{k_1}{s+1} + \frac{k_2}{s+2} + \frac{k_3}{s}\right)$$

where $k_1 = \lim_{s\to -1}\left((s+1)G(s)U(s)\right) = \left[\frac{3(-1)+1}{(-1+2)}\left(\frac{1}{-1}\right)\right] = 2$

$$k_2 = \lim_{s\to -2}\left((s+2)G(s)U(s)\right) = \left[\frac{3(-2)+1}{(-2+1)}\left(\frac{1}{-2}\right)\right] = -5/2 = -2.5$$

$$k_3 = \lim_{s\to 0}\left((s)G(s)U(s)\right) = \left[\frac{3(0)+1}{(0+1)(0+2)}\right] = 1/2 = 0.5$$

so that

$$y(t) = 2e^{-t} - 2.5e^{-2t} + 0.5$$

11.5 **a)** The system's steady state gain is

$$G(s=0) = (3*0+1)/((0+1)(0+2)) = 1/(1*2) = 1/2 = 0.5$$

11.6 **a)** The frequency response of the system $Y(\omega)/F(\omega)$ at $\omega = 2\pi f = 2\pi(3) = 18.85 \text{ rad / sec}$ is

$$\frac{Y(\omega)}{F(\omega)} = \frac{(3)18.85j+1}{(18.85j+1)(18.85j+2)} = 0.0222 - 0.1565j = 0.1581e^{-1.4298j}$$

The frequency response has a magnitude ratio, $\|Y(\omega)/F(\omega)\| = 0.1581$ and a phase angle,

$$\text{angle}[Y(\omega)/F(\omega)] = -1.4298 \text{ rad} = -81.92 \text{ deg}$$

11.7 **b)** The system's order is 2 based on the order of the denominator polynomial, choice.

11.8 **a)** The system's two poles are both "negative real" so the system is "stable".

11.9 **c)** The (2) poles are at $s = -1$ and -2.

11.10 **d)** The single zero is at $s = -1/3$.

11.11 **a)** To determine closed-loop stability, sketch the root-locus:

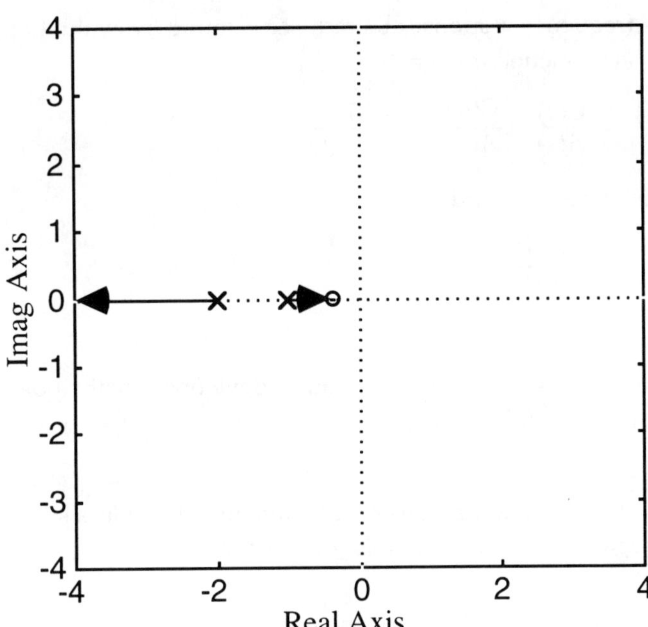

The system is closed loop stable for all positive values of k_p. To examine the stability for negative values of k_p, write the closed-loop characteristic equation, $1 + k_p G_c G_p H = 0$ as

$$1 + k_p \frac{3s+1}{(s+1)(s+2)} = 0 \Rightarrow (s+1)(s+2) + k_p(3s+1) = 0 \quad \text{or}$$

$$s^2 + \left(3 + 3k_p\right)s + \left(2 + k_p\right) = 0$$

which has negative real roots when

$$\left(3 + 3k_p\right) > 0 \Rightarrow k_p > -1 \quad \text{and} \quad \left(2 + k_p\right) > 0 \Rightarrow k_p > -2$$

Simultaneous application of the two conditions requires

$$k_p > -1$$

11.12 **e)** The root locus and stability analysis above indicates the system's least stable positive control gain is $K_p = \infty$.

11.13 **b)** For the plant model differential equation given, the Laplace transform is $sC(s) = -3C(s) + 5A(s)$ yielding the plant transfer function

$$C(s)/A(s) = 5/(s+3)$$

11.14 **d)** The steady-state gain of the plant is the value of the plant transfer function at $s = 0$,

$$C(0)/A(0) = 5/(0+3) = 5/3$$

11.15 **c)** The frequency response at 4 rad/sec is found by evaluating the plant transfer function for $s = 4j$:

$$\frac{C(s)}{A(s)} = \frac{C(4j)}{A(4j)} = \frac{5}{(4j+3)}$$

which has magnitude,

$$\left\|\frac{C(4j)}{A(4j)}\right\| = \frac{5}{\sqrt{4^2+3^2}} = 1$$

11.16 **b)** The plant's slowest time constant is developed by the pole at $s = -3$, $\tau = 1/3 \text{ sec} = 0.33 \text{ sec}$.

11.17 **a)** Based on the plant's transfer function, the plant has a single pole which is negative real. The plant is therefore stable.

11.18 **a)** Using the plant transfer function above, the system closed-loop transfer function is

$$T(s) = \frac{C(s)}{R(s)} = \frac{KG}{1+KGH} = \left[\frac{5K}{(s+3)}\right] \Big/ \left[1 + \frac{25K}{(s+3)(s+5)}\right]$$

which simplifies to

$$T(s) = \frac{5K(s+5)}{\left(s^2+8s+15\right)+25K}$$

11.19 **a)** The closed-loop system steady-state gain is the value of the closed-loop transfer function at $s = 0$,

$$T(0) = 5K(0+5) \Big/ \left[\left(0^2 + 8\times 0 + 15\right) + 25K\right] = 25K/[15+25K],$$

11.20 **c)** The closed-loop frequency response at 4 rad/sec is found by evaluating the closed-loop transfer function for $s = 4j$,

$$T(4j) = \frac{5K(4j+5)}{\left[\left(-16+8(4j)+15\right)+25K\right]} = \frac{(25+20j)}{[24+32j]}$$

which has magnitude,

$$\|T(4j)\| = \sqrt{\left(25^2+20^2\right)} \Big/ \sqrt{24^2+32^2} = 1025/1600 = 0.64$$

11.21 **c)** The closed-loop system's slowest time constant can be identified through a root locus. First the closed-loop characteristic equation is written in the standard form as $1 + KGH = 0 \Rightarrow 1 + \frac{25K}{(s+3)(s+5)} = 0$ where $N(s) = 25$ and $D(s) = (s+3)(s+5) = s^2 + 8s + 15$ The sketch below shows the resulting Root Locus.

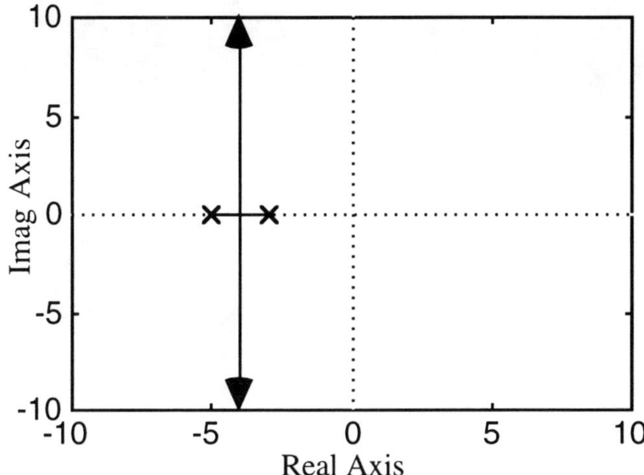

To calibrate the Root Locus, select values of the Laplace variable "s" and compute the Gain, $K = -D(s)/N(s)$ on the indicated paths of the Root Loci. At $K = 1$, both closed-loop poles are equidistant from the imaginary axis at $s \cong -4 \pm 5j$ and contribute equally slow components to the time response.

s	K
-3.0+ 0.0j	0
-3.5+ 0.0j	0.03
-4.0+ 0.0j	0.04
-4.0 + 0.5j	0.05
-4.0 + 1.0j	0.08
-4.0 + 1.5j	0.13
-4.0 + 2.0j	0.20
-4.0 + 2.5j	0.29
-4.0 + 3.0j	0.40
-4.0 + 3.5j	0.53
-4.0 + 4.0j	0.68
-4.0 + 4.5j	0.85
-4.0 + 5.0j	1.04

11.22 **c)** The Root Locus indicates that the closed-loop system has negative-real poles for all positive gains K.

Heat Transfer

by Keith Woodbury

"Heat transfer" is the transmission of energy due only to a temperature difference. There are three modes of heat transfer: *conduction* in solids, *convection at solid-fluid interfaces,* and *radiation* through vacuum and gas-filled spaces. All three modes are dependent on various properties of the solids and fluids involved in the heat transfer process. Many of these properties are tabulated in the Appendices at the end of this chapter after the problems.

12.1 Conduction

The flow of heat in solids is proportional to the local temperature gradient through the material conductivity according to Fourier's Law:

$$\vec{q} = -k\nabla T \quad \textbf{(12.1.1)}$$

The temperature field in the solid is found by solution of the appropriate form of the heat conduction equation:

$$\rho c_p \frac{\partial T}{\partial t} = \nabla \cdot (k\nabla T) + q''' \quad \textbf{(12.1.2)}$$

where q''' accounts for heat generation. This equation must be combined with the appropriate boundary and initial conditions to determine the temperature distribution in the solid.

For one-dimensional problems, the temperature distribution is a function of only one space coordinate. The appropriate forms of Eq. 12.1.2 for one-dimensional heat flow are listed in Table 12.1.

Table 12.1 One-Dimensional Heat Conduction Equation in various coordinate systems.

System	Equation (k = constant)
Cartesian	$\rho c_p \frac{\partial T}{\partial t} = k\frac{\partial^2 T}{\partial x^2}$
Cylindrical	$\rho c_p \frac{\partial T}{\partial t} = \frac{k}{r}\frac{\partial}{\partial r}\left(r\frac{\partial T}{\partial r}\right)$
Spherical	$\rho c_p \frac{\partial T}{\partial t} = \frac{k}{r^2}\frac{\partial}{\partial r}\left(r^2\frac{\partial T}{\partial r}\right)$

12.2 Steady One-Dimensional Conduction

The steady-state temperature distribution in a solid in each of the three coordinate systems is found by solving the steady form of the appropriate equation from Table 12.1. For constant temperature boundaries T_1 (at x_1 or r_1) and T_2 (at x_2 or r_2), these distributions are shown in Table 12.2.

The heat flux through any of these systems follows from the Fourier Law (Eq 12.1.1) and heat flow results from multiplying by the area. These results are summarized in Table 12.3 for each of the three systems. Note that for both the cylindrical and spherical systems the heat *flux* is a function of the coordinate r, but the total heat flow is constant (independent of the radial coordinate).

Table 12.2 Steady One-Dimensional Temperature Distributions

System	Distribution
Cartesian	$T(x) = T_1 + \frac{(T_2 - T_1)}{(x_2 - x_1)}(x - x_1)$
Cylindrical	$T(r) = \frac{(T_1 - T_2)}{\ln(r_1 / r_2)}\ln\left(\frac{r}{r_2}\right) + T_2$
Spherical	$T(r) = \frac{(T_2 - T_1)}{(1/r_1 - 1/r_2)}\left(\frac{1}{r_1} - \frac{1}{r}\right) + T_1$

Table 12.3 Heat Flux and Heat Flow for Steady 1-D Systems

System	Heat Flux	Heat Flow
Cartesian	$q'' = k\frac{(T_1 - T_2)}{(x_2 - x_1)}$	$q = kA\frac{(T_1 - T_2)}{(x_2 - x_1)}$
Cylindrical	$q'' = \frac{k}{r}\frac{(T_1 - T_2)}{\ln(r_2 / r_1)}$	$q = 2\pi kL\frac{(T_1 - T_2)}{\ln(r_2 / r_1)}$
Spherical	$q'' = \frac{k}{r^2}\frac{(T_1 - T_2)}{(1/r_1 - 1/r_2)}$	$q = 4\pi k\frac{(T_1 - T_2)}{(1/r_1 - 1/r_2)}$

Thermal Resistance

The fact that the heat flow through any of the one-dimensional systems is constant gives rise to the *thermal resistance analogy*. The steady heat flow through a one-dimensional body is proportional to the overall temperature difference by the inverse of the thermal resistance. The thermal resistances for the different coordinate systems are shown in Table 12.4.

Table 12.4 Thermal Conduction Resistances for Different Coordinate Systems

System	Thermal Resistance
Cartesian	$R_{th} = \dfrac{\Delta x}{kA}$
Cylindrical	$R_{th} = \dfrac{\ln(r_o / r_i)}{2\pi kL}$
Spherical	$R_{th} = \dfrac{(1/r_i - 1/r_o)}{4\pi k}$

Thus, the heat flow under steady conditions between two constant temperatures T_1 and T_2 is

$$q = \frac{(T_1 - T_2)}{\sum R_{th}} \qquad \textbf{(12.2.1)}$$

A further resistance can be defined for a surface subjected to a constant heat transfer coefficient h as

$$R_{th} = \frac{1}{hA} \qquad \textbf{(12.2.2)}$$

where A is the area available for convection heat transfer.

Example 12.1

A plane wall is comprised of a 4" layer of masonary brick (k = 0.416 Btu/hr-ft-F) and a 0.75" layer of gypsum plaster board (k = 0.098 Btu/hr-ft-F). The air inside is still (h_i = 1.0 Btu/hr-ft^2-F) at T_i=75 F, while the outside air at T_o=-10 F is blowing at 5 mph (h_o = 5.0 Btu/hr-ft^2-F). Compute the heat flux through the wall and the inside wall temperature.

Solution. The heat flux (heat flow through the wall per unit area) is given by Eq.12.2.1 divided by the total wall area:

$$q'' = \frac{T_1 - T_2}{A\sum R_{th}} = \frac{(75-(-10))}{A(1/1.0A + (0.75/12)/0.098A + (4/12)/0.416A + 1/5.0A)}$$

$$= \frac{85}{1.0 + 0.638 + 0.801 + 0.200} = \frac{85}{2.639} = 32.2 \text{ Btu / hr - ft}^2$$

The inner surface temperature can be found, since the heat flow at steady state must be the same through each and every resistance in the series:

$$q'' = \frac{T_i - T_w}{A(1/h_i A)} = \frac{75 - T_w}{A(1/1.0A))} = 32.2 \text{ Btu/hr - ft}^2$$

Solving for T_w yields:

$$T_w = 42.8 \text{ F}$$

Example 12.2

A 2" diameter (I.D.) copper pipe ($k = 210$ Btu/hr-ft-F) with a 0.125" wall thickness carries condensing steam at 400 F. The pipe has a 0.5" layer of insulation ($k = 0.025$ Btu/hr-ft-F), and the pipe runs through a room where the temperature is 75 F. The heat transfer coefficient on the inside of the pipe is very high, and on the outside is $h_o = 2.0$ Btu/hr-ft^2-F. Calculate the heat loss from the steam per foot of pipe.

Solution. The heat loss per unit length is given by Eq. 12.2.1 divided by the length of the pipe L:

$$q' = \frac{q}{L} = \frac{T_1 - T_2}{L\left(\sum R_{th}\right)}$$

$$= \frac{400 - 75}{\dfrac{L\ln(1.125/1.0)}{2\pi \cdot 210L} + \dfrac{L\ln(1.625/1.125)}{2\pi \cdot 0.025L} + \dfrac{L}{2\pi(1.625/12)2.0L}}$$

$$= \frac{325}{8.93\times 10^{-5} + 2.34 + 0.588} = 111 \text{ Btu / hr - ft}$$

Heat Generation

In the presence of uniform heat generation, the steady temperature distributions in each of the three coordinate systems is changed, and the corresponding thermal resistances are invalidated. However, it can be recognized that all of the heat generated within the body must be conducted out of the body. Frequently, this bit of information can be exploited to simplify an analysis.

Cartesian. In the presence of uniform internal heat generation and constant temperature boundary conditions $T(0) = T_1$ and $T(L) = T_2$, the steady temperature distribution in a plane slab is

$$T(x) = T_1 + (T_2 - T_1)\frac{x}{L} + \frac{q'''L^2}{2k}\left(\frac{x}{L} - \frac{x^2}{L^2}\right)$$

Solid Cylinder. In the presence of uniform internal heat generation and constant temperature boundary condition $T(r_1) = T_1$, the steady temperature distribution in a solid cylinder is

$$T(r) = T_1 + \frac{q'''{r_1}^2}{4k}\left(1 - \frac{r^2}{{r_1}^2}\right)$$

This result is applicable to a solid wire subjected to ohmic (electrical) heating.

Hollow Cylinder. In the presence of uniform internal heat generation and constant temperature boundary conditions $T(r_1) = T_1$ and $T(r_2) = T_2$ the steady temperature distribution in a hollow cylinder is

$$T(r) = T_2 + \frac{q'''{r_2}^2}{4k}\left(1 - \frac{r^2}{{r_2}^2}\right) + C_1 \ln\left(\frac{r}{r_2}\right)$$

where

$$C_1 = \frac{(T_1 - T_2) + q'''(r_1^2 - r_2^2)/4k}{\ln(r_1/r_2)}$$

Solid Sphere. In the presence of uniform internal heat generation and constant temperature boundary condition $T(0) = T_1$ and $T(r_2) = T_2$, the steady temperature distribution in a solid sphere is

$$T(r) = T_1 + \frac{q'''r_2^2}{6k}\left(\frac{r_2}{r} - \frac{r^2}{r_2^2}\right) + \frac{r_2}{r}(T_2 - T_1)$$

Fins (Extended Surfaces)

Extended surfaces are used to enhance heat transfer from a surface, primarily by enlarging the available surface area for heat exchange. For fins of *constant cross sectional area* (such as rectangular fins or circular spines/pins), the temperature distribution along the fin and the heat loss from the fin depend only on the tip condition and the physical parameters of the problem. The relevant fin parameters are:

h	heat transfer coefficient
P	perimeter of fin
A_c	cross-sectional area of fin
k	thermal conductivity of fin material
L	length of fin
T_b	temperature of the base of the fin
T_∞	temperature of surrounding fluid
θ	excess temperature $\theta(x) = T(x) - T_\infty$

The temperature distribution and heat loss from constant area fins for different tip conditions are summarized in Table 12.5.

Table 12.5 Temperature distribution and heat flux for constant area fins

Tip Condition	Temperature Distribution	Heat Loss
Insulated (adiabatic)	$\frac{\cosh m(L-x)}{\cosh mL}$	$M \tanh mL$
Very long fin ($x \to \infty$)	$\exp(-mx)$	M
Fixed Temperature	$\frac{(\theta_L/\theta_b)\sinh mx + \sinh m(L-x)}{\sinh mL}$	$M\frac{\cosh mL - \theta_L/\theta_b}{\sinh mL}$
where	$M = \sqrt{hPkA_c}\,\theta_b$ $m = \sqrt{hP/kA_c}$	

Mathematical analysis of fins can be quite complicated, especially when the fin cross section is not constant, but fortunately engineering analysis is facilitated through the concept of *fin efficiency*. The fin efficiency is the ratio of the actual heat transferred q_{act} with the fin to the heat which would be transferred q_{max} if the entire fin were at the base temperature of the fin:

$$\eta_{fin} = \frac{q_{act}}{q_{\max}}$$

Expressions for fin efficiencies of various geometries are given by Gardner (1945) and are depicted in Figure 12.1.

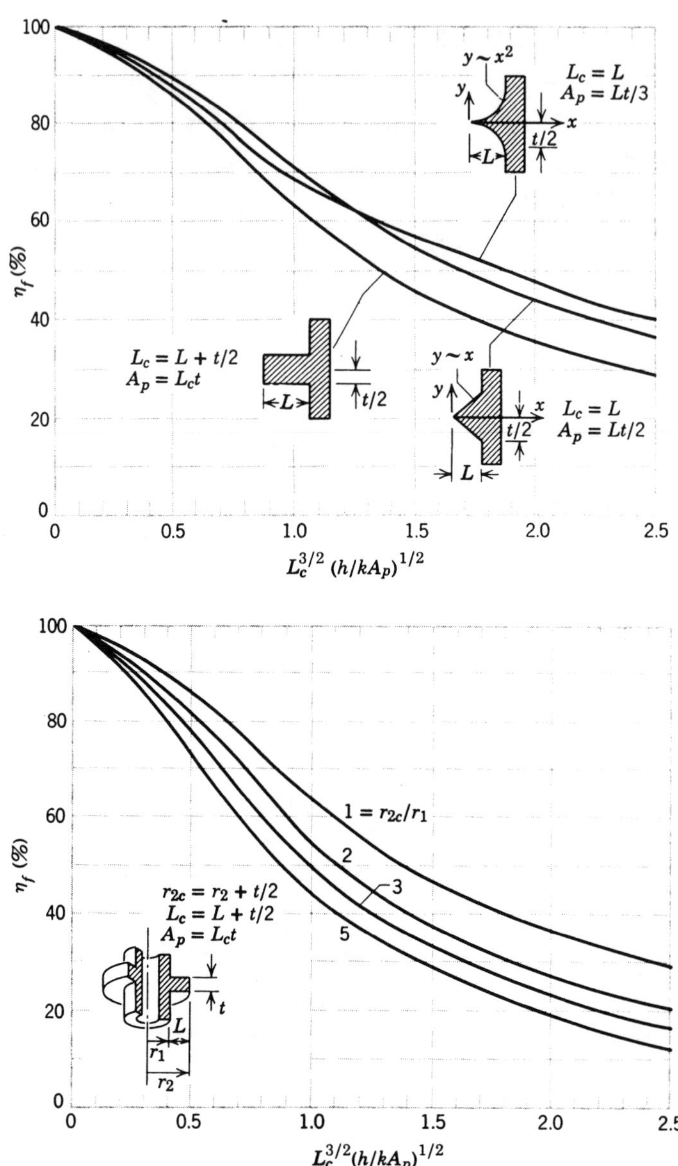

Figure 12.1 Efficiencies for Straight and Annular Fins (reproduced from Incropera and DeWitt, with permission)

When a surface is covered with an array of fins, it is sometimes convenient to use the total array efficiency:

$$\eta_{total} = 1 - \frac{A_{fin}}{A_{total}}(1 - \eta_{fin})$$

where A_{fin}/A_{total} is the ratio of the finned area to the total area available for heat transfer (including the fins). With this definition, the resistance of the finned surface can be expressed as

$$R_{th} = \frac{1}{hA_{total}\eta_{total}}$$

12.3 Transient

Lumped Capacitance

When the external resistance to heat flow is much higher that the internal conduction resistance, the internal temperature may be considered uniform and the transient analysis is sometimes referred to as a lumped capacitance analysis. Generally, the internal resistance to heat flow can be considered negligible when the Biot modulus is "small". Thus, if

$$\mathrm{Bi} = \frac{hL_c}{k} < 0.1(x) = T(x) - T_\infty$$

then the internal resistance is negligible. Here L_c is the characteristic length of the body, which is the ratio of the volume to the surface area:

$$L_c = \frac{V}{A}$$

Of course, V is the total volume, but A is the surface area available for heat transfer.

Under these conditions, the cooling follows a first order (exponential) response:

$$\frac{T(t) - T_\infty}{T_i - T_\infty} = \frac{\theta(t)}{\theta_0} = \exp\left(-\frac{hA}{\rho c_p V} t\right)$$

where ρc_p is the volumetric heat capacity of the material.

Example 12.3

A steel billet 5 cm by 10 cm by 70 cm initially at 900 C is cooled by a stream of cold air at 20 C. If the average heat transfer coefficient is 250 W/m^2·C, estimate the time required for the billet to cool to 100 C.

Solution. First check the Biot modulus to see if the lumped capacitance approximation is valid. The thermal conductivity of mild steel (1% carbon) is about 60 W/m·C, so

$$L_c = V/A = (0.05 \times 0.10 \times 0.70)\mathrm{m}^3 / (2 \times (0.05 \times 0.10 + 0.10 \times 0.70 + 0.05 \times 0.70))\mathrm{m}^2$$

$$= 0.016 \text{ m}$$

$$\mathrm{Bi} = \frac{hL_c}{k} = \frac{250 \text{ W}/\mathrm{m}^2 \cdot \mathrm{C} \times 0.016 \text{ m}}{60 \text{ W}/\mathrm{m} \cdot \mathrm{C}} = 0.07$$

As Bi < 0.1, the lumped approximation will be valid. Therefore,

$$t = \frac{-\rho c_p V}{hA} \ln\left(\frac{\theta(t)}{\theta_0}\right) = \frac{-(7850 \text{ kg}/\mathrm{m}^3)(434 \text{ J}/\mathrm{kg} \cdot \mathrm{C})(0.016 \text{ m})}{250 \text{ W}/\mathrm{m}^2 \cdot \mathrm{C}} \ln\left(\frac{100 - 20}{900 - 20}\right)$$

$$= 523 \text{ secs} = 8.7 \text{ min}$$

One-dimensional Charts

In cases where the Biot number is large (> 0.1), the temperature gradients within the body must be considered. For one-dimensional geometries (symmetrically cooled slab, infinitely long cylinder, or symmetrically cooled sphere) subjected to cooling with a constant heat transfer coefficient h, the transient temperature distribution at the center of the object will be determined by the Biot number and the Fourier modulus (defined below). The characteristic length in the Biot number in this case will be the slab half thickness, the cylinder radius, or the sphere radius for each of the respective geometries:

$$\mathrm{Bi} = \begin{cases} \dfrac{hL}{k} & \text{slab of thickness } 2L \\ \dfrac{hR}{k} & \text{cylinder or sphere of radius } R \end{cases} \tag{12.3.1}$$

The Fourier modulus is a dimensionless time parameter based on the thermal diffusivity of the material and the characteristic dimension of the body:

$$\mathrm{Fo} = \begin{cases} \dfrac{\alpha t}{L^2} & \text{slab of thickness } 2L \\ \dfrac{\alpha t}{R^2} & \text{cylinder or sphere of radius } R \end{cases} \tag{12.3.2}$$

The solution for the object's center temperature can be expressed analytically in terms of an infinite series, but these equations can be represented graphically, as depicted in Figure 12.2, Figure 12.4, and Figure 12.6 for the slab, cylinder, and sphere, respectively. These charts are applicable to all but the smallest times, specifically, they are applicable in cases where the $\mathrm{Fo} \geq 0.2$

Example 12.4

A low carbon steel slab 10 cm thick initially at 900 C is cooled by quenching in oil at 50 C. If the average heat transfer coefficient is 5000 W/m^2·C, estimate the time required for the billet to cool to 100 C.

Solution. With heat loss from the major faces only, this becomes a case of a symmetrically cooled slab. Taking the thermal conductivity of mild steel as $k = 60$ W/m· C, the Biot number is computed as:

$$\mathrm{Bi} = \frac{hL}{k} = \frac{5000 \text{ W/m}\cdot\text{C} \times 0.05 \text{ m}}{60 \text{ W/m}^2\cdot\text{C}} = 4.17$$

The final temperature ratio $\theta(t)/\theta_0 = (100 - 50)/(900 - 50) = 0.058$. Consulting Figure 12.2, looking across along $\theta(t)/\theta_0 = 0.058$ until it intersects $\mathrm{Bi}^{-1} = 0.24$, then reading from the lower axis, the value of Fo = 1.85 when the slab is 100 C. Using the thermal diffusivity of mild steel $\alpha = 1.76 \times 10^{-5}$ m^2/s, the dimensional time can be found as

$$t = \frac{\mathrm{Fo}\, L^2}{\alpha} = \frac{1.85 \times 0.05^2 \text{ m}^2}{1.76 \times 10^{-5} \text{ m}^2/\text{s}} = 263 \text{ s} = 4.4 \text{ min}$$

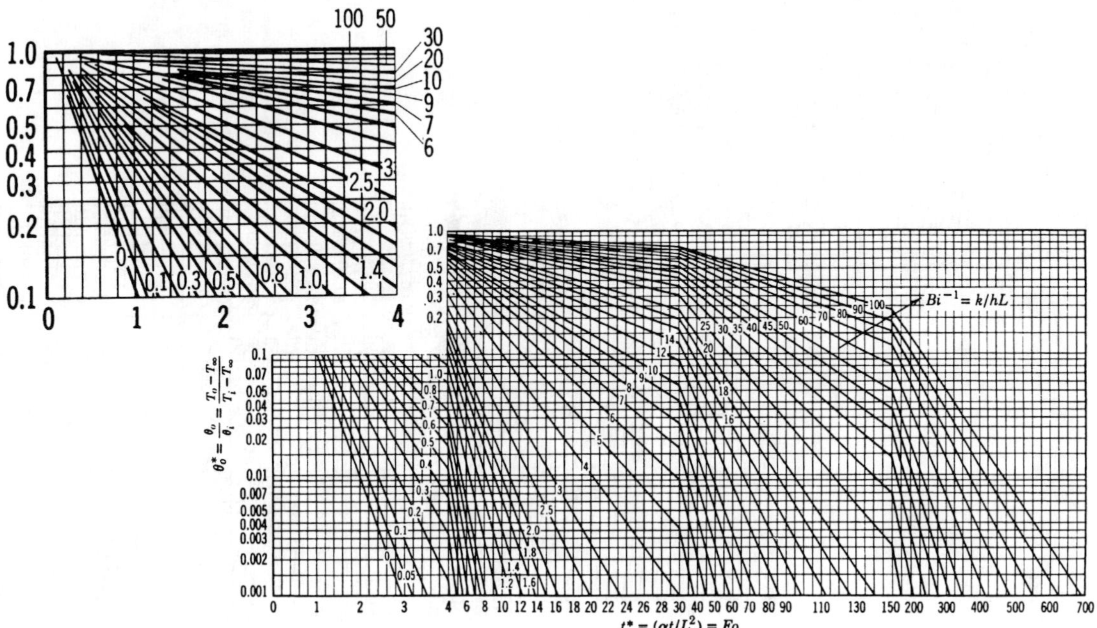

Figure 12.2 Centerline temperature for a symmetrically cooled slab

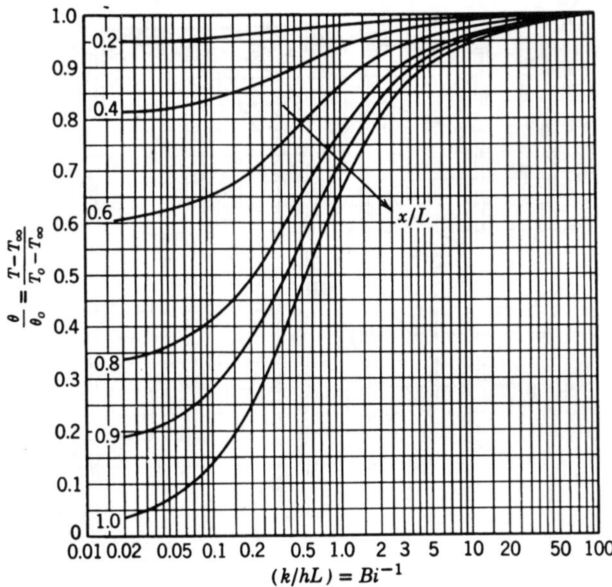

Figure 12.3 Ratio of off-centerline to centerline temperature for a slab

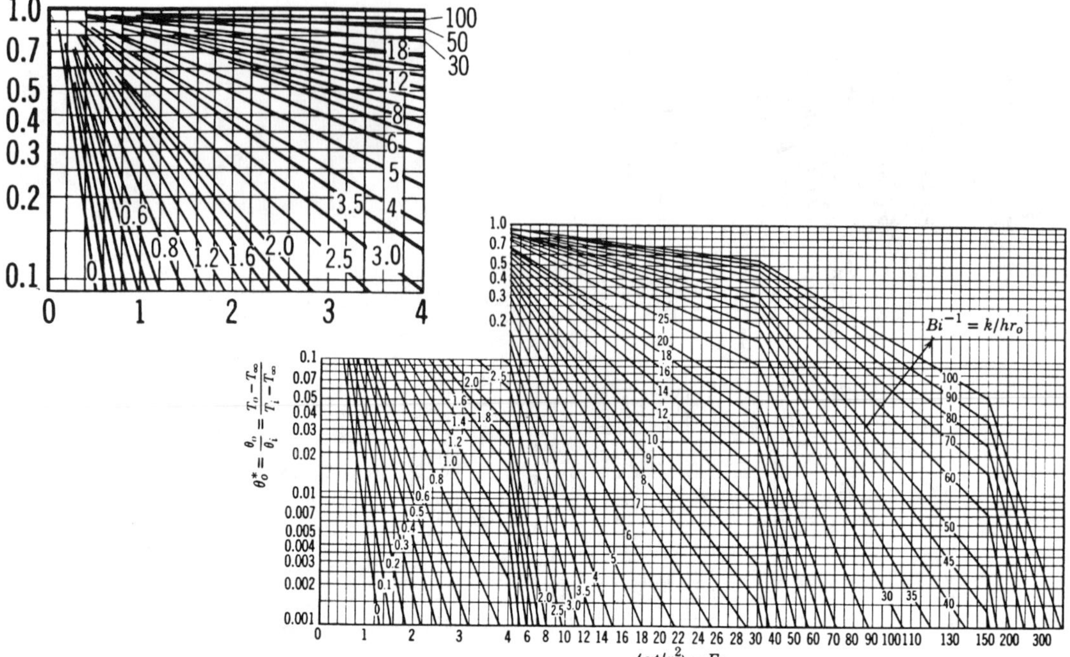

Figure 12.4 Center temperature for a cylinder

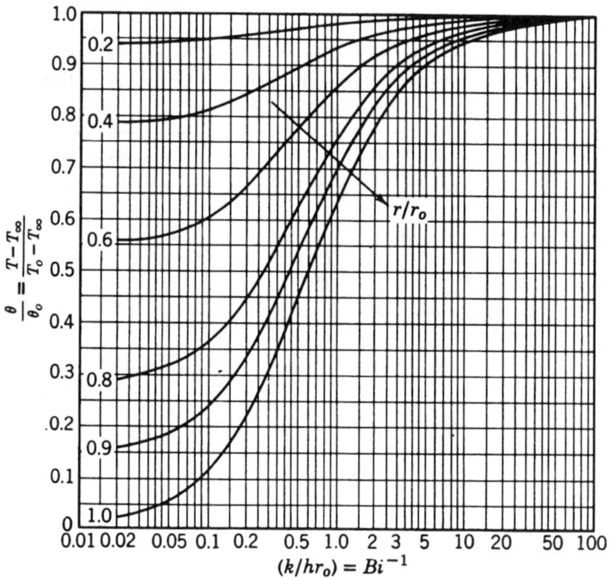

Figure 12.5 Ratio of off-center to center temperature for a cylinder

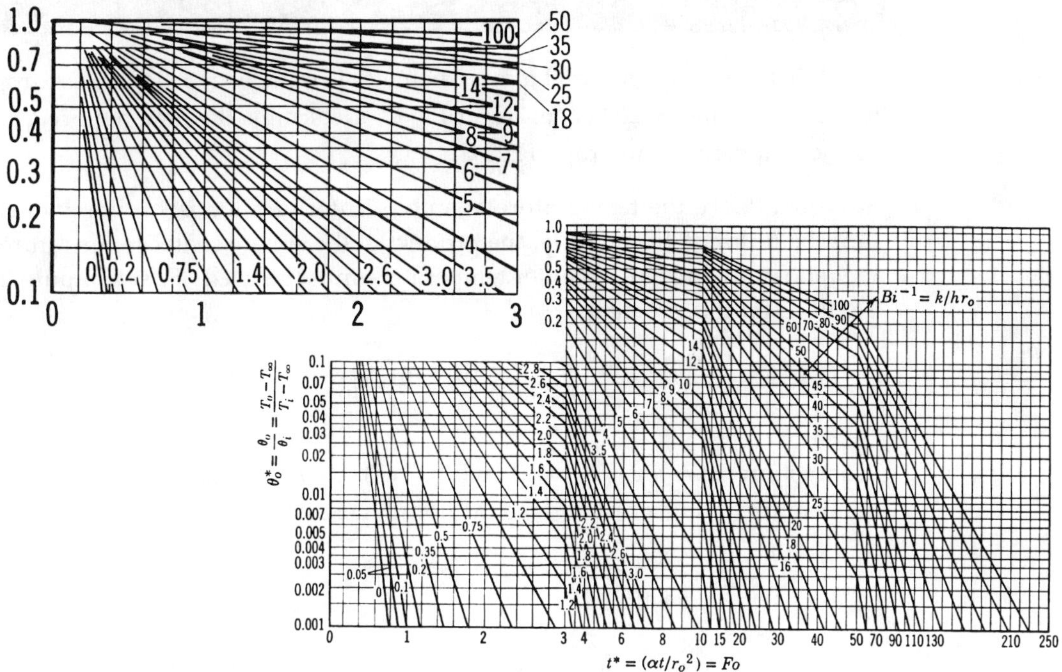

Figure 12.6 Center temperature for a sphere

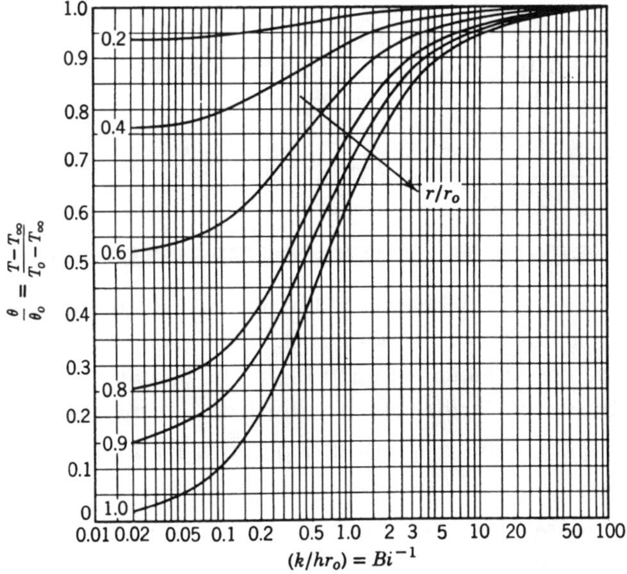

Figure 12.7 Ratio of off-center to centerline temperature for a sphere

If the temperature is to be determined at a location away from the center, it can be found as a ratio of the center temperature for a slab, cylinder, or sphere with the aid of, Figure 12.3, Figure 12.5, and Figure 12.7. Each of these figures shows the ratio of the temperature at any x/L (or r/R) location away from the center to the temperature at the center as a function of Bi^{-1}.

Example 12.5

A 304 Stainless sphere 50 mm in diameter is initially at 800 C and is dropped into a vat of oil (h = 200 W/m^2·C) at 100 C. Determine the time required for the surface temperature to drop to 200 C.

Solution. Since the temperature at an off-center location is known, two charts must be utilized. Specifically, the product of the sphere center temperature and the ratio of the surface to center temperature must be used. First compute the Bi modulus for this problem:

$$\text{Bi} = \frac{hR}{k} = \frac{200\ \text{W}/\text{m}^2\cdot\text{C}\times 0.050/2\ \text{m}}{15\ \text{W}/\text{m}\cdot\text{C}} = 0.333$$

From Figure 12.7, for r/r_0 = 1.0 and Bi^{-1} = 3.0, θ/θ_0= 0.84. The given surface temperature ratio corresponds to the non-dimensional value of

$$\frac{\theta}{\theta_i} = \frac{T - T_\infty}{T_i - T_\infty} = \frac{200-100}{800-100} = \frac{1}{7} = 0.143$$

Now the ratio of the center to initial temperature can be found algebraically as

$$\frac{\theta_0}{\theta_i} = \frac{\theta/\theta_i}{\theta/\theta_0} = \frac{0.143}{0.84} = 0.170.$$

Now, from Figure 12.6, the Fo at which q_0/q_i = 0.170 and Bi^{-1} = 3.0 is Fo = 2.10. The time required is then

$$t = \frac{\text{Fo}L^2}{\alpha} = \frac{2.10\times(0.050/2)^2}{3.95\times 10^{-6}} = 332\ \text{s} = 5.5\ \text{min}$$

Another case which logically belongs to the present category is that of a semi-infinite solid. This is the case of a "very thick" media; one whose temperature far below the convective surface is undisturbed. The temperature response in this case is quantified in terms of the unnamed dimensionless quantities $h\sqrt{\alpha t}/k$and $x/2\sqrt{\alpha t}$. This response is depicted in Figure 12.8.

Multidimensional Charts

In cases where the transient response is multidimensional, solutions can be constructed by computing the product of the appropriate component one-dimensional solutions. For example, the solution for the cooling of a very long bar of cross-sectional dimensions $W \times H$ is found by multiplying the solutions for two one-dimensional bars of half thicknesses $L_1 = W/2$ and $L_2 = H/2$:

$$P(x,y,t) = X_1(x,t)\times X_2(y,t)$$

If the component solutions are identified as:

$X(x,t)$	Symmetric Plane wall
$R(r,t)$	Infinite Cylinder
$S(x,t)$	Semi-infinite Plane Slab

then the multicomponent solution can be computed according to the prescriptions given in Table 12.6.

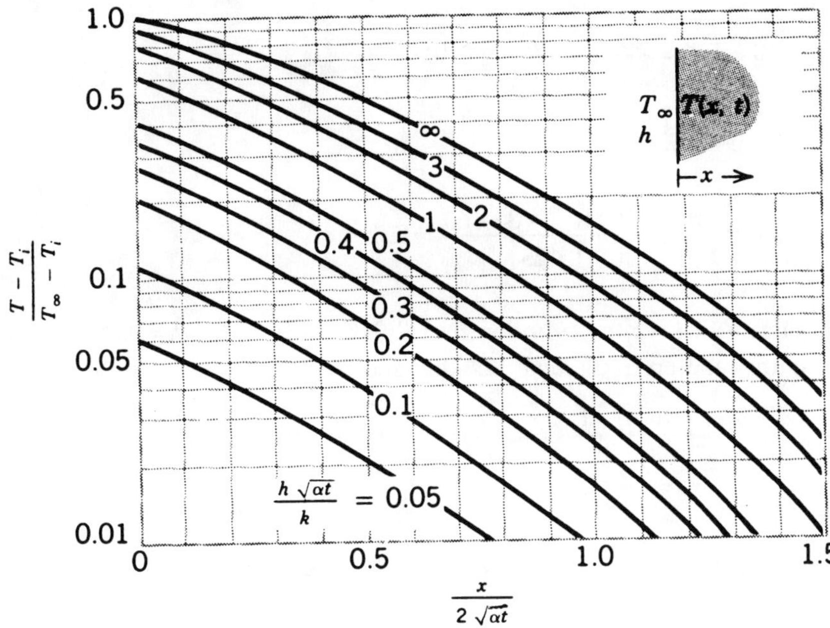

Figure 12.8 Temperature history in a semi-infinite media

Table 12.6 Multidimensional Solution Components

Semi-infinite plate (2D)	$X(x,t) \times S(y,t)$
Semi-infinite cylinder (2D)	$R(r,t) \times S(y,t)$
Very long rectangular bar (2D)	$X_1(x,t) \times X_2(y,t)$
Finite Cylinder (2-D)	$R(r,t) \times X(x,t)$
Semi-infinite Rectangular bar (3-D)	$X_1(x,t) \times X_2(y,t) \times S(z,t)$
Finite rectangular bar (3-D)	$X_1(x,t) \times X_2(y,t) \times X_3(z,t)$

Example 6.6

An aluminum extrusion billet (0.6 m long and 0.25 m in diameter) initially at 20 C is to be heated in a 700 C oven (h = 200 W/m^2·C). What is the temperature in the center of the billet after 30 minutes? Assume that all surfaces of the billet are exposed to the convection environment.

Solution. From Table 12.6, the finite cylinder requires information from both the plane slab $X(x,t)$ of half thickness 0.6 /2 = 0.3 m and the cylinder $R(r,t)$ of radius 0.25/2 = 0.125 m. For the slab,

$$\text{Bi} = \frac{hL}{k} = \frac{200 \times 0.3}{237} = 0.253$$

$$\therefore \text{Bi}^{-1} = 3.95$$

$$\text{Fo} = \frac{\alpha t}{L^2} = \frac{97 \times 10^{-6} \times 30 \times 60}{0.3^2} = 1.94$$

From Figure 12.2, the ratio $X(0, 30 \text{ min}) = \theta_0 / \theta_i = 0.68$ is found. For the cylinder,

$$\text{Bi} = \frac{hr_0}{k} = \frac{200 \times 0.125}{237} = 0.105$$

$$\therefore \text{Bi}^{-1} = 9.48$$

$$\text{Fo} = \frac{\alpha t}{r_0^2} = \frac{97 \times 10^{-6} \times 30 \times 60}{0.125^2} = 11.2$$

From Figure 12.4 for the cylinder, find $R(0{,}30 \text{ min}) = \theta_0 / \theta_i = 0.102$. Now, employing the relation from Table 12.6 the ratio

$$\frac{T(0,0,30\,\text{min}) - T_\infty}{T_i - T_\infty} = X(x,t)R(r,t) = 0.68 \times 0.10 = 0.069$$

Thus,

$$T(0,0,30\,\text{min}) = 0.065 \times (-680\text{ C}) + 700\text{ C} = 652\text{ C}$$

12.4 Convection

Convection heat transfer is the science of determining the heating or cooling provided by a moving stream of fluid. As such, it generally requires fluid dynamic and thermodynamic analysis in addition to heat transfer analysis. This exceedingly complex phenomenon is most often reduced to the following simple expression to compute the heat loss from a solid surface exposed to a moving fluid stream:

$$q_{conv} = hA(T_s - T_\infty) \tag{12.4.1}$$

Here q_{conv} is the heat removed from the surface (Btu/hr or Watts), T_s is the temperature of the surface, T_∞ is the temperature of the flowing fluid stream, A is the area of the surface exposed to the stream, and h is the *heat transfer coefficient*.

The heat transfer coefficient is sometimes called the film coefficient, because it is the action in the thin boundary layer ("film") near the surface which is responsible for the energy exchange. Another term derived from the same concept is the *film temperature*; this is the average temperature of the fluid in the boundary layer. The film temperature is routinely computed as the arithmetic average of the surface and free-stream temperatures:

$$T_{film} = \frac{(T_s + T_\infty)}{2} \tag{12.4.2}$$

The energy removed by the fluid is conducted into the fluid stream at the interface where the viscous forces in the fluid bring it to rest relative to the solid ("no-slip" condition). Thus, the energy crossing the interface is conducted from the solid into the fluid and must obey the energy balance:

$$\frac{q_{conv}}{A} = h(T_s - T_\infty) = -k_{solid}\frac{\partial T}{\partial n}\bigg|_{solid} = -k_{fluid}\frac{\partial T}{\partial n}\bigg|_{fluid}$$

For many practical purposes, convection heat transfer analysis requires determining an appropriate value of the heat transfer coefficient to use in a particular situation. The heat transfer coefficient will generally be provided by a correlation for the Nusselt number:

$$\mathrm{Nu}_L = \frac{hL}{k} \tag{12.4.3}$$

where h is the heat transfer coefficient, L is an appropriate reference length, and k is the thermal conductivity of the fluid.

Tube Flow

Flow inside tubes or ducts accounts for a large portion of practical problems of interest, including air flow in conditioning ducts and flow in tubes in heat exchangers (condensers and evaporators).

As flow enters a pipe or duct, the boundary layers begin to grow from the walls. Only after these boundary layers have grown together to fill the passage is the flow considered *fully developed.* The flow up to this point is considered *developing flow.*

The heat transferred to the fluid in the passage gives rise to an increase in its temperature. The bulk or mean temperature (also called the "mixing-cup" temperature) is the velocity-averaged temperature across the flow passage. This temperature will increase between two streamwise stations based on the thermodynamic energy balance. For two important cases, constant wall temperature T_s and constant wall heating q_w'', this variation in T_m is

$$T_m(x) - T_s = (T_m(0) - T_s)\exp\left(-\frac{hPx}{\dot{m}c_p}\right) \qquad \text{constant } T_s \tag{12.4.4}$$

$$T_m(x) = T_m(0)_s + \frac{q_s''P}{\dot{m}c_p}x \qquad \text{constant } q_s'' \tag{12.4.5}$$

Here P is the perimeter around the passage, and $\dot{m} = \rho AV$ is the mass flow rate.

The fluid dynamics are drastically different for laminar and turbulent flow, and of course the energy transfer is correspondingly different. Hence, it is important to know which flow regime is applicable in any situation. The flow regime is customarily determined on the basis of the Reynolds number; for a pipe, it is

$$\mathrm{Re}_D = \frac{V_{mean}D}{\mu/\rho} = \frac{V_{mean}D}{\nu} = \begin{cases} < 2300 & \text{laminar} \\ \geq 2300 & \text{turbulent} \end{cases} \tag{12.4.6}$$

Ordinarily, properties required in Eq.12.4.6 are evaluated at the film temperature (Eq. 12.4.2).

Laminar

For fully-developed laminar flow in smooth tubes, the Nusselt number is constant. For circular tubes, there are two possible wall conditions: constant wall temperature (with varying heat flux) or constant heat flux (with varying wall temperature).

For the case of constant wall temperature:

$$\mathrm{Nu}_D = 3.66 \qquad T_w = \text{constant} \tag{12.4.7}$$

For the case of constant wall heat flux:

$$\mathrm{Nu}_D = 4.36 \qquad q''_w = \text{constant} \tag{12.4.8}$$

Turbulent

The most widely-used correlation for turbulent flow in tubes is the Colburn equation:

$$\mathrm{Nu}_D = 0.023\,\mathrm{Re}_D{}^{4/5}\,\mathrm{Pr}^{1/3} \tag{12.4.9}$$

In this relation, the properties of the fluid should be evaluated at the film temperature. Slightly better results are obtained by modifying the exponent on the Prandtl number to account for heating or cooling of the fluid. The resulting relation is known as the Dittus-Boelter relation:

$$\mathrm{Nu}_D = 0.023\,\mathrm{Re}_D{}^{4/5}\,\mathrm{Pr}^{n} \qquad n = \begin{cases} 0.3 & T_m > T_s \\ 0.4 & T_m < T_s \end{cases} \tag{12.4.10}$$

Example 12.7

Diesel fuel is supplied to a small engine at a rate of 50 cm^3/s. The fuel is to be preheated from 0 C to 40 C by flowing through a 5 mm diameter tube of constant wall temperature T_s = 50 C. What is the heat transfer coefficient inside the tube?

Solution. The properties of diesel fuel at the film temperature are needed. Since this is changing along the length, the simple average film temperature is used:

$$T_{f1} = \frac{T_{m1} + T_s}{2} = \frac{0 + 50}{2} = 25\ \mathrm{C}$$

$$T_{f2} = \frac{T_{m2} + T_s}{2} = \frac{40 + 50}{2} = 45\ \mathrm{C}$$

$$\overline{T}_f = \frac{T_{f1} + T_{f2}}{2} = \frac{25 + 45}{2} = 35\ \mathrm{C}$$

From tables of properties of oil (see "Unused Engine Oil" in the Appendix at the end of this chapter), the following properties are obtained:

$$\nu = 360 \times 10^{-6}\ \mathrm{m^2/s}$$

$$k = 0.145\ \mathrm{W/m \cdot C}$$

$$\mathrm{Pr} = 4235$$

So the Reynolds number is

$$\mathrm{Re}_D = \frac{VD}{\nu} = \frac{(\dot{V}/A)D}{\nu}$$

$$= \frac{(50\ \mathrm{cm^3/s})/(\pi 0.5^2\,\mathrm{cm^2}/4) \times 0.005\ \mathrm{m}}{360 \times 10^{-6}\ \mathrm{m^2/s}} \times \frac{1\ \mathrm{m}}{100\ \mathrm{cm}} = 35$$

Since the Reynolds number is less than the turbulent threshold (2300), the flow is laminar, and the heat transfer coefficient can be found from Eq. 12.4.7:

$$h = \frac{\text{Nu}\, k}{D} = \frac{3.66 \times 0.145}{0.005} = 106 \text{ W}/\text{m}^2 \cdot \text{C}$$

Example 12.8

Air for combustion in a boiler is preheated by flowing through tubes in an exhaust gas stack. The tubes are 4 inches in diameter, and the average velocity of the air in the tubes is 15 ft/s. If the average tube wall temperature is 500 F, and air enters at atmospheric conditions (70 F), estimate the heat transfer coefficient inside the tubes. Also estimate the exiting air temperature if the tubes are 15 feet long.

Solution. The average film temperature is not known, so an engineering estimate must be made. The exit temperature of the air cannot exceed 500 F, or the second law of thermodynamics will be violated. Estimate the exit temperature as 400 F. This yields an average film temperature of (285+450)/2 ≈ 370 F. From tables of air properties, find:

$$\rho = 0.0483 \text{ lb/ft}^3$$
$$\nu = 1.24 \text{ ft}^2\text{/hr}$$
$$k = 0.0214 \text{ Btu/hr - ft - F}$$
$$c_p = 0.244 \text{ Btu/lb - F}$$
$$\text{Pr} = 0.69$$

Then the Reynolds number is

$$\begin{aligned} \text{Re}_D &= \frac{VD}{\nu} \\ &= \frac{(15 \text{ ft/s})(4/12) \text{ ft}}{1.24 \text{ ft}^2/\text{hr}} \times \frac{3600 \text{ s}}{1 \text{ hr}} = 14{,}500 \end{aligned}$$

So the flow is turbulent. Then the Nusselt number can be estimated from Eq. 12.4.10:

$$\begin{aligned} \text{Nu} &= 0.023(14{,}500)^{0.8}(0.69)^{0.4} \\ &= 42.3 \end{aligned}$$

and the heat transfer coefficient is:

$$\begin{aligned} h &= \text{Nu}\, k/D = (42.3)(0.0214) \;/\; (4/12) \\ &= 2.7 \text{ Btu}/\text{hr} \cdot \text{ft}^2 \cdot \text{F} \end{aligned}$$

The exiting fluid temperature now follows from Eq. 12.4.4 as

$$T_m(15 \text{ ft}) = 70\text{F} + (500 - 70)\text{F} \times$$

$$\exp\left(-\frac{2.7 \times \pi(4/12)/3600 \times 15}{0.0483(15)\left(\pi(4/12)^2/4\right)(0.244)} \right)$$

$$= 270\text{F}$$

Now it can be seen that the estimated exit temperature of 400 F was too high; to obtain better results, another calculation using this new temperature (i.e., using an average film temperature of (285+385)/2 = 335 F) should be made. Such a "corrector" step will produce slightly better values for the results.

Reynolds/Chilton-Colburn Analogy

The preceding results for flow in tubes are valid only for smooth tubes. In cases where rough surfaces are encountered, the Reynolds analogy for tube flow, also known as the Chilton-Colburn analogy, may be used.
The analogy is based on the mathematical similarity of the transport of momentum and heat in the boundary layer. Mathematically, it relates the frictional effect to the heat transfer effect through the following relation:

$$\frac{f}{8} = \mathrm{St}\,\mathrm{Pr}^{2/3} = \frac{\mathrm{Nu}}{\mathrm{Re}\,\mathrm{Pr}}\mathrm{Pr}^{2/3} = \frac{\mathrm{Nu}}{\mathrm{Re}\,\mathrm{Pr}^{1/3}} \tag{12.4.11}$$

Here f is the friction factor (from the Moody diagram), St is the Stanton number, and the group $\mathrm{StPr}^{2/3}$ is the Colburn j factor.

Example 12.9

Water at 70 F flows through an 8 inch diameter concrete pipe (absolute roughness $e = 0.00005$ ft) at a velocity of 20 ft/s. Estimate the heat transfer coefficient.

Solution. The wall temperature is not known so the film temperature cannot be evaluated. As a first approximation, properties of the fluid at the given temperature are found:

$$\rho = 62.3\ \mathrm{lb/ft^3}$$
$$k = 0.354\ \mathrm{Btu/hr\text{-}ft\text{-}F}$$
$$\nu = 0.033\ \mathrm{ft^2/hr}$$
$$\mathrm{Pr} = 5.8$$

The Reynolds number is Re = (20 ft/s)(8/12)ft/(0.033 $\mathrm{ft^2}$/hr × 1hr/3600s) = 1.45×10^6. Thus, the flow is highly turbulent and, consulting the Moody diagram at a relative roughness of e/D = 0.00005/(8/12) = 0.000075, the friction factor f = 0.0138. Employing the Chilton-Colburn analogy (Eq.12.4.11), the Nusselt number is first computed as:

$$\mathrm{Nu} = \frac{f}{8}\mathrm{Re}\,\mathrm{Pr}^{1/3}$$
$$= \frac{0.0138}{8}\left(1.45\times10^6\right)(5.8)^{1/3} = 4500$$

Then the heat transfer coefficient is computed as

$$h = \frac{\mathrm{Nu}\,k}{D} = \frac{(4500)0.354}{(8/12)} = 2400\frac{\mathrm{Btu}}{\mathrm{hr\cdot ft^2\cdot F}}$$

Non-Circular Ducts

In the case of noncircular flow passages (rectangular, triangular, or polygonal-shaped) flow conditions are generally similar to smooth pipe flow and can be handled with simple modifications. In all cases, the dimensionless parameters (such as Re and Nu) are computed using the *hydraulic diameter* concept. The hydraulic diameter is the defined as:

$$D_H = \frac{4A_{cs}}{P}$$

where A_{cs} is the cross sectional area of the duct and P is its perimeter. Note that for a circular duct, $D_H = D$, that is, the hydraulic diameter equals the physical diameter.

Laminar Flow

For laminar flows, the Nusselt number is constant, and the values for different cross sections are presented in Table 12.7. These values are in terms of the hydraulic diameter as defined above.

Table 12.7 Nusselt Numbers for Laminar Flow in Non-Circular Ducts

Shape	Width/Height	Constant Wall Flux	Constant T_s
Rectangular	1.0	3.61	2.98
Rectangular	1.43	3.73	3.08
Rectangular	2.0	4.12	3.39
Rectangular	3.0	4.79	3.96
Rectangular	4.0	5.33	4.44
Rectangular	8.0	6.49	5.60
Parallel Plates	---	8.23	7.54
Triangular	---	3.00	2.35

Turbulent Flow

The case of turbulent flow in a non-circular pipe is handled by simply replacing the physical diameter with the hydraulic diameter and then proceeding as though the duct is circular. Thus, the Reynolds number is computed on the basis of D_H, and the regime of the flow dictated by Eq. 12.4.6. The Nusselt number is computed via an appropriate correlation, then the heat transfer coefficient extracted using D_H.

Example 12.10

Air at 20 C flows in a rectangular air conditioning duct 20 cm by 40 cm at a velocity of 5 m/s. Estimate the heat transfer coefficient. Assume the walls of the duct are at 30 C.

Solution. The film temperature is 25 C, and the properties of the air are approximately

$$\nu = 15.9 \times 10^{-6} \text{ m}^2/\text{s}$$

$$k = 0.0263 \text{ W/m} \cdot \text{C}$$

$$\text{Pr} = 0.707$$

The hydraulic diameter for this duct is

$$D_H = \frac{4\times(0.2\text{m}\times 0.4\text{m})}{2\times(0.2\text{m}+0.4\text{m})} = 0.267 \text{ m}$$

and the Reynolds number is

$$\text{Re} = \frac{5\times 0.267}{15.9\times 10^{-6}} = 84\,000$$

Thus the flow is turbulent, and the Dittus-Boelter equation (Eq. 12.4.10) can be used to find Nu:

$$\begin{aligned}\text{Nu} &= 0.023(84000)^{0.8}(0.707)^{0.4}\\ &= 174\end{aligned}$$

Then the heat transfer coefficient is

$$h = \frac{\text{Nu}\,k}{D_H} = \frac{174\times 0.0263}{0.267} = 17 \text{ W}/\text{m}^2\cdot\text{C}$$

External Flow

External flow over plates, cylinders, and banks of cylinders accounts for the next largest section of convection heat transfer. Flow over spheres and non-circular tubes occur much less frequently and are not included in this review.

Convection heat transfer in external flow continues to be a phenomenon associated with the fluid boundary layer interaction with a solid surface. In cases of cylinders, this boundary layer generally separates at some point behind the cylinder, giving a decrease in the heat transfer effectiveness and an increase in friction (drag).

Plates

For flow over plates, the boundary layer near the leading edge is laminar. This laminar boundary layer conditions persists for some distance until the viscous forces lead to instabilities, and the boundary layer becomes turbulent. For flat plates aligned with the flow, the *local* Reynolds number at which the boundary layer become turbulent can be taken as 500,000. Thus,

$$\text{Re}_{crit} = \frac{Vx_{crit}}{\nu} = 5\times 10^5$$

Generally, the heat transfer analyst is only interested in the average heat transfer coefficient over the surface. If the flow is laminar over the entire distance up to a point x, the average Nusselt number over the distance is

$$\overline{\text{Nu}_x} = \frac{\overline{h_x}x}{k} = 0.664\,\text{Re}_x^{\,0.5}\,\text{Pr}^{1/3} \qquad \text{Laminar}$$

If the flow is turbulent over most of the length (negligibly short laminar region), then the average Nusselt number over the distance is

$$\overline{\text{Nu}_L} = \frac{\overline{h}L}{k} = 0.037\,\text{Re}_L^{\,0.8}\,\text{Pr}^{1/3} \qquad \text{Turbulent}$$

If, as is generally the case, the flow is laminar up to a point, then becomes turbulent for the rest of the length, the average Nusselt number over the length is

$$\overline{Nu}_L = \frac{\bar{h}L}{k} = 0.037\left(Re_L^{0.8} - 871\right)Pr^{1/3} \qquad \text{Mixed, } Re_{crit} = 5\times10^5$$

Cylinders

Flow over single cylinders is complex, as the flow regime continually changes as the free stream velocity increases, owing to the continual shift in the boundary layer separation point. As a result, the correlations for the average heat transfer coefficient are broken down into a number of ranges bracketed by the Reynolds number computed based on the cylinder diameter. The empirical correlation of Hilpert is of the form:

$$\overline{Nu}_D = \frac{\bar{h}D}{k} = C\,Re_D{}^m\,Pr^{1/3} \tag{12.4.12}$$

where the constants C and m are given in Table 6.8.

Table 12.8 Constants C and m for Flow over a Cylinder (Eq.12.4.12)

Re_D	C	m
0.4 - 4	0.989	0.330
4 - 40	0.911	0.385
40 - 4000	0.683	0.466
4000 - 40,000	0.193	0.618
40,000 - 400,000	0.027	0.805

Tube Banks

When a fluid flows over a group of tubes arranged in a bank, the complex interaction of the boundary layers results in heat exchange which differs from that for a single cylinder. Two common possible arrangements for the tube bank are the inline and staggered arrangements, which are illustrated in Figure 12.9.

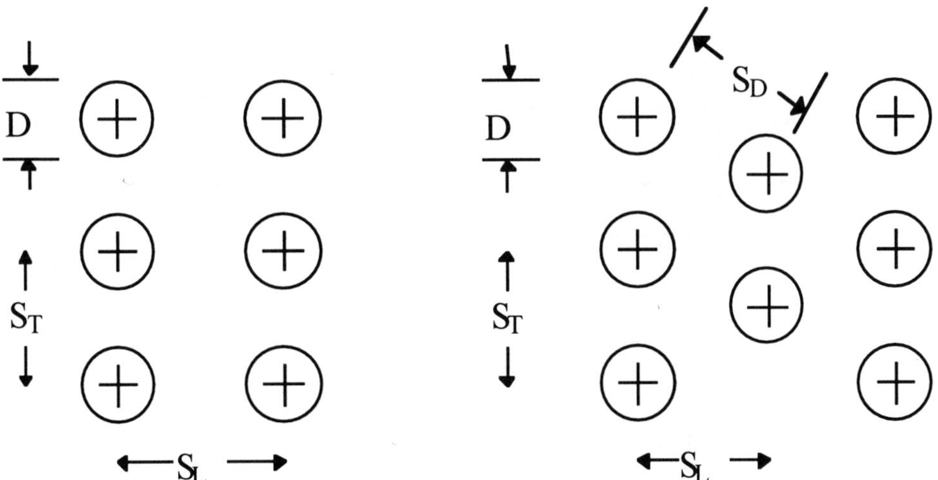

Figure 12.9 Inline and Staggered Tube Banks

The correlation for these two arrangements are in terms of the maximum value of the Reynolds number and the tube diameter. Additionally, the correlations are for the average Nusselt number (heat transfer coefficient) and are valid when the fluid flows over 10 or more rows in the streamwise direction.

The maximum value of Re will of course occur at the minimum cross sectional area for the flow. Thus,

$$\mathrm{Re}_{max} = \frac{V_{max} D}{\nu}$$

If the fluid approaches the bank with a velocity V_∞, then the maximum velocity generally will occur at the transverse section and

$$V_{max} = \frac{S_T}{S_T - D} V_\infty$$

However, for the staggered arrangement, the maximum velocity may occur at S_D and it will when $S_D < (S_T + D)/2$. In this case

$$V_{max} = \frac{S_T}{2(S_D - D)} V_\infty$$

The correlation for the average Nusselt number is of the form:

$$\overline{\mathrm{Nu}}_D = \frac{\bar{h}D}{k} = 1.13C\,\mathrm{Re}^m{}_{max}\,\mathrm{Pr}^{1/3} \tag{12.4.13}$$

where the constants C and m depend only on the relative spacing of the tubes. These constants are listed in Table 12.9. All fluid properties in Equation 12.4.13 should be evaluated at the film temperature.

Table 12.9 Constants C and m for use in Eq.12.4.13 (10 or more rows of tubes)

	S_T/D							
Aligned	1.25		1.5		2.0		3.0	
S_L/D	C	m	C	m	C	m	C	m
1.25	0.348	0.592	0.275	0.608	0.100	0.704	0.0633	0.752
1.50	0.367	0.586	0.250	0.620	0.101	0.702	0.0678	0.744
2.00	0.418	0.570	0.299	0.602	0.229	0.632	0.198	0.648
3.00	0.290	0.601	0.357	0.584	0.374	0.581	0.286	0.608
Staggered								
S_L/D	C	m	C	m	C	m	C	m
0.600	---	---	---	---	---	---	0.213	0.636
0.900	---	---	---	---	0.446	0.571	0.401	0.581
1.000	---	---	0.497	0.558	---	---	---	---
1.125	---	---	---	---	0.478	0.565	0.518	0.560
1.500	0.518	0.556	0.505	0.554	0.519	0.556	0.522	0.562
2.000	0.404	0.572	0.416	0.568	0.482	0.556	0.449	0.570
3.000	0.310	0.592	0.356	0.580	0.440	0.562	0.428	0.574

When there are fewer than 10 rows of tubes in the streamwise direction, a correction factor must be applied to the result computed from Equation 12.4.13. These correction factors are presented in Table 12.10.

Table 12.10 Correction factor for Eq.12.4.13 for less than 10 rows of tubes

N	1	2	3	4	5	6	7	8	9
Aligned	0.64	0.80	0.87	0.90	0.92	0.94	0.96	0.98	0.99
Staggered	0.68	0.75	0.83	0.89	0.92	0.95	0.97	0.98	0.99

Example 12.11

Air flows over a bank of tubes in a heat recovery unit. The air enters at 600 F with a velocity of 25 ft/s. The 0.5 in. diameter tubes have an average surface temperature of 300 F and are spaced in a staggered fashion with $S_T = 1.0$ in. and $S_L = 1.5$ in. What is the average heat transfer coefficient on the air side?

Solution. From the geometry of the staggered arrangement, S_D is the hypotenuse of the right triangle with sides of length S_L and $S_T/2$. Thus

$$S_D = \sqrt{S_L{}^2 + S_T{}^2/4} = \sqrt{2.25 + 0.25} = 1.58''$$

Since $S_D > (S_T + D)/2 = 0.75$ in., the minimum flow section in the bank is at S_T, and the maximum velocity is

$$V_{max} = \frac{S_T}{S_T - D} V_\infty = \frac{1.0}{0.5} 25 = 50 \text{ ft/s}$$

The properties of the air evaluated at the film temperature (450 F) are

$$\nu = 1.50 \text{ ft}^2/\text{hr}$$

$$k = 0.0235 \text{ Btu/hr - ft - F}$$

$$\text{Pr} = 0.684$$

and the Re_{max} is

$$\text{Re}_{max} = \frac{50 \text{ ft/s } (0.5/12) \text{ ft}}{1.50 \text{ ft}^2/\text{hr} \times 1\text{hr}/3600 \text{ s}} = 5000$$

With $S_T/D = 1.0/0.5 = 2.0$ and $S_L/D = 1.5/0.5 = 3.0$, Table 12.9 gives the constants $C = 0.440$ and $m = 0.562$. Thus, Eq.12.4.13 gives the average Nu as

$$\overline{\text{Nu}}_D = 1.13 \times 0.440 \times 5000^{0.562} \times 0.684^{1/3}$$
$$= 52.5$$

Then the heat transfer coefficient for 10 or more rows of tubes is

$$h = \frac{\text{Nu}k}{D} = \frac{52.5 \times 0.0235}{0.5/12} = 30 \text{ Btu / hr - ft}^2 \text{ - F}$$

Heat Exchangers

A heat exchanger is a device designed to transfer heat from one fluid to another. There are two major categories of heat exchangers: shell and tube heat exchangers and cross-flow heat exchangers (sometimes called compact heat exchangers). The simplest type of shell and tube exchanger has a single tube housed concentrically within the outer shell. This type of exchanger is called a tube-in-tube exchanger, or double-pipe exchanger, and is shown schematically in Figure 12.10. The tube-in-tube exchanger is frequently used to qualitatively discuss general heat exchanger performance and analysis.

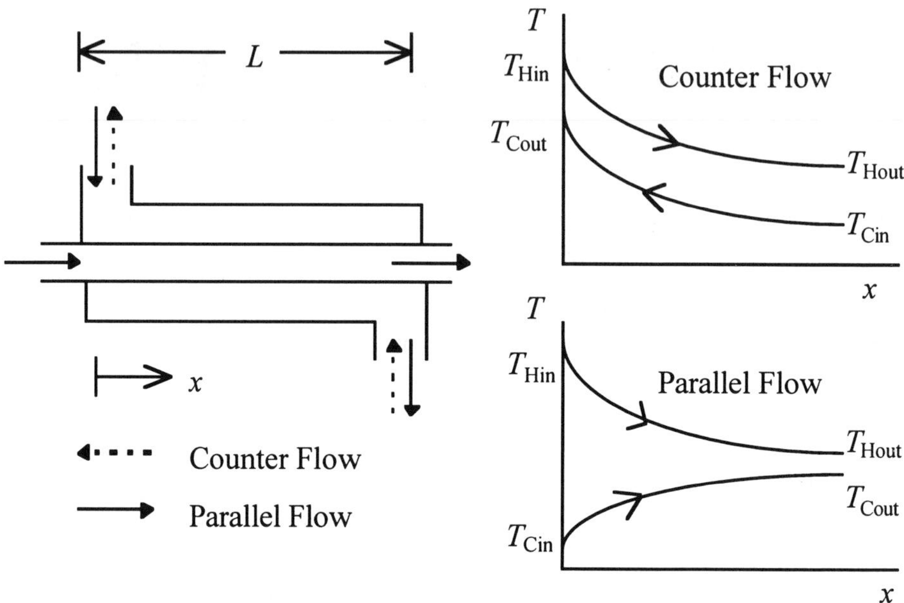

Figure 12.10 Parallel and Counter Flow Heat Exchangers

Shell and tube heat exchangers in general and tube-in-tube exchangers in particular are classified as parallel flow or counter flow according to whether the two fluid streams flow in the same or opposite directions through the device. These two possibilities are illustrated in Figure 12.10 along with typical fluid temperature variations for the two types. Note from the temperature distribution schematics that 1) there is a larger average ΔT between the two fluids throughout the exchanger for the counter flow arrangement and 2) the maximum possible exit temperature (in the limit of $L \to \infty$) of the cold fluid T_{Cout} is equal to the entering temperature of the hot fluid T_{Hin} for the counter flow arrangement (by contrast the limiting exit temperature of the cold fluid in the parallel flow arrangement is something between T_{Hin} and T_{Cin}). For these reasons, *the counter-flow arrangement is more effective than the parallel flow arrangement.*

Cross flow heat exchangers are categorized according to whether or not each fluid stream is free to mix with itself. If the flow is constrained between walls or in tubes, then the flow is said to be unmixed. On the other hand, if the flow is unconstrained (as in the case of fluid flowing over tubes with no fins) then the flow is said to be mixed. Thus mixed-unmixed and unmixed-unmixed (referring to the conditions of the two streams) cross-flow exchangers are possible.

A well designed heat exchanger will have no external losses. That is, all the heat lost by the hot fluid will be gained by the cold fluid:

$$q_H = -q_C$$

If the fluids are incompressible (such as water, oil, or air at moderate velocities), then

$$\left[\dot{m}c_p\right]_H\left(T_{Hin}-T_{Hout}\right)=\left[\dot{m}c_p\right]_C\left(T_{Cout}-T_{Cin}\right)$$

or **(12.4.14)**

$$\dot{C}_H\left(T_{Hin}-T_{Hout}\right)=\dot{C}_C\left(T_{Cout}-T_{Cin}\right)$$

where the notation $\dot{C}=\dot{m}c_p$ has been introduced.

Overall Heat Transfer Coefficient

In addition to the thermodynamic relation given by Equation 12.4.14, a heat transfer relation quantifying the heat exchange is necessary to analyze the heat exchanger. This heat transfer relation specifies the rate of energy exchange between the two fluid streams to be proportional to the local temperature difference by the overall heat transfer coefficient U. Thus, at any x-location, the heat exchange per unit area is

$$q''(x)=U\left[T_H(x)-T_C(x)\right]=U\Delta T(x)$$

The overall heat transfer coefficient can be visualized with the aid of Figure 12.11, which illustrates the heat flow path between fluids at a cross section in a tube-in-tube exchanger. The overall heat transfer coefficient is related to the thermal resistances in the heat flow path as

$$UA=\frac{1}{\sum R_{th}}=\frac{1}{R_{th,i}+R_{th,cond}+R_{th,o}} \quad \textbf{(12.4.15)}$$

where A is the appropriate area for heat flow. In fact, either the inner or outer area of the pipe may be chosen as a basis for the overall heat transfer coefficient.

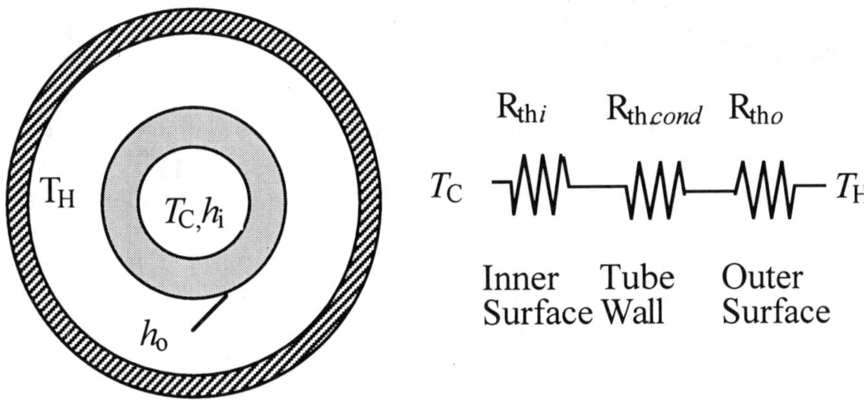

Figure 12.11 Overall Heat Transfer Coefficient for Tube-in-Tube Exchanger

Example 6.12

Hot water flows through a 75 mm ID copper pipe (k_{cu} = 400 W/m·C) with a 3 mm wall thickness. The water has a heat transfer coefficient of 600 W/m^2·C at the pipe wall. The pipe is covered with a 10 mm layer of insulation (k_{ins} = 0.15 W/m·C), and still air on the outside of the insulation produces a heat transfer coefficient of 10 W/m^2·C. What is the overall heat transfer coefficient per unit length of pipe?

Solution. Equation 12.4.15 gives the overall heat transfer coefficient. In this case, there are two conductive resistances, and two convective resistances:

$$UA = \left[\frac{1}{h_i A_i} + \frac{\ln(d_o / d_i)}{2\pi k_{cu} L} + \frac{\ln((d_o + 2\Delta r) / d_o)}{2\pi k_{ins} L} + \frac{1}{h_o A_o}\right]^{-1}$$

$$= \left[\frac{1}{h_i \pi d_i L} + \frac{\ln(d_o / d_i)}{2\pi k_{cu} L} + \frac{\ln((d_o + 2\Delta r) / d_o)}{2\pi k_{ins} L} + \frac{1}{h_o \pi (d_o + 2\Delta r) L}\right]^{-1}$$

So the overall heat transfer coefficient, multiplied by the area for heat flow, per unit length is

$$\frac{UA}{L} = \left[\frac{1}{600\pi \times 0.075} + \frac{\ln(81/75)}{2\pi \times 400} + \frac{\ln(101/81)}{2\pi \times 0.15} + \frac{1}{10\pi \times 0.101}\right]^{-1}$$

$$= [0.0071 + 0.000031 + 0.234 + 0.315]^{-1} = 1.80 \text{ W / m} \cdot \text{C}$$

Note that the effect of the inside heat transfer coefficient and the thermal resistance of the pipe wall are insignificant in comparison to the other terms and could easily be neglected in the computation.

Effectiveness-NTU method

The effectiveness-NTU (ε-NTU) method of heat exchanger analysis allows easy computation of the exit stream temperatures if the exchanger characteristics and the fluid entering conditions are known. This method and the LMTD method to be presented below give consistent results and either method may be used in a particular situation. Often, the e-NTU is easier to apply.

The heat exchanger's *effectiveness*, ε, is defined as the actual heat transfer for the exchanger compared to the maximum possible heat exchange for the given parameters. This maximum possible heat transfer would be realized if one of the fluids underwent the largest temperature change in the exchanger ($\Delta T_{max} = T_{Hin} - T_{Cin}$). According to second law arguments, the fluid with the minimum volumetric heat capacity $\dot{C}_{min}$ must be the one to undergo this maximum temperature change. Thus,

$$q_{max} = \dot{C}_{min} \Delta T_{max} = \dot{C}_{min}(T_{Hin} - T_{Cin}) \quad \textbf{(12.4.16)}$$

where

$$\dot{C}_{min} = \min\left([\dot{m}c_p]_H, [\dot{m}c_p]_C\right)$$

The Number of Transfer Units, or NTUs, is defined as the ratio

$$\text{NTU} = \frac{UA}{\dot{C}_{min}}$$

It can be shown that for any heat exchanger the effectiveness is a function of NTU and the ratio of $\dot{C}_{min} / \dot{C}_{max}$:

$$\varepsilon = f(\text{NTU}, \dot{C}_{min} / \dot{C}_{max})$$

These relationships are depicted graphically in Figure 12.12 for several common heat exchanger configurations. In these figures, the ratio $C_r = \dot{C}_{min} / \dot{C}_{max}$.

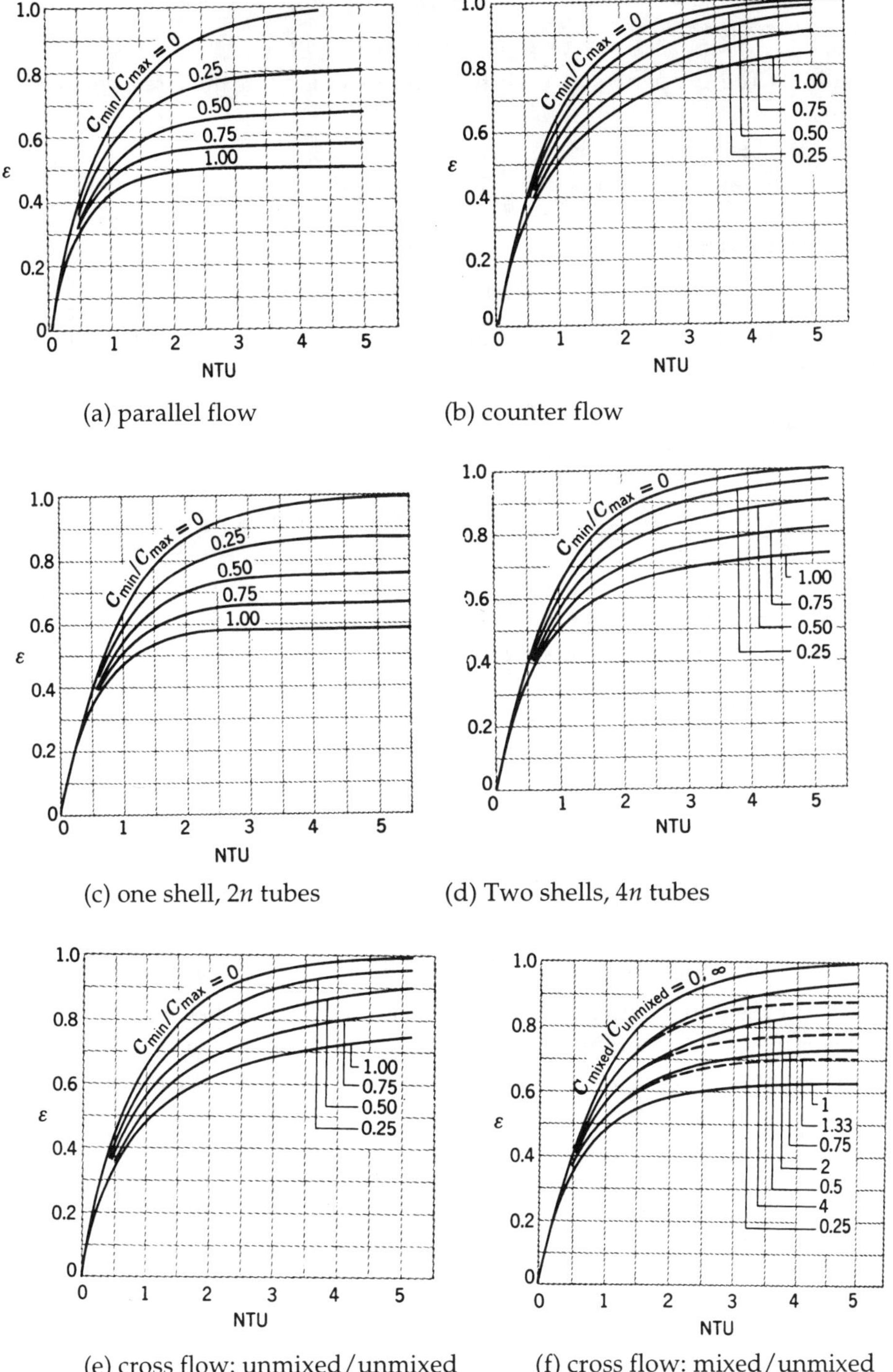

Figure 12.12 Effectiveness-NTU relations for different exchangers

Example 12.13

A shell and tube heat exchanger with $U_i = 350\ \text{W/m}^2\cdot\text{C}$ (based on inside tube area) is to be designed to heat 0.5 kg/s of oil from 20 C to 90 C using hot water available at 150 C.

a) If the water exits at 30 C, what flow rate of water is needed?
b) If the tubes are 25 mm ID and are 1 m long, how many tubes are required?

Solution. The properties of each fluid are evaluated at their respective mean values:

For oil at 60 C,

$$c_{poil} = 2.05\ \text{kJ/kg}\cdot\text{K}$$

For water at 90 C,

$$c_{pw} = 4.20\ \text{kJ/kg}\cdot\text{K}$$

Then, from the thermodynamic energy balance, Equation 12.4.14, the required flow rate of water is

$$\dot{m}_w = \frac{\dot{m}_{oil} c_{poil} \Delta T_{oil}}{c_{pw} \Delta T_w}$$

$$= \frac{0.5 \times 2.05 \times (90 - 20)}{4.20 \times (150 - 30)} \quad = 0.142\ \text{kg/s}$$

To find the number of tubes, the total area of the exchanger must first be found. The NTU parameter will contain this information. First compute the required exchanger effectiveness. Note that:

$$\frac{\dot{m}_w c_{pw}}{\dot{m}_{oil} c_{poil}} = \frac{\Delta T_{oil}}{\Delta T_w} = \frac{90 - 20}{150 - 30} = 0.583$$

Thus, water is the minimum fluid (the one with the lowest value of $\dot{m}c_p$), and $C_r = 0.583$. So

$$\varepsilon = \frac{q}{q_{\max}} = \frac{m_{oil} c_{poil} \Delta T_{oil}}{\dot{m}_w c_{pw} \Delta T_{\max}} = \frac{1}{C_r} \frac{\Delta T_{oil}}{(T_{win} - T_{oin})}$$

$$= \frac{1}{0.583} \times \frac{90 - 20}{150 - 20} = 0.924$$

Now, consulting Figure 12.12 note that the only type of shell and tube heat exchanger that can deliver this level of effectiveness is the simple counterflow type. For $C_r \approx 0.583$ and $\varepsilon \approx 0.924$, find NTU = 4.3 for the counterflow configuration. Now

$$A_i = \frac{\text{NTU}\ \dot{C}_{\min}}{U_i} = \frac{4.3 \times (0.142\text{kg/s} \times 4.20\text{kJ/kg}\cdot\text{K})}{350\ \text{W/m}^2\cdot\text{C}} \times \frac{1000\ \text{W}}{\text{kW}} = 7.33\ \text{m}^2$$

Since the inside area of one tube is $A_i/N = \pi d_i L = \pi(0.025\ \text{m})(1\ \text{m}) = 0.0785\ \text{m}^2$, the number of tubes required is

$$N = \frac{A_i}{A_i/N} = \frac{7.33}{0.0785} = 93$$

Special Case: One Fluid at constant temperature. This arises approximately in many circumstances, but is especially applicable when one fluid is undergoing a change of phase (evaporation or condensation). In this case, the heat capacity of the phase change fluid is virtually infinite, and the *other* fluid will always be the minimum fluid. This has the further consequence that the ratio $C_r = 0$, and the NTU-ε relationship simplifies for all exchangers to:

$$NTU = -\ln(1-\varepsilon) \qquad \text{All Exchangers, } C_r = 0$$

LMTD Method

In the LMTD method, the heat transfer relationship used to supplement the thermodynamic relations is:

$$q = UAF\Delta T_{LM} \tag{12.4.17}$$

where ΔT_{LM} is the *log mean temperature difference* (or LMTD), and is defined as the temperature difference at one end of the exchanger minus the temperature difference at the other end of the exchanger divided by the ratio of the same two temperature differences. So

$$\Delta T_{LM} = \frac{(T_H - T_C)_1 - (T_H - T_C)_2}{\ln\left[\frac{(T_H - T_C)_1}{(T_H - T_C)_2}\right]} = \frac{(T_H - T_C)_2 - (T_H - T_C)_1}{\ln\left[\frac{(T_H - T_C)_2}{(T_H - T_C)_1}\right]} \tag{12.4.18}$$

In Equation 12.4.17, F is a correction factor applicable to all but counterflow shell and tube exchangers (for which F = 1). These correction factors are shown in Figure 12.13 for different exchanger configurations.

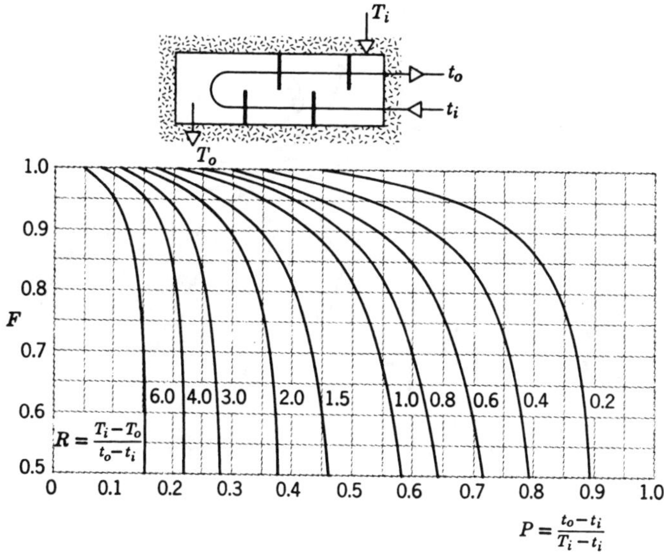

(a) one shell, 2*n* tubes

Figure 12.13 (a) Correction factors *F* for Eq.12.4.17 for different exchangers

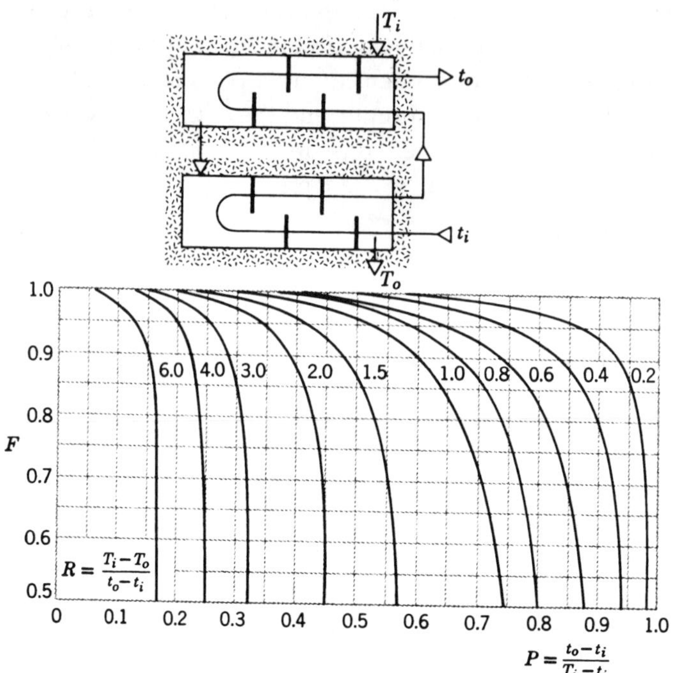

(b) two shells, 4*n* tubes

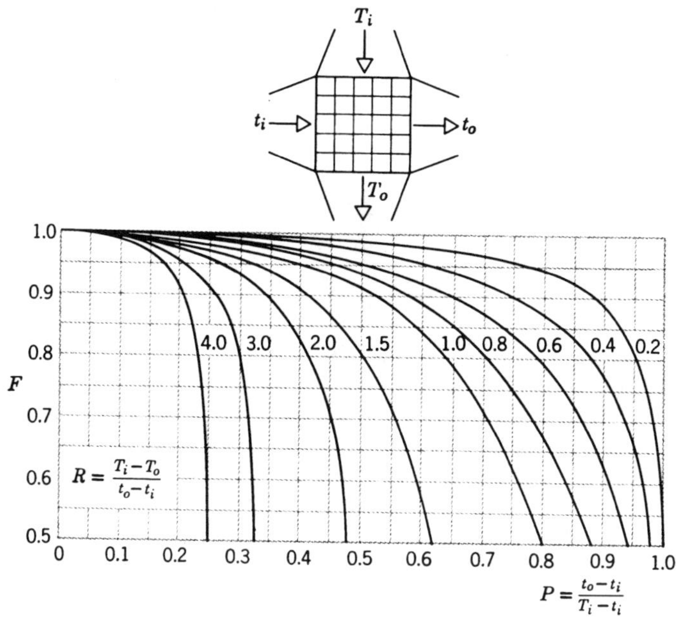

(c) unmixed/unmixed

Figure 12.13 (b,c) Correction factors *F* for Eq.12.4.17 for different exchangers

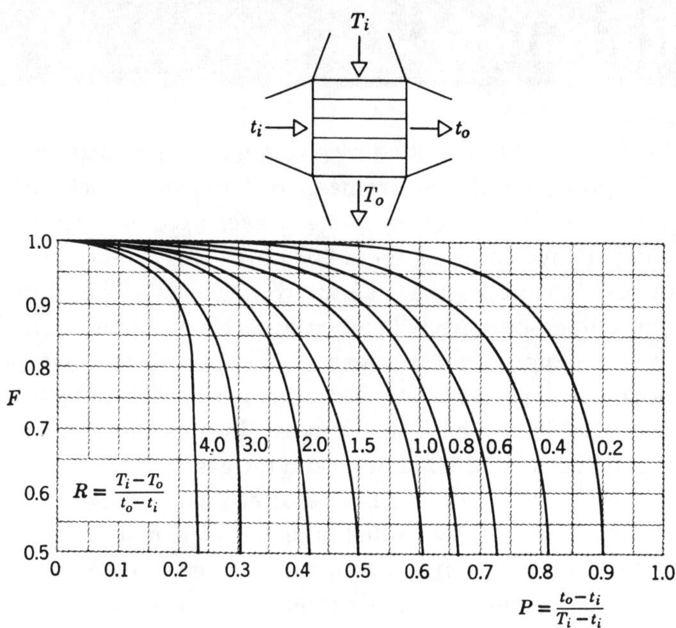

(d) mixed/unmixed

Figure 12.13 (d) Correction factors *F* for Eq.12.4.17 for different exchangers

Special Case: One Fluid at constant temperature. This arises approximately in many circumstances, but is especially applicable when one fluid is undergoing a change of phase (evaporation or condensation). In this case, either *P* or *R* in Figure 12.13 will be zero, and the factor *F* will be equal to 1.0. Then,

$$q = UA\Delta T_{LM} \qquad \text{All Exchangers, } F = 1.0$$

Example 12.14

Repeat the calculations of Example 12.13 using the LMTD method.

Solution. Of course, the required mass flow rate of water is unchanged as it is determined from thermodynamic considerations alone:

$$\dot{m}_w = 0.142 \text{ kg/s}$$

Now the LMTD is

$$\Delta T_{LM} = \frac{(150-90)-(30-20)}{\ln\left(\frac{150-90}{30-20}\right)} = 27.9 \text{ C}$$

Assuming $F = 1.0$ (counterflow operation), then the area required is:

$$A = q/U\Delta T_{LM} = \frac{\dot{m}c_p\Delta T_{oil}}{U_i\Delta T_{LM}}$$

$$= \frac{0.5 \times 2.05(90-20)}{350(27.9)} = 7.35 \text{ m}^2$$

This compares favorably with the value found previously.

12.5 Radiation

Thermal radiation is heat transmitted by electromagnetic radiation. Thermal radiation covers a portion of the electromagnetic spectrum which includes visible light, stretching from the ultraviolet range (wavelength of about 0.1 μm) to the microwave range (wavelength of about 100 μm). This type of heat transfer occurs without resistance across a vacuum, but may also be important when transmitted through gases, especially when surface temperatures are high. Many gases (such as air, oxygen, and nitrogen) do not absorb thermal radiation and may be considered non-participating; however, some gases (notably carbon dioxide and water vapor) have a high absorptivity. In such cases an analysis should include interaction between the radiation and the gas.

Generally, thermal radiation has both a directional and wavelength-dependent (spectral) character. *Diffuse* radiation is the same in all directions. *Integral* or *total* radiation accounts for energy across the entire wavelength spectrum.

The ideal source for emission of thermal radiation is called a *blackbody*. A blackbody is any object for which the emitted thermal radiation is

$$E = E_b = \sigma T^4$$

where σ is the Stefan-Boltzman constant ($\sigma = 5.67 \times 10^{-8} \mathrm{W/m^2 \cdot K^4} = 0.1714 \times 10^{-8} \mathrm{Btu/hr\text{-}ft^2\text{-}R^4}$), and the temperature is expressed in absolute (Kelvin = C + 273 or Rankine = F +460). For objects less than ideal, the amount of energy emitted is characterized by the emissivity ε:

$$E = \varepsilon \sigma T^4 = \varepsilon E_b$$

When thermal radiation passes through a medium (or strikes a solid surface) some of it may be absorbed, some reflected, and some transmitted through the medium. Three properties are defined which characterize these: absorptivity α, reflectivity ρ, and transmissivity τ. These individual properties must sum to unity:

$$\alpha + \rho + \tau = 1.$$

In general, these properties are also wavelength-dependent, however they are expressed here as integral (total) quantities.

A *gray surface* is one for which the absorptivity is equal to the emissivity:

$$\alpha = \varepsilon \qquad \text{(gray surface)}$$

A *diffuse surface* is one that does not emit or reflect radiation preferentially in any direction.

A wide class of radiation problems may be attacked by assuming thermal radiation exchange between diffuse gray surfaces with nonparticipating gases.

Gray Two-Body Problems

The simplest type of radiation problem involves one (relatively small) object at temperature T_1 which is completely enclosed by a second surface at T_2. If the surfaces are diffuse and gray, then the heat flow from surface 1 to surface 2 by thermal radiation is

$$q_{12} = \varepsilon_1 A_1 \sigma (T_1^{\,4} - T_2^{\,4}) \qquad \textbf{(12.5.1)}$$

Example 12.15

Small painted parts (A = 0.3 m^2, ε = 0.87) are passed through an oven on a conveyor to dry. The walls of the oven are at 400 C, and the parts are initially at 20 C. What is the initial rate of radiant heating of the parts?

Solution. The temperatures must be converted to absolute, then Equation 12.5.1 applies. Thus,

$$q_{12} = 0.87 \times 0.3 \times 5.67 \times 10^{-8} \times \left(293^4 - 673^4\right)$$
$$= -2930 \text{ W}$$

The minus sign indicates that the net heat flows *to* surface 1.

Radiation in enclosures

The total amount of radiant energy arriving at a surface per unit area from all sources is called the *irradiation, G*. The total amount of radiation leaving a surface (emitted plus reflected) per unit area is called the *radiosity, J*. Thus, for surface *i*:

$$J_i = E_i + \rho_i G_i = \varepsilon_i \sigma T_i^4 + \rho G_i$$

The difference between J_i and G_i is the net energy which must be supplied to the surface (by conduction and convection) to maintain its surface temperature. That is, the net heat transfer to the surface is

$$q_i = A_i(J_i - G_i)$$

which can be expressed as

$$q_i = \frac{(\sigma T^4 - J_i)}{(1-\varepsilon_i)/\varepsilon_i A_i} = \frac{E_{b_i} - J_i}{R_{surface}} \quad \textbf{(12.5.2)}$$

View Factors

To determine the total amount of radiant energy arriving at a surface, the energy leaving all the other surfaces and arriving at the current one needs to be known. The geometric *view factor*, or *configuration factor*, conveys this information:

$$F_{ij} = \frac{\text{Radiation reaching surface } j}{\text{Radiation leaving surface } i}$$

The reciprocity relation relates F_{ij} and F_{ji}:

$$A_i F_{ij} = A_j F_{ji}$$

The *enclosure relation* applies for any surface which is part of an *N*-surface enclosure, and reflects conservation of radiant energy leaving a surface:

$$\sum_{i=1}^{N} F_{ij} = 1$$

The view factors depend only on geometry, and are given for some simple two-dimensional geometries in Figure 12.14.

The net radiant exchange between surface *i* and surface *j* can be expressed as

$$q_{ij} = A_i F_{ij}\left(J_i - J_j\right) = \frac{J_i - J_j}{1/(A_i F_{ij})} = \frac{J_i - J_j}{R_{spatial}} \quad \textbf{(12.5.3)}$$

Resistance Networks

Equations 12.5.2 and 12.5.3 form the basis for a simple network analogy for radiation problems. The denominator of Equation 12.5.2 can be considered a *surface resistance* relating the surface radiosity to its blackbody emissive power. Similarly, the denominator of Equation 12.5.3 can be considered the *spatial resistance* relating the radiosities of the two surfaces. By combining these two concepts, a network representation of a complex radiation interchange in an enclosure can be drawn, and conventional DC electric circuit concepts can be used to solve them.

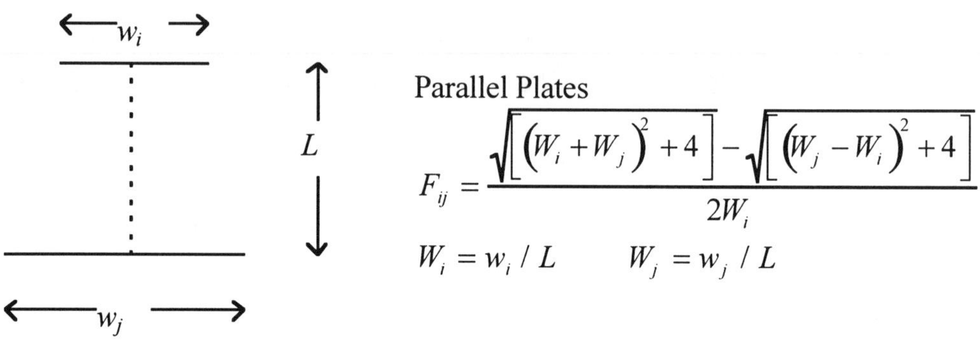

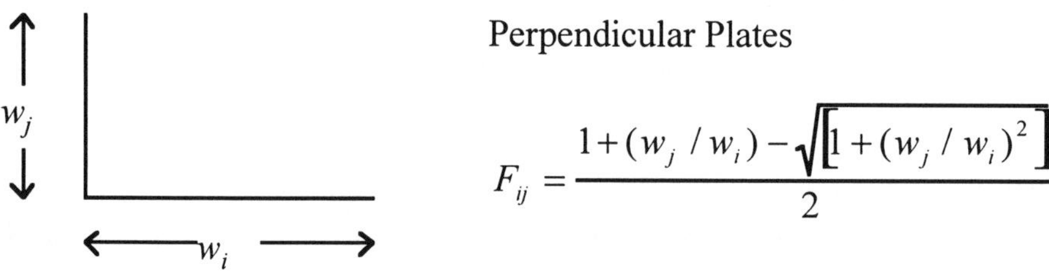

Figure 12.14 View Factors for some two dimensional geometries

Re-radiating (adiabatic) surfaces

An adiabatic surface is one which has no external heat supply. Accordingly, any incident radiation is either reflected or absorbed and re-emitted or a combination of both. The key feature is that $q_i = 0$, and the surface's temperature will be that required to attain this equilibrium condition. Also, for a re-radiating (adiabatic) surface, the surface radiosity J is equal to the blackbody emissive power E_b, since all of the energy arriving at the surface must be emitted or reflected away:

$$J_i = E_{b_i} = \sigma T_i^4 \qquad \text{(Adiabatic surface)}$$

Example 12.16

A three surface enclosure, as shown in the figure has surface 1 maintained at 27 C, surface 2 maintained at 377 C, and surface 3 is adiabatic.

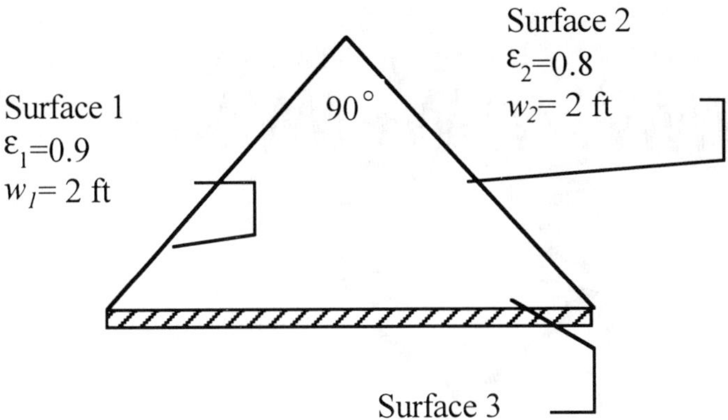

a) What is the view factor F_{31}?
b) What is the temperature of surface 3?
c) What is the net heat which must be supplied to surface 2 to maintain its temperature?

Solution. The view factor from 1-2 (or 2-1) can easily be computed using the relation from Figure 12.14 for perpendicular planes:

$$F_{12} = F_{21} = \frac{1+(2/2)-\sqrt{1+(2/2)^2}}{2} = 0.293$$

Now F_{13} can be found from the enclosure rule

$$F_{13} = 1 - F_{12} - F_{11} = 1 - 0.293 - 0 = 0.707$$

Finally, F_{31} can be found by reciprocity:

$$F_{31} = \frac{A_1 F_{13}}{A_{31}} = \frac{(2)(1)(0.707)}{2.83} = 0.50$$

which might have been apparent to the alert reader due to symmetry.

The remainder of the analysis can be completed with the aid of the network constructed in the figure on the next page.

Note that the total heat flow from surface 1 to surface 2 can be found by combining the intervening resistances in series and parallel:

$$\frac{1}{R_{eq}} = \frac{1}{1/A_1F_{12}} + \frac{1}{1/A_1F_{13} + 1/A_2F_{23}}$$

$$= (2)(0.293) + \frac{1}{1/(2)(0.707) + 1/(2)(0.707)} = 1.293$$

So, the R_{eq} which could replace the three resistors in the center is $R_{eq} = 0.773$. Now the heat flow from 2 to 1 is

$$q_{21} = \frac{E_{b2} - E_{b1}}{(1-\varepsilon_1)/\varepsilon_1 A_1 + R_{eq} + (1-\varepsilon_2)/\varepsilon_2 A_2}$$

$$= \frac{\left(5.67\times10^{-8}\right)\left(650^4 - 100^4\right)}{(1-0.9)/(0.9\times2) + 0.773 + (1-0.8)/(0.8\times2)} = 10600 \text{ W/m}$$

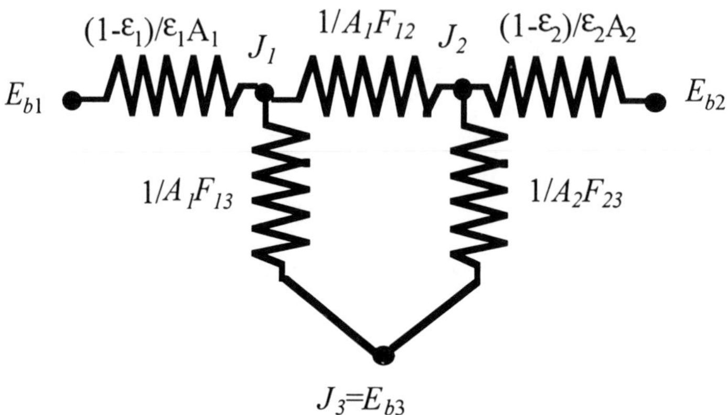

Now the two radiosities J_1 and J_2 can be found using voltage divider concepts:

$$J_1 = E_{b1} + q_{21} \times (1-\varepsilon_1)/\varepsilon_1 A_1$$

$$= \left(5.67\times10^{-8}\right)\left(100^4\right) + 10600\times(0.0556) = 595 \text{ W/m}^2\text{/m}$$

and

$$J_2 = E_{b2} - q_{21} \times (1-\varepsilon_2)/\varepsilon_2 A_2$$

$$= \left(5.67\times10^{-8}\right)\left(650^4\right) - 10600\times(0.125) = 8800 \text{ W/m}^2\text{/m}$$

The value J_3 is similarly found from the adiabatic condition ($q_{2\text{-}3} = q_{3\text{-}1}$) using network algebra

$$\frac{J_2 - J_3}{1/A_2F_{23}} = \frac{J_3 - J_1}{1/A_1F_{13}}$$

Since $A_2F_{23} = A_1F_{13}$ (due to the symmetry of the system), then

$$E_{b3} = J_3 = \frac{(J_1 + J_2)}{2} = 4700 \text{ W/m}^2\text{/m}$$

and

$$T_3 = \sqrt[4]{E_{b3}/\sigma} = 537 \text{ K} = 264 \text{ C}$$

Note that all the radiosities were computed to solve this problem. This is usually required to solve a general radiation problem.

Non-Gray Behavior

An important situation involving non-gray behavior of surfaces arises when considering solar radiation. Because solar radiation is emitted from a high temperature source (the sun!), it has a wavelength character much different from any radiation emitted from a terrestrial source. Specifically , the sun's radiation energy is concentrated at shorter wavelengths. Although it is the case that the surface emissivity and the absorptivity are equal in each wavelength band, they can be dramatically different for "high-temperature" (solar) radiation and for "low-temperature" (terrestrial) radiation. This behavior is, in fact, desirable for cases where the surface is to be heated by (or kept cool in spite of) direct solar radiation.

Example 12.17

An absorber plate for a solar collector has an absorptivity for solar radiation of $\alpha_{solar} = 0.8$ and an emissivity for long wavelength (terrestrial) emission of $\varepsilon = 0.2$. The plate receives an incident solar flux of $q_{solar} = 1100$ W/m^2, and exchanges long wavelength radiation with the sky, which is at an effective temperature of $T_{sky} = 285$ K. Determine the adiabatic temperature of the plate.

Solution. The net energy absorbed from the incident flux is $\alpha_{solar} q_{solar}$ (the amount $(1-\alpha_{solar})q_{solar}$ is reflected away). If adiabatic, then the amount absorbed must be transferred to the sky at the effective temperature T_{sky}:

$$\alpha_{solar} q_{solar} = \sigma\varepsilon({T^4}_{plate} - {T^4}_{sky})$$

So,

$$T_{plate} = \left({T^4}_{sky} + \frac{q_{solar}}{\sigma}\frac{\alpha_{solar}}{\varepsilon}\right)^{1/4}$$

$$= \left(285^4 + \frac{1100}{5.67\times10^{-8}}\frac{0.8}{0.2}\right)^{1/4} = 539 \text{ K} = 266 \text{ C}$$

Practice Problems

Essay Questions

12.1 The wind chill temperature is the temperature the air *feels* like on a cold windy day. Consider that the body temperature beneath a 3 mm thick layer of fatty tissue is maintained at about 36 C, and that the heat transfer coefficient on a calm day is nominally 10 $W/m^2 \cdot K$, while on a windy day it may be 50 $W/m^2 \cdot K$. Assume the air temperature is –5 C in both cases.

a) What is the surface temperature of the tissue on the calm day?
b) What is the surface temperature of the tissue on the windy day?
c) What temperature will the air need to be on the calm day to have the same heat loss as that on the windy day?

12.2 An electronic chip must dissipate 0.75 W of power in an environment where $T_\infty = 30$ C and $h = 20$ $W/m^2 \cdot C$. The chip is 30 mm × 30 mm, and is mounted on a circuit board that effectively insulates the bottom of the chip.

a) What is the surface temperature of the chip under steady conditions?
b) It is proposed to cool the chip and limit its surface temperature to 50 C by adding a heat sink with 4 mm diameter aluminum pins 15 mm in length. How many pins are needed?

12.3 A uranium fuel rod 15 mm in diameter is clad with 1 mm thick stainless steel skin. The uranium experiences uniform internal heat generation at a rate of 700 MW/m^3, while the heat is convected away from the outer surface of the steel with $h = 5000$ $W/m^2 \cdot K$ and $T_\infty = 90$ C.

a) What is the maximum temperature of the stainless steel?
b) What is the maximum temperature of the uranium?

12.4 Large aluminum ingots (60 cm × 150 cm × 500 cm) at 30 C are placed in a furnace for preheating. The average velocity of the air in the furnace is 10 m/s which blows across both of the faces in the 150 cm direction. The air in the furnace is at 600 C.

a) What is the heat transfer coefficient?
b) How long will it take for all parts of the slab to reach at least 500 C?

12.5 Exhaust gases from an industrial process are discharged through a thin-walled cylindrical stack. The stack is 0.5 m in diameter and 6.0 m tall. The gases, which have approximately the same thermophysical properties as air, enter the stack at 600 C and a mass flow rate of 0.5 kg/s. Atmospheric air flows over the cylinder at 5 C and a velocity of 5 m/s.

a) What is the temperature of the gas at the exit?
b) What is the temperature of the stack (cylinder) at the exit?

12.6 The feedwater heater for a boiler supplies 8,000 kg/h of water at 60 C from an inlet temperature of 20 C. The water is to be heated in a single-shell, double pass heat exchanger by condensing steam at 1.30 bars. The overall heat transfer coefficient is 2000 $W/m^2{\cdot}K$.

a) What is the required heat transfer area?
b) What is the rate of condensation of the steam?

12.7 Oil at 200 C is cooled by heating a large room at 27 C. The oil flows at 1.5 kg/s in a 3 cm ID steel pipe which has a 3 mm wall thickness. To enhance the cooling, annular fins (also made of steel) 3 mm thick and 3.5 cm long are attached to the outside of the pipe and are spaced with a 5 mm gap between fins. The heat transfer coefficient over the bare pipe and also the fins may be taken as 200 $W/m^2{\cdot}K$. It is desired to reduce the oil temperature to 120 C.

a) If the pipe is bare (without fins), how long must the pipe be to achieve the desired cooling?
b) With the fins in place, how long must the pipe be to achieve the desired cooling?

12.8 Waste heat is recovered from industrial exhaust gases by mounting a bank of unfinned tubes in the stack. Water at 25 C and a mass flow rate of 0.025 kg/s flows through each tube, while the exhaust gas approaches the bank at 550 C and a velocity of 5 m/s. The total gas mass flow rate is 2.25 kg/s. The bank is a 10 × 10 square array with 100 thin walled tubes 25 mm in diameter and 4 m long in an aligned configuration with a 50 mm transverse pitch. Assume the gas thermophysical properties to be similar to those of air.

a) What is the overall heat transfer coefficient?
b) What is the exit temperature of each stream?

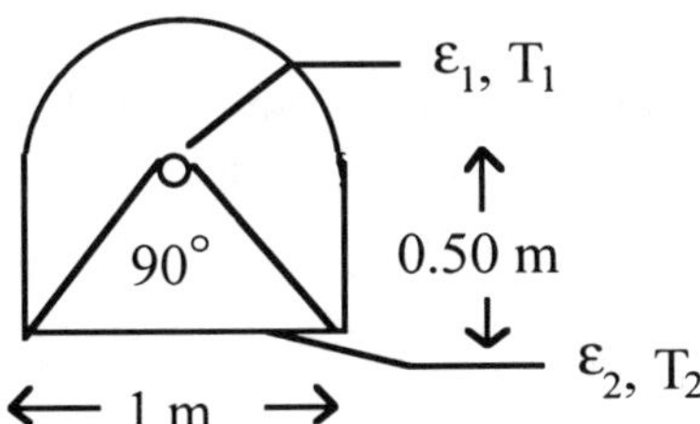

12.9 A long heater rod ($D = 1$ cm) is used to maintain conditions is a furnace as shown in the sketch. The walls of the furnace are refractory brick ($\varepsilon = 0.9$), while the floor is diffuse grey with $\varepsilon_2 = 0.6$. The heater element may be treated as black, and its temperature is maintained at 1200 C.

a) Determine the heat power (per unit length of furnace) required to maintain the floor at 300 C.
b) What is the temperature of the furnace walls under these conditions?

12.10 The temperature of the gas in a exhaust stack is measured by a cylindrical thermocouple probe. The probe axis is oriented normal to the gas flow, and the thermocouple indicates the equilibrium temperature T_t of the outside of the tube. This equilibrium temperature is the result of radiation exchange with the walls of the stack at T_s and convection from the hot gases at T_g. Given the following information, estimate the true gas temperature T_g.

D_s = stack duct diameter = 0.8 m
D_t = thermocouple probe diameter = 5 mm
$\dot{m}_g$ = mass flow rate of gas = 1 kg/s
T_s = stack wall temperature = 300 K
T_t = indicated gas temperature = 300 C
ε_t = emissivity of thermocouple probe = 0.9

Multiple Choice Questions

Questions 12.11-12.20 pertain to the following scenario:

Hot water flows through a 3-cm ID steel pipe which has a 3 mm wall thickness. The pipe passes through a large room.
For questions 12.11 through 12.16, use the following information:

h_i = water side heat transfer coefficient = 200 W/m^2·C
h_o = air side heat transfer coefficient = 50 W/m^2·C
T_{air} = 30 C
T_{surr} = temperature of surrounding walls = 25 C
ε = emissivity of the outside pipe surface = 0.8
k_{steel} = 60 W/m·C

12.11 At a particular x-location along the pipe the water temperature is 120 C. Neglecting radiation effects, the heat loss per unit length is:

a) 1695 W/m
b) 509 W/m
c) 391 W/m
d) 156 W/m

12.12 If the heat loss per unit length (neglecting radiation) at a particular location is 500 W/m, the temperature of the outside of the bare pipe at that location is:

a) 89 C
b) 99 C
c) 104 C
d) 118 C

12.13 Assume the outside pipe wall temperature is 80 C. The fraction of the total energy loss due to radiation alone is:

a) 26%
b) 55%
c) 43%
d) 66%

12.14 If a 0.5 cm layer of insulation (k_{ins}=0.05 W/m·C) is added and the water temperature is 120 C, the heat loss per unit length, neglecting radiation effects, is:

a) 89 W/m c) 107 W/m

b) 94 W/m d) 124 W/m

12.15 If the pipe is 10 m long and the water enters at 120 C, with $\dot{m}c_p = 84$ W/C, what is the exit temperature of the water? Neglect radiation effects.

a) 54 C c) 84 C

b) 69 C d) 99 C

12.16 Circumferential steel fins 8 mm wide and 25 mm long are added to the outside of the pipe with 8 mm of space between the fins. The efficiency of a single fin is 0.88 (88%). Neglecting radiation effects, the heat loss per unit length of pipe at a location where the water temperature is 120 C is:

a) 204 W/m c) 1100 W/m

b) 354 W/m d) 1730 W/m

__**Questions 12.17-20:**__ Use the following properties. Ignore the previously provided heat transfer coefficients:

	ρ (kg/m^3)	c_p (kJ/kg·K)	$\nu \times 10^6$ (m^2/s)	$k \times 10^3$ (W/m·K)	Pr
water	983	4.18	0.456	654	3.00
air	1.08	1.00	18.0	28.0	0.70

12.17 For a mass flow rate of 0.02 kg/s of water in the pipe, the heat transfer coefficient on the inside of the pipe is:

a) 10 W/m^2·C c) 80 W/m^2·C

b) 50 W/m^2·C d) 110 W/m^2·C

12.18 For a mass flow rate of 0.08 kg/s of water in the pipe, the heat transfer coefficient on the inside of the pipe is:

a) 560 W/m^2·C c) 1100 W/m^2·C

b) 920 W/m^2·C d) 1200 W/m^2·C

12.19 If the inside of the pipe is artificially roughened to enhance heat transfer so that the absolute surface roughness is ε = 0.125 mm, then the heat transfer coefficient for a mass flow rate of 0.08 kg/s will be nearly:

a) 560 W/m^2·C
b) 920 W/m^2·C
c) 1100 W/m^2·C
d) 1200 W/m^2·C

12.20 If the air flows over the cylinder with an approach velocity of 2.5 m/s, the exterior heat transfer coefficient will be:

a) 12 W/m^2·C
b) 26 W/m^2·C
c) 56 W/m^2-C
d) 86 W/m^2·C

Solutions to Practice Problems

Essay Questions

12.1 a) $q_{overall} = q_{surface}$ $\quad \dfrac{36-(-5)}{\dfrac{0.003}{0.2}+\dfrac{1}{10}} = \dfrac{T_s-(-5)}{\dfrac{1}{10}}$ $\quad \Rightarrow T_s = 30.7$ C

b) $T_s = 18.4$ C

c) $q_{windy} = \dfrac{36-T_\infty}{\left(\sum R_{th}\right)_{calm}}$ $\quad \dfrac{36-(-5)}{\dfrac{0.003}{0.2}+\dfrac{1}{50}} = \dfrac{36-T_\infty}{\dfrac{0.003}{0.2}+\dfrac{1}{10}}$ $\quad \Rightarrow T_\infty = -99$ C

12.2 a) $P_i = hA(T_s - T_\infty)$ $\quad T_s = 0.75/(20\times 0.03\times 0.03)+30 = 72$ C

b) Different assumptions about the tip condition are possible. Assume no heat loss from the tip (insulated tip). Then the heat dissipated by one fin follows from the first case in Table 12.5:

$q_{fin} = M\tanh(mL)$

$$M = \sqrt{20\cdot(0.004\pi)\cdot 177\cdot\left(\pi 0.004^2/4\right)}(50-30) = 0.4729$$

$$m = \sqrt{20\cdot(0.004\pi)/\left[177\cdot\left(\pi 0.004^2/4\right)\right]} = 10.63$$

$$q_{fin} = 0.4729\times\tanh(10.63\times 0.015) = 0.0748 \text{ W}$$

$$q_{total} = q_{fins} + q_{no\,fins} = N\cdot q_{fin} + (w^2 - N\cdot\pi d^2/4)h\Delta T = 0.75 \qquad \Rightarrow N = 5.6 \approx 6$$

12.3 a) All of the heat generated must be conducted through the steel:

$$q'''A\cdot L=\frac{T_{\max}-T_{\infty}}{\frac{\ln((d_i+2\Delta r)/d_i)}{2\pi kL}+\frac{1}{h\pi d_o L}}=\frac{T_{\max}-90}{\frac{\ln(17/15)}{2\pi(14.9)L}+\frac{1}{5000\cdot\pi\cdot(0.017)L}}$$

$$\Rightarrow T_{\max,st}=719\text{ C}$$

b) Maximum temperature occurs at centerline. Use equation for solid cylinder with constant internal heat generation:

$$T(0)=T_{\max,st}+\frac{700\times 6\cdot(0.015^2)}{4\cdot 43.9}=1616\text{ C}$$

12.4 a) Assume the average surface temperature is 300 C, then the average film temperature is T_{film} = (600+300)/2 = 450 C. Then the air properties are:

$$\nu=71.9\times10^{-6}\text{ m}^2/\text{s}\qquad k=0.0536\text{ W/m}\cdot\text{C}\qquad \Pr=0.698$$

Now compute Reynolds number to determine the flow regime:

$$\text{Re}_L=\frac{V\cdot L}{\nu}=\frac{10\times1.50}{71.9\times10^{-6}}=208{,}600<500{,}000\text{ so laminar over entire length}$$

Then

$$\overline{\text{Nu}}_L=0.664\cdot208{,}600^{0.5}\cdot0.698^{1/3}=269\qquad\Rightarrow\bar{h}=269\cdot0.0536/1.50=9.6\text{ W/m}^2\text{-C}$$

b) The center of the slab will take the longest to heat. Assume all faces of slab convect heat with same h just computed. Check for lumped capacitance assumption:

$$L_c=V/A=(1.50\cdot5.0\cdot0.6)/2/(1.5\cdot5+1.5\cdot0.6+5.0\cdot0.6)=0.197\text{ m}$$

$$\text{Bi}=\frac{h(V/A)}{k_{Al}}=\frac{9.6\cdot(0.197)}{177}=0.01<0.1\text{ Lumped Capacitance O.K.}$$

$$t=\frac{-\rho c_p L_c}{h}\ln\left(\frac{\theta(t)}{\theta_0}\right)=\frac{-2770\cdot875\cdot0.197}{9.6}\ln\left(\frac{500-600}{30-600}\right)=86{,}600\text{ s}=24\text{ h}$$

12.5 a) The heat transfer coefficient on each side must be found from appropriate correlations. Assume the average wall temperature is about 100 C. Then the approximate film temperature on the inside of the tube is (600 + 100)/2 = 350 C. Properties of air are

$$\nu=56.15\times10^{-6}\text{ m}^2/\text{s}\qquad k=0.0482\text{ W/m-C}\qquad \Pr=0.687$$

$$\rho=0.5598\text{ kg/m}^3\qquad c_p=1057\text{ J/kg-C}$$

Then

$$Re_D = \frac{VD}{\nu} = \frac{\dot{m}}{\rho\pi D^2/4}\cdot\frac{D}{\nu} = \frac{4\cdot 0.5}{0.5598\cdot\pi\cdot 0.5\cdot 56.15\times 10^{-6}} = 40,500 > 2300$$

$$Nu_D = 0.023\times 40,500^{0.8}\times 0.687^{1/3} = 98.5 \quad \Rightarrow h = \frac{98.5\cdot 0.0482}{0.5} = 9.5\ \text{W/m}^2\text{-C}$$

For the outside of the cylinder, the film temperature is (5 + 100)/2 ≈ 50 C.

$\nu = 18.20\times 10^{-6}\ \text{m}^2/\text{s}$ $\quad k = 0.0280\ \text{W/m-C}$ $\quad \Pr = 0.704$

Then

$$Re_D = \frac{VD}{\nu} = \frac{5\cdot 0.5}{18.2\times 10^{-6}} = 137,360$$

From Table 12.8, C = 0.027 and m = 0.805, and

$$Nu_D = 0.027\cdot 137,360^{0.805}\times 0.704^{1/3} = 329 \quad \Rightarrow h = \frac{329\cdot 0.0280}{0.5} = 18.4\ \text{W/m}^2\text{-C}$$

Now, the overall heat transfer coefficient is

$$UA = \frac{1}{\sum R_{th}} = \left[\frac{1}{h_i} + \frac{1}{h_o}\right]^{-1} A = \left[\frac{1}{9.5} + \frac{1}{18.4}\right]^{-1} A = 6.3A\ \ (\text{W/C})$$

And

$$\dot{m}c_p(T_{out} - T_{in}) = -UA\frac{((T_{out} - T_\infty) - (T_{in} - T_\infty))}{\ln\left(\frac{T_{out} - T_\infty}{T_{in} - T_\infty}\right)}$$

$$\Rightarrow T_{out} = (T_{in} - T_\infty)\times\exp\left[-\frac{UA}{\dot{m}c_p}\right] + T_\infty = 536\ \text{C}$$

b) At the exit:

$$\frac{(T_{out} - T_\infty)}{A\sum R_{th}} = (T_{wall} - T_\infty)h_o A \quad \Rightarrow T_{wall} = \frac{U}{h_o}(536 - 5) + 5 = 187\ \text{C}$$

12.6 a) The steam will maintain constant temperature as long as it undergoes phase change. The temperature will be the saturation temperature at the given pressure. From steam tables, $T_{steam} \approx 106.9$ C. Since the steam temperature is constant, $C_r = 0$, and the water is the minimum fluid. For the exchanger, the effectiveness is

$$\varepsilon = \frac{q}{q_{max}} = \frac{\dot{C}_{water}(T_{water,out} - T_{water,in})}{\dot{C}_{min}(T_{steam,in} - T_{water,in})} = \frac{(60 - 20)}{(106.9 - 20)} = 0.460$$

When one fluid temperature is constant,

$$NTU = -\ln(1-\varepsilon) = -\ln(1 - .460) = 0.616 = UA/\dot{C}_{min}$$

$$\Rightarrow A = NTU\cdot\dot{C}_{min}/U = 0.616\cdot\left(\frac{8000}{3600}4179\right)/2000 = 2.86\ \text{m}^2$$

b) $\dot{m}_{steam}h_{fg} = \dot{C}_{water}(T_{out} - T_{in})$ From steam tables $h_{fg} \approx 2238$ kJ/kg

$$\Rightarrow \dot{m}_{steam} = \frac{1}{2238}\frac{8000}{3600}\cdot 4.179(60 - 20) = 0.165\ \text{kg/s} = 600\ \text{kg/hr}$$

12.7 a) The heat transfer coefficient on the inside of the pipe must be found. Assume the pipe wall temperature is about 90 C, then the average film temperature of the oil is 100 C. For unused engine oil:

$$\nu = 20.5\times10^{-6}\ \text{m}^2/\text{s} \qquad k = 0.1367\ \text{W/m-C} \qquad \text{Pr} = 280$$

$$\rho = 840.1\ \text{kg/m}^3 \qquad c_p = 2219\ \text{J/kg-C}$$

Then the Reynolds number is

$$\text{Re}_D = \frac{VD}{\nu} = \frac{\dot{m}D}{\rho\pi\left(D^2/4\right)\nu} = \frac{4\cdot1.5}{840.1\cdot\pi\cdot0.030\cdot20.5\times10^{-6}} = 3696 > 2300$$

And the Nusselt number is

$$\text{Nu}_D = 0.023\times3700^{0.8}\times280^{1/3} = 108$$

$$\Rightarrow h = 108\times0.1367/0.03 = 491\ \text{W/m}^2\text{-C}$$

Then the overall heat transfer coefficient is

$$UA = \left[\frac{1}{\pi\cdot0.03\cdot L\cdot491} + \frac{\ln(36/30)}{2\pi\cdot60\cdot L} + \frac{1}{\pi\cdot0.036\cdot L\cdot200}\right]^{-1}$$

$$= L[0.0216 + 0.00048 + 0.442]^{-1} = 15.1L$$

The heat exchanger factor F is unity since one fluid temperature is constant, so

$$L = \dot{m}c_p\Delta T_{oil} = \Delta T_{LM}(UA/L)L$$

$$\Delta T_{LM} = (200-120)/\ln((200-27)/(120-27)) = 128.9$$

$$L = 1.5\times2219\times80/128.9/15.1 = 137\ \text{m}$$

b) Need the efficiency for one of the fins. Use Figure 12.1:

$$L_c^{3/2}(h/kA_p)^{1/2} = (0.035+0.003/5)^{3/2}(200/(60\cdot0.00011))^{1/2} = 1.217$$

$$r_{2c}/r_1 = (2\cdot(3.5+0.3)+3.0)/3.6 = 2.94$$

Entering Figure 12.1, find h_{fin}= 0.44. Now find the resistance of the bank of fins:

$$A_{fin} = 2\cdot\pi\left(D_o^2 - d_o^2\right)/4 + \pi D_o t = 166\ \text{cm}^2$$

$$A_{space} = \pi d_o s = 5.7\ \text{cm}^2$$

$$\frac{A_{fin}}{A_{total}} = \frac{A_{fin}}{A_{fin}+A_{space}} = 0.967$$

$$\eta_{total} = 1 - \frac{A_{fin}}{A_{total}}(1-\eta_{fin}) = 0.459$$

Now the overall fin bank resistance is found. The key is to express the total area for heat transfer per unit length:

$$A_{total}/L = (A_{fin}+A_{space})/(s+t) = 2.15\ \text{m}^2/\text{m}$$

$$R_{bank} = L[h(A_{total}/L)\eta_{total}]^{-1} = 0.00507L\ \ \text{C-m}^2/\text{W}$$

Assuming the interior heat transfer coefficient is unchanged, and neglecting the thermal resistance of the pipe wall itself, the overall heat transfer coefficient per unit length is

$UA/L = [1/\pi d_i h_i + R_{bank}/L]^{-1} = 37.5$

And

$L = L_{no\,fins} \times (15.1/37.5) = 44$ m

12.8 a) Need heat transfer coefficients for both sides of the tubes. Assume the tube wall temperature is about 100 C, then the average film temperature of the water could be about 75 C:

$\nu = 0.3842 \times 10^{-6}$ m^2/s $\quad k = 0.668$ W/m-C $\quad \Pr = 2.35$

$\rho = 974.9$ kg/m^3 $\quad c_p = 4193$ J/kg-C

Reynolds and Nusselt numbers:

$$\mathrm{Re}_D = \frac{4\dot{m}}{\rho\pi D\nu} = \frac{4(0.025)}{974.9\cdot\pi\cdot 0.025\cdot 0.3842\times 10^{-6}} = 3,400 > 2300$$

$$\mathrm{Nu}_D = 0.023\cdot 3400^{0.8} 2.35^{1/3} = 20.4$$

$$\Rightarrow h_i = 20.4\cdot 0.668/0.025 = 545 \text{ W/m}^2\text{-C}$$

For the air on the exterior of the tubes, the film temperature may be 300 C:

$\nu = 48.85 \times 10^{-6}$ m^2/s $\quad k = 0.0453$ W/m-C $\quad \Pr = 0.684$

$\rho = 0.6088$ kg/m^3 $\quad c_p = 1045$ J/kg-C

Since the bank is an aligned arrangement, the maximum Re occurs at the transverse section:

$$\mathrm{Re}_{\max} = \frac{S_T}{S_T - D}\frac{V_\infty D}{\nu} = \frac{50}{50-25}\frac{5\cdot 0.025}{48.85\times 10^{-6}} = 5,125$$

In Table 12.9, find $C = 0.229$ and $m = 0.632$ (since $S_T/D = S_L/D = 2.0$). Then

$$\overline{\mathrm{Nu}}_D = 1.13\times 0.229\times 5,125^{0.632}\times 0.684^{1/3} = 50.4$$

$$\Rightarrow \bar{h}_o = 50.4\times 0.0453/0.025 = 91.3 \text{ W/m}^2\text{-C}$$

Then

$$UA = [1/h_i A + 1/h_o A]^{-1} = 78.2A = 78.2\cdot N_{tubes}\pi DL = 2,460 \text{ W/C}$$

b) Find the effectiveness from the NTU:

$\dot{C}_{air} = 2.25\cdot 1045 = 2350$ W/C $\quad \dot{C}_{water} = 100\cdot 0.025\cdot 4193 = 10,500$

So the air is the minimum fluid and $C_r = 0.224$. Also

$$NTU = \frac{UA}{\dot{C}_{\min}} = \frac{2460}{2350} = 1.05$$

For a cross flow exchanger, with one fluid mixed (air) and one unmixed (water), the effectiveness $\varepsilon = 0.6$. So

$$\Delta T_{air} = \frac{\varepsilon\dot{C}_{\min}\Delta T_{\max}}{\dot{C}_{air}} = 0.6\times(550-25) = 315 \text{ C} \quad \Rightarrow T_{air,out} = 235 \text{ C}$$

$$\Delta T_{water} = \frac{\varepsilon\dot{C}_{\min}\Delta T_{\max}}{\dot{C}_{water}} = 0.6\times 0.224\times(550-25) = 71 \text{ C} \quad \Rightarrow T_{water,out} = 96 \text{ C}$$

12.9 a) Key is recognition that the view factor $F_{1\text{-}2} = 90/360 = 0.25$ (25% of the energy leaving the heater rod is intercepted by the floor). The analogous network for this problem is the same as that in Figure 12.16. The equivalent resistance of the spatial resistors is:

$$\frac{1}{R_{eq}} = \left[A_1F_{1-2} + \frac{1}{1/A_1F_{1-3} + 1/A_2F_{2-3}}\right]$$

$$A_1F_{1-2} = A_2F_{2-1}; \; F_{2\text{-}3} = 1 - F_{2-1} = 1 - 0.00785 = 0.992$$

$$\frac{1}{R_{eq}} = \left[0.00785 + \frac{1}{42.4 + 1.01}\right] = 0.0309 \quad \Rightarrow R_{eq} = 32.4$$

Then

$$\frac{\dot{P}}{L} = \frac{q_2}{L} = -\frac{q_1}{L} = \frac{E_{b_1} - E_{b_2}}{R_{eq} + (1-\varepsilon_2)/\varepsilon_2 A_2} = \frac{5.67E{-}8(1473^4 - 573^4)}{32.4 + 0.4/(0.6 \cdot (1))} = 7.89 \text{ kW/m}$$

b) The radiosities must be found:

$$J_2 = E_{b_1} - \frac{q_2}{L} R_{eq} = 11{,}290 \text{ W/m}^2$$

$$\frac{q_{1-2}}{L} = \frac{A_1}{L} F_{1-2}\left(E_{b_1} - J_2\right) = 2{,}010 \text{ W/m} \quad \Rightarrow q_{1-3} = q_2 - q_{1-2} = 5{,}880 \text{ W/m}$$

$$J_3 = E_{b_1} - \frac{q_{1-3}}{L}\frac{1}{A_1F_{1-3}} = 17{,}370 \text{ W/m}^2$$

Since surface 3 is refractory (adiabatic):

$$J_3 = E_{b_3} \quad \Rightarrow T_3 = 744 \text{ K} \approx 470 \text{ C}$$

12.10 The heat transfer coefficient on the outside of the probe must be found. Assume the depth of immersion of the probe is sufficient so that end effects are insignificant. Assume the film temperature is not much different from the indicated 300C:

$$\nu = 48.85 \times 10^{-6} \text{ m}^2/\text{s} \qquad k = 0.0453 \text{ W/m-C} \qquad \text{Pr} = 0.684$$

$$\rho = 0.6088 \text{ kg/m}^3 \qquad c_p = 1045 \text{ J/kg-C}$$

Then the Reynolds number:

$$\text{Re}_D = \frac{VD}{\nu} = \frac{\dot{m}_g}{\rho\pi D_s^2/4}\frac{D_t}{\nu} = 334$$

Then from Table 12.8 $C = 0.683$ and $m = 0.466$, and $\text{Nu}_D = C\,\text{Re}_D{}^m\,\text{Pr}^{1/3} = 9.03 \quad \Rightarrow h = 81.8 \text{ W/m}^2\text{-C}$

At equilibrium, the energy gained by convection from the gas is equal to the energy radiated to the walls (neglecting conduction along the probe):

$$q_{conv} = q_{rad}$$

$$hA(T_g - T_t) = \varepsilon_t \sigma A(T_t^4 - T_s^4)$$

$$T_g = \frac{\varepsilon_t \sigma}{h}(T_t^4 - T_s^4) + T_t = 363 \text{ C}$$

Multiple Choice Questions

12.11 **c)**

$$q' = \frac{\Delta T}{L\sum R_{th}}; \; L\sum R_{th} = \left(\frac{1}{h_i \pi d_i} + \frac{\ln((d_i + 2\Delta r)/d_i)}{2\pi k_{steel}} + \frac{1}{h_o \pi (d_i + 2\Delta r)}\right)$$

$$q' = \frac{120 - 30}{0.0531 + 0.00048 + 0.177} = 391 \text{ W/m}$$

12.12 **d)**

$$q' = \frac{T_o - T_{air}}{1/(h_o \pi d_o)} \quad \Rightarrow T_o = \frac{500}{50 \cdot \pi \cdot 0.036} + 30 = 118 \text{ C}$$

12.13 **b)**

$$q'_{conv} = \frac{T_o - T_{air}}{1/(h_o \pi d_o)} = 283 \text{ W/m};$$

$$q'_{rad} = \varepsilon\sigma\left(T_o^4 - T_{surr}^4\right) = 0.8\left(5.67 \times 10^{-8}\right)\left(353^4 - 298^4\right) = 347 \text{ W/m}$$

12.14 **a)**

$$L\sum R_{th} = 0.2306 + \frac{\ln((0.036 + 0.01)/0.036)}{2\pi \cdot 0.05} = 1.011$$

$$q' = 90/1.011 = 89 \text{ W/m}$$

12.15 **c)**

$$UA = \left(\sum R_{th}\right)^{-1} = 1/0.02306 = 43.4 \text{ W/C}$$

$$NTU = UA/\dot{C}_{\min} = 43.4/84 = 0.517$$

$$\varepsilon = 1 - \exp(-NTU) = 0.403 = q_{act}/q_{\max} = \dot{C}_{water}\Delta T_{water}/\dot{C}_{\min}\Delta T_{\max}$$

$$T_{out} = T_{in} - \varepsilon\Delta T_{\max} = 84 \text{ C}$$

12.16 **c)**

$$A_{1fin} = \left[2 \cdot \pi\left(86^2 - 36^2\right)/4 + \pi \cdot 86 \cdot 8\right] \times 10^{-6} = 0.0117 \text{ m}^2$$

$$A_{space} = \pi \cdot 0.036 \cdot 0.008 = 0.000905 \text{ m}^2$$

$$\eta_{total} = 1 - \frac{A_{fin}}{A_{total}}(1 - \eta_{fin}) = 1 - \frac{0.0117}{0.0126}(1 - 0.88) = 0.889$$

$$A_{total}/L = (0.0126/0.016) = 0.788 \text{ m}^2/\text{m}$$

$$LR_{fins} = 1/\left(h \cdot (A_{total}/L)\eta_{total}\right) = 0.0286$$

$$L\sum R_{th} = 0.0531 + +0.00048 + 0.0286 = 0.0822$$

$$q' = \Delta T/\left(L\sum R_{th}\right) = 90/(0.0822) = 1095 \text{ W/m}$$

12.17 **c)**

$$\text{Re} = \frac{4\dot{m}}{\rho\pi D\nu} = 1890 \quad \Rightarrow \text{Laminar}$$

Nu = 3.66 (assuming constant wall temperture)

$$h = 80 \text{ W/m}^2\text{-C}$$

12.18 **b)**

$$\text{Re} = \frac{4\dot{m}}{\rho\pi D\nu} = 7575 \quad \Rightarrow \text{Turbulent}$$

$$\text{Nu} = 0.023\,\text{Re}^{0.8}\,\text{Pr}^{1/3} = 42.1$$

$$h = 918 \text{ W/m}^2\text{-C}$$

12.19 **d)**

use Chilton-Colburn Analogy.

$$\text{Re} = 7575; \quad \varepsilon / D = 0.0042 \quad \Rightarrow \text{from Moody diagram } f \approx 0.040$$

$$\text{Nu} = \frac{f}{8}\text{Re}\,\text{Pr}^{1/3} = 55$$

$$h = 1190 \text{ W / m}^2\text{-C}$$

12.20 **b)**

$$\text{Re} = \frac{VD}{\nu} = \frac{2.5 \cdot 0.036}{18 \times 10^{-6}} = 5000$$

$$\text{Nu} = 0.193 \cdot 5000^{0.618} \cdot 0.70^{1/3} = 33$$

$$h = 26 \text{ W / m}^2\text{-C}$$

References

Gardner, K. A., "Efficiency of Extended Surfaces,", *Trans. ASME*, 67, 1945, 621.
Heisler, M. P., "Temperature Charts for Induction and Constant Temperature Heating," *Trans. ASME*, 69, 227-236, 1947.
Incropera, F. P., and DeWitt, D. P., *Fundamentals of Heat and Mass Transfer*, Third Edition, John Wiley and Sons.
Holman, J. P., *Heat Transfer*, Seventh Edition, McGraw-Hill.

Appendices

Courtesy of Incropera, F. P., and DeWitt, D. P., *Fundamentals of Heat and Mass Transfer*, Third Edition, John Wiley and Sons.

Table 12.11 Properties of Some Solids

	T C	ρ kg/m^3	c_p J/kg·K	k W/m·K
Metals:				
Aluminum (pure)	27	2702	903	237
Aluminum (2024-T6)	27	2770	875	177
Copper (pure)	27	8933	385	401
Brass (70%Cu30%Zn)	27	8530	380	110
Iron (pure)	27	7870	447	80.2
Carbon Steel	27	7854	434	60.5
Chrome (1%) Steel	27	7858	442	42.3
Stainless (304) Steel	27	7900	477	14.9
Building Materials:				
Gypsum Board		800		0.17
Plywood		545	1215	0.12
Common Brick		1920	835	0.72
Hardwood		1010	1380	0.15
Other Materials:				
Glass		2300	835	1.4
Ice		920	2040	1.88
Sand		1515	800	0.27
Fatty Tissue				0.2
Uranium	727	19070	180	43.9

Table 12.12 Properties of Liquid Water (SI)

T	ρ	c_p	$\nu \times 10^6$	$k \times 10^3$	Pr
C	kg/m^3	kJ/kg·K	m^2/s	W/m·K	
0	1000.0	4.217	1.7500	569	12.99
5	1000.0	4.203	1.5140	579	11.04
10	1000.0	4.193	1.3038	587	9.39
15	999.4	4.186	1.1386	595	8.06
20	998.4	4.182	1.0090	603	7.00
25	997.4	4.180	0.8989	610	6.15
30	995.8	4.178	0.8067	617	5.45
35	993.8	4.178	0.7291	625	4.85
40	991.9	4.179	0.6620	632	4.34
45	989.9	4.180	0.6047	638	3.93
50	987.9	4.181	0.5543	643	3.56
55	985.4	4.183	0.5120	648	3.26
60	983.1	4.185	0.4754	654	2.99
65	980.6	4.187	0.4418	658	2.75
70	977.7	4.190	0.4105	665	2.53
75	974.9	4.193	0.3842	668	2.35
80	972.0	4.197	0.3619	670	2.20
85	968.6	4.201	0.3423	673	2.07
90	964.9	4.207	0.3246	676	1.95
95	961.7	4.212	0.3076	678	1.84
100	957.9	4.217	0.2913	680	1.76
105	958.4	4.219	0.2883	681	1.72
110	955.5	4.224	0.2782	682	1.65

Table 12.13 Properties of Air at Atmospheric Pressure (SI)

T	ρ	c_p	$\nu \times 10^6$	$k \times 10^3$	Pr
C	kg/m^3	kJ/kg·K	m^2/s	W/m·K	
0	1.2874	1.006	13.49	24.1	0.714
50	1.0849	1.008	18.20	28.0	0.704
100	0.9380	1.011	23.45	31.7	0.695
150	0.8264	1.017	29.16	35.4	0.688
200	0.7383	1.025	35.33	38.9	0.685
250	0.6672	1.035	41.91	42.2	0.684
300	0.6088	1.045	48.85	45.3	0.684
350	0.5598	1.057	56.15	48.2	0.687
400	0.5181	1.069	63.84	50.9	0.692
450	0.4822	1.081	71.90	53.6	0.698
500	0.4510	1.093	80.31	56.0	0.705
550	0.4236	1.104	89.01	58.4	0.712
600	0.3992	1.115	97.99	60.7	0.718
650	0.3775	1.126	107.18	63.1	0.721
700	0.3581	1.136	116.66	65.4	0.724
750	0.3409	1.145	126.48	67.8	0.726

Table 12.14 Properties of (Unused) Engine Oil (SI)

T	ρ	c_p	$\nu \times 10^6$	$k \times 10^3$	Pr
C	kg/m^3	kJ/kg·K	m^2/s	W/m·K	
0	899.1	1.796	4280	147.0	47000
10	893.7	1.839	2037	144.3	23120
20	888.2	1.880	949	145.0	10950
30	882.2	1.922	471	145.0	5500
40	876.1	1.964	250	144.4	2970
50	870.0	2.006	142	142.4	1737
60	864.0	2.047	86.1	140.4	1081
70	858.1	2.089	55.7	138.7	719
80	852.1	2.131	38.1	138.0	501
90	846.0	2.175	27.4	137.7	367
100	840.1	2.219	20.5	136.7	280
110	834.4	2.263	15.8	135.7	219
120	829.0	2.307	12.5	134.7	177
130	823.2	2.350	9.98	133.7	144
140	816.9	2.395	8.05	133.0	118
150	810.4	2.440	6.61	132.7	99

Heating Ventilation, Air Conditioning and Refrigeration

by Bassem H. Ramadan

A Brief Introduction to HVAC

In order for people, animals, and plants to live and work comfortably in closed spaces such as residences, offices, and shopping malls, the proper thermal environment must be provided. Temperature and humidity of the air can influence the body's skin and interior temperature. Hence, the temperature and humidity of the air should be controlled to maintain a comfortable and safe environment. Heating, ventilation, and air-conditioning (HVAC) systems are used to control the thermal environment in enclosed spaces. Heating is employed during the winter season to heat the inside air when the outside air conditions are unsafe. Likewise, air-conditioning or cooling of the inside air is required when humidity and temperature of the outside air are intolerable for humans. Moreover, the air inside an occupied space should be fresh and free of toxic fumes. Fresh air should be supplied through proper ventilation of the space. Control of the thermal environment has also helped maintain the proper conditions for food storage. Now various food products can be stored for long periods of time by the use of refrigeration systems. In such a system heat is removed from the food product and maintained at the proper temperature for storage.

All pertinent figures and tables are located at the end of this chapter.

13.1 Psychrometrics

In air conditioning systems, moist air is always a working substance, but associated systems may involve refrigerants, steam, water or other fluids. Hence, to study a heating or air conditioning system, the state of the working substance should be defined in terms of its thermodynamic properties.

Atmospheric Air

The earth's atmosphere is a mixture of several gases such as, oxygen, nitrogen, carbon dioxide, argon, water vapor, and traces of other gases. Atmospheric air usually contains various particulate matter and additional vapors. Water vapor and other vapors are usually concentrated in the atmosphere only within a few thousand feet of the earth's surface. Above an altitude of about 20,000 feet the atmosphere consists essentially of dry air.

Moist Air

In air conditioning problems the working substance is moist air. Moist air is defined as a binary mixture of dry air and water vapor. It may contain variable amounts of water vapor from zero (dry air) to that of saturated moist air. Moist air is considered to be saturated when it may coexist in neutral equilibrium with associated condensed water.

The humidity ratio W (often ω in thermodynamics) defined as the weight of water vapor associated with unit weight of dry air. Humidity ratio is related to the mol-fraction of water vapor by the expression:

$$W = \frac{18.016}{28.966}\frac{x_w}{(1-x_w)} = 0.622\frac{x_w}{(1-x_w)} \tag{13.1.1}$$

where x_w is the mol-fraction of water vapor, and the molecular weights for dry air and water vapor are 28.966 and 18.016, respectively.

The degree of saturation μ is defined as

$$\mu = \frac{W}{W_s} \tag{13.1.2}$$

where W_s is the humidity ratio at saturation for the same temperature and pressure as those for the actual state.

Relative humidity ϕ is defined as

$$\phi = \frac{x_w}{x_{w,s}} \tag{13.1.3}$$

where $x_{w,s}$ is the mol-fraction of water vapor at saturation for the same pressure and temperature of the actual state. Alternatively, ϕ may be written as

$$\phi = \mu\frac{0.622 + W_s}{0.622 + W} \tag{13.1.4}$$

Other moist-air properties that are associated with temperature are:

Dry-bulb temperature T is the true temperature of moist-air at rest.

Dew-point temperature T_d is the saturation temperature corresponding to the humidity ratio and pressure of a given moist-air state.

Wet-bulb temperature T_w is the temperature at which water, by evaporating into air, can bring the air to saturation adiabatically at the same temperature. To help understand this definition consider the process shown in the sketch below.

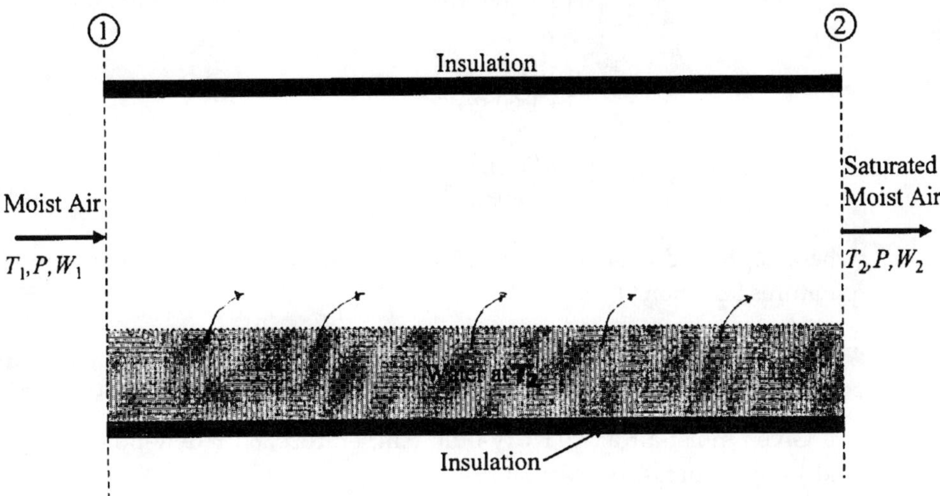

Assuming steady-state conditions the energy equation for the above system is

$$h_1 + (W_{s,2} - W_1)\, h_{f,2} = h_{s,2} \tag{13.1.5}$$

where h_1 is the enthalpy of moist-air at state (1)

$h_{s,2}$ is the enthalpy of saturated air at state (2)

$h_{f,2}$ is the enthalpy of saturated liquid water at state (2)

Since this is a constant pressure process we may deduce that the temperature T_2 is a function of h_1, W_1, and P. In other words, T_2 is a function of state (1).

The Goff and Gratch Tables for Moist Air

J.A. Goff and S. Gratch calculated accurate thermodynamic properties of moist-air for standard sea-level pressure. These properties are listed in Table 13.1. (All tables and figures are located at the end of this chapter.) Calculations for volume, enthalpy, and entropy of moist-air are given by

$$v = v_a + \mu v_{as} + \overset{=}{v} \tag{13.1.6}$$

$$h = h_a + \mu h_{as} + \overset{=}{h} \tag{13.1.7}$$

$$s = s_a + \mu s_{as} + \overset{=}{s} \tag{13.1.8}$$

where v is specific volume

h is specific enthalpy

s is specific entropy

Subscript a indicates properties for dry air, and subscript s indicates properties for saturated air.

Hence

$$v_{as} = v_s - v_a \tag{13.1.9}$$

$$h_{as} = h_s - h_a \tag{13.1.10}$$

$$s_{as} = s_s - s_a \tag{13.1.11}$$

Moreover,

$$\bar{\bar{v}} = \frac{\mu(1-\mu)A}{1+1.608\mu W_s} \tag{13.1.12}$$

$$\bar{\bar{h}} = \frac{\mu(1-\mu)B}{1+1.608\mu W_s} \tag{13.1.13}$$

$$\bar{\bar{s}} = \frac{\mu(1-\mu)C}{1+1.608\mu W_s} \tag{13.1.14}$$

where, A, B, and C are constants. However, they may be taken as zero for temperatures less than 96°F.

Example 13.1

Given moist-air at 75°F dry-bulb temperature, 60°F dew-point temperature, and 14.7 psia pressure, calculate:

a) the humidity ratio

b) degree of saturation

c) relative humidity

d) specific enthalpy

e) specific volume

Solution:

a) $W = W_s$ at $T = 60°F$, $P = 14.7$ psia from Table 13.1:

$W = 0.01108$ lbw/lba

b) For $T = 75°F$ and $P = 14.7$ psia we find $W_s = 0.01882$ lbw/lba. Hence,

$$\mu = \frac{W}{W_s} = \frac{0.01108}{0.01882} = 0.5887 \text{ or } 58.87\%$$

c) $$\phi = \mu\frac{0.662+W_s}{0.662+W} = 0.5887\frac{0.662+0.01882}{0.662+0.01108} = 0.596 \text{ or } 59.6\%$$

d) Since $T = 75°F < 96°F$, then $A = B = C = 0$, or $\bar{\bar{v}} = \bar{\bar{h}} = \bar{\bar{s}} = 0$.

$h = h_a + \mu(h_s - h_a)$

From Table 13.1:

$h_a = 18.018$ Btu/lba $\qquad v_a = 13.474$ ft^3/lba

$h_{as} = 20.59$ Btu/lba $\qquad v_{as} = 0.407$ ft^3/lba

$h = 18.018 + 0.5887(20.59) = 30.14$ Btu/lba

$v = 13.474 + 0.5887(0.407) = 13.71$ ft^3/lba

Perfect Gas Equations

Since Table 13.1 is restricted to a pressure of 14.7 psia, we use perfect gas relations to calculate properties of moist-air at pressures other than 14.7 psia. We will assume that circumstances permit assumption of Dalton's Rule. Thus, $x_w = P_w/P$, which allows writing

$$W = 0.622\frac{P_w}{P_{w,s}}$$

where P_w is the partial pressure of water vapor, and

$$W_s = 0.622\frac{P_{w,s}}{P - P_{w,s}} \quad \textbf{(13.1.15)}$$

Also,

$$h = 0.240T + Wh_g$$

or,

$$h = c_{P,a}T + 1061W \quad \textbf{(13.1.16)}$$

where

$$c_{P,a} = 0.240 + 0.45W \quad \textbf{(13.1.17)}$$

is the specific heat of moist-air. The above gives h_g = 1075 Btu/lba at T = 32°F. The volume of moist-air, using Dalton's Rule, can be written as

$$v = \frac{R_a T}{P - P_w} \quad \textbf{(13.1.18)}$$

where R_a = 53.35 (ft-lb)/(lba-°R) and T is in degrees Rankine. In addition, ϕ is written as

$$\phi = \frac{P_w}{P_{w,s}} \quad \textbf{(13.1.19)}$$

where $P_{w,s}$ is the pressure of water vapor at saturation. By Eqs.13.1.5 and 13.1.6 we write

$$W = \frac{W_s^* h_{fg}^* - 0.240\left(T - T^*\right)}{h_g - h_f^*} \quad \textbf{(13.1.20)}$$

where the superscript star refers to the saturated state.

Example 13.2

Repeat Example 13.1, using perfect-gas relations and steam tables. The steam tables are in Chapter 5.

Solution:

a) Use Eq.13.1.15. Note, $P_w = P_s$ at $T = T_d$ = 60°F. From the steam tables P_w = 0.25618 psia; hence,

$$W = 0.622\frac{0.256}{14.7 - 0.256} = 0.01103 \text{ lbw/lba}$$

b) At T = 75°F, $P_{w,s}$ = 0.42979 psia. Therefore,

$$W_s = 0.622\frac{0.430}{14.7 - 0.430} = 0.01875 \text{ lbw/lba}$$

$$\therefore \mu = W/W_s = 0.01103/0.01875 = 0.588 \text{ or } 58.8\%$$

c) The relative humidity is

$$\phi = \frac{P_w}{P_{w,s}} = \frac{0.25618}{0.42979} = 0.596 \text{ or } 59.6\%$$

d) Use Eq.13.1.16 where

$$c_{p,a} = 0.240 + 0.45W = 0.240 + 0.45(0.01103) = 0.245 \text{ Btu/lba-°F}$$

Hence,

$$h = c_{p,a}T + 1061W$$

$$= 0.245(75) + 1061(0.01103) = 30.08 \text{ Btu/lba}$$

e) The ideal gas law provides

$$v = \frac{R_a T}{P - P_w} = \frac{(53.35)(535)}{(14.7 - 0.256)144} = 13.72 \text{ ft}^3/\text{lba}$$

Psychrometric Charts

A psychrometric chart is a plot of the thermodynamic properties of moist air, and the various relations between them. It allows graphical reading of the various properties, and provides a graphical method for solving problems. It provides quick but less accurate results compared with the previous equations.

A Mollier Diagram is a psychrometric chart that uses enthalpy and humidity ratio as the coordinates. A psychrometric chart is constructed for some single value of barometric pressure. Typically, standard sea-level pressure has been used to construct these charts. Three *ASHRAE* psychrometric charts are widely used. They are of the Mollier type and are constructed for sea-level pressure. The three charts cover a wide range of temperatures. The low temperature chart covers temperatures from −40 to 50°F. The second one, the one included at the end of this chapter, is for the normal range of temperatures from 32°F to 120°F where most air conditioning problems occur. And the third chart covers temperatures from 60 to 250°F. The three charts were constructed using data from the Goff and Gratch tables for moist air.

Example 13.3

Solve Example 13.1 using the psychrometric chart found at the end of this chapter, Fig.13.1.

Solution:

a) Using $T_2 = 60$°F and the saturation curve we read $W = 0.0111$ lbw/lba

b) For T=75°F, read W_s=0.01882 lbw/lba, then $\mu = \frac{W}{W_s} = \frac{0.0111}{0.01882} = 0.5898$

c) For $T = 75$°F and $W = 0.0111$ lbw/lba, read $\phi = 59\%$

d) Using $T = 75°F$ and $W = 0.0111$ lbw/lba we get $h = 30.20$ Btu/lba

e) Similarly, we read $v = 13.70\ ft^3$/lba

Elementary Psychrometric Processes

In the following processes the fundamental problems are independent of heat and/or mass transfer rates in equipment. The problems in these sections are considered to be at steady-state, and the barometric pressure is assumed to remain constant throughout the process.

Sensible heating or cooling of moist air

Sensible heating is the process in which heat is added to moist air with no addition of moisture. Sensible cooling is the process in which heat is removed from moist air with no moisture removed. The surface temperature of the cooling element must be higher than the air dew-point temperature to prevent moisture condensation. For these processes, the energy equation gives

$$\dot{Q} = \dot{m}_a(h_2 - h_1) \quad \textbf{(13.1.21)}$$

or, from Eq. 13.1.17 and 13.1.18

$$\dot{Q} = \dot{m}_a(0.24 + 0.45W)(T_2 - T_1) \quad \textbf{(13.1.22)}$$

Here, $\dot{m}_a$ = mass flow rate in lba/hr or lba/min, and $\dot{Q}$ is the heat transfer rate in Btu/hr or Btuh.

Dehumidification of moist air by cooling

When moist air is cooled below its dew-point temperature, then condensation of moisture will occur. In this process we assume that all of the condensed moisture is brought to the final temperature T_2. Writing the conservation equations for this process we get

energy: $$\dot{m}_{a,1}h_1 = \dot{m}_{a,2}h_2 + \dot{Q} + \dot{m}_f h_{f,2} \quad \textbf{(13.1.23)}$$

water: $$\dot{m}_{a,1}W_1 = \dot{m}_{a,2}W_2 + \dot{m}_f \quad \textbf{(13.1.24)}$$

air: $$\dot{m}_{a,1} = \dot{m}_{a,2} \quad \textbf{(13.1.25)}$$

Combining, we find

$$\dot{Q} = \dot{m}_a(h_1 - h_2) - \dot{m}_f h_{f,2} \quad \textbf{(13.1.26)}$$

$$\dot{m}_f = \dot{m}_a(W_1 - W_2) \quad \textbf{(13.1.27)}$$

where $\dot{m}_f$ is the mass flow rate of condensate.

Example 13.4

Moist air at 80°F dry-bulb temperature, 50% relative humidity, enters a refrigeration coil at a rate of 200 lb/min and a pressure of 14.7 psia. The air leaves the coil as saturated moist air at 50°F. Calculate the tons of refrigeration needed.

Solution: First, we determine the enthalpies from the psychrometric charts: h_1 = 31.2 Btu/lba, h_2 = 20.3 Btu/lba; the humidity ratios are W_1 = 0.01096 lbw/lba and W_2 = 0.00766 lbw/lba. From the steam tables at 50°F we find h_f = 18.07 Btu/lbw.

From Eq.13.1.27 we calculate the mass flux of condensed moisture to be

$$\dot{m}_f = 200\,(0.01096 - 0.00766) = 0.66 \text{ lbw/min}$$

Using Eq.13.1.26 we find

$$\dot{Q} = 200(31.2 - 20.3) - 0.66(18.07) = 2168 \text{ Btu/min}$$

Thus tons of refrigeration needed

$$\dot{Q} = \frac{2168 \text{ Btu/min}}{200 \text{ Btu/min/ton}} = 10.8 \text{ tons}$$

Note, one ton of refrigeration = 200 Btu/min.

Combined heating and humidification of moist air

In this process heat and moisture are added to the air as it circulates through the processing element, during the winter, or as the air passes through a cooled space, during the summer. In this case we write

$$\dot{m}_{a,1}h_1 + \dot{Q} + \dot{m}_w h_w = \dot{m}_{a,2}h_2 \quad \textbf{(13.1.28)}$$

$$\dot{m}_{a,1}W_1 + \dot{m}_w = \dot{m}_{a,2}W_2 \quad \textbf{(13.1.29)}$$

Hence,

$$\dot{Q} = \dot{m}_a(h_2 - h_1) - \dot{m}_w h_w \quad \textbf{(13.1.30)}$$

and

$$\dot{m}_w = \dot{m}_a(W_2 - W_1) \quad \textbf{(13.1.31)}$$

Example 13.5

Atmospheric air at 46°F dry-bulb temperature and 30°F wet-bulb temperature enters a heating chamber with a volume flow rate of 3000 ft^3/min. The air is heated at a rate of 153,000 Btu/hr and picks up 88 lbw/hr of saturated steam at 230°F. Determine the dry-bulb temperature and thermodynamic wet-bulb temperature of the leaving air.

Solution: For T_d = 46°F and T_w = 30°F we have from the psychrometric chart, W_1 = 0.0007 lbw/lba. We also find h_1 = 11.4 Btu/lba, v_1 = 12.75 ft^3/lba. Consequently,

$$m_a = \frac{3000}{12.75} = 235.3 \text{ lba/min}$$

From Eq.13.1.31 we have

$$W_2 = W_1 + \frac{\dot{m}_w}{\dot{m}_a} = 0.0007 + \frac{88}{235.3} = 0.00693 \text{ lbw/lba}$$

From Eq.13.1.30 we can write:

$$h_2 = \frac{\dot{Q}}{\dot{m}_a} + \frac{\dot{m}_w}{\dot{m}_a} h_w + h_1$$

$$= \frac{153{,}000/60}{235.3} + \frac{88/60}{235.3}(1157) + 11.4 = 29.45 \text{ Btu/lba}$$

For W_2 = 0.00693 lbw/lba and h_2 = 29.45 But/lba we determine (using the *ASHRAE* psychrometric chart),

$$T_{d2} = 91.2°\text{F} \text{ and } T_{w2} = 64.3°\text{F}$$

Humidification of moist air

In this process moisture, in vapor or liquid form, is added to air as it flows through a chamber. It is assumed that all of the added moisture is retained by the passing air. In this case $\dot{Q}$ is equal to zero. Hence,

$$\dot{m}_w h_w = \dot{m}_a (h_2 - h_1) \quad \textbf{(13.1.32)}$$

$$\dot{m}_w = \dot{m}_a (W_2 - W_1) \quad \textbf{(13.1.33)}$$

The humidification process can be achieved in several ways. If air is humidified at constant dry-bulb temperature, then the steam added must have a specific enthalpy equal to that of saturated steam at the air dry-bulb temperature. If water at the air wet-bulb temperature is added, the entering and leaving air wet-bulb temperatures must be equal; in this case, if $h_w > h_g$, then air may be sensibly heated as well as humidified. However, if $h_w < h_g$ then the air may be sensibly cooled during humidification.

Evaporative Cooling

Evaporative cooling can be used when the temperature is relatively high and the humidity of the atmospheric air is relatively low. This is an adiabatic heat exchange process, in which heat is supplied by the air, into which water is evaporated. Since this occurs adiabatically, the amount of heat removed from the air must equal the amount of heat absorbed by the water. When this process is occurring in an evaporative cooling apparatus, the water will reach equilibrium, and its temperature will approach the wet-bulb temperature of the entering air. On a psychrometric chart this process is represented by a line of constant wet-bulb temperature (essentially, a line of constant enthalpy). In a 100% efficient process the air is cooled to the wet-bulb temperature. However, in practice this process could be designed to be about 90% efficient. In this case, the air leaves the apparatus at a temperature higher than the wet-bulb temperature.

Example 13.6

Air at 100°F and 30% relative humidity is supplied to a humidifier using recirculated water. The humidifier has adequate capacity to bring the air to within 65% of the original wet-bulb depression. Find the condition of the air as it leaves the humidifier.

Solution: For $T = 100°F$ and $\phi = 30\%$ we determine from the psychrometric chart, $T_w = 74.3°F$.

The original wet-bulb depression is 100 – 74.3 = 25.7°F. However, the air is brought to within 65% of this depression. Hence, the actual depression is (100 – 0.65 × 25.7= 83.3°F). Therefore, the air leaves the humidifier at $T_d = 83.3°F$ and T_w= 74.3°F which gives $\phi = 68\%$.

Adiabatic mixing of two streams of moist air

In some air conditioning systems mixing of two air streams (1) and (2) often occurs. This mixing usually occurs under adiabatic conditions. In this case we write:

energy:
$$\dot{m}_{a,1}h_1 + \dot{m}_{a,2}h_2 = \dot{m}_{a,3}h_3 \quad \textbf{(13.1.34)}$$

water:
$$\dot{m}_{a,1}W_1 + \dot{m}_{a,2}W_2 = \dot{m}_{a,3}W_3 \quad \textbf{(13.1.35)}$$

air:
$$\dot{m}_{a,1} + \dot{m}_{a,2} = \dot{m}_{a,3} \quad \textbf{(13.1.36)}$$

Eliminating $\dot{m}_{a,3}$ from Eqs. 13.1.34 and 13.1.35 we get

$$\frac{\dot{m}_{a,1}}{\dot{m}_{a,2}} = \frac{h_2 - h_3}{h_3 - h_1} = \frac{W_2 - W_3}{W_3 - W_1} \quad \textbf{(13.1.37)}$$

which indicates that state(3) must lie on the line connecting state(1) to state(2) on the psychrometric chart. It should be noted that segments of this line are proportional to the masses of dry air, so we can write

$$\frac{\dot{m}_{a,1}}{\dot{m}_{a,2}} = \frac{\overline{32}}{\overline{13}} \quad \text{or} \quad \frac{\dot{m}_{a,1}}{\dot{m}_{a,3}} = \frac{\overline{32}}{\overline{12}} \quad \text{or} \quad \frac{\dot{m}_{a,2}}{\dot{m}_{a,3}} = \frac{\overline{13}}{\overline{12}} \quad \textbf{(13.1.38)}$$

Example 13.7

A stream of moist air at 65°F dry-bulb temperature, 60°F wet-bulb temperature, and volume flow rate of 900 ft^3/min mixes with a second stream of air at 82°F dry-bulb temperature, 68°F wet-bulb temperature, and 450 ft^3/min. Determine the dry-bulb temperature of the mixture.

Solution: Given:

$\dot{m}_{a,1} = 900\ ft^3/min$ (cfm) $\qquad \dot{m}_{a,2} = 450\ ft^3/min$

$T_{d1} = 65°F \qquad T_{d2} = 82°F$

$T_{w1} = 60°F \qquad T_{w2} = 68°F$

We find from the psychrometric chart the following:

$v_1 = 13.4\ \text{ft}^3/\text{lba}$ $v_2 = 13.9\ \text{ft}^3/\text{min}$

$W_1 = 0.0100\ \text{lbw/lba}$ $W_2 = 0.0115\ \text{lbw/lba}$

Then,

$$\dot{m}_{a,1} = (900\ \text{cfm})/(13.4\ \text{ft}^3/\text{lba}) = 67.2\ \text{lb}$$

$$\dot{m}_{a,2} = (450\ \text{cfm})/(13.9\ \text{ft}^3/\text{lba}) = 32.4\ \text{lba/min}$$

From Eq.13.1.36

$$\dot{m}_{a,3} = \dot{m}_{a,1} + \dot{m}_{a,2} = 67.2 + 32.4 = 99.6\ \text{lba/min}$$

From Eq.13.1.35

$$W_3 = \frac{\dot{m}_{a,1}}{\dot{m}_{a,3}} W_3 + \frac{\dot{m}_{a,2}}{\dot{m}_{a,3}} W_2$$

or,

$$W_3 = \frac{67.2}{99.6}(0.0102) + \frac{32.4}{99.6}(0.0115) = 0.01062\ \text{lbw/lba}$$

State(3) is determined by the intersection of $W = W_3$ line and the line connecting state(1) to state(2). Hence,

$T_{d3} = 70.2°\text{F}$.

13.2 Heating Load Calculation

In this section we discuss the procedure for calculating the heating load in a conditioned space. The procedure includes estimating the heat loss through the walls, roof or ceiling, and floor of the conditioned space, that are exposed to the outside. In addition, heat loss through glass windows and doors is also calculated. Moreover, the heat required to heat the air that enters by infiltration through window and door cracks, or by a ventilation system must be included in the heat load calculations.

The equation used for calculating wall heat loss is

$$\dot{Q}_w = U_w A_w (T_i - T_o) \tag{13.2.1}$$

where $\dot{Q}_w$ is the heat loss in Btu/hr, and U_w is the overall heat transfer coefficient in Btu/hr-ft^2-°F, found either by direct experiment or by calculation or selected from suitably tabulated values, such as those found in the *ASHRAE Heating, Ventilating, Air Conditioning Guide.* The area A_w is the wall, roof or floor surface area in ft^2, excluding glass and door areas. The temperature T_i is the inside design temperature of the heated space, which can be different for different types of space, but usually a value of 72°F is selected. The design outside temperature T_o is selected based on the geographic location of the heated space. This temperature is usually calculated using temperature data from previous years, available from *U.S. Weather Bureau* records. The design temperature T_0 is often taken as the temperature that occurred during 97.5% of the time during the heating season, such as December, January and February.

For the windows and doors, we use a similar relationship:

$$\dot{Q}_g = U_g A_g (T_i - T_o) \tag{13.2.2}$$

where A_g is the total area of windows and doors.

Air infiltration occurs through cracks and clearances around windows and doors. It depends on the wind velocity, the tightness of the construction of the building, and the crack length. The crack length is the perimeter of each window sash, counting only once the crack at the meeting rail of a double-hung sliding sash. Infiltration values in cubic feet per hour per feet of crack, for various window and door types, under different wind speeds are listed in Table 13.2.

For almost any type of space a volume of air, usually over half to three times the volume of the heated space, enters by infiltration each hour, and consequently an equivalent volume of air is ejected. This results in the air being changed several times per hour. This is referred to as the number of air changes per hour. At least 0.5 air changes per hour should be considered for any conventional space. Very tight spaces can have as low as 0.02 air changes per hour. Health considerations are then of special concern.

The equation for calculating heat loss due to air infiltration is

$$\dot{Q}_{\text{inf}} = 0.244 \dot{m}_a (T_i - T_o) \tag{13.2.3}$$

where $\dot{m}_a$ is the mass flow rate of air, in lb/hr, entering by infiltration. In terms of the volume flow rate in ft^3/min (*cfm*) as

$$\dot{Q}_{\text{inf}} = 1.08 \times cfm \times (T_i - T_o) \tag{13.2.4}$$

In some cases, spaces may contain heat producing items, such as equipment, lights and people. Except for unusual layouts, it is not common practice to deduct the heat produced by equipment, lights and people from the heat load calculations. However, in some cases, such as buildings containing large equipment, huge heat producing lights, or large numbers of people, care should be taken in determining the heating or cooling load.

The above is a simplified procedure used to estimate the heating load in buildings. More extensive calculations are usually used to determine a more accurate value of the heating load. Computer codes are available and are usually employed when calculating the heating load in large commercial buildings with varying operating schedules.

Example 13.8

A room in a building is 11 ft long, 10 ft wide and 9 ft high, has one double-hung window 4 ft by 6 ft, in each of the two outside walls. Calculate the heating load for the room based on an inside design temperature of 72°F, outside air temperature at 0°F and 10 mph wind speed. Use the following *U*-values:

Item	Wall	Roof	Floor	Glass
U-(Btu/hr-ft^2-F)	0.35	0.28	0.34	1.10

Solution: Wall heat loss:

$A_g = 2\,(4\times 6) = 48\ \text{ft}^2$ $A_w = (10+11)\times 9 - 48 = 141\ \text{ft}^2$

$\dot{Q}_w = 0.35\times 141(72-0) = 3553\ \text{Btu/hr}$

Roof: $A_R = 10\times 11 = 110\ \text{ft}^2$, $\dot{Q}_R = 0.28\times 110\times(72-0) = 2218\ \text{Btu/hr}$

Floor: $A_F = 10\times 11 = 110\ \text{ft}^2$, $\dot{Q}_F = 0.34\times 110\times(72-0) = 2693\ \text{Btu/hr}$

Windows: $\dot{Q}_g = 1.1\times 48\times(72-0) = 3802\ \text{Btu/hr}$

Infiltration: Crack length = 2[2(4+6)+6] = 52 ft

From Table 13.2 for a 10 mph wind velocity $q_h = 21.4\ \text{ft}^3/\text{hr/ft}$. Assuming the number of air changes per hour to be 0.5, the total air infiltration is $0.5\times 52\times 21.4 = 556.4\ \text{ft}^3/\text{hr}$. Hence,

$$\dot{Q}_{\text{inf}} = 0.244\times(0.075\ \text{lb/ft}^3)(556.4\ \text{ft}^3/\text{hr})(72\text{–}0) = 733\ \text{Btu/hr}$$

Total:

$$\dot{Q} = 3553+2218+2693+3802+733 = 13{,}000\ \text{Btu/hr}$$

13.3 Cooling Load Calculations

When an air-conditioned space is to be cooled, such as by supplying chilled air to the space, heat must be removed from the conditioned space to provide a comfortable thermal environment. The cooling load calculations are based on the amount of heat to be removed to offset all heat gains to the space. Sources of heat gains include:

1. Heat gain through exposed walls, roof, windows, doors and floor, due to temperature difference between inside and outside air.
2. Solar heat gain. This includes solar radiation through glass windows, and solar energy absorbed by exposed surfaces of the conditioned space.
3. Heat gain from occupants
4. Heat gain from equipment, lights and machinery.
5. Heat gain from infiltration air.

The procedure for calculating the heat transmitted through walls, roofs, floors, windows and doors due to the temperature difference existing on both sides of the particular surface, is the same as Eq. 13.2.1. However, in this case we usually have $T_i<T_0$. The outside temperature T_0 during the cooling season depends on the geographic location of the space being considered; T_i can again be taken as 72°F or even higher in some cases. Hence, Eq. 13.2.1 is rewritten as

$$\dot{Q}_w = U_w A_w (T_o - T_i) \qquad \textbf{(13.3.1)}$$

Solar Load Calculations

Computations of the solar load involved in calculating the cooling load are complex and extensive. Many variables such as location, time of day, building orientation, cloud cover, shading, type of construction and glass types are involved in calculating heat gain through buildings due to solar energy. Hence, computer programs are usually used to handle these complex calculations. In this section, however, a simplified approach is presented. The procedure involves using tables obtained from the *ASHRAE Handbook of Fundamentals*. Table 13.3 gives the equivalent temperature differential for calculating the total heat gain due to both the air-temperature difference and solar effects, through sunlit and shaded walls. Table 13.4 is for sunlit and shaded roofs. Table 13.5 gives the overall heat transfer coefficient *U* for flat roofs. Similarly heat gain through glass windows can be determined by using Tables 13.6 to 13.10. These tables are based on an air-temperature difference of 15°F. A correction must be made to the values in the tables for air-temperature differences other than 15°F.

Example 13.9

Estimate the heat gain from a 3 ft by 5 ft single-sheet plate glass window in a northwest wall at 3 p.m. in 40°N Latitude location. The inside air is maintained at 70°F and the outside air is at 85°F.

Solution: From Table 13.6 we find the instantaneous heat gain as 79 Btu/hr-ft^2. Table 13.8 gives the application factors as 0.87 and 1.0(*X*)+0.25(*Y*). The value of X is obtained from Table 13.7 as 19 Btu/ hr-ft^2, and Y is obtained from Table 13.9 as 15 Btu/hr-ft^2. Hence,

$$\dot{q}_{glass} = 0.87 \times 79 + 1.0(19) + 0.25(15) = 91.5 \text{ Btu/hr-ft}^2$$

The heat gain is

$$\dot{Q}_{glass} = A_g \dot{q}_{glass} = (3 \times 5) \times 91.5 = 1372 \text{ Btu/hr}$$

Example 13.10

Estimate the solar heat gain at 4 p.m. through a 6 inch thick concrete roof with a 2 inch thick insulation layer with an inside temperature of 72°F and outside temperature equal to 90°F.

Solution: We use Table 13.4 to get $\Delta T_e = 42$°F. We determine from Table 13.5 $U = 0.13$ Btu/hr-ft^2-F. However, since $(T_o - T_i) = (90–72) = 18$°F which is greater than 15°F, we must correct ΔT_e by adding the difference (18–15 =3) to it. Hence,

$$\dot{q}_{roof} = U\Delta T_e = 0.13(42+3) = 5.85 \text{ Btu/hr-ft}^2$$

Heat Gain from Infiltration Air

Infiltration air enters the cooled space through cracks or leakage and brings with it high temperature air and moisture, both of which contribute to the cooling load. The sensible load is calculated as before:

$$\dot{Q}_{inf} = 0.244 \dot{m}_a (T_o - T_i) \quad \textbf{(13.3.2)}$$

The latent load is calculated using

$$\dot{Q}_L = 4880(cfm)(W_o - W_i) \quad \textbf{(13.3.3)}$$

where W_o and W_i are the humidity ratios of outside and inside air, respectively.

Infiltration air can also enter through doors when opened. This can be significant if a door is opened many times, such as in restaurants and office buildings. Table 13.11 lists the infiltration air per person for some commercial establishments.

Heat Gain from Occupants

Heat gain arising from occupants consists of sensible and latent heat, and depends on the type of activity of an individual. Values of heat gain from occupants are shown in Table 13.12.

Heat Gain from Equipment

Equipment such as motors, heaters, appliances, lights and burners can contribute to the cooling load requirement, if such equipment is being operated in the conditioned space. The heat generated by these devices is absorbed by the air in the conditioned space, and hence can contribute to the cooling load requirements. Table 13.13 lists the heat gain from various components.

Example 13.11

A restaurant building in Philadelphia, Pennsylvania is 25 ft wide, 50 ft long, and 12 ft high. The longest side of the building faces east. The south wall has two glazed 7 ft by 7 ft swinging doors, and a plate glass window 12 ft wide and 8 ft high with awning. The east wall has three 4 ft by 4 ft vertically pivoted metal windows with awnings. The north wall, which is a partition, separates the dining area from an unconditioned kitchen where the peak temperature could reach 100°F. It has two swinging hollow wooden doors, each 4 ft by 7 ft. The ceiling is made of concrete 4 in thick and has a 2 in insulation board. All walls are 12 in thick concrete with 0.5 in. plaster on the outside. The restaurant can accommodate 50 people and 5 employees during peak occupancy at noon. It has 2500 W of lighting, and two 3-gallon electrically heated coffee urns. Estimate the cooling load required to maintain the air inside the restaurant at 75°F during peak occupancy, for an outside dry-bulb temperature of 95°F, and a wet-bulb temperature of 78°F. The relative humidity in the restaurant is to be maintained at 60%. Neglect the heat gain through the floor and use for the glass a U-value of 0.50 Btu/hr-ft^2-F.

Solution:

1. First we calculate the walls, roof, doors and windows areas:

Location	Gross Area	Window Area	Door Area	Net Area
East Wall	600	48	--	552
West Wall	600	--	--	600
North Wall	300	--	56	244
South Wall	300	96	98	106
Roof	1250	--	--	1250

2. Next calculate the solar heat gain, through roof, walls, windows and doors.

From Table 13.5 we determine U= 0.13 Btu/hr-ft^2-F, and from Table 13.4 we read ΔT_e= 38°F for the roof.

Similarly, we obtain ΔT_e for the walls from Table 13.3. ΔT_e (south)= 6°F, ΔT_e (east)= 2°F, and ΔT_e(west)= 6°F. The heat gain is calculated using $\dot{Q} = UA\Delta T_e$. The results are shown below.

Location	U-Value	ΔT_e	$\dot{Q} = UA\,\Delta T_e$
East Wall	0.53	6	1755
West Wall	0.53	6	1908
South Wall	0.53	2	112
Roof	0.13	38	6175
Total			9950

To determine the solar gain through glass windows and doors we use $\dot{Q} = (\dot{q}_s \times F_s \times F_t + C_1X + C_2Y)A$ where F_t, C_1 and C_2 are obtained from Table 13.8, X is obtained from Table 13.7 and Y from Table 13.9. The shading coefficient F_s is obtained from Table 13.10. The heat gain $\dot{q}_s$ is found in Table 13.6.

For the windows in the east wall we determine:

$$\dot{q}_s = 19 \text{ Btu/hr-ft}^2, F_s = 0.25, F_t = 1.0, C_1 = 1.0, C_2 = 0.0, \text{ and } X = 12$$

This gives

$$\dot{Q} = (19 \times 1.0 \times 0.25 + 1.0 \times 12) \times 48 = 804 \text{ Btu/hr}$$

For the south wall windows and doors we determine:

$$\dot{q}_s = 98, F_s = 0.25, F_t = 0.87, C_1 = 1.0, C_2 = 0.25, X = 14, \text{ and } Y = 19$$

Hence,

$$\dot{Q} = (98 \times 0.87 \times 0.25 + 1.0 \times 14 + 0.25 \times 19) \times 194 = 7773 \text{ Btu/hr}$$

Therefore, the total solar gain is = 9950+804+7773 = 18,527 Btu/hr.

3. To determine the heat gain from occupants we use Table 13.12. For people eating we select sensible heat/person = 250 Btu/hr. For the employees, we choose $\dot{Q}_{sensible}$ = 305 Btu/hr and $\dot{Q}_{latent}$ = 545 Btu/hr per person. Hence, total heat gain from occupants is:

$$\dot{Q}_{sensible} = 50 \times 250 + 5 \times 305 = 14{,}025 \text{ Btu/hr}$$

$$\dot{Q}_{latent} = 50 \times 250 + 5 \times 545 = 15{,}225 \text{ Btu/hr}$$

4. Heat gain from equipment is obtained from Table 13.13. For the lighting,

$$\dot{Q}_{sensible} = 2500W \times 3.413 \text{Btu/hr/W} = 8533 \text{ Btu/hr}$$

The coffee urns produce

$$\dot{Q}_{sensible} = 2 \times 2200 = 4400 \text{ Btu/hr, and } \dot{Q}_{latent} = 2 \times 1500 = 3000 \text{ Btu/hr}$$

5. The infiltration air can enter through the windows in the east wall, swinging doors in the south wall, and swinging doors in the partition. The glass window in the south wall is of the tight plate-glass type which has no leakage. We assume a 10 mph outside wind velocity and we use Table 13.2 to determine the infiltration for vertically pivoted metal windows as 88 ft^3/hr/ft of crack length. For the east windows we calculate the crack length as 3[(4+4)2+4] = 60 ft; the air infiltration is 88 ft^3/hr/ft×60 ft = 5280 ft^3/hr or 88 cfm. For the south wall the door crack length is 2[(7+7)+7] = 42 ft. From Table 13.2 the infiltration for poorly fitted doors is 138 ft^3/hr/ft of crack length. Hence the infiltration air is 138×42 = 5796 ft^3/hr or 96.6 cfm. The north partition has a swinging door to the unconditioned kitchen. Hence, infiltration can occur due to the opening of the doors. Likewise, some air can enter due to opening of the swinging doors in the south wall. Table 13.11 gives the infiltration air for a swinging door as 2.5 cfm/person. For the south wall we assume 50 persons entering the restaurant. So, the infiltration air is 50×2.5 = 125 cfm. For the north partition, we assume 5 employees. Hence, the infiltration air is 5×2.5 = 12.5 cfm. The sensible heat gain due to infiltration is calculated as follows:

$$\dot{Q}_{inf} = 0.244 \times \dot{m}_a (T_o - T_i), \text{ where } \dot{m}_a = \rho_a \times cfm \times 60$$

For the east and south walls T_o=95°F and T_i=75°F. Hence,

$$\dot{Q}_{inf} = 0.244 \times \left(\frac{88+125}{14.35} \right) \times 60(95-75) = 4346 \text{ Btu/hr}$$

where $\rho_a = \frac{1}{v_a} = \frac{1}{14.35}$ lb/ft^3. For the north partition we use T_o = 100°F. So,

$$\dot{Q}_{inf} = 0.244 \times \left(\frac{12.5}{14.35} \right) \times 60(100-75) = 319 \text{ Btu/hr}$$

The latent heat gain is calculated using Eq.13.3.3. First, we determine W_o using a dry-bulb temperature of 95°F and a wet-bulb temperature of 75°F; W_i is determined using a dry-bulb temperature of 75°F and a relative humidity of 60% for comfort reasons. Using the psychrometric chart in Fig.13.1 we determine:

$$W_o = 0.0169 \text{ lbw/lba}$$

$$W_i = 0.0112 \text{ lbw/lba}$$

Hence,

$$\dot{Q}_L = 4880(225)(0.0169\text{-}0.0112) = 6259 \text{ Btu/hr}$$

The total cooling load is summarized in the table below.

Item	Sensible Load	Latent Load
Transmission	18,527	—
People	14,025	15,225
Equipment	12,933	3000
Infiltration	4665	6259
Total	50,150 Btu/hr	24,484 Btu/hr

13.4 Ventilation and Air Distribution

Air flows into and out of a building due to either natural ventilation, infiltration or exfiltration. The rate at which the infiltration air flows through depends on the pressure and temperature differences between the inside and outside of the building, and the tightness of the construction. In some buildings, the infiltration air and/or the natural ventilation of the building are sufficient to provide adequate ventilation of the building. However, in some cases where windows and doors or cracks in the building are not enough to allow infiltration air to enter, ventilation air should be provided by the heating or cooling system.

In some heating systems and almost all cooling systems, heating/cooling is supplied by forcing and circulating hot/cold air to the conditioned space. In both cases the proper flow rate of air at a comfortable temperature and speed should be selected. The air is distributed to the conditioned space or zones through a ducting system. The ducts sizes and shapes should be selected to reduce friction loss as the air flows through the ducts.

Ventilation Air

The air in a space that is occupied by people should be free from toxic, unhealthy, or disagreeable fumes. Enough fresh air or ventilation air must be supplied to a space in order to prevent the air from becoming contaminated with toxins. Fresh air should also help keep odors and dust concentrations to a low level. On average a person requires about 8 cfm of fresh air in a smoke free environment. Table 13.14 lists suggested values of ventilation air for various space applications. Some of the ventilation air is supplied by infiltration air. But, to ensure proper ventilation of a conditioned space, fresh air is often brought in from the outside.

In most cooling systems the outside air is introduced directly into the conditioned space at the outside temperature. Since the air's temperature is considered to be unaltered as it flows through the system, it is referred to as by-pass air. Hence, manufacturers of cooling equipment describe their cooling equipment in terms of the cooling coil efficiency which is related to a by-pass factor. For example, a coil which is 80% efficient has 20% of the air as by-pass air. That is 80% of the air flowing through the coil is cooled down to the coil temperature, while the remaining 20% remains unchanged. Sometimes, the infiltration air is sufficient to provide adequate amount of ventilation to a conditioned space. Any additional ventilation air introduced into a building should be treated as infiltration air when calculating the system load.

Example 13.12

Determine the required ventilation air for the restaurant in Example 13.11.

Solution: From Table 13.14 we get the suggested ventilation air as 15 cfm per person. The restaurant during peak occupancy has 55 people inside. Hence, the total ventilation air required is 55x15=825 cfm. The mass flow rate is

$$\dot{m}_a = \frac{825 \text{ cfm} \times 60 \text{ min/hr}}{14.35 \text{ ft}^3/\text{lb}} = 3450 \text{ lb/hr}$$

Cooling Load Air Supply

After calculating the cooling load the system that will provide the cooling of the conditioned space must be selected. Cooling systems supply cooled air to the space in order to offset the cooling load of the space. The amount of air to be supplied must be calculated from the cooling load. The temperature of the supplied air must be chosen so as to provide a comfortable thermal environment to the occupants. The air quantity is determined from

$$\dot{Q}_s = 0.244\dot{m}_a(T_i - T_s) \quad \textbf{(13.4.1)}$$

where $\dot{Q}$ is the space internal sensible cooling load in Btu/hr, excluding the ventilation air load, T_i is the inside air temperature to be maintained, and T_s is the supply air temperature in degrees Fahrenheit. To determine $\dot{m}_a$ we write Eq.13.4.1 in the form

$$\dot{m}_a = \frac{\dot{Q}_s}{0.244(T_i - T_s)} \quad \textbf{(13.4.2)}$$

The volume flow rate in cfm is determined from $\dot{m}_a$ as

$$cfm = \frac{\dot{m}_a \times v_a}{60} \quad \textbf{(13.4.3)}$$

Sometimes, it is convenient to add the by-pass ventilation air load to the space load. The sensible and latent loads of the ventilation air are calculated using

$$\dot{Q}_{vs} = 0.244\dot{m}_a(T_o - T_i)(b) \quad \textbf{(13.4.4)}$$

$$\dot{Q}_{vL} = 1100\dot{m}_a(W_o - W_i)(b) \quad \textbf{(13.4.5)}$$

where $\dot{Q}_{vs}$ and $\dot{Q}_{vL}$ are the sensible and latent heat gains from the ventilation air, and b is the by-pass factor.

Example 13.13

Calculate the amount of supply air to be delivered at 60°F to the restaurant in Example 13.11.

Solution: From Example 13.11, $\dot{Q}_{vs}$ = 50,150 Btuh, T_i = 75°F, and T_s = 60°F. Hence, from Eq.13.4.2

$$\dot{m}_a = \frac{50,150}{0.244(75-60)} = 13,702 \text{ lb/hr}$$

The humidity ratio for the supply air is determined using Eq.13.4.5 with $\dot{Q}_{vL}$ = 24,484 Btuh, and W_i = 0.0112 lbw/lba from Example 13.11. Hence, we calculate W_s = 0.00958 lbw/lba. Then, for T_s = 60°F and W_s = 0.00958 lbw/lba we determine v_a = 13.3 ft^3/lba. Hence, the supply air in cfm is

$$\frac{13702 \times 13.3}{60} = 2969 \text{ cfm}$$

Example 13.14

Determine the sensible and latent loads of the ventilation air for Example 13.12 if the by-pass factor is 40%.

Solution: Using Eqs.13.4.4 and 13.4.5:

$$\dot{Q}_{vs} = 0.244(3450)(95-75)(0.4) = 6734 \text{ Btuh}$$

and,

$$\dot{Q}_{vL} = 1100(3450)(0.0169-0.0112)(0.4) = 8653 \text{ Btuh}$$

Air Flow

In air conditioning systems, air is supplied and distributed to the conditioned space through ducts. As the air flows through the ducts the fluid sticks to the wall resulting in a wall shearing stress. The velocity of the fluid increases as its distance from the wall increases. The maximum velocity occurs at the center. In normal flow of air in a duct a pressure drop occurs due to the wall shear. The pressure drop per unit length of duct is approximated as

$$\Delta p = f\left(\frac{L}{D_{hc}}\right)\frac{\rho V^2}{2g_c} \tag{13.4.6}$$

where Δp is the pressure drop, L is the length of duct, ρ is density, V is the average velocity in the duct, f is a friction factor, D_{hc} is the hydraulic diameter, and g_c = 1 for SI units and 32.2 for English units. D_{hc} is defined as

$$D_{hc} = \frac{\text{cross}-\text{sectional area of duct}}{\text{wetted perimeter of duct}} \tag{13.4.7}$$

The friction factor f is a function of the Reynolds number R_e and the relative roughness $\varepsilon = e/D_{hc}$ of the surface of the duct; e is the roughness of the duct wall, $R_e = \frac{\rho V D_{hc}}{\mu}$ and μ = viscosity of air.

The above discussion applies to fluid flowing in straight runs of ducting. Losses through fittings, elbows, and bends are obtained from experimental data. But, usually these losses are stated as equivalent to a certain additional length of straight run of duct. For air flowing in a duct we can write the energy equation as

$$p_1 + \frac{\rho V_1^2}{2g_c} = p_2 + \frac{\rho V_2^2}{2g_c} + h_L \rho g \tag{13.4.8}$$

where potential energy change and heat transfer are neglected. The head loss h_L represents the frictional losses in the duct. Results of tests on friction losses for air flowing in galvanized ducts have been tabulated and are presented in Fig.13.2. This chart applies to standard air, but could also be used for air temperatures in the range of 50 to 90°F. Values taken from this chart should be corrected for density variations as,

$$(h_L)_c = h_L \frac{\rho_a}{0.075} \tag{13.4.9}$$

and for temperature variations by using,

$$\left(h_L\right)_c = h_L \frac{530}{460+T} \tag{13.4.10}$$

where T is in degrees Fahrenheit.

To account for surface roughness we use a correction factor in Fig.13.3. Since these charts were constructed for circular ducts an equation was developed by Huebscher to determine the diameter of a circular duct having the same frictional losses as a rectangular duct delivering the same flow rate and having the same length. The equation is

$$D = 1.3\frac{(H\times W)^{5/8}}{(H+W)^{1/4}} \tag{13.4.11}$$

where, D is the diameter of the circular duct and H and W are the height and width of the rectangular duct. Table 13.15 lists values and circular equivalents for rectangular and circular ducts.

Frictional losses in elbows can be obtained from Fig.13.4 The loss is expressed in equivalent length of duct (L) and is measured in duct diameter (D) or duct width (W).

Example 13.15

(a) Determine the friction loss in a 50 ft long pipe carrying 400 cfm of air at 70°F. The pipe diameter is 20 inches.

(b) Repeat (a) if the pipe has a medium rough surface.

(c) Find an equivalent size of rectangular duct with a width not to exceed 16 inches.

(d) What is the friction loss in the pipe if the air is at 150°F?

Solution:

(a) From Fig.13.2 for $\dot{Q}$ = 4000 cfm and D = 20 in we determine the friction loss h_L = 0.23 in of water/100 ft. Hence,

$$h_L = 0.23\times 50/100 = 0.46 \text{ in. water}$$

(b) Using Fig.13.3 for a medium rough surface and $V = \frac{cfm}{A} = \frac{4000}{\pi(10/12)^2}$ $= 1834$ fpm, we read a correction factor of 1.4. Therefore, the corrected head loss is

$$h_L = 1.4\times 0.46 = 0.644 \text{ in water}$$

(c) From Table 13.15 we find, for W = 16 in and H = 20, D = 19.5, and for W = 16 and H = 22, D = 20.4 in, by interpolation we find

$$H = 21 \text{ in.}$$

(d) In this case we employ Eq.13.4.10:

$$h_L = 0.644\left(\frac{460+70}{460+150}\right) = 0.56 \text{ in water}$$

Example 13.16

The rectangular duct shown has an elbow at BC with $H/W = 2$, $R/W = 1$, and $H = 50$ in.

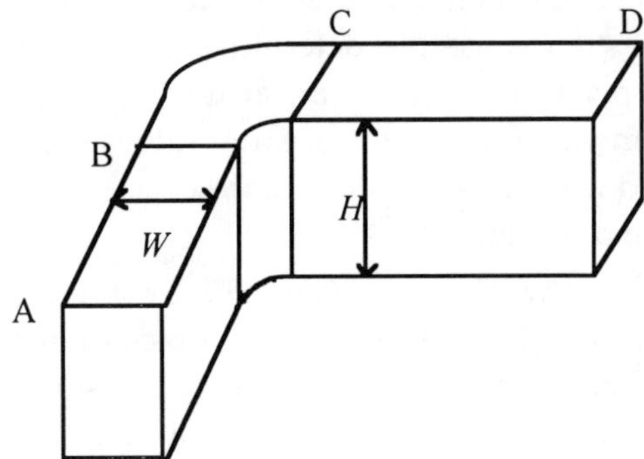

(a) Find the pressure loss in an equivalent length of duct for the elbow.

(b) If the velocity in the duct is 1100 fpm, calculate the loss in in. of water for the elbow.

Solution:

(a) From Fig.13.4 for $R/W = 1.0$ and $H/W = 2.0$ we find $L/W = 13.2$. Hence, $L = 13.2 \times W$, where $W = 50/2 = 25$ in. Hence,

$$L = 13.2 \times 25 / 12 = 27.5 \text{ ft}$$

(b) The loss in elbow BC can be calculated by finding

$$Q = VA = V \times (H \times W) = 1100 \times \left(\frac{50}{12} \times \frac{25}{12}\right) = 9549 \text{ cfm},$$

and determining the equivalent diameter circular duct from Table 13.15. It is $D = 38.1$ in. From Fig.13.2 for $Q = 9549$ cfm and $D = 38.1$ in. (do not use the velocity lines) we determine

$$h_L = 0.042 \times \frac{27.5}{100} = 0.0116 \text{ in. of water}$$

Note: the head loss over relatively short distances is quite small.

Duct Design

When designing an air flow distribution system for an air-conditioning system, the air delivered by the fan must be supplied at a static pressure slightly greater than the total resistance generated by the ducting system. Several parameters influence the duct design procedure. The designer must take into consideration noise and vibration in the system due to high velocities, and frictional losses due to bends and elbows in the system. The designer should always try to avoid sharp bends and the duct cross section for rectangular ducts should have an aspect ratio less than 10. The design procedure follows.

Step 1 - Determine the air requirements at each duct outlet for each zone or division of the conditioned space.

Step 2 -Lay out the simplest or most suitable duct system for proper air distribution, considering space limitations, and ease of construction and installation of the system.

Step 3 - Determine the sizes of the various duct branches, and air outlets based on velocity or pressure drop calculations.

Step 4 - Determine, after using any of the following methods, the circuit with the maximum frictional resistance, note, this is not necessarily the longest circuit. The maximum resistance offered by the duct system is used to determine the static pressure the fan must supply to deliver the air at the required flow rates. Also, the fan must be selected to account for leakage, heat transfer, and added resistance due to the presence of grilles louvers, humidifiers, and heaters.

Four methods of duct design will be presented. These methods are:

1. *Assumed Velocity Method*: The velocity in each branch or section of the duct system is assumed. The total pressure loss is then calculated based on this assumption. In a modified version of this method, called the velocity-reduction method, the assumed velocities are gradually reduced in the various duct sections. The highest velocity is assumed at the fan outlet, and velocities are decreased in the main after each branch is taken off. This method is, in general, suitable for designing simple duct systems. Dampers are used to control the flow in the various branches.
2. *Constant-Pressure-Drop Method*: The friction loss per foot of duct is considered constant. However, the constant pressure drop is usually established on the basis of the desirable velocity in the duct main behind the fan. Dampers must be used to control the flow.
3. *Balanced-Pressure-Loss Method*: The static pressure required for the flow in any branch from its point of attachment is made equal to the static pressure of the main system at the juncture point. Theoretically, dampers are not required for this system.
4. *Static-Pressure Regain Method*: The velocity after each branch is reduced by an amount sufficient to compensate for the frictional pressure loss in the succeeding section. Ideally, pressure regain is highly dependent on the physical configuration.

Static-Pressure-Regain

Pressure loss in ducts can occur due to friction, abrupt geometry changes, and increased turbulence. Bernoulli's equation suggests that pressure decreases when the velocity increases. Conversely, pressure increases when the velocity decreases. This pressure rise can occur in a diverging duct, a duct whose cross-sectional area is increasing with flow direction. This pressure rise is called regain. Regain can also occur in ducts of constant cross-sectional area, if the velocity of the flow decreases due to branching. Static-pressure-regain *SPR* can be calculated, in inches of water, as follows:

$$SPR = 0.5\left[\left(\frac{V_i}{4005}\right)^2 - \left(\frac{V_f}{4005}\right)^2\right] \qquad \textbf{(13.4.12)}$$

where, V_i and V_f are the inlet and final velocity, in fpm, of the transformation.

Example 13.17

Design the duct system shown below for a school building, using the constant-pressure-drop method. The duct depth should not exceed 16 in. The supply air flow rates are shown. Assume $R/W = 1.0$ for all elbows.

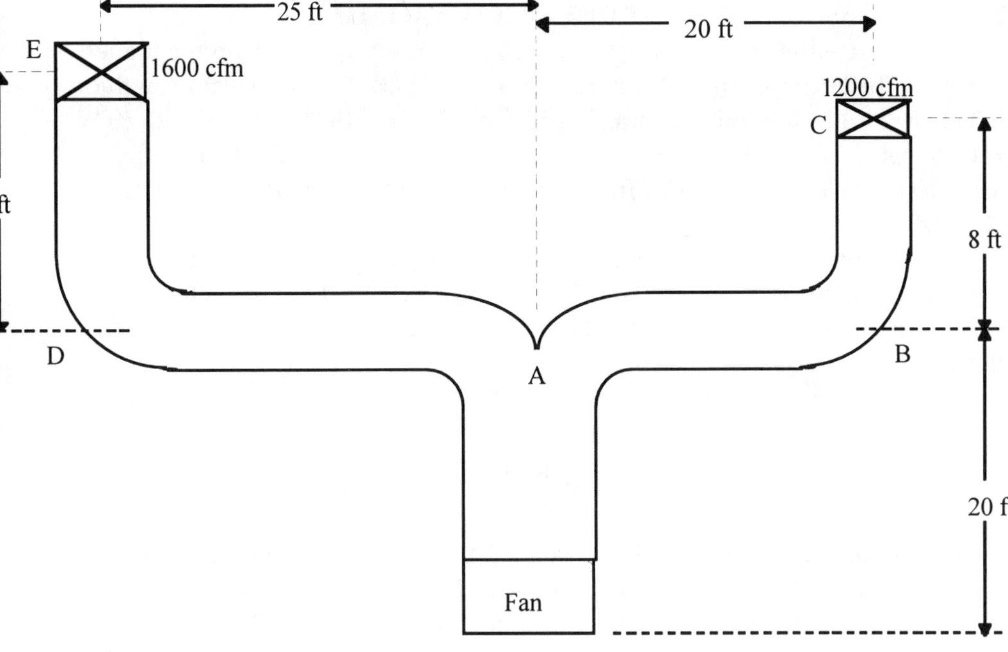

Solution: The inlet duct must carry (1600 + 1200) cfm = 2800 cfm. Using a velocity of 1200 fpm (see Table 13.15) we can write,

$$Q = VA = V \times H \times W$$

$$2800 \text{ cfm} = 1200 \times (10/12) \times W, \quad \text{so} \quad W = 21 \text{ in}$$

The equivalent diameter of a circular duct is found from Table 13.15 to be 19.95 in, or 20 in. From Fig.13.2 for $Q = 2800$ cfm and $D = 20$ in we find the friction loss to be 0.13 in. water/100 ft.

Branch *ABC* carries 1200 cfm and for a loss of 0.13 in. water we determine from Fig.13.2 an equivalent duct diameter of 14 inches which from Table 13.15 gives a 10 in. by 16 in. rectangular duct.

Branch *ADE* has $Q = 1600$ cfm and a loss of 0.13 in. water. From Fig. 13.2 we get $D = 15.8$ in. Table 13.15 gives the dimensions of a 13 in. by 16 in. rectangular duct.

The right run from the fan outlet has a total length equal to 20 + 20 + 8 = 48 ft. To calculate additional lengths due to elbows we take $R/W = 1.0$. At *A* we determine from Fig.13.4 for $H/W = 16/12 = 1.33$, $L/W = 12.25$, which gives $L_A = 12.25 \times 10/12 = 10.2$ ft. At *B* we also have $L/W = 12.25$. Hence, $L_B = 10.2$ ft. The total length of right run is 48+10.2+10.2 = 68.4 ft. The total right duct and elbow frictional loss is $0.12 \times 72/100 = 0.0864$ in. water. We apply a factor of safety of 10% to this loss to get $0.0889 \times 1.10 = 0.0978$ in. water.

Some overall static pressure regain occurs in each branch of the duct system. The starting velocity is 1200 fpm. Branch ABC has a final velocity

$$V_f = \frac{Q}{A} = \frac{1200}{16 \times 10/144} = 1080 \text{ fpm}$$

Hence, from Eq.13.4.12

$$SPR = 0.5\left[\left(\frac{1200}{4005}\right)^2 + \left(\frac{900}{4005}\right)^2\right] = 0.0196$$

$$SPR = 0.5\left[\left(\frac{1200}{4005}\right)^2 + \left(\frac{1080}{4005}\right)^2\right] = 0.0085 \text{ in. water}$$

The net pressure loss is 0.0978 – 0.0085 = 0.0893 in. water.

Next we use the same procedure to calculate the pressure drop for the left branch: The length of run *ADE* is 20 + 25 + 10 = 55 ft. The equivalent length of left elbow at *A* is determined using Fig.13.4 and $H/W = 16/13 = 1.23$, with $R/W = 1.0$ to get $L/W = 11.3$. Then $L_A = 11.3 \times 13/12 = 12.2$ ft. Similarly, the equivalent length of elbow at *B* is 12.2 ft. The total length of branch *ADE* is 55 + 12.2 + 12.2 = 79.4 ft.

The pressure drop is $0.13 \times 79.4/100 = 0.103$ in. water. Again we add a 10% factor of safety to get a total pressure loss of $0.103 \times 1.10 = 0.113$ in. water.

The final velocity is $V_f = Q/A$, where Q = 1600 cfm and $A = (16 \times 13)/(144)$, gives

$$V_f = 1108 \text{ fpm}$$

The static pressure regain is

$$SPR = 0.5\left[\left(\frac{1200}{4005}\right)^2 - \left(\frac{1108}{4005}\right)^2\right] = 0.0066 \text{ in. water}$$

The net pressure loss is 0.113 – 0.0066 = 0.106 in. water. Hence, the pressure drop in branch *ADE* is greater than in branch *ABC*. In addition to this static pressure loss in the duct system, the fan is required to overcome the losses in the accessory equipment such as grilles, louvers, filters, heater, humidifier, etc. Assuming the pressure loss through the accessory equipment to be 0.5 in. water, the fan selected must have a static pressure greater than (0.5 + 0.106) = 0.61 in. water.

Example 13.18

Size the duct system shown below. Use the static-regain method, and use a maximum depth of duct of 15 inches. The outlet pressure for the branches should average about 0.2 in. water. Use *R/W* = 1.5 for all elbows.

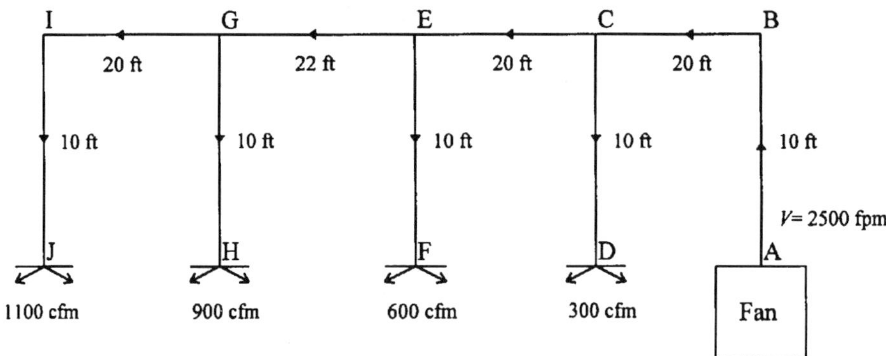

Solution: For duct *ABC* we have Q = 2900 cfm and V = 2500 fpm. Hence, $A = Q/V = 2900/2500 = 1.16$ ft^2. For a circular cross section $A = \pi D^2/4$, which gives D = 14.6 inches. From Fig.13.2 we get the friction loss as 0.6 in. water per 100 ft. From Table 13.15 we get the size of the equivalent rectangular duct as 12×15 in.

From Fig.13.4 we get the additional equivalent length of elbow at B, with R/W= 1.5 and H/W= 1.25, we find L/W= 4.7. So $L_B = 4.7 \times 12 = 56$ in. Loss in ABC= $(10 + 20 + 56/12) \times 0.6/100 = 0.208$ in. water.

For section CE: Q = 2600 cfm and we keep the size of duct the same as for section ABC. From Fig.13.2 we determine the friction loss as 0.46 in. water/100 ft. The velocity is

$$V = Q/A = (2600 \text{ cfm})/(1.16 \text{ ft}^2) = 2241 \text{ fpm}$$

The static pressure regain is

$$SPR = 0.5\left[\left(\frac{2500}{4005}\right)^2 - \left(\frac{2241}{4005}\right)^2\right] = 0.038 \text{ in. water}$$

The friction loss in $CE = 0.46 \times 20 / 100 = 0.092$ in. water. The net loss is = 0.092 – 0.038 = 0.054 in. water. In order to achieve an average of 0.2 in. of water at the branches, we assume a static pressure of 0.24 in. water at C, the static pressure at E is = 0.24 – 0.054 = 0.186 in. water; this is close enough to the 0.2 static pressure requirement of 0.2 in.

Section EG: Q = 2000 cfm. If the size of the duct of section EG is kept the same as before, then $V = Q/A$= 2000/1.16 = 1724 fpm.

The static pressure regain in this case is

$$SPR = 0.5\left[\left(\frac{2241}{4005}\right)^2 - \left(\frac{1724}{4005}\right)^2\right] = 0.064 \text{ in water}$$

From Fig.13.2 the loss is 0.30 in. water/100 ft. The loss in $EG = 0.30 \times 22 / 100$= 0.066 in. water. Hence, the net pressure drop = 0.066 – 0.064 = 0.002 in. water. For this case, the static pressure regain almost offsets the pressure drop in section EG. So, the static pressure at G is = 0.186 – 0.002 = 0.184 in. water. This static pressure at G is close enough.

Section GI: In this case the decrease in Q could result in significant static pressure regain if we keep the duct size the same. Hence, a smaller duct size will be selected. Choose D = 12 in. so $A = \pi D^2/4 = 0.7854$ ft^2. Hence, $V = Q/A$= 1100/.7854 = 1401 fpm.

From Fig.13.2 we find the loss as 0.22 in. water/100 ft. The friction loss is $0.22 \times 20 / 100 = 0.044$ in. water.

The static pressure regain is

$$SPR = 0.5\left[\left(\frac{1724}{4005}\right)^2 - \left(\frac{1401}{4005}\right)^2\right] = 0.0315 \text{ in water}$$

Net loss = 0.044 – 0.0315 = 0.0125 in. water. Static pressure at I = 0.184 – 0.0125 = 0.172 in. water. So the average pressure at C, E, G, and I is 0.195 in. water which is close to the required value of 0.2. If this is too low (or too high) it is necessary to repeat the calculations with a different assumption for the pressure at C.

13.5 Refrigeration

When selecting refrigeration equipment the heat load of the refrigerated space must be calculated. The heat load consists of heat gain from walls, air changes, product to be refrigerated, people, and equipment.

The wall heat gain is calculated as before using

$$\dot{Q}_w = U_w A_w (T_o - T_i) \tag{13.5.1}$$

The value of the overall heat transfer coefficient U_w can be determined as before. However, the refrigeration industry has tended to standardize the minimum insulation of a refrigerator. Hence, heat transfer factors can be obtained in Btu/ft^2 for a 24-hour period for various temperature differences. The heat gain from walls over 24 hours is calculated as

$$\dot{Q} = F \times A_w \tag{13.5.2}$$

where F is the heat transfer factor obtained from Table 13.17, and A_w is the surface area of the wall.

The product load is calculated by knowing the properties of the product to be refrigerated, and the initial and final temperatures. The procedure for calculating the product load is as follows:

1. Calculate the heat removed from the product to reduce its temperature to some lower temperature above freezing as

$$Q = w \times c(T_1 - T_2) \tag{13.5.3}$$

where w = weight of product in pounds and c = specific heat of product above freezing, in Btu/°F-lb.

2. Calculate heat removed to freeze product using

$$Q = w \times h_{if} \tag{13.5.4}$$

where h_{if} = latent heat of fusion, in Btu/lb.

3. Calculate heat removed from freezing point to final temperature below freezing by

$$Q = w \times c_i (T_f - T_3) \tag{13.5.5}$$

where c_i = specific heat below freezing, T_f = freezing temperature of product, and T_3 = is the final temperature.

Values of specific heats above and below freezing, and latent heats of fusion for various products can be found in Table 13.18. Most food products have a freezing temperature in the range of 26 to 31°F. If the exact freezing temperature of a product is unknown it may be assumed to be 28°F.

Outside air enters a refrigerated space each time the door of the space is opened to the outside. The air enters at a warm temperature and is then cooled to the temperature of the refrigerated space. The heat removed from one cubic foot of air is listed in Table 13.19. The number of air changes due to door openings for a 24 hour period is listed in Table 13.20. These values are based on practical experience.

Heat gain from people entering or working in the refrigerated space depends on the temperature of the space, type of work, size, clothing, and time duration a person is inside. Average hourly loads from occupants have been compiled and are shown in Table 13.21.

Example 13.19

A storage room $9 \times 10 \times 11$ ft is to be maintained at 40°F. It has 4 in. cork insulation and 250 watts of electrical lighting. 1500 lb of beef are to be cooled from 50°F to 40°F from 6 p.m. to 6 a.m. Two men work in the room from 8 a.m. to 4 p.m. The outside air is at 90°F and 50% relative humidity. Calculate the refrigeration load for a 24 hour period.

Solution: The wall surface area is $2[9 \times 10 + 9 \times 11 + 10 \times 11] = 598 \text{ft}^2$. From Table 13.17 we determine the heat transfer factor for a temperature difference of 90 – 40 = 50°F and 4 in. cork insulation to be F = 90 Btu/ft^2/day. The heat gain is

$$\dot{Q} = (90)(598) = 53{,}820 \text{ Btu/day}$$

The beef is to be cooled from 50 to 40°F. From Table 13.18 the average freezing temperature for beef is 29°F, the specific heat above freezing is 0.77 Btu/lb-°F. The product heat gain is (1500)(0.77)(50–40) = 11,550 Btu. However, this is the load to cool the beef in 24 hours. But, the beef is to be cooled in 12 hours from 6 p.m. to 6 a.m. The actual product load is

$$\dot{Q} = 11{,}550 \times 24 / 12 = 23{,}100 \text{ Btu/day.}$$

To determine the heat gain from air infiltration we calculate the volume of the room as $9 \times 10 \times 11 = 990 \text{ ft}^3$. From Table 13.20 the number of air changes is 17.5/day. We determine the heat removed from the air from Table 13.19 to be 2.0 Btu/ft^3. The heat gain is

$$\dot{Q} = (2.0 \text{ Btu/ft}^3)(990 \text{ ft}^3)(17.5/\text{day}) = 34{,}650 \text{ Btu/day}$$

Since two men work in the room from 8 a.m. to 4 p.m., we obtain from Table 13.21 the heat equivalent/person to be 840 Btu/hr. For two men the heat gain is

$$\dot{Q} = 2(840)(24) = 40{,}320 \text{ Btu/day}$$

The electrical load is

$$\dot{Q} = (250 \text{ W})(3.413 \text{ Btu/W})(24 \text{ hr}) = 20478 \text{ Btu/day}$$

In order to determine the hourly load requirement, we compute the day and night loads as follows:

Item	**Based on 24 hr**	
	Day	**Night**
Walls	53,820	53,820
Product	--	23,100
Air Changes	34,650	34,650
People	40,320	--
Electrical	20,500	--
Total	**149,000 Btu/day**	**112,000 Btu/day**

The day hourly load is 149,000/24 = 6200 Btu/hr. The night hourly load is 112,000/24 = 4700 Btu/hr. The day hourly load is higher and should be used in selecting the equipment.

13.6 HVAC/R Control

HVAC Systems are designed to operate daily during the heating/cooling season for a predetermined set of operating conditions. In order for the system to supply heating/cooling whenever needed , automatic controls are employed. These may include temperature control and/or humidity, pressure, and flow rate control. Control devices can be either mechanical or electrical, and are used to perform the automatic control with limited human intervention.

Control systems can be of the open-loop type or the closed-loop type. In the open-loop control system, the effect of external disturbances is corrected by changing the control system variables. For example, using an outdoor thermostat to control the operation of a heating system of a building in proportion to the load changes caused by the variations in the outdoor temperature is an open-loop control system. In this case, the controller actions are independent of the actual temperature inside the conditioned space. Instead, a controller that measures the delivered air temperature and adjusts the system variables based on the measured temperature, can bring the system to deliver the air at the desired temperature or set point. However, the controller may have to perform this task several times before the correct result is achieved. This system is said to use a feedback signal that closes the control loop, and hence the name closed-loop control system is used.

The performance of a control system is usually evaluated based on its stability, proportional band, and response. A system is considered to have stability if it is free of excessive periodic oscillations in the controlled variable. The proportional band of a control system is the range over which the controlled variable changes to move the controlled device from one extreme to another. System response is determined by the system's ability to rapidly correct for any disturbances.

The elements of a control system consist of sensors, controllers, and auxiliary devices. The sensing elements measure changes in the controlled variables. These sensors include temperature, humidity, pressure, and water flow sensing elements. The controllers take the information from the sensors, compare it with the desired condition or set-point, and regulate an output signal to cause various control actions. A room thermostat houses both a sensor and a controller. However, controllers and sensors can be in separate devices. Moreover, a control system may require auxiliary devices such as transformers, electric relays, potentiometers, manual switches, and auxiliary switches.

When designing a control system the designer should limit the area of an individually controlled space to less than 5000 ft^2, to achieve good distribution, and for ease of finding a representative location for the controls. A sensor usually can accurately measure conditions at its location, and not far away where conditions may be very different. A large area exceeding 5000 ft^2 should be divided into zones, each zone having its own controls. The smaller the controlled area, the easier it is to achieve proper control and optimal system performance.

For proper operation of HVAC/R systems trained operators should always be employed. However, sometimes this task is turned over to untrained custodial staff in small to medium- sized buildings. To compensate for this, the control system designer can resort to making the system as simple and automatic as possible.

Practice Problems (Essay)

13.1 Moist air at 85°F dry-bulb temperature, 14.7 psia, has a humidity ratio of 0.0148 lbw/lba. Using the psychrometric charts determine: specific volume, enthalpy, relative humidity and dew-point temperature.

13.2 Moist air at a pressure of 14.0 psia and 60% relative humidity has a dew-point temperature of 63°F. Determine: the humidity ratio and the specific volume of the air.

13.3 Determine the relative humidity and dew-point temperature of moist air at 92°F dry-bulb temperature, 72°F wet-bulb temperature and 13.0 psia pressure.

13.4 Moist air at 85°F dry-bulb temperature, 68°F wet-bulb temperature and 14.7 psia, enters a perfect contact cooling coil at a rate of 3500 cfm. Determine the tons of refrigeration required to cool the air to 50°F.

13.5 A vertical 10-by 6-ft high plate glass window is facing south at 40°N latitude. Compute the instantaneous heat gain through the glass at noon, if the outside temperature is 95°F, and the inside is at 80°F.

13.6 A theater is to be maintained at 80°F db and 65° F wb. At 4 p.m. it has a transmission and solar load of 50,000 Btuh, and 500 people. At 8 p.m. the solar and transmission heat gain is 25,000 Btuh and 1000 people are present. Neglecting the heat gain from lights and equipment, determine the peak load and time at which it occurs.

13.7 For the theater in Problem 13.6 determine the mass flow rate of air to be supplied at 60°F db under maximum sensible load conditions.

13.8 Air at 70°F db and 50% relative humidity flows at a rate of 10,000 cfm in a 40-in diameter pipe. If the pipe is 150 ft long determine the friction loss in the pipe.

13.9 Air at standard conditions flows in the duct system shown below. The air enters the duct at a velocity of 1200 fpm and a static pressure of 1 in. water. The duct is 24 in high and 46 in wide. After the branch at B the duct continues with the same dimensions.

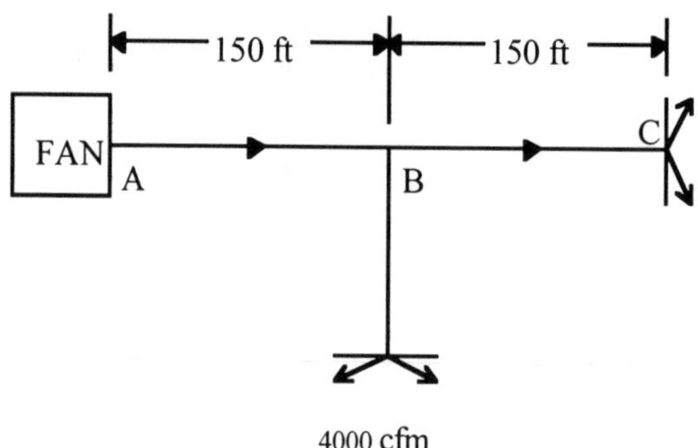

Determine:

a) The static pressure at B.
b) The velocity in branch BC.
c) The pressure at C.

13.10 A refrigerated storage room $10 \times 10 \times 12$ ft has 4 in cork insulation. 1,000 lb of beef are to be cooled from 65°F to 40°F from 6 p.m. to 6 a.m. One man works in the room from 8 a.m. to 4 p.m. If the outside air temperature is 85°F find the total load required to maintain the room at 40°F

Practice Problems (Multiple Choice)

Questions 13.11 – 13.20

A building is 50 ft long, 20 ft wide, and 12 ft high is to be maintained at 72°F and 60% relative humidity when the outside has a dry-bulb temperature of 95°F and a wet-bulb temperature of 78°F. The building is located near 40N latitude. The long side of the building face west and east. The east wall has a 15-by-7 ft plate-glass window, and the south wall has a 5-by-7 ft plate-glass window.

13.11 Calculate the transmission and solar heat gains, in Btu/hr, at 2:00 p.m. through the roof and walls of the building. The roof is 6 in. concrete with no insulation, and the walls are made of 12 in. thick concrete and have a light exterior with $U_w = 0.34$ Btu/hr-ft^2-°F.

a) 23,566 b) 23,947 c) 22,586 d) 25,806 e) 15,966

13.12 Calculate the heat gain, in Btu/hr, through the glass windows at 2 p.m.

a) 16,231 b) 15,750 c) 13,703 d) 16,013 e) 18,060

13.13 If the building has 20 people doing light office work during peak occupancy, calculate the heat gain, in Btu/hr, from the occupants.

a) 8000 b) 9000 c) 4900 d) 3100 e) 2200

13.14 Calculate the infiltration air load , in Btu/hr, for the building assuming a 15 mph wind velocity. The building has a 8-by-4 ft poorly fitted wooden door in the south wall.

a) 2065 b) 2186 c) 4820 d) 2920 e) 5106

13.15 If the total sensible load for the building is 50,000 Btuh, and the total latent load is 7,000 Btuh, determine the amount of supply air, in cfm, at 60°F that is required.

a) 4107 b) 3858 c) 4398 d) 4315 e) 3785

13.16 If 350 cfm of fresh air is required for proper ventilation of the building, calculate the sensible load, in Btu/hr, for a by-pass factor of 40%.

a) 4286 b) 1814 c) 2722 d) 1704 e) 2572

13.17 Calculate the latent load, in Btu/hr, for the ventilation air in Problem 13.17.

a) 6969 b) 4646 c) 4379 d) 6568 e) 10,946

13.18 The building has two electric coffee urns of the 3.0 gal. size, and 2000 W of lighting. Calculate the equipment load, in Btu/hr.

a) 6826 b) 11,226 c) 4400 d) 14,260 e) 7400

13.19 Moist air enters the refrigeration coil of the building at 95°F dry-bulb temperature and 78°F wet-bulb temperature, at a rate of 300 lba/min. The air leaves the coil at 60°F and 60% relative humidity. Calculate the tons of refrigeration required.

a) 12.8 b) 29.8 c) 30.4 d) 30.3 e) 29.6

13.20 If the air in problem 19 leaves the coil at 65F and 60% relative humidity, calculate the tons of refrigeration required.

a) 26.5 b) 27.2 c) 25.8 d) 27.1 e) 11.0

Solutions to Practice Problems

13.1 Given: $T_d = 85°\text{F}$, $P = 14.7$ psia, and $W = 0.0148$ lbw / lba. From the psychrometric chart:

$T_d = 68°\text{F}$, $W_s = 0.0265$ lbw / lba, $v = 14.1\ \text{ft}^3/\text{lba}$, $h = 36.9$ Btu / lba

$$\therefore \phi = \frac{W}{W_s} = \frac{0.0148}{0.0265} = 0.558 \quad \text{or} \quad 55.8\%$$

13.2 Given: $P = 14.0$ psia, $\phi = 60\%$, and $T_d = 63°\text{F}$. From the steam tables for $T = 63°\text{F}$, we find $P_w = 0.28488$ psia. Hence, $P_{w,s} = P_w / \phi = 0.28488/).6 = 0.4748$ psia. For $P_{w,s} = 0.4748$ we find $T_d = 78°\text{F}$:

$$W = 0.622\frac{P_w}{P - P_w} = \frac{0.285}{14 - 0.285} = 0.021$$

$$v = \frac{R_a T}{P - P_w} = \frac{53.35 \div (78 + 460)}{(14 - 0.285) \times 144} = 14.53\ \text{ft}^3/\text{lba}$$

13.3 Given: $T_d = 92°\text{F}$, $T_w = 72°\text{F}$, and $P = 13$ psia. From the steam tables we get for $T_1 = 92°\text{F}$, $P_{w,s} = 0.7434$ psia, $h_g = 1101.3$ Btu / lba. For $T_w = 72°\text{F}$ we get $P^*_{w,s} = 0.3356$ psia, $h^*_{fg} = 1052.6$ Btu/lba:

$$W^*_s = 0.622\frac{P^*_{w,s}}{P - P^*_{w,s}} = 0.622\frac{0.38856}{13 - 0.38856} = 0.0308\ \text{lbw / lba}$$

$$W = \frac{W^*_s h^*_{fg} - 0.24(T - T^*)}{h_g - h^*_f} = \frac{.0308 \times 1052.6 - .24(92 - 72)}{1101.3 - 40.07} = 0.026$$

Rearranging Eq.13.1.15,

$$P_w = \frac{PW}{0.622 + W} = \frac{13 \times 0.026}{0.622 + 0.026} = 0.5216\ \text{psia}$$

By interpolation from the steam tables $T_d = 80.9°\text{F}$:

$$\phi = \frac{P_w}{P_{w,s}} = \frac{0.526}{0.7434} = 0.702 \quad \text{or} \quad 70.2\%$$

13.4 Given: $T_d = 85°\text{F}$, $T_w = 68°\text{F}$, $P = 14.7$ psia, and $m_a = 3500$ cfm. Use $\dot{Q}_{sensible} = 0.244\dot{m}_a(T_i - T_s)$ where $\dot{m}_a = cfm \times 60 / v_a$. For $T_d = 85°\text{F}$ and $T_w = 68°\text{F}$, we find $v_a = 13.9\text{ft}^3 / \text{lba}$. Hence,

$$\dot{Q}_{sensible} = 0.244\frac{3500 \times 60}{13.9}(85 - 50) = 129,000 \text{ Btu / hr}$$

For latent heat gain use

$$\dot{Q}_{latent} = 4880 \times cfm(W_i - W_s)$$

To determine W_s use $T_d = 50°\text{F}$ and $\phi = 60\%$ for comfort reasons; this gives $W_s = 0.0046$ lbw / lba. For $T_d = 85°\text{F}$ and $T_w = 68°\text{F}$ we find $W_i = 0.0108$ lbw/lba. Thus,

$$\dot{Q}_{latent} = 4880 \times 3500(0.0108 - 0.0046) = 105,900 \text{ Btu / hr}$$

Finally,

$$\dot{Q}_{total} = 129,000 + 105,900 = 234,900 \text{ Btu / hr}$$

$$\text{Tons of refrigeration} = 234,900 / 12,000 = 19.6 \text{ tons}$$

13.5 To the instantaneous heat gain:

$$\dot{Q} = A[\dot{q}_s \times F_s \times F_t + c_1 X + c_2 Y]$$

For $T_o = 95°\text{F}$, $T_i = 80°\text{F}$, $40°$ north latitude, we find at noon,

$$\dot{Q} = 10 \times 6[98 \times 0.87 + 1.0 \times 14 + 0.25 \times 19] = 6240 \text{ Btu / hr}$$

13.6 At 4:00 p.m. the total heat gain is

$$\dot{Q} = 50,000 + 500 \times 350 = 225,000 \text{ Btu / hr}$$

At 8:00 p.m. the total heat gain is

$$\dot{Q} = 25,000 + 1000 \times 350 = 375,000 \text{ Btu / hr}$$

Hence, the peak load of 375,000 Btu/hr occurs at 8:00 p.m.

13.7 First, find the sensible heat load:

$$\begin{aligned} \dot{Q}_{sensible} &= 25,000 + 1000 \times 245 = 270,000 \text{ Btu / hr} \\ &= 0.244\dot{m}_a(T_i - T_s) \\ \therefore \dot{m}_a &= \frac{270,000}{0.244(80 - 60)} = 55,300 \text{ lba / hr} \end{aligned}$$

13.8 Given: $L = 150$ ft, $D = 40$ in., $T_d = 70°\text{F}$, $\phi = 50\%$, and $Q = 10,000$ cfm. From Fig.13.2, the friction loss is 0.0355 in. water/100 ft. Hence, the friction loss in the pipe is

$$0.0355\frac{150}{100} = 0.0533 \text{ in. water}$$

Note: in this case, no corrections are needed since the air is at standard conditions.

13.9 The cross-sectional area of the duct at the inlet is

$$A = W \times H = 46 \times 24 / 144 = 7.67 \text{ ft}^2$$

The volume flow rate is

$$Q = 1200 \text{ fpm} \times 7.67 \text{ ft}^2 = 9200 \text{ cfm}$$

The equivalent diameter of the rectangular duct is $D = 35.9$ ". The friction loss is 0.055"water/100 ft:

$$\Delta P_{AB} = 0.055 \frac{150}{100} = 0.083 \text{ in. water}$$

a)

$$P_B = P_A - \Delta P_{AB} = 1.0 - 0.083 = 0.917 \text{ in. water}$$

$$\text{At B, Q} = 9200 - 4000 = 5200 \text{ cfm}$$

b)

$$V_B = \frac{Q_B}{A} = \frac{5200}{7.67} = 680 \text{ fpm}$$

c)

$$SPR = 0.5\left[\left(\frac{V_i}{4005}\right)^2 - \left(\frac{V_B}{4005}\right)^2\right] = 0.5\left[\left(\frac{1200}{4005}\right)^2 - \left(\frac{680}{4005}\right)^2\right] = 0.03 \text{ in. water}$$

The friction loss in branch BC (for $Q = 5200$ cfm and $D = 35.9$") is 0.018 in. water/100 ft. The total friction loss in branch BC is

$$\Delta P = 0.018 \frac{150}{100} = 0.027 \text{ in. water}$$

$$\therefore P_C = P_B - \Delta P + SPR = 0.917 + 0.03 + 0.027 = 0.92 \text{ in. water}$$

13.10 First, calculate the surface area of the walls:

$$A = 2(10 \times 12 + 10 \div 12 + 10 \times 10) = 680 \text{ ft}^2$$

The heat gain factor from Table 13.17 is 81 Btu/ft^2/day. For the walls the heat gain is $81 \times 680 = 55{,}100$ Btu/day. Next, the internal volume of the room is $10 \times 10 \times 12 = 1200 \text{ ft}^3$. From Table 13.20 the number of air changes are 15.4/day. The heat gain from infiltration air is 1.92 Btu/ft^3. Hence,

$$\dot{Q}_{\text{inf}} = 1.92 \times 15.4 \times 1200 = 35{,}500 \text{ Btu / day}$$

The product load is

$$Q_{pro} = 1000 \text{ lb} \times 0.77 \times 25 \times 24 / 12 = 38{,}500 \text{ Btu / day}$$

The load due to the people is

$$\dot{Q}_{people} = 750 \times 24 = 18{,}000 \text{ Btu / day}$$

The loads are summarized below:

Item	Day	Night
Product	—	38,500
People	18,000	—
Infiltration	35,500	35,500
Transmitted	55,100	55,100
Total ------	108,600 --	129,100

The day hourly load is 108,600/24 = 4530 Btu/hr, and the night hourly load is 129,100/24 = 5380 Btu/hr. The night load is higher and should be used when selecting refrigeration equipment.

13.11 **a)** Use $\dot{Q} = UA\Delta T_e$. The solution is summarized as follows:

Surface	Area	U	ΔT_e	$\dot{Q}$
roof	1000	0.54	38	$\dot{Q}_1$
west wall	600	0.34	6	$\dot{Q}_2$
east wall	495	0.34	10	$\dot{Q}_3$
south wall	205	0.34	2	$\dot{Q}_4$

$$\begin{aligned}\dot{Q}_{total} &= \dot{Q}_1 + \dot{Q}_2 + \dot{Q}_3 + \dot{Q}_4 \\ &= 10000.54 \times 38 + 600 \times 0.34 \times 6 + 495 \times 0.34 \times 10 + 205 \times 0.34 \times 2 \\ &= 23{,}600 \text{ Btu/hr}\end{aligned}$$

13.12 **a)** Use $\dot{Q} = A(\dot{q}_s \times F_s \times F_t + c_1 X + c_2 Y)$. For the east facing window,

$$\begin{aligned}\dot{Q}_{east} &= 15 \times 7(127 \times 0.87 + 16 + 0.25 \times 3) \\ &= 13{,}360 \text{ Btu / hr}\end{aligned}$$

For the south facing window,

$$\begin{aligned}\dot{Q}_{south} &= 5 \times 7(69 \times 0.87 + 18 + 0.25 \times 16) \\ &= 16{,}200 \text{ Btu / hr}\end{aligned}$$

13.13 **a)** Heat gain per person doing light office work is 400 Btu/hr. Hence, for 20 people the heat gain is

$$\dot{Q} = 20 \times 400 = 8000 \text{ Btu / hr}$$

13.14 **a)** Assume leakage through the door only; plate glass windows have negligible leakage. The crack length is 2(8+4) = 24 ft. Hence, from Table 13.2,

$$\dot{q}_{\text{inf}} = 220 \text{ ft}^3 \text{ / hr / ft}$$

The volume flow rate of infiltration air is 220(24) = 5280 ft^3/hr. Hence,

$$\dot{m}_a = \frac{5280 \text{ ft}^3 \text{ / hr}}{14.35 \text{ ft}^3 \text{ / lba}} = 368 \text{ lba / hr}$$

$$\dot{Q}_{sensible} = 0.244\dot{m}_a(T_o - T_i) = 0.244 \times 368(95 - 72) = 2070 \text{ Btu / hr}$$

$$\dot{Q}_{latent} = 1100\dot{m}_a(W_o - W_i) = 1100 \times 368(.0169 - .0101) = 2750 \text{ Btu / hr}$$

$$\therefore \dot{Q}_{total} = 4820 \text{ Btu / hr}$$

13.15 **e)** The mass flow rate is

$$\dot{m}_a = \frac{Q_{sensible}}{0.244(T_i - T_o)} = \frac{50,000}{0.244(72-60)} = 17,077 \text{ lba / hr}$$

To find the volume rate of flow we proceed as follows:

$$\dot{Q}_{latent} = 1100\dot{m}_a(W_i - W_s)$$

$$7000 = 1100 \times 17,077(0.0101 - W_s). \quad \therefore W_s = 0.0097 \text{ lbw / lba}$$

Now, at $T_s = 60°\text{F}$, and $W_s = 0.0097$, we get $v_a = 13.3 \text{ ft}^3/\text{lba}$.

Hence,

$$cfm = \frac{\dot{m}_a v_a}{60} = \frac{17,077 \times 13.3}{60} = 3785$$

13.16 **d)** The sensible load is

$$\dot{Q} = 0.244\dot{m}_a(T_o - T_i)b$$

$$= 0.244\left(\frac{350 \times 60}{14.35}\right)(95 - 72) = 1700 \text{ Btu / hr}$$

13.17 **c)** The latent heat load is

$$\dot{Q} = 1100\dot{m}_a(W_o - W_i)b$$

$$= 1100\left(\frac{350 \times 60}{14.35}\right)(0.0169 - 0.0101) = 4380 \text{ Btu / hr}$$

13.18 **d)** The equipment load is found as follows:

$$\dot{Q}_{sensible} = 2 \times 2200 + 2000 \div 3.413 = 11,230 \text{ Btu / hr}$$

$$\dot{Q}_{latent} = 2 \times 1500 = 3000 \text{ Btu / hr}$$

$$\therefore \dot{Q}_{total} = 14,200 \text{ Btu / hr}$$

13.19 **e)** Given: $T_o = 95°\text{F}$, $T_w = 78°\text{F}$, $T_s = 60°\text{F}$, and $\phi = 60\%$. The load is calculated as follows:

For $T_o = 95°, T_w = 78°$, $h_o = 42$ Btu / lba and $W_o = 0.0169$ lbw / lba

For $T_s = 60°, \phi = 60\%$, $h_s = 22$ Btu / lba and $W_s = 0.0066$ lbw / lba

$$\therefore \dot{Q} = \dot{Q}_S + \dot{Q}_L = 300[(42 - 22) + (0.0169 - 0.0066)28.08]$$

$$= 5913 \text{ Btu / min}$$

$$\text{Tons of refrigeration} = \frac{5913}{200} = 29.6 \text{ tons}$$

13.20 **b)** For this case, $T_s = 65°\text{F}$ and $\phi = 60\%$. $\therefore h_s = 24.1$ Btu / lba and $W_s = 0.0079$ lb. Then

$$\dot{Q} = 300[(42 - 24,1) + (0.0169 - 0.0079) \times 33.08]$$

$$= 5460 \text{ Btu / min}$$

$$\therefore \quad \text{Tons of refrigeration} = \frac{5460}{200} = 27.3$$

Cited Figures and Tables for HVAC

Figure 13.1 Psychrometric Chart No. 1
Normal Temperature
Barometric pressure 29.921 inches of mercury
Copyright 1963
American Society of Heating, Refrigerating and Air-Conditioning Engineers, Inc.
SEA LEVEL

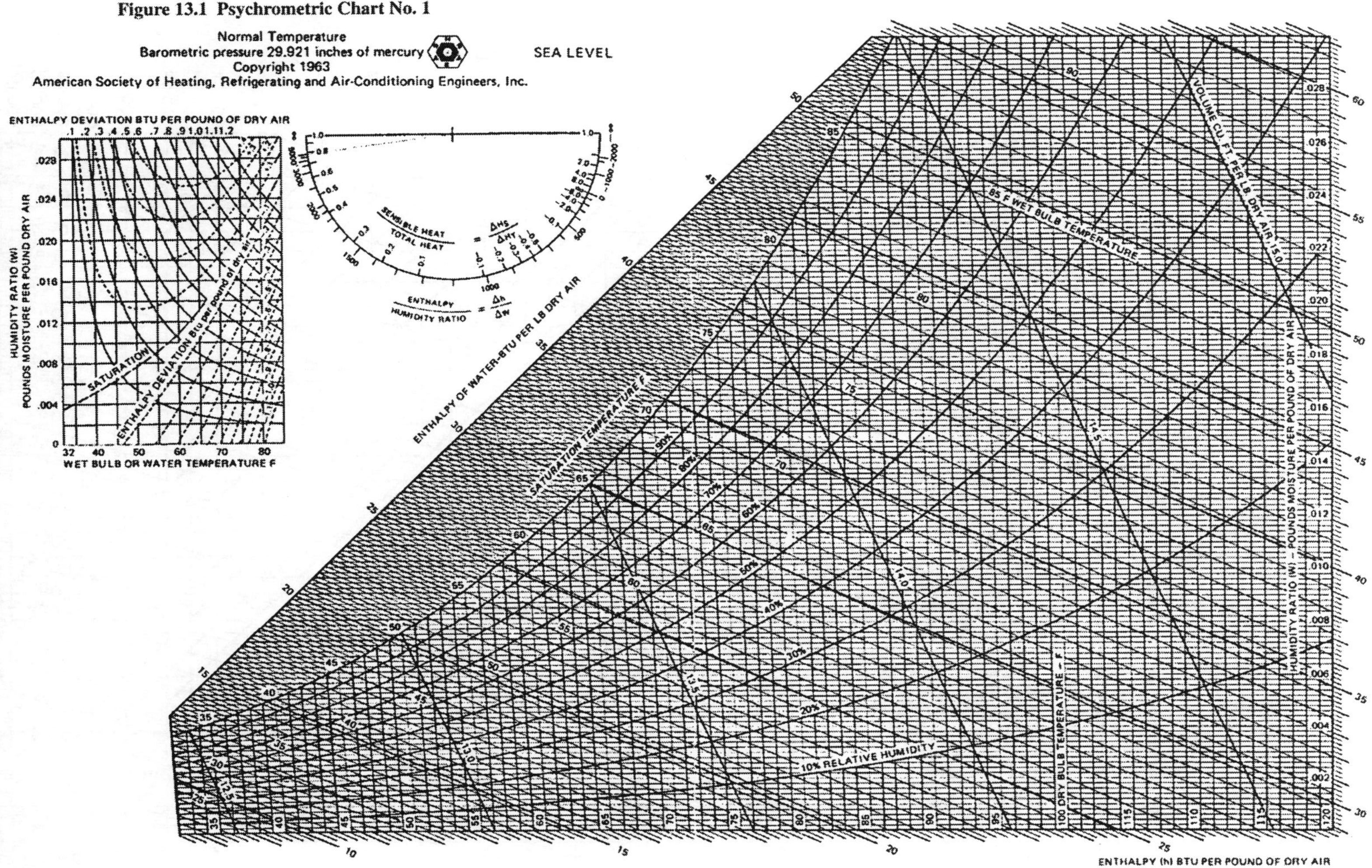

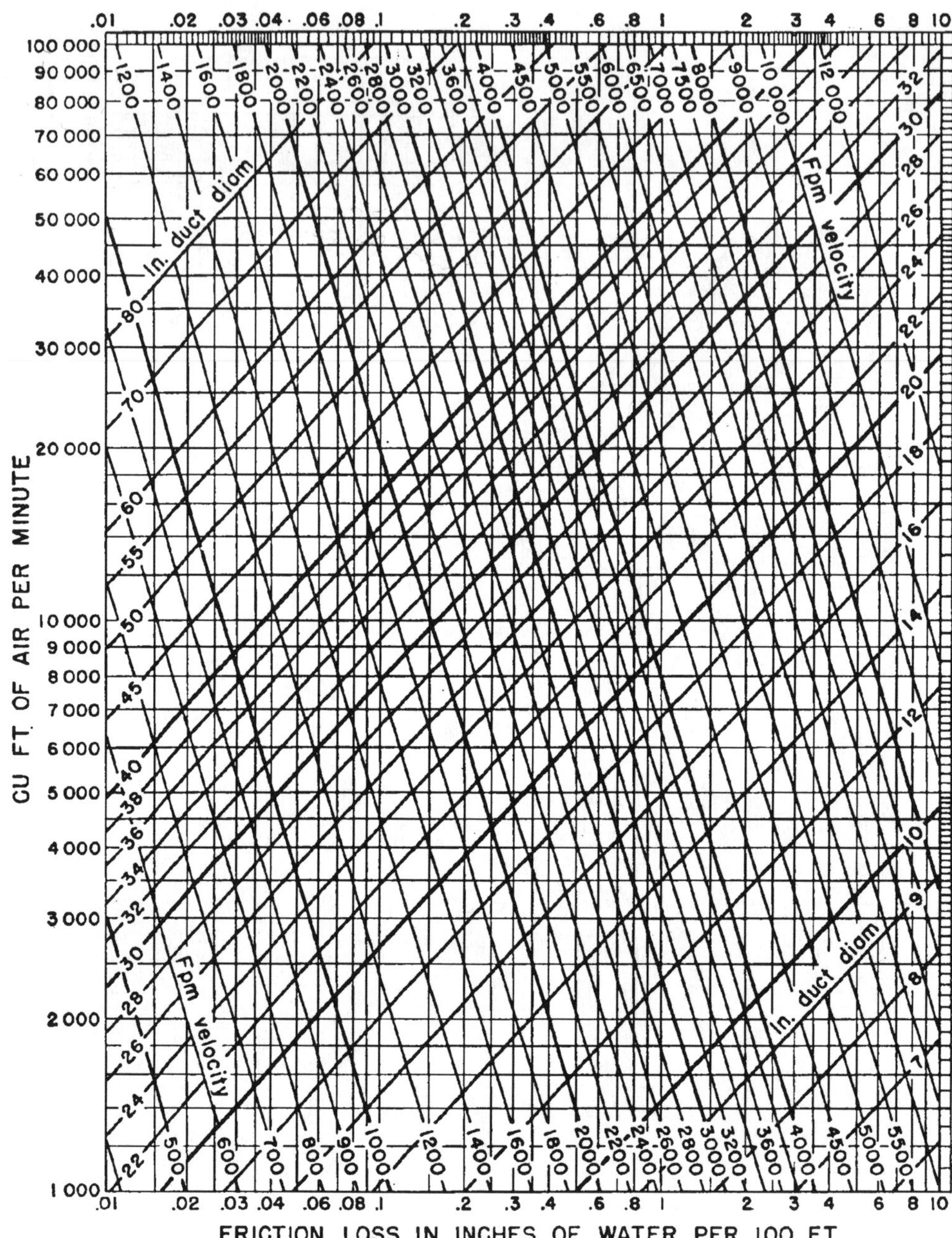

Figure 13.2 Friction of Air in Straight Ducts for Volumes of 1000 to 100,000 cfm

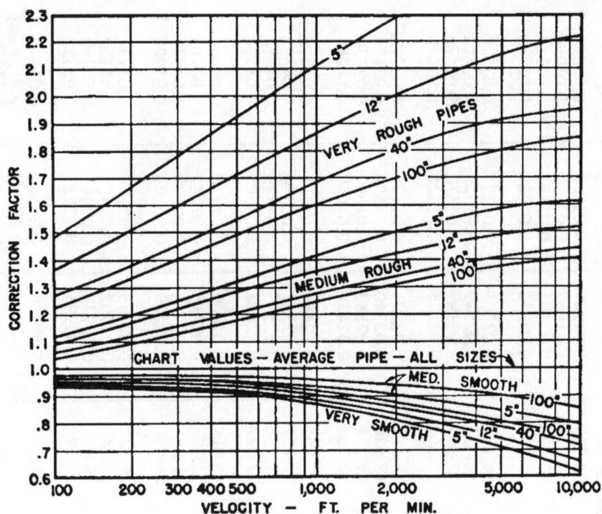

Figure 13.3 Correction Factors for Pipe Roughness

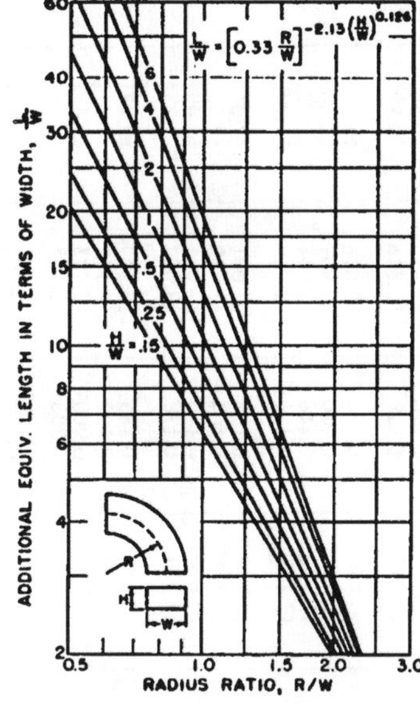

Figure 13.4 Loss in 90-Deg Elbows of Rectangular Cross-Section

TABLE 13.1 Thermodynamic Properties of Moist Air. (P = 14.7 psia)

Fahr. Temp. t(F)	Humidity Ratio $W_s \times 10^3$	Volume cu ft/lb dry air			Enthalpy Btu/lb dry air			Entropy Btu per (R)(lb dry air)		
		v_a	v_{as}	v_s	h_a	h_{as}	h_s	s_a	s_{as}	s_s
38	4.818	12.540	0.097	12.637	9.128	5.191	14.319	0.01909	0.01097	0.03006
39	5.012	12.565	0.101	12.666	9.368	5.403	14.771	0.01957	0.01139	0.03096
40	5.213	12.590	0.105	12.695	9.608	5.622	15.230	0.02005	0.01183	0.03188
41	5.421	12.616	0.109	12.725	9.848	5.849	15.697	0.02053	0.01228	0.03281
42	5.638	12.641	0.114	12.755	10.088	6.084	16.172	0.02101	0.01275	0.03376
43	5.860	12.666	0.119	12.785	10.329	6.328	16.657	0.02149	0.01323	0.03472
44	6.091	12.691	0.124	12.815	10.569	6.580	17.149	0.02197	0.01373	0.03570
45	6.331	12.717	0.129	12.846	10.809	6.841	17.650	0.02245	0.01425	0.03670
46	6.578	12.742	0.134	12.876	11.049	7.112	18.161	0.02293	0.01478	0.03771
47	6.835	12.767	0.140	12.907	11.289	7.391	18.680	0.02340	0.01534	0.03874
48	7.100	12.792	0.146	12.938	11.530	7.681	19.211	0.02387	0.01591	0.03978
49	7.374	12.818	0.151	12.969	11.770	7.981	19.751	0.02434	0.01650	0.04084
50	7.658	12.843	0.158	13.001	12.010	8.291	20.301	0.02481	0.01711	0.04192
51	7.952	12.868	0.164	13.032	12.250	8.612	20.862	0.02528	0.01774	0.04302
52	8.256	12.894	0.170	13.064	12.491	8.945	21.436	0.02575	0.01839	0.04414
53	8.569	12.919	0.178	13.097	12.731	9.289	22.020	0.02622	0.01906	0.04528
54	8.894	12.944	0.185	13.129	12.971	9.644	22.615	0.02669	0.01976	0.04645
55	9.229	12.970	0.192	13.162	13.211	10.01	23.22	0.02716	0.02047	0.04763
56	9.575	12.995	0.200	13.195	13.452	10.39	23.84	0.02762	0.02121	0.04883
57	9.934	13.020	0.208	13.228	13.692	10.79	24.48	0.02809	0.02197	0.05006
58	10.30	13.045	0.216	13.261	13.932	11.19	25.12	0.02855	0.02276	0.05131
59	10.69	13.071	0.224	13.295	14.172	11.61	25.78	0.02902	0.02357	0.05259
60	11.08	13.096	0.233	13.329	14.413	12.05	26.46	0.02948	0.02441	0.05389
61	11.49	13.121	0.242	13.363	14.653	12.50	27.15	0.02994	0.02527	0.05521
62	11.91	13.147	0.251	13.398	14.893	12.96	27.85	0.03040	0.02616	0.05656
63	12.35	13.172	0.261	13.433	15.134	13.44	28.57	0.03086	0.02708	0.05794
64	12.80	13.197	0.271	13.468	15.374	13.94	29.31	0.03132	0.02803	0.05935
65	13.26	13.222	0.282	13.504	15.614	14.45	30.06	0.03177	0.02901	0.06078
66	13.74	13.247	0.292	13.539	15.855	14.98	30.83	0.03223	0.03002	0.06225
67	14.24	13.273	0.303	13.576	16.095	15.53	31.62	0.03269	0.03106	0.06375
68	14.75	13.298	0.315	13.613	16.335	16.09	32.42	0.03314	0.03213	0.06527
69	15.28	13.323	0.327	13.650	16.576	16.67	33.25	0.03360	0.03323	0.06683

TABLE 13.1 (*Continued*)

THERMODYNAMIC PROPERTIES OF MOIST AIR (14.696 PSIA)

Fahr. Temp. t(F)	Humidity Ratio $W_s \times 10^2$	Volume cu ft/lb dry air			Enthalpy Btu/lb dry air			Entropy Btu per (R)(lb dry air)		
		v_a	v_{as}	v_s	h_a	h_{as}	h_s	s_a	s_{as}	s_s
70	1.582	13.348	0.339	13.687	16.816	17.27	34.09	0.03405	0.03437	0.06842
71	1.639	13.373	0.351	13.724	17.056	17.89	34.95	0.03450	0.03554	0.07004
72	1.697	13.398	0.364	13.762	17.297	18.53	35.83	0.03495	0.03675	0.07170
73	1.757	13.424	0.377	13.801	17.537	19.20	36.74	0.03540	0.03800	0.07340
74	1.819	13.449	0.392	13.841	17.778	19.88	37.66	0.03585	0.03928	0.07513
75	1.882	13.474	0.407	13.881	18.018	20.59	38.61	0.03630	0.04060	0.07690
76	1.948	13.499	0.422	13.921	18.259	21.31	39.57	0.03675	0.04197	0.07872
77	2.016	13.525	0.437	13.962	18.499	22.07	40.57	0.03720	0.04337	0.08057
78	2.086	13.550	0.453	14.003	18.740	22.84	41.58	0.03765	0.04482	0.08247
79	2.158	13.575	0.470	14.045	18.980	23.64	42.62	0.03810	0.04631	0.08441
80	2.233	13.601	0.486	14.087	19.221	24.47	43.69	0.03854	0.04784	0.08638
81	2.310	13.626	0.504	14.130	19.461	25.32	44.78	0.03899	0.04942	0.08841
82	2.389	13.651	0.523	14.174	19.702	26.20	45.90	0.03943	0.05105	0.09048
83	2.471	13.676	0.542	14.218	19.942	27.10	47.04	0.03987	0.05273	0.09260
84	2.555	13.702	0.560	14.262	20.183	28.04	48.22	0.04031	0.05446	0.09477
85	2.642	13.727	0.581	14.308	20.423	29.01	49.43	0.04075	0.05624	0.09699
86	2.731	13.752	0.602	14.354	20.663	30.00	50.66	0.04119	0.05807	0.09926
87	2.824	13.777	0.624	14.401	20.904	31.03	51.93	0.04163	0.05995	0.10158
88	2.919	13.803	0.645	14.448	21.144	32.09	53.23	0.04207	0.06189	0.10396
89	3.017	13.828	0.668	14.496	21.385	33.18	54.56	0.04251	0.06389	0.10640
90	3.118	13.853	0.692	14.545	21.625	34.31	55.93	0.04295	0.06596	0.10890
91	3.223	13.879	0.716	14.595	21.865	35.47	57.33	0.04339	0.06807	0.11146
92	3.330	13.904	0.741	14.645	22.106	36.67	58.78	0.04382	0.07025	0.11407
93	3.441	13.929	0.768	14.697	22.346	37.90	60.25	0.04426	0.07249	0.11675
94	3.556	13.954	0.795	14.749	22.587	39.18	61.77	0.04469	0.07480	0.11949
95	3.673	13.980	0.822	14.802	22.827	40.49	63.32	0.04513	0.07718	0.12231
96	3.795	14.005	0.851	14.856	23.068	41.85	64.92	0.04556	0.07963	0.12519
97	3.920	14.030	0.881	14.911	23.308	43.24	66.55	0.04600	0.08215	0.12815
98	4.049	14.056	0.911	14.967	23.548	44.68	68.23	0.04643	0.08474	0.13117
99	4.182	14.081	0.942	15.023	23.789	46.17	69.96	0.04686	0.08741	0.13427
100	4.319	14.106	0.975	15.081	24.029	47.70	71.73	0.04729	0.09016	0.13745
101	4.460	14.131	1.009	15.140	24.270	49.28	73.55	0.04772	0.09299	0.14071
102	4.606	14.157	1.043	15.200	24.510	50.91	75.42	0.04815	0.09591	0.14406
103	4.756	14.182	1.079	15.261	24.751	52.59	77.34	0.04858	0.09891	0.14749
104	4.911	14.207	1.117	15.324	24.991	54.32	79.31	0.04900	0.1020	0.1510

Table 13.2 Infiltration through Window and Door Crack in Cubic Feet per Hour per Foot of Crack ($ft^3/h/ft$)
$ft^3/h/ft \times 0.0929 = m^3/h/m$

Type of Aperture	Remarks	Wind Velocity (mph) 5	10	15	20	25
		(km/h) 8.0	16.1	24.1	32.2	40.2
Double-hung wood-sash windows (unlocked)	Average; non-weather-stripped	7	21.4	39	59	80
	Average; weather-stripped	4	13	24	36	49
	Poorly fitted; non-weather-stripped	27	69	111	154	199
	Poorly fitted; weather-stripped	6	19	34	51	71
	Around window frame: masonry wall, uncalked	3	8	14	20	27
	Around window frame: masonry wall, calked	1	2	3	4	5
	Around window frame: wood frame structure	2	6	11	17	23
Double-hung metal windows	Non-weather-stripped; unlocked	20	47	74	104	137
	Non-weather-stripped; locked	20	45	70	96	125
	Weather-stripped; unlocked	6	19	32	46	60
Single-sash metal windows	Industrial; horizontally pivoted	52	108	176	244	304
	Residential casement	14	32	52	76	100
	Vertically pivoted	30	88	145	186	221
Doors	Well-fitted	27	69	110	154	199
	Poorly fitted	54	138	220	308	398

Data based on research papers in *Trans. ASHVE*, vols. 30, 34, 36, 37, and 39.

TABLE 13.3 Total Equivalent Temperature Differentials for Calculating Heat Gain Through Sunlit and Shaded Walls

North Latitude Wall Facing	Sun Time																		South Latitude Wall Facing
	A.M.						P.M.												
	8		10		12		2		4		6		8		10		12		
	Exterior color of wall—D = dark, L = light																		
	D	L	D	L	D	L	D	L	D	L	D	L	D	L	D	L	D	L	
	Frame																		
NE	22	10	24	12	14	10	12	10	14	14	14	14	10	10	6	4	2	2	SE
E	30	14	36	18	32	16	12	12	14	14	14	14	10	10	6	6	2	2	E
SE	13	6	26	16	28	18	24	16	16	14	14	14	10	10	6	4	2	2	NE
S	−4	−4	4	0	22	12	30	20	26	20	16	14	10	10	6	6	2	2	N
SW	−4	−4	0	−2	6	4	26	22	40	28	42	28	24	20	6	4	2	2	NW
W	−4	−4	0	0	6	6	20	12	40	28	48	34	22	22	8	8	2	2	W
NW	−4	−4	0	−2	6	4	12	10	24	20	40	26	34	24	6	4	2	2	SW
N (Shade)	−4	−4	−2	−2	4	4	10	10	14	14	12	12	8	8	4	4	0	0	S (Shade)
	4 In. Brick or Stone Veneer + Frame																		
NE	−2	−4	24	12	20	10	10	6	12	10	14	14	12	12	10	10	6	4	SE
E	2	0	30	14	31	17	14	14	12	12	14	14	12	12	10	8	6	6	E
SE	2	−2	20	10	28	16	26	16	18	14	14	14	12	12	10	8	6	6	NE
S	−4	−4	−2	−2	12	6	24	16	26	18	20	16	12	12	8	8	4	4	N
SW	0	−2	0	−2	2	2	12	8	32	22	36	26	34	24	10	8	6	6	NW
W	0	−2	0	0	4	2	10	8	26	18	40	28	42	28	16	14	6	6	W
NW	−4	−4	−2	−2	2	2	8	6	12	12	30	22	34	24	12	10	6	6	SW
N (Shade)	−4	−4	−2	−2	0	0	6	6	10	10	12	12	12	12	8	8	4	4	S (Shade)
	8 In. Hollow Tile or 8 In. Cinder Block																		
NE	0	0	0	0	20	10	16	10	10	6	12	10	14	12	12	10	8	8	SE
E	4	2	12	4	24	12	26	14	20	12	12	10	14	12	14	10	10	8	E
SE	2	0	2	0	16	8	20	12	20	14	14	12	14	12	12	10	8	6	NE
S	0	0	0	0	2	0	12	6	24	14	26	16	20	14	12	10	8	6	N
SW	2	0	2	0	2	0	6	4	12	10	26	18	30	20	26	18	8	6	NW
W	4	2	4	2	4	2	6	4	10	8	18	14	30	22	32	22	18	14	W
NW	0	0	0	0	2	0	4	2	8	6	12	10	22	18	30	22	10	8	SW
N (Shade)	−2	−2	−2	−2	−2	−2	0	0	6	6	10	10	10	10	10	10	6	6	S (Shade)
	8 In. Brick or 12 In. Hollow Tile or 12 In. Cinder Block																		
NE	2	2	2	2	10	2	16	8	14	8	10	6	10	8	10	10	10	8	SE
E	8	6	8	6	14	8	18	10	18	10	14	8	14	10	14	10	12	10	E
SE	8	4	6	4	6	4	14	10	18	12	16	12	12	10	12	10	12	10	NE
S	4	2	4	2	4	2	4	2	10	6	16	10	16	12	12	10	10	8	N
SW	8	4	6	4	6	4	8	4	10	6	12	8	20	12	24	16	20	14	NW
W	8	4	6	4	6	6	8	6	10	6	14	8	20	16	24	16	24	16	W
NW	2	2	2	2	2	2	4	2	6	4	8	6	10	8	16	14	18	14	SW
N (Shade)	0	0	0	0	0	0	0	0	2	2	6	6	8	8	8	8	6	6	S (Shade)
	12 In. Brick																		
NE	8	6	8	6	8	4	8	4	10	4	12	6	12	6	10	6	10	6	SE
E	12	8	12	8	12	8	10	6	12	8	14	10	14	10	14	8	14	8	E
SE	10	6	10	6	10	6	10	6	10	6	12	8	14	10	14	10	12	8	NE
S	8	6	8	6	6	4	6	4	6	4	8	4	10	6	12	8	12	8	N
SW	10	6	10	6	10	6	10	6	10	6	10	8	10	8	12	8	14	10	NW
W	12	8	12	8	12	8	10	6	10	6	10	6	10	6	12	8	16	10	W
NW	8	6	8	6	8	4	8	4	8	4	8	4	8	6	10	6	10	6	SW
N (Shade)	4	4	2	2	2	2	2	2	2	2	2	2	2	2	4	4	6	6	S (Shade)

TABLE 13.4 Total Equivalent Temperature Differentials for Calculating Heat Gain Through Sunlit and Shaded Roofs

Description of Roof Construction[a]	Sun Time								
	A.M.			P.M.					
	8	10	12	2	4	6	8	10	12
Light Construction Roofs—Exposed to Sun									
1″ Wood[b] or 1″ Wood[b] + 1″ or 2″ insulation	12	38	54	62	50	26	10	4	0
Medium Construction Roofs—Exposed to Sun									
2″ Concrete or 2″ Concrete + 1″ or 2″ insulation or 2″ Wood[b]	6	30	48	58	50	32	14	6	2
2″ Gypsum or 2″ Gypsum + 1″ insulation 1″ Wood[b] or, 2″ Wood[b] or, 2″ Concrete or, 2″ Gypsum } + 4″ rock wool in furred ceiling	0	20	40	52	54	42	20	10	6
4″ Concrete or 4″ Concrete with 2″ insulation	0	20	38	50	52	40	22	12	6
Heavy Construction Roofs—Exposed to Sun									
6″ Concrete	4	6	24	38	46	44	32	18	12
6″ Concrete + 2″ insulation	6	6	20	34	42	44	34	20	14
Roofs Covered with Water—Exposed to Sun									
Light construction roof with 1″ water	0	4	16	22	18	14	10	2	0
Heavy construction roof with 1″ water	−2	−2	−4	10	14	16	14	10	6
Any roof with 6″ water	−2	0	0	6	10	10	8	4	0
Roofs with Roof Sprays—Exposed to Sun									
Light construction	0	4	12	18	16	14	10	2	0
Heavy construction	−2	−2	2	8	12	14	12	10	6
Roofs in Shade									
Light construction	−4	0	6	12	14	12	8	2	0
Medium construction	−4	−2	2	8	12	12	10	6	2
Heavy construction	−2	−2	0	4	8	10	10	8	4

[a] Includes ⅜ in. felt roofing with or without slag. May also be used for shingle roof.
[b] Nominal thickness of the wood.

Table 13.5 Summer Coefficients of Heat Transmission U of Flat Roofs Covered with Built-up Roofing*
(Btu/h · ft^2 of F Temperature Difference between the Air on the Two Sides)

Type of Roof Deck (ceiling not shown)	Thickness of Roof Deck (in.)	Insulation on Top of Deck (covered with built-up roofing)									
		No Ceiling—Underside of Roof Exposed					Furred Ceiling with Air Space, Metal Lath, and Plaster				
		No Insulation	Insulating Board † Thickness (in.) ½	1	1½	2	No Insulation	Insulating Board † Thickness (in.) ½	1	1½	2
Flat metal roof deck (Roofing, Insulation, Metal deck)	4-ply felt roof	0.73	0.35	0.23	0.17	0.13	0.40	0.25	0.18	0.14	0.12
	Ditto + ½-in. slag	0.54	0.30	0.20	0.16	0.13	0.34	0.22	0.16	0.13	0.11
Precast cement tile (Roofing, Insulation, Cast tile)	4-ply felt roof 1⅝	0.67	0.33	0.22	0.17	0.13	0.38	0.24	0.18	0.14	0.12
	Ditto + ½-in. slag 1⅝	0.50	0.28	0.20	0.15	0.12	0.32	0.21	0.17	0.13	0.11
Concrete (Roofing, Insulation, Concrete)	4-ply felt roof 2	0.65	0.33	0.22	0.16	0.13	0.37	0.24	0.18	0.14	0.12
	4	0.59	0.31	0.21	0.16	0.13	0.36	0.23	0.17	0.13	0.12
	6	0.54	0.30	0.20	0.16	0.13	0.33	0.22	0.17	0.13	0.11
	Ditto + ½-in. slag 2	0.49	0.28	0.20	0.15	0.12	0.31	0.21	0.16	0.13	0.11
	4	0.46	0.27	0.19	0.15	0.12	0.30	0.21	0.16	0.13	0.11
	6	0.42	0.26	0.19	0.14	0.12	0.29	0.20	0.16	0.13	0.10
Gypsum and wood fiber‡ on ½-in. gypsum board (Roofing, Insulation, Gypsum, Plaster board)	4-ply felt roof 2½	0.34	0.23	0.17	0.13	0.12	0.25	0.18	0.14	0.12	0.097
	3½	0.28	0.20	0.15	0.12	0.11	0.21	0.16	0.13	0.11	0.094
	Ditto + ½-in. slag 2½	0.29	0.20	0.16	0.13	0.11	0.22	0.16	0.13	0.11	0.093
	3½	0.25	0.18	0.14	0.12	0.10	0.19	0.15	0.13	0.10	0.090
Wood§ (Roofing, Insulation, Sheathing)	4-ply felt roof 1	0.43	0.26	0.19	0.15	0.12	0.29	0.20	0.15	0.13	0.11
	1½	0.33	0.22	0.17	0.13	0.11	0.24	0.18	0.14	0.12	0.097
	2	0.29	0.20	0.16	0.13	0.11	0.22	0.16	0.13	0.11	0.094
	3	0.22	0.16	0.13	0.11	0.09	0.17	0.13	0.12	0.10	0.085
	Ditto + ½-in. slag 1	0.35	0.23	0.17	0.14	0.11	0.25	0.18	0.14	0.12	0.10
	1½	0.29	0.20	0.15	0.12	0.10	0.21	0.17	0.13	0.11	0.093
	2	0.26	0.19	0.14	0.12	0.10	0.20	0.15	0.13	0.10	0.090
	3	0.20	0.15	0.12	0.10	0.09	0.16	0.13	0.11	0.09	0.081

TABLE 13.6 Instantaneous Rates of Heat Gain Due to Transmitted Direct and Diffuse or Sky Solar Radiation by a Single Sheet of Unshaded Common Window Glass

Latitude	Sun Time AM → ↓		Instantaneous Heat Gain in Btu per (hr) (sq ft) N	NE	E	SE	S	SW	W	NW	Horiz.
30 Deg north	6 a.m.	6 p.m.	25	98	108	52	*5*	*5*	*5*	*5*	17
	7	5	23	155	190	110	*10*	*10*	*10*	*10*	71
	8	4	16	148	205	136	*14*	*13*	*13*	*13*	137
	9	3	*16*[a]	106	180	136	21	*15*	*15*	*15*	195
	10	2	*17*	54	128	116	34	*17*	*16*	*16*	241
	11	1	*18*	20	59	78	45	*19*	*18*	*18*	267
	12		*18*	*19*	*19*	35	49	35	*19*	*19*	276
40 Deg north	5 a.m.	7 p.m.	3	7	6	2	*0*	*0*	*0*	0	1
	6	6	26	116	131	67	*7*	*6*	*6*	6	25
	7	5	16	149	195	124	*11*	*10*	*10*	10	77
	8	4	*14*	129	205	156	18	*12*	*12*	12	137
	9	3	*15*	79	180	162	42	*14*	*14*	14	188
	10	2	*16*	31	127	148	69	*16*	*16*	16	229
	11	1	*17*	*18*	58	113	90	23	*17*	17	252
	12		*17*	*17*	19	64	98	64	*19*	17	259
50 Deg north	5 a.m.	7 p.m.	20	54	54	20	*3*	*3*	*3*	*3*	6
	6	6	25	128	149	81	*8*	*7*	*7*	*7*	34
	7	5	*12*	139	197	136	*12*	*10*	*10*	*10*	80
	8	4	*13*	107	202	171	32	*12*	*12*	*12*	129
	9	3	*14*	54	176	183	72	*14*	*14*	*14*	173
	10	2	*15*	18	124	174	110	*16*	*15*	*15*	206
	11	1	*16*	*16*	57	143	136	42	*16*	*16*	227
	12		*16*	*16*	*18*	96	144	96	*18*	*16*	234
	↑ PM →		N	NW	W	SW	S	SE	E	NE	Horiz.

[a] Italics indicate shaded glass.

TABLE 13.7 Instantaneous Rates of Heat Gain by Convection and Radiation from a Single Sheet of Unshaded Common Window Glass

Sun Time	Dry-Bulb F Deg	Instantaneous Heat Gain in Btu per (hr) (sq ft) N	NE	E	SE	S	SW	W	NW	Horiz.
5 a.m.	74	−6	−6	−6	−6	−6	−6	−6	−6	−6
6	74	−5	−4	−4	−5	−5	−6	−6	−6	−5
7	75	−5	−2	−2	−3	−5	−5	−5	−5	−3
8	77	−3	0	1	0	−2	−3	−3	−3	0
9	80	0	2	4	3	1	0	0	0	3
10	83	3	4	6	6	5	3	3	3	8
11	87	8	8	10	11	10	9	8	8	13
12	90	12	12	12	13	14	13	12	12	16
1 p.m.	93	15	15	15	16	17	17	17	15	20
2	94	16	16	16	16	18	19	19	17	21
3	95	17	17	17	17	19	21	21	19	21
4	94	16	16	16	16	17	20	20	19	19
5	93	15	15	15	15	15	18	19	18	17
6	91	13	13	13	13	13	14	15	15	13
7	87	8	8	8	8	8	8	8	8	8
8	85	6	6	6	6	6	6	6	6	6
9	83	3	3	3	3	3	3	3	3	3

Table 13.8 Application Factors to Apply to Tables 13.6, 13.7, 13.9 to Obtain Instantaneous Rates of Heat Gain for Various Types of Single Flat Glass and Combinations of Two Sheets of Flat Glass Spaced at ¼ In.

Glass[a]	Normal Incidence Transmittance	Factor to Apply to Table 13.6	Factor to Apply to Table 13.7
Single common window	0.87	1.00	1.0(X)[c] + 0.0(Y)[d]
Single regular plate	0.77	0.87	1.0(X) + 0.25(Y)
Single heat absorbing plate	0.41	0.46[b]	1.0(X) + 1.00(Y)
Double common window	0.76	0.85	0.6(X) + 0.10(Y)
Double regular plate	0.60	0.66[b]	0.6(X) + 0.55(Y)
Heat absorbing plate outdoors, Regular plate indoors	0.35	0.37[b]	0.6(X) + 0.75(Y)

[a] Common window glass ⅛ in. thick. Plate glass ¼ in. thick.
[b] For better precision, increase factors 10 percent when glass is in the shade.
[c] X values are Table 13.7 values.
[d] Y values are Table 13.9 values.

Table 13.9 Heat Absorbed in Glass. Values of Y to be Used with Factors in Table 13.8 in the Determination of Instantaneous Rates of Heat Gain Due to Convection and Radiation for Various Types of Single Glass and Combinations of Two Sheets of Glass Spaced at ¼ In.

For Clear Atmospheres and 18 Deg Declination, North (August 1)

Sun Time	Latitude	Values of Y in Btu per (hr) (sq ft)[a]								
		N	NE	E	SE	S	SW	W	NW	Horiz.
5 a.m.		0	0	1	0	0	0	0	0	0
6		4	16	18	9	1	1	1	1	3
7		2	24	30	20	2	2	2	2	11
8		2	22	33	25	2	2	2	2	21
9		2	16	30	29	8	3	3	3	32
10		3	5	25	27	14	3	3	3	37
11	40 Degrees North	3	3	12	21	18	3	3	3	42
12		3	3	3	15	19	12	3	3	45
1 p.m.		3	3	3	3	19	22	10	3	44
2		3	3	3	3	16	27	24	4	41
3		3	3	3	3	10	30	31	15	35
4		3	3	3	3	4	29	36	23	26
5		2	2	2	2	2	23	34	27	17
6		4	1	1	1	1	14	24	21	6
7		0	0	0	0	0	2	3	3	1

Sun Time	Latitude	SE	S	SW	Sun Time	Latitude	SE	S	SW
5 a.m.		0	0	0	5 a.m.		2	0	0
6		7	1	1	6		13	1	1
7		18	2	2	7		22	2	2
8		22	2	2	8		28	3	2
9		24	3	3	9		30	13	3
10		22	5	3	10		31	20	3
11	30[b] Degrees North	16	7	3	11	50[b] Degrees North	27	25	5
12		6	9	4	12		20	27	17
1 p.m.		3	9	14	1 p.m.		9	25	26
2		3	6	21	2		3	22	32
3		3	5	27	3		3	16	33
4		3	3	26	4		2	7	31
5		2	2	21	5		2	2	26
6		1	1	11	6		1	1	17
7		0	0	0	7		0	0	7

[a] Values of Y for 8 and 9 p.m. are zero.
[b] For N, NE, E, W, NW, and horizontal use 40 deg North Latitude values.

TABLE 13.10 Shade Factors for Various Types of Shading

Type of Shading	Finish on Side Exposed to Sun	Shade Factor	
Canvas awning sides open	Dark or medium	0.25	
Canvas awning top and sides tight against building	Dark or medium	0.35	
Inside roller shade, fully drawn[a]	White, cream	0.41[f]	
Inside roller shade, fully drawn[a]	Medium	0.62	
Inside roller shade, fully drawn[a]	Dark	0.81	
Inside roller shade, half drawn[a]	White, cream	0.71	
Inside roller shade, half drawn[a]	Medium	0.81	
Inside roller shade, half drawn[a]	Dark	0.91	
Inside venetian blind, slats set at 45 deg[b]	White, cream	0.56[f]	
Inside venetian blind, slats set at 45 deg[b]	Diffuse reflecting aluminum metal	0.45[f]	
Inside venetian blind, slats set at 45 deg[b]	Medium	0.65[f]	
Inside venetian blind, slats set at 45 deg[b]	Dark	0.75[f]	
Outside venetian blind, slats set at 45 deg[b]	White, cream	0.15[f]	
Outside venetian blind, slats set at 45 deg[b, c] extended as awning fully covering window	White, cream	0.15[f]	
Outside venetian blind, slats set at 45 deg, extended as awning covering ⅔ of window[c]	White, cream	0.43	
		Dark[d, f]	Green tint[e, f]
Outside shading screen, solar altitude 10 deg		0.52	0.46
Outside shading screen, solar altitude 20 deg		0.40	0.35
Outside shading screen, solar altitude 30 deg		0.25	0.24
Outside shading screen, solar altitude, above 40 deg		0.15	0.22

Table 13.11 Door Infiltration in Summer for Commercial Establishments

Revolving and Swinging Doors Opening to Outside			Average Occupancy (Patrons and Employees) on Which Values Are Based (min.)
Application	Infiltration per Person in Room (cfm)		
	72-In. Revolving Door	36-In. Swinging Door	
Bank	7.5	10.0	20
Barber shop	3.5	4.5	45
Broker's office	5.0	6.5	30
Candy and soda store	5.0	6.5	30
Cigar store	15.0	20.0	10
Department store (small)	5.0	6.5	30
Dress shop	2.0	2.5	75
Drugstore	10.0	13.0	15
Furrier	2.0	2.0	90
Lunchroom	5.0	6.5	30
Men's shop	3.5	4.5	45
Office (professional)	2.5	3.0	60
Restaurant	2.0	2.5	75
Shoe store	3.5	4.5	45

When Doors Are Left Open Continuously	
72-in. revolving door (panels open)	1200 cfm
36-in. swinging door (standing open)	800 cfm

Table 13.12 Heat Gain from Occupants

Type of Activity	Typical Application	Total Heat Dissipation, Adult Male (Btuh)	Total Adjusted* Heat Dissipation (Btuh)	Sensible Heat (Btuh)	Latent Heat (Btuh)
Seated at rest	Theater				
	Matinee	390	330	225	105
	Evening	390	350	245	105
Seated; very light work	Offices, hotels, apartments, restaurants	450	400	245	155
Moderately active office work	Offices, hotels, apartments	475	450	250	200
Standing; light work; walking slowly	Department store, retail store	550	450	250	250
Walking; seated	Drug store	550	500	250	250
Standing; walking slowly	Bank	550	500	250	250
Sedentary work	Restaurant	590	550	275	275
Light bench work	Factory	800	750	275	475
Moderate work	Small-parts assembly	900	850	305	545
Moderate dancing	Dance hall	900	850	305	545
Walking, 3 mph; moderately heavy work	Factory	1000	1000	375	625
Bowling (participant)	Bowling alley	1500	1450	580	870
Heavy work	Factory	1500	1450	580	870

TABLE 13.13 Rate of Heat Gain from Applainces WITHOUT HOODS

Appliance	Capacity	Overall Dimensions (Less Legs and Handles; Last Dimension is Height) Inches	Control A—Automatic M—Manual	Miscellaneous Data	Manufacturer's Rating Watts	Btu/Hr	Maintaining Rate Btu per Hour	Recommended Rate of Heat Gain Btu per Hour		
								Sensible	Latent	Total
Restaurant Electrical Appliances										
Coffee brewer and warmer	½ gal		M M	Brewer 660 w Warmer 90 w	600 90	2000 300	 306	900 230	220 60	1120 290
Coffee brewer unit with tank	½ gal	20 x 30 x 26		2000 w Water heater, 2960 w brewer	4960	17000		4800	1200	6000
Coffee urn	3 gal 5 gal	12 x 23 x 21 18 (Diam.) x 37	A A	Nickel plated Nickel plated	4500 5000	15000 17000	2600 3600	2200 3400	1500 2300	3700 5700
Doughnut machine		22 x 22 x 57	A	Exhaust system	4700	16000		5000	0	5000
Egg boiler	2 cups	10 x 13 x 25	M		1100	3750		1200	800	2000
Food warmer, with plate warmer, per sq ft of top surface			A	Insulated, separate heat unit for each pot; plate warmer in base	400	1350	500	350	350	700
Food warmer, alone, per sq ft of top surface			A		300	1000	400	200	350	550
Fry kettle	11½ lb fat	12 (Diam.) x 14	A		2600	8900	1100	1600	2400	4000
Fry kettle	25 lb fat	16 x 18 x 12	A	Area 12 x 14 in.	7000	24000	2000	3800	5700	9500
Griddle, frying		18 x 18 x 8	A	Area 18 x 14 in.	2350	8000	2800	3100	1700	4800
Griddle, frying		24 x 20 x 10	A	Area 23 x 18 in.	4000	13500	5000	5300	2900	8200
Grill, meat		14 x 14 x 10	A	Area 10 x 12 in.	3000	10250	1900	3900	2100	6000
Grill, sandwich		13 x 14 x 10	A	Area 12 x 12 in.	1650	5600	1900	2700	700	3400
Roll warmer		23 x 23 x 29	A	Three drawers	1000	3400	900	2400	300	2700
Toaster, continuous	360 slices/hr	15 x 15 x 28	A	2 slices wide	2200	7500	5000	5100	1300	6400
Toaster, continuous	720 slices/hr	20 x 15 x 28	A	4 slices wide	3000	10250	6000	6100	2600	8700
Toaster, pop-up	216 slices/hr	12 x 11 x 9	A	4 slice	2450	8400	2000	4900	900	5800
Waffle iron	20 waffles/hr	12 x 13 x 10	A	7 in. diam. waffle	750	2500	600	1100	750	1850
Restaurant Gas-Burning Appliances										
Coffee brewer and warmer	½ gal		M M	Brewer Warmer		3400 500	 500	1350 400	350 100	1700 500
Coffee brewer unit with tank	4½ gal Tank	19 x 30 x 26		4 Brewers and tank				7200	1800	9000
Coffee urn	3 gal 5 gal	12 x 23 x 21 18 (Diam.) x 37	A A	Nickel plated Nickel plated			3400 4700	2500 3900	2500 3900	5000 7800
Food Warmer, per sq ft of top surface			M	Water bath		2000	900	850	430	1280
Fry kettle	15 lb fat	12 x 20 x 18	A	Area 10 x 10		14250	3000	4200	2800	7000
Fry kettle	28 lb fat	15 x 35 x 11	A	Area 11 x 16		24000	4500	7200	4800	12000
Grill		22 x 14 x 17	M	Insulated, grill surface of 1.4 sq ft Top burner 22,000 Btu/hr Bottom burner 15,000 Btu/hr		37000		14400	3600	18000

Table 13.14 Ventilation Standards

Application	Smoking	Air per Person (cfm)		Minimum Air per Sq Ft of Floor (cfm)
		Recommended	Minimum	
Apartment				
Average	Some	20	15	...
De luxe	Some	30	25	.33
Banking space	Occasional	10	7½	...
Barber shops	Considerable	15	10	...
Beauty parlors	Occasional	10	7½	...
Brokers' board rooms	Very heavy	50	30	...
Cocktail bars	Heavy	30	25	...
Corridors (supply or exhaust)	.....	...	...	.25
Department stores	None	7½	5	.05
Directors' room	Extreme	50	30	...
Drug stores	Considerable	10	7½	...
Factories	None	10	7½	.10
Five-and-ten-cent stores	None	7½	5	...
Funeral parlors	None	10	7½	...
Garage	.....	...	...	1.0
Hospitals				
Operating rooms *	None	...	...	2.0
Private rooms	None	30	25	.33
Wards	None	20	15	...
Hotel rooms	Heavy	30	25	.33
Kitchen				
Restaurant	.....	...	...	4.0
Residence	.....	...	...	2.0
Laboratories	Some	20	15	...
Meeting rooms	Very heavy	50	30	1.25
Office				
General	Some	15	10	...
Private	None	25	15	.25
Private	Considerable	30	25	.25
Restaurant				
Cafeteria	Considerable	12	10	...
Dining rooms	Considerable	15	12	...
Schoolrooms	None	...	...	...
Shop, retail	None	10	7½	...
Theater	None	7½	5	...
	Some	15	10	...
Toilets (exhaust)	.....	...	...	2.0

* Use outside air only, to overcome explosion hazard of anesthetics.

TABLE 13.15 Circular Equivalents of Rectangular Ducts for Equal Friction and Capacity
All dimensions in inches.

Side Rectangular Duct	4.0	4.5	5.0	5.5	6.0	6.5	7.0	7.5	8.0	9.0	10.0	11.0	12.0	13.0	14.0	15.0	16.0
3.0	3.8	4.0	4.2	4.4	4.6	4.8	4.9	5.1	5.2	5.5	5.7	6.0	6.2	6.4	6.6	6.8	7.0
3.5	4.1	4.3	4.6	4.8	5.0	5.2	5.3	5.5	5.7	6.0	6.3	6.5	6.8	7.0	7.2	7.4	7.6
4.0	4.4	4.6	4.9	5.1	5.3	5.5	5.7	5.9	6.1	6.4	6.8	7.1	7.3	7.6	7.8	8.1	8.3
4.5	4.6	4.9	5.2	5.4	5.6	5.9	6.1	6.3	6.5	6.9	7.2	7.5	7.8	8.1	8.4	8.6	8.9
5.0	4.9	5.2	5.5	5.7	6.0	6.2	6.4	6.7	6.9	7.3	7.6	8.0	8.3	8.6	8.9	9.1	9.4
5.5	5.1	5.4	5.7	6.0	6.3	6.5	6.8	7.0	7.2	7.6	8.0	8.4	8.7	9.0	9.4	9.6	9.8

Side Rectangular Duct	6	7	8	9	10	11	12	13	14	15	16	17	18	19	20	22	24	26	28	30	Side Rectangular Duct
6	6.6																				6
7	7.1	7.7																			7
8	7.5	8.2	8.8																		8
9	8.0	8.6	9.3	9.9																	9
10	8.4	9.1	9.8	10.4	10.9																10
11	8.8	9.5	10.2	10.8	11.4	12.0															11
12	9.1	9.9	10.7	11.3	11.9	12.5	13.1														12
13	9.5	10.3	11.1	11.8	12.4	13.0	13.6	14.2													13
14	9.8	10.7	11.5	12.2	12.9	13.5	14.2	14.7	15.3												14
15	10.1	11.0	11.8	12.6	13.3	14.0	14.6	15.3	15.8	16.4											15
16	10.4	11.4	12.2	13.0	13.7	14.4	15.1	15.7	16.3	16.9	17.5										16
17	10.7	11.7	12.5	13.4	14.1	14.9	15.5	16.1	16.8	17.4	18.0	18.6									17
18	11.0	11.9	12.9	13.7	14.5	15.3	16.0	16.6	17.3	17.9	18.5	19.1	19.7								18
19	11.2	12.2	13.2	14.1	14.9	15.6	16.4	17.1	17.8	18.4	19.0	19.6	20.2	20.8							19
20	11.5	12.5	13.5	14.4	15.2	15.9	16.8	17.5	18.2	18.8	19.5	20.1	20.7	21.3	21.9						20
22	12.0	13.1	14.1	15.0	15.9	16.7	17.6	18.3	19.1	19.7	20.4	21.0	21.7	22.3	22.9	24.1					22
24	12.4	13.6	14.6	15.6	16.6	17.5	18.3	19.1	19.8	20.6	21.3	21.9	22.6	23.2	23.9	25.1	26.2				24
26	12.8	14.1	15.2	16.2	17.2	18.1	19.0	19.8	20.6	21.4	22.1	22.8	23.5	24.1	24.8	26.1	27.2	28.4			26
28	13.2	14.5	15.6	16.7	17.7	18.7	19.6	20.5	21.3	22.1	22.9	23.6	24.4	25.0	25.7	27.1	28.2	29.5	30.6		28
30	13.6	14.9	16.1	17.2	18.3	19.3	20.2	21.1	22.0	22.9	23.7	24.4	25.2	25.9	26.7	28.0	29.3	30.5	31.6	32.8	30
32	14.0	15.3	16.5	17.7	18.8	19.8	20.8	21.8	22.7	23.6	24.4	25.2	26.0	26.7	27.5	28.9	30.1	31.4	32.6	33.8	32
34	14.4	15.7	17.0	18.2	19.3	20.4	21.4	22.4	23.3	24.2	25.1	25.9	26.7	27.5	28.3	29.7	31.0	32.3	33.6	34.8	34
36	14.7	16.1	17.4	18.6	19.8	20.9	21.9	23.0	23.9	24.8	25.8	26.6	27.4	28.3	29.0	30.5	32.0	33.0	34.6	35.8	36
38	15.0	16.4	17.8	19.0	20.3	21.4	22.5	23.5	24.5	25.4	26.4	27.3	28.1	29.0	29.8	31.4	32.8	34.2	35.5	36.7	38
40	15.3	16.8	18.2	19.4	20.7	21.9	23.0	24.0	25.1	26.0	27.0	27.9	28.8	29.7	30.5	32.1	33.6	35.1	36.4	37.6	40
42	15.6	17.1	18.5	19.8	21.1	22.3	23.4	24.5	25.6	26.6	27.6	28.5	29.4	30.4	31.2	32.8	34.4	35.9	37.3	38.6	42
44	15.9	17.5	18.9	20.2	21.5	22.7	23.9	25.0	26.1	27.2	28.2	29.1	30.0	31.0	31.9	33.5	35.2	36.7	38.1	39.5	44
46	16.2	17.8	19.2	20.6	21.9	23.2	24.3	25.5	26.7	27.7	28.7	29.7	30.6	31.6	32.5	34.2	35.9	37.4	38.9	40.3	46
48	16.5	18.1	19.6	20.9	22.3	23.6	24.8	26.0	27.2	28.2	29.2	30.2	31.2	32.2	33.1	34.9	36.6	38.2	39.7	41.2	48
50	16.8	18.4	19.9	21.3	22.7	24.0	25.2	26.4	27.6	28.7	29.8	30.8	31.8	32.8	33.7	35.5	37.3	38.9	40.4	42.0	50
52	17.0	18.7	20.2	21.6	23.1	24.4	25.6	26.8	28.1	29.2	30.3	31.4	32.4	33.4	34.3	36.2	38.0	39.6	41.2	42.8	52
54	17.3	19.0	20.5	22.0	23.4	24.8	26.1	27.3	28.5	29.7	30.8	31.9	32.9	33.9	34.9	36.8	38.7	40.3	42.0	43.6	54
56	17.6	19.3	20.9	22.4	23.8	25.2	26.5	27.7	28.9	30.1	31.2	32.4	33.4	34.5	35.5	37.4	39.3	41.0	42.7	44.3	56
58	17.8	19.5	21.1	22.7	24.2	25.5	26.9	28.2	29.3	30.5	31.7	32.9	33.9	35.0	36.0	38.0	39.8	41.7	43.4	45.0	58
60	18.1	19.8	21.4	23.0	24.5	25.8	27.3	28.7	29.8	31.0	32.2	33.4	34.5	35.5	36.5	38.6	40.4	42.3	44.0	45.8	60
62	18.3	20.1	21.7	23.3	24.8	26.2	27.6	29.0	30.2	31.4	32.6	33.8	35.0	36.0	37.1	39.2	41.0	42.9	44.7	46.5	62
64	18.6	20.3	22.0	23.6	25.2	26.5	27.9	29.3	30.6	31.8	33.1	34.2	35.5	36.5	37.6	39.7	41.6	43.5	45.4	47.2	64
66	18.8	20.6	22.3	23.9	25.5	26.9	28.3	29.7	31.0	32.2	33.5	34.7	35.9	37.0	38.1	40.2	42.2	44.1	46.0	47.8	66
68	19.0	20.8	22.5	24.2	25.8	27.3	28.7	30.1	31.4	32.6	33.9	35.1	36.3	37.5	38.6	40.7	42.8	44.7	46.6	48.4	68
70	19.2	21.	22.8	24.5	26.1	27.6	29.1	30.4	31.8	33.1	34.3	35.6	36.8	37.9	39.1	41.3	43.3	45.3	47.2	49.0	70

Table 13.16 Recommended and Maximum Duct Velocities for Low-Velocity Systems

Designation	Recommended Velocities (fpm)			Maximum Velocities (fpm)		
	Residences	Schools, Theaters, Public Buildings	Industrial Buildings	Residences	Schools, Theaters, Public Buildings	Industrial Buildings
Outside air intakes[a]	500	500	500	800	900	1200
Filters[a]	250	300	350	300	350	350
Heating coils[a]	450	500	600	500	600	700
Cooling coils[a]	450	500	600	450	500	600
Air washers	500	500	500	500	500	500
Suction connections	700	800	1000	900	1000	1400
Fan outlets	1000–1600	1300–2000	1600–2400	1700	1500–2200	1700–2800
Main ducts	800–900	1000–1300	1200–1800	800–1000	1100–1600	1300–2200
Branch ducts	600	600–900	800–1000	700–1000	800–1300	1000–1800
Branch risers	500	600–700	800	650–800	800–1200	1000–1600

[a] These velocities are for total face area, not the net free area.
Adapted by permission from *ASHRAE Handbook of Fundamentals*, 1972, chap. 25.

TABLE 13.17 Heat Gain Factors (Walls, Floor and Ceiling)

Insulation	*Temp difference (ambient temp minus storage temp), F deg*																	
Cork or equivalent in.	1	40	45	50	55	60	65	70	75	80	85	90	95	100	105	110	115	120
3	2.4	96	108	120	132	144	156	168	180	192	204	216	228	240	252	264	276	288
4	1.8	72	81	90	99	108	117	126	135	144	153	162	171	180	189	198	207	216
5	1.44	58	65	72	79	87	94	101	108	115	122	130	137	144	151	159	166	173
6	1.2	48	54	60	66	72	78	84	90	96	102	108	114	120	126	132	138	144
7	1.03	41	46	52	57	62	67	72	77	82	88	93	98	103	108	113	118	124
8	0.90	36	41	45	50	54	59	63	68	72	77	81	86	90	95	99	104	108
9	0.80	32	36	40	44	48	52	56	60	64	68	72	76	80	84	88	92	96
10	0.72	29	32	36	40	43	47	50	54	58	61	65	68	72	76	79	83	86
11	0.66	26	30	33	36	40	43	46	50	53	56	60	63	66	69	73	76	79
12	0.60	24	27	30	33	36	39	42	45	48	51	54	57	60	63	66	69	72
13	0.55	22	25	28	30	33	36	39	41	44	47	50	52	55	58	61	63	66
14	0.51	20	23	26	28	31	33	36	38	41	43	46	49	51	54	56	59	61
Single glass	27.0	1080	1220	1350	1490	1620	1760	1890	2030	2160	2290	2440	2560	2700	2840	2970	3100	3240
Double glass	11.0	440	500	550	610	660	715	770	825	880	936	990	1050	1100	1160	1210	1270	1320
Triple glass	7.0	280	320	350	390	420	454	490	525	560	595	630	665	700	740	770	810	840

Note: Where wood studs are used multiply the above values by 1.1

TABLE 13.18 Food Products Data

Product	Average Freezing Point F	Percent Water	Sp ht, Btu/(lb)(F deg) Above Freezing	Sp ht, Btu/(lb)(F deg) Below Freezing	Latent Heat of Fusion Btu/lb	Heat of Respiration Btu per (24 hr) (ton) (F deg)	
Corn (dried)		10.5	0.28	0.23	15		
Cucumbers	30.5	96.1	0.97	0.49	137		
Eggplant	30.4	92.7	0.94	0.47	132		
Endive (escarole	30.9	93.3	0.94	0.48	132		
Horseradish	26.4	73.4	0.78	0.42	104		
Kale	30.7	86.6	0.89	0.46	124		
Kohlrabi	30	90	0.92	0.47	128		
Lettuce	31.2	94.8	0.96	0.48	136	32	638
						40	7392
Mushrooms	30.2	91.1	0.93	0.47	130	32	6160
						50	22,000
Olives	28.5	75.2	0.80	0.42	108		
Onions	30.1	87.5	0.91	0.46	124	32	880
						50	1870
Parsnips	28.9	78.6	0.84	0.46	112		
Peas (green)	30	74.3	0.79	0.42	106	40	14,000
Peas (dried)		9.5	0.28	0.22	14		
Peppers (sweet)	30.1	92.4	0.94	0.47	132	40	1518
Potatoes (white)	28.9	77.8	0.82	0.43	111	32	660
						40	1430
Potatoes (sweet)	28.5	68.5	0.75	0.40	97	40	1710
Pumpkin	30.1	90.5	0.92	0.47	130		
Radishes	30.1	93.6	0.95	0.48	124		
Rhubarb	28.4	94.9	0.96	0.48	134		
Sauerkraut	26	89	0.92	0.47	129		
Spinach	30.3	92.7	0.94	0.48	132	40	8000
Squash	30.1	90.5	0.92	0.47	130		
Tomatoes (green)	30.4	94.7	0.95	0.48	134	60	6230
Tomatoes (ripening)	30.4	94.1	0.95	0.48	134	40	1260
Turnips	30.5	90.9	0.93	0.40	137	32	66
						40	572
Vegetables (mixed)	30	90	0.90	0.45	130		
MEATS AND FISH							
Bacon		20	0.50	0.30	29		
Beef (dried)		5–15	0.22–0.34	0.19–0.26	7–22		
Beef (fresh-lean)	29	68	0.77	0.40	100		
Beef (fresh-fat)	28		0.60	0.35	79		
Brined Meats			0.75				
Cod fish (fresh)	28		0.90	0.49	119		
Cut meats	29	65	0.72	0.40	95		
Fish (frozen)	28	70	0.76	0.41	101		
Fish (iced)		70	0.76	0.41	101		
Fish (dried)			0.56	0.34	65		
Hams and loins	27	60	0.68	0.38	86.5		
Lamb	29	58	0.67	0.30	83.5		
Livers	29	65.5	0.72	0.40	93.3		
Oyster (shell)	27	80.4	0.83	0.44	116		
Oysters (tub)	27	87	0.90	0.46	125		
Pork (fresh)	28	60	0.68	0.38	86.5		
Pork (smoked)		57	0.60	0.32			
Poultry (fresh)	27	74	0.79	0.37	106		
Poultry (frozen)	27	74	0.79	0.37	106		
Sausage (casings)			0.60				
Sausage (drying)	26	65.5	0.89	0.56	93		
Sausage (franks)	29	60	0.86	0.56	86		
Sausage (fresh)	26	65	0.89	0.56	93		
Sausage (smoked)	25	60	0.86	0.56	86		
Scallops	28	80.3	0.89	0.48	116		
Shrimp	28	70.8	0.83	0.45	119		
Veal	29	63	0.71	0.39	91		
MISCELLANEOUS							
Beer	28	92	1.0				
Bread		32–37	0.70	0.34	46–53		
Bread (dough)		58	0.75				
Butter	30–0	15	0.64	0.34	15		
Candy			0.93				
Caviar (tub)	20	55				40	3820
Cheese (American)	17	60	0.64	0.36	79	40	4680
Cheese (Camembert)	18	60	0.70	0.40	86	40	4920

TABLE 13.19 Heat Removed in Cooling Air to Storage Room Conditions

Storage room temp F	Temperature of outside air, F							
	85		90		95		100	
	Relative humidity, Percent							
	50	60	50	60	50	60	50	60
65	0.65	0.85	0.93	1.17	1.24	1.54	1.58	1.95
60	0.85	1.03	1.13	1.37	1.44	1.74	1.78	2.15
55	1.12	1.34	1.41	1.66	1.72	2.01	2.06	2.44
50	1.32	1.54	1.62	1.87	1.93	2.22	2.28	2.65
45	1.50	1.73	1.80	2.06	2.12	2.42	2.47	2.85
40	1.69	1.92	2.00	2.26	2.31	2.62	2.67	3.06
35	1.86	2.09	2.17	2.43	2.49	2.79	2.85	3.24
30	2.00	2.24	2.26	2.53	2.64	2.94	2.95	3.35

Storage room temp F	Temperature of outside air, F							
	40		50		90		100	
	Relative humidity, Percent							
	70	80	70	80	50	60	50	60
30	0.24	0.29	0.58	0.66	2.26	2.53	2.95	3.35
25	0.41	0.45	0.75	0.83	2.44	2.71	3.14	3.54
20	0.56	0.61	0.91	0.99	2.62	2.90	3.33	3.73
15	0.71	0.75	1.06	1.14	2.80	3.07	3.51	3.92
10	0.85	0.89	1.19	1.27	2.93	3.20	3.64	4.04
5	0.98	1.03	1.34	1.42	3.12	3.40	3.84	4.27
0	1.12	1.17	1.48	1.56	3.28	3.56	4.01	4.43
−5	1.23	1.28	1.59	1.67	3.41	3.69	4.15	4.57
−10	1.35	1.41	1.73	1.81	3.56	3.85	4.31	4.74
−15	1.50	1.53	1.85	1.92	3.67	3.96	4.42	4.86
−20	1.63	1.68	2.01	2.09	3.88	4.18	4.66	5.10
−25	1.77	1.80	2.12	2.21	4.00	4.30	4.78	5.21
−30	1.90	1.95	2.29	2.38	4.21	4.51	4.90	5.44

TABLE 13.20 Average Air Changes per 24 hr for Storage Rooms due to Door Openings and Infiltration (above 32 F)

Volume cu ft	Air changes per 24 hr	Volume cu ft	Air changes per 24 hr
200	44.0	6,000	6.5
300	34.5	8,000	5.5
400	29.5	10,000	4.9
500	26.0	15,000	3.9
600	23.0	20,000	3.5
800	20.0	25,000	3.0
1,000	17.5	30,000	2.7
1,500	14.0	40,000	2.3
2,000	12.0	50,000	2.0
3,000	9.5	75,000	1.6
4,000	8.2	100,000	1.4
5,000	7.2		

Note: For heavy usage multiply the above values by 2.
For long storage multiply the above values by 0.6.

TABLE 13.21 Heat Equivalent of Occupancy

Cooler temperature F	Heat equivalent/person Btu/hr
50	720
40	840
30	950
20	1,050
10	1,200
0	1,300
−10	1,400

Power Plants

by Craig W. Somerton

Introduction

Electric power generation in an essential element in a modern industrialized society. In fact many of the indicators of such a society –longevity, per capita income, infant mortality– are directly correlated with electricity production. Most electric power generation is achieved through the conversation of fossil fuel energy in either a steam power system or a gas turbine system. Nuclear power and hydroelectric power sources are significant, but still minor contributors to world wide electricity production. The focus of this chapter will be the fossil fuel based power plants. The chapter will begin with a brief review of thermodynamics, followed by a treatment on the components found in a fossil fueled power plant. Next, the combustion process will be explored which will include a discussion on fuels. Finally, system analysis will be introduced to determine plant efficiencies and operating conditions.

14.1 Review of Thermodynamics

Thermodynamics forms the basis of the tools used to analyze and design power plants. A brief review of basic thermodynamics should prove helpful in the power plant calculations that will follow. A review of Chapter 5 on Thermodynamics may be helpful here. The various components found in power plants are thermodynamically modeled as control volumes. Neglecting potential energy and kinetic energy changes, the first law of thermodynamics, which represents energy conservation, may be written for a control volume as follows:

$$\sum_{outlets} \dot{m}_k h_k - \sum_{inlets} \dot{m}_i h_i = \dot{Q}_{in} - \dot{W}_{out} \qquad (14.1.1)$$

where

$\sum_{outlets} \dot{m}_k h_k$ is the sum of the enthalpy flows leaving the control volume

$\sum_{inlets} \dot{m}_i h_i$ is the sum of the enthalpy flows entering the control volume

$\dot{Q}_{in}$ is the heat transfer rate into the control volume (negative for heat transfer rate out of the control volume)

$\dot{W}_{out}$ is the shaft power out of the control volume (negative for shaft power into the control volume)

With this expression we will be able to analyze the energy flows associated with all of the devices present in a power plant. A major facet of utilizing Eq.14.1.1 is the evaluation of the enthalpies. To properly evaluate the enthalpies we must determine which substance model will be used for the working fluids. There are three principal substance models utilized: ideal gas (e.g., air), compressible substance (e.g., steam), incompressible substance (e.g., water).

Most gases at normal pressures can be modeled as ideal gases. The enthalpy of an ideal gas can be determined two different ways. Quite often the enthalpy can be read from tables, such as the air tables. For an ideal gas the enthalpy only depends on the temperature, so it can be a simple matter of turning to the ideal gas table for the specific gas of interest and reading the enthalpy that corresponds to the temperature specified. The only care that needs to be exercised deals with the units of the enthalpy. Quite often ideal gas tables will provide enthalpy on a per mole basis rather than a per mass basis. To make the conversion to a per mass basis, one simply divides by the molecular weight of the gas. A second way to determine enthalpy, or more correctly enthalpy differences, is to use the average specific heat at constant pressure for the gas and write

$$h_2 - h_1 = c_{p,avg}(T_2 - T_1) \qquad (14.1.2)$$

where $c_{p,avg}$ is the constant pressure specific heat for the specified gas evaluated at the average temperature $(T_2 + T_1)/2$; we usually write $c_{p,avg} = c_p$. Quite often a value for the specific heat at only one temperature is known. It is a not an unreasonable approximation to use this constant value for the specific heat, especially if the temperature difference is relatively small.

A compressible substance model is used for substances for which phase change may play a role for the situation provided. Steam and refrigerants are the most common of the compressible substances. Properties for compressible substances are normally obtained from one of the four tables provided for the substance: saturation pressure table, saturation temperature table, superheated vapor table, compressed liquid table. The key is to determine the appropriate table to use which depends on the phase of the substance. Normally, five different phases are identified:

1. Subcooled or compressed liquid
2. Saturated liquid
3. Two phase mixture
4. Saturated vapor
5. Superheated vapor

It is known from the state principle that two independent properties must be known to fix the state of the system, so that if two properties are known the remaining properties can be determined. Consider that temperature T and pressure P are the two known properties. Then to determine the phase, one can go to the saturation pressure table and obtain the saturation temperature at the specified pressure. Recall that this is really the boiling point of the substance at that pressure, so the following conditions will indicate the fluid phase:

- Specified temperature less than saturation temperature
  —sub-cooled liquid
- Specified temperature greater then saturation temperature
  —superheated vapor
- Specified temperature equal to saturation temperature
  —phase indeterminate

Quite often the two known properties are either the temperature or the pressure and one other property such as the enthalpy or entropy. For these cases the phase may be identified by going to the appropriate saturation table, temperature or pressure depending on what has been specified, and obtaining the saturated liquid and vapor values for the other property specified. For example, if the enthalpy has been specified then the values for the enthalpy of saturated liquid h_f and saturated vapor h_g would be taken from the table. Comparing the specified value of the property to the saturated values will determine the phase as shown below (let z stand for the property):

$z_{specified} < z_f$ — subcooled liquid

$z_f = z_{specified}$ — saturated liquid

$z_f < z_{specified} < z_g$ — two phase mixture with quality,

$$x = \frac{z_{specified} - z_f}{z_g - z_f} = \frac{z_{specified} - z_f}{z_{fg}} \tag{14.1.3}$$

$z_{specified} = z_g$ — saturated vapor

$z_{specified} > z_g$ — superheated vapor

Most common liquids, such as water and oil, are modeled as incompressible liquids. For an incompressible liquid the specific volume (or density) depends only on the temperature, not the pressure. If the liquid also exists as a compressible substance, such as water, we can then use the saturation temperature table and evaluate the specific volume as v_f at the substance temperature. The enthalpy change for an incompressible liquid is given by

$$h_2 - h_1 = c_p(T_2 - T_1) + v_f(P_2 - P_1) \tag{14.1.4}$$

The second law of thermodynamics also has ramifications in power plant analyses. First, is the observation that energy delivered to a work engine by the heat transfer mechanism cannot be converted 100% into work. That is, there must always be some heat rejected by the system. The second observation is that energy which is moved by heat transfer can only move from a region of high temperature to a region of low temperature. That is, if the heat rejection process interacts with a heat reservoir, the system temperature at that point can be no less than the heat reservoir temperature. This condition also applies for any part of the system that has heat addition. The third observation is that the maximum thermal efficiency any work producing system can have is bounded by that produced by the equivalent Carnot cycle system. Finally , several devices used in power plant systems are reasonably approximated as isentropic or constant entropy devices. Hence, it is useful to review the determination of entropy and entropy changes for our three types of substances.

For an ideal gas, the entropy change can be written, assuming a constant specific heat

$$s_2 - s_1 = c_p \ln\left(\frac{T_2}{T_1}\right) - R\ln\left(\frac{P_2}{P_1}\right) \tag{14.1.5}$$

For an isentropic process it is often convenient to rewrite this equation in the following form

$$\frac{T_2}{T_1} = \left(\frac{P_2}{P_1}\right)^{(k-1)/k} \tag{14.1.6}$$

where k is the ratio of specific heats, c_p / c_v.

For a compressible substance, the entropy may be found in the tables using the procedure laid out for the enthalpy. For an incompressible substance the entropy change is given by

$$s_2 - s_1 = c_p \ln\left(\frac{T_2}{T_1}\right) \tag{14.1.7}$$

Note for an incompressible substance, an isentropic process is also an isothermal process.

14.2 Components

Several different components or devices operate within a power system to achieve the conversion of fossil fuel energy to electricity. These components may be formed into two groupings: devices involved with work addition or extraction (pumps, compressors, turbines), and devices involved with heat addition, extraction or transfer (boilers, reheaters, condensers, intercoolers, regenerators, feedwater heaters).

Work Devices

Work devices such as pumps, compressors, or turbines are modeled thermodynamically as control volume devices with no heat transfer. The thermodynamic model begins by assuming that the devices are isentropic (constant entropy). Since the exit pressure is normally given, the exit state for the ideal device is fixed and the exit enthalpy can be determined. The ideal work or power is then calculated using the first law equation. Next, the adiabatic efficiency is used to calculate the actual work or power associated with the device. The adiabatic efficiency is defined as

$$\eta_s = \frac{\dot{W}_{ideal}}{\dot{W}_{actual}} \quad \text{for turbines} \tag{14.2.1}$$

$$\eta_s = \frac{\dot{W}_{actual}}{\dot{W}_{ideal}} \quad \text{for pumps and compressors}$$

Once the actual work or power is determined using the adiabatic efficiency, the actual exit enthalpy can be determined form the first law equation. At this point, two properties are known at the exit state, so that any other property of interest can be determined.

Example 14.1

Steam at 250 psia and 600°F enters a turbine at 75,000 lb_m/hr. The steam exits at 20 psia and a quality of 85%. Determine the power output of the turbine.

Solution: It is useful to set up the following table.

State	T(°F)	P(psia)	Fluid Phase	h(BTU/lb_m)	$\dot{m}$ (lb_m/hr)
Inlet	600	250	Superheat	*1318*	75,000
Outlet	*228*	20	2 Phase x=0.85	1012	75,000

Note: *Italicized* values from steam tables in Chapter 5. Underlined values are calculated.

Since both the pressure and temperature are known at the inlet, the steam tables can provide the remaining information. At the outlet a two phase mixture exists, so that the temperature is the saturation temperature and the enthalpy are calculated from

$$h_{out} = h_{f,out} + x_{out}h_{fg,out}$$

$$= 196.3 + 0.85 \times 960.1 = 1012 \text{ BTU / lb}_m$$

The power output is calculated from the first law

$$\dot{W}_{out} = \dot{m}_{stm}\left(h_{in} - h_{out}\right) = (75{,}000)(1318 - 1012)$$

$$= 23 \text{ MBTU / hr} \quad \text{or} \quad 6.73 \text{ MW}$$

Example 14.2

Water at 100 kPa and 25°C enters a pump of adiabatic efficiency 73% at 4 kg/s. The pump provides a pressure boost to 250 kPa. Determine the pumping power required and the actual outlet temperature.

Solution: We begin by entering the operating data on the table below.

State	T(°C)	P(kPa)	$\dot{m}$ (kg/s)
Inlet	25	100	4
Outlet(ideal)	25	250	4
Outlet(actual)	25.013	250	4

Note: Underlined values are calculated.

To determine the actual power given an adiabatic efficiency requires us to first determine the ideal power. For the ideal pumping process, we have an isentropic process. Treating water as incompressible liquid tells us we also have an isothermal process. Then the ideal power is given by

$$\dot{W}_{ideal} = \dot{m}\left(h_{out} - h_{in}\right) = \dot{m}v_f\left(P_{out} - P_{in}\right)$$

$$= (4)(.001003)(250 - 100) = 0.60 \text{ kW}$$

Now applying the adiabatic efficiency

$$\dot{W}_{act} = \frac{\dot{W}_{ideal}}{\eta_s} = \frac{0.60}{0.73}$$
$$= 0.82 \text{ kW}$$

The actual exit temperature is then calculated from the first law:

$$\dot{m}\Delta h_{act} = \dot{W}_{act}$$
$$\dot{m}(c_p\Delta T + v\Delta P) = \dot{W}_{act}$$

This is written as

$$T_{out,act} = T_{in} + \frac{\left(\dot{W}_{act} / \dot{m}\right) - v_f\left(P_{out} - P_{in}\right)}{c_p}$$
$$= 25 + \frac{0.82 / 4 - (0.001003)(250 - 100)}{4.177} = 25.01°\text{C}$$

We usually ignore the temperature rise in liquids due to losses.

Heat Devices

Heat devices found in power systems include combustion chambers, boilers, feedwater heaters, and condensers. From a thermodynamics perspective, they are modeled as control volume devices with no work. If no specification is made otherwise, they are also assumed to be at constant pressure. Two types of problems are typical for these devices: (i) given the inlet and exit conditions, calculate the heat transfer required and (ii) given the inlet conditions and the heat transfer, determine the exit conditions. For the first type of problem the first law equation is used directly to calculate the heat transfer rate once the inlet and exit enthalpies are determined. For the second type of problem the exit enthalpy is calculated from the first law equation and with one other known exit property (normally the pressure) the exit state is fixed and all the remaining exit properties can be determined.

Example 14.3

A boiler operating at 160 psia provides a heat transfer rate of 40 MBTU/hr to steam with a flow rate of 35,000 lb_m/hr. If the inlet temperature is 250°F, determine the exit temperature of the steam.

Solution: We begin by entering the operating data on the following table:

State	T(ºF)	P(psia)	Fluid Phase	h(BTU/lb_m)	$\dot{m}$ (lb_m/hr)
Inlet	250	160	Sub.Liq.	*219*	35,000
Outlet	*673*	160	*Superheat*	1362	35,000

Since the state is fixed for the inlet, the steam tables can be used to determine the inlet enthalpy. Now applying the first law to solve for the enthalpy at the outlet

$$h_{out} = h_{in} + \frac{\dot{Q}_{in}}{\dot{m}_{steam}} = 219 + \frac{40 \times 10^6}{35{,}000}$$
$$= 1362 \text{ BTU / lb}_m$$

From the steam tables we find that at 160 psia and an enthalpy of 1362 BTU/lb_m, we have superheated vapor at 673°F.

Example 14.4

After a first stage of compression, air at 400 kPa and 400 K enters an intercooler with a flow rate of 3 kg/s. Determine the heat transfer rate required to cool the air to 320 K.

Solution: We begin by entering the operating data on the following table:

State	T(K)	P(kPa)	$\dot{m}$ (kg/s)
Inlet	400	400	3
Outlet	320	160	3

The device simply rejects heat; there is no work involved. So, apply the first law to determine the required heat transfer:

$$\dot{Q}_{out} = \dot{m}_{air} c_p (T_{in} - T_{out}) = (3)(1.0035)(400 - 320)$$

$$= 241 \text{ kW}$$

There are two other devices that must be treated somewhat differently than shown above: regenerator and open feedwater heaters. In a regenerator two fluid streams enter and all of the heat transfer occurs between the two fluid streams; the streams do not mix. The first law equation then becomes

$$\dot{q}_{internal} = (\dot{m} \cdot \Delta h)_{stream\ \#1} = (\dot{m} \cdot \Delta h)_{stream\ \#2} \qquad \textbf{(14.2.2)}$$

A regenerator is used in a gas turbine power system to recover some waste heat that would normally be rejected to the surrounds. Both fluid streams have the same mass flow rates and it is also assumed that both streams are ideal gases. The performance of a regenerator is then characterized by the effectiveness which is defined by

$$\varepsilon_{reg} = \frac{(T_{out} - T_{in})_C}{T_{H,in} - T_{C,in}} = \frac{(T_{in} - T_{out})_H}{T_{H,in} - T_{C,in}} \qquad \textbf{(14.2.3)}$$

where the subscript H denotes the hot fluid stream and C denotes the cold fluid stream. With the specification of an effectiveness the exit temperatures can be calculated. For an ideal regenerator the effectiveness is unity since $T_{H,in} = T_{C,out}$ and $T_{H,out} = T_{C,in}$.

An open feedwater heater is a direct contact heat exchanger in which subcooled liquid is heated to saturated liquid by mixing with a certain amount of steam. It is the amount of steam needed that is calculated for the device. Applying the first law to an open feedwater heater gives

$$\dot{m}_{out} h_f = (\dot{m} \cdot h)_{liquid,in} + (\dot{m} \cdot h)_{steam,in} \qquad \textbf{(14.2.4)}$$

Mass conservation gives

$$\dot{m}_{out} = \dot{m}_{liquid,in} + \dot{m}_{steam,in} \qquad \textbf{(14.2.5)}$$

So that the mass flow rate of steam required is

$$\dot{m}_{steam,in} = \frac{\dot{m}_{liquid,in}(h_f - h_{liquid,in})}{h_{steam,in} - h_f} \qquad \textbf{(14.2.6)}$$

Example 14.5

An open feedwater heater operates at 20 psia. The feedwater (FW) enters at 200°F and 100,000 lb_m/hr and is to be heated to saturated liquid with steam at 400°F. Determine the mass flow rate required.

Solution: We begin by entering the operating data on the following table:

State	T(°F)	P(psia)	Fluid Phase	h(BTU/lb_m)	$\dot{m}$ (lb_m/hr)
FW in	200	20	Sub.Liq.	*168*	100,000
FW out	*228*	20	Sat.Liq.	*196*	102,685
Steam in	400	20	Superheat	*1239*	2685

Since the state is fixed for each condition, the steam tables are used to complete the thermodynamic information on the table. The mass flow rate of steam required is then calculated from

$$\dot{m}_{steam,in} = \frac{\dot{m}_{FW,in}\left(h_f - h_{FW,in}\right)}{h_{steam,in} - h_f}$$

$$= \frac{(100,000)(196 - 168)}{1239 - 196} = 2685 \text{ lb}_m / \text{hr}$$

Then the mass flow rate of the feedwater leaving the heater is

$$\dot{m}_{FW,out} = 100,000 + 2685 = 102,685 \text{ lb}_m / \text{hr}$$

14.3 Fuels and Combustion Processes

The energy input for fossil fuel power plants derives from the combustion of hydrocarbon fuels with air. During the combustion of a hydrocarbon fuel, considerable energy is released as the chemical bonds between the carbon and hydrogen atoms are broken. An analysis of these combustion processes first requires a specification of the chemical reaction occurring. Next a first law analysis must be performed to determine the heat released, the combustion temperature, and the effect of the amount of air present on the process.

Fuels and their Chemical Reactions

In general, fossil fuel power plants burn one of three types of hydrocarbon fuels: natural gas, coal, or fuel oil. All three of these fuel types can be represented by an effective molecular formula, $C_xH_yO_z$. To affect the combustion, or chemical reaction, oxygen must be supplied. Normally, this oxygen is supplied by air. For oxygen supplied in this manner it is important to remember that for every mole of O_2 supplied, 3.76 moles of N_2 must accompany it. For simplicity, it is often assumed that complete combustion occurs which results in the products of combustion being CO_2 and H_2O. To describe the combustion process a chemical reaction equation must be written as follows:

$$C_xH_yO_z + aO_2 + a(3.76)N_2 \rightarrow bCO_2 + dH_2O + a(3.76)N_2 \qquad \textbf{(14.3.1)}$$

The above equation represents the *stoichiometric combustion* of the fuel as all of the oxygen is consumed. It is also important to note that the equation is written relative to the combustion of one mole of fuel. To fully describe the process the moles a of O_2 consumed, the moles b of CO_2 produced, and the moles d of H_2O produced must be determined. To determine these mole values, conservation of the various chemical elements is applied. Beginning with carbon, the right and left side of the reaction equation are balanced:

$$\text{C: } x = b \tag{14.3.2}$$

Next the hydrogen is balanced

$$\text{H: } y = 2d \tag{14.3.3}$$

Finally the oxygen is balanced

$$\text{O: } z + 2a = 2b + d \tag{14.3.4}$$

Solving yields

$$\begin{aligned} b &= x \\ d &= y/2 \\ a &= (2x + y/2 - z)/2 \end{aligned} \tag{14.3.5}$$

Quite often a fossil fuel is represented as a combination of two or more hydrocarbon compounds. In these cases the balancing of the chemical reaction equation still proceeds as outlined above.

Example 14.6

A fossil fuel consisting of 80% octane, C_8H_{18}, and 20% dodecane, $C_{12}H_{26}$, by volume, is burned in air. If the reaction is stoichiometric, determine the moles of air required to burn one mole of fuel.

Solution: Considering that one mole of fuel reacts, the general reaction equation may be written as:

$$(0.8)C_8H_{18} + (0.2)C_{12}H_{26} + aO_2 + a(3.76)N_2 \rightarrow bCO_2 + dH_2O + a(3.76)N_2$$

Now proceeding with the element balance,

$$\text{C: } (0.8)8 + (0.2)12 = b \quad => b = 8.8$$

$$\text{H: } (0.8)18 + (0.2)26 = 2d \quad => d = 9.8$$

$$\text{O: } 2a = 2b + d \quad => a = 13.7$$

Then the number of moles of air required are

$$a + a(3.76) \quad \text{or} \quad a = 65.21 \text{ moles.}$$

In actual combustion processes, it is not unusual to have air amounts other than stoichiometric. As will be seen later, increasing the air amount above stoichiometric is used to control the combustion temperature. To quantify the amount of air provided, the terms percent theoretical air (TA) or percent excess air (EA) are used. Theoretical air of 100% is equal to stoichiometric air. Excess air of 0% is also equal to stoichiometric air. Hence, a relationship exists between percent theoretical air and percent excess air,

$$EA = TA - 100 \tag{14.3.6}$$

Then, for example, 150% theoretical air is equal to 50% excess air.

The final parameter used to describe the combustion process is the air to fuel ratio, AF. This may be represented on either a mole basis or mass basis and is defined by

$$(AF)_{mole} = \frac{\text{moles of air burned}}{\text{moles of fuel burned}} \tag{14.3.7}$$

$$(AF)_{mass} = \frac{\text{mass of air burned}}{\text{mass of fuel burned}}$$

Recall that to convert moles to mass the molecular weight of the substance is used.

Example14.7

Propane, C_3H_8, is burned with 130% theoretical air. Determine the air to fuel ratio on both a per mole and per mass basis.

Solution: The chemical reaction equation is written

$$C_3H_8 + a(1.30)O_2 + a(1.30)(3.76)N_2 \rightarrow bCO_2 + dH_2O + a(0.30)O_2 + a(4.888)N_2$$

Now proceeding with the element balance,

$$\text{C: } 3 = b \quad => b = 3$$

$$\text{H: } 8 = 2d \quad => d = 4$$

$$\text{O: } 2(1.3)a = 2b + d + 2(0.3)a \quad => a = 5$$

First computing $(AF)_{mole}$:

$$(AF)_{mole} = \frac{\text{moles of air burned}}{\text{moles of fuel burned}} = \frac{(4.76)(1.3)(5)}{1} = 30.94$$

Now on a mass basis:

$$(AF)_{mass} = \frac{\text{moles of air burned} \times MW_{\text{air}}}{\text{moles of fuel burned} \times MW_{\text{fuel}}} = \frac{(4.76)(1.3)(5)(28.97)}{(1)(44.097)} = 20.33$$

First Law Analysis for Combustion Processes

The first law for a combustion process is written

$$H_{reactants} = H_{products} + Q_{released} \tag{14.3.8}$$

For a standard combustion process the enthalpies $H_{\text{reactants}}$ and H_{products} are defined as

$$H_{reactants} = h_{fuel} + a(TA)h_{O_2} + a(TA)(3.76)h_{N_2} \tag{14.3.9}$$

$$H_{products} = b\,h_{CO_2} + d\,h_{H_2O} + a(TA - 1)h_{O_2} + a(TA)(3.76)h_{N_2} \tag{14.3.10}$$

The enthalpies of specific compounds are composed of two parts. A base enthalpy associated with the chemical energy the compound has at a specified base temperature, called the enthalpy of formation, and a sensible enthalpy associated with the energy due to being at a temperature other than the base temperature. In general, it is written

$$h = h_f^o + c_P(T - T_o) \tag{14.3.11}$$

The base temperature, T_O, is normally taken at 25°C or 77°F. The enthalpy of formation for a number of compounds may be found in Table 14.1. Rather than an enthalpy of formation, quite often a heating value for the fuel is provided. The *heating value HV* is defined as the energy released when the fuel is completely burned, with the temperature of the reactants and the products being equal. Using the general chemical reaction equation the heating value is related to the enthalpies of formation as

$$HV = b\left(h_f^o\right)_{CO_2} + d\left(h_f^o\right)_{H_2O} - \left(h_f^o\right)_{fuel} \tag{14.3.12}$$

where the enthalpies of formation for N_2 and O_2 of zero have been employed. Two different heating values are often used. The higher heating value HHV is used when the water in the products is in the liquid phase, while the lower heating value LHV is used when the water in the products is in vapor form. The two heating values are related by

$$HHV = LHV + d\left(h_{fg}\right)_{H_2O} \tag{14.3.13}$$

Using the first law and enthalpy concepts given above in conjunction with the chemical reaction equation, one of three types of problems can be solved:

(i) determine the heat release given the product and reactant temperatures and the percent theoretical air.

(ii) determine the product temperature given the heat release, the reactant temperature, and the percent theoretical air.

(iii) determine the percent theoretical air given the heat release and the product and reactant temperatures.

Example 14.8

Methane, CH_4, is burned with 15% excess air. The reactants enter the combustion chamber at 85°F and the products exit the chamber at 300°F. Determine the heat transfer in the combustion chamber in BTU/lbmole of fuel.

Solution: We begin by writing the chemical reaction equation for the process:

$$CH_4 + a(1.15)O_2 + a(1.15)(3.76)N_2 \rightarrow bCO_2 + dH_2O + a(0.15)O_2 + a(4.324)N_2$$

Now proceed with the element balance:

$$\text{C:}\quad 1 = b \quad => b = 1$$

$$\text{H:}\quad 4 = 2d \quad => d = 2 \qquad \text{O:}\quad 2(1.15)a = 2b + d + 2(0.15)a \quad => a = 2$$

Table 14.1 Enthalpy of Formations and Specific Heats for Several Gases

Substance	Formula	Molecular Weight	SI Units	
			h_f^o (kJ/kgmole)	c_p (kJ/kgmole·°C)
Air		29	0	29.1
Oxygen	O_2	32	0	29.5
Nitrogen	N_2	28	0	29.2
Carbon Dioxide	CO_2	44	-393,520	37.0
Steam	H_2O	18	-241,820	33.7
Methane	CH_4	16	-74,850	36.1
Acetylene	C_2H_2	26	226,730	44.3
Ethane	C_2H_6	30	-84,680	53.0
Propane	C_3H_8	44	-103,850	73.9
Butane	C_4H_{10}	58	-126,150	99.6
Octane	C_8H_{18}	114	-208,450	195.1
Dodecane	$C_{12}H_{26}$	170	-291,010	

Substance	Formula	Molecular Weight	USCS Units	
			h_f^o (BTU/lbmole)	c_p (BTU/lbmole·°F)
Air		29	0	6.96
Oxygen	O_2	32	0	7.01
Nitrogen	N_2	28	0	6.94
Carbon Dioxide	CO_2	44	-169,300	8.93
Steam	H_2O	18	-104,040	8.01
Methane	CH_4	16	-32,210	8.51
Acetylene	C_2H_2	26	97,540	10.53
Ethane	C_2H_6	30	-36,420	12.81
Propane	C_3H_8	44	-44,680	17.91
Butane	C_4H_{10}	58	-54,270	24.07
Octane	C_8H_{18}	114	-107,530	46.63
Dodecane	$C_{12}H_{26}$	170	-125,190	

Now the chemical reaction equation may be rewritten as:

$$CH_4 + 2.3O_2 + 8.648N_2 \rightarrow CO_2 + 2H_2O + 0.30O_2 + 8.648N_2$$

The first law is written as

$$Q_{released} = H_{reactants} - H_{products}$$

The enthalpy of the products is given by

$$H_{products} = h_{CO_2} + 2h_{H_2O} + 0.3h_{O_2} + 8.694h_{N_2}$$

$$= [h_f^o + c_p(T_{prod} - T_o)]_{CO_2} + 2[h_f^o + c_p(T_{prod} - T_o)]_{H_2O}$$

$$+ 0.3[c_p(T_{prod} - T_o)]_{O_2} + 8.694[c_p(T_{prod} - T_o)]_{N_2}$$

$$= [(-169{,}300) + 8.934(300 - 77)] + 2[(-104{,}040) + 8.017(300 - 77)]$$

$$+ 0.3[7.006(300 - 77)] + 8.694[6.947(300 - 77)]$$

$$= -357{,}875 \text{ BTU / lbmole of fuel}$$

The enthalpies of formation and specific heats have been obtained from Table 14.1. For the reactants it is similarly written

$$H_{reactants} = h_{CH_4} + 2.3h_{O_2} + 8.694h_{N_2}$$

$$= [h_f^o + c_P(T_{reac} - T_o)]_{CH_4} + 2.3[c_P(T_{reac} - T_o)]_{O_2}$$

$$+ 8.694[c_P(T_{reac} - T_o)]_{N_2}$$

$$= [(-32{,}210) + 8.535(85 - 77)] + 2.3[7.006(85 - 77)]$$

$$+ 8.694[6.947(85 - 77)]$$

$$H_{reactants} = h_{CH_4} + 2.3h_{O_2} + 8.694h_{N_2}$$

$$= [h_f^o + c_p(T_{reac} - T_o)]_{CH_4} + 2.3[c_p(T_{reac} - T_o)]_{O_2}$$

$$+ 8.694[c_p(T_{reac} - T_o)]_{N_2}$$

$$= [(-32{,}210) + 8.535(85 - 77)] + 2.3[7.006(85 - 77)]$$

$$+ 8.694[6.947(85 - 77)]$$

$$= -31{,}530 \text{ BTU / lbmole of fuel}$$

Then the heat transfer in the combustion chamber must be

$$Q_{released} = -31{,}530 - (-357{,}875) = 326{,}345 \text{ BTU / lbmole of fuel}$$

Example 14.9

A specific coal type with effective molecular weight, $C_{54}H_4O_2$, and higher heating value of 28,500 kJ/kg is used in a combustion chamber of a steam power plant. Determine the maximum temperature of the combustion products, if the reactants enter the chamber at 60°C.

Solution: We begin by writing the chemical reaction equation for the process:

$$C_{54}H_4O_2 + a(1.00)O_2 + a(1.00)(3.76)N_2 \rightarrow bCO_2 + dH_2O + a(3.76)N_2$$

The theoretical air has been set to 100% as this will give the maximum product temperature. Now proceeding with the element balance,

$$\text{C: } 54 = b \quad => b = 54$$

$$\text{H: } 4 = 2d \quad => d = 2$$

$$\text{O: } 2 + 2a = 2b + d \quad => a = 54$$

Now the chemical reaction equation may be rewritten as:

$$C_{54}H_4O_2 + 54O_2 + 203N_2 \rightarrow 54CO_2 + 2H_2O + 203N_2$$

The maximum product temperature will also occur when the heat transfer is zero. Then the first law is written as

$$H_{products} = H_{reactants}$$

Using the heating value representation, this may be written in terms of temperature as

$$54c_{p,CO_2}(T_{prod} - T_o) + 2c_{p,H_2O}(T_{prod} - T_o) + 203c_{p,N_2}(T_{prod} - T_o)$$
$$= HHV + 54c_{p,O_2}(T_{reac} - T_o) + 203c_{p,N_2}(T_{reac} - T_o)$$

Solving for the product temperature gives

$$T_{prod} = T_o + \frac{HHV + 54c_{p,O_2}(T_{reac} - T_o) + 203c_{p,N_2}(T_{reac} - T_o)}{54c_{p,CO_2} + 2c_{p,H_2O} + 203c_{p,N_2}}$$

Using the specific heat values from Table 14.1 and a base state temperature of 25°C, yields

$$T_{prod} = 25 + \frac{(28,500 \times 684) + 54 \times 29.38(35) + 203 \times 29.11(35)}{54 \times 37.23 + 2 \times 33.73 + 203 \times 29.11} = 2500°\text{C}$$

14.4 Power System Analysis

The analysis of a power plant system can be done on two different levels. The first level, called an overall analysis, focuses on mass and energy flows in and out of the system and the various overall efficiencies of the system. The second level, called cycle analysis, focuses on the working fluids and their state changes due to the various devices within the system. From the cycle analysis results, overall efficiencies and energy and mass outputs can be determined. Two specific types of power systems will be considered: steam power systems, based on the Rankine cycle, and gas turbine power systems, based on the Brayton cycle.

Overall Analysis

The premise of an overall system analysis is to use definitions of the efficiencies of a power system and the first law applied on the complete system in conjunction with specified mass and energy inflows and outflows; this allows us to asses the viability of a system. We begin with some basic definitions.

First Law Thermal Efficiency: the net power output divided by the rate of heat transfer input to the system:

$$\eta_{th} = \frac{\dot{W}_{net}}{\dot{Q}_{in,total}} \quad (14.4.1)$$

Net Power Output: the power output of the work producing devices (turbines) minus the power input required for the work consuming devices (pumps and compressors):

$$\dot{W}_{net} = \dot{W}_{turbines} - \dot{W}_{pumps\ \&\ compressors} \quad (14.4.2)$$

Heat Transfer Rate Input: the energy gathered by the working fluid of the system in the furnace:

$$\dot{Q}_{in,total} = \sum \dot{m}_{WF} \cdot \Delta h_{WF} \quad (14.4.3)$$

Cogeneration Thermal Efficiency: Often in steam power systems some fraction of the steam produced by the boiler is extracted out of the system and used for heating, air conditioning, or process applications. It is defined as the ratio of total power output plus steam enthalpy flows out of the system to the heat transfer rate input for the system:

$$\eta_{cogen} = \frac{\dot{W}_{net} + \dot{H}_{steam,out}}{\dot{Q}_{in,total}} \quad (14.4.4)$$

Furnace Efficiency: the percent of fuel energy supplied in the furnace of the system that results in the heat transfer rate input to the working fluid of the system:

$$\eta_{furn} = \frac{\dot{Q}_{in,total}}{\dot{Q}_{combust}}$$

Fuel Energy Supplied: the mass flow rate of fuel times the fuel heating value:

$$\dot{Q}_{combust} = \dot{m}_{fuel} \cdot HV_{fuel} \quad (14.4.5)$$

Carnot Cycle Efficiency: the maximum possible thermal efficiency a system can possess:

$$\eta_{Carnot} = 1 - \frac{T_L}{T_H} \quad (14.4.6)$$

The temperature T_L is the low temperature of the system and T_H is the high temperature of the system.

First Law for the Working Fluid for Closed System (steam power system):

$$\dot{Q}_{in,total} = \dot{W}_{net} + \dot{Q}_{out,total} \quad (14.4.7)$$

First Law for the Working Fluid for a Cogeneration System (steam power system):

$$\dot{Q}_{in,total} = \dot{W}_{net} + \dot{Q}_{out,total} + \dot{H}_{steam,out} \quad (14.4.8)$$

First Law for the Working Fluid for an Open System (gas power system):

$$\dot{Q}_{in,total} + \dot{H}_{gas,in} = \dot{W}_{net} + \dot{Q}_{out,total} + \dot{H}_{gas,out} \quad (14.4.9)$$

With these definitions and relationships a variety of information concerning a power system may be determined. The following example demonstrates some of these calculations.

Example 14.10

A steam power system operates under the following conditions:

- -Water enters the boiler at 1400 psia and 400°F.
- -350,000 lb_m/hr of steam exits the boiler at 1400 psia and 900°F.
- -Coal with heating value of 11,000 BTU/lb_m is burned in an 83% efficient furnace.
- -Condenser rejects heat to cooling tower water at 65°F.

Determine

- a.) Coal consumption rate in tons per day.
- b.) Maximum power output possible in MW.
- c.) Minimum operating pressure of the condenser.
- d.) Minimum required heat rejection from the condenser in MBTU per hour.

Solution: a.) The coal consumption rate can be determined from an overall energy balance on the furnace. Assuming no reheating in the system,

$$\eta_{furn}(\dot{m} \times HV)_{coal} = \dot{m}_{stm}(h_{out} - h_{in})_{H_2O}$$

Solving for $\dot{m}_{coal}$ gives

$$\dot{m}_{coal} = \frac{\dot{m}_{stm}(h_{out} - h_{in})_{H_2O}}{\eta_{furn} \times HV_{coal}}$$

where h_{in} and h_{out} refer to the boiler. From the steam tables we find

$$h_{out,H_2O} = 1,322 \text{ BTU / lb}_\text{m} \text{ and } h_{in,H_2O} = 375 \text{ BTU / lb}_\text{m}$$

The consumption rate of coal required can then be calculated as

$$\dot{m}_{coal} = \frac{350,000 \times (1322 - 375)}{0.83 \times 11,000} = 36,303 \text{ lb}_\text{m} \text{ / hr or } 426 \text{ tons / day}$$

b.) The maximum power output would occur for a plant that had the maximum possible efficiency which is set by the Carnot cycle efficiency. Calculating the Carnot cycle efficiency gives

$$\eta_{Carnot} = 1 - \frac{T_L}{T_H} = 1 - \frac{65 + 460}{900 + 460} = 1 - \frac{525}{1360} = 0.614$$

The maximum power output is then calculated from

$$\dot{W}_{net,\max} = \eta_{Carnot} \cdot \dot{Q}_{in,total} = \eta_{Carnot} \cdot \dot{m}_{stm}(h_{out} - h_{in})_{H_2O}$$

$$= 0.614 \times 350,000 \times (1322 - 375)$$

$$= 203.5 \text{ MBTU / hr or } 59.6 \text{ MW}$$

c.) So as not to violate the second law of thermodynamics the lowest operating temperature possible in the condenser must be equal to the temperature of the reservoir to which the heat is rejected, or 65°F. The lowest possible operating pressure for the condenser must then be the saturation pressure at this temperature. From the steam tables, we find this pressure to be

$$P_{cond} = 0.319 \text{ psia}$$

d.) The minimum heat rejection required in the condenser will occur when the power output is maximized. The first law for the overall system yields

$$\dot{Q}_{out,\min} = \dot{Q}_{in} - \dot{W}_{net,\max} = \dot{m}_{stm}\left(h_{out} - h_{in}\right)_{H_2O} - \dot{W}_{net,\max}$$

$$= 350{,}000 \cdot (1322 - 375) \; - \; 203.5 = \; 128 \text{ MBTU / hr}$$

Cycle Analysis

As shown above, considerable useful information may be determined about a power system through an overall system analysis, however to obtain detailed information concerning the system it is necessary to determine the state changes of the working fluid as it flows through the system. A methodology for this sort of calculation has been developed and is presented below. It will be used in the next two sections to perform cycle analysis calculations.

Procedure For Cycle Analysis

1. The system layout is sketched. The devices representing the various processes are placed and connected according to the system description.
2. The nodes between the devices are numbered. These nodes represent locations within the system where the state of the working fluid is of interest.
3. A table is constructed with the following headings:

   For a compressible substance (Steam Turbine Power System):

   Node, Temperature, Pressure, Fluid Phase, Enthalpy, Entropy, Mass Flow Rate

   For an ideal gas (Gas Turbine Power System):

   Node, Temperature, Pressure, Mass Flow Rate
4. With the given operating conditions and system description, all known thermodynamic information is entered in the table.
5. Using the working fluid property tables or the appropriate equations, all obtainable thermodynamic information is added to the table.
6. The system is traversed, device by device, analyzing the fluid as it passes through each device. This analysis provides additional fluid properties, which when used in conjunction with step #5 systematically completes the table.
7. If there is insufficient information to fix the state at the node, continue on to the next node. After the cycle has been traversed, repeat the traversing until the state at each node is fixed.
8. With the completed table, system information (such as thermal efficiency and work produced) is calculated.

14.5 Steam Power Systems

All steam power systems are based upon the ideal Rankine cycle shown below.

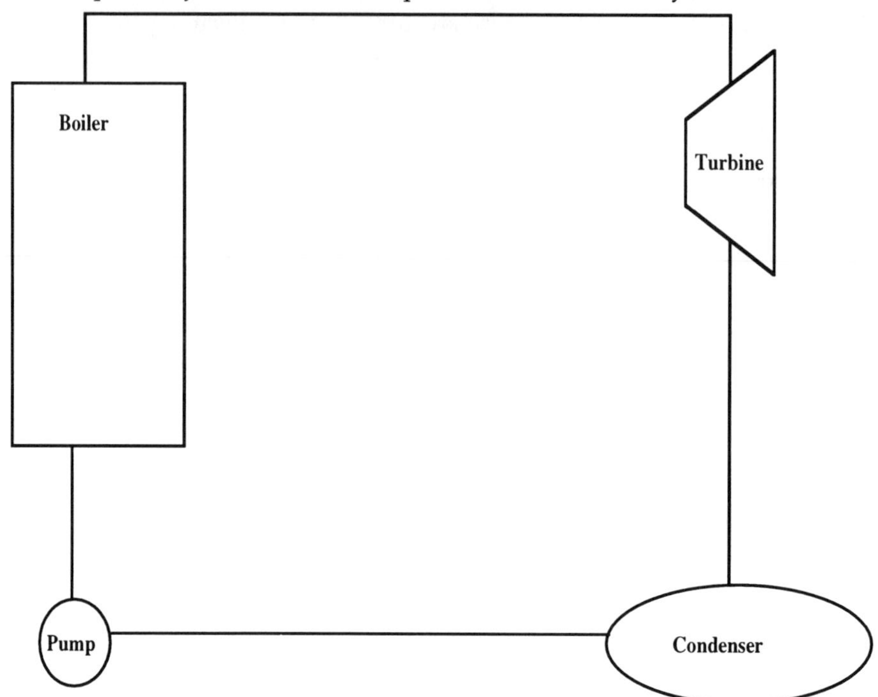

Figure 14.1 An Ideal Rankine Cycle

Compressed liquid enters the boiler, is heated, converted to steam, and normally leaves superheated. Ideally, the heat addition that occurs within the boiler is assumed to be at constant pressure. The high pressure, superheated steam then enters the turbine in which it expands to a lower pressure while work is extracted. Ideally, the turbine is an isentropic device. According to the second law of thermodynamics, no system can convert heat added completely into work output, that is, there must be some heat rejected; the condenser serves this purpose. In an ideal condenser a low pressure steam, close to saturation, is converted to a saturated liquid by heat rejection at constant pressure. Finally, the pressure of this saturated liquid is raised to the boiler pressure as it passes through the pump. An ideal pump is an isentropic device.

There are several ways to improve the efficiency of a Rankine cycle and some of these lead to the more extensive power plant layouts. We first consider adjusting the operating conditions of the ideal Rankine cycle to improve efficiency. Considering the Carnot cycle efficiency, it is apparent that to increase the efficiency we would like to maximize the boiler exit temperature and pressure and minimize the condenser temperature (and consequently its pressure). The minimum condenser temperature is fixed by the effectiveness of the condenser and, most importantly, the temperature of the heat rejection reservoir. The maximum temperature at the boiler exit is fixed by material/structural limits, the effectiveness of the boiler, and the combustion temperature of the fuel. The upper bound for the boiler pressure is fixed by material/structural limits.

The efficiency of a steam power system can also be improved by the addition of other devices. As the thermal efficiency of the Rankine cycle is increased by decreasing the condenser pressure, the quality of the steam leaving the turbine will also decrease. Wet steam can be physically very damaging to a turbine. Imagine wet steam consisting as a gas /liquid droplet suspension. As the steam pushes against the blades of the turbine, these liquid droplets traveling at rather large velocities collide with the blades in almost a sand blasting action. This sort of action is sure to lead to blade erosion and eventually mechanical failure of the turbine. Some way of maintaining the higher thermal efficiencies that result from lower condenser pressures, yet also maintaining high turbine exit steam qualities, normally around 95%, is sought. The solution to this problem is the use of reheating. In a reheat cycle, multiple turbines are used. In the first turbine, usually called the high pressure turbine, steam directly from the boiler is expanded down to a pressure which yields steam at the target exit quality. The steam exiting the high pressure turbine is then reheated in the boiler to a temperature close to the boiler exit temperature. The reheated steam is then expanded through a second turbine, again down to a pressure corresponding to the target exit quality. This process is continued until the expansion to the low condenser pressure produces a steam at or above the target quality. The vast majority of steam power systems operate with only one reheat leg and two turbines, a high pressure turbine and a low pressure turbine. An example of such a reheat cycle system is shown in Fig.14.2.

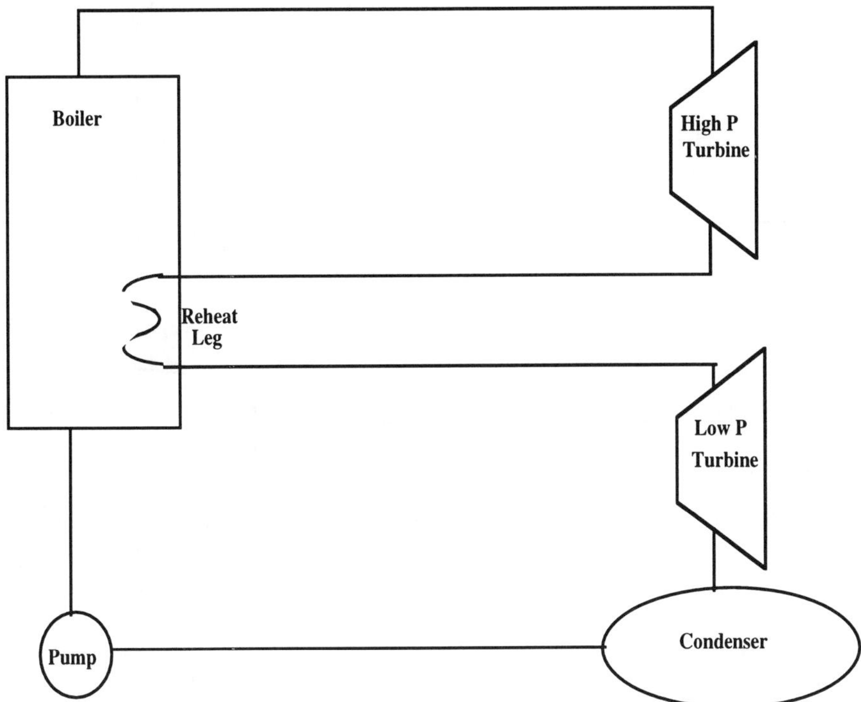

Figure 14.2 An Ideal Rankine Cycle with Reheat

Example 14.11

It is desired to compare the performance of a steam power plant that operates on the ideal Rankine cycle to one that operates as a reheat cycle. The following conditions apply to both power plants:

-Boiler exit conditions: 10 MPa and 600°C
-Ideal pumps and turbines
-Water at 20°C available for heat rejection
-Turbine exit quality must be 95% or greater

Part of the solution should involve specifying a condenser pressure for the Rankine cycle and reheat pressure and reheat exit temperature for the reheat cycle.

Solution: A cycle analysis will be performed for each system and the thermal efficiencies will then be compared.

Rankine Cycle

We begin with a sketch of the system noded appropriately.

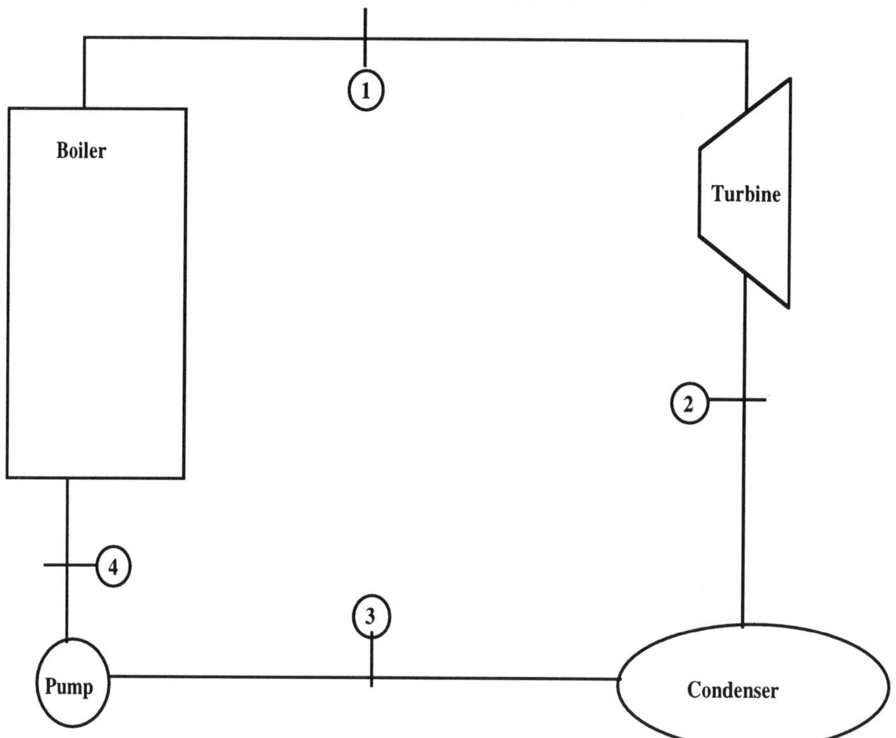

Now set up the node table and enter the operating conditions. Note that for simplicity the turbine exit quality (node #2) has been set equal to 100% (saturated vapor), which will satisfy the turbine constraints. The mass flow rate has also been set equal to 1.0 kg/s for ease of calculations. Since these calculations are of a linear nature due to the ideal pumps and turbines, heat transfer rate and power scale up accordingly.

Node	T(°C)	P(MPa)	Fluid Phase	h(kJ/kg)	s(kJ/kg·K)	$\dot{m}$ (kg/s)
1	600	10	Superheat	*3625*	*6.90*	1.0
2	*141.8*	*0.380*	Sat.Vap.	*2736*	6.90	1.0
3	*141.8*	0.380	Sat.Liq.	*597*	*1.76*	1.0
4	*141.8*	10	*Sub.Liq.*	607	1.76	1.0

The next step is use the steam tables to specify h and s at node #1. Now traverse the cycle:

1-2 Ideal Turbine For an ideal turbine we have

$$s_2 = s_1 = 6.90 \text{ kJ/kg·K}$$

With the entropy known at #2 and the condition that the steam is saturated vapor, the remaining properties can be obtained from the steam tables

$$P_2 = 0.380 \text{ MPa} \qquad T_2 = 141.8°\text{C} \qquad h_2 = 2736 \text{ kJ/kg}$$

Note: If we let $T_2 = 20°C$, the quality of state 2 would be much less than unity and hence unacceptable.

2-3 Condenser For an ideal condenser we have

$$P_3 = P_2 = 0.380 \text{ MPa}$$

With the pressure known and the condition of saturated liquid, the remaining properties can be obtained from the steam tables:

$$T_3 = 141.8°\text{C} \qquad h_3 = 597 \text{ kJ/kg} \qquad s_3 = 1.76 \text{ kJ/kg·K}$$

3-4 Ideal Pump For an ideal pump we have, assuming an incompressible liquid,

$$s_4 = s_3 = 1.76 \text{ kJ / kg·K}$$

$$h_4 = h_3 + v_3(P_4 - P_3) = 597 + (0.001082)(10\,000 - 380) = 607 \text{ kJ / kg}$$

With the node table now complete, system information can be calculated.

Net Power Output:

$$\begin{aligned} \dot{W}_{net} &= \dot{m}_1(h_1 - h_2) - \dot{m}_3(h_4 - h_3) \\ &= (1)(3625 - 2736) - (1)(607 - 597) = 879 \text{ kW} \end{aligned}$$

Total Heat Transfer Rate In:

$$\begin{aligned} \dot{Q}_{in} &= \dot{m}_1(h_1 - h_4) \\ &= (1)(3625 - 607) = 3018 \text{ kW} \end{aligned}$$

Thermal Efficiency:

$$\eta_{th} = \frac{\dot{W}_{net}}{\dot{Q}_{in}} = \frac{879}{3018} = 0.29$$

Note: The pump energy could have been neglected (i.e., $h_3 = h_4 = 597 \text{ kJ / kg}$) with no significant change in the thermal efficiency.

Reheat Cycle

We begin with a sketch of the system noded appropriately.

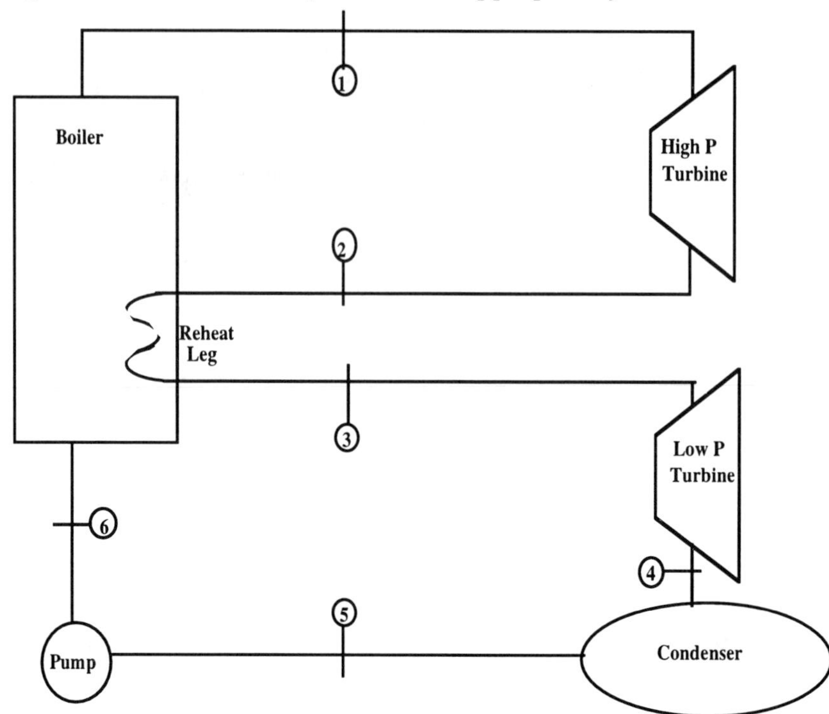

Now set up the node table and enter the operating conditions. Note that for simplicity the turbine exit qualities (nodes #2 and #4) have been set equal to 100% (saturated vapor), which will satisfy the turbine constraints. The mass flow rate has also been set equal to 1.0 kg/s for ease of calculations. Since we will want to take advantage of the cooling water available at 20°C, we will have our condenser operate at this temperature, if possible.

Node	T(°C)	P(MPa)	Fluid Phase	h(kJ/kg)	s(kJ/kg·K)	$\dot{m}$ (kg/s)
1	600	10	Superheat	*3625*	*6.90*	1.0
2	*141.8*	*0.380*	Sat.Vap.	*2736*	6.90	1.0
3	*677*	0.380	Superheat	*3876*	8.67	1.0
4	20	*0.00234*	Sat.Vap.	*2538*	*8.67*	1.0
5	20	*0.00234*	Sat.Liq.	*84*	*0.30*	1.0
6	*20*	10	Sub.Liq.	94	0.30	1.0

The next step is to use the stream tables and specify h and s at node #1. Also P, h, and s can be specified at nodes #4 and #5. Now traverse the cycle.

1-2 **Ideal Turbine** (High Pressure) For an ideal turbine we have

$$s_2 = s_1 = 6.90 \text{ kJ / kg} \cdot \text{K}$$

With the entropy known at #2 and the condition that the steam is saturated vapor, the remaining properties can be obtained from the steam tables:

$$P_2 = 0.380 \text{ MPa} \quad T_2 = 141.8°\text{C} \quad h_2 = 2736 \text{ kJ/kg}$$

2-3 **Reheat Leg** For an ideal reheat leg, the heating occurs at constant pressure so that,

$$P_3 = P_2 = 0.380 \text{ MPa}$$

Unfortunately, at this point we have insufficient information to continue. As we will see, the condenser conditions and the turbine exit quality constraint will give us the reheat leg exit temperature.

3-4 **Ideal Turbine** (Low Pressure) For an ideal turbine we use $x_4 = 1.0$ at 20° C. Then,

$$s_3 = s_4 = 8.67 \text{ kJ / kg} \cdot \text{K}$$

Knowing the pressure at node #3 allows us to use the steam tables and find

$$T_3 = 677°\text{C} \qquad h_3 = 3876 \text{ kJ/kg}$$

In practice, we may wish to have the maximum value for the exit reheat leg temperature to be the exit boiler temperature. If we wanted to apply this constraint, we would let node #3 be at 600°C and 380 kPa and then calculate whether the exit quality of the low pressure turbine would be acceptable. Carrying this calculation out we would have

$$s_4 = s_3 = 8.48 \text{ kJ / kg} \cdot \text{K}$$

which would lead to a quality at #4 of

$$x_4 = \frac{s_4 - s_{f,4}}{s_{fg,4}} = \frac{8.48 - 0.30}{8.37} = 0.98$$

which is still very acceptable.

5-6 **Ideal Pump** For an ideal pump, the enthalpy can be calculated by applying the first law:

$$\begin{aligned} h_6 &= h_5 + v_5(P_6 - P_5) \\ &= 84 + 0.001002 \times (10000 - 2.34) = 94 \text{ kJ / kg} \end{aligned}$$

With the node table now complete, the system information can be calculated.

Net Power Output:

$$\begin{aligned} \dot{W}_{net} &= \dot{m}_1(h_1 - h_2) + \dot{m}_3(h_3 - h_4) - \dot{m}_5(h_6 - h_5) \\ &= (1)(3625 - 2736) + (1)(3876 - 2538) - (1)(94 - 84) \\ &= 2217 \text{ kW} \end{aligned}$$

Total Heat Transfer Rate In:

$$\begin{aligned} \dot{Q}_{in} &= \dot{m}_1(h_1 - h_4) + \dot{m}_3(h_3 - h_2) \\ &= (1)(3625 - 94) + (1)(3876 - 2736) \\ &= 4671 \text{ kW} \end{aligned}$$

Thermal Efficiency:

$$\eta_{th} = \frac{\dot{W}_{net}}{\dot{Q}_{in}} = \frac{2217}{4671} = 0.47$$

Note the significant improvement in the thermal efficiency, from 29% to 47%, that reheating provides. This is due primarily to the reduced pressure (temperature) in the condenser allowed by the reheat process. Of course any decision as to what system to choose must also incorporate the additional equipment cost associated with the reheat cycle system.

Feedwater Heaters

The last modification to the Rankine cycle used to increase efficiency is the incorporation of feedwater heaters. It is observed that a good fraction of the heat transfer in at the boiler is used to bring the feedwater up to its boiling point. It is proposed to preheat the feedwater prior to its entry into the boiler utilizing some steam extracted from a turbine. Hence the pressure boost needed between the condenser and boiler is achieved in stages, with heating by extraction steam occurring after every stage. An example of a power system with one feedwater heater, or regeneration, is shown in Fig.14.3.

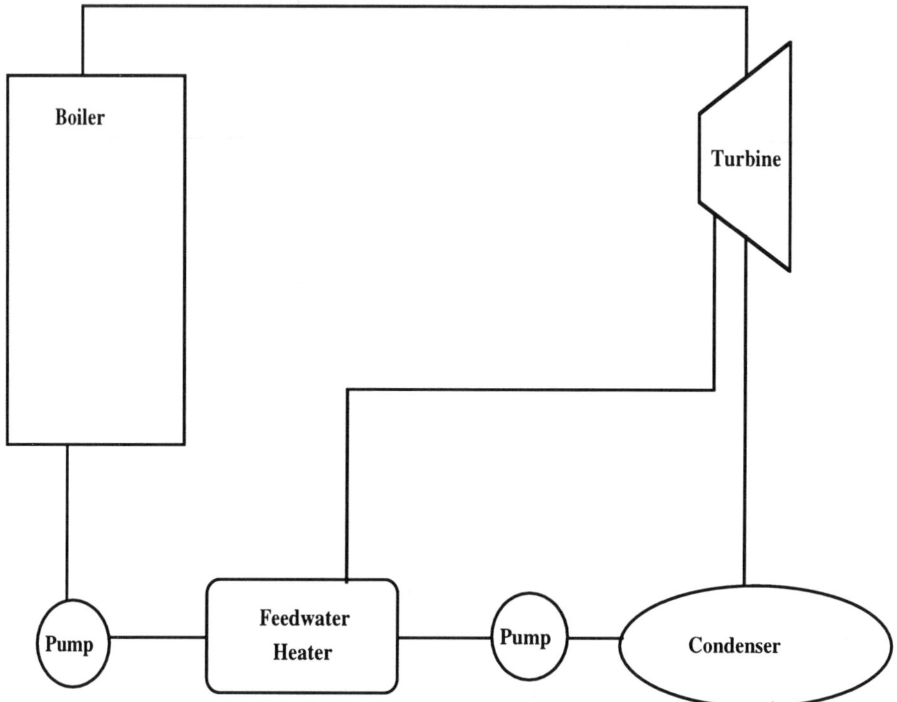

Figure 14.3 A Rankine Cycle with Regeneration

In general, there are two types of feedwater heaters. Open feedwater waters have direct mixing between the feedwater and extracted steam. The mass flow rate of extracted steam is calculated so as to produce saturated liquid leaving the heater. Open feedwater heaters are usually used at lower pressures. A closed feedwater heater is the more conventional heat exchanger in which the two fluids are not allowed to mix. The steam extraction is returned to the feedwater stream either upstream of the heater by the use of a steam trap or downstream of the heater by use of a pump.

In practice a combination of reheating, regeneration, maximizing boiler temperature and pressure, and minimizing condenser temperature are used in the thermal hydraulic design of a steam power system. The exact layout of the system depends mainly on the required power output, any plans for cogeneration, environmental constraints, and economics.

Gas Turbine Systems

The ideal Brayton cycle is the basis for the gas turbine power system. The ideal Brayton cycle is shown in Fig.14.4. The Brayton cycle is an open cycle in which the working fluid is taken in from the surrounds and exhausted to the surrounds. It consists of a compressor followed by a combustion chamber, and finally a turbine. Atmospheric air is sucked into the compressor where its pressure is significantly increased. In the combustion chamber fuel is added to the highly compressed air, the air-fuel mixture is ignited and combustion occurs which leads to a dramatic increase in the gas temperature. This energy is then extracted as the gas expands through a turbine back to the atmospheric pressure. The net power output is the turbine power minus that needed to power the compressor.

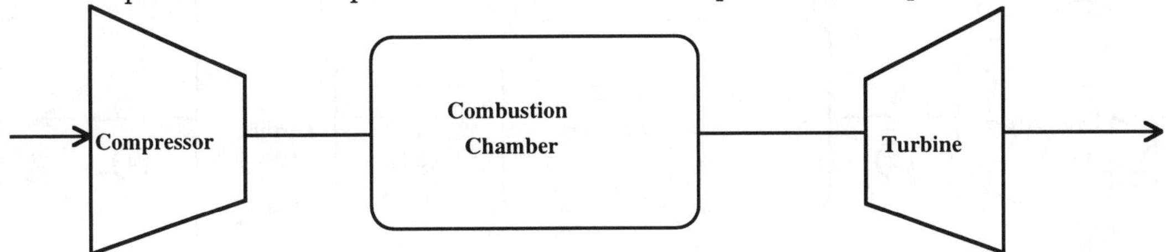

Figure 14.4 The Brayton Cycle

Three modifications of this simple system are employed in various ways to improve the efficiency and operation of the system: compressor staging and intercooling, regeneration, and turbine staging and reheating. All three of these modifications are shown in Fig.14.5.

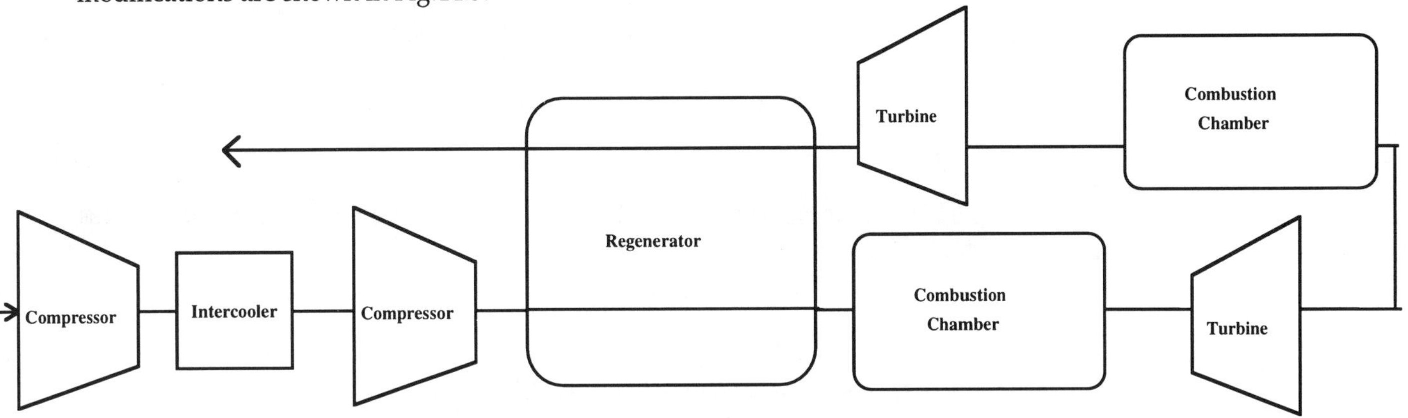

Figure 14.5 The Brayton Cycle with Modifications

It can be shown that multistage compression with intercooling reduces the work required for a specified compression ratio. Likewise, multistage expansion with reheating increases the power output from a specified expansion ratio. Both of these modifications are intended to make an isentropic process look more like an isothermal process, thereby approaching the Carnot cycle. The regenerator is used to recover some of the waste heat that would be rejected to the surrounds. The following example demonstrates some of these principles.

Example 14.12

Compare the efficiency of a simple Brayton cycle gas turbine system to a gas turbine system with intercooling and regeneration. The systems have the following operating conditions:

-Air enters the systems at 14.7 psia and 75°F
-The overall compression ratio is 9:1
-The exit temperature of the combustion chamber is 3000°F
-Intercooler exit temperature is 80°F
-Turbines and compressors are ideal
-Regenerator is 90% effective

Solution: A cycle analysis will be performed for each system and the thermal efficiencies will be compared.

Brayton Cycle

We begin with a sketch of the system noded appropriately.

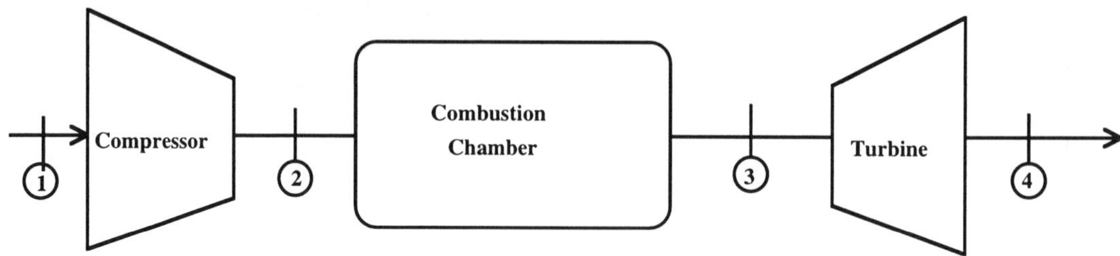

Now set up the node table and enter the operating conditions.

Node	*T*(°F)	*P*(psia)
1	75	14.7
2	542	132.3
3	3000	132.3
4	1387	14.7

(Underlined entries are from calculations.)

Now traverse the cycle.

1-2 **Ideal Compressor** An ideal compressor operates on an isentropic process, so that we may use

$$\left(\frac{T_2}{T_1}\right) = \left(\frac{P_2}{P_1}\right)^{(k-1)/k}$$

Using a specific heat ratio of 1.4 and solving for T_2 gives

$$T_2 = T_1\left(\frac{P_2}{P_1}\right)^{(k-1)/k} = (75+460)(9)^{(1.4-1)/1.4}$$

$$= 1002°\text{R} = 542°\text{F}$$

2-3 **Combustion Chamber** For an ideal combustion chamber we have

$$P_3 = P_2 = 132.3 \text{ psia}$$

3-4 **Ideal Turbine** For an ideal turbine we have an isentropic process, so that

$$T_4 = T_3\left(\frac{P_4}{P_2}\right)^{(k-1)/k} = (3000+460)\left(\frac{1}{9}\right)^{(1.4-1)/1.4}$$

$$= 1847°\text{R} = 1387°\text{F}$$

With the node table now complete, system information can be calculated.

Net Power Output:

$$W_{net} = W_T - W_C = c_p(T_3 - T_4) - c_p(T_2 - T_1)$$

$$= (0.24)(3000-1387) - (0.24)(542-75)$$

$$= 275 \text{ BTU / lb}_\text{m}$$

Total Heat Transfer Rate In:

$$Q_{in} = c_p(T_3 - T_2) = (0.24)(3000 - 542)$$
$$= 590 \text{ BTU / lb}_m$$

Thermal Efficiency:

$$\eta_{th} = \frac{W_{net}}{Q_{in}} = \frac{275}{590} = 0.47$$

Brayton Cycle with Intercooling, and Regeneration

We begin with a sketch of the system noded appropriately.

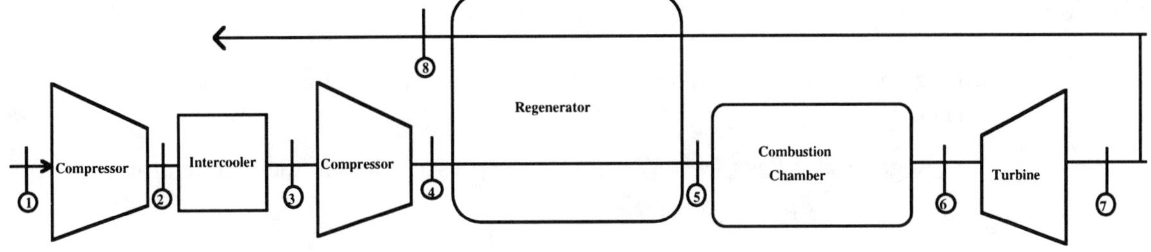

Now set up the node table and enter the operating conditions.

Node	*T*(°F)	*P*(psia)
1	75	14.7
2	272	44.1
3	80	44.1
4	279	132.3
5	1276	132.3
6	3000	132.3
7	1387	14.7
8	390	14.7

Now traverse the cycle.

1-2 **Ideal Compressor** To make the most of the staging and intercooling it has been found that the intercooling pressure should be given by

$$P_2 = \sqrt{P_4 \cdot P_1} = \sqrt{132.3 \times 14.7} = 44.1 \text{ psia}$$

For an ideal compressor we have an isentropic process:

$$T_2 = T_1\left(\frac{P_2}{P_1}\right)^{(k-1)/k} = (75 + 460)(3)^{(1.4-1)/1.4}$$
$$= 732°\text{R} = 272°\text{F}$$

2-3 **Intercooler** For an ideal intercooler we have

$$P_3 = P_2 = 44.1 \text{ psia}$$

3-4 **Ideal Compressor** For an ideal compressor we have an isentropic process:

$$T_4 = T_3\left(\frac{P_4}{P_3}\right)^{(k-1)/k} = (80 + 460)(3)^{(1.4-1)/1.4}$$
$$= 739°\text{R} = 279°\text{F}$$

4-5 Regenerator Pass One We assume no pressure loss in the regenerator, so that

$$P_5 = P_4 = 132.3 \text{ psia}$$

Using the regenerator effectiveness

$$\varepsilon_{reg} = \frac{T_5 - T_4}{T_7 - T_4}$$

But this equation has two unknowns, T_5 and T_7, at this point in our cycle analysis, so we will need to continue and return to this process later in our calculations.

6-7 Ideal Turbine For an ideal turbine we have an isentropic process, so that

$$T_7 = T_6\left(\frac{P_7}{P_6}\right)^{(k-1)/k}$$

But again we have two unknowns, T_7 and P_7 so we must move on and return to this later.

7-8 Regenerator Pass Two We assume no pressure loss in the regenerator, so that

$$P_7 = P_8 = 14.7 \text{ psia}$$

Unfortunately, we cannot proceed further without knowing T_7.

We now proceed around the system a second time in an attempt to complete our node table.

4-5 Regenerator Pass Two We still do not know T_7, so we must continue.

6-7 Ideal Turbine We now know P_7, so that we can calculate T_7 by

$$T_7 = (3000 + 460)\left(\frac{1}{9}\right)^{(1.4-1)/1.4} = 1847°\text{R} = 1387°\text{F}$$

7-8 Regenerator Pass Two Now applying our effectiveness, we obtain

$$T_8 = T_7 - \varepsilon_{reg}(T_7 - T_4)$$
$$= 1387 - (0.9)(1387 - 279) = 390°\text{F}$$

4-5 Regenerator Pass Three Finally we can obtain

$$T_5 = T_4 + \varepsilon_{reg}(T_7 - T_4)$$
$$= 279 - (0.9)(1387 - 279) = 1276°\text{F}$$

With the node table now complete, system information can be calculated.

Net Power Output:

$$W_{net} = c_p(T_6 - T_7) - c_p(T_2 - T_1) - c_p(T_4 - T_3)$$
$$= (0.24)(3000 - 1387) - (0.24)(272 - 75) - (0.24)(279 - 80)$$
$$= 292 \text{ BTU / lb}_m$$

Total Heat Transfer Rate In:

$$Q_{in} = c_p(T_6 - T_5) = (0.24)(3000 - 1276)$$
$$= 414 \text{ BTU / lb}_m$$

Thermal Efficiency:

$$\eta_{th} = \frac{W_{net}}{Q_{in}} = \frac{292}{414} = 0.71$$

We see a significant improvement in the efficiency, from 47% to 71%. However, any decision between the two systems must also include the efficiencies of the compressors and turbines and the additional equipment cost for compressor staging, intercooling, and regeneration.

Practice Problems (Essay)

14.1 Air enters a compressor with a pressure ratio of 5:1 and an adiabatic efficiency of 93%. The inlet conditions are 40 psia and 240°F. Determine the exit temperature.

14.2 Steam at 15 MPa and 500°C enters a two stage steam turbine at 150 kg/s. The first stage has an efficiency of 90% and an exit pressure of 1 MPa. After the first stage 50 kg/s of steam is bled off for process use. The second stage has an efficiency of 85% and an exit pressure of 400 kPa. Determine the power output of the turbine.

14.3 Two streams of air both flowing at 50,000 lb_m/hr enter a regenerator of effectiveness 0.83. One air stream enters at 220°F, while the other enters at 1500°F. Determine the two exit temperatures and the heat transfer in BTU/hr that is moved between the two streams.

14.4 Coal with a heating value of 25,000 kJ/kg is consumed in a furnace at a rate of 10 kg/s to produce steam at a flow rate of 75 kg/s. If the steam enters at 20 MPa and 340°C, what is the exit temperature of the steam for a furnace efficiency of 73%.

14.5 A natural gas composed of 70% methane, CH_4, and 30% propane, C_3H_8, is burned in the combustion chamber of a gas turbine system to produce a hot gas at 2800°F. Determine the percent excess air required and the required molar air to fuel ratio. The fuel and air may be assumed to enter the combustion chamber at 77°F.

14.6 What is the heat release available in BTU/lb_m when a coal with effective molecular formula $C_{65}H_{22}O_{12}$ and heating value 6,500 BTU/lb_m is burned with 110% theoretical air? The coal and air enter the combustion chamber at 150°F and the combustion gases leave at 500°F.

14.7 A gas turbine system with intercooling operates under the following conditions:

-Air enters at 25°C and 5 kg/s
-Combustion gases leave the combustion chamber at 1500°C
-Heat transfer to the gas in the combustion chamber is 5 MW
-Net power output is 3 MW
-Air leaves the system at 200°C

Determine

a) Thermal Efficiency
b) Maximum Efficiency
c) Heat rejected by the intercoolers

Practice Problems (Multiple Choice)

Questions 14.8 – 14.15

A steam cogeneration facility is laid out as shown below. The following operating conditions exist.

- -Steam at 800°F and 800 psia is produced in the boiler at 375,000 lb_m
- -Steam for heating is extracted at 100 psia and 125,000 lb_m/hr
- -The condenser pressure is 10 psia
- -The reheater pressure is 100 psia and its exit temperature is 750°F
- -The high pressure open feedwater heater operates at 300 psia
- -The low pressure open feedwater heater operates at 60 psia
- -The process steam returns as saturated liquid
- -All devices are assumed to be ideal

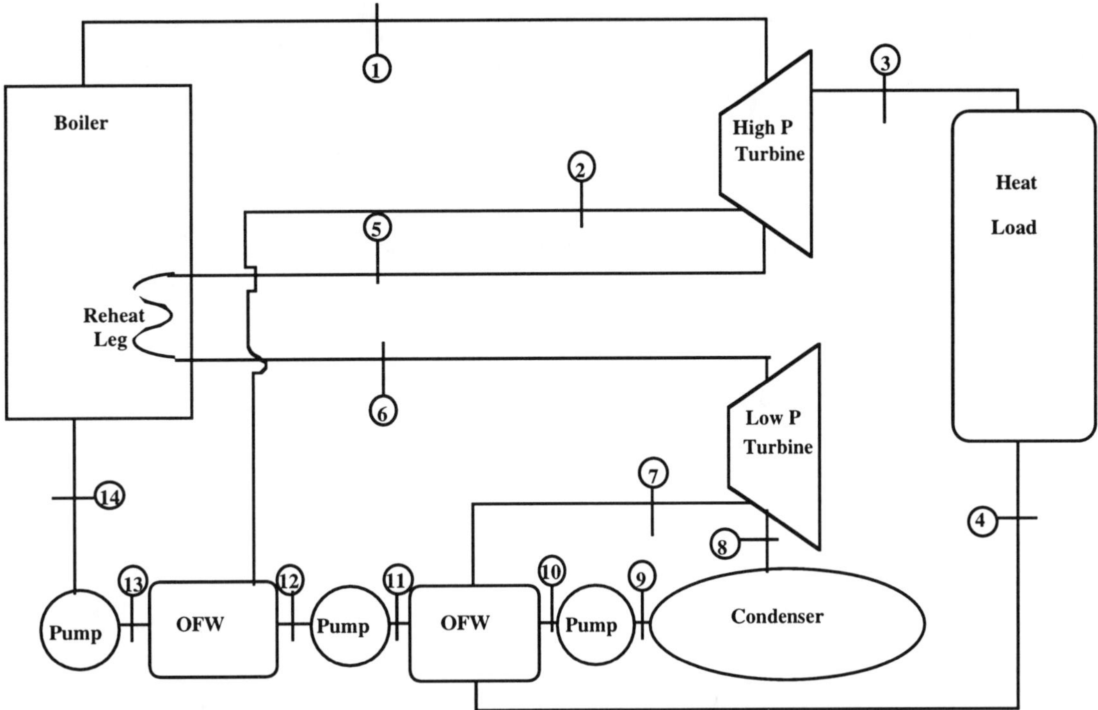

14.8 The first law thermal efficiency is

A) 32%
b) 17%
c) 40%
d) 28%

14.9 The exit quality of the low pressure turbine is

a) Not Defined
b) 90%
c) 98%
d) 95%

14.10 The net power output is (in MBTU/hr)

a) 74
b) 90
c) 118
d) 200

14.11 Percent of the boiler heat transfer used to produce the process steam is

a) 44%
b) 40%
c) 27%
d) 35%

14.12 Steam consumed by the feedwater heaters is (in lb_m/hr)

a) 71,418
b) 48,000
c) 17,334
d) 65,300

14.13 Total pump power is (in MBTU/hr)

a) 1.02
b) 0.94
c) 0.66
d) 0.83

14.14 Heat transfer in heat load is (in MBTU/hr)

a) 157
b) 96
c) 116
d) 202

14.15 For a furnace efficiency of 87% and a natural gas with energy content of 22,000 BTU/lb_m, the amount of natural gas required is (in lb_m/hr)

a) 16,000
b) 22,000
c) 25,000
d) 19,000

Solutions to Problems

14.1 We begin by entering the operating data on the table below:

State	T(°F)	P(psia)
Inlet	240	40
Outlet(ideal)	649	200
Outlet(actual)	680	200

The outlet pressure is given by

$$P_{out} = 5 \times P_{in} = 5 \times 40 = 200 \text{ psia}$$

We use our isentropic relationship to calculate the ideal exit temperature

$$T_{2s} = T_1\left(\frac{P_2}{P_1}\right)^{(k-1)/k} = (240+460)(5)^{(1.4-1)/1.4}$$

$$= 1109°\text{R} = 649°\text{F}$$

The actual work is given by

$$W_{act} = \frac{W_{ideal}}{\eta_s} = \frac{c_p(T_{2s} - T_1)}{h_s} = \frac{0.24(649-240)}{0.93}$$

$$= 105.5 \text{ BTU / lb}_\text{m}$$

and the actual exit temperature is then calculated as

$$T_2 = T_1 + \frac{W_{act}}{c_p} = 240 + \frac{105.5}{0.24} = 680°\text{F}$$

14.2 We begin by entering the operating data on the table below.

State	T(°C)	P(MPa)	Fluid Phase	h(kJ/kg)	s(kJ/kg K)	$\dot{m}$ (kg/s)
Inlet-1	500	15	*Superheat*	*3309*	*6.34*	150
Outlet-1s	*180*	1	*2 Phase* x=0.94	2657	6.34	150
Outlet-1a	*180*	1	*2 Phase* x=0.97	2722	6.46	150
Inlet-2	*180*	1	2 Phase x=0.97	2722	6.46	100
Outlet-2s	*144*	0.4	2 Phase x=0.91	2546	6.46	100
Outlet-2a	*144*	0.4	2 Phase x=0.92	2572	6.49	100

Italicized values from steam tables. Underlined values are calculated.

At the inlet to stage 1, we know the temperature and pressure , so that the remaining properties can be obtained from the steam tables. The ideal outlet conditions for stage 1 are given by

$$s_{O1s} = s_{I1} = 6.34 \text{ kJ / kg} \cdot \text{K}$$

Then from the steam tables we find

$$x_{O1s} = \frac{6.34 - 2.14}{4.45} = 0.94$$

which gives a temperature of 180°C and an enthalpy of 2657 kJ/kg for the ideal outlet to stage 1. For the actual enthalpy for the outlet of stage 1, we use the adiabatic efficiency to yield

$$h_{O1a} = h_{I1} - \eta_s(h_{I1} - h_{O1s}) = 3309 - 0.90(3309 - 2657)$$
$$= 2722 \text{ kJ / kg}$$

From the steam tables we can determine this is a two phase mixture with quality 0.97, entropy of 6.46 kJ/kg·K, and temperature of 180°C. The inlet conditions for stage 2 are the same as for the actual exit of stage 1 and they are entered onto the table appropriately. For the ideal outlet for stage 2 we have

$$s_{O2s} = s_{I2} = 6.46 \text{ kJ / kg} \cdot \text{K}$$

Then from the steam tables we find

$$x_{O2s} = \frac{6.46 - 1.78}{5.12} = 0.91$$

which gives a temperature of 144°C and an enthalpy of 2546 kJ/kg for the ideal outlet to stage 2. For the actual enthalpy for the outlet of stage 2 we use the adiabatic efficiency to yield

$$h_{O2a} = h_{I2} - \eta_s(h_{I2} - h_{O2s}) = 2722 - 0.85(2722 - 2546)$$
$$= 2572 \text{ kJ / kg}$$

From the steam tables we can determine this is a two phase mixture with quality 0.92, entropy of 6.49 kJ/kg·K, and temperature of 144°C. The total power output is then

$$\dot{W}_{tot} = \dot{m}_{I1}(h_{I1} - h_{O1a}) + \dot{m}_{I2}(h_{I2} - h_{O2a})$$
$$= 150(3309 - 2722) + 100(2722 - 2572)$$
$$= 103 \text{ MW}$$

14.3 We begin by entering the operating data on the table below

State	***T*(°F)**	**$\dot{m}$ (lb$_m$/hr)**
Cold In	220	50,000
Cold Out	1282	50,000
Hot In	1500	50,000
Hot Out	438	50,000

Using the definition of our regenerator effectiveness, we calculate

$$T_{c,out} = T_{c,in} + \varepsilon_{reg}(T_{h,in} - T_{c,in}) = 220 + 0.83(1500 - 220)$$
$$= 1282°\text{F}$$

and

$$T_{h,out} = T_{h,in} - \varepsilon_{reg}(T_{h,in} - T_{c,in}) = 1500 - 0.83(1500 - 220)$$
$$= 438°\text{F}$$

The internal heat transfer rate is

$$\dot{Q}_{int} = \dot{m}_h c_p \left(T_{h,in} - T_{h,out}\right) = 50000 \times 0.24(1500 - 438)$$
$$= 12.7 \text{ MBTU / hr}$$

14.4 The energy delivered to the steam is given by

$$\dot{Q}_{stm} = h_{furn} \dot{m}_{coal} HV = (0.73)(10)(25\,000)$$
$$= 182\,500 \text{ kW}$$

The steam exit enthalpy is then calculated

$$h_{out} = h_{in} + \frac{\dot{Q}_{stm}}{\dot{m}_{stm}} = 1594 + \frac{182\,500}{75} = 4027 \text{ kJ / kg}$$

Then from the steam tables we find an exit temperature of 784°C.

14.5 Considering that one mole of fuel reacts, the general reaction equation may be written as:

$$(0.7)\mathrm{CH_4} + (0.3)\mathrm{C_3H_8} + a(TA)\mathrm{O_2} + a(TA)(3.76)\mathrm{N_2}$$
$$\rightarrow b\mathrm{CO_2} + d\mathrm{H_2O} + a(TA-1)\mathrm{O_2} + a(TA)(3.76)\mathrm{N_2}$$

Now proceeding with the element balance,

$$\text{C:} \quad (0.7) + (0.3)3 = b \quad => b = 1.6$$

$$\text{H:} \quad (0.7)4 + (0.3)8 = 2d \quad => d = 2.6$$

$$\text{O:} \quad 2a = 2b + d \quad => a = 2.9$$

The chemical reaction equation is then rewritten as

$$(0.7)\mathrm{CH_4} + (0.3)\mathrm{C_3H_8} + 2.9(TA)\mathrm{O_2} + 2.9(TA)(3.76)\mathrm{N_2}$$
$$\rightarrow 1.6\mathrm{CO_2} + 2.6\mathrm{H_2O} + 2.9(TA-1)\mathrm{O_2} + 2.9(TA)(3.76)\mathrm{N_2}$$

To determine the percent excess air, or equivalently the percent theoretical air, we must apply the first law and solve for the percent theoretical air required for the exit temperature. The first law is written

$$H_{products} = H_{reactants}$$

The above equation assumes adiabatic combustion. We also assume that the fuel and air enter at 77°F and write

$$1.6\left\{h_f^o + c_P(T_{prod} - T_o)\right\}_{CO_2} + 2.6\left\{h_f^o + c_P(T_{prod} - T_o)\right\}_{H_2O}$$
$$+ 2.9(EA)c_{P,O_2}(T_{prod} - T_o) + 10.9(1 + EA)c_{P,N_2}(T_{prod} - T_o)$$
$$= 0.70h^o_{f,CH_4} + 0.30h^o_{f,C_3H_8}$$

Solving for EA:

$$EA = \left\{0.70h^o_{f,CH_4} + 0.30h^o_{f,C_3H_8} - 1.6\left[h_f^o + c_P(T_{prod} - T_o)\right]_{CO_2}\right.$$
$$\left. - 2.6\left[h_f^o + c_P(T_{prod} - T_o)\right]_{H_2O} - 10.9c_{P,N_2}(T_{prod} - T_o)\right\}$$
$$\div \left\{10.9c_{P,N_2}(T_{prod} - T_o) + 2.9c_{P,O_2}(T_{prod} - T_o)\right\}$$

Substituting in numerical values:

$$EA = \{0.70(-32,210) + 0.30(-44,680) - 1.6[(-169,300) + (8.93)(2723)] - 2.6[(-104,040) + (8.01)(2723)] - 10.9(6.94)(2723)\} \div \{10.9(6.94)(2723) + 2.9(7.01)(2723)\}$$

which gives a required excess air of 78%. The molar air to fuel ratio is found to be 24.6.

14.6 We begin by writing the chemical reaction equation for the process.

$$C_{65}H_{22}O_{12} + a(1.10)O_2 + a(1.10)(3.76)N_2 \rightarrow bCO_2 + dH_2O + a(0.10)O_2 + a(1.10)(3.76)N_2$$

The theoretical air has been set to 100% as this will give the maximum product temperature. Now proceeding with the element balance,

$$C:\ 65 = b \quad \Rightarrow b = 65$$

$$H:\ 22 = 2d \quad \Rightarrow d = 11$$

$$O:\ 12 + 2a = 2b + d \quad \Rightarrow a = 64.5$$

Now the chemical reaction equation may be rewritten as:

$$C_{65}H_{22}O_{12} + 71O_2 + 266.8N_2 \rightarrow 65CO_2 + 11H_2O + 6.5O_2 + 266.8N_2$$

The first law can be solved for the heat transfer to give

$$Q_{released} = H_{reactants} - H_{products}$$

The enthalpy of the products is given by

$$Q_{released} = HHV \cdot MW_{fuel} + 71c_{P,O_2}(T_{reac} - T_o) + 266.8c_{P,N_2}(T_{reac} - T_o) - 65c_{P,CO_2}(T_{prod} - T_o) - 11c_{P,H_2O}(T_{prod} - T_o) - 6.5c_{P,O_2}(T_{prod} - T_o) - 266.8c_{P,N_2}(T_{prod} - T_o)$$

Substituting numerical values

$$Q_{released} = 6500 \times 994 + 71(7.01)(73) + 266.8(6.94)(73) - 65(8.93)(423) - 11(8.01)(423) - 6.5(7.01)(423) - 266.8(6.94)(423)$$

for a heat release of 5.55 MBTU/lb_m of coal.

14.7 a) The thermal efficiency is given by

$$\eta_{th} = \frac{\dot{W}_{net}}{\dot{Q}_{in,total}} = \frac{3}{5} = 0.60$$

b) The maximum efficiency is given by the Carnot cycle efficiency

$$\eta_{max} = 1 - \frac{T_L}{T_H} = 1 - \frac{25 + 273}{1500 + 273} = 0.83$$

c) Applying the first law overall on the system gives

$$
\begin{aligned}
\dot{Q}_{intclr} &= \dot{Q}_{in,total} + \dot{H}_{gas,in} - \dot{W}_{net} - \dot{H}_{gas,out} \\
&= \dot{Q}_{in,total} + \dot{m}_{gas} c_P (T_{in} - T_o) - \dot{W}_{net} - \dot{m}_{gas} c_P (T_{in} - T_o) \\
&= 5000 + (5)(1.00)(25-25) - 3000 - (5)(1.00)(200-25) \\
&= 1.125 \text{ MW}
\end{aligned}
$$

Questions 14.8–14.15

We begin by performing a cycle analysis which results in the following table:

Node	T(°F)	P(psia)	Phase	h(BTU/lb$_m$)	s(BTU/lb$_m$·°R)	$\dot{m}$ (lb$_m$/hr)
1	800	800	Superheat	*1398*	*1.60*	375,000
2	*553*	300	*Superheat*	*1288*	1.60	48,000
3	*328*	100	*Sat.Vap.*	*1188*	1.60	125,000
4	*293*	60	Sat.Liq.	*262*	*0.43*	125,000
5	*328*	100	Sat.Vap.	*1188*	1.60	202,000
6	*750*	100	*Superheat*	*1404*	*1.82*	202,000
7	*608*	60	*Superheat*	*1336*	1.82	17,334
8	*241*	10	*Superheat*	*1166*	1.82	184,666
9	*193*	10	Sat.Liq.	*161*	*0.28*	184,666
10	193	60	Sub.Liq.	161.19	0.28	184,666
11	293	60	Sat.Liq.	262	*0.43*	327,000
12	293	300	Sub.Liq.	262.77	0.43	327,000
13	*417*	300	Sat.Liq.	*394*	*0.59*	375,000
14	417	800	Sub.Liq.	395.75	0.59	375,000

Traverse the cycle.

1-2 **Ideal Turbine** (Stage 1):

$$s_2 = s_1 = 1.60 \text{ BTU / lb}_m \cdot {}^\circ\text{R}$$

The remaining properties can be obtained from the steam tables. The mass flow rate cannot be calculated at this point.

2-3 **Ideal Turbine** (Stage 2):

$$s_3 = s_2 = 1.60 \text{ BTU / lb}_m \cdot {}^\circ\text{R}$$

The remaining properties can be obtained from the steam tables.

3-4 **Heat Load**:

All information known

4-5 **Ideal Turbine** (Stage 3):

$$s_5 = s_3 = 1.60 \text{ BTU / lb}_m \cdot {}^\circ\text{R}$$

The remaining properties can be obtained from the steam tables.

5-6 **Reheater**:

All information known

6-7 **Ideal Turbine** (Stage 1):

$$s_5 = s_3 = 1.60 \text{ BTU / lb}_m \cdot {}^\circ\text{R}$$

The remaining properties can be obtained from the steam tables. The mass flow rate can not be calculated at this point.

7-8 **Ideal Turbine** (Stage 2):

$$s_8 = s_7 = 1.82 \text{ BTU / lb}_m \cdot {}^\circ\text{R}$$

The remaining properties can be obtained from the steam tables.

8-9 **Condenser**:

All information known

9-10 **Ideal Pump**:

$$s_{10} = s_9 = 0.28 \text{ BTU / lb}_m \cdot {}^\circ\text{R}$$

Assuming an incompressible liquid we have

$$T_{10} = T_9 = 193 \cdot {}^\circ\text{F}$$

$$\begin{aligned} h_{10} &= h_9 + v_9(P_{10} - P_9) \\ &= 161 + (0.0166)(60 - 10)\left(144\frac{\text{in}^2}{\text{ft}^2}\right)\left(\frac{1}{778.16}\frac{\text{BTU}}{\text{ft}\cdot\text{lb}_f}\right) \\ &= 161.19 \text{ BTU / lb}_m \end{aligned}$$

11-12 **Ideal Pump**:

$$s_{12} = s_{11} = 0.43 \text{ BTU / lb}_m \cdot {}^\circ\text{R}$$

Assuming an incompressible liquid we have

$$T_{12} = T_{11} = 293 \cdot {}^\circ\text{F}$$

$$\begin{aligned} h_{12} &= h_{11} + v_{11}(P_{12} - P_{11}) \\ &= 262 + (0.0174)(300 - 60)\left(144\frac{\text{in}^2}{\text{ft}^2}\right)\left(\frac{1}{778.16}\frac{\text{BTU}}{\text{ft}\cdot\text{lb}_f}\right) \\ &= 262.77 \text{ BTU / lb}_m \end{aligned}$$

13-14 **Ideal Pump**:

$$s_{14} = s_{13} = 0.59 \text{ BTU / lb}_m \cdot {}^\circ\text{R}$$

Assuming an incompressible liquid we have

$$T_{14} = T_{13} = 417^\circ\text{F}$$

$$\begin{aligned} h_{14} &= h_{13} + v_{13}(P_{14} - P_{13}) \\ &= 394 + (0.0189)(800 - 300)\left(144\frac{\text{in}^2}{\text{ft}^2}\right)\left(\frac{1}{778.16}\frac{\text{BTU}}{\text{ft}\cdot\text{lb}_f}\right) \\ &= 395.75 \text{ BTU / lb}_m \end{aligned}$$

We can now determine the mass flow rates. Starting with the high pressure feedwater heater, we write an energy balance:

$$\dot{m}_{13}h_{13} = \dot{m}_{12}h_{12} + \dot{m}_2 h_2$$

We also note that

$$\dot{m}_{12} = \dot{m}_{13} - \dot{m}_2$$

So we eliminate $\dot{m}_{12}$ from our energy balance and solve for $\dot{m}_2$ as

$$\dot{m}_2 = \frac{\dot{m}_{13}(h_{13} - h_{12})}{h_2 - h_{12}} = \frac{(375{,}000)(394 - 262.77)}{1288 - 262.77}$$

$$= 48{,}000 \text{ lb}_m / \text{hr}$$

Similarly,

$$\dot{m}_{12} = \dot{m}_{13} - \dot{m}_2 = 375{,}000 - 48{,}000 = 327{,}000 \text{ lb}_m / \text{hr} = \dot{m}_{11}$$

A mass balance on the high pressure turbine gives

$$\dot{m}_5 = \dot{m}_1 - \dot{m}_2 - \dot{m}_3 = 375{,}000 - 48{,}000 - 125{,}000 = 202{,}000 = \dot{m}_6$$

The final mass flow rates are determined form the analysis of the low pressure feedwater heater. We have

$$\dot{m}_7 = \frac{\dot{m}_{11}(h_{11} - h_{10}) - \dot{m}_4(h_{10} - h_4)}{h_7 - h_{10}}$$

$$= \frac{(327{,}000)(262 - 161.19) + (125{,}000)(161.19 - 262)}{1336 - 161.19}$$

$$= 17{,}334 \text{ lb}_m / \text{hr}$$

A mass balance on the low pressure turbine gives

$$\dot{m}_8 = \dot{m}_6 - \dot{m}_7 = 202{,}000 - 17{,}334 = 184{,}466 \text{ lb}_m / \text{hr} = \dot{m}_9 = \dot{m}_{10}$$

With the table filled out, the questions can now be answered.

14.8 **d)** The first law thermal efficiency is defined as

$$\eta_{th} = \frac{\dot{W}_{net}}{\dot{Q}_{in,tot}}$$

$$\begin{aligned}\dot{W}_{net} &= \dot{m}_1 h_1 - \dot{m}_2 h_2 - \dot{m}_3 h_3 - \dot{m}_5 h_5 \\ &\quad + \dot{m}_6 h_6 - \dot{m}_7 h_7 - \dot{m}_8 h_8 \\ &\quad - \dot{m}_{14}(h_{14} - h_{13}) - \dot{m}_{12}(h_{12} - h_{11}) \\ &\quad - \dot{m}_{10}(h_{10} - h_9) \\ &= 118.14 \text{ MBTU / hr}\end{aligned}$$

$$\begin{aligned}\dot{Q}_{in,tot} &= \dot{m}_1(h_1 - h_{14}) + \dot{m}_6(h_6 - h_5) \\ &= 419.48 \text{ MBTU / hr}\end{aligned}$$

which gives a thermal efficiency of 28%.

14.9 **a)** The cycle analysis table shows that the steam leaving the low pressure turbine is superheated vapor so that the quality is not defined.

14.10 **c)** As shown in the efficiency calculation it is 118 MBTU/hr

14.11 **d)** We find

$$\frac{\dot{m}_3 h_3}{\dot{Q}_{in,tot}} = \frac{(125,000)(1188)}{419.48 \times 10^6} = 35\%$$

14.12 **d)** We simply sum the mass flows at nodes 2 and 7 to obtain 65,334 lb_m/hr

14.13 **b)** We sum the enthalpy change times the mass flow rate for each pump to obtain 0.94 MBTU/hr.

14.14 **c)** The heat load is heat transfer is

$$\dot{Q}_{HL} = \dot{m}_3(h_4 - h_3) = (125,000)(1188 - 262)$$

$$= 116 \text{ MBTU / hr}$$

14.15 **b)** The natural gas flow rate can be calculated from

$$\dot{m}_{gas} = \frac{\dot{Q}_{in,tot}}{\eta_{furn} e_{coal}} = \frac{419.48 \times 10^6}{(0.87)(22,000)}$$

$$= 22,000 \text{ lb}_m \text{ / hr}$$

Fluid Flow

by David C. Wiggert

Introduction

The design and analysis of piping systems and associated turbomachinery are common occurrences for mechanical engineers. The complexity of piping ranges from a single pipeline providing fluids to various device to an extensive network transporting fluids in a large manufacturing plant. Piping networks are analyzed using computer algorithms, and they will not be considered in this chapter. We will focus on simpler systems that can be solved by hand calculation: pipe elements either operating singly, or joined to form series, parallel, or branching configurations. The analysis of a single pipeline is presented first, followed by the more complex systems that are treated using ad hoc solutions suitable for use with calculators. The basic idea behind the ad hoc method of solution presented herein is to identify all of the unknowns and write an equivalent number of independent equations to be solved. Subsequently, the system is simplified by eliminating as many unknowns as possible and reducing the problem to a series of single pipe problems. It is possible that the resulting equation or grouping of equations will be nonlinear, so that an iterative solution will be required; either a trial-and-error or numerical procedure could be utilized. A brief introduction to pumps is provided, focusing primarily on centrifugal, or radial flow pumps operating in piping systems. Also, momentum principles applied to vanes encountered in turbomachines will be reviewed.

15.1 Losses in Piping

Losses are divided into two categories: (1) those due to wall friction in the piping, and (2) those created by piping components such as valves, bends, orifices, etc. The former are distributed along the length of the pipe, and the latter, termed "minor losses", are concentrated at prescribed locations in the system. They are defined separately and then introduced into the system energy equation, allowing one to determine the discharge, pressure drop, or pipe diameter.

Frictional Losses

Frictional pipe loss is normally expressed using either the Darcy-Weisbach or the Hazen-Williams relation. The *Darcy-Weisbach* equation is

$$h_L = \frac{fL}{D}\frac{V^2}{2g} = \frac{8fL}{g\pi^2 D^5}Q^2 \tag{15.1.1}$$

in which h_L = head loss, f = friction factor, L = pipe length, D = pipe diameter, V = mean velocity, and Q = discharge. The behavior of the friction factor is described by the Moody diagram, shown in Fig. 15.2. Generally, f is a function of the Reynolds number $Re = VD/\nu$ and the relative roughness $\varepsilon = e/D$, where ν = kinematic viscosity of the liquid, and e = absolute roughness of the pipe wall. An empirical formula for f that accurately reproduces the turbulent portion of the Moody diagram is

$$f = 1.325\left[\ln(0.27\varepsilon + 5.74Re^{-0.9})\right]^{-2} \tag{15.1.2}$$

Either Eq. 15.1.2 or the Moody diagram can be used to determine the friction factor. Equation 15.1.2 is valid over the ranges $0.01 > \varepsilon > 10^{-6}$, and $10^8 > Re > 3000$. For values of Re and ε outside of those ranges, the Moody diagram must be used directly. Numerical values of e are provided in Table 15.1 and on the Moody diagram.

In many instances it is acceptable to assume that the friction factor is independent of the Reynolds number. It can be observed in Eq. 15.1.2 that the second term in brackets becomes relatively smaller as Re increases, so that it can be used in the form $f = 1.325[\ln 0.27\varepsilon]^{-2}$, either to estimate f as a constant for a given problem, or to obtain an estimate for an iterative solution in which initially Re is unknown. In many engineering situations this simplified form is acceptable.

TABLE 15.1 Values of the Average Wall Roughness Height *e*

Type of pipe	e (ft)
Riveted steel	~0.01
Concrete	~0.001-0.01
Wood	~0.001
Cast iron	0.00085
Galvanized iron	0.0005
Wrought iron	0.00015
Drawn tubing	0.000005

The *Hazen-Williams* formula applies only to water as the flowing liquid. It normally expresses the head loss as a function of the discharge Q:

$$h_L = \frac{4.72L}{C^{1.85}D^{4.87}}Q^{1.85} \qquad \textbf{(15.1.3)}$$

in which C is the *Hazen-Williams loss coefficient*. Values of C are given in Table 15.2.

TABLE 15.2 Values of the Hazen-Williams Coefficient ***C***

Type of pipe	C
Extremely smooth; asbestos-cement	140
New or smooth cast iron; concrete	130
Wood stave; newly welded steel	120
Average cast iron; newly riveted steel	110
Cast iron or riveted steel after some years of use	95-100
Deteriorated old pipes	60-80

Note that in the form given the Hazen-Williams formula is applicable only with English units. It is neither as universally applicable, nor as accurate, as is the Darcy-Weisbach equation. It is particularly useful when analyzing series piping systems.

Example 15.1

Water is flowing at $Q = 20$ gal/min in a 500-ft-long, 2-inch-diameter cast iron steel pipe. The water temperature is 50° F. Determine the head loss h_L using (a) the Darcy-Weisbach formula and (b) the Hazen-Williams formula.

Solution. From a table of properties of water, the kinematic viscosity of water at the given temperature is found to be $\nu = 1.41 \times 10^{-5}\ \text{ft}^2/\text{sec}$. Using Table 15.1, one finds the roughness height to be e = 0.00085 ft. The diameter is $D = 2/12 = 0.167$ ft and the pipe length is $L = 500$ ft. The pipe area, discharge and velocity are:

$$A = (\pi/4)\times(2/12)^2 = 0.0218\ \text{ft}^2$$
$$Q = (20\ \text{gal/min})\times(1\ \text{ft}^3/7.48\ \text{gal})\times(1\ \text{min}/60\ \text{sec}) = 0.0446\ \text{ft}^3/\text{sec}$$
$$V = Q/A = 0.0446/0.0218 = 2.05\ \text{ft/sec}$$

The Reynolds number is $Re = 2.05\times0.167/1.41\times10^{-5} = 24{,}300$, and ε = 0.00085/0.167 = 0.00509. Substitute the data into Eq. 15.1.2 to determine the friction factor:

$$f = 1.325\left[\ln(0.27\times0.00509 + 5.74\times24300^{-0.9})\right]^{-2} = 0.034$$

a) The head loss is computed with Eq. 15.1.1:

$$h_L = 0.034\times\frac{500}{0.167}\times\frac{2.05^2}{2\times32.2} = 6.64\ \text{ft}$$

b) From Table 15.2, the Hazen-Williams coefficient is $C = 110$. Application of

Eq. 15.1.3 yields the head loss:

$$h_L = \frac{4.72 \times 500}{110^{1.85} \times 0.167^{4.87}} \times 0.0446^{1.85} = 7.64 \text{ ft}$$

These two results, although numerically different, are sufficiently close for most design problems.

Minor losses

Minor losses are expressed proportional to the kinetic energy in the manner

$$h_L = K\frac{V^2}{2g} = \frac{8K}{g\pi^2 D^4}Q^2 \quad \textbf{(15.1.4)}$$

in which K is the loss coefficient. Representative values of K can be found in Table 15.4. Application of Eq. 15.1.4 is relatively straight-forward and will be shown in subsequent examples.

It is often convenient to express a loss coefficient as an equivalent length L_e of pipe. This is done by equating Eq. 15.1.4 to Eq. 15.1.1 and eliminating like terms to produce the relation

$$L_e = K\frac{D}{f} \quad \textbf{(15.1.5)}$$

Collectively, frictional losses and minor losses make up the head loss term. Equations 15.1.1 to 15.1.5 are used in the following sections to express h_L in the energy equation.

15.2 Energy and Continuity Equations

The two fundamental control volume equations for piping problems are presented here without derivation. We assume steady, incompressible flow with uniform velocity profiles and constant pressures at the control surfaces, Fig. 15.1. Between two control surfaces 1 and 2 the equations take the form:

Continuity: $$Q = A_1V_1 = A_2V_2 \quad \textbf{(15.2.1)}$$

Energy: $$\frac{p_1}{\gamma} + \frac{V_1^2}{2g} + z_1 + H_P = \frac{p_2}{\gamma} + \frac{V_2^2}{2g} + z_2 + h_L \quad \textbf{(15.2.2)}$$

in which p = pressure, γ = specific weight of the fluid, z = elevation, and H_P = head rise across a pump. If

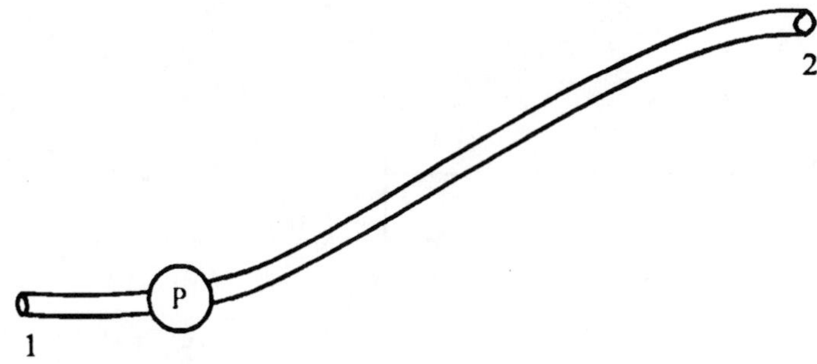

Figure 15.1 System control volume.

no pump is present, then $H_P = 0$. It is convenient in many applications to make use of the hydraulic grade line (HGL), defined in the manner

$$H = \frac{p}{\gamma} + z \tag{15.2.3}$$

in which H is the distance from a selected datum to the HGL. If the kinetic energy terms $V^2/2g$ in Eq. 15.2.2 are relatively small, they can be neglected, and the energy equation becomes

$$H_1 + H_P = H_2 + h_L \tag{15.2.4}$$

In this form, the energy equation is useful for analyzing parallel and branching systems.

The power $\dot{W}_f$ delivered to the fluid by a pump is designated as

$$\dot{W}_f = \gamma Q H_P \tag{15.2.5}$$

and the power $\dot{W}_P$ delivered to the pump, called the brake power, is

$$\dot{W}_P = \frac{\dot{W}_f}{\eta} = \frac{\gamma Q H_P}{\eta} \tag{15.2.6}$$

where η = the pump efficiency. The units for $\dot{W}_f$ and $\dot{W}_P$ are (ft-lb/sec); to convert to horsepower, divide the expressions by 550 ft-lb/sec. Equations 15.2.5 and 15.2.6 are used in conjunction with the energy and continuity equations when a pump is present in the system.

Single Pipe Systems

A single pipe system consists of one pipe connected to one or more reservoirs, Fig. 15.2. The system may include a pump, or the flow may be induced by gravity forces. There are three categories of single pipe problems:

Category 1: Determining the pressure drop, or head loss
Category 2: Determining the discharge
Category 3: Determining the pipe diameter

In conjunction with the Darcy-Weisbach formulation, the first category can be solved directly by substitution into the energy equation, and the second and third categories require either trial and error solutions using the Moody diagram, or use of empirically derived formulas. Table 15.3 lists the various alternative solutions.

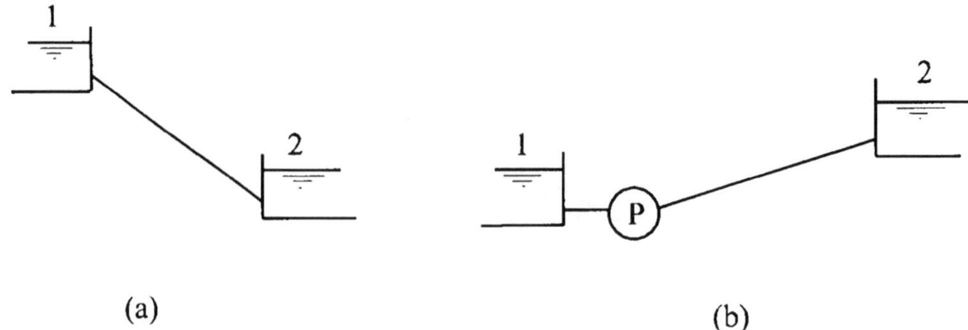

Figure 15.2 Single pipe systems: (a) gravity flow; (b) pump-driven flow.

TABLE 15.3 Solutions for Single Pipe Problems using the Darcy-Weisbach Formulation

Category	Variables	Moody diagram $f=f(Re,\varepsilon)$	Empirical formula*
1	Known: Q, D, e, ν Unknown: h_L	Direct solution	$h_L = 1.07\frac{Q^2 L}{gD^5}\left\{\left[\ln\frac{e}{3.7D} + 4.62\left(\frac{\nu D}{Q}\right)^{0.9}\right]\right\}^{-2}$ Range: $10^{-6} < \varepsilon < 10^{-2}, \quad 3000 < Re < 3\times10^8$
2	Known: D, e, ν, h_L Unknown: Q	Trial and error: $fQ^2 = \text{Constant}$	$Q = -0.965\left(\frac{gD^5 h_L}{L}\right)^{0.5}\ln\left[\frac{e}{3.7D} + \left(\frac{3.17\nu^2 L}{gD^3 h_L}\right)^{0.5}\right]$ Range: $Re > 2000$
3	Known: Q, e, ν, h_L Unknown: D	Trial and error: $\frac{f}{D^5} = \text{Constant}$	$D = 0.66\left[e^{1.25}\left(\frac{LQ^2}{gh_L}\right)^{4.75} + \nu Q^{9.4}\left(\frac{L}{gh_L}\right)^{5.2}\right]^{0.04}$ Range: $10^{-6} < \varepsilon < 10^{-2}, \quad 5000 < Re < 3\times10^6$

*No pumps or components with minor losses are incorporated in the empirical formulas.

Example 15.2

A pumping scheme requires that water be transported from a river at elevation $z_1 = 610$ ft to a reservoir at elevation $z_2 = 795$ ft. The discharge $Q = 1{,}030$ gallons per minute, the pipe has length $L = 1{,}500$ ft, and diameter $D = 6$ inches. The pipe is made of concrete and contains a gate valve half open, two standard elbows, and a pipe exit loss. Determine the following: (a) pump head, (b) fluid power, and (c) brake power, if the pump efficiency is η = 0.75. Use the Hazen-Williams formula to estimate pipe frictional loss.

Solution. This is a category 1 problem. The discharge, pipe area and velocity are:

$$Q = (1{,}030\ \text{gal/min})/(7.48\ \text{gal/ft}^3 \times 60\ \text{sec/min}) = 2.30\ \text{ft}^3/\text{sec}$$
$$A = (\pi/4)\times 0.5^2 = 0.196\ \text{ft}^2$$
$$V = 2.30/0.196 = 11.7\ \text{ft/sec}$$

From Table 15.2, C = 130, and the sum of the minor loss coefficients from Table 15.4 is 5.6 + 2 × 0.9 + 1 = 8.4. The system head loss is

$$h_L = \frac{4.72\times 1{,}500}{130^{1.85}\times 0.5^{4.87}}\times 2.30^{1.85} + 8.4\times\frac{11.7^2}{2\times 32.2} = 118.7 + 17.9 = 137\ \text{ft}$$

The energy equation, Eq. 15.2.2, is solved for H_P, noting that the kinetic energy terms $V^2/2g$ and the pressure terms p/γ at the two reservoirs are zero:

(a) $H_P = z_2 - z_1 + h_L = 795 - 610 + 137 = 322\ \text{ft}$

(b) $\dot{W}_f = \gamma Q H_P$
$= 62.4\times 2.30\times 322 = 46{,}200\ \text{ft-lb/sec},\ \text{or 84 horsepower}$

(c) $\dot{W}_P = \dfrac{\dot{W}_f}{\eta} = \dfrac{46{,}200}{0.75} = 61{,}600\ \text{ft-lb/sec},\ \text{or 112 horsepower}$

Note: The information given does not merit more than two significant numbers. However, it is common practice to provide three significant numbers.

It is customary to neglect minor loss terms if the pipe length is sufficiently large: one "rule-of-thumb" is to ignore them when the ratio $L/D > 1{,}000$. Caution is urged using such a rule; in the previous example, the minor losses, primarily dominated by the gate valve, amounted to 13 percent of the head loss and 6 percent of the required brake power. It is recommended to include the minor losses when sufficient information is provided to estimate the terms.

Example 15.3

A pump is situated between sections 1 and 2 of a horizontal pipeline. The upstream section has dimensions D_1 = 2 in, with p_1 = 50 lb/in², and the downstream section has dimensions D_2 = 3 in, with p_2 = 110 lb/in². Determine the required fluid power if Q = 25 gal/min, h_L = 20 ft, and γ = 62.4 lb/ft³.

Solution. This is also a category 1 problem. Substitute $V = Q/A$ and $H_P = \dot{W}_f/(\gamma Q)$ into the energy relation, Eq. 15.2.2, noting that $z_1 = z_2$,

$$\frac{p_1}{\gamma} + \frac{Q^2}{2gA_1^2} + \frac{\dot{W}_f}{\gamma Q} = \frac{p_2}{\gamma} + \frac{Q^2}{2gA_2^2} + h_L$$

Substitute $Q = 25/(7.48 \times 60) = 0.0557\ \text{ft}^3/\text{sec}$, $A_1 = (\pi/4) \times (2/12)^2 = 0.0218\ \text{ft}^2$, $A_2 = (\pi/4) \times (3/12)^2 = 0.0491\ \text{ft}^2$, $p_1/\gamma = 50 \times 144/62.4 = 115.4$ ft, and $p_2/\gamma = 110 \times 144/62.4 = 253.8$ ft into the relation:

$$115.4 + \frac{(0.0557)^2}{2 \times 32.2 \times 0.0218^2} + \frac{\dot{W}_f}{62.4 \times 0.0557} = 253.8 + \frac{(0.0557)^2}{2 \times 32.2 \times 0.0491^2} + 20$$

Reducing and solving for the fluid power, one finds $\dot{W}_f = 551$ ft-lb/sec, or approximately 1 horsepower. The kinetic energy terms are quite small relative to the head loss term, and could justifiably be neglected.

Example 15.4

A liquid with a specific gravity of 0.68 is pumped from a storage tank to a discharge station through a 1,500-ft-long, 8-in-diameter pipe. The pump provides 15 horsepower of useful power to the liquid. Determine the discharge in the system. Assume a constant friction factor of $f = 0.015$ and include minor losses.

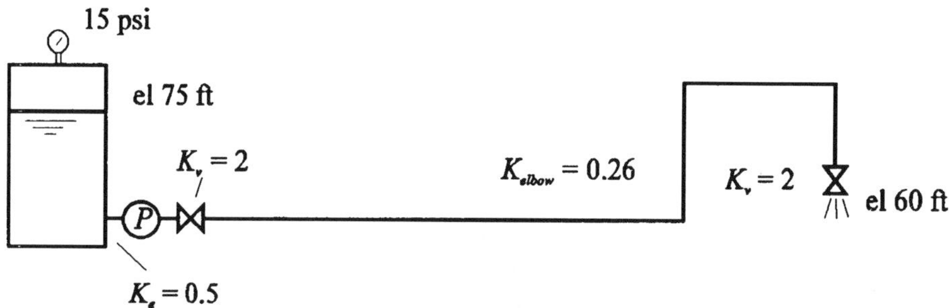

Solution. This is a category 2 problem. Write the energy equation from the liquid surface in the tank (location 1) to the valve exit (location 2):

$$\frac{p_1}{\gamma} + z_1 + \frac{\dot{W}_f}{\gamma Q} = \left(\sum K + \frac{fL}{D} + 1\right)\frac{Q^2}{2gA^2} + z_2$$

Rearrange and substitute in known data:

$$\frac{p_1}{\gamma} + z_1 - z_2 + \left(\frac{\dot{W}_f}{\gamma}\right)\frac{1}{Q} - \left(\sum K + \frac{fL}{D} + 1\right)\frac{Q^2}{2gA^2} = 0$$

$$\frac{15 \times 144}{0.68 \times 62.4} + 75 - 60 + \frac{15 \times 550}{0.68 \times 62.4} \times \frac{1}{Q} - \frac{(0.5 + 3 \times 0.26 + 2 + 0.015 \times 1500/0.667 + 1) \times Q^2}{2 \times 32.2 \times (\pi/4)^2 \times 0.667^4} = 0$$

The equation reduces to $F(Q) = 65.9 + 194.4/Q - 4.83Q^2 = 0$. A trial and error solution is used to solve for Q. The first value of Q is selected by noting that $F(4)$ is positive and $F(5)$ is negative, so that the root lies between those two values. After a couple tries, the discharge is

$$Q = 4.7\ \text{cfs}.$$

Example 15.5

Select a pipe to convey 50° F water delivering at least 250 gallons per minute from a higher reservoir to a lower one. The distance between the reservoirs is 750 ft, and the difference in water surface elevations is 20 ft. The pipe is made of concrete, with e = 0.001 ft, and minor losses can be neglected. The flow is gravity driven.

Solution. This is a category 3 problem, where we need to find the diameter. The following data is provided: $Q = 250/(7.48 \times 60) = 0.557$ ft^3/sec, $\nu = 2.74 \times 10^{-5}$ ft^2/sec, $z_1 - z_2 = 20$ ft, and $L = 750$ ft. Substitute the data into the energy equation and reduce:

$$z_1 - z_2 = \frac{8fLQ^2}{g\pi^2 D^5}, \quad 20 = \left(\frac{8 \times 750 \times 0.557^2}{32.2 \times \pi^2}\right)\frac{f}{D^5}, \quad \text{or} \quad D = 0.782 f^{1/5} \qquad \textbf{(A)}$$

Equation 15.1.2 is expressed in terms of f and D, recognizing that $Re = 4Q/(\pi \nu D)$:

$$\begin{aligned} f &= 1.325\left\{\ln\left[\frac{0.27 \times 0.001}{D} + 5.74 \times \left(\frac{4 \times 0.557}{\pi \times 2.74 \times 10^{-5}}\right)^{-0.9} D^{0.9}\right]\right\}^{-2} \\ &= 1.325\left\{\ln\left[\frac{2.7 \times 10^{-4}}{D} + 6.13 \times 10^{-4} D^{0.9}\right]\right\}^{-2} \end{aligned} \qquad \textbf{(B)}$$

A trial and error solution will yield the diameter. To begin the iteration, guess a value of f, say, 0.02 (this is an intermediate value that is commonly used), and then solve for D with Eq. A and f with Eq. B. The results are tabulated as follows:

f (Eq B)	0.02 (guess)	0.028	0.026	0.0275
D (Eq A)	0.358	0.382	0.378	0.381

With $D = 0.381$ ft, the computed friction factor is $f = 0.0275$. Hence a diameter of $0.381 \times 12 = 4.57$ inches is to be used. Likely, a 6-in-diameter pipe available "off the shelf" would be specified. This problem can also be solved using Eq. A and the Moody diagram, Fig. 15.2, or solved directly with the empirical equation given in Table 15.3, using $h_L = z_1 - z_2 = 20$ ft.

Series and Parallel Piping Systems

Multiple pipe systems can be generalized by considering combinations of series and parallel elements, Fig. 15.3. If several pipe segments are connected end to end, then the system is labeled series piping. The discharge is the same in each element, and the head loss is the sum of the loss in each element:

$$\begin{aligned} &\text{(a)} \quad Q_{sys} = Q_1 = Q_2 = Q_3 = \cdots \\ &\text{(b)} \quad (h_L)_{sys} = (h_L)_1 + (h_L)_2 + (h_L)_3 + \cdots = \Sigma(h_L)_i \end{aligned} \qquad \textbf{(15.2.7)}$$

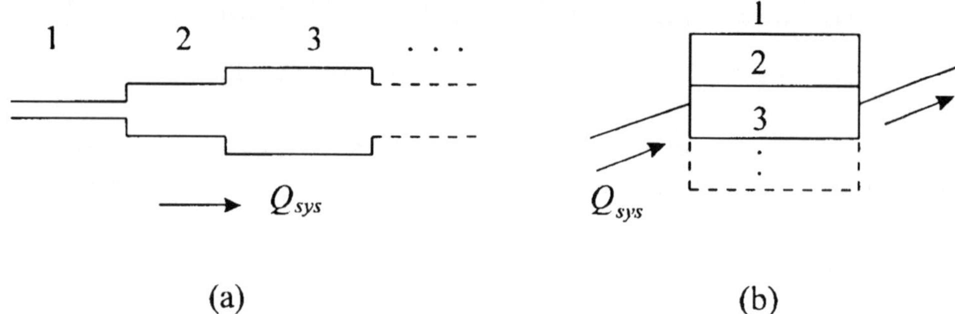

Figure 15.3 Two piping system configurations: (a) series; (b) parallel.

For pipes in parallel, Fig. 15.3b, the system discharge is cumulative, and the system head loss is the same for each element:

$$\begin{aligned}&\text{(a)}\quad Q_{sys} = Q_1 + Q_2 + Q_3 + \cdots = \sum Q_i \\ &\text{(b)}\quad (h_L)_{sys} = (h_L)_1 = (h_L)_2 = (h_L)_3 = \cdots = (h_L)_i\end{aligned} \tag{15.2.8}$$

The standard problem associated with parallel piping is category 2, i.e., determining the discharge for a given head loss across the system. The methodology for a trial and error solution is as follows. The head loss for any pipe i can be expressed as

$$Q_i = \sqrt{\frac{(h_L)_{sys}}{R_i}} \tag{15.2.9}$$

$$R_i = \frac{1}{2gA_i^2}\left[\frac{fL}{D} + \sum K\right]_i = \left[\frac{8f(L+L_e)}{g\pi^2 D^5}\right]_i \tag{15.2.10}$$

where L_e comes from Eq. 15.1.5. Combining Eqs. 15.2.9 and 15.2.8b, and eliminating Q_i results in

$$(h_L)_{sys} = \left[\frac{Q_{sys}}{\sum\sqrt{1/R_i}}\right]^2 \tag{15.2.11}$$

An iterative procedure to solve for the discharges Q_i and system head loss $(h_L)_{sys}$, given the system discharge Q_{sys} is as follows:

Step 1: Estimate or guess the initial friction factor in each element, provided they are not given.

Step 2: Compute R_i for each element with Eq. 15.2.10 and estimate $(h_L)_{sys}$ with Eq. 15.2.11.

Step 3: Compute Q_i in each element using Eq. 15.2.9.

Step 4: With the current values of Q_i, update estimates of f for each element using Eq. 15.1.2 or the Moody diagram.

Step 5: Repeat steps 2 to 4 until $(h_L)_{sys}$ and Q_i do not vary significantly.

Quite often in problems of this nature the friction factor is known. In that case R_i is found using Eq. 15.2.10, steps 4 and 5 are unnecessary, and a solution results with the first iteration.

Example 15.6

Three pipes in series deliver flow between two reservoirs under the action of gravity. The upper reservoir elevation is 820 ft, and the elevation at the lower reservoir is 351 ft. Minor losses are proportional to the square of the discharge, and the Hazen-Williams formula is used to account for frictional losses. Determine the system discharge Q.

Pipe	L (ft)	D (in)	K	C
1	600	8	2	100
2	300	10	3	120
3	900	12	0	90

Solution. The energy equation between the two reservoirs is

$$H_A - H_B = (h_L)_1 + (h_L)_2 + (h_L)_3, \quad \text{with } h_L = \frac{4.72L}{C^{1.85}D^{4.87}}Q^{1.85} + \frac{\Sigma K}{2gA^2}Q^2.$$

Substituting in known data and reducing the equations results in

$$820 - 351 = (4.07 + 0.49 + 1.03)Q^{1.85} + (0.255 + 0.157)Q^2$$

or

$$469 = 5.59Q^{1.85} + 0.41Q^2.$$

To compute Q we may employ a trial-and-error solution. For the first iteration, neglect the minor losses (the Q^2 term) so that $Q = (469/5.59)^{1/1.85} = 10.96$. After a couple tries, the discharge is

$$Q = 10.4 \text{ cfs}$$

Example 15.7

Determine the flow distribution of water in a three-pipe parallel system, as shown in Fig. 15.3b. The system flow is $Q_{sys} = 0.35 \text{ ft}^3/\text{sec}$, and the friction factors are specified.

Pipe	L (ft)	D (in)	f	K
1	90	2	0.020	3
2	120	3	0.025	5
3	180	2.5	0.022	1

Solution. For each pipe we first compute the equivalent lengths and the coefficients R_i:

$$\text{Pipe 1:} \quad L_e = \frac{2/12}{0.02} \times 3 = 25, \quad R_1 = \frac{8 \times 0.02 \times 115}{32.2 \times \pi^2 \times (2/12)^5} = 450.2$$

Pipe 2: $L_e = \dfrac{3/12}{0.025} \times 5 = 50$, $R_2 = \dfrac{8 \times 0.025 \times 170}{32.2 \times \pi^2 \times (3/12)^5} = 109.6$

Pipe 3: $L_e = \dfrac{2.5/12}{0.022} \times 1 = 9.5$, $R_3 = \dfrac{8 \times 0.022 \times 189.5}{32.2 \times \pi^2 \times (2.5/12)^5} = 267.4$

Since the friction factors are constant, no iterations are required. The solution for the system head loss and the discharges are now computed:

$$(h_L)_{sys} = \frac{0.35^2}{\left[1/\sqrt{450.2} + 1/\sqrt{109.6} + 1/\sqrt{267.4}\right]^2} = 2.95 \text{ ft}$$

$$Q_1 = \sqrt{2.95/450.2} = 0.081 \text{ ft}^3/\text{sec}$$

$$Q_2 = \sqrt{2.95/109.6} = 0.164 \text{ ft}^3/\text{sec}$$

$$Q_3 = \sqrt{2.95/267.4} = 0.105 \text{ ft}^3/\text{sec}$$

A check of the sum of the individual discharges reveals that $Q_{sys} = 0.35 \text{ ft}^3/\text{sec}$.

Branching Pipe Systems

Consider the branching pipe system illustrated in Fig. 15.4a, which is made up of three pipe elements, each connected to a reservoir, and joined at a single junction. As in the previous section, the most common situation is to determine the flow distribution, a category 2 problem. First assume a flow direction in each pipe as shown; then, neglecting the kinetic energy terms, the energy equations can be written in the manner

$$H_A - H_B = (h_L)_1 , \quad H_B - H_C = (h_L)_2 , \quad H_B - H_D = (h_L)_3 \tag{15.2.12}$$

and the continuity balance at location B is

$$\Sigma Q = Q_1 - Q_2 - Q_3 = 0 \tag{15.2.13}$$

The hydraulic grade lines at A, C, and D (that is, the reservoir elevations) are considered known, and the unknowns are H_B, Q_1, Q_2, and Q_3. A convenient ad hoc solution is outlined below.

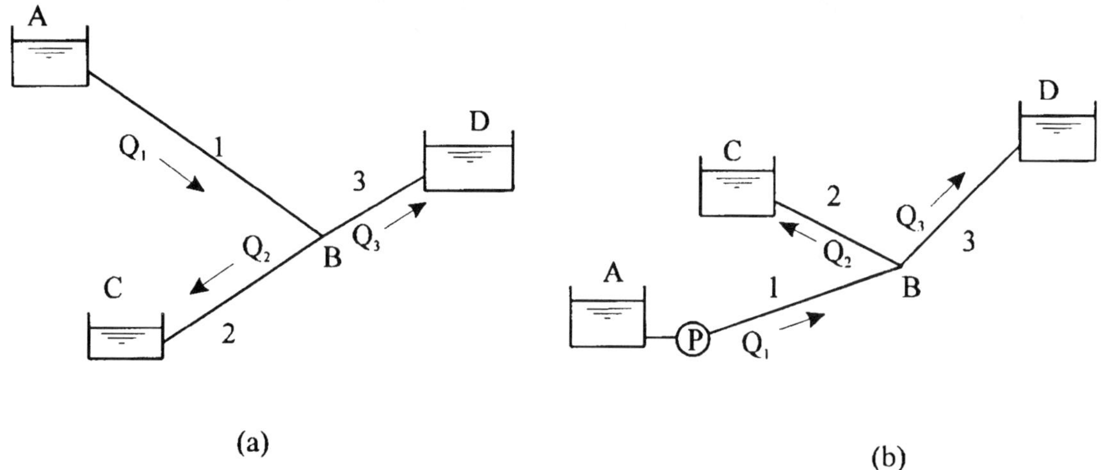

Figure 15.4 Branch piping system: (a) gravity flow; (b) pump-driven flow.

Step 1: Assume Q_1 and compute the hydraulic grade line H_B at location B.

Step 2: Compute Q_2 and Q_3 in branches 2 and 3.

Step 3: Substitute Q_1, Q_2, and Q_3 into the continuity equation. Generally there will be a flow imbalance, i.e., $\Sigma Q \neq 0$.

Step 4: Adjust the flow Q_1 and repeat steps 2 and 3 until ΣQ is within desired limits.

If a pump exists in pipe one (see Figure 15.4b) the energy equation for that pipe is altered as

$$H_A - H_B + H_P = (h_L)_1 \tag{15.2.14}$$

An additional unknown H_P is introduced. Another necessary relationship is the pump characteristic curve. It is often convenient to follow the solution graphically by plotting the assumed discharge in pipe 1 versus either the hydraulic grade line at B or Q at B. Such a procedure is helpful to determine in what manner the discharge in pipe 1 should be altered for the next iteration.

Example 15.8

Consider the system shown in Figure 15.4b, with the data tabulated below. Determine the water distribution and the piezometric head at the junction. Assume constant friction factors and a pump characteristic curve given by $H_P = 55 - 0.1Q^2$, with units of H_P in ft and Q in ft³/sec. The reservoir elevations are $H_A = 20$ ft, $H_C = 50$ ft, and $H_D = 45$ ft.

Pipe	L (ft)	D (in).	f	K
1	100	10	0.020	2
2	200	8	0.015	0
3	300	6	0.025	0

Solution. The losses for each pipe are expressed using Eqs. 15.1.5 and 15.1.1:

Pipe 1: $L_e = \dfrac{10/12}{0.02} \times 2 = 83.3$, $(h_L)_1 = \dfrac{8 \times 0.02 \times 183.3}{32.2 \times \pi^2 \times (10/12)^5} Q_1^2 = 0.23Q_1^2$

Pipe 2: $(h_L)_2 = \dfrac{8 \times 0.015 \times 200}{32.2 \times \pi^2 \times (8/12)^5} Q_2^2 = 0.57Q_2^2$

Pipe 3: $(h_L)_3 = \dfrac{8 \times 0.025 \times 300}{32.2 \times \pi^2 \times (6/12)^5} Q_3^2 = 6.04Q_3^2$

The energy relations for the system are Eq. 15.2.14 for pipe 1 and the latter two of Eq. 15.2.12 for pipes 2 and 3:

Pipe 1: $20 + 55 - 0.1Q_1^2 = H_B + 0.23Q_1^2$, or $H_B = 75 - 0.33Q_1^2$

Pipe 2: $H_B = 50 + 0.57Q_2^2$, or $Q_2 = \sqrt{(H_B - 50)/0.57}$

Pipe 3: $H_B = 45 + 6.04Q_3^2$, or $Q_3 = \sqrt{(H_B - 45)/6.04}$

The flow balance at the junction is $\Sigma Q = Q_1 - Q_2 - Q_3$. It is assumed that the hydraulic grade line at B lies below the water surfaces at C and D. If during the iterations, it becomes apparent that the flow direction will reverse in either pipe 2 or pipe 3, it will be necessary to reformulate the appropriate energy equation and the flow balance. The solution is tabulated below. The first estimate of Q_1 is based on assuming that H_B = 65 ft, so that $Q_1 = \sqrt{(75-65)/0.33} = 5.5$.

Q_1 (ft^3/sec)	H_B(ft)	Q_2 (ft^3/sec)	Q_3 (ft^3/sec)	ΣQ
5.5	65	5.13	1.82	-1.45
6	63.12	4.80	1.73	-0.53
6.5	61.10	4.41	1.63	+0.46
6.27*	62.03	4.59	1.68	0.00

*Estimated by linear interpolation between the two previous values of Q_1 and Q.

Hence the solution is H_B = 62 ft, Q_1 = 6.3 ft^3/sec, Q_2 = 4.6 ft^3/sec, and Q_3 = 1.7 ft^3/sec.

15.3 Use of Centrifugal Pumps in Piping Systems

Thus far, we have included pumps only by making simplified assumptions regarding their behavior. Let us now focus our attention on centrifugal pumps, which include radial-flow, mixed-flow and axial-flow pumps. If a centrifugal pump is included in the piping system and the flow rate or useful power is specified, the solution is straight-forward, employing techniques we have already developed. On the other hand, if the flow rate is not specified, a situation that is often the case (category 2 problem), a trial-and-error solution is required. The reason is that the head produced by a centrifugal pump depends on the discharge in a unique relationship, termed the *pump characteristic curve.* Sometimes the characteristic curve is approximated by a polynomial equation (refer to Example 15.8), but it is more appropriate and accurate to make use of one provided by the pump manufacturer. The characteristic curve provides a relationship between the discharge Q and the pump head H_P; a representative one is shown in Fig. 15.5 along with the pump efficiency. The pump characteristic curve and the appropriate energy equation(s) for the piping system are solved simultaneously to determine the desired discharge.

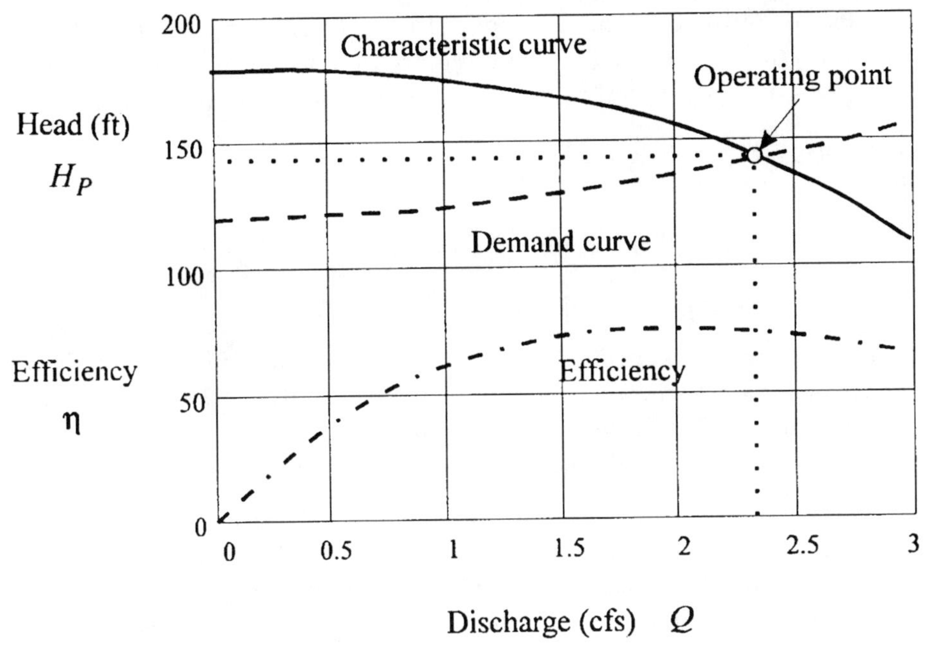

Figure 15.5 Pump characteristic curve and system demand curve.

Matching pump to system demand

Consider a single pipeline that contains a pump to deliver fluid between two reservoirs, Fig. 15.2b. The energy equation for the system is

$$H_P = z_2 - z_1 + \left(\frac{fL}{D} + \Sigma K\right)\frac{Q^2}{2gA^2} \qquad \textbf{(15.3.1)}$$

and is termed the *system demand curve*. Note that z_1 can be less than z_2 and that the friction factor can vary with the discharge as shown earlier. Equation 15.3.1 is represented by the dashed curve in Fig. 15.5; on the right-hand-side, z_1 - z_2 is the static head, and the remaining term is the head loss. The steep nature of the demand curve is dependent on the magnitude of the losses: as the losses increase, the pumping head required for a given discharge increases. Piping systems may experience short-term changes in the demand such as throttling of valves, and long-term changes due to aging of pipes. The intersection of the characteristic curve with the demand curve is termed the *operating point*; it provides the unknown head and discharge. When selecting a pump for a given application, it is desirable to have the operating point occur as close as possible to the point of maximum efficiency.

Example 15.9

The piping system shown in Fig. 15.2b has the following data: z_1 = 100 ft, z_2 = 220 ft, D = 8 in (wrought iron pipe), L = 1,500 ft. The entrance and exit loss coefficients are $K_{ent} = 0.5$, and $K_{exit} = 1.0$. Using the characteristic pump curve shown in Fig. 15.5, determine the operating head, system discharge, and required horsepower.

Solution. To generate the system demand curve, we assume that the Reynolds number is sufficiently large, so that it can be neglected when evaluating the friction factor with Eq. 15.1.2. From Table 15.1, e = 0.00015 ft, and $\varepsilon = 0.00015/(8/12) = 0.000225$. Then the friction factor is computed to be $f = 1.325[\ln(0.27 \times 0.000225)]^{-2} = 0.014$. The system demand curve is thus

$$H_P = z_2 - z_1 + \left(\frac{fL}{D} + \Sigma K\right)\frac{Q^2}{2gA^2}$$

$$= 220 - 100 + \left[\frac{0.014 \times 1500}{8/12} + 1.5\right]\frac{Q^2}{2 \times 32.2 \times (\pi/4)^2 \times (8/12)^4}$$

$$= 120 + 4Q^2$$

The demand curve is plotted in Fig. 15.5. At the operating point the desired head, discharge, and efficiency are observed to be $H_P \cong 140$ ft, $Q \cong 2.3$ ft³/sec, or 1030 gal/min, and $\cong 70\,\%$. The required input power is

$$\dot{W}_P = \gamma Q H_P/\eta = 62.4 \times 2.3 \times 140/0.70 = 28{,}700 \text{ ft-lb/sec, or 52 horsepower}$$

For more complex systems, one can work the pump characteristic curve into the system equations in an ad hoc fashion. One example is the branching system presented in Section 15.2.3, where in Eq. 15.2.14 the pump head H_P is considered a function of the discharge Q_1, with values provided by the characteristic curve.

Cavitation and Net Positive Suction Head

Cavitation refers to conditions within a pump where the local pressures reduce to vapor pressure. Consequently, vapor cavities are formed, and when transported into regions of greater pressure they collapse, creating high localized pressures near solid boundaries. After repeated collapse cycles, the pump impeller and other components may become materially damaged. The proper selection of a pump for a given piping system requires that a calculation be made to eliminate the possibility of cavitation at the design flow condition.

A convenient parameter used to designate the potential for cavitation is the *net positive suction head, NPSH*. Consider a pump operating as shown in Figure 15.6; location 1 is at the liquid surface on the inlet, or suction, reservoir, and location 2 is the center of the pump. The *NPSH* design requirement for the pump is established by the inequality

$$NPSH \le \frac{p_{atm} - p_{vap}}{\gamma} - \Delta z - h_L \qquad \textbf{(15.3.2)}$$

in which p_{atm} = atmospheric pressure, p_{vap} = vapor pressure of the liquid, $\Delta z = z_2 - z_1$, the difference in elevation between locations 1 and 2, and h_L = head loss on the suction pipe between locations 1 and 2. The performance data provided by the pump manufacturer usually include the *NPSH* curve. Application of Eq. 15.3.2 is demonstrated in the following example.

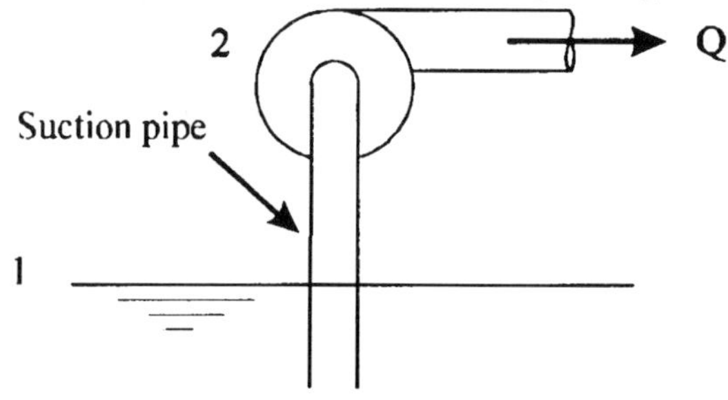

Figure 15.6 Cavitation setting for a pump.

Example 15.10

Water at 80° F is being pumped through a 20-ft-long, 4-inch-diameter suction pipe at a rate of 500 gal/min. The suction pipe includes two vaned bends (K=0.2) and a reentrant entrance (K = 0.8). At the given flow rate the friction factor is 0.015. If the allowable $NPSH$ at that flow rate is 20 ft, determine the height z above the inlet water surface that the pump can be located to operate without cavitating.

Solution. Assume p_{atm} = 14.7 psi. From a table of properties of water one finds that at a temperature of 80° F, p_{vap} = 0.507 psi, and γ = 62.1 lb/ft³. The suction pipe head loss is

$$h_L = \left(\frac{fL}{D} + \Sigma K\right)\frac{Q^2}{2gA^2}$$

$$= \left(\frac{0.015\times 20}{4/12} + 2\times 0.2 + 0.8\right)\frac{500/(7.48\times 60)}{2\times 32.2\times(\pi/4)^2\times(4/12)^4} = 4.77 \text{ ft}$$

Substituting the accumulated data into Eq. 15.3.2 results in

$$20 \le \frac{(14.7-0.507)\times 144}{62.1} - \Delta z - 4.77\text{ , or } \Delta z \le 8.1 \text{ ft}$$

Hence, the pump should be placed no higher than about 8 ft above the suction reservoir surface.

Pump Similarity Rules

Following the principles of similitude presented in Section 4.3, relationships for two pumps that are geometrically similar can be developed. Let $\dot{W}$ = pump power, N = pump speed, and D = impeller diameter, with subscripts 1 and 2 designating the two pumps. Then the pump similarity relationships are:

$$\frac{\dot{W}_2}{\dot{W}_1} = \left(\frac{N_2}{N_1}\right)^3\left(\frac{D_2}{D_1}\right)^5 \quad \textbf{(15.3.3)}$$

$$\frac{H_2}{H_1} = \left(\frac{N_2}{N_1}\right)^2\left(\frac{D_2}{D_1}\right)^2 \quad \textbf{(15.3.4)}$$

$$\frac{Q_2}{Q_1} = \frac{N_2}{N_1}\left(\frac{D_2}{D_1}\right)^3 \quad \textbf{(15.3.5)}$$

These relations are used to design or select a pump from a family of geometrically similar units. Another use of them is to examine the effects of changing speed on a given unit.

Example 15.11

A pump is designed to operate at 1,500 rev/min with an efficiency of 0.70 when delivering water at 600 gal/min against a head of 65 ft. If the same pump is now required to deliver water at a head of 100 ft at the same efficiency, determine the required rotational speed, the discharge and power delivered to the pump.

Solution. The given data are the following:

$H_1 = 65$ ft; $Q_1 = 600$ gal/min, or $1.34\ \text{ft}^3/\text{sec}$; $N_1 = 1{,}500$ rev/min;
$\dot{W}_1 = 62.4 \times 1.34 \times 65 / 0.7 = 7{,}760$ ft - lb/sec, or 14.1 horsepower; $H_2 = 100$ ft

Solve for N_2, Q_2, and $\dot{W}_2$ using Eqs. 15.3.4, 15.3.5, and 15.3.3, respectively:

$$N_2 = \sqrt{\frac{H_2}{H_1}}\left(\frac{D_2}{D_1}\right)N_1 = \sqrt{\frac{100}{65}} \times 1 \times 1{,}500 = 1{,}860 \text{ rev/min}$$

$$Q_2 = \frac{N_2}{N_1}\left(\frac{D_2}{D_1}\right)^3 Q_1 = \frac{1{,}860}{1{,}500} \times 1^3 \times 600 = 744 \text{ gal/min}$$

$$\dot{W}_2 = \left(\frac{N_2}{N_1}\right)^3\left(\frac{D_2}{D_1}\right)^5 \dot{W}_1 = \left(\frac{1{,}860}{1{,}500}\right)^3 \times 1^5 \times 14.1 = 26.9 \text{ horsepower}$$

Note that for this problem the ratio $D_2 / D_1 = 1$ since the same pump is used for both flow conditions.

15.4 Impeller Analysis for Turbomachines

When analyzing the flow of a fluid around the blades of a turbomachine, we assume the pressure to be constant. Then, with no elevation change, Bernoulli's equation allows us to conclude that the magnitude of the velocity, relative to a moving blade, remains constant. Consequently, the relative velocities $v_{r1} = v_{r2}$, or the magnitude of the inlet relative velocity equals the magnitude of the outlet relative velocity. In addition, blades are designed so the inlet relative velocity is tangent to the blade when the machine is operating at design speed. The outlet relative velocity is obviously tangent to the blade. A typical blade is shown in Fig.15.7.

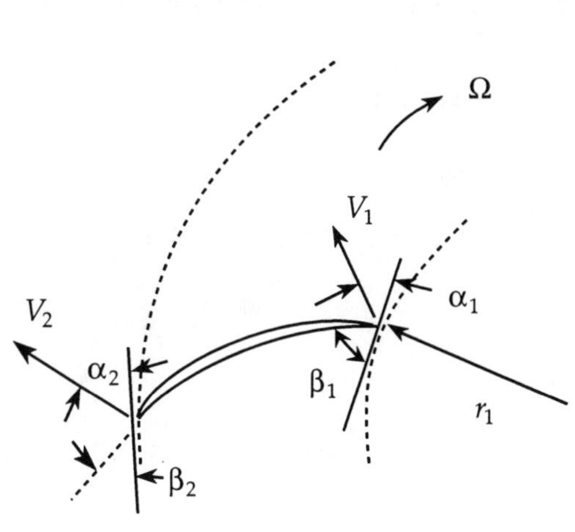

a) absolute velocities

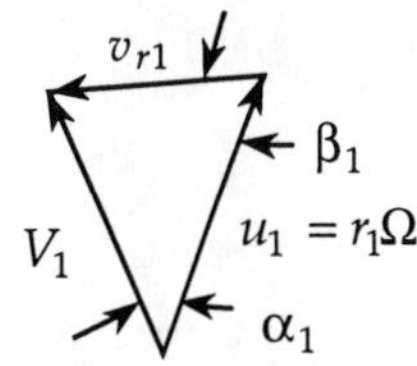

b) inlet velocity polygon

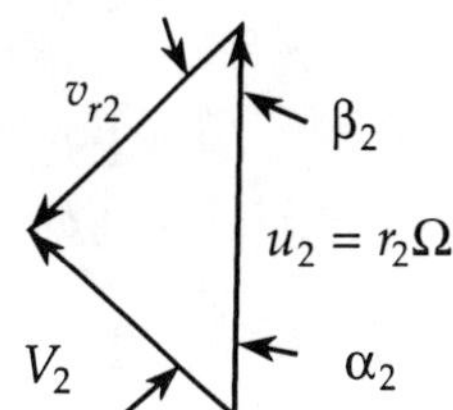

c) outlet velocity polygon

Figure 15.7 A blade in a turbomachine

The torque exerted on the fluid is given by

$$T = \dot{m}(r_2 V_2 \cos\alpha_2 - r_1 V_1 \cos\alpha_1) \quad \textbf{(15.4.1)}$$

so that the power transfer to the fluid is

$$\dot{W} = T\Omega = \dot{m}\Omega(r_2 V_2 \cos\alpha_2 - r_1 V_1 \cos\alpha_1) \quad \textbf{(15.4.2)}$$

Turbines are often designed so there is no post swirl, i.e., $\alpha_2 = 90°$ and pumps so that there is no preswirl, i.e., $\alpha_1 = 90°$, for maximum efficiency. If an efficiency is specified for a turbomachine, the output of a turbine is $\dot{W}\eta_T$ and the input required is $\dot{W} / \eta_P$.

Example 15.12

The 16-mm-deep impeller of a centrifugal pump is designed without inlet swirl and with $r_1 = 20$ mm, $r_2 = 160$ mm, $\beta_1 = 55°$ and $\beta_2 = 40°$ with an impeller rotational speed of 1800 rpm. Estimate a) the discharge, b) the power requirement if the efficiency is 85%, and c) the maximum possible pressure rise in the water.

Solution

a) If there is no inlet swirl, $\alpha_1 = 90°$ in Fig.15.7b and, from the inlet velocity polygon, we find

$$u_1 = r_1\Omega = 0.02 \times 1800 \times 2\pi / 60 = 3.77 \text{ m/s}$$

$$V_1 = u_1 \tan\beta_1 = 3.77 \tan 55° = 5.38 \text{ m/s}$$

The discharge is found using the velocity component normal to the inlet area:

$$Q = A_1 V_1 = 2\pi \times 0.02 \times 0.016 \times 5.38 = 0.0108 \text{ m}^3/\text{s}$$

b) We must now find V_2. Refer to Fig.15.7c:

$$u_2 = r_2\Omega = 0.16 \times 1800 \times 2\pi / 60 = 30.16 \text{ m/s}$$

Continuity: $Q = A_2 V_2 \sin\alpha_2$

$0.0108 = (2\pi \times 0.16 \times 0.016) V_2 \sin\alpha_2. \quad \therefore V_2 \sin\alpha_2 = 0.671 \text{ m/s}$

With $\beta_2 = 40°$, the velocity polygon provides the following:

$$v_{r2} \sin 40° = 6.71. \quad \therefore v_{r2} = 1.044 \text{ m/s}$$

$$V_2 \cos \alpha_2 + 1.044 \cos 40° = 30.16$$

Consequently,

$$\left. \begin{array}{l} V_2 \sin \alpha_2 = 0.671 \\ V_2 \cos \alpha_2 = 29.36 \end{array} \right\} \therefore V_2 = 29.37 \text{ m/s}, \ \alpha_2 = 1.31°$$

The power requirement is

$$\dot{W} = \dot{m}\Omega(r_2 V_2 \cos \alpha_2 - r_1 V_1 \cos \alpha_1) / \eta_P$$

$$= 1000 \times 0.0108 \times \frac{1800 \times 2\pi}{60}(0.16 \times 29.37 \cos 1.31° - 0.02 \times 5.38 \cos 90°) / 0.85$$

$$= 11300 \text{ W}$$

c) Apply the energy equation, Eq.15.2.2, between the inlet and the outlet recognizing that input energy is negative:

$$-\frac{\dot{W}}{\gamma Q} = \frac{\Delta p}{\gamma} + \frac{V_2^2 - V_1^2}{2g}$$

$$\frac{11300 \times 0.85}{9800 \times 0.0108} = \frac{\Delta p}{9800} + \frac{29.37^2 - 5.38^2}{2 \times 9.8} \qquad \therefore \Delta p = 473000 \text{ Pa}$$

Practice Problems (Essay)

15.1 Water at 60° F flows in a 1.5-in-diameter cast iron pipe at flow rate of 27 gal/min. Determine the head loss in a 350-ft-section of pipe using (a) the Darcy-Weisbach relation, and (b) the Hazen-Williams formula.

15.2 A pressure drop of 5 psi is not to be exceeded over a 600-ft length of horizontal 4-ft-diameter concrete pipe transporting water at 60° F. What flow rate can be accommodated?

15.3 Find the size of 1,000 ft of plastic tubing that can transport 20 gal/min of water at 80° F if the maximum allowable drop in the hydraulic grade line is 210 ft.

15.4 Determine the flow rate in the pipe shown. Discuss the nature of the hydraulic grade line.

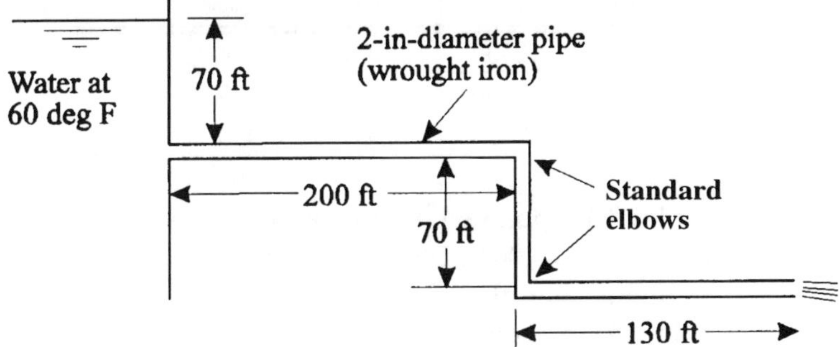

15.5 Water at 60° F is being pumped through the three pipes in series as shown. The power delivered to the pump is 2,500 horsepower, and the pump efficiency is 0.82. Compute the discharge. Assume constant friction factors.

Pipe	L (ft)	D (ft)	f	K
1	600	5	0.018	2
2	1,000	3	0.020	0
3	400	4	0.019	10

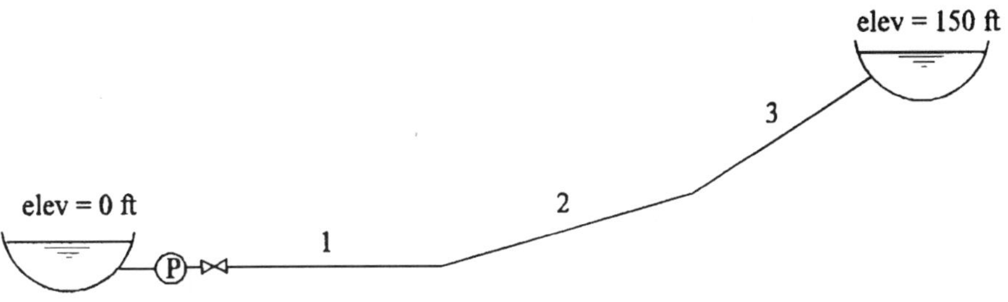

15.6 The system discharge for the four-pipe parallel system is $Q_{sys} = 5\ \text{ft}^3/\text{sec}$. Compute the flow distribution. Assume $f = 0.020$ for all four pipes.

Pipe	L (ft)	D (ft)	K
1	200	0.333	2
2	300	0.500	0
3	180	0.167	4
4	260	0.333	1

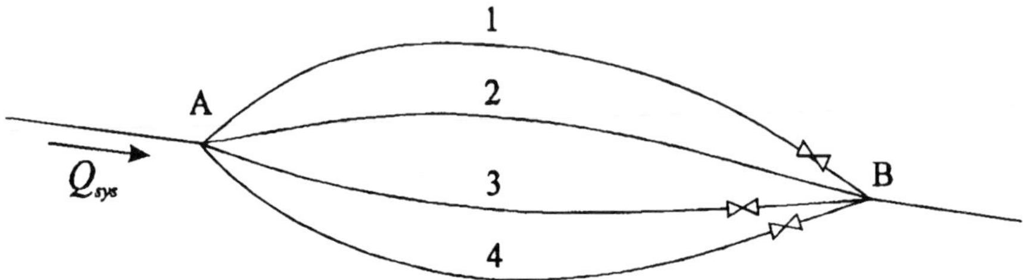

15.7 Consider the three branch piping system shown in Fig. 15.4a with the data tabulated below. The reservoir elevations are $z_A = 65$ ft, $z_C = 5$ ft, and $z_D = 43$ ft. Find the water flow distribution assuming constant friction factors.

Pipe	L (ft)	D (ft)	f	K
1	2,500	0.500	0.02	2
2	1,600	0.333	0.025	3
3	3,300	0.500	0.018	7

15.8 An oil ($\gamma = 53\ \text{lb/ft}^3$) is pumped through a single pipe system with the following constraints: $L = 8{,}000$ ft, $D = 8$ in, $f = 0.02$, $K = 12.5$. The upper reservoir is 105 ft higher than the lower one. Using the pump characteristic curve shown in Fig. 15.5, determine (a) the oil discharge in the pipe, and (b) the power requirement.

15.9 During a test on a water pump, no cavitation is detected when the atmospheric pressure is 15 psi, the water temperature is 80° F, and the discharge is 13 gal/sec. The inlet pipe diameter is 6 in, and the pump is situated 16 ft above the suction reservoir.

(a) Compute the *NPSH* at standard atmospheric pressure conditions. (b) If the pump is to produce the same head and discharge at a location where the atmospheric pressure is 12 psi, what is the required change in elevation of the pump relative to the inlet reservoir to avoid cavitation?

15.10 A pump at design conditions delivers 230 ft^3/sec of water at a head of 38 ft. The pump speed is 970 rev/min, and its efficiency is 0.85. A geometrically similar pump that is one-half the size is to run at a speed of 1,200 rev/min. Find the resulting head, discharge, and required power.

Practice Problems (Multiple Choice)

A pump, whose performance curve is given below, has been chosen to deliver water in a piping system consisting of four pipes arranged as shown. The pumping system consists of four pipes arranged as shown. Water at 60° F is being pumped from reservoir A and exits at either reservoir B or at location D. The atmospheric pressure is p_{atm} = 14.7 psi, the vapor pressure is p_{vap} = 0.256 psi, and the specific weight is γ = 62.4 lb/ft^3. The pipe characteristics are shown in the accompanying table. All of the pipe diameters are 4 inches, and the friction factor in each pipe is assumed to be constant, f = 0.02.

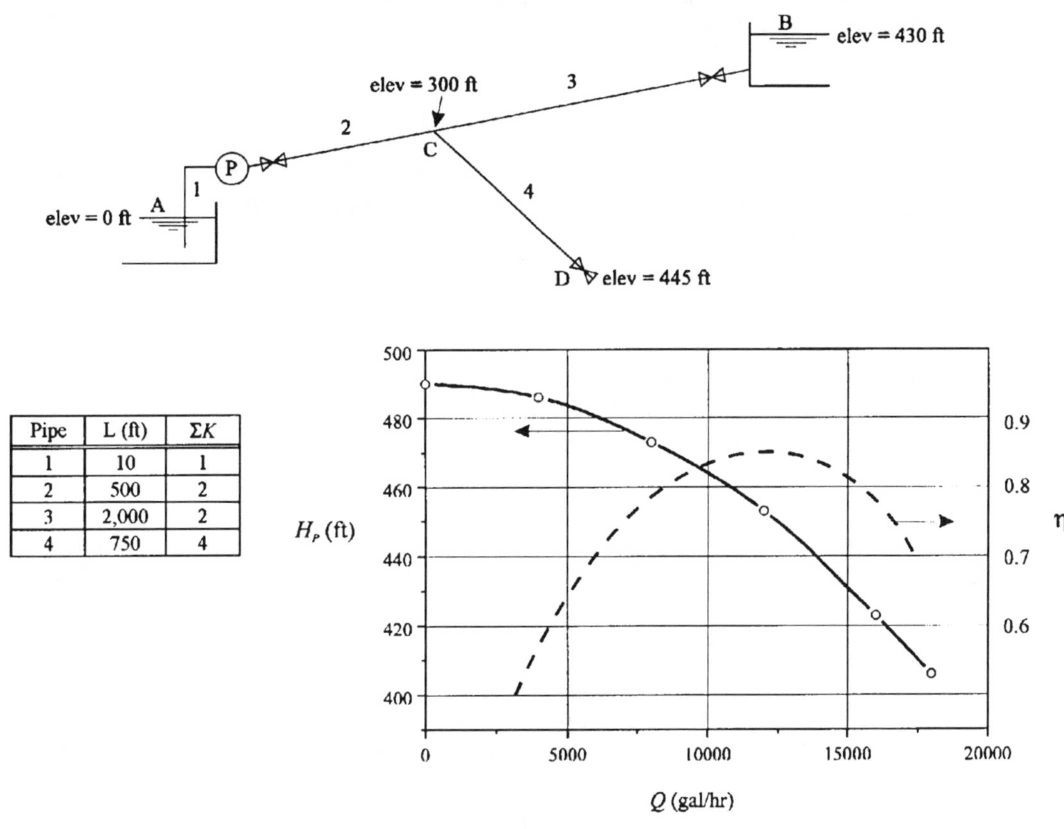

Pipe	L (ft)	ΣK
1	10	1
2	500	2
3	2,000	2
4	750	4

Questions 15.11 - 15.20

15.11 If the discharge through the pump is 5,000 gal/hr, then the head loss across pipe 2, in feet, is

a) 11.2 b) 300 c) 485 d) 2.3 e) 8.6

15.12 Assume that the valve at location D is closed, then the discharge in the system is, in gal/hr

a) 5,000 b) 9,000 c) 11,000 d) 7,000 e) 13,000

15.13 If the valve at D is open and the discharge through the pump is 11,000 gal/hr, then the discharge in pipe 4, in gal/hr, is

a) 3,600 b) 11,000 c) 7,400 d) 6,000 e) 2,400

15.14 Assume that the discharge through the pump is 8,000 gal/hr. Then the required power for the pump, in horsepower, is approximately

a) 148 b) 16 c) 118 d) 20 e) 27

15.15 At a given operating state, the *NPSH* for the pump is 25 ft, and the discharge is 10,000 gal/hr. The elevation of the pump above the water surface at location A, in ft, is about

a) 33 b) 25 c) 8 d) 0.4 e) -8

15.16 Assume that the pump delivers 12,000 gal/hr. The pressure at location C, in pounds per square inch, is

a) -2 b) 61 c) 195 d) -63 e) 130

15.17 With the same assumption as Problem 16, if the valve at location D is closed, the gage pressure at that location, in pounds per square inch, is approximately

a) 195 b) -2 c) 130 d) 61 e) -63

15.18 If the valve at location D is open and the head rise across the pump is 460 ft, then the discharge in pipe 3, in gal/hr, is

a) 3,600 b) 11,000 c) 2,400 d) 7,400 e) 6,000

15.19 If the valve at location B were closed, and pipe 4 was open, the head across the pump, in ft, would be about

a) 485 b) 450 c) 470 d) 490 e) 435

15.20 Assume that the pump is delivering 12,000 gal/hr to the system at a rotational speed of 900 rev/min. If a geometrically similar pump 2/3 the original size were used to deliver the same discharge, the speed of the new pump, in rev/min, would be

a) 2,020 b) 276 c) 400 d) 900 e) 3,040

Solutions to Problems

15.1 This is a category 1 problem.

a) The kinematic viscosity of 60° F water is 1.22×10^{-5} ft²/sec. From Table 15.1 we find

$e = 0.00085$ ft. Compute the velocity, Reynolds number, and relative roughness:

$$D = 1.5/12 = 0.125 \text{ ft}, \quad V = \frac{27/(7.48 \times 60)}{\pi \times 0.125^2/4} = 4.9 \text{ ft/sec},$$

$$Re = \frac{4.9 \times 0.125}{1.22 \times 10^{-5}} = 3.2 \times 10^6, \quad \varepsilon = \frac{0.00085}{0.125} = 0.0068$$

Compute the friction factor:

$$f = 1.325\left[\ln\left(0.27 \times 0.0068 + 5.74 \times (3.2 \times 10^6)^{-0.9}\right)\right]^{-2} = 0.033$$

(One could also use the Moody Diagram to estimate the friction factor.) Substitute into the Darcy-Weisbach equation:

$$h_L = \frac{0.033 \times 350 \times 4.9^2}{0.125 \times 2 \times 32.2} = 34.4 \text{ ft}, \quad \text{or} \quad 34.4 \times \frac{62.4}{144} = 14.9 \text{ psi}$$

b) From Table 15.2, assume $C = 110$. Then substitute known data into the Hazen-Williams formula:

$$h_L = \frac{4.72 \times 350 \times [27/(7.48 \times 60)]^{1.85}}{110^{1.85} \times 0.125^{4.87}} = 38.1 \text{ ft}, \quad \text{or} \quad 38.1 \times \frac{62.4}{144} = 16.5 \text{ psi}$$

Note the difference in the two results; both are acceptable answers for most problems. Solution (a), however, is more accurate.

15.2 This is a category 2 problem. The head loss due to friction is $h_L = 5 \times 144/62.4 = 11.5$ ft. Substitute into the Darcy-Weisbach equation:

$$11.5 = f\frac{600}{4} \times \frac{V^2}{2 \times 32.2}, \quad \text{or} \quad fV^2 = 4.94$$

The Moody diagram is used to estimate f. The kinematic viscosity is $\nu = 1.22 \times 10^{-5}$ ft²/sec, and the relative roughness is $\varepsilon = 0.01/4 = 0.0025$. First assume $f = 0.03$. Then $V = \sqrt{4.94/0.03} = 12.8$ ft/sec, $Re = 12.8 \times 4/1.22 \times 10^{-5}$ $= 4.2 \times 10^6$, and from the Moody diagram, we find the friction factor to be $f = 0.025$. Hence the new estimate of the velocity is $V = 14.1$ ft/sec, and $Re = 14.1 \times 4/1.22 \times 10^{-5} = 4.6 \times 10^6$. The friction factor remains at 0.025, so that the final velocity is $V = 14.1$ ft/sec, and the discharge is $Q = 14.1 \times \pi \times 4^2/4 = 177$ ft³/sec.

As an alternative to the trial and error approach, the empirical formula for a category 2 problem, Table 15.3, can be used since no pump is involved:

$$Q = -0.965 \times \left(\frac{32.2 \times 4^5 \times 11.5}{600} \right)^{0.5} \ln \left[\frac{0.01}{3.7 \times 4} + \left(\frac{3.17 \times (1.22 \times 10^{-5})^2 \times 600}{32.2 \times 4^3 \times 11.5} \right)^{0.5} \right]$$

$$= 177 \text{ ft}^3 / \text{sec}$$

The empirical formula gives an answer that is as precise as the trial and error solution.

15.3 This is a category 3 pipe problem. The following data are given: L = 1000 ft, Q = 20/(7.48 × 60) = 0.0446 ft³/sec, and h_L = 210 ft. Substitute into the Darcy-Weisbach equation and simplify:

$$210 = \frac{8 \times f \times 1000 \times 0.0446^2}{32.2 \times \pi^2 \times D^5}, \quad \text{or} \quad D = 0.189 f^{1/5}$$

This relation will be used in conjunction with the Moody diagram to estimate the friction factor. For drawn tubing, $e = 5 \times 10^{-6}$ ft from Table 15.1, and the kinematic viscosity for water at 80° F is $\nu = 0.93 \times 10^{-5} \text{ ft}^2/\text{sec}$. The two parameters required for using the diagram are

$$\varepsilon = \frac{e}{D} = \frac{5 \times 10^{-6}}{D}, \quad \text{and} \quad Re = \frac{4Q}{\pi D \nu} = \frac{4 \times 0.0446}{\pi \times D \times 0.93 \times 10^{-5}} = \frac{6110}{D}$$

Begin the iterative solution by assuming that $f = 0.015$. Then

$$D = 0.189 \times 0.015^{1/5} = 0.0816 \text{ ft}, \quad \varepsilon = \frac{5 \times 10^{-6}}{0.0816} = 6 \times 10^{-5}, \quad Re = \frac{6110}{0.0816} = 75{,}000$$

The updated friction factor read from the Moody diagram is f = 0.019. Repeating the calculations one finds that $D = 0.0855$ ft, $\varepsilon = 6 \times 10^{-5}$, and Re = 71,000. From the Moody diagram, the friction factor again is f = 0.019, unchanged from the previous estimate. Hence D = 0.0855 × 12 = 1.03 in. One could use either a 1-inch-diameter pipe or a 2-inch-diameter pipe. A 1-inch-diameter pipe will yield a head loss of h_L = 237 ft, and a 2-inch-diameter pipe will yield h_L = 74 ft.

The solution using the formula for a category 3 problem in Table 15.3 is

$$D = 0.66 \left[(5 \times 10^{-6})^{1.25} \left(\frac{1{,}000 \times 0.0446^2}{32.2 \times 210} \right)^{4.75} + 0.93 \times 10^{-5} \times 0.0446^{9.4} \left(\frac{1{,}000}{32.2 \times 210} \right)^{5.2} \right]^{0.04}$$

$$= 0.0868 \text{ ft}, \text{ or } 1.04 \text{ in}$$

Note that the solution is as accurate as using the Moody diagram with the trial and error procedure.

15.4 The kinematic viscosity is $\nu = 1.22 \times 10^{-5} \text{ ft}^2/\text{sec}$, the relative roughness is $\varepsilon = 0.00015/0.167 = 0.0009$, and the pipe diameter is $D = 2/12 = 0.167$ ft.

Assume a completely turbulent flow regime, so that the friction factor can be approximated by $f = 1.325[\ln(0.27 \times 0.0009)]^{-2} = 0.019$. Next, write the energy equation from the reservoir surface to the pipe outlet:

$$140 = \frac{V^2}{2g} + \left[\frac{fL}{D} + 2K_{elbow} + K_{entrance}\right]\frac{V^2}{2g}$$

Assume standard elbows so that $K_{elbow} = 0.9$, and with $K_{entrance} = 0.5$ from Table 15.4 substitute known data into the energy equation:

$$140 = \left[1 + \frac{0.019 \times 400}{0.167} + 2 \times 0.9 + 0.5\right]\frac{V^2}{2 \times 32.2} . \quad \therefore \ V = 13.6 \text{ ft/sec}$$

We perform one more iteration to determine a better estimate of the velocity:

$$Re = \frac{13.6 \times 0.167}{1.22 \times 10^{-5}} = 1.9 \times 10^5 ,$$

$$f = 1.325\left[\ln\left(0.27 \times 0.0009 + 5.74(1.9 \times 10^5)\right)^{-0.9}\right]^{-2} = 0.021$$

Substituting the new value of f into the energy equation and solving we find that V = 13.0 ft/sec. Hence $Q = 13.0 \times \pi \times 0.167^2 / 4 = 0.285 \text{ ft}^3/\text{sec}$. The hydraulic grade line has sudden drops at the entrance and elbows, and a linear drop over the length of the pipe.

15.5 The energy equation for the system is $H_P = (h_L)_{sys} + 150$, or

$$\frac{\dot{W}_P \eta}{\gamma Q_{sys}} = (R_1 + R_2 + R_3)Q_{sys}^2 + 150 , \text{ in which } R_i = \frac{1}{2gA_i^2}\left(\frac{fL}{D} + \Sigma K\right)_i$$

Compute the R-values:

$$A_1 = \frac{\pi}{4} \times 5^2 = 19.6 \text{ ft}^2, \quad R_1 = \frac{1}{2 \times 32.2 \times 19.6^2}\left(\frac{0.018 \times 600}{5} + 2\right) = 0.00017$$

$$A_2 = \frac{\pi}{4} \times 3^2 = 7.1 \text{ ft}^2, \quad R_2 = \frac{1}{2 \times 32.2 \times 7.1^2}\left(\frac{0.020 \times 1000}{3} + 0\right) = 0.00205$$

$$A_3 = \frac{\pi}{4} \times 4^2 = 12.6 \text{ ft}^2, \quad R_3 = \frac{1}{2 \times 32.2 \times 12.6^2}\left(\frac{0.019 \times 400}{4} + 10\right) = 0.00116$$

Substituting the known data into the energy equation we have

$$\frac{2500 \times 550 \times 0.82}{62.4 Q_{sys}} = (0.00017 + 0.00205 + 0.00116)Q_{sys}^2 + 150$$

which reduces to $18{,}070 = 0.00338Q_{sys}^3 + 150Q_{sys}$. Solving the equation by trial and error gives

$$Q_{sys} = 99 \text{ ft}^3/\text{sec}.$$

15.6 Compute the R-values:

$$\text{Pipe 1:}\ \ L_e = \frac{2\times 0.333}{0.02} = 33\,,\ \ R_1 = \frac{8\times 0.02\times 233}{32.2\times \pi^2 \times 0.333^5} = 29$$

$$\text{Pipe 2:}\ \ L_e = 0\,,\qquad R_2 = \frac{8\times 0.02\times 300}{32.2\times \pi^2 \times 0.5^5} = 5$$

$$\text{Pipe 3:}\ \ L_e = \frac{2\times 0.167}{0.02} = 33\,,\ \ R_3 = \frac{8\times 0.02\times 213}{32.2\times \pi^2 \times 0.167^5} = 826$$

$$\text{Pipe 4:}\ \ L_e = \frac{1\times 0.333}{0.02} = 17\,,\ \ R_4 = \frac{8\times 0.02\times 277}{32.2\times \pi^2 \times 0.333^5} = 34$$

We next find the system head loss, and then the discharge in each pipe:

$$(h_L)_{sys} = \frac{5^2}{\left(1/\sqrt{29} + 1/\sqrt{5} + 1/\sqrt{826} + 1/\sqrt{34}\right)^2} = 35.5\text{ ft}$$

$$Q_1 = \sqrt{35.5/29} = 1.11\text{ ft}^3/\text{sec}\,,\quad Q_2 = \sqrt{35.5/5} = 2.66\text{ ft}^3/\text{sec}$$

$$Q_3 = \sqrt{35.5/826} = 0.21\text{ ft}^3/\text{sec}\,,\quad Q_4 = \sqrt{35.5/34} = 1.02\text{ ft}^3/\text{sec}$$

The sum of the four discharges is equal to the system discharge.

15.7 Compute the loss terms for each pipe:

$$\text{Pipe 1:}\ \ L_e = \frac{2\times 0.5}{0.02} = 50\,,\ \ (h_L)_1 = \frac{8\times 0.02\times 2550}{32.2\times \pi^2 \times 0.5^5} Q_1^2 = 41Q_1^2$$

$$\text{Pipe 2:}\ \ L_e = \frac{3\times 0.333}{0.025} = 40\,,\ \ (h_L)_2 = \frac{8\times 0.025\times 1640}{32.2\times \pi^2 \times 0.333^5} Q_2^2 = 252Q_2^2$$

$$\text{Pipe 3:}\ \ L_e = \frac{7\times 0.5}{0.018} = 190\,,\ \ (h_L)_3 = \frac{8\times 0.018\times 3490}{32.2\times \pi^2 \times 0.5^5} Q_3^2 = 51Q_3^2$$

Assume the flow directions shown in Fig. 15.4a, and write the energy relations for each pipe:

$$\text{Pipe 1:}\ \ H_B = 65 - 41Q_1^2\,,\ \ \text{or } Q_1 = \sqrt{(65 - H_B)/41}$$

$$\text{Pipe 2:}\ \ H_B = 5 + 252Q_2^2\,,\ \ \text{or } Q_2 = \sqrt{(H_B - 5)/252}$$

$$\text{Pipe 3:}\ \ H_B = 43 + 51Q_3^2\,,\ \ \text{or } Q_3 = \sqrt{(H_B - 43)/51}$$

The solution is found by iterating values of H_B, computing Q_1, Q_2, and Q_3, and checking the continuity balance at the junction: $\Sigma Q = Q_1 - Q_2 - Q_3$. The results are tabulated below; five iterations are performed until the continuity balance is satisfied within a tolerable limit.

Iteration	H_B	Q_1	Q_2	Q_3	ΣQ
1	43	0.732	0.388	0	+0.344
2	45	0.698	0.398	0.198	+0.102
3	46	0.681	0.403	0.242	+0.036
4	47	0.663	0.408	0.280	-0.025
5	415.6*	0.670	0.406	0.266	-0.002

* Estimated by linear interpolation between the two previous values of H_B and Q.

Hence the discharges are $Q_1 = 0.67$ ft^3/sec, $Q_2 = 0.40$ ft^3/sec, and $Q_3 = 0.27$ ft^3/sec.

15.8 With $D = 8/12 = 0.668$ ft, assuming a constant friction factor, and recognizing that $z_2 - z_1 = 105$ ft, the demand curve can be constructed:

$$H_P = 105 + \left(\frac{0.02 \times 8,000}{0.667} + 12.5\right)\frac{Q^2}{2 \times 32.2 \times \left(\pi \times 0.667^2 / 4\right)^2}$$

$$= 105 + 32Q^2$$

a) Assume Q = 1.0 ft^3/sec, then $H_p = 105 + 32 \times 1^2 = 137$ ft. On the pump characteristic curve of Fig. 15.5, $H_P \cong 175$ ft at $Q = 1.0$ ft^3/sec. Try another discharge, say $Q = 1.5$ ft^3/sec; then from the demand curve $H_P = 177$ ft, and from the pump curve $H_P \cong 170$ ft. Finally, with $Q = 1.4$ ft^3/sec, from the demand curve $H_P = 168$ ft, and from the pump curve $H_P \cong 170$ ft. Thus $Q = 1.4$ ft^3/sec and $H_P = 170$ ft.

b) From Fig 15.4, the pump efficiency at $Q = 1.4$ ft^3/sec is approximately 0.7. Hence, the required power input to the pump is $\dot{W}_P = 53 \times 1.4 \times 170 / 0.7$ $= 18,000$ ft - lb/sec, or 32.8 horsepower.

15.9 First compute the head loss in the suction pipe. Neglecting minor losses (no information is given) we have

$$h_L = \frac{8 \times 0.02 \times 16 \times (13/7.48)^2}{32.2 \times \pi^2 \times 0.5^5} = 0.78 \text{ ft}$$

a) For water at 80° F, γ = 62.2 lb/ft^3, and $p_{vap} = 0.507$ psi. Substituting into Eq. 15.3.2, the condition for imminent cavitation is

$$NPSH = \frac{(15 - 0.507) \times 144}{62.2} - 16 - 0.78 = 16.8 \text{ ft}$$

b) With $p_{atm} = 12$ psi and the same flow conditions for part a), use Eq. 15.3.2 to determine the required z:

$$\Delta z = \frac{p_{atm} - p_{vap}}{\gamma} - h_L - NPSH = \frac{(12 - 0.507) \times 144}{62.2} - 0.78 - 16.8 = 9.0 \text{ ft}$$

Hence the change in elevation is $16 - 9 = 7$ ft.

15.10 The following parameters are given: $N_1 = 970$ rpm, $N_2 = 1{,}200$ rpm, $H_1 = 38$ ft, $Q_1 = 230$ ft^3/sec, and $D_2 : D_1 = 1 : 2$. Using the similitude relations, Eqs. 15.3.4 and 15.3.5, compute Q_2 and H_2:

$$H_2 = 38 \times \left(\frac{1,200}{970}\right)^2 \times \left(\frac{1}{2}\right)^2 = 14.5 \text{ ft}$$

$$Q_2 = 230 \times \frac{1,200}{970} \times \left(\frac{1}{2}\right)^3 = 35.6 \text{ ft}^3 / \text{sec}$$

The power is $\dot{W}_2 = 62.4 \times 35.6 \times 14.5 / 0.85 = 37{,}900$ ft - lb/sec, or 70 horsepower.

Questions 15.11-15.20

Before answering the specific questions, it is expedient to compute the R-values for each pipe (recall that $h_L = RQ^2$):

$$\text{Pipe 1:}\quad Le = \frac{1 \times 0.333}{0.02} = 17 \text{ ft}, \quad R_1 = \frac{8 \times 0.02 \times 27}{32.2 \times \pi^2 \times 0.333^5} = 3$$

$$\text{Pipe 2:}\quad Le = \frac{2 \times 0.333}{0.02} = 33 \text{ ft}, \quad R_2 = \frac{8 \times 0.02 \times 533}{32.2 \times \pi^2 \times 0.333^5} = 66$$

$$\text{Pipe 3:}\quad Le = \frac{2 \times 0.333}{0.02} = 33 \text{ ft}, \quad R_3 = \frac{8 \times 0.02 \times 2{,}033}{32.2 \times \pi^2 \times 0.333^5} = 250$$

$$\text{Pipe 4:}\quad Le = \frac{4 \times 0.333}{0.02} = 67 \text{ ft}, \quad R_4 = \frac{8 \times 0.02 \times 1{,}817}{32.2 \times \pi^2 \times 0.333^5} = 224$$

15.11 **d)** Compute the discharge and head loss in pipe 2:

$$Q_2 = 5{,}000 \frac{\text{gal}}{\text{hr}} \times \frac{1 \text{ ft}^3}{7.48 \text{ gal}} \times \frac{1 \text{ hr}}{3600 \text{ sec}} = 0.186 \frac{\text{ft}^3}{\text{sec}},$$

$$(h_L)_2 = R_2 Q_2^2 = 66 \times 0.186^2 = 2.28 \text{ ft}$$

15.12 **b)** The flow path is pipe 1 - pipe 2 - pipe 3; there is no flow in pipe 4. The demand curve from A to B is

$$H_P = z_B - z_A + (h_L)_1 + (h_L)_2 + (h_L)_3 = 430 + (3 + 66 + 250)Q^2 = 430 + 319Q^2$$

A trial and error solution is employed: assume Q, compute H_P from the demand curve, and compare it with H_P from the pump characteristic curve.

Trial 1: $Q = 10{,}000$ gal/hr $= 0.371$ ft³/sec,

$$H_P \text{ (demand)} = 430 + 319 \times 0.371^2 = 474 \text{ ft},$$
$$H_P \text{ (pump)} = 465 \text{ ft}$$

Trial 2: $Q = 9{,}000$ gal/hr $= 0.334$ ft³/sec,

$$H_P \text{ (demand)} = 430 + 319 \times 0.334^2 = 467 \text{ ft},$$
$$H_P \text{ (pump)} = 466 \text{ ft}$$

15.13 **a)** $Q_2 = 11{,}000$ gal/min $= 0.409$ ft^3/sec, and from the pump curve, $H_P =$ 460 ft. Write the energy equation from A to C:

$$H_C = H_P - (R_1 + R_2)Q_2^2 = 460 - 69 \times 0.409^2 = 449 \text{ ft}$$

The energy equation from C to D is $H_C = z_D + R_4 Q_4^2$, where H_C is the hydraulic grade line at location C. Therefore the discharge in pipe 4 is

$$Q_4 = \sqrt{\frac{H_C - z_D}{R_4}} = \sqrt{\frac{449-445}{224}} = 0.134 \text{ ft}^3/\text{sec}, \text{ or } 3{,}600 \text{ gal/hr.}$$

15.14 **d)** The discharge is $Q = 8{,}000/(7.48\times 3600) = 0.297 \text{ ft}^3/\text{sec}$. From the pump characteristic curve,

$H_P = 470$ ft, and $\eta = 0.8$. Therefore the required power is

$\dot{W}_P = \gamma Q H_P/\eta = 62.4\times 0.297 \quad \times 470/0.8 = 10{,}900 \text{ ft-lb/sec}$, or 19.8 horsepower.

15.15 **c)** First determine the head loss in the suction pipe:

$$Q = \frac{10{,}000}{7.48\times 3600} = 0.371 \text{ ft}^3/\text{sec}, \quad (h_L)_1 = 3\times 0.371^2 = 0.41 \text{ ft}$$

Using Eq. 15.3.2, the required pump height is:

$$\Delta z = \frac{p_{atm} - p_{vap}}{\gamma} - (h_L)_1 - NPSH = \frac{(14.7-0.256)\times 144}{62.4} - 0.41 - 25 = 7.9 \text{ ft}$$

15.16 **b)** The discharge through the pump is $Q = 12{,}000/(7.48\times 3600)$ $= 0.446 \text{ ft}^3/\text{sec}$, and from the characteristic curve the pump head is H_P = 454 ft. The energy equation from A to C is $H_P = z_C + p_C/\gamma - z_A + (h_L)_1 + (h_L)_2$. Therefore

$$\frac{p_C}{\gamma} = 454 - 300 - (3+66)\times 0.446^2 = 140 \text{ ft}, \quad \text{and } p_C = \frac{62.4\times 140}{144} = 61 \text{ lb/in}^2$$

15.17 **b)** Write the energy equation from C to D and solve for the pressure at location D:

$p_C/\gamma + z_C = p_D/\gamma + z_D$. (Since there is no flow in pipe 4, the energy equation reduces to the law of hydrostatics.) From Problem 16, the pressure at location C is 61 lb/in^2. Therefore,

$$p_D = 61 + 62.4\times\frac{300-445}{144} = -1.83 \text{ lb/in}^2$$

15.18 **d)** From the pump curve, for H_P = 460 ft, Q = 11,000 gal/hr, or 0.409 ft^3/sec. Write the energy equation from A to C and evaluate the hydraulic grade line:

$$H_C = H_P - (h_L)_1 - (h_L)_2 = 460 - 69\times 0.409^2 = 449 \text{ ft}$$

Thus the discharge in pipe 3, between location C and location B is

$$Q_3 = \sqrt{\frac{(H_C - H_B)}{R_3}} = \sqrt{\frac{449-430}{250}} = 0.276 \text{ ft}^3/\text{sec}, \text{ or } 7{,}400 \text{ gal/hr}$$

15.19 **c)** The flow path is pipe 1 - pipe 2 - pipe 4; there is no flow in pipe 3. Write the energy equation from location A to location D:

$$H_P = z_D - z_A + (h_L)_1 + (h_L)_2 + (h_L)_4 = 445 + (3 + 66 + 224)Q^2 = 445 + 293Q^2$$

A trial and error solution is employed: assume Q, compute H_P from the demand curve, and compare it with H_P from the pump characteristic curve.

Trial 1: $Q = 10{,}000 \text{ gal/hr} = 0.371 \text{ ft}^3/\text{sec}$,

$$H_P \text{ (demand)} = 445 + 293 \times 0.371^2 = 485 \text{ ft},$$
$$H_P \text{ (pump)} = 465 \text{ ft}$$

Trial 2: $Q = 7{,}500 \text{ gal/hr} = 0.278 \text{ ft}^3/\text{sec}$,

$$H_P \text{ (demand)} = 445 + 293 \times 0.278^2 = 468 \text{ ft},$$
$$H_P \text{ (pump)} = 475 \text{ ft}$$

Trial 3: $Q = 8{,}000 \text{ gal/hr} = 0.297 \text{ ft}^3/\text{sec}$,

$$H_P \text{ (demand)} = 445 + 293 \times 0.297^2 = 471 \text{ ft},$$
$$H_P \text{ (pump)} = 472 \text{ ft}$$

15.20 **e)** Given data are $Q_1 = Q_2$, $N_1 = 900$ rev/min. $D_1 : D_2 = 2 : 3$. Use the similitude relation, Eq. 15.3.5 to compute the new pump speed N_2:

$$N_2 = N_1\left(\frac{Q_2}{Q_1}\right)\left(\frac{D_1}{D_2}\right)^3 = 900 \times 1 \times \left(\frac{3}{2}\right)^3 = 3{,}040 \text{ rev/min}$$

Engineering Economics

by Frank Hatfield

Engineering designs are intended to produce good results. In general, the good results are accompanied by undesirable effects including the costs of manufacturing or construction. Selecting the best design requires the engineer to anticipate and compare the good and bad outcomes. If outcomes are evaluated in dollars, and if "good" is defined as positive monetary value, then design decisions may be guided by the techniques known as *engineering economy*. Decisions based solely on engineering economy may be guaranteed to result in maximum goodness only if all outcomes are anticipated and can be monetized (measured in dollars).

16.1 Value and Interest

"Value" is not synonymous with "amount." The value of an amount of money depends on when the amount is received or spent. For example, the promise that you will be given a dollar one year from now is of less value to you than a dollar received today. The difference between the anticipated amount and its current value is called *interest* and is frequently expressed as a time rate. If an interest rate of 10% per year is used, the expectation of receiving \$1.00 one year hence has a value now of about \$0.91. In engineering economy, interest usually is stated in percent per year. If no time unit is given, "per year" is assumed.

Example 16.1

What amount must be paid in two years to settle a current debt of $1,000 if the interest rate is 6%?

Solution. Value after one year $= 1000 + 1000 \times 0.06$

$$= 1000(1 + 0.06)$$

$$= \$1060$$

Value after two years $= 1060 + 1060 \times 0.06$

$$= 1000(1 + 0.06)^2$$

$$= \$1124$$

Hence, $1,124 must be paid in two years to settle the debt.

16.2 Cash Flow Diagrams

As an aid to analysis and communication, an engineering economy problem may be represented graphically by a horizontal time axis and vertical vectors representing dollar amounts. The cash flow diagram for Example 16.1 is sketched in Fig. 16.1. Income is up and expenditures are down. It is important to pick a point of view and stick with it. For example, the vectors in Fig. 16.1 would have been reversed if the point of view of the lender had been adopted. It is a good idea to draw a cash flow diagram for every engineering economy problem that involves amounts occurring at different times.

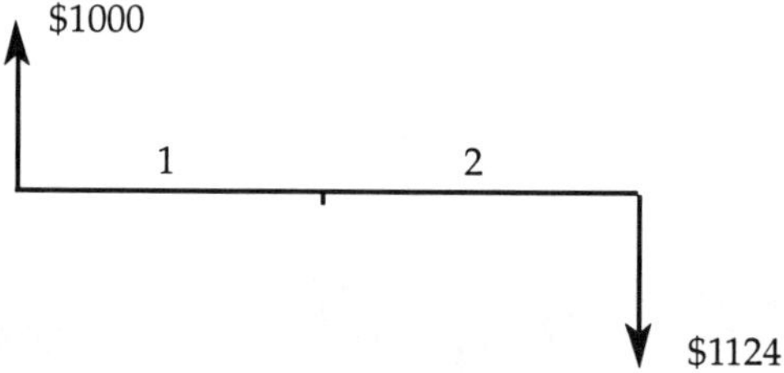

Figure 16.1 Cash flow diagram for Example 16.1.

In engineering economy, amounts are almost always assumed to occur at the ends of years. Consider, for example, the value today of the future operating expenses of a truck. The costs probably will be paid in varied amounts scattered throughout each year of operation, but for computational ease the expenses in each year are represented by their sum (computed without consideration of interest) occurring at the end of the year. The error introduced by neglecting interest for partial years is usually insignificant compared to uncertainties in the estimates of future amounts.

16.3 Cash Flow Patterns

Engineering economy problems involve the following four patterns of cash flow, both separately and in combination:

P-pattern: A single amount P occurring at the beginning of n years. P frequently represents "present" amounts.

F-pattern: A single amount F occurring at the end of n years. F frequently represents "future" amounts.

A-pattern: Equal amounts A occurring at the ends of each of n years. The A-pattern frequently is used to represent "annual" amounts.

G-pattern: End-of-year amounts increasing by an equal annual gradient G. Note that the first amount occurs at the end of the second year. G is the abbreviation of "gradient."

The four cash flow patterns are illustrated in Fig. 16.2.

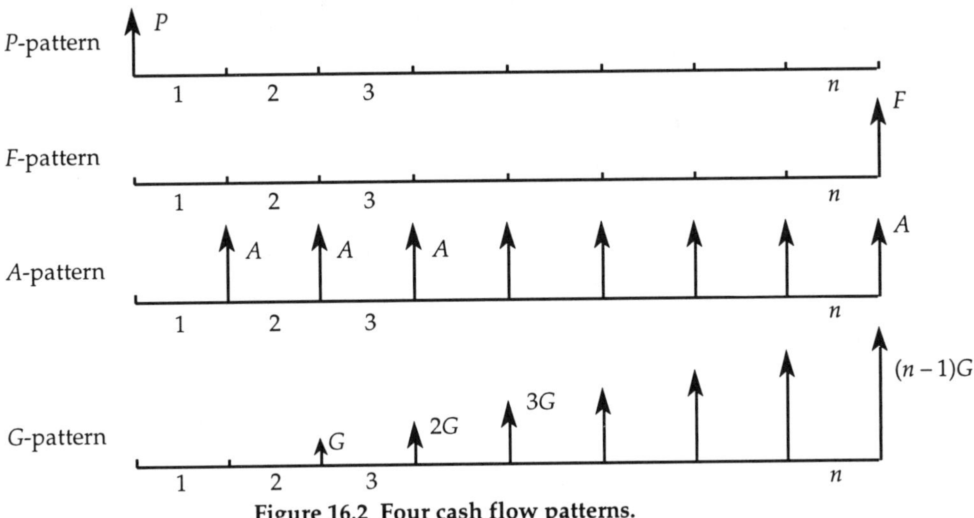

Figure 16.2 Four cash flow patterns.

16.4 Equivalence of Cash Flow Patterns

Two cash flow patterns are said to be equivalent if they have the same value. Most of the computational effort in engineering economy problems is directed at finding a cash flow pattern that is equivalent to a combination of other patterns. Example 16.1 can be thought of as finding the amount in an F-pattern that is equivalent to \$1,000 in a P-pattern. The two amounts are proportional, and the factor of proportionality is a function of interest rate i and number of periods n. There is a different factor of proportionality for each possible pair of the cash flow patterns defined in Section 16.3. To minimize the possibility of selecting the wrong factor, mnemonic symbols are assigned to the factors. For Example 16.1, the proportionality factor is written $(F/P)_n^i$ and solution is achieved by evaluating

$$F = (F/P)_n^i P$$

To analysts familiar with the canceling operation of algebra, it is apparent that the correct factor has been chosen. However, the letters in the parentheses together with the sub- and super-scripts constitute a single symbol; therefore, the canceling operation is not actually performed. Table 16.1 lists symbols and formulas for commonly used factors. Table 16.2, located at the end of this chapter, presents a convenient way to find numerical values of interest factors. Those values are tabulated for selected interest rates i and number of interest periods n; linear interpolation for intermediate values of i and n is acceptable for most situations.

TABLE 16.1 Formulas for Interest Factors

Symbol	To Find	Given	Formula
$(F/P)_n^i$	F	P	$(1+i)^n$
$(P/F)_n^i$	P	F	$\frac{1}{(1+i)^n}$
$(A/P)_n^i$	A	P	$\frac{i(1+i)^n}{(1+i)^n-1}$
$(P/A)_n^i$	P	A	$\frac{(1+i)^n-1}{i(1+i)^n}$
$(A/F)_n^i$	A	F	$\frac{i}{(1+i)^n-1}$
$(F/A)_n^i$	F	A	$\frac{(1+i)^n-1}{i}$
$(A/G)_n^i$	A	G	$\frac{1}{i}-\frac{n}{(1+i)^n-1}$
$(F/G)_n^i$	F	G	$\frac{1}{i}\left[\frac{(1+i)^n-1}{i}-n\right]$
$(P/G)_n^i$	P	G	$\frac{1}{i}\left[\frac{(1+i)^n-1}{i(1+i)^n}-\frac{n}{(1+i)^n}\right]$

Example 16.2

Derive the formula for $(F/P)_n^i$.

Solution. For $n = 1$,

$$F = (1+i)P$$

that is,

$$(F/P)_1^i = (1+i)^1$$

For any n,

$$F = (1+i)(F/P)_{n-1}^i P$$

that is,

$$(F/P)_n^i = (1+i)(F/P)_{n-1}^i$$

By induction,

$$(F/P)_n^i = (1+i)^n$$

Example 16.3

A new widget twister, with a life of six years, would save $2,000 in production costs each year. Using a 12% interest rate, determine the highest price that could be justified for the machine. Although the savings occur continuously throughout each year, follow the usual practice of lumping all amounts at the ends of years.

Solution. First, sketch the cash flow diagram.

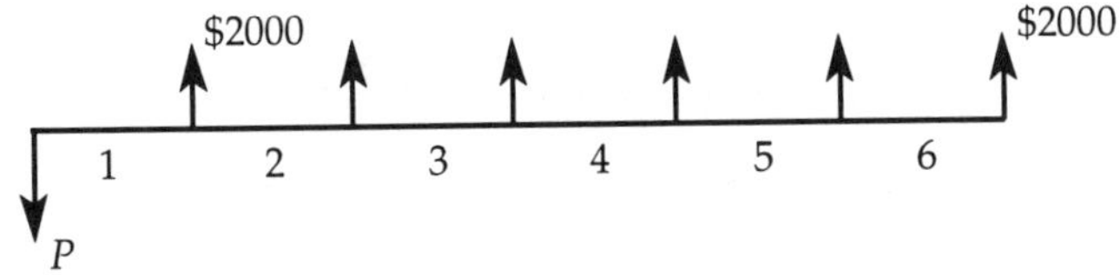

The cash flow diagram indicates that an amount in a P-pattern must be found that is equivalent to $2,000 in an A-pattern. The corresponding equation is

$$\begin{aligned} P &= (P/A)_n^i A \\ &= (P/A)_6^{12\%} 2000 \end{aligned}$$

Table 16.2 is used to evaluate the interest factor for $i = 12\%$ and $n = 6$:

$$\begin{aligned} P &= 4.1114 \times 2000 \\ &= \$8223 \end{aligned}$$

Example 16.4

How soon does money double if it is invested at 8% interest?

Solution. Obviously, this is stated as

$$F = 2P$$

Therefore,

$$(F/P)_n^{8\%} = 2$$

In the 8% interest table, the tabulated value for (F/P) that is closest to 2 corresponds to $n = 9$ years.

Example 16.5

Find the value in 1987 of a bond described as "Acme 8% of 2000" if the rate of return set by the market for similar bonds is 10%.

Solution. The bond description means that the Acme company has an outstanding debt that it will repay in the year 2000. Until then, the company will pay out interest on that debt at the 8% rate. Unless otherwise stated, the principal amount of a single bond is $1,000. If it is assumed that the debt is due December 31, 2000, interest is paid every December 31, and the bond is purchased January 1, 1987, then the cash flow diagram, with unknown purchase price P, is:

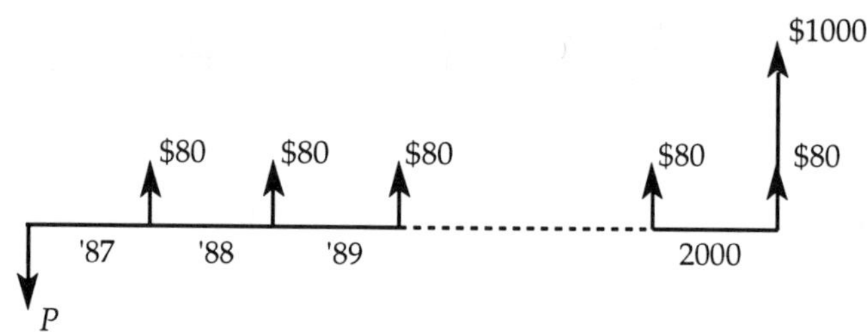

The corresponding equation is

$$\begin{aligned} P &= (P/A)_{14}^{10\%}\, 80 + (P/F)_{14}^{10\%}\, 1000 \\ &= 7.3667 \times 80 + 0.2633 \times 1000 \\ &= \$853 \end{aligned}$$

That is, to earn 10% the investor must buy the 8% bond for $853, a "discount" of $147. Conversely, if the market interest rate is less than the nominal rate of the bond, the buyer will pay a "premium" over $1,000.

The solution is approximate because bonds usually pay interest semiannually, and $80 at the end of the year is not equivalent to $40 at the end of each half year. But the error is small and is neglected.

Example 16.6

You are buying a new appliance. From past experience you estimate future repair costs as:

First Year	$ 5
Second Year	15
Third Year	25
Fourth Year	35

The dealer offers to sell you a four-year warranty for $60. You require at least a 6% interest rate on your investments. Should you invest in the warranty?

Solution. Sketch the cash flow diagram.

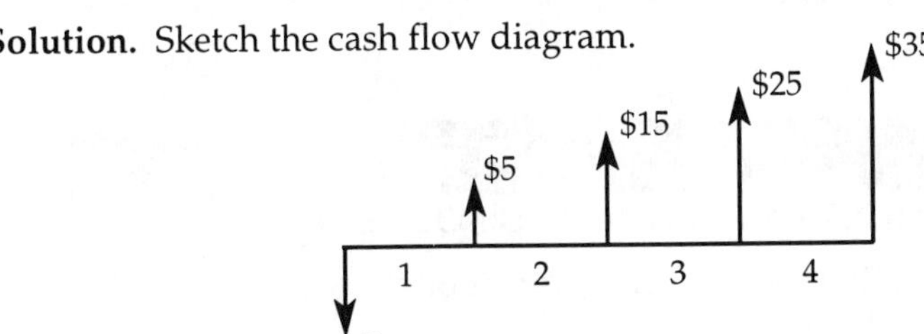

The known cash flows can be represented by superposition of a $5 A-pattern and a $10 G-pattern. Verify that statement by drawing the two patterns. Now it is clear why the standard G-pattern is defined to have the first cash flow at the end of the second year. Next, the equivalent amount P is computed:

$$\begin{aligned} P &= (P/A)_4^{6\%} A + (P/G)_4^{6\%} G \\ &= 3.4651 \times 5 + 4.9455 \times 10 \\ &= \$67 \end{aligned}$$

Since the warranty can be purchased for less then $67, the investment will earn a rate of return greater than the required 6%. Therefore, you should purchase the warranty .

If the required interest rate had been 12%, the decision would be reversed. This demonstrates the effect of a required interest rate on decision making. Increasing the required rate reduces the number of acceptable investments.

Example 16.7

Compute the annual equivalent maintenance costs over a 5-year life of a laser printer that is warranted for two years and has estimated maintenance costs of $100 annually. Use $i = 10\%$.

Solution. The cash flow diagram appears as:

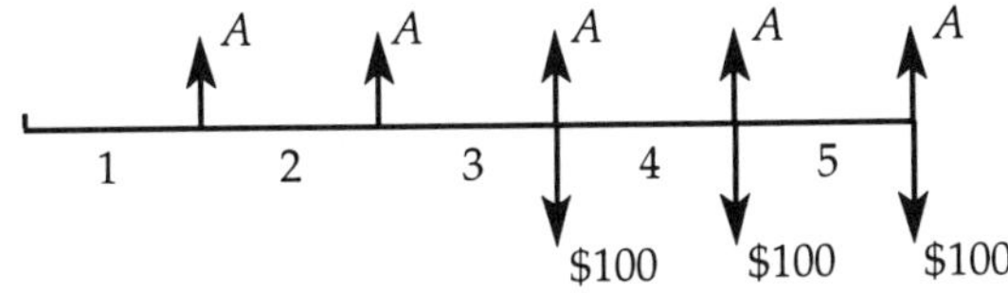

There are several ways to find the 5-year A-pattern equivalent to the given cash flow. One of the more efficient methods is to convert the given 3-year A-pattern to an F-pattern, and then find the 5-year A-pattern that is equivalent to that F-pattern. That is,

$$\begin{aligned} A &= (A/F)_5^{10\%} (F/A)_3^{10\%} 100 \\ &= \$54. \end{aligned}$$

16.5 Unusual Cash Flows and Interest Periods

Occasionally an engineering economy problem will deviate from the year-end cash flow and annual compounding norm. The examples in this section demonstrate how to handle these situations.

Example 16.8

PAYMENTS AT BEGINNINGS OF YEARS

Using a 10% interest rate, find the future equivalent of:

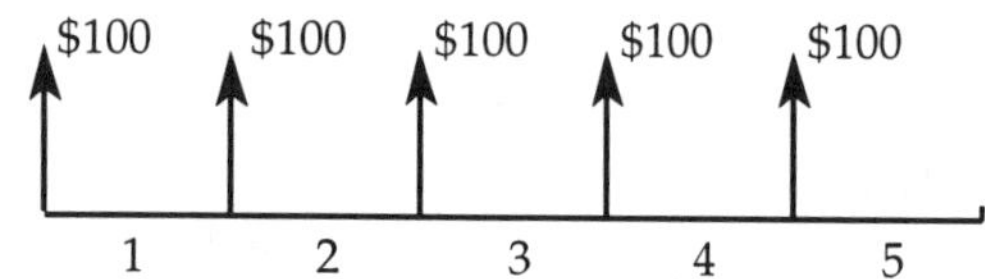

Solution. Shift each payment forward one year. Therefore,

$$A = (F/P)_1^{10\%} 100 = \$110.$$

This converts the series to the equivalent A-pattern:

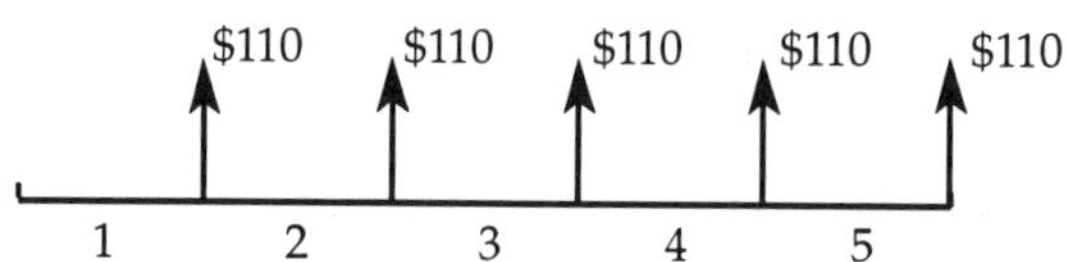

and the future equivalent is found to be

$$F = (F/A)_5^{10\%} 110 = \$672.$$

Alternative Solution. Convert to a six-year series:

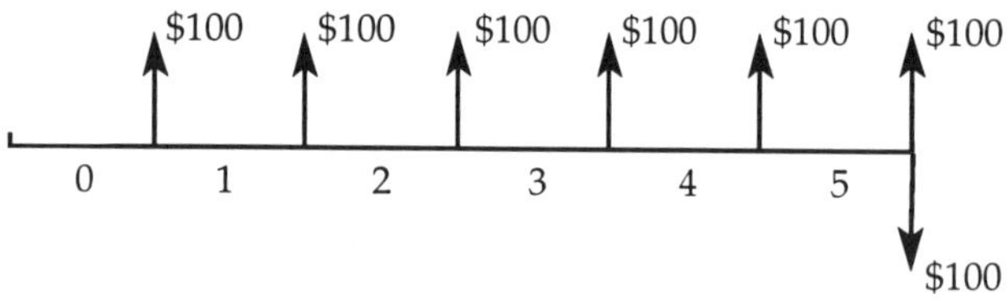

The future equivalent is

$$F = (F/A)_6^{10\%} 100 - 100 = \$672.$$

Example 16.9

SEVERAL INTEREST AND PAYMENT PERIODS PER YEAR

Compute the present value of eighteen monthly payments of \$100 each, where interest is 1/2% per month.

Solution. The present value is computed as

$$P = (P/A)_{18}^{1/2\%} 100 = \$1717$$

Example 16.10

ANNUAL PAYMENTS WITH INTEREST COMPOUNDED *m* TIMES PER YEAR

Compute the effective annual interest rate equivalent to 5% nominal annual interest compounded daily. (There are 365 days in a year.)

Solution. The legal definition of nominal annual interest is

$$i_n = mi$$

where i is the interest rate per compounding period. For the example,

$$\begin{aligned} i &= i_n/m \\ &= 0.05/365 = 0.000137 \text{ or } 0.0137\% \text{ per day.} \end{aligned}$$

Because of compounding, the effective annual rate is greater than the nominal rate. By equating (F/P)-factors for one year and m periods, the effective annual rate i_e may be computed as follows:

$$\begin{aligned} (1+i_e)^1 &= (1+i)^m \\ i_e &= (1+i)^m - 1 \\ &= (1.000137)^{365} - 1 = 0.05127 \text{ or } 5.127\% \end{aligned}$$

Example 16.11

CONTINUOUS COMPOUNDING

Compute the effective annual interest rate i_e equivalent to 5% nominal annual interest compounded continuously.

Solution. As m approaches infinity, the value for i_e is found as follows:

$$\begin{aligned} i_e &= e^{mi} - 1 \\ &= e^{0.05} - 1 \\ &= 0.051271 \text{ or } 5.1271\% \end{aligned}$$

Example 16.12

ANNUAL COMPOUNDING WITH m PAYMENTS PER YEAR

Compute the year-end amount equivalent to twelve end-of-month payments of $10 each. Annual interest rate is 6%.

Solution. The usual simplification in engineering economy is to assume that all payments occur at the end of the year, giving an answer of $120. This approximation may not be acceptable for a precise analysis of a financial agreement. In such cases, the agreement's policy on interest for partial periods must be investigated.

Example 16.13

ANNUAL COMPOUNDING WITH PAYMENT EVERY m YEARS

With interest at 10% compute the present equivalent of

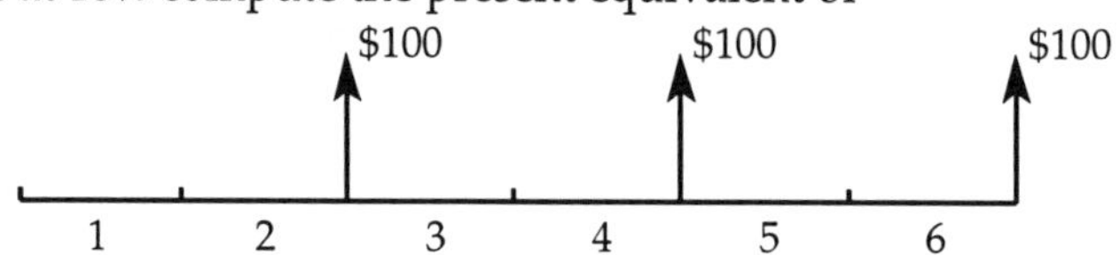

Solution. First convert each payment to an A-pattern for the m preceding years. That is,

$$A = (A/F)_2^{10\%} 100$$

$$= \$47.62$$

Then, convert the A-pattern to a P-pattern:

$$P = (P/A)_6^{10\%} 47.62$$

$$= \$207$$

16.6 Evaluating Alternatives

The techniques of engineering economy assume the objective of maximizing net value. For a business, "value" means after-tax cash flow. For a not-for-profit organization, such as a government agency, value may include non-cash benefits, such as clean air, improved public health, or recreation to which dollar amounts have been assigned.

This section concerns strategies for selecting alternatives such that net value is maximized. The logic of these methods will be clear if the following distinctions are made between two different types of interest rates, and between two different types of relationships among alternatives.

Types of Interest Rates

•*Rate of Return* (*ROR*): The estimated interest rate produced by an investment. It may be computed by finding the interest rate such that the estimated income and non-cash benefits (positive value), and the estimated expenditures and non-cash costs (negative value), sum to a net equivalent value of zero.

•*Minimum Attractive Rate of Return (MARR)*: The lowest rate of return that the organization will accept. In engineering economy problems, it is usually a given quantity and may be called, somewhat imprecisely, "interest," "interest rate," "cost of money," or "interest on capital."

Types of Alternative Sets

•*Mutually Exclusive Alternatives*: Exactly one alternative must be selected. Examples: "Shall Main Street be paved with concrete or asphalt?" "In which room will we put the piano?" If a set of alternatives is mutually exclusive, it is important to determine whether the set includes the null (do nothing) alternative. Serious consequences can arise from failure to recognize the null alternative.

•*Independent Alternatives*: It is possible (but not necessarily economical) to select any number of the available alternatives. Examples: "Which streets should be paved this year?" "Which rooms shall we carpet?"

Annual Equivalent Cost Comparisons

The estimated income and benefits (positive) and expenditures and costs (negative) associated with an alternative are converted to the equivalent A-pattern using an interest rate equal to $MARR$. The A-value is the *annual net equivalent value* ($ANEV$) of the alternative. If the alternatives are mutually exclusive, the one with the largest $ANEV$ is selected. If the alternatives are independent, all that have positive $ANEV$ are selected.

Example 16.14

A new cap press is needed. Select the better of the two available models described below. $MARR$ is 10%.

Model	Price	Annual Maintenance	Salvage Value	Life
Reliable	11,000	1,000	1,000	10 years
Quicky	4,000	1,500	0	5 years

Solution. The $ANEV$ is calculated for each model:

Reliable: $$ANEV = -(A/P)_{10}^{10\%} 11000 - 1000 + (A/F)_{10}^{10\%} 1000$$
$$= -\$2730$$

Quicky: $$ANEV = -(A/P)_{5}^{10\%} 4000 - 1500$$
$$= -\$2560$$

Negative $ANEV$ indicates a rate of return less than $MARR$. However, these alternatives are mutually exclusive and the null is not available. The problem is one of finding the less costly way to perform a necessary function. Therefore, *Quicky* is selected. If $MARR$ had been much lower, *Reliable* would have been selected. By setting the $MARR$ relatively high, the organization is indicating that funds are not available to invest now in order to achieve savings in the future.

Present Equivalent Cost Comparisons

The estimated income and benefits (positive), and expenditures and costs (negative), associated with an alternative are converted to the equivalent P-pattern using an interest rate equal to $MARR$. The P-value is the *present net equivalent value* ($PNEV$) of the alternative. If the alternatives are mutually exclusive, the one with the largest $PNEV$ is selected. $PNEV$ is also called "life cycle cost," "present worth," "capital cost," and "venture worth." If the alternatives are independent, all that have positive $PNEV$ are selected.

The present equivalent cost method requires that alternatives be evaluated over the same span of time. If their lives are not equal, the lowest common multiple of the lives is used for the time span, with each alternative repeated to fill the span. A variation, called the *capitalized cost method*, computes the $PNEV$ for repeated replacement of the alternatives for an infinite time span. The capitalized cost P of an infinite series of equal amounts A is given by

$$P = A(P/A)_{\infty}^{i} = A/i\,.$$

Example 16.15

Repeat Example 16.14 using the present equivalent cost method.

Solution. The $PNEV$ is calculated for each model:

$$\textit{Reliable:}\quad PNEV = -11000 - (P/A)_{10}^{10\%}1000 + (P/F)_{10}^{10\%}1000$$

$$= -\$16{,}800$$

$$\textit{Quicky:}\quad PNEV = -4000 - (P/F)_{5}^{10\%}4000 - (P/A)_{10}^{10\%}1500$$

$$= -\$15{,}700$$

Note that *Quicky* was replaced in order to fill the ten-year time span. As in Example 16.14, *Quicky* is selected. The two methods will always give the same decision if used correctly. Observe that for both alternatives

$$PNEV = (P/A)_{10}^{10\%}ANEV\,.$$

Incremental Approach

For a set of mutually exclusive alternatives, only the differences in amounts need to be considered. Compute either the $ANEV$ or the $PNEV$ and base the decision on the sign of that value.

Example 16.16

Repeat Example 16.14 using an incremental present net equivalent value approach.

Solution. *Reliable* costs \$7,000 more than *Quicky* but saves \$500 each year in maintenance expenses and eliminates the need for a \$4,000 replacement after five years. In addition, *Reliable* has a \$1,000 salvage value whereas *Quicky* has none.

Reliable – *Quicky*:

$$PNEV = -7000 + (P/A)_{10}^{10\%} 500 + (P/F)_{5}^{10\%} 4000 + (P/F)_{10}^{10\%} 1000$$

$$= -\$1060$$

The negative result dictates selection of *Quicky*. That is, the additional initial cost required to purchase *Reliable* is not justified.

Rate of Return Comparisons

The expression for *ANEV* or *PNEV* is formulated and then solved for the interest rate that will give a zero *ANEV* or *PNEV*. This interest rate is the rate of return (*ROR*) of the alternative. To apply the rate of return method to mutually exclusive alternatives requires incremental comparison of each possible pair of alternatives; increments of investment are accepted if their rates of return exceed *MARR*. For independent alternatives, all those with *ROR* exceeding *MARR* are accepted. The rate of return method permits conclusions to be stated as functions of *MARR*, which is useful if *MARR* has not been determined precisely.

Example 16.17

A magazine subscription costs $50 for one year or $80 for two years. If you want to receive the magazine for at least two years, which alternative is better?

Solution. The two-year subscription requires an additional initial investment of $30 and eliminates the payment of $50 one year later. The rate of return formulation is:

$$PNEV = 0$$

$$-30 + 50(P/F)_{1}^{i} = 0$$

The solution for i is as follows:

$$-30 + 50\frac{1}{(1+i)} = 0$$

$$i = 0.67 \text{ or } 67\%$$

Therefore, if your *MARR* is less than 67%, subscribe for two years.

Example 16.18

Repeat Example 16.14 using the rate of return method.

Solution. Use the incremental expression derived in Example 16.16, but set *PNEV* equal to zero and use the interest rate as the unknown:

$$-7000 + (P/A)_{10}^{i} 500 + (P/F)_{5}^{i} 4000 + (P/F)_{10}^{i} 1000 = 0$$

By trial and error, the interest rate is found to be 6.6%. Therefore, *Reliable* is preferred if, and only if, *MARR* is less than 6.6%.

Benefit/Cost Comparisons

The benefit/cost ratio is determined from the formula:

$$\frac{B}{C} = \frac{\text{Uniform net annual benefits}}{\text{Annual equivalent of initial cost}}$$

where *MARR* is used in computing the A-value in the denominator. As with the rate of return method, mutually exclusive alternatives must be compared incrementally, the incremental investment being accepted if the benefit/cost ratio exceeds unity. For independent alternatives, all those with benefit/cost ratios exceeding unity are accepted.

Note that the only pertinent fact about a benefit/cost ratio is whether it exceeds unity. This is illustrated by the observation that a project with a ratio of 1.1 may provide greater net benefit than a project with a ratio of 10 if the investment in the former project is much larger than the investment in the latter. It is incorrect to rank mutually exclusive alternatives by their benefit/cost ratios as determined by comparing each alternative to the null (do nothing) alternative.

The benefit/cost ratio method will give the same decision as the rate of return method, present equivalent cost method, and annual equivalent cost method if the following conditions are met:

1. Each alternative is comprised of an initial cost and uniform annual benefit.
2. The form of the benefit/cost ratio given above is used without deviation.

Example 16.19

A road resurfacing project costs \$200,000, lasts five years, and saves \$100,000 annually in patching costs. *MARR* is 10%. Should the road be resurfaced?

Solution. The benefit/cost ratio is

$$\frac{B}{C} = \frac{100{,}000}{(A/P)_5^{10\%}\,200{,}000} = 1.9$$

Since the ratio exceeds unity, the resurfacing is justified.

A Note on *MARR*

In engineering economy examination problems, *MARR* is a given quantity. However, the following discussion of the determination of *MARR* will help clarify the logic underlying the various comparison methods.

In general, an organization will be able to identify numerous opportunities to spend money now that will result in future returns. For each of these independent investment opportunities, an expected rate of return can be estimated. Similarly, the organization will be able to find numerous sources of funds for investment. Associated with each source of funds is an interest rate. If the source is a loan, the associated interest rate is simply that charged by the lender. Funds generated by operations of the organization, or provided by its owners (if the organization is a business), or extracted from taxpayers (if the organization is a government agency) can be thought of as being borrowed from the owners or taxpayers. Therefore, such funds can be assigned a fictitious interest rate, which should not be less than the maximum rate of return provided by other opportunities in which the owners or taxpayers might invest.

Value will be maximized if the rates of return of all the selected investments exceed the highest interest rate charged for the money borrowed, and if every opportunity has been taken to invest at a rate of return exceeding that for which money can be borrowed. That rate is the Minimum Attractive Rate of Return. No investments should be made that pay rates of return less than *MARR*, and no loans should be taken that charge interest rates exceeding *MARR*. Furthermore, the organization should exploit all opportunities to borrow money at interest rates less than *MARR* and invest it at rates of return exceeding *MARR*.

To estimate *MARR* precisely would require the ability to foresee the future, or at least to predict all future investment and borrowing opportunities and their associated rates. A symptom of *MARR* being set too low is insufficient funds for all the investments that appear to be acceptable. Conversely, if *MARR* has been set too high, some investments will be rejected that would have been profitable.

Replacement Problems

How frequently should a particular machine be replaced? This type of problem can be approached by varying the life n. For each value of n, the annual costs and salvage value are estimated, and then the *ANEV* is computed. The value of n resulting in the smallest annual equivalent cost is the optimum, or economic, life of the machine. This approach is complicated by technological improvements in replacement machinery, which may make it advantageous to replace a machine before the end of its economic life. In practice, technological advances are difficult to anticipate.

Another form of the replacement problem asks if an existing asset should be replaced by a new (and possibly different) one. Again, the annual equivalent cost method is recommended. The *ANEV* of the replacement is computed, using its economic life for n. However, the annual cost of the existing asset is simply the estimated expense for one more year of operation. This strategy is based on the assumption that the annual costs of the existing asset increase monotonically as it ages.

Always Ignore the Past

Engineering economy, and decision making in general, deals with alternatives. But there is only one past and it affects all future alternatives equally. Therefore, past costs and income associated with an existing asset should not be included in computations that address the question of replacing the asset. Only the estimated cash flows of the future are relevant.

The mistake of counting past costs is common in everyday affairs. For example, a student may say, "I paid $90 for this textbook so I will not sell it for $20." A more rational approach would be to compare the highest offered price to the value of retaining the text.

Example 16.20

Yesterday a machine was bought for $10,000. Estimated life is ten years, with no salvage value at the end of its useful life. Current book value is $10,000. Today a vastly improved model was announced. It costs $15,000, has a ten-year life and no salvage value, but reduces operating costs by $4,000 annually. The current resale value of the older machine has dropped to $1,000 due to this stunning technological advance. Should the old model be replaced with a new model at this time?

Solution. The purchase price of the old machine, its book value, and the loss on the sale of the old machine are irrelevant to the analysis. The incremental cost of the new machine is $14,000 and the incremental income is $4,000 annually. A rate of return comparison is formulated as follows:

$$-14,000+(P/A)_{10}^{i}\,4000=0$$

Solving for rate of return gives $i = 26\%$, indicating that the older machine should be replaced immediately if *MARR* is less than 26%.

Break-Even Analysis

A break-even point is the value of an independent variable such that two alternatives are equally attractive. For values of the independent variable above the break-even point, one of the alternatives is preferred; for values of the independent variable below the break-even point, the other alternative is preferred. Break-even analysis is particularly useful for dealing with an independent variable that is subject to change or uncertainty since the conclusion of the analysis can be stated as a function of the variable. The rate of return method, as applied to mutually exclusive alternatives, is an example of break-even analysis. The independent variable is *MARR*.

Example 16.21

An item can be manufactured by hand for $15. Alternatively, the item can be produced by a machine at a fixed annual equivalent cost of $4,000 plus a variable cost of $1 per item. Assume that the cost of laying off and hiring workers is zero. For each of the two manufacturing processes, answer the following questions:

a)For what production rate is one method more economical than the other?

b)If the item is sold for $6, how many must be sold to make a profit?

c)How low must the price fall, in the short term, before production is discontinued?

Solution.

a) Let P be production rate in units per year. Production costs for the two processes are equated:

$$\text{Cost by machine} = \text{Cost manually}$$

$$4000+1P=5P$$

$$\therefore P=1000$$

If annual production is expected to be less than 1,000 units, the manual process is more economical. For production rates exceeding 1,000 units per year, the machine process is preferred.

b) Setting profit equal to zero is expressed as:

$$\text{gross income} - \text{cost} = 0$$

$$\text{Manual production: } 6P - 5P = 0$$

$$\therefore P = 0$$

$$\text{Machine production: } 6P - (4000 + 1P) = 0$$

$$\therefore P = 800$$

With price maintained at $6, the mechanized operation will be unprofitable if production rate is less than 800 units per year, but the manual operation is profitable at all production rates.

c) Manual production becomes unprofitable if the price drops below $5, and production will cease at that level. For the machine, the $4,000 cost continues whether or not the machine is running. Incremental income is generated so long as the price stays above the variable (per item) cost. Therefore, production will continue at any price over $1, even though a net loss may be sustained. Of course, if it appears that the price and production rate will not soon increase sufficiently to provide a profit, then the operation will be terminated.

16.7 Income Tax and Depreciation

Business pays to the federal government a tax that is a proportion of taxable income. Taxable income is gross revenue less operating costs (wages, cost of materials, etc.), interest payments on debts, and depreciation. Depreciation is different from the other deductions in that it is not a cash flow.

Depreciation is an accounting technique for charging the initial cost of an asset against two or more years of production. For example, if you buy a $50,000 truck for use in your construction business, deducting its total cost from income during the year of purchase gives an unrealistically low picture of income for that year, and an unrealistically high estimate of income for the succeeding years during which you use the truck. A more level income history would result if you deducted $10,000 per year for five years. In fact, the Internal Revenue Service (IRS) requires that most capital assets used in business be depreciated over a number of years rather than being deducted as expenses during the year of purchase.

An asset is depreciable if it is used to produce income, has a determinable life greater than one year, and decays, wears out, becomes obsolete, or gets used up. Examples include tools, production machinery, computers, office equipment, buildings, patents, contracts, franchises, and livestock raised for wool, eggs, milk or breeding. Non-depreciable assets include personal residence, land, natural resources, annual crops, livestock raised for sale or slaughter, and items intended primarily for resale such as stored grain and the merchandise in a department store.

Since depreciation is not a cash flow, it will not directly enter an engineering economy analysis. However, depreciation must be considered when estimating future income taxes, which are cash flows.

The IRS requires that the Modified Accelerated Cost Recovery System (MACRS) be applied to most tangible property placed in service after 1986. In general, MACRS is based on computing depreciation using a declining balance method or the straight line method treating the property as being placed in service and retired from service at midpoints of tax years, and setting salvage value equal to zero. Older methods, such as the Accelerated Cost Recovery System (ACRS) or the straight line method with a non-zero salvage value, may still show up in engineering economy problems, and therefore are included in this discussion. The following notation will be used in defining methods for computing depreciation:

B — The installed first cost, or basis.
n — Recovery period in years.
D_x — Depreciation in year x. at age n.
V_x — Undepreciated balance at the end of year x, also called book value.
V_n — Estimated salvage at age n.

In computing depreciation there is no attempt to equate book value with resale value, productive worth, or any other real figure. A business is not obliged to keep an asset for exactly n years, nor to sell it for exactly its book value or estimated salvage value. These, then, are the depreciation methods:

1. *Declining Balance*: Depreciation is taken as a proportion of book value:
$$D_x = V_{x-1}C/n$$
where n is the recovery period. For values of C equaling 1.25, 1.5 and 2 the method is called, respectively: 125% declining balance, 150% declining balance, and double declining balance.
2. *Straight Line Depreciation*: Depreciation is the same for every full year and is calculated as
$$D_x = (B - V_n)/n.$$
3. *Accelerated Cost Recovery System (ACRS)*: An asset is classed as having a recovery period n of 3, 5, 10, or 15 years using IRS guidelines. For each class, a set of annual rates R_x is specified by the IRS. With 3-year property, for example, $R_1 = 0.25,\ R_2 = 0.38,\ R_3 = 0.37$. Depreciation is calculated as follows:
$$D_x = R_x B.$$
By definition, the salvage value using *ACRS* is zero.

Example 16.22

The purchase price of an over-the-road tractor unit is \$100,000, its recovery period is three years, and it can be sold for an estimated \$20,000 at that time. Compute the depreciation schedules using each of the methods described.

Solution.

Double Declining Balance (MACRS)

Year	Depreciation	Book Value
		$100,000
1	0.5 × ($100,000×2/3) = $33,333	$ 66,667
2	$66,667 × 2/3 = $44,444	$ 22,222
3	$22,222 × 2/3 = $14,815	$ 7,407
4	$7407	$ 0

*The book value must be zero after three years of service, so the formula was not used for the last year.

Straight Line (MACRS)

Year	Depreciation	Book Value
		$100,000
1	$100,000/3 × 0.5 = $16,667	$ 83,333
2	$100,000/3=$33,333	$ 50,000
3	$100,000/3=$33,333	$ 16,667
4	$100,000/3 × 0.5 = $16,667	$ 0

Straight Line (General)

Year	Depreciation	Book Value
		$100,000
1	(100,000 – 20,000)/3 = $26,667	$ 73,333
2	$26,667	$ 46,667
3	$26,667	$ 20,000

Accelerated Cost Recovery System (ACRS)

Year	Depreciation	Book Value
		$100,000
1	0.25 × $100,000 = $25,000	$ 75,000
2	0.38 × $100,000 = $38,000	$ 37,000
3	0.37 × $100,000 = $37,000	$ 0

Note: The factor 0.5 in the MACRS results from the requirement to use the midpoint of a tax year.

16.8 Inflation

The "buying power" of money changes with time. A decline in "buying power" is experienced due to a general increase in prices, called "inflation."

Inflation, if it is anticipated, can be exploited by fixing costs and allowing income to increase. A manufacturing business can fix its costs by entering long-term contracts for materials and wages, by purchasing materials long before they are needed, or by stockpiling its product for later sale. Income is allowed to respond to inflation by avoiding long-term contracts for the product. Borrowing becomes more attractive if inflation is expected since the debt will be paid with the less valuable cash of the future.

MARR may be adjusted for anticipated uniform inflation using the formula

$$d = i + f + if$$

where d is inflation-adjusted *MARR*, i is unadjusted *MARR*, and f is the rate of inflation. This formula facilitates solution of some types of engineering economy problems.

Example 16.23

A machine having a five-year life can replace a worker who is compensated $20,000 per year with 5% annual "cost of living" increases. Operating and maintenance costs for the machine are negligible. *MARR* is 10%. Find the maximum price that can be justified for the machine if:

a) general price inflation is 5%

b) general price inflation is zero

Solution.

a) Although the worker gets a larger amount of money each year, her raises are exactly matched by increased prices, including those of her employer's product. "Buying power" of her annual compensation remains equal to the current value of $20,000. Hence, the maximum justifiable price for the machine is

$$P = (P/A)_5^{10\%} 20,000 = \$75,816$$

b) The maximum justifiable price of the machine is equal to the present equivalent value of the annual amounts of compensation:

$$(P/F)_1^{10\%}(1.05)\ 20,000 = \$19,090$$

$$(P/F)_2^{10\%}(1.05)^2 20,000 = \$18,224$$

$$(P/F)_3^{10\%}(1.05)^3 20,000 = \$17,394$$

$$(P/F)_4^{10\%}(1.05)^4 20,000 = \$16,604$$

$$(P/F)_5^{10\%}(1.05)^5 20,000 = \underline{\$15,850}$$

$$\text{therefore,} \qquad P = \$86,162$$

Example 16.24

Recompute the value, in terms of 1987 "buying power," of the "Acme 8% of 2000" bond discussed in Example 16.5, but assume 6% annual inflation.

Solution. The cash flow for each year must be divided by an inflation factor, as well as multiplied by an interest factor, and then the factored cash flows are added:

$$
\begin{aligned}
&(P/F)_1^{10\%} \quad 80/(1.06) = \$\ 69 \\
&(P/F)_2^{10\%} \quad 80/(1.06)^2 = \$\ 59 \\
&(P/F)_3^{10\%} \quad 80/(1.06)^3 = \$\ 50 \\
&\qquad \vdots \\
&(P/F)_{13}^{10\%} \quad 80/(1.06)^{13} = \$\ 11 \\
&(P/F)_{14}^{10\%} \quad 80/(1.06)^{14} = \$\ 9 \\
&(P/F)_{14}^{10\%} \quad 1000/(1.06)^{14} = \$\ \underline{116} \\
&\text{therefore,} \qquad P = \$\ 541
\end{aligned}
$$

Note that investors can account for anticipated inflation simply by using increased values of *MARR*. A *MARR* of 16.6% gives the same conclusions as a *MARR* of 10% with 6% inflation.

Alternative Solution. Using inflation-adjusted *MARR* we have

$$
\begin{aligned}
d &= i + f + if \\
&= .10 + .06 + .10 \times .06 \\
&= .166 \quad \text{or} \quad 16.6\%
\end{aligned}
$$

The value of the bond is

$$
\begin{aligned}
P &= (P/A)_{14}^{16.6\%} 80 + (P/F)_{14}^{16.6\%} 1000 \\
&= 5.3225 \times 80 + .1165 \times 1000 \\
&= \$542
\end{aligned}
$$

Formulas from Part 5.4 (a table in the NCEES Handbook) were used to evaluate the interest factors.

Practice Problems

(If you attempt only a few, select those with a star.)

16.1 Which of the following would be most difficult to monetize?

a) maintenance cost b) selling price c) fuel cost
d) prestige e) interest on debt

VALUE AND INTEREST

*16.2 If $1,000 is deposited in a savings account that pays 6% annual interest and all the interest is left in the account, what is the account balance after three years?

a) $840 b) $1,000 c) $1,180 d) $1,191 e) $3,000

*16.3 Your perfectly reliable friend, Merle, asks for a loan and promises to pay back $150 two years from now. If the minimum interest rate you will accept is 8%, what is the maximum amount you will loan him?

a) $119 b) $126 c) $129 d) $139 e) $150

16.4 $12,000 is borrowed now at 12% interest. The first payment is $4000 and is made 3 years from now. The balance of the debt immediately after the payment is

a) $4000 b) $8000 c) $12,000 d) $12,860 e) $16,860

16.5 An alumnus establishes a perpetual endowment fund to help Saint Louis University. What amount must be invested now to produce income of $100,000 one year from now and at one-year intervals forever? Interest rate is 8%.

a) $8000 b) $100,000 c) $1,250,000 d) $10,000,000 e) Can't be done.

EQUIVALENCE OF CASH FLOW PATTERNS

*16.6 The annual amount of a series of payments to be made at the end of each of the next twelve years is $500. What is the present worth of the payments at 8% interest compounded annually?

a) $500 b) $3,768 c) $6,000 d) $6,480 e) $6,872

*16.7 Consider a prospective investment in a project having a first cost of $300,000, operating and maintenance costs of $35,000 per year, and an estimated net disposal value of $50,000 at the end of thirty years. Assume an interest rate of 8%.

What is the present equivalent cost of the investment if the planning horizon is thirty years?

a) $670,000 b) $689,000 c) $720,000 d) $791,000 e) $950,000

If the project replacement will have the same first cost, life, salvage value, and operating and maintenance costs as the original, what is the capitalized cost of perpetual service?

a) $670,000 b) $689,000 c) $720,000 d) $765,000 e) infinite

*16.8 Maintenance expenditures for a structure with a twenty-year life will come as periodic outlays of $1,000 at the end of the fifth year, $2,000 at the end of the tenth year, and $3,500 at the end of the fifteenth year. With interest at 10%, what is the equivalent uniform annual cost of maintenance for the twenty-year period?

a) $200 b) $262 c) $300 d) $325 e) $342

16.9 An alumnus has given Michigan State University ten million dollars to build and operate a laboratory. Annual operating cost is estimated to be one hundred thousand dollars. The endowment will earn 6% interest. Assume an infinite life for the laboratory and determine how much money may be used for its construction.

a) $\$5.00\times10^6$ b) $\$8.33\times10^6$ c) $\$8.72\times10^6$

d) $\$9.90\times10^6$ e) $\$9.98\times10^6$

16.10 An investment pays $6000 at the end of the first year, $4000 at the end of the second year, and $2000 at the end of the third year. Compute the present value of the investment if a 10% rate of return is required.

a) $8333 b) $9667 c) $10,300 d) $12,000 e) $14,000

16.11 An amount F is accumulated by investing a single amount P for n compounding periods with interest rate of i. Select the formula that relates P to F.

a) $P=F(1+i)^{n-1}$ b) $P=F(1+i)^n$ c) $P=F(1+n)^{-i}$

d) $P=F(1+ni)^{-1}$ e) $P=F(1+i)^{-n}$

16.12 At the end of each of the next ten years, a payment of $200 is due. At an interest rate of 6%, what is the present worth of the payments?

a) $27 b) $200 c) $1472 d) $2000 e) $2636

16.13 The purchase price of an instrument is $12,000 and its estimated maintenance costs are $500 for the first year, $1500 for the second and $2500 for the third year. After three years of use the instrument is replaced; it has no salvage value. Compute the present equivalent cost of the instrument using 10% interest.

a) $14,070 b) $15,570 c) $15,730 d) $16,500 e) $20,720

16.14 If an amount invested five years ago has doubled, what is the annual interest rate?

a) 5% b) 6% c) 10% d) 12% e) 15%

16.15 After a factory has been built near a stream, it is learned that the stream occasionally overflows its banks. A hydrologic study indicates that the probability of flooding is about 1 in 8 in any one year. A flood would cause about $20,000 in damage to the factory. A levee can be constructed to prevent flood damage. Its cost will be $54,000 and its useful life is thirty years. Money can be borrowed at 8% interest. If the annual equivalent cost of the levee is less than the annual expectation of flood damage, the levee should be built. The annual expectation of flood damage is $(1/8)\times20{,}000=\$2{,}500$. Compute the annual equivalent cost of the levee.

a) $1,261 b) $1,800 c) $4,320 d) $4,800 e) $6,750

16.16 If $10,000 is borrowed now at 6% interest, how much will remain to be paid after a $3,000 payment is made four years from now?

a) $7,000 b) $9,400 c) $9,625 d) $9,725 e) $10,700

*16.17 A piece of machinery costs $20,000 and has an estimated life of eight years and a scrap value of $2,000. What uniform annual amount must be set aside at the end of each of the eight years for replacement if the interest rate is 4%?

a) $1,953 b) $2,174 c) $2,250 d) $2,492 e) $2,898

*16.18 The maintenance costs associated with a machine are $2,000 per year for the first ten years, and $1,000 per year thereafter. The machine has an infinite life. If interest is 10%, what is the present worth of the annual disbursements?

a) $16,145 b) $19,678 c) $21,300 d) $92,136 e) $156,600

*16.19 A manufacturing firm entered into a ten-year contract for raw materials which required a payment of $100,000 initially and $20,000 per year beginning at the end of the fifth year. The company made unexpected profits and asked that it be allowed to make a lump sum payment at the end of the third year to pay off the remainder of the contract. What lump sum is necessary if the interest rate is 8%?

a) $85,600 b) $92,700 c) $122,300 d) $196,700 e) $226,000

UNUSUAL CASH FLOWS AND INTEREST PAYMENTS

16.20 A bank currently charges 10% interest compounded annually on business loans. If the bank were to change to continuous compounding, what would be the effective annual interest rate?

a) 10% b) 10.517% c) 12.5% d) 12.649% e) 12.92%

*16.21 Terry bought a CD-ROM drive for $50 down and $30 per month for 24 months. The same drive could have been purchased for $675 cash. What nominal annual interest rate is Terry paying?

a) 7.6% b) 13.9% c) 14.8% d) 15.2% e) 53.3%

16.22 How large a contribution is required to endow perpetually a research laboratory which requires $500,000 for original construction, $200,000 per year for operating expenses, and $100,000 every three years for new and replacement equipment? Interest is 4%.

a) $700,000 b) $6,400,000 c) $7,900,000 d) $10,000,000 e) $12,490,000

16.23 A set of speakers may be purchased now for $400 or by making a down payment of $35 and additional payments of $45 at the end of each of the next ten months. Compute the nominal annual interest rate for the time payment plan.

a) 11.2% b) 21.2% c) 23.3% d) 48.0% e) 60.1%

16.24 Same as preceding question except find the effective annual interest rate.

a) 11.2% b) 21.2% c) 23.3% d) 48.0% e) 60.1%

16.25 A $1000 debt is to be repaid in four weekly payments of $300 each, beginning one week after the debt is incurred. Compute the nominal annual interest rate.

a) 8% b) 20% c) 240% d) 400% e) 5000%

ANNUAL EQUIVALENT COST COMPARISONS

*16.26 One of the two production units described below must be purchased. The minimum attractive rate of return is 12%. Compare the two units on the basis of equivalent annual cost.

	Unit A	Unit B
Initial Cost	$16,000	$30,000
Life	8 years	15 years
Salvage value	$ 2,000	$ 5,000
Annual operating cost	$ 2,000	$ 1,000

a) *A*—$5,058; *B*—$5,270
b) *A*—$4,916; *B*—$4,872
c) *A*—$3,750; *B*—$2,667
d) *A*—$1,010; *B*—$1,010
e) *A*—$2,676; *B*—$4,250

16.27 Tanks to hold a corrosive chemical are now being made of material *A*, and have a life of eight years and a first cost of $30,000. When these tanks are four years old, they must be relined at a cost of $10,000. If the tanks could be made of material *B* , their life would be twenty years and no relining would be necessary. If the minimum rate of return is 10%, what must be the first cost of a tank made of material *B* to make it economically equivalent to the present tanks?

a) $38,764 b) $42,631 c) $51,879 d) $58,760 e) $92,361

PRESENT EQUIVALENT COST COMPARISONS

16.28 Compute the life cycle cost of a reciprocating compressor with first cost of $120,000, annual maintenance cost of $9,000, salvage value of $25,000 and life of six years. The minimum attractive rate of return is 10%.

a) $120,000 b) $145,000 c) $149,000 d) $153,280 e) $167,900

16.29 A punch press costs $100,000 initially, requires $10,000 per year in maintenance expenses, and has no salvage value after its useful life of ten years. With interest of 10%, the capitalized cost of the press is:

a) $100,000 b) $161,400 c) $197,300 d) $200,000 e) $262,700

16.30 A utility is considering two alternatives for serving a new customer. Both plans provide twenty years of service, but plan *A* requires one large initial investment, while plan *B* requires additional investment at the end of ten years. Neglect salvage value, assume interest at 8%, and determine the present cost of both plans.

	Plan A	Plan B
Initial Investment	$50,000	$30,000
Investment at end of 10 years	none	$30,000
Annual property tax and maintenance, years 1–10	$ 800	$ 500
Annual property tax and maintenance, years 11–20	$ 800	$ 900

a) *A*—$48,780; *B*—$49,250
b) *A*—$50,000; *B*—$30,000
c) *A*—$50,000; *B*—$60,000
d) *A*—$57,900; *B*—$50,000
e) *A*—$66,000; *B*—$74,000

*16.31 The heat loss of a bare stream pipe costs $206 per year. Insulation *A* will reduce heat loss by 93% and can be installed for $116; insulation *B* will reduce heat loss by 89% and can be installed for $60. The insulations require no additional expenses and will have no salvage value at the end of the pipe's estimated life of eight years. Determine the present net equivalent value of the two insulations if the interest rate is 10%.

a) *A*—$116; *B*—$90
b) *A*—$906; *B*—$918
c) *A*—$1,022; *B*—$978
d) *A*—$1,417; *B*—$1,406
e) *A*—$1,533; *B*—$1,467

INCREMENTAL APPROACH

16.32 A desalinator is needed for six years. Cost estimates for two are:

	The Life of Brine	The Salty Tower
Price	$95,000	$120,000
Annual maintenance	3,000	9,000
Salvage value	12,000	25,000
Life in years	3	6

With interest at 10%, what is the annual cost advantage of the Salty Tower?

a) 0 b) $4,260 c) $5,670 d) $5,834 e) $56,000

16.33 A motor costs $20,000 and has an estimated life of six years. By the addition of certain auxiliary equipment, an annual savings of $300 in operating costs can be obtained, and the estimated life of the motor extended to nine years. Salvage value in either case is $5,000. Interest on capital is 8%. Compute the maximum expenditure justifiable for the auxiliary equipment.

a) $1,149 b) $1,800 c) $2,700 d) $7,140 e) $13,300

*16.34 An existing electrical power line needs to have its capacity increased, and this can be done in either of two ways. The first method is to add a second conductor to each phase wire, using the same poles, insulators and fittings, at a construction cost of $15,000. The second method for increasing capacity is to build a second line parallel to the existing line, using new poles, insulators and fittings, at a construction cost of $23,000. At some time in the future, the line will require another increase in capacity, with the first alternative now requiring a second line at a cost of $32,500, and the second alternative requiring added conductors at a cost of $23,000. If interest rate is 6%, how many years between the initial expenditure and the future expenditure will make the two methods economically equal?

a) 1 b) 3 c) 5 d) 10 e) 25

REPLACEMENT PROBLEMS

16.35 One year ago machine *A* was purchased at a cost of $2,000, to be useful for five years. However, the machine failed to perform properly and costs $200 per month for repairs, adjustments and shut-downs. A new machine *B* designed to perform the same functions is quoted at $3,500, with the cost of repairs and adjustments estimated to be only $50 per month. The expected life of machine *B* is five years. Except for repairs and adjustments, the operating costs of the two machines are substantially equal. Salvage values are insignificant. Using 8% interest rate, compute the incremental annual net equivalent value of machine *B*.

a) – $877 b) $923 c) $1,267 d) $1,800 e) $2,677

BREAK-EVEN ANALYSIS

16.36 Bear Air, an airline serving the Arctic, serves in-flight snacks on some routes. Preparing these snacks costs Bear Air \$5000 per month plus \$1.50 per snack. Alternatively, prepared snacks may be purchased from a supplier for \$4.00 per snack. What is the maximum number of snacks per month for which purchasing from the supplier is justified economically?

a) 769 b) 1250 c) 2000 d) 3333 e) 4000

*16.37 Bear Air has been contracting its overhaul work to Aleutian Aeromotive for \$40,000 per plane per year. Bear estimates that by building a \$500,000 maintenance facility with a life of 15 years and a salvage value of \$100,000, they could handle their own overhauls at a variable cost of only \$30,000 per plane per year. The maintenance facility could be financed with a secured loan at 8% interest. What is the minimum number of planes Bear must operate in order to make the maintenance facility economically feasible?

a) 5 b) 6 c) 10 d) 40 e) 50

16.38 It costs Bear Air \$1,200 to run a scheduled flight, empty or full, from Coldfoot to Frostbite. Moreover, each passenger generates a cost of \$40. The regular ticket costs \$90. The plane holds 65 people, but it is running only about 20 per flight. The sales director has suggested selling introductory tickets for \$50 to people who have never flown Bear Air.

What is the minimum number of introductory tickets that must be sold in order for a flight to produce a profit?

a) 5 b) 10 c) 15 d) 20 e) 45

16.39 What would be the total profit on the flight from Coldfoot to Frostbite of Problem 16.38 if all 65 passengers claimed introductory tickets?

a) – \$800 b) – \$550 c) 0 d) \$400 e) \$500

16.40 Two electric motors are being considered for an application in which there is uncertainty concerning the hours of usage. Motor *A* costs \$4,500 and has an efficiency of 90%. Motor *B* costs \$3,000 and has an efficiency of 89%. Each motor has a ten-year life and no salvage value. Electric service costs \$18.70 per year per kW of demand and \$0.10 per kW of energy. The output of the motors is to be 75 kW, and interest rate is 8%. At how many hours usage per year would the two motors be equally economical? If the usage is less than this amount, which motor is preferable?

a) 1800, *A* b) 1800, *B* c) 2200, *A* d) 2200, *B* e) 2500, *A*

INCOME TAX AND DEPRECIATION

16.41 A drill press is purchased for \$10,000 and has an estimated life of twelve years. The salvage value at the end of twelve years is estimated to be \$1,300. Using general straight-line depreciation, compute the book value of the drill press at the end of eight years.

a) \$1,300 b) \$3,333 c) \$3,475 d) \$4,200 e) \$4,925

16.42 Excelsys Inc. purchased a desk top computer for \$4500. Using the general straight line method, compute the depreciation for the first year if the recovery period is three years and the salvage value is \$1500.

a) \$500 b) \$750 c) \$1000 d) \$1500 e) \$2250

16.43 A grading contractor owns earth-moving equipment that costs \$300,000 and is classed as 7-year property. After seven years of use, its salvage value will be \$50,000. Using the general straight line method, compute the first two depreciation deductions and the book value at the end of four years.

a) \$35,714; \$35,714; \$157,143
b) \$85,714; \$85,714; \$0
c) \$21,429; \$42,857; \$150,000
d) \$42,857; \$73,469; \$93,711
e) \$85,714; \$61,224; \$78,092

16.44 Rework Prob. 16.43 using the MACRS straight line method to compute the first two depreciation deductions and the book value at the end of the fourth tax year.

a) \$35,714; \$35,714; \$157,143
b) \$85,714; \$85,714; \$0
c) \$21,429; \$42,857; \$150,000
d) \$42,857; \$73,469; \$93,711
e) \$85,714; \$61,224; \$78,092

16.45 Rework Prob. 16.43 using the general double-declining balance method to compute the first two depreciation deductions and the book value at the end of the four years.

a) \$35,714; \$35,714; \$157,143
b) \$85,714; \$85,714; \$0
c) \$21,429; \$42,857; \$150,000
d) \$42,857; \$73,469; \$93,711
e) \$85,714; \$61,224; \$78,092

16.46 Rework Prob. 16.43 using the MACRS double-declining balance method to compute the first two depreciation deductions and the book value at the end of the fourth tax year.

a) \$35,714; \$35,714; \$157,143
b) \$85,714; \$85,714; \$0
c) \$21,429; \$42,857; \$150,000
d) \$42,857; \$73,469; \$93,711
e) \$85,714; \$61,224; \$78,092

GENERAL

__Questions 16.47–16.50__

The market for laser ignition units is estimated to be 1000 per month if the selling price is \$400 per unit. The monthly cost of production is

$$C = 35,000 + 0.5S^2$$

where S is the number of units produced each month. Assume that monthly sales and production rates are equal.

16.47 Compute the monthly profit if production is 1000 units per month.

a) –\$135,000 b) \$112 c) \$135 d) \$35,800 e) \$45,000

16.48 Compute the profitable range of production rates.

a) 100 – 400 units/mo b) 100 – 700 units/mo c) 265 – 400 units/mo
d) 265 – 700 units/mo e) 400 – 700 units/mo

16.49 Compute the maximum possible monthly profit.

a) –$135,000 b) $112 c) $135 d) $35,800 e) $45,000

16.50 Compute the maximum possible profit per unit.

a) –$135,000 b) $112 c) $135 d) $35,800 e) $45,000

Questions 16.51–16.53:

As a byproduct of its manufacturing process, Technocel Inc. generates 100 kg of highly toxic waste per year. Currently Technocel pays Enviroserv, a waste management service, $15 per kg to remove and treat the waste. Alternatively, Technocel could treat and dispose of the waste for $5 per kg if it installed a $4000 on-site treatment reactor. The estimated life of the reactor is five years, and it would have no salvage value. Technocel's before-tax minimum attractive rate of return is 20%.

16.51 Use a before-tax analysis to compute the difference in annual equivalent costs of the two alternatives, and to select the more economic alternative.

a) ±$200, prefer Enviroserv d) ±$338, prefer on-site treatment

b) ±$200, prefer on-site treatment e) equal costs, no preference

c) ±$338, prefer Enviroserv

16.52 Compute the annual mass of toxic waste above which on-site treatment is the more economic alternative.

a) 80 kg/yr b) 100 kg/yr c) 123 kg/yr d) 134 kg/yr e) 165 kg/yr

16.53 Use the general straight line method to compute depreciation of the on-site treatment reactor for the second year, and its book value at the end of that year. The recovery period is three years and there is no salvage value.

a) $1000, $2000 b) $1333, $1333 c) $1333, $2000

d) $1333, $2667 e) $2000, 0

Questions 16.54-16.55:

A new bus route was added last week, but the average number of passengers on that bus has been only 20; the capacity is 60. The fare is $0.50 and the cost of operating the bus is $0.05 per passenger plus $25 per trip. The marketing director suggests offering a $0.25 fare to people who have never before taken the bus.

16.54 Assume that the director's suggestion is implemented, and that 15 additional passengers per trip are gained. The total profit per trip will be

a) –$26.75 b) –$13 c) –$3.75 d) $10 e) $13.75

16.55 Assume that the director's suggestion is implemented, and that 15 additional passengers per trip are gained, but that all passengers claim the $0.25 fare. The total profit per trip will be

a) –$25 b) –$23 c) –$18 d) $7 e) $8.75

Questions 16.56-16.60:

A corporation evaluates all capital investments using a 20% annual rate of return before taxes. The corporation must purchase a new tangent scanner. The following estimates pertain to the two models available:

	Scanx	Holo-Scan
First cost	$90,000	$170,000
Life (years)	5	5
Salvage value	$15,000	$ 50,000
Annual cost	$44,000	$ 70,000
Generated income (annual)	$100,000	$160,000

16.56 Using general straight line depreciation, compute the book value of the Scanx at the end of two years.

a) $15,000 b) $30,000 c) $45,000 d) $60,000 e) $75,000

16.57 Suppose the Scanx is depreciated by the general double declining balance method. At the end of which year will book value be reduced to salvage value?

a) 1 b) 2 c) 3 d) 4 e) 5

16.58 If income is disregarded, an annual cost comparison indicates that the preferred model is

a) Scanx by about $22,000/yr.
b) Scanx by about $26,000/yr.
c) Scanx by about $48,000/yr.
d) Holo-Scan by about $2700/yr.
e) Holo-Scan by about $53,000/yr.

16.59 The present worth of costs and income for the two models indicates that Holo-Scan is worth about how much more than Scanx?

a) $80,000 b) $65,000 c) $45,000 d) $36,000 e) $7600

16.60 The scanner is served by high-voltage electric lines. Total cost of the lines is $F + tI + L/t$ where t is the thickness of insulation on the lines, F is a fixed cost independent of insulation, tI is the cost of insulation, and L/t is the cost of leakage through the insulation. The thickness of the insulation that will give the lowest cost is

a) $\sqrt{L/I}$ b) $F+L/I$ c) $\sqrt{I/L}$ d) $[(I+F)/L]^2$ e) $[L/(I+F)]^2$

Solutions to Practice Problems

16.1 **d)** Prestige.

16.2 **d)** $1000 \times 1.06^3 = \$1191$.

16.3 **c)** $150/1.08^2 = \$129$.

16.4 **d)** $F = 12,000(F/P)_3^{12} - 4000 = \$12,860$.

16.5 **c)** $P = 100,000(P/A)_\infty^8 = \$1,250,000$.

16.6 **b)** $500(\mathrm{P/A})_{12}^8 = 500 \times 7.536 = \$3768$.

16.7 **b), d)** A: $300,000 + 35,000(P/A)_{30}^8 - 50,000(P/F)_{30}^8 = \$689,000$

B: $689,000(A/P)_{30}^8(P/A)_\infty^8 = \$765,000$.

16.8 **b)** $\left[1000(P/F)_5^{10} + 2000(P/F)_{10}^{10} + 3500(P/F)_{15}^{10}\right](A/P)_{20}^{10} = \$262$.

16.9 **b)** $10 \times 10^6 = X + 0.1 \times 10^6(P/A)_\infty^6$. $\therefore X = 8.33 \times 10^6$.

16.10 **c)** $P = 6000(P/A)_3^{10} - 2000(P/G)_3^{10} = 10,300$.

16.11 **e)** By inspection.

16.12 **c)** $P = 200(P/A)_{10}^6 = \$1472$.

16.13 **b)** $P = 12,000 + 500(P/A)_3^{10} + 1000(P/G)_3^{10} = \$15,570$.

16.14 **e)** $2P = (F/P)_5^i P$. $(F/P)_5^i = (1+i)^5 = 2$. $\therefore i = 15\%$.

16.15 **d)** $54,000(A/P)_{30}^8 = \$4800$.

16.16 **c)** $10,000(F/P)_4^6 - 3000 = \$9625$.

16.17 **a)** $18,000(A/F)_8^4 = \$1953$.

16.18 **a)** $1000(P/A)_\infty^{10} + 1000(P/A)_{10}^{10} = \$16,145$. Note: for the first 10 years, this accounts for \$2000/yr.

16.19 **a)** $20,000(P/A)_6^8(P/F)_1^8 = \$85,600$.

16.20 **b)** $e^{0.1} - 1 = 0.10517$ or 10.517%.

16.21 **b)** $675 = 50 + 30(P/A)^i_{24}$. $\therefore (P/A)^i_{24} = 20.833$.
$\therefore$ by trial and error $i = 0.0116$. $\therefore 12i = 0.139$ or 13.9%.

16.22 **b)** $500{,}000 + \left[200{,}000 + 100{,}000(A/P)^4_3\right](P/A)^4_\infty = \$6{,}400{,}000$.

16.23 **d)** $400 = 35 + 45(P/A)^i_{10}$. $(P/A)^i_{10} = 8.11$. $i = 4\%$. $i_n = 12i$. $\therefore i_n = 48\%$.

16.24 **e)** From preceding solution $i = 4\%$.
$i_e = (1+0.04)^{12} - 1 = 0.601$. $\therefore i_e = 60.1\%$.

16.25 **d)** $1000 = 300(P/A)^i_4$. $(P/A)^i_4 = 3.33$. $i = 7.7\%$. $\therefore i_n = 52 \times 7.7 = 400\%$.

16.26 **a)** $A:\ -16{,}000(A/P)^{12}_8 - 2000 + 2000(A/F)^{12}_8 = -\$5058$.
$B:\ -30{,}000(A/P)^{12}_{15} - 1000 + 5000(A/F)^{12}_{15} = -\$5270$

16.27 **d)** $P(A/P)^{10}_{20} = \left[30{,}000 + 10{,}000(P/F)^{10}_4\right](A/P)^{10}_8$. $\therefore P = \$58{,}760$.

16.28 **b)** $120{,}000 + 9000(P/A)^{10}_6 - 25{,}000(P/F)^{10}_6 = \$145{,}000$.

16.29 **e)** $\left[100{,}000(A/P)^{10}_{10} + 10{,}000\right](P/A)^{10}_\infty = \$262{,}700$.

16.30 **d)** $A:\ 50{,}000 + 800(P/A)^8_{20} = \$57{,}900$
$B:\ 30{,}000 + 500(A/P)^8_{20} + \left[30{,}000 + 400(P/A)^8_{10}\right](P/F)^8_{10} = \$50{,}000$.

16.31 **b)** $A:\ -116 + 0.93 \times 206(P/A)^{10}_8 = \$906$
$B:\ -60 + 0.89 \times 206(P/A)^{10}_8 = \$918$

16.32 **b)** $A:\ \left[-25{,}000 + 83{,}000(P/F)^{10}_3\right](A/P)^{10}_6 - 6000 + 13{,}000(A/F)^{10}_6 = \$4260$.

16.33 **d)** $(20{,}000 + P)(A/P)^8_9 - 300 - 5000(A/F)^8_9 = 20{,}000(A/P)^8_6 - 5000(A/F)^8_6$.
$\therefore P = \$7140$.

16.34 **b)** $(23{,}000 - 15{,}000) + (23{,}000 - 32{,}500)(P/F)^6_N = 0$. $(1.06)^{-N} = 0.84$.
$\therefore N = 3$ yrs.

16.35 **b)** $-3500(A/P)^8_5 + 12(200 - 50) = \$923$.

16.36 **c)** $5000 + 1.50n = 4.00n$. $\therefore n = 2000$.

16.37 **b)** $40{,}000x = 500{,}000(A/P)^8_{15} - 100{,}000(A/F)^8_{15} + 30{,}000x$.
$\therefore x = 5.47$. Use $x = 6$.

16.38 **d)** $1200 + 40(20 + x) = 90(20) + 50x$. $\therefore x = 20$

16.39 **b)** $65(50)-65(40)-1200=-550$

16.40 **d)** $4500(A/P)_{10}^{8}+(18.7+.1x)75/.9=3000(A/P)_{10}^{8}+(18.7+.1x)75/.89$.
$\therefore\ x=2200$ hr., B

16.41 **d)** $10,000-(10,000-1300)8/12=4200$

16.42 **c)** $D=(4500-1500)/3=1000$

16.43 **a)** $(300,000-50,000)/7=35,714$
$300,000-4(35,714)=157,143$

16.44 **c)** $(300,000/7)0.5=21,419$
$300,000/7=42,857$
$300,000-3.5(42,857)=150,000$

16.45 **e)** $300,000\times 2/7=85,714$
$(300,000-85,714)2/7=61,224$
$300,000-85,714-61,224-43,732-31,237=78,092$

16.46 **d)** $(300,000\times 2/7)0.5=42,857$
$(300,000-42,857)2/7=73,469$
$300,000-42,857-73,469-52,478-37,484=93,711$

16.47 **a)** $P=400S-(35000+0.5S^2)$. $S=1000$ gives $P=-\$135,000$

16.48 **b)** $P=0.\ S=\left(-400\pm\sqrt{400^2-4\times 0.5\times 35000}\right)\Big/(-2\times 0.5)$.
$\therefore S=400\pm 300$ units/mo

16.49 **e)** $dP/dS=400-2\times 0.5S=0$. $S=400$. $P=\$45,000$ per month.

16.50 **c)** $U=P/S=400-(35000S^{-1}+0.5S)$. $dU/dS=35,000S^{-2}-0.5=0$.
$S=265$. $U=\$135/\text{unit}$.

16.51 **c)** $\text{ANEV}=\left[-4000(A/P)_{5}^{20}-500\right]-[-1500]=-\$338$.

16.52 **d)** $\text{ANEV}=0=\left[-4000(A/P)_{5}^{20}-5X\right]-[-15X]$. $\therefore X=134$ kg/yr.

16.53 **b)** $D=4000/3=\$1333$. $V_2=4000-2\times 1333=\$1333$.

16.54 **b)** $P=20\times(0.50-0.05)+15(0.25-0.05)-25=-\$13$.

16.55 **c)** $P=(20+15)\times(0.25-0.05)-25=-\$18$.

16.56 **d)** $D=\dfrac{(90,000-15,000)}{5}=15,000;\ V_2=90,000-2\times 15,000=\$60,000$

16.57 **d)** $D_1 = \frac{90,000 \times 2}{5} = 36,000. \quad V_1 = 90,000 - 36,000 = 54,000$

$D_2 = \frac{54,000 \times 2}{5} = 21,600. \quad V_2 = 54,000 - 21,600 = 32,400$

$D_3 = \frac{32,400 \times 2}{5} = 12,960. \quad V_3 = 32,400 - 12,960 = 19,440$

$D_4 = \frac{19,440 \times 2}{5} = \ 7,776. \quad V_4 = 19,440 - \ 7,776 = 11,664 < 15,000$

16.58 **c)** $\text{ANEV} = -(170,000 - 90,000)(A/P)_5^{20} - (70,000 - 44,000)$

$+ (50,000 - 15,000)(A/F)_5^{20} = -\$48,000$

16.59 **d)** $\text{PNEV} = (160,000 - 100,000)(P/A)_5^{20} - (170,000 - 90,000)$

$-(70,000 - 44,000)(P/A)_5^{20} + (50,000 - 15,000)(P/F)_5^{20} = \$36,000.$

16.60 **a)** $dC/dt = I - \left(L/t^2\right) = 0; \quad t = \sqrt{L/I}\,.$

This page is intentionally blank.

TABLE 16.2 Compound Interest Factors

$i = \frac{1}{2}\%$

n	(P/F)	(P/A)	(P/G)	(F/P)	(F/A)	(A/P)	(A/F)	(A/G)	n
1	.9950	0.995	0.000	1.005	1.000	1.0050	1.0000	0.000	1
2	.9901	1.895	0.990	1.010	2.005	0.5038	0.4988	0.499	2
3	.9851	2.970	2.960	1.015	3.015	0.3367	0.3317	0.997	3
4	.9802	3.950	5.901	1.020	4.030	0.2531	0.2481	1.494	4
5	.9754	4.926	9.803	1.025	5.050	0.2030	0.1980	1.990	5
6	.9705	5.896	14.655	1.030	6.076	0.1696	0.1646	2.485	6
7	.9657	6.862	20.449	1.036	7.106	0.1457	0.1407	2.980	7
8	.9609	7.823	27.176	1.041	8.141	0.1278	0.1228	3.474	8
9	.9561	8.779	34.824	1.046	9.182	0.1139	0.1089	3.967	9
10	.9513	9.730	43.386	1.051	10.228	0.1028	0.0978	4.459	10
11	.9466	10.677	52.853	1.056	11.279	0.0937	0.0887	4.950	11
12	.9419	11.619	63.214	1.062	12.336	0.0861	0.0811	5.441	12
13	.9372	12.556	74.460	1.067	13.397	0.0796	0.0746	5.930	13
14	.9326	13.489	86.583	1.072	14.464	0.0741	0.0691	6.419	14
15	.9279	14.417	99.574	1.078	15.537	0.0694	0.0644	6.907	15
16	.9233	15.340	113.424	1.083	16.614	0.0652	0.0602	7.394	16
17	.9187	16.259	128.123	1.088	17.697	0.0615	0.0565	7.880	17
18	.9141	17.173	143.663	1.094	18.786	0.0582	0.0532	8.366	18
19	.9096	18.082	160.036	1.099	19.880	0.0553	0.0503	8.850	19
20	.9051	18.987	177.232	1.105	20.979	0.0527	0.0477	9.334	20
21	.9006	19.888	195.243	1.110	22.084	0.0503	0.0453	9.817	21
22	.8961	20.784	214.061	1.116	23.194	0.0481	0.0431	10.299	22
23	.8916	21.676	233.677	1.122	24.310	0.0461	0.0411	10.781	23
24	.8872	22.563	254.082	1.127	25.432	0.0443	0.0393	11.261	24
25	.8828	23.446	275.269	1.133	26.559	0.0427	0.0377	ll.N1	25
26	.8784	24.342	297.228	1.138	27.692	0.0411	0.0361	12.220	26
28	.8697	26.068	343.433	1.150	29.975	0.0384	0.0334	13.175	28
30	.8610	27.794	392.632	1.161	32.280	0.0360	0.0310	14.126	30
∞	0	200.000	40000.0	∞	∞	.0050	0	200.00	∞

$i = 2.00\%$

n	(P/F)	(P/A)	(P/G)	(F/P)	(F/A)	(A/P)	(A/F)	(A/G)	n
I	.9804	0.9804	0.0000	1.0200	1.0000	1.0200	1.0000	0.0000	1
2	.9612	1.9416	0.9612	1.0404	2.0200	0.5150	0.4950	0.4950	2
3	.9423	2.8839	2.8458	1.0612	3.0604	0.3468	0.3268	0.9868	3
4	.9238	3.8077	5.6173	1.0824	4.1216	0.2626,	0.2426	1.4752	4
5	.9057	4.7135	9.2403	1.1041	5.2040	0.2122	0.1922	1.9604	5
6	.8880	5.6014	13.6801	1.1262	6.3081	0.1785	0.1585	2.4423	6
7	.8706	6.4720	18.905	1.1487	7.4343	0.1545	0.1345	2.9208	7
8	.8535	7.3255	24.8779	1.1717	8.5830	0.1365	0.1165	3.3961	8
9	.8368	8.1622	31.5720	1.1951	9.7546	0.1225	0.1025	3.8681	9
10	.8203	8.9826	38.9551	1.2190	10.9497	0.1113	0.0913	4.3367	10
11	.8043	9.7868	46.9977	1.2434	12.1687	0.1022	0.0822	4.8021	11
12	.7885	10.5753	55.6712	1.2682	13.4121	0.0946	0.0746	5.2642	12
13	.7730	11.3484	64.9475	1.2936	14.6803	0.0881	0.0681	5.7231	13
14	.7579	12.1062	74.7999	1.3195	15.9739	0.0826	0.0626	6.1786	14
15	.7430	12.8493	85.2021	1.3459	17.2934	0.0778	0.0578	6.6309	15
16	.7284	13.5777	96.1288	1.3728	18.6393	0.0737	0.0537	7.0799	16
17	.7142	14.2919	107.5554	1.4002	20.0121	0.0700	0.0500	7.5256	17
18	.7002	14.9920	119.4581	1.4282	21.4123	0.0667	0.0467	7.9681	18
19	.6864	15.6785	131.8139	1.4568	22.8406	0.0638	0.0438	8.4073	19
20	.6730	16.3514	144.6003	1.4859	24.2974	0.0612	0.0412	8.8433	20
21	.6598	17.0112	157.7959	1.5157	25.7833	0.0588	0.0388	9.2760	21
22	.6468	17.6580	171.3795	1.5460	27.2990	0.0566	0.0366	9.7055	22
23	.6342	18.2922	185.3309	1.5769	28.8450	0.0547	0.0347	10.1317	23
24	.6217	18.9139	199.6305	1.6084	30.4219	0.0529	0.0329	10.5547	24
25	.6095	19.5235	214.2592	1.6406	32.0303	0.0512	0.0312	10.9745	25
26	.5976	20.1210	229.1987	1.6734	33.6709	0.0497	0.0297	11.3910	26
28	.5744	21.2813	259.9392	1.7410	37.0512	0.0470	0.0270	12.2145	28
30	.5521	22.3965	291.7164	1.8114	40.5681	0.0446	0.0246	13.0251	30
∞	.0000	50.0000	2500.0000	∞	∞	0.0200	0.0000	50.0000	∞

TABLE 16.2 Compound Interest Factors (continued)

$i = 4.00\%$

n	(P/F)	(P/A)	(P/G)	(F/P)	(F/A)	(A/P)	(A/F)	(A/G)	*n*
1	.9615	0.9615	-0.0000	1.0400	1.0000	1.0400	1.0000	-0.0000	1
2	.9246	1.8861	0.9246	1.0816	2.0400	0.5302	0.4902	0.4902	2
3	.8890	2.7751	2.7025	1.1249	3.1216	0.3603	0.3203	0.9739	3
4	.8548	3.6299	5.2670	1.1699	4.2465	0.2755	0.2355	1.4510	4
5	.8219	4.4518	8.5547	1.2167	5.4163	0.2246	0.1846	1.9216	5
6	.7903	5.2421	12.5062	1.2653	6.6330	0.1908	0.1508	2.3857	6
7	.7599	6.0021	17.0657	1.3159	7.8983	0.1666	0.1266	2.8433	7
8	.7307	6.7327	22.1806	1.3686	9.2142	0.1485	0.1085	3.2944	8
9	.7026	7.4353	27.8013	1.4233	10.5828	0.1345	0.0945	3.7391	9
10	.6756	8.1109	33.8814	1.4802	12.0061	0.1233	0.08333	4.1773	10
11	.6496	8.7605	40.3772	1.5395	13.4864	0.1141	0.0741	4.6090	11
12	.6246	9.3851	47.2477	1.6010	15.0258	0.1066	0.0666	5.0343	12
13	.6006	9.9856	54.4546	1.6651	16.6268	0.1001	0.0601	5.4533	13
14	.5775	10.5631	61.9618	1.7317	18.2919	0.0947	0.0547	5.8659	14
15	.5553	11.1184	69.7355	1.8009	20.0236	0.0899	0.0499	6.2721	15
16	.5339	11.6523	77.7441	1.8730	21.8245	0.0858	0.0458	6.6720	16
17	.5134	12.1657	85.9581	1.9479	23.6975	0.0822	0.0422	7.0656	17
18	.4936	12.6593	94.3498	2.0258	25.6454	0.0790	0.0390	7.4530	18
19	.4746	13.1339	102.8933	2.1068	27.6712	0.0761	0.0361	7.8342	19
20	.4564	13.5903	111.5647	2.1911	29.7781	0.0736	0.0336	8.2091	20
21	.4388	14.0292	120.3414	2.2788	31.9692	0.0713	0.0313	8.5779	21
22	.4220	14.4511	129.2024	2.3699	34.2480	0.0692	0.0292	8.9407	22
23	.4057	14.8568	138.1284	2.4647	36.6179	0.0673	0.0273	9.2973	23
24	.3901	15.2470	147.1012	2.5633	39.0826	0.0656	0.0256	9.6479	24
25	.3751	15.6221	156.1040	2.6658	41.6459	0.0640	0.0240	9.9925	25
26	.3607	15.9828	165.1212	2.7725	44.3117	0.0626	0.0226	10.3312	26
28	.3335	16.6631	183.1424	2.9987	49.9676	0.0600	0.0200	10.9909	28
30	.3083	17.2920	201.0618	3.2434	56.0849	0.0578	0.0178	11.6274	30
∞	.0000	25.000	625.0000	∞	∞	0.0400	0.0000	25.0000	∞

$i = 6.00\%$

n	(P/F)	(P/A)	(P/G)	(F/P)	(F/A)	(A/P)	(A/F)	(A/G)	*n*
1	.9434	0.9434	–0.0000	1.0600	1.0000	1.0600	1.0000	-0.0000	1
2	.8900	1.8334	0.8900	1.1236	2.0600	0.5454	0.4854	0.4854	2
3	.8396	2.6730	2.5692	1.1910	3.1836	0.3741	0.3141	0.9612	3
4	.7921	3.4651	4.9455	1.2625	4.3746	0.2886	0.2286	1.4272	4
5	.7473	4.2124	7.9345	1.3382	5.6371	0.2374	0.1774	1.8836	5
6	.7050	4.9173	11.4594	1.4185	6.9753	0.2034	0.1434	2.3304	6
7	.6651	5.5824	15.4497	1.5036	8.3938	0.1791	0.1191	2.7676	7
8	.6274	6.2098	19.8416	1.5938	9.8975	0.1610	0.1010	3.1952	8
9	.5919	6.8017	24.5768	1.6895	11.4913	0.1470	0.0870	3.6133	9
10	.5584	7.3601	29.6023	1.7908	13.1808	0.1359	0.0759	4.0220	10
11	.5268	7.8869	34.8702	1.8983	14.9716	0.1268	0.0668	4.4213	11
12	.4970	8.3838	40.3369	2.0122	16.8699	0.1193	0.0593	4.8113	12
13	.4688	8.8527	45.9629	2.1329	18.8821	0.1130	0.0530	5.1920	13
14	.4423	9.2950	51.7128	2.2609	21.0151	0.1076	0.0476	5.5635	14
15	.4173	9.7122	57.5546	2.3966	23.2760	0.1030	0.0430	5.9260	15
16	.3936	10.1059	63.4592	2.5404	25.6725	0.0990	0.0390	6.2794	16
17	.3714	10.4773	69.4011	2.6928	28.2129	0.0954	0.0354	6.6240	17
18	.3503	10.8276	75.3569	2.8543	30.9057	0.0924	0.0324	6.9597	18
19	.3305	11.1581	81.3062	3.0256	33.7600	0.0896	0.0296	7.2867	19
20	.3118	11.4699	87.2304	3.2071	36.7856	0.0872	0.0272	7.6051	20
21	.2942	11.7641	93.1136	3.3996	39.9927	0.0850	0.0250	7.9151	21
22	.2775	12.0416	98.9412	3.6035	43.3923	0.0830	0.0230	8.2166	22
23	.2618	12.3034	104.7007	3.8197	46.9958	0.0813	0.0213	8.5099	23
24	.2470	12.5504	110.3812	4.0489	50.8156	0.0797	0.0197	1.87951	24
25	.2330	12.7834	115.9732	4.2919	54.8645	0.0782	0.0182	9.0722	25
26	.2198	13.0032	121.4684	4.5494	59.1564	0.0769	0.0169	9.3414	26
28	.1956	13.4062	132.1420	5.1117	68.5281	0.0746	0.0146	9.8568	28
30	.1741	13.7648	142.3588	5.7435	79.0582	0.0726	0.0126	10.3422	30
∞	.0000	16.6667	277.7778	∞	∞	0.0600	0.0000	16.667	∞

TABLE 15.2 Compound Interest Factors (continued)

$i = 8.00\%$

n	(P/F)	(P/A)	(P/G)	(F/P)	(F/A)	(A/P)	(A/F)	(A/G)	*n*
1	.9259	0.9259	-0.0000	1.0800	1.0000	1.0800	1.0000	-0.0000	1
2	.8573	1.7833	0.8573	1.1664	2.0800	0.5608	0.4808	0.4808	2
3	.7938	2.5771	2.4450	1.2597	3.2464	0.3880	0.3080	0.9487	3
4	.7350	3.3121	4.6501	1.3605	4.5061	0.3019	0.2219	1.4040	4
5	.6806	3.9927	7.3724	1.4693	5.8666	0.2505	0.1705	1.8465	5
6	.6302	4.6229	10.5233	1.5869	7.3359	0.2163	0.1363	2.2763	6
7	.5835	5.2064	14.0242	1.7138	8.9228	0.1921	0.1121	2.6937	7
8	.5403	5.7466	17.8061	1.8509	10.6366	0.1740	0.0940	3.0985	8
9	.5002	6.2469	21.8081	1.9990	12.4876	0.1601	0.0801	3.4910	9
10	.4632	6.7101	25.9768	2.1589	14.4866	0.1490	0.0690	3.8713	10
11	.4289	7.1390	30.2657	2.3316	16.6455	0.1401	0.0601	4.2395	11
12	.3971	7.5361	34.6339	2.5182	18.9771	0.1327	0.0527	4.5957	12
13	.3677	7.9038	39.0463	2.7196	21.4953	0.1265	0.0465	4.9402	13
14	.3405	8.2442	43.4723	2.9372	24.2149	0.1213	0.0413	5.2731	14
15	.3152	8.5595	47.8857	3.1722	27.1521	0.1168	0.0368	5.5945	15
16	.2919	8.8514	52.2640	3.4259	30.3243	0.1130	0.0330	5.9046	16
17	.2703	9.1216	56.5883	3.7000	33.7502	0.1096	0.0296	6.2037	17
18	.2502	9.3719	60.8426	3.9960	37.4502	0.1067	0.0267	6.4920	18
19	.2317	9.6036	65.0134	4.3157	41.4463	0.1041	0.0241	6.7697	19
20	.2145	9.8181	69.0898	4.6610	45.7620	0.1019	0.0219	7.0369	20
21	.1987	10.0168	73.0629	5.0338	50.4229	0.0998	0.0198	7.2940	21
22	.1839	10.2007	76.9257	5.4365	55.4568	0.0980	0.0180	7.5412	22
23	.1703	10.3711	80.6726	5.8715	60.8933	0.0964	0.0164	7.7786	23
24	.1577	10.5288	84.2997	6.3412	66.7648	0.0950	0.0150	8.0066	24
25	.1460	10.6748	87.8041	6.8485	73.1059	0.0937	0.0137	8.2254	25
26	.1352	10.8100	91.1842	7.3964	79.9544	0.0925	0.0125	8.4352	26
28	.1159	11.0511	97.5687	8.6271	95.3388	0.0905	0.0105	8.8289	28
30	.0994	11.2578	103.4558	10.0627	113.2832	0.0888	0.0088	9.1897	30
∞	.0000	12.500	156.2500	∞	∞	0.0800	0.0000	12.5000	∞

$i = 10.00\%$

n	(P/F)	(P/A)	(P/G)	(F/P)	(F/A)	(A/P)	(A/F)	(A/G)	*n*
1	.9091	0.9091	-0.0000	1.1000	1.0000	1.1000	1.0000	-0.0000	1
2	.8264	1.7355	0.8264	1.2100	2.1000	0.5762	0.4762	0.4762	2
3	.7513	2.4869	2.3291	1.3310	3.3100	0.4021	0.3021	0.9366	3
4	.6830	3.1699	4.3781	1.4641	4.6410	0.3155	0.2155	1.3812	4
5	.6209	3.7908	6.8618	1.6105	6.1051	0.2638	0.1638	1.8101	5
6	.5645	4.3553	9.6842	1.7716	7.7156	0.2296	0.1296	2.2236	6
7	.5132	4.8684	12.7631	1.9487	9.4872	0.2054	0.1054	2.6216	7
8	.4665	5.3349	16.0287	2.1436	11.4359	0.1874	0.0874	3.0045	8
9	.4241	5.7590	19.4215	2.3579	13.5795	0.1736	0.0736	3.3724	9
10	.3855	6.1446	22.8913	2.5937	15.9374	0.1627	0.0627	3.7255	10
11	.3505	6.4951	26.3963	2.8531	18.5312	0.1540	0.0540	4.0641	11
12	.3186	6.8137	29.9012	3.1384	21.3843	0.1468	0.0468	4.3884	12
13	.2897	7.1034	33.3772	3.4523	24.5227	0.1408	0.0408	4.6988	13
14	.2633	7.3667	36.8005	3.7975	27.9750	0.1357	0.0357	4.9955	14
15	.2394	7.6061	40.1520	4.1772	31.7725	0.1315	0.0315	5.2789	15
16	.2176	7.8237	43.4164	4.5950	35.9497	0.1278	0.0278	5.5493	16
17	.1978	8.0216	46.5819	5.0545	40.5447	0.1247	0.0247	5.8071	17
18	.1799	8.2014	49.6395	5.5599	45.5992	0.1219	0.0219	6.0526	18
19	.1635	8.3649	52.5827	6.1159	51.1591	0.1195	0.0195	6.2861	19
20	.1486	8.5136	55.4069	6.7275	57.2750	0.1175	0.0175	6.5081	20
21	.1351	8.6487	58.1095	7.4002	64.0025	0.1156	0.0156	6.7189	21
22	.1228	8.7715	60.6893	8.1403	71.4027	0.1140	0.0140	6.9189	22
23	.1117	8.8832	63.1462	8.9543	79.5430	0.1126	0.0126	7.1085	23
24	.1015	8.9847	65.4813	9.8497	88.4973	0.1113	0.0113	7.2881	24
25	.0923	9.0770	67.6964	10.8347	98.3471	0.1102	0.0102	7.4580	25
26	.0839	9.1609	69.7940	11.9182	109.1818	0.1092	0.0092	7.6186	26
28	.0693	9.3066	73.6495	14.4210	134.2099	0.1075	0.0075	7.9137	28
30	.0573	9.4269	77.0766	17.4494	164.4940	0.1061	0.0061	8.1762	30
∞	.0000	10.0000	100.0000	∞	∞	0.1000	0.0000	10.0000	∞

TABLE 16.2 Compound Interest Factors (continued)

i= 12.00%

n	(P/F)	(P/A)	(P/G)	(F/P)	(F/A)	(A/P)	(A/F)	(A/G)	*n*
1	.8929	0.8929	-0.0000	1.1200	1.0000	1.1200	1.0000	–0.0000	1
2	.7972	1.6901	0.7972	1.2544	2.1200	0.5917	0.4717	0.4717	2
3	.7118	2.4018	2.2208	1.4049	3.3744	0.4163	0.2963	0.9246	3
4	.6355	3.073	4.1273	1.5735	4.7793	0.3292	0.2092	1.3589	4
5	.5674	3.6048	6.3970	1.7623	6.3528	0.2774	0.1574	1.7746	5
6	.5066	4.1114	8.9302	1.9738	8.1152	0.2432	0.1232	2.1720	6
7	.4523	4.5638	11.6443	2.2107	10.0890	0.2191	0.0991	2.5515	7
8	.4039	4.9676	14.4714	2.4760	12.2997	0.2013	0.0813	2.9131	8
9	.3606	5.3282	17.3563	2.7731	14.7757	0.1877	0.0677	3.2574	9
10	.3220	5.6502	20.2541	3.1058	17.5487	0.1770	0.0570	3.5847	10
11	.2875	5.9377	23.1288	3.4785	20.6546	0.1684	0.0484	3.8953	11
12	.2567	6.1944	25.9523	3.8960	24.1331	0.1614	0.0414	4.1897	12
13	.2292	6.4235	28.7024	4.3635	28.0291	0.1557	0.0357	4.4683	13
14	.2046	6.6282	31.3624	4.8871	32.3926	0.1509	0.0309	4.7317	14
15	.1827	6.8109	33.9202	5.4736	37.2797	0.1468	0.0268	4.9803	15
16	.1631	6.9740	36.3670	6.1304	42.7533	0.1434	0.0234	5.2147	16
17	.1456	7.1196	38.6973	6.8660	48.8837	0.1405	0.0205	5.4353	17
18	.1300	7.2497	40.9080	7.6900	55.7497	0.1379	0.0179	5.6427	18
19	.1161	7.3658	42.9979	8.6128	63.4397	0.1358	0.0158	5.8375	19
20	.1037	7.4694	44.9676	9.6463	72.0524	0.1339	0.0139	6.0202	20
21	.0926	7.5620	46.8188	10.8038	81.6987	0.1322	0.0122	6.1913	21
22	.0826	7.6446	48.5543	12.1003	92.5026	0.1308	0.0108	6.3514	22
23	.0738	7.7184	50.1776	13.5523	104.6029	0.1296	0.0096	6.5010	23
24	.0659	7.7843	51.6929	15.1786	118.1552	0.1285	0.0085	6.6406	24
25	.0588	7.8431	53.1046	17.0001	133.3339	0.1275	0.0075	6.7708	25
26	.0525	7.8957	54.4177	19.0401	150.3339	0.1267	0.0067	6.8921	26
28	.0419	7.9844	56.7674	23.8839	190.6989	0.1252	0.0052	7.1098	28
30	.0334	8.0552	58.7821	29.9599	241.3327	0.1241	0.0041	7.2974	30
∞	.0000	8.333	69.4444	∞	∞	0.1200	0.0000	8.3333	∞

i = 20.00%

n	(P/F)	(P/A)	(P/G)	(F/P)	(F/A)	(A/P)	(A/F)	(A/G)	*n*
1	.8333	0.8333	-0.0000	1.2000	1.0000	1.2000	1.0000	-0.0000	1
2	.6944	1.5278	0.6944	1.4400	2.2000	0.6545	0.4545	0.4545	2
3	.5787	2.1065	1.8519	1.7280	3.6400	0.4747	0.2747	0.8791	3
4	.4823	2.5887	3.2986	2.0736	5.3680	0.3863	0.1863	1.2742	4
5	.4019	2.9906	4.9061	2.4883	7.4416	0.3344	0.1344	1.6405	5
6	.3349	3.3255	6.5806	2.9860	9.9299	0.3007	0.1007	1.9788	6
7	.2791	3.6046	8.2551	3.5832	12.9159	0.2774	0.0774	2.2902	7
8	.2326	3.8372	9.8831	4.2998	16.4991	0.2606	0.0606	2.5756	8
9	.1938	4.0310	11.4335	5.1598	20.7989	0.2481	0.0481	2.8364	9
10	.1615	4.1925	12.8871	6.1917	25.9587	0.2385	0.0385	3.0739	10
11	.1346	4.3271	14.2330	7.4301	32.1504	0.2311	0.0311	3.2893	11
12	.1122	4.4392	15.4667	8.9161	39.5805	0.2253	0.0253	3.4841	12
13	.0935	4.5327	16.5883	10.6993	48.4966	0.2206	0.0206	3.6597	13
14	.0779	4.6106	17.6008	12.8392	59.1959	0.2169	0.0169	3.8175	14
15	.0649	4.6755	18.5095	15.4070	72.0351	0.2139	0.0139	3.9588	15
16	.0541	4.7296	19.3208	18.4884	87.4421	0.2114	0.0114	4.0851	16
17	.0451	4.7746	20.0419	22.1861	105.9306	0.2094	0.0094	4.1976	17
18	.0376	4.8122	20.6805	26.6233	128.1167	0.2078	0.0078	4.Z975	18
19	.0313	4.8435	21.2439	31.9480	154.7400	0.2065	0.0065	4.3861	19
20	.0261	4.8696	21.7395	38.3376	186.6880	0.2054	0.0054	4.4643	20
21	.0217	4.8913	22.1742	46.0051	225.0256	0.2044	0.0044	4.5334	21
22	.0181	4.9094	22.5546	55.2061	271.0307	0.2037	0.0037	4.5941	22
23	.0151	4.9245	22.8867	66.2474	326.2369	0.2031	0.0031	4.6475	23
24	.0126	4.9371	23.1760	79.4968	392.4842	0.2025	0.0025	4.6943	24
25	.0105	4.9476	23.4276	95.3962	471.9811	0.2021	0.0021	4.7352	25
26	.0087	4.9563	23.6460	114.4755	567.3773	0.2018	0.0018	4.7709	26
28	.0061	4.9697	23.9991	164.8447	819.2233	0.2012	0.0012	4.8291	28
30	.0042	4.9789	24.2628	237.3763	1181.8816	0.2008	0.0008	4.8731	30
∞	.o000	5.0000	25.0000	∞	∞	0.2000	0.0000	5.5000	∞

Index

D

E

F

Reader Remarks

Yes, you! Your suggestions and comments help us continuously improve this review.

Also, we survive and grow by word-of-mouth. Check the box below and let us tell others your opinion!

Comments: ______________________________

Name: ______________________________

Address: ______________________________

______________ Position: ______________

"You can tell them I said so!" (check here) ☐

Errata...

I noticed the following errors in your new *ME/PE*.....

error	proposed correction	pg. #

Name & Address ______________________________

Try our Civil Review!

CATALOG of titles by GREAT LAKES PRESS

Editor and Principal Author: Dr. Merle C. Potter, Ph.D., P.E.

FUNDAMENTALS OF ENGINEERING, 6th edition. Retail: $49.95

New edition of the most effective review for the FE/EIT exam. Designed for the easiest pass on the complex new exam format. Written by *eight Ph.D. teaching professors* for the General Tests, both a.m. and p.m. Our strategy advises you to choose the General DS Test in the p.m. for the easiest pass! One title contains all you need to pass—no more, no less...and no need to study another dozen subjects for the afternoon! •Includes hardcover reference book which reviews all 12 general subjects; tables and charts; 1100 problems and full solutions; two 8-hour practice exams; NCEES equation summaries. 560 pages; 8.5"x11".

NEW 6th!

PROFESSIONAL CIVIL ENGINEERING EXAM REVIEW

2nd edition. Retail: $69.95

New second edition of this highly-effective review for the Principles & Practice of Engineering (PE) Civil Exam. Also an excellent desk reference for civil engineers. •The **only** review written by ***Ph.D. professors***—ten experts in their fields. •Includes reviews of all 16 major subject areas tested; hundreds of excellent exam-simulating practice problems with full solutions; equation summaries. The only review with all you need in one volume. 693 pages; hardcover.

NEW 2nd!

GRE TIME•SAVER™, 2nd edition. Retail: $17.95

The most authoritative study guide for the GRE. All authors are *Ph.D. teaching professors* from major universities. Praised by the national Psychology honor society, Psi Chi. •Study same subjects as in other reviews...in *half* the time! •Includes diagnostic test and evaluation to identify your strengths and weaknesses, two practice tests and score-analysis charts, hundreds of problems, full solutions, complete reviews of all 3 subjects, and 3000-word vocabulary list with quick-phrases. Paperback, 510 pages.

GMAT TIME•SAVER™, 2nd edition. Retail: $17.95

The most authoritative study guide for the GMAT Test—authors are *Ph.D. teaching professors* from major universities. Now with fax-in writing evaluation offer lets you maximize the new writing portion of the GMAT. •Study the same material as in other reviews...in *half* the time! •Includes a diagnostic test and evaluation to identify your skills, two practice tests and score-analysis charts, hundreds of problems and full solutions, complete reviews of all subjects. Paperback, 397 pages.

GRE ENGINEERING REVIEW, 1st edition. Retail: $19.95

The only full review for the difficult GRE Engineering Subject Test. •The *only* prep that reviews all the subjects covered in the exam—and *fully solves all problems.* •The *only* prep written by nine expert *Ph.D. professors.* •Includes a diagnostic test, two practice tests and score-analysis charts, complete reviews of all subjects, and a self-assessment scale. 297 pages; hardcover.

ORDER FORM

Name (print): ______________________

Street Address: ______________________

____

____

Day ph.: ______________________

Credit Card: VISA MC (circle one)

Number: ______________________ Exp.: __________

Name of Title	Price	Quan.	Multi-Discount!	Subtotal
Civil PE Exam Review	$69.95 x		– Take 20% off each book on orders of 3 or more!	=
Fundamentals of Engr	$49.95 x		–	=
GRE Time•Saver	$17.95 x		–	=
GMAT Time•Saver	$17.95 x		–	=
GRE/Engr Review	$19.95 x		–	=
Shipping	—		x $3 per item	=
			(Mich. residents add 6% sales tax.) TOTAL	$

Mail this order form, credit info., check or money order to:

Great Lakes Press POB550 Grover, MO 63040

Ph. 1-800-837-0201/fax 1-314-273-6086

www.glpbooks.com